Advanced Transmission Electron Microscopy

Jian Min Zuo · John C.H. Spence

Advanced Transmission Electron Microscopy

Imaging and Diffraction in Nanoscience

 Springer

Jian Min Zuo
Frederick-Seitz Materials Research
    Laboratory, Department of Materials
    Science and Engineering
University of Illinois, Urbana-Champaign
Urbana, IL
USA

John C.H. Spence
Department of Physics
Arizona State University
Tempe, AZ
USA

ISBN 978-1-4939-8249-3   ISBN 978-1-4939-6607-3   (eBook)
DOI 10.1007/978-1-4939-6607-3

This Springer imprint is published by Springer Nature
The registered company is Springer Science+Business Media LLC
The registered company address is: 233 Spring Street, New York, NY 10013, U.S.A.

*To my family, Xiurong, Yuan and Ling,
and in memory of my parents,
Yijun and Qiaozhen*

—Jian Min Zuo

To my family, in gratitude

—John C.H. Spence

Preface

This book is written and organized around three topics: electron diffraction, electron optics, and electron crystallography. It is intended as an advanced undergraduate- or graduate-level text in support of course materials in materials science, physics, or chemistry departments. High-resolution transmission electron microscope imaging and scanning transmission electron microscopy are treated as major applications of electron optics, as well as powerful electron crystallographic techniques for structure determination. The emphasis here is on the fundamentals and applications of electron diffraction and imaging in materials research, especially in the study of nanoscience. For this purpose, we have included theory for electron wave propagation, electron diffraction and imaging, and a detailed treatment of electron optics, aberration correction, and instrument techniques, on a level that can be followed by a materials science or physics graduate student. For crystallography, we have emphasized the fundamentals of symmetry, structure and bonding, diffuse scattering, imaging of defects, strain measurement, and determination of nanostructures. Structure determination of large crystals, including polymeric and biological samples, is not discussed specifically in this book, although the electron diffraction and imaging theories presented here and instrumental techniques apply equally to these topics.

Transmission electron microscopy (TEM) traditionally refers to electron diffraction and imaging techniques that are enabled by a transmission electron microscope (with the same TEM acronym). Scanning transmission electron microscopy (STEM) embodies a separate set of techniques. The development of modern TEMs that function as both TEM and STEM has brought them together, as complementary techniques, often abbreviated as S/TEM. For this reason, we have simply used TEM in the book's title. STEM, more than TEM, is associated with powerful analytical techniques, such as electron energy loss spectroscopy and energy-dispersive X-ray spectroscopy. This aspect of TEM is not covered here, and readers are referred to the excellent books on these subjects by Egerton (2011), Hawkes and Spence (2007), and Pennycook and Nellist (2011).

The materials included here come from multiple sources. Firstly, we have updated our previous book on "Electron Microdiffraction" (Plenum, New York, 1992, by J.C.H. Spence and J.M. Zuo). The previous Chaps. 2–4, 7, and 9 are now parts of Chaps. 3, 5, 12, 13, and 10, respectively. The new Chap. 10 on instrumental techniques also incorporates the previous Chap. 6. The previous Chap. 8 is now separated into Chap. 14, which discusses atomic-resolution imaging and Chap. 15 on the characterization of defects. Secondly, we have incorporated much new teaching material throughout the book, such as waves and wave properties (Chap. 2), kinematical theory of electron diffraction (Chap. 4), electron optics (Chaps. 6 and 7), diffuse scattering (Chap. 13), and electron imaging (Chaps. 14 and 15). This material is based on the lectures given to graduate students at University of Illinois, Urbana-Champaign in two courses: diffraction physics and advanced electron microscopy. The writing of Chaps. 6 and 7 has benefitted from the special invited lectures given by Prof. Harald Rose in 2011 to the advanced electron microscopy class.

In writing this book, we have also relied on the original research work by many graduate students, post-docs, and our collaborators. To them, we owe special thanks, especially to Profs. Michael O'Keeffe, Ragnvald Hoier (1938–2009), Miyoung Kim, Randi Holmestad, Jerome Pacuad, Jean-Paul Morniroli, Syo Matsumura, Yoshitsugu Tomokiyo and Drs. Bin Jiang, Weijie Huang, Jing Tao, Jiong Zhang, Min Gao, Celik Ayten, Shankar Sivaramakrishnan, Amish Shah, Ke Ran, Wenpei Gao, and Honggyu Kim. The work at University of Illinois was funded by the Department of Energy, Basics Science and Division of Materials Research, National Science Foundation. Especially, JMZ wishes to thank Dr. Jane Zhu at the Department of Energy for the support of the electron nanocrystallography project.

On reading the literature, one is struck by the enormous variety of applications of TEM/STEM. These include studies of various defects, grain boundaries and interfaces in a broad range of materials, analyses of the symmetry changes which accompany phase transitions, polarization and charge ordering including charge-density waves in layered structures, accurate mapping of the distribution of valence electrons in crystals, phase identification and strain measurement around defects, precipitates and interfaces in alloys or semiconductors, in addition to the characterization of all sorts of nanostructures. To review all this work, published in a vast number of papers, and draw out its implications for materials physics would be a Herculean task. Our aim has been a limited one, to explain the principles of TEM, to provide the theory in a consistent format and to convey enough understanding to students and researchers to let them get started with modern TEM for materials characterization. Thus, to experts in the field, some examples in this book may seem somewhat oversimplified. Also, we have cited references that are directly related to our discussions. We offer our apologies to many of our colleagues whose works were not covered or cited here. With regret, for reasons of space, we have not been able to include the topics of structure determination (see Zou et al. 2011), electron tomography, or coherent diffractive imaging.

Several chapters were written during the sabbatical stay of JMZ at CEA, Grenoble, France, in the fall of 2014. He is therefore grateful to Drs. Jean-Luc Rouviere and Alain Fontaine for their hospitality and also to the Nanoscience Foundation, Grenoble, for the Chair of Excellence position which made his visit possible.

The study of electron diffraction and imaging can be significantly helped by computer simulations. For this purpose, we have made available of computer programs listed in the "Electron Microdiffraction" book on the website http://cbed.matse.illinois.edu/, as well as links to other online resources.

Urbana, USA   Jian Min Zuo
Tempe, USA   John C.H. Spence
                                                                                                           ForMemRS

References

Egerton, R.F.: Electron Energy-Loss Spectroscopy in the Electron Microscope, 2nd edn. Springer, New York (2011)

Pennycook, S., Nellist, P. (eds.): Scanning Transmission Electron Microscopy, Imaging and Analysis. Springer, New York (2011)

Rose, H.: Electron Optics. University of Illinois, Urbana-Champaign (2011) http://cbed.matse.illinois.edu/download/Rose_optics_of_magnetic_lenses.pdf

Hawkes, P.W. and Spence, J.C.H. (eds) Science of Microscopy. Springer, New York (2007) , and Springer Handbook of Microscopy, (2017) to follow.

Spence, J.C.H., Zuo, J.M.: Electron Microdiffraction. Plenum, New York (1992)

Zou, X., Hovmöller, S., Oleynikov, P.: Electron crystallography, electron microscopy and electron diffraction. Oxford University Press (2011)

Preface to "Electron Microdiffraction," Plenum, New York, 1992

Much of this book was written during a sabbatical visit by J.C.H.S. to the Max Planck Institute in Stuttgart during 1991. We are therefore grateful to Profs. M. Ruhle and A. Seeger for acting as hosts during this time and to the Alexander von Humboldt Foundation for the Senior Scientist Award which made this visit possible. The Ph.D. work of one of us (J.M.Z.) has also provided much of the background to the book, together with our recent papers with various collaborators. Of these, perhaps the most important stimulus to our work on convergent beam electron diffraction resulted from a visit to the National Science Foundation's Electron Microscopy Facility at Arizona State University by Prof. R. Hoier in 1988 and from a return visit to Trondheim by J.C.H.S. in 1990. We are therefore particularly grateful to Prof. R. Hoier and his students and coworkers for their encouragement and collaboration. At ASU, we owe a particular debt of gratitude to Prof. M. O'Keeffe for his encouragement. The depth of his understanding of crystal structures and his role as passionate skeptic have frequently been invaluable. Professor John Cowley has also been an invaluable sounding board for ideas and was responsible for much of the experimental and theoretical work on coherent nanodiffraction. The sections on this topic derive mainly from collaborations by J.C.H.S. with him in the seventies. Apart from that, we have tried to review the literature as impartially as possibly and at the same time bring out the underlying concepts in a clear and unified manner, so that the book will be useful for graduate students. We are particularly grateful to Dr. J.A. Eades for his critical review of Chap. 7. We apologize to those authors whose work may have been overlooked among the many hundreds of papers. In order to make the book more practically useful, we have included some FORTRAN source listings, together with POSTSCRIPT code which allows the direct printing of Kikuchi and HOLZ line patterns on modern laser printers from the programs. Support from NSF award DMR-9015867 ("Electron Crystallography") and the facilities of the NSF-ASU National Center for High Resolution Electron Microscopy is gratefully acknowledged.

Tempe, USA   John C.H. Spence
                                                                  Jian Min Zuo

Contents

Symbols and Abbreviations

Symbols

$a, b, c, \alpha, \beta, \gamma$	Lattice constants and angles
$\vec{a}, \vec{b}, \vec{c}$	Unit cell vectors
$\vec{a}^*, \vec{b}^*, \vec{c}^*$	Reciprocal space unit cell vectors
$a_{\vec{k}}$ and $\vec{e}$	Vibrational amplitude and direction
A	Amplitude, structure matrix, scalar magnetic potential, lattice matrix
A_i	Astigmatism
A_o	Cross-sectional area, lattice matrix
$A\left(\vec{k}\right)$	Objective aperture function
$\vec{b}$	Burger's vector
B	Magnetic field, Debye–Waller factor
B_i	Coma
B_g	Normalized eigenvector coefficient
$B^j(x, y)$	Electron lateral eigen function
c_{ijkl}	Stiffness tensor, fourth-rank
C	Eigenvector matrix, speed of light, centered
$\vec{C}$	Chiral vector
C_g	Eigenvector
C_1	Defocus
C_c	Chromatic aberration constant. Positive
C_s	Spherical aberration constant. Positive for round magnetic lenses
C_{IJ}	The 6×6 elastic stiffness matrix
d	d-spacing, distance, diameter
d_{hkl}	d-spacing of hkl lattice plane
$d_{lm\pm}$	Real spherical harmonic function
D	Diameter, size, aberration coefficient
$D\left(\vec{K}_t\right)$	STEM detector response function

e	Electron charge
E	Electron energy, Young's modulus
$E_c(\Delta_f, K_t)$, $E_K(\alpha_s, K_t)$	Envelope functions in contrast transfer function
EI	Bending stiffness
f	Atomic scattering factor, focal length
f^x	*X-ray* atomic scattering factor
$f(\vec{k}, \vec{k}_o)$	Scattering amplitude as function of the scattered and incident wave vectors
$f'(s)$	Atomic absorptive coefficient
$f_r = \dfrac{1 + e\Phi/m_e c^2}{1 + 2e\Phi/m_e c^2}$	Relativistic factor
F	Fano noise, deformation gradient matrix, face-centered, functional
$\vec{F}$	Force
F_g, $F(hkl)$, F_{hkl}	Structure factor
F_g^B	Electron structure factor
F_g^x	X-ray structure factor
$\vec{g}$ or g	Reciprocal lattice vector, reflection or reflected beam
G	$2\pi g$, gain, group, G matrix, Gibbs energy
$G(\vec{r}, \vec{r}')$, $G(\vec{r} - \vec{r}')$	Green's function
h	Planck's constant
$h(z)$	Multipole characteristic function
$h(\vec{r}, E)$	Lens resolution function
$\vec{h}$ or h	Reciprocal lattice vector, reflection or reflected beam
$\hbar$	$h/2\pi$, reduced Planck's constant
H	Hamiltonian, Eikonal, principle plane, a subset of a group, parameter
H_χ	2×2 matrix
$H(\vec{K}_t, E)$	Fourier transform of $h(\vec{r}, E)$
h, k, l	Miller indices
I	Intensity, body-centered, unit diagonal matrix
J	Current density
j_n	Spherical Bessel function of nth order
J	Joule, a unit of energy
J_n	Bessel function of nth order
k	Kilo, one thousand, wave number
k_B	Boltzmann constant
$\vec{k}$	Electron wave vector, phonon wave vector
K	Wave number inside the crystal
$\vec{K}$	Scattered wave vector
$\vec{K}_t$	Tangential component of wave vector
$\vec{K}_o$	Incident wave vector inside the crystal
L	Length, camera length

L_c	Coherence length
L_T	Transverse coherence length
$L(\vec{r})$	Lattice function
$L(\vec{S})$	Reciprocal lattice function
m_e and m	Electron mass and relativistic mass
m	Mirror symmetry, mixing factor
M	Magnification, number, molecule
$M = B(\sin\theta/\lambda)^2$	Thermal damping factor
M_A	Atomic mass
n	Normal, integer number, n-fold rotation axis
N	Integer number, nodal point
$\vec{n}$	Surface normal vector
$\vec{n}$	Roto-inversion axis
o	Zero order, origin
p	momentum, point
P	Perturbation function, primitive, population
$P_z(x,y)$, $P(\vec{r}-\vec{r}')$	Fresnel propagator
$P_g(\vec{r})$	Phase map
q	$2\pi k$, charge, quantum efficiency
$q(\vec{r})$	Object transmission function
$Q(\vec{K}_t)$	Fourier transform of $q(\vec{r})$
$\vec{Q}$	$2\pi\vec{g}$
r	Radius
$\vec{r}$	Real space vector
$\vec{r}^*$	Reciprocal space vector
$\vec{R}$	Lattice vector
s	$s = \frac{\sin\theta}{\lambda}$, path length
S	Symmetry, symmetry operation, surface area, pole-piece gap size, signal
S_i	Aberration coefficient
S_{IJ}	The 6×6 elastic compliance matrix
$\vec{S}$	Scattering vector
$S(\vec{q})$	Diffuse scattering
S_g	Excitation error of reflection g
$S_2(\vec{r}_1,\vec{r}_2)$	Ursell function
$S = C\left\{\mathrm{e}^{2\pi i\gamma^i t}\right\}C^{-1}$	Scattering matrix
t	Thickness
T	Temperature, translation, transformation matrix, time, transmittance
$\vec{T}$	Translation vector
$T(x,y)$	Objective lens resolution function
$T(z)$	Paraxial lens function

u	Electron ray path
u, $\vec{u}$	Atomic displacement amplitude and vector
U	Electron interaction potential, object distance
U'	Imaginary electron interaction potential or absorption potential
U_g	Fourier coefficient of electron interaction potential, or electron structure factor
U_o	Mean electron interaction potential
$U^C(\vec{r})$	Real crystal potential
$\bar{U}(\vec{R})$	Projected electron interaction potential
$U_{jk} = \frac{\beta_{jk}}{2\pi^2 a_j^* a_k^*}$	Averaged, squared, thermal displacements
v	Electron velocity
V	Electrostatic potential, image distance, volume, electron velocity
V_o	Mean electrostatic potential, electron speed
$\vec{V}$ or V_x, V_y, V_z	Electron velocity
V_g	Fourier coefficient of electron potential, or electron structure factor
$\bar{V}(\vec{R})$	Projected electron potential
W_{adh}	Work of adhesion
x, y, z	Cartesian coordinates, variable
X	Electronegativity
X, Y, Z	Cartesian coordinates
Z	Atomic number
α	Angle, phase
β	Angle, variable
β_s	Source brightness
$\beta_{jk} = 2\pi^2 \langle u_j u_k \rangle$	Averaged, squared, thermal displacements
χ	Phase shift due to lens aberrations, parameter, wave function
χ^2	Least-square function
δ	δ function, small number, phase
ε	Small number, coefficient matrix, dielectric constant, energy
ε_{ij}	Strain, i, $j = x$, y, z
ϕ	Wave function
γ	Relativistic constant, fundamental ray, correlation, eigenvalue, dispersion, surface energy
η	Parameter
$\varphi(\vec{r})$	Bloch wave function
κ	Curvature
λ	Wavelength, Lamè constant
μ	Lamè constant

μ_o	Vacuum permeability
v	Poisson's ratio
θ	Angle
θ_{B}	Bragg's angle
ρ	Charge density, radius in the cylindrical coordinates, density
σ	Electron interaction constant, scattering cross section, standard deviation, line charge density
σ_{ij}	Stress
$\sigma_{jk} = \langle u_j u_k \rangle$	Averaged, squared, thermal displacements
τ	Time delay, eigenvalue
υ	Visibility
ϖ	Complex coordinate $(x - iy)$
ω	Frequency, complex coordinate $(x + iy)$, parameter
$\omega\!\left(\vec{k}\right)$	Phonon frequency
ξ_{g}	Extinction distance
ψ	Wave function
Δ	Interval, distance
Δf	Defocus
Φ	Electron acceleration voltage
Φ_g	Wave function
Γ	Correlation, dispersion rate
Λ	Lattice shape function
$\Lambda_{\min} = \frac{1}{K_{t\max}}$	Information transfer limit
Θ_{D}	Debye temperature
Σ	Density of coincidence sites, in its reciprocal form
Υ	Eigenvalue matrix
Ω	Solid angle
Ψ	Phase invariant, multipole potential

Abbreviations

ADF	Annular dark field
BCC	Body-centered cubic
CA	Condenser aperture
CBED	Convergent beam electron diffraction
CBIM	Convergent beam imaging
CCD	Charge-coupled device
CL	Cathodoluminescence
CTF	Contrast transfer function
DQE	Detector quantum efficiency
DR	Dynamic range
DWBA	Distorted wave Born approximation

EBSD	Electron backscattered diffraction
fcc or FCC	Face-centered cubic
FFT	Fast Fourier transform
FT	Fourier transform
G–M	Gjønnes–Moodie
HAADF	High-angle annular dark field
HCP	Hexagonal close packed
HIO	Hybrid input and output
HREM	High-resolution electron microscopy
ICSD	Inorganic Crystal Structure Database
IP	Imaging plates
LACBED	Large-angle convergent beam electron diffraction
LEED	Low-energy electron diffraction
MPB	Morphotropic phase boundary
MTF	Modulated transfer function
NAED	Nanoarea electron diffraction
NBD	Nanobeam diffraction
NBED	Nanobeam electron diffraction
PED	Precession electron diffraction
PMN-xPT	$(1-x)Pb(Mg_{1/3}Nb_{2/3})O_3 - xPbTiO_3$
PMT	Photomultiplier tube
PSF	Point spread function
RHEED	Reflection high-energy electron diffraction
SAED	Selected area electron diffraction
SEM	Scanning electron microscopy or scanning electron microscope
SEND	Scanning electron nanodiffraction
SNR	Signal-to-noise ratio
STEM	Scanning transmission electron microscopy
TEM	Transmission electron microscopy or transmission electron microscope
WPO	Weak phase object
XAFS	X-ray absorption fine structure

Chapter 1
Introduction and Historical Background

1.1 Electrons and the Electron Wavelength

The electron is a negatively charged subatomic particle with elemental charge $e = 1.60218 \times 10^{-19}$ C and mass $m_e = 9.10938 \times 10^{-31}$ kg. The electron was discovered by British physicist J.J. Thomson in 1897. At the Cavendish Laboratory, Cambridge University, J.J. Thomson was experimenting with a cathode ray tube (CRT), investigating a long-standing puzzle known as "cathode rays." The CRT consists of a negatively charged cathode and a positively charged anode sealed inside a vacuum tube (Fig. 1.1). Electrons flow from the cathode and are accelerated toward the anode. A beam of electrons is formed by letting through some of the electrons through a small hole in the anode. Using a deflection plate and a phosphor coating on the other side of the tube that glows when struck by the electrons, Thomson was able to demonstrate that the electrons are negatively charged, and to measure the electron charge/mass ratio.

Electrons have wave-like properties, according to the particle–wave duality in quantum mechanics. The electron wavelength λ in vacuum may be derived using the relationship between λ, the electron momentum p, and the kinetic energy E_o as proposed by Louis de Broglie in 1924:

$$\lambda = \frac{h}{p} = \frac{h}{\sqrt{2mE_o}}. \qquad (1.1)$$

Here, m is the electron relativistic mass and h is Planck's constant ($h = 6.62607 \times 10^{-34}$ J). Inside a TEM, electrons are accelerated to tens or hundreds of keV in energy ($E_o = e\Phi$ with Φ standing for the acceleration voltage in volts) and the electron velocity approaches the speed of light. The electron momentum p in Eq. (1.1) is relativistic (see Appendix A), and this relativistic correction gives

© Springer Science+Business Media New York 2017
J.M. Zuo and J.C.H. Spence, *Advanced Transmission Electron Microscopy*,
DOI 10.1007/978-1-4939-6607-3_1

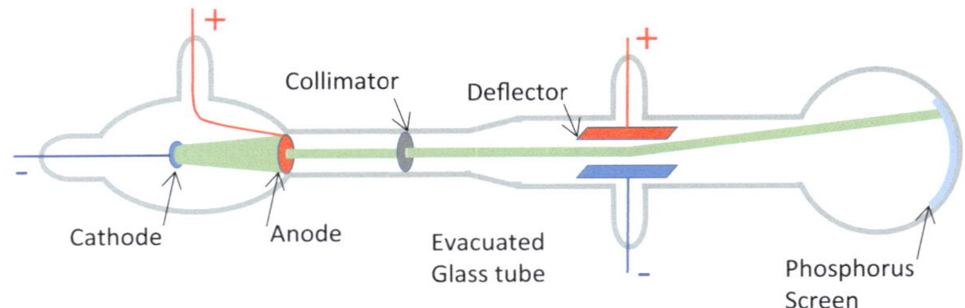

Fig. 1.1 Diagram of cathode ray tube used by J.J. Thomson for his discovery of electrons

$$\lambda = 12.2643/\sqrt{\Phi(1 + 0.978476 \times 10^{-6}\Phi)} \qquad (1.2)$$

where Φ is given in volts (V) and λ is obtained in Angstroms (Å). (1 Å = 0.1 nm). At the commonly used electron acceleration voltages of 100 and 200 kV, λ = 0.037015 and 0.025079 Å, and v/c = 0.5482 and 0.6953, respectively (additional values for many common microscope voltages are tabulated in Appendix B).

The electron matter wave was confirmed by electron diffraction in the experiments reported first by Davisson and Germer of USA (in reflection), and then by G.P. Thompson and Reid of Scotland (in transmission) in 1927, who independently demonstrated crystal electron diffraction and electron interference patterns similar to the patterns formed by light waves scattered by a diffraction grating. So it may be said that while J.J. Thompson showed that an electron was a particle, his son G.P. Thompson showed that it was a wave.

1.2 Electron and Sample Interaction

In an electron diffraction experiment, an incident electron beam of energy E impinges on a specimen and the scattered electrons are observed some distance away. Various diffraction patterns are obtained, as determined by the incident beam energy, scattering geometry, and sample.

The incident electrons interact with the electrons and nuclei within the sample through Coulomb forces. Electrons are scattered as the result of these interactions. Figure 1.2 summarizes various electron scattering effects in a thin, electron transparent sample, where the sample is represented as made of atoms. The electron scattering is separated according to whether it involves electron energy loss or not. Inelastic scattering involves a measurable amount of electron loss and a loss of coherence, whereas elastic scattering from neighboring atoms is normally coherent and suffers no detectable energy loss. The elastic scattering events are further separated into those scattered into relatively low scattering angles by both atomic electrons and the nuclei, or into high angles by the nuclei alone, or backscattered, or

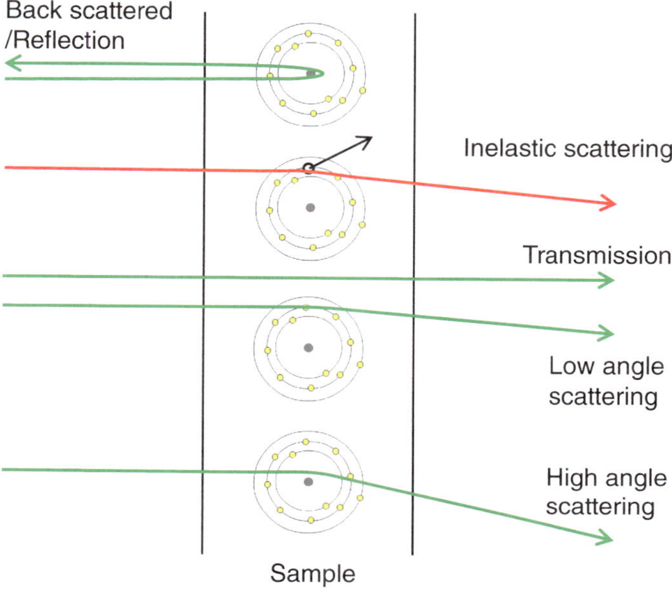

Fig. 1.2 Electron scattering by a thin sample

those reflected by the sample. The direct transmitted beam involves no change in the beam direction. Other types of signals are also generated from the electron and sample interactions; these include secondary electrons, Auger electrons, characteristic and continuum X-rays, long-wavelength radiation in the visible, IR and UV spectral range (cathodoluminescence), lattice vibrations or phonons, and plasmon. The inelastic scattering events can be element-specific or related to specific materials properties. This type of scattering is especially useful in analytical electron microscopy. For electron diffraction, elastic scattering is important as it is used to probe atomic structure as well as chemical bonding.

In a thick sample, each inelastic scattering event results in a loss of energy and a change in the electron momentum. The electron's energy is steadily lost, and the direction of flight is modified. The three-dimensional space in which the incident electrons have sufficient energy to interact with the sample is known as the interaction volume.

The electron beam intensity after passing through a thin sample can be written as follows:

$$I_o - I_B = I_t + I_e + I_i = I_{zero} + I_i \tag{1.3}$$

where I_o, I_B, I_t, I_e, and I_i are the incident beam, backscattered, transmitted, elastic, and inelastic scattering intensities, respectively. The zero-loss intensity, i.e., these electrons which suffer no energy loss, is defined by $I_{zero} = I_t + I_e$. For high-energy electrons, the backscattered intensity I_B is very small and can be neglected. The terms I_{zero} and I_i can be measured using an electron energy loss spectrometer (EELS); however, the amount of current measured depends on the detector collection angle β as well as the smallest energy loss that can be detected. Experimentally, the following relationship is found between I_i and the total intensity $(I_{tot} = I_{zero} + I_i)$

$$I_i(\beta) = I_{tot}(\beta)\left[1 - e^{-t/\lambda_i(\beta)}\right], \tag{1.4}$$

where λ_i is the electron inelastic mean free path (MFP). Appendix F lists the electron inelastic MFP for elemental crystals and selected binary crystals. Most of the MFP values were obtained by EELS. Some were measured using the energy filtering capability of electron holography (McCartney and Gajdardziska-Josifovska 1994). These two measurements give different values since they differ by the energy resolution.

The electron diffraction experiment performed by Davisson and Germer used low-energy electrons, and the diffraction patterns were observed on the back side of the sample, which eventually gave birth to the field of low-energy electron diffraction (LEED) that is still used today for surface analysis. Reflection high-energy electron diffraction (RHEED) was first reported by Nishikawa and Kikuchi in Japan in 1928. In RHEED, the electron beam impinges on the sample at a glancing angle. The electrons are reflected as well as diffracted. The diffraction angles do not follow Bragg's law exactly, because of the effects of the mean crystal potential (Ichimiya and Cohen 2004). This effect was first analyzed in 1927 by H. Bethe during his PhD work under A. Sommerfeld on multiple scattering in this new field of electron diffraction as discussed below in Sect. 1.6. At a small incidence angle, the electron penetration depth is very small and thus very sensitive to surface structure. For this reason as well as its compatibility with the vacuum-based growth techniques, RHEED performed with electron energies of 5–30 keV is a major in situ characterization technique for molecular beam epitaxy (MBE) as well as pulsed laser deposition (PLD).

The experiment carried out by Thomson and Reid used high-energy electrons at normal incidence to a very thin sample (Thomson and Reid 1927). In their experimental setup, cathode rays were generated by an air induction coil and collimated using a fine tube that also shielded the electron beam from external magnetic fields. The diffraction camera was evacuated to a low vacuum using a three-stage mercury vapor pump in order to minimize the electron scattering by gas molecules, and electron diffraction patterns were recorded on film. The thin films of aluminum, gold, celluloid and an unknown substance were used as samples. The necessary components of the electron source, illumination, thin samples, electron detector, and vacuum constitute the essential requirements for transmission electron diffraction in which a high-energy, collimated, beam of electrons traverses a thin slab of sample.

1.3   Transmission Electron Microscope

Within an external field, an electron is subject to a total force

$$\vec{F} = -e\big(\vec{E} + \vec{v} \times \vec{B}\big), \tag{1.5}$$

or

$$\vec{F} = -e\big(\vec{v} \times \vec{B}\big), \tag{1.6}$$

in the absence of electric field ($\vec{E} = 0$). The magnetic force in Eq. (1.6) is known as the Lorentz force. Since it is described by the vector product, the resulting force is perpendicular to the electron velocity $\vec{v}$ and the magnetic field $\vec{B}$.

This force can be used to focus electrons. In 1926, Busch showed theoretically that the axially symmetric electric and magnetic fields could act as a lens for charged particles (Busch 1927). Figure 1.3 illustrates the case of a magnetic coil. The magnetic lens action is equivalent to that of an optical lens, which is characterized by its focal length f and other lens parameters. For a thin lens, the object and image positions (U and V) are related through the thin lens equation

$$\frac{1}{U} + \frac{1}{V} = \frac{1}{f}. \tag{1.7}$$

The first magnetic lens was built by the German scientists Knoll and Ruska (1932). A key innovation was the introduction of the ironclad coils with a small gapped pole piece (Fig. 1.3). Ruska realized that the short focus required for a large lens magnification could be achieved by concentrating the magnetic fields. For example, in 1933, Ruska, together with Bodo von Borries, demonstrated a focal length of 3 mm for electrons of 75 keV in energy (Ruska 1987). This design is still used today.

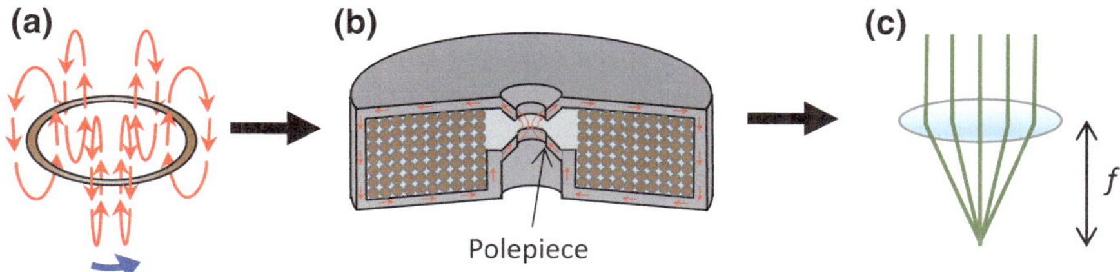

Fig. 1.3 Operation of an electron magnetic lens. **a** Magnetic field generated by a circular coil, showing magnetic field lines (*red arrows*) and electrical current flow (*blue arrow*), **b** magnetic lens with electric coils, soft iron shielding, and the pole piece, and **c** electron beam focusing represented by a lens diagram (from Sanchez et al. 2010)

The construction of the first magnetic lens by Ruska and his collaborator Knoll led to the development of the first TEM (Knoll and Ruska 1932). They were interested in improving the resolution of light microscopes using electrons, although they were initially unaware of the matter waves proposed by de Broglie. In a light microscope, the resolution d_{min} is diffraction-limited according to the Rayleigh criterion

$$d_{min} = 0.61 \frac{\lambda}{n \sin \theta}, \tag{1.8}$$

where $n\sin\theta$ is the numerical aperture (NA) of the lens and θ is half the acceptance angle of the lens. This result applies to the ability to separate images of neighboring incoherent point emitters, such as a binary star or a pair of independently fluorescencing molecules in solution. It does not apply to high-resolution coherent TEM imaging, but is a good approximation for dark-field STEM. From Eq. 1.8, d_{min} is 0.61λ with $NA = 1$. For electrons, $n = 1$ in vacuum, $d_{min} = 2.3$ Å at 100 kV with $\lambda = 0.037$ Å and $\theta = 10$ mrad. A similar calculation by Ruska and Knoll convinced them that it was possible to reach a better-than-light-microscope resolution. The small acceptance angle is due to the aberrations of electron magnetic lenses, which increase with θ to the third-order power. Experimentally, the 2 Å resolution level was reached only 40 years after Ruska and Knoll's invention, after many improvements in TEM technology.

The TEM functions similarly to an optical microscope (Fig. 1.4). It uses electrons instead of light and electron magnetic lenses instead of glass lenses. Electrons accelerated from the electron source are focused into a small, uniform, beam by the condenser lens. A small condenser aperture is used to exclude the electrons travelling at large angles to the optical axis. The beam then traverses the sample at, or close to, normal incidence. Parts of the beam are diffracted, depending upon the thickness and electron transparency of the sample. The transmitted beams are then focused by the objective lens to form an image. This image is subsequently magnified by the intermediate and projector lenses onto a phosphor screen or electron camera. Optionally, an objective aperture can be used to enhance the image contrast by letting through only parts of the transmitted or diffracted beams, or both.

After passing through the sample, beams in the same direction are focused onto the same point at the back focal plane of the objective lens, which generates the electron diffraction pattern. All electrons scattered from the same sample point are ideally focused into the same point at the image plane. In practice, because of lens aberrations and the small lens acceptance angle, the point is more like a blurred disk, albeit a very small disk. For electron diffraction, the intermediate lens focus is changed (by changing the current in the lens), so that the objective lens back focal plane becomes the object plane for the intermediate lens. This method then allows both the electron image and diffraction patterns to be recorded from the same sample area, just by changing the strength of the lenses, which is one of the major advantages of TEM.

Fig. 1.4 Geometric optics of
a basic TEM for imaging

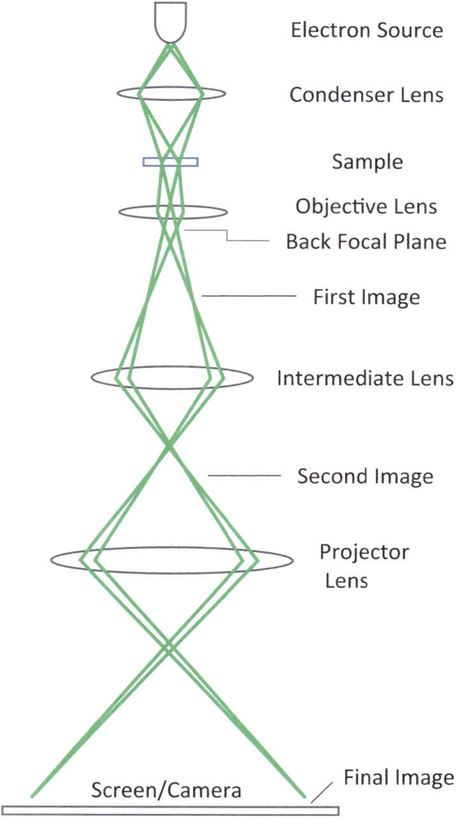

Early TEMs had a single condenser lens and illuminated a sample area of about
1 mm. This large illumination area caused sample heating. This problem was
solved using the two condenser lens arrangement. This solution was installed in the
first mass produced TEM, the Siemens "Elmiskop," which gave this microscope the
"small region radiation" capability. A beam of 1 micron could be formed on the
sample for an image size of 10 cm at a magnification of 100,000 (Ruska 1987).

1.4 Electron Microdiffraction and STEM

Electron diffraction patterns obtained from the early TEMs are spot patterns
because the beam is approximately parallel, with the second condenser lens far
away from the sample.

The origins of microdiffraction using a convergent beam lie in the discovery of
Kossel patterns in X-ray diffraction in 1937 and of Kikuchi patterns in 1928. Both
may be thought of as due to internal sources of radiation from atomic sites in the

crystal. In order to understand this process in more detail, the young G. Möllenstedt was asked in 1937 by W. Kossel in Danzig to build a 45 kV convergent beam electron diffraction camera [see Mollenstedt (1989) for a review]. This would supply an external source of diverging radiation. Möllenstedt's design (for his diploma thesis (M.Sc.) project) used a plasma discharge in a wine bottle as the electron source. A two-magnetic-lens illumination setup was used to form a beam which converged onto the sample. The second condenser lens was positioned close to the sample, and a beam with larger convergence angles could also be formed. The vacuum at the specimen was 10^{-3} torr, and the probe size was about 40 μm. With such a large probe, contamination was not a problem, despite the poor vacuum. Using flakes of mica as a sample, CBED patterns such as that shown in Fig. 1.5 were obtained, exactly like the patterns obtainable on the most modern machines.

The principles underlying scanning electron microscopy (SEM) and scanning transmission electron microscopy (STEM) were first put forward by von Ardenne in 1938 (von Ardenne 1985). To obtain a fine electron probe, von Ardenne proposed to demagnify the electron source crossover using two magnetic lenses. For SEM or STEM, the scanning coils were placed between the two magnetic lenses. The apparatus built by von Ardenne produced a focused electron probe of 50–100 nm in diameter, and this was first used to examine the surface of a specimen by collecting secondary electrons using a specially made low-capacitance detector system. The later SEM images were reproduced on a television tube with afterglow.

SEM went on to become a popular technique for surface imaging, when the first commercial instrument became available in 1960s. von Ardenne's own effort shifted to STEM. His innovations included the first annular aperture for dark-field

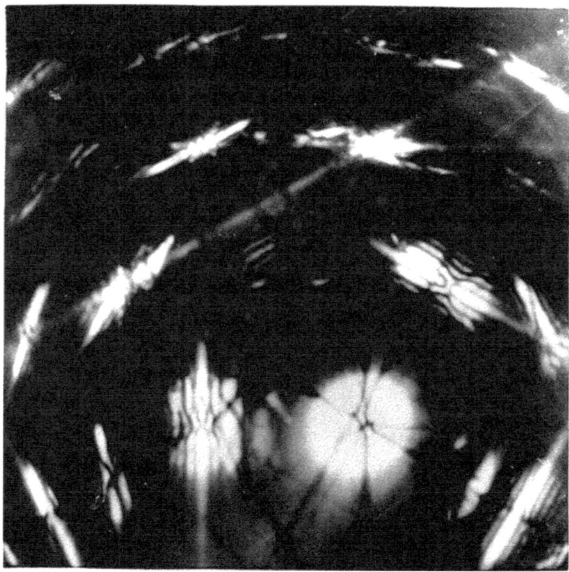

Fig. 1.5 The first CBED pattern, obtained from mica in 1937 at 45 kV using a diffraction camera with two condenser lenses. The pattern was analyzed using two-beam theory by MacGillavry and others in 1940 (Mollenstedt 1989)

imaging, complementary dark- and bright-field imaging, and the idea of stereomicroscopy, suggesting that it was "the ultimate tool for future structure investigations" (von Ardenne 1940). Unfortunately, this line of research stopped in 1944, when von Ardenne's STEM instrument was destroyed in an air raid. The next STEM was developed after over 20 years later by Albert Crewe when he introduced the field emission gun.

1.5   Analytical TEM

A major development for analytical TEM was the adoption of the single-field condenser objective lens, which was first developed by Riecke and Ruska (1966) (for details of objective lenses, see Chap. 6). The specimen is located at the center of the pole-piece gap in the middle of the objective lens's magnetic field. This natural position allows maximum space for specimen translation and tilt as well as access by energy-dispersive X-ray (EDX) detectors. Inside the lens, the field above the sample serves as the final condenser lens for electron probe formation, while the field after the sample serves as the first magnifying lens for electron imaging. The field symmetry above and below the sample provides the same lens parameters for high-resolution TEM and STEM.

Modern commercial TEM/STEM instruments offer various options for analytical TEMs, combined TEM/STEM instruments, and dedicated STEMs, as shown in Fig. 1.6. These instruments share some common features as well as having unique features. Figure 1.7 shows a schematic diagram of the common features found in a modern analytical TEM/STEM. The microscope is shown to be configured with a C_s probe corrector for high-resolution STEM and an electron biprism for electron holography. Inside the microscope column, we have three optical systems: (1) the illumination system, (2) the objective lens, and (3) the projection system; each can be configured to serve multiple functions in a TEM or STEM mode of operation. The properties of individual lenses that make up the illumination and projection systems will be discussed in Chap. 6. Since illumination is critical to STEM and electron nanodiffraction, we will have separate sections in Chap. 10 to discuss the different aspects of the illumination system.

A range of magnifications or camera lengths are provided by the projection system. In a modern TEM, the projection system has three intermediate lenses and one projector lens placed above the electron detector. The first intermediate lens is also called the diffraction lens, since it enables the switch from imaging to diffraction. The final image recorded on the electron detector is a product of the magnifications of individual lenses

$$M_{\text{total}} = M_{\text{obj}} \times M_{\text{int1}} \times M_{\text{int2}} \times M_{\text{int3}} \times M_{\text{proj}}. \tag{1.9}$$

At medium and high magnifications, the objective lens is operated at nearly constant strength with both its image and object fixed at the predesigned positions.

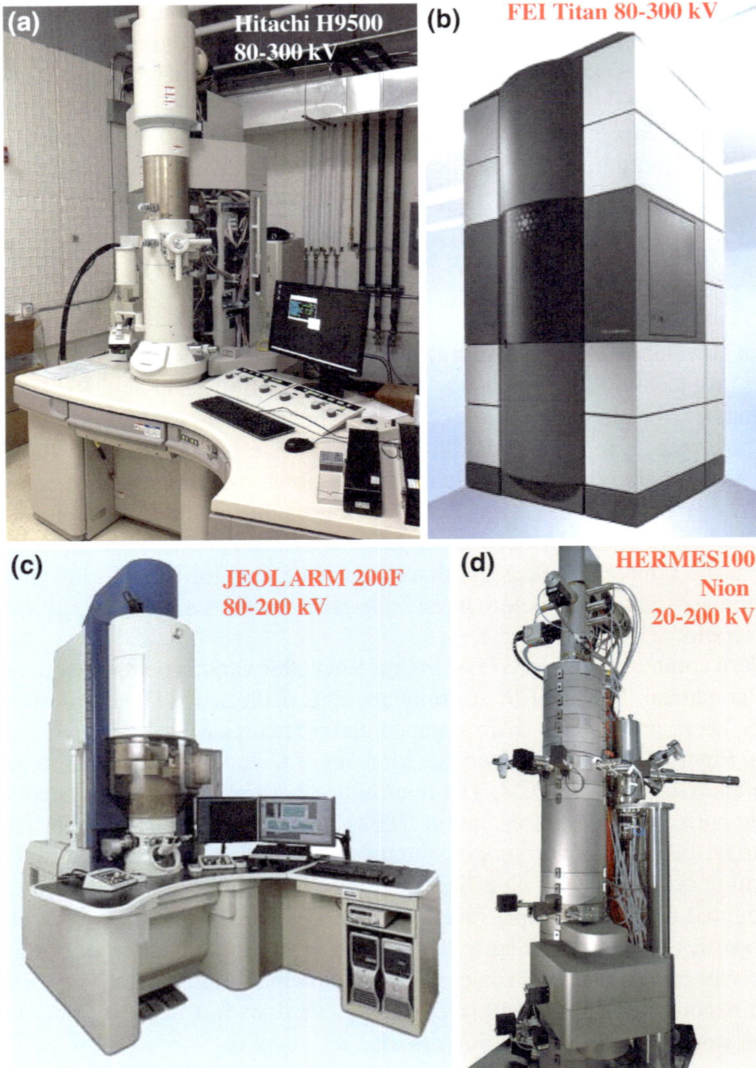

Fig. 1.6 Modern TEM, TEM/STEM, and dedicated STEM. **a** Hitachi H9500 TEM, **b** FEI Titan TEM/STEM, **c** JEOL ARM 200F TEM/STEM, and **d** Nion 100, dedicated TEM (images in **b, c,** and **d** are reproduced with permission)

The image of the projector lens is also fixed at the position of the electron detector. Thus, it is the change in the intermediate lenses that provide the magnification from several thousand to several million. At a magnification of low hundreds to several thousand, the objective lens is switched off and a minilens (objective minilens) placed below the objective lens is sometimes used for the first-stage magnification.

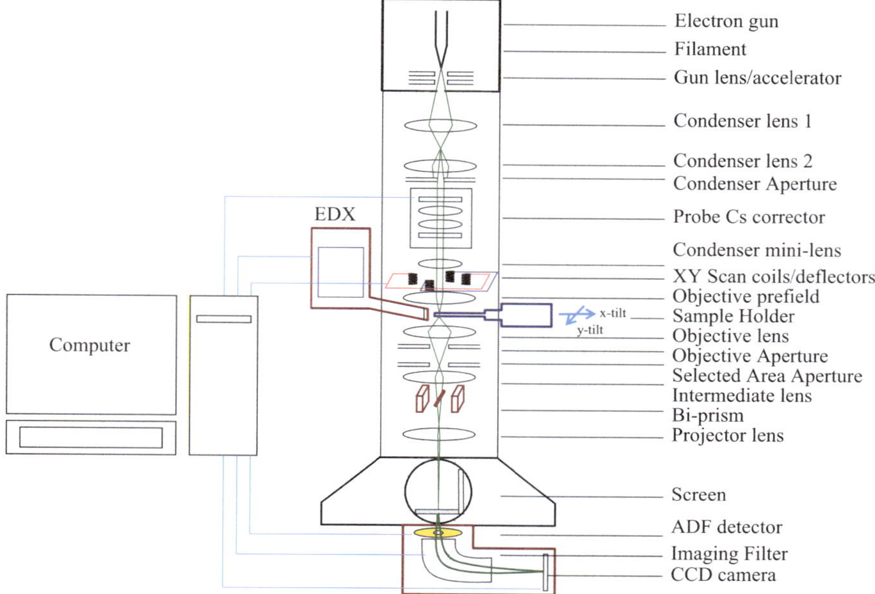

Fig. 1.7 Schematic illustration of a modern analytical S/TEM, its major optical components, and attachments

Some microscopes are also equipped with a special lens called a Lorentz lens, which can provide intermediate magnification with the objective lens turned off. This lens is used for imaging magnetic materials or other materials that cannot be studied in the presence of the objective lens' strong magnetic field.

In diffraction mode, instead of using magnification, we use the effective distance from the sample to the detector, or camera length, to measure the distances recorded in diffraction patterns. The total camera length in a projection system with four lenses is given by

$$L_{total} = f_{obj} \times M_{int1} \times M_{int2} \times M_{int3} \times M_{proj}, \tag{1.10}$$

Here, f_{obj} is the focal length of the objective lens, where the first diffraction pattern is formed.

TEM optical alignment is achieved by deflecting the electron beam, since the electron magnetic lenses are fixed in their positions. Beam deflectors are installed strategically inside a TEM, in the electron gun for gun tilt/shift, in the illumination system for beam shift/tilt and in the projector system for image or diffraction shift. Together, these deflectors are used to steer the electron beam to the optical axis of the magnetic lenses, and the sample and the detector.

TEM apertures are metal diaphragms of different opening sizes. They are mounted on linear translation stages that allow both x and y motion at high precision. Apertures are used to limit the number of electrons as well their direction, which are allowed to enter the magnetic lens or pass onto the specimen. The apertures accessible to the TEM user are the condenser aperture, the objective aperture, and the selected area aperture after the objective lens.

The microscope has a vacuum pumping/handling system installed to keep the electron gun, sample area, and detectors as separate vacuum spaces. A major development in electron microscope instrumentation was the use of differential pumping, which separates the vacuum of the electron gun from that of the specimen region and the camera chamber. In a conventional TEM, the camera chamber is maintained at a lower vacuum, and differential pumping thus enables the TEM to be fitted with a field-emission gun, which requires ultrahigh vacuum (Tonomura et al. 1979). In a dedicated STEM, which is maintained at ultrahigh vacuum, special electron detectors were developed in order to work with the vacuum (Cowley 1993).

Other major components of a TEM or STEM are specimen holders and the mechanisms designed for sample translation and rotation. The analytical TEMs use the so-called side entry holders, which work with the TEM sample stage or goniometer to provide x and y motion over ±1 mm range and z motion of about 0.5 mm along the beam direction. Additionally, the sample stage can be rotated to tens of degrees in both directions depending on the pole-piece design. The double-tilt holders or tilt-rotate holders provide an additional rotation in two directions normal to the sample stage rotation axis. Other special holders are designed to work with mechanical forces for in situ mechanical testing (Hysitron Inc. MN), liquids and gases (Protochips Inc. NC and Hummingbird Inc. WA), heating and cooling (Gatan, CA, Danssolution, Netherland, Protochips, NC), and electrical measurements (Kim et al. 2008).

Additional functionalities on an analytical TEM are obtained by adding optical components, such as aberration correctors, monochromator and an energy filter, and attachments to the electron column. The attachments include 2D pixel array detectors for electron imaging and diffraction, bright-field/annular dark-field detectors for STEM, an electron energy loss spectrometer for EELS, an energy-dispersive X-ray detector for EDX, and an electron biprism for electron holography. The addition of EDX and EELS provides valuable composition information that is critical for diffraction data analysis (Botton 2007; Egerton 2011; Pennycook and Nellist 2011).

1.6   A Brief History of Electron Microdiffraction

The history of the development of electron diffraction to today's state of the art is a fascinating blend of elegantly derived results from the basic physics of electron scattering theory, together with their application to subtle problems in understanding real crystalline solids. The following is a very brief review of some of the main historical developments in transmission electron diffraction using a focused beam, following the initial discovery of G.P. Thomson. This focused-beam mode produces the so-called convergent beam electron diffraction patterns. The literature references are given elsewhere in the relevant section of this book, but an excellent set of historical essays by the pioneers can be found in Goodman (1981).

In 1928, the results of Hans Bethe's "Ph.D." thesis (under the direction of A. Sommerfeld) were published, describing the use of "Bloch waves" to solve the high-energy electron diffraction problem for the reflection case, in order to account for the observations of Davisson and Germer and Thomson and Reid. Bethe's work was based on Ewald's earlier treatment of the multiple scattering problems in X-ray diffraction. A perturbation method (the "Bethe potentials," on which three-beam theory is based) was also given. This work was contemporaneous with, and apparently independent of, Bloch's introduction of "Bloch waves" into band theory. Bethe's theory was extended to the transmission case by Blackman in 1939.

Following the work of G. Möllenstedt on electron microdiffraction, deviations from the kinematic theory were immediately noted. Improved 65 and 750 kV machines were also built, before the war brought developments to an end. In 1940, MacGillavry used two-beam theory to fit experimental CBED patterns in the first attempt to measure structure factors using dynamical electron diffraction theory (MacGillavry 1940).

Theoretical work on the dynamical theory with emphasis on the symmetry properties of the scattering was continued throughout the 1950s by researchers such as Niehrs, Fukahara, Fues, Howie and Whelan, Fujimoto, Miyake, Tournarie, Sturkey, and Cowley and Moodie. In 1957, K. Kambe, in his study of three-beam theory, showed that the intensity depends on a certain sum of three structure-factor phases (the three-phase invariant), which is independent of the choice of origin and so might be measured. Throughout the sixties, the CBED method was developed almost solely by Lehmpfuhl in Berlin, and by Goodman and Moodie in Melbourne using the unsatisfactory (but modifiable) instruments available to them. In 1965, Gjønnes and Moodie (building on earlier work) explained the occurrence of forbidden reflections in the presence of strong multiple scattering (as previously observed by Goodman and Lehmpfuhl). These could then be used to identify translational symmetry elements. The implications of combining results from the reciprocity theorem with crystal symmetry elements were first appreciated in 1968 by Pogeny and Turner, working in John Cowley's group. Work by Uyeda and Høier during this period showed how the position of Kikuchi lines may be used to determine accelerating voltages and lattice constants, and the importance of dynamical corrections to the line positions understood. [This would later be studied in great detail for the closely related problem of the higher-order Laue zone (HOLZ) lines used for strain measurement.] The critical voltage effect on Kikuchi lines was discovered at about this time by Uyeda, Watanabe and coworkers in Japan. In their 1971 three-beam analysis of this effect, Gjønnes and Høier showed that an eigenvalue degeneracy existed in three-beam theory, so that the absence of intensity at certain points in these patterns may be used to determine the three-phase invariant for centrosymmetric crystals. Thus, with the dynamical problem "solved," the implications of crystal symmetry and reciprocity understood, dynamical corrections to the Bragg law understood, and the three-phase invariants defined (with their effects elucidated for centric crystals), we might say that the heroic age of CBED theory came to an end.

By the early seventies, then, systematic procedures for point-group and space-group determination by CBED had begun to emerge from the groups in Melbourne (Goodman, Moodie, et al.), Bristol (Steeds, Buxton, and coworkers), and Tohoku (Tanaka) and a lively debate ensued on the possible symmetry-breaking effects of boundary conditions. The theoretical foundations for point-group determination were firmly established by Buxton and coworkers in the context of group theory; they went on to develop the perturbation theory for HOLZ interactions (with later elegant contributions from Portier and Gratias, Tinnable, Kogiso, Kastner, and others). The Bristol group then embarked on a systematic application of the CBED method to a wide variety of problems in materials science and condensed matter physics. This focused effort in Bristol over many years (resulting among other things in the publication of an atlas of CBED patterns for alloy phases in 1984) was perhaps responsible more than anything else for establishing the success of that method.

As one example among many, the work of the Bristol group on phase transformations in layer compounds supporting charge-density waves brought the CBED technique to the attention of a much larger audience of solid-state physicists for the first time. The most comprehensive attack on the problem of structure determination by CBED has also been described by this group in their successful determination of the structure of AuGeAs. In the USA, the subsequent popularity of the method owed most to Eades' work at Illinois. Throughout the late 1970s and early eighties, Cowley (together with one of the authors) in Arizona were developing the coherent CBED method using subnanometer probes. The application of the superlattice method within the multislice algorithm was developed for coherent CBED patterns from defects in 1977, and the theory of STEM lattice imaging developed as a result. In 1981, Cowley produced some remarkable CBED patterns using the Vacuum Generators HB5 instrument from regions of crystal smaller than the unit cell, which were seen to repeat with the period of the lattice as the probe was moved across the crystal. Similar work on nanodiffraction, including the use of imaging energy filters and novel detectors, was later developed by Brown and coworkers in Cambridge, UK. Both groups subsequently developed techniques for the study of defects in crystals using subnanometer probes. A nonscanning alternative to the Eades/Tanaka scanning method for avoiding overlap of adjacent orders was developed by Tanaka, whose group also produced (starting in 1985) two invaluable volumes of beautiful CBED patterns covering a wide range of applications and "case studies." This group has since produced much of the highest quality work in the field. The CBED method had otherwise been slow to develop in Japan prior to Tanaka's efforts.

In Berlin, quantitative work on structure factor measurement continued under Lehmpfuhl in the early eighties. The use of CBED patterns to study line and planar defects also first began to be studied at about this time. Large-angle CBED methods (LACBED) for HOLZ and ZOLZ reflections were then developed in 1986 by Taftø, Vincent, and coworkers. HOLZ effects from artificial superlattices appeared first in the work of Cherns in 1987. The persistent rediscovery of the value of shadow imaging in CBED (producing various hybrid modes such as CBIM and LACBED) and of the value of HOLZ lines (because their intensities are frequently kinematic, allowing simple rules to be derived) is a feature of work during this period. Research on structure-factor phase measurement in noncentrosymmetric crystals was begun in earnest in the mid-1980s by Marthinsen, Høier, and later Bird and others, resulting finally in experimental structure-factor phase measurements with an accuracy of better than one degree by Zuo and coworkers in 1989. During the same period, many measurements of local strains began to appear, based on measurements of HOLZ line positions with various approximate dynamical correction schemes (summarized by Zuo 1992) in development of the earlier work by Gjønnes, Høier, and Olsen on dynamical shifts on Kikuchi lines.

Only recently have the computing times for whole CBED patterns, including three-dimensional multiple scattering effects, been reduced to less than an hour or so for simple crystals, and this progress in computer hardware explains much of the recent renewed interest in CBED. The use of elastic energy filtering has greatly improved the accurate quantification of these data. This, together with the use of cooled CCD cameras, online work stations, and field-emission guns, brings our subject to the threshold of its most exciting era, in which the techniques of quantitative electron microcrystallography will be applied to a wide range of problems in materials science, solid-state chemistry, mineralogy, and condensed matter physics.

More recent developments of electron diffraction are entwined with the development of electron microscope technologies, and they are further helped with the development of computer algorithms for data analysis. The development of field-emission guns (FEGs) in the 1970s and their adoption in conventional transmission electron microscopes (TEM) brought high source brightness, smaller probe size, and improved coherence to electron diffraction. Electron energy filters, such as the in-column Ω-energy filter, allow the inelastic background from plasmon, or higher electron energy losses, to be removed from recorded diffraction patterns with an energy resolution of a few eV. The development of array detectors, such as CCD cameras or imaging plates, enables parallel recording of diffraction patterns and quantification of diffraction intensities over a large dynamic range that was not available to electron microscopy before.

These developments in the electron diffraction hardware were accompanied by the development of efficient and accurate algorithms to simulate electron diffraction patterns and modeling structures on a first-principle basis using fast computers, which has significantly improved our ability to interpret experimental electron diffraction patterns. The more recent developments are time-resolved electron diffraction at the time resolution approaching femtoseconds (Zewail 2006; LaGrange et al. 2008) and scanning electron nanodiffraction and coherent nanoarea electron diffraction for the study of nanostructures (Chaps. 16 and 17). Further developments of these techniques will significantly improve our ability to interrogate structures at high spatial and time resolution that hitherto has not been available.

1.7 A Note to Students and Lecturers

In this book, we have attempted to summarize and develop most of the useful knowledge which has been gained over the years from the study of the multiple electron scattering problem, quantitative electron diffraction, and microscopy and from their applications to materials structure characterization. This book is intended for advanced undergraduate and graduate students and professional research workers in materials science, chemistry, and condensed matter physics. A physics background, knowledge of crystallography, and some familiarity with electron microscopy at the undergraduate level are assumed.

Students embarking on a course in materials research at the graduate level will want to know where this book fits into the overall scheme of modern materials science and engineering. It is often said that materials science and engineering is the study of the structure, properties, synthesis, and processing of materials. The study of relationships between structure and properties is what differentiate materials science and engineering from chemistry, physics, or mechanical engineering. Characterization forms an essential part of this endeavor. Transmission electron microscopy, more than any other techniques, has contributed to our knowledge of materials microstructure because of the large electron elastic scattering cross section and electron imaging. In particular, the combination of the large scattering cross sections and a small probe (down to subnanometer dimensions) makes TEM the essential characterization tool for nanoscience and nanotechnology, which gives the strongest signal from the smallest volume of matter of any analysis technique. The field of TEM has grown tremendously as considerable efforts are now underway in many countries aimed at nanostructured materials. This book presents theory of electron diffraction, optics, microscopy techniques, and their practice in relationship to materials structure characterization.

Working in several electron microscopy laboratories for many years has convinced us that most of the enormous amount of information contained in electron diffraction patterns and images is simply thrown away. This trend was reversed only recently, with the help of digital detectors and development of quantitative analysis techniques. This book is an attempt to further this trend and to promote TEM as truly quantitative characterization techniques. For this purpose, rather than writing a book exclusively about TEM, we have emphasized the fundamentals and applications of TEM.

This book starts with a brief introduction to TEM and its history in this chapter. After this, this book is organized into three parts: Part I, Electron Diffraction; Part II, Electron Optics; and Part III, Electron Crystallography. In Part I, we first develop the wave theory of electrons and wave properties in Chap. 2. This is followed by Chaps. 3 and 4 on kinematical diffraction theory. Chapter 3 concerns mainly with diffraction geometry, while Chap. 4 deals with kinematical diffraction intensity. Electron dynamic theory, including two-, three- and many-beam theories, is described in Chap. 5. In Part II, Chaps. 6–8 introduce electron optics, magnetic lens, aberrations, and aberration correction and electron sources. This is followed by Chap. 9 on electron detectors. The last part, Part III, starts with Chap. 10 on instrumentation for electron diffraction and imaging. Here, we included the sections on the optics of magnetic sectors and the principles of energy filters. We then move on to Chaps. 11–17 on the following topics: crystal symmetry, crystal structure and bonding, diffuse scattering, atomic resolution imaging, defects, strain measurements, and determination of nanostructures.

Experiences in teaching graduate students in materials science and engineering showed that the materials included in this book can be taught in two semester-based courses. One is on diffraction physics, where selected materials from Chaps. 1–5 and 11–13 can be combined with materials on X-ray diffraction, such as the excellent book by Warren (1990) on "X-ray diffraction." The second course is advanced electron microscopy, which can be offered to students after they finished an introductory course on TEM based on the book of Williams and Carter (2009). This course could be designed based on materials covered in Chaps. 6–8, 14, and 15 and additional materials on inelastic scattering and electron spectroscopy such as these covered in the books by Williams and Carter (2009), Reimer and Kohl (2008), and Botton (2007).

References

Botton G (ed) (2007) Analytical electron microscopy. Science of microscopy, Springer, New York

Busch H (1927) On the operation of the concentration coil in a Braun tube. Arch Electrotech 18:583

Cowley JM (1993) Configured detectors for STEM imaging of thin specimens. Ultramicroscopy 49:4–13

Egerton RF (2011) Electron energy-loss spectroscopy in the electron microscope, 2nd edn. Springer, New York

Goodman P (ed) (1981) Fifty years of electron diffraction. D. Reidel. Dordrecht, Holland, IUCr

Ichimiya A, Cohen PI (2004) Reflection high energy electron diffraction. Cambridge University Press

Kim T, Kim S, Olson E, Zuo JM (2008) In situ measurements and transmission electron microscopy of carbon nanotube field-effect transistors. Ultramicroscopy 108:613–618

Knoll M, Ruska E (1932) Das elektronenmikroskop. Zeitschrift Fur Physik 78:318–339

LaGrange T, Campbell GH, Reed B, Taheri M, Pesavento JB, Kim JS, Browning ND (2008) Nanosecond time-resolved investigations using the in situ of dynamic transmission electron microscope (DTEM). Ultramicroscopy 108:1441–1449

MacGillavry CH (1940) Examination of the dynamic theory of electron diffraction on lattice. Physica 7:329–343

McCartney MR, Gajdardziskajosifovska M (1994) Absolute measurement of normalized thickness, t/λi, from off-axis electron holography. Ultramicroscopy 53:283–289

Mollenstedt G (1989) My early work on convergent-beam electron-diffraction. Phys Status Solidi A 116:13–22

Pennycook S, Nellist P (eds) (2011) Scanning transmission electron microscopy, imaging and analysis. Springer, New York

Reimer L, Kohl H (2008) Transmission electron microscopy (4th). Springer, Berlin

Riecke WD, Ruska E (1966) A 100 kV transmission electron microscope with single-field condenser objective. VI. Int. Congress for Electron Microscopy, Kyoto, Japan

Ruska E (1987) The development of the electron-microscope and of electron-microscopy. Rev Mod Phys 59:627–638

Sanchez SI, Small MW, Sivaramakrishnan S, Wen JG, Zuo JM, Nuzzo RG (2010) Visualizing materials chemistry at atomic resolution. Anal Chem 82:2599

Thomson GP, Reid A (1927) Diffraction of cathode rays by a thin film. Nature 119:890

Tonomura A, Matsuda T, Endo J, Todokoro H, Komoda T (1979) Development of a field-emission electron-microscope. J Electron Microsc 28:1–11

von Ardenne M (1940) About a universal electron microscope for brightfield, darkfield and stereo operation. Z Physik 115:339–368

von Ardenne, M. (1985). On the history of scanning electron-microscopy, of the electron-microprobe, and of early contributions to transmission electron-microscopy. In: Hawkes PW (ed) The beginnings of electron microscopy, Elsevier

Warren BE (1990) X-ray diffraction, Reprint edn. Dover Publications

Williams DB, Carter BC (2009) Transmission electron microscopy, a textbook for materials science (2nd Editiom). Springer, New York

Zewail AH (2006) 4d ultrafast electron diffraction, crystallography, and microscopy. Annu Rev Phys Chem 57:65–103

Zuo JM (1992) Automated lattice-parameter measurement from HOLZ lines and their use for the measurement of oxygen-content in $YBa_2Cu_3O_{7-\Delta}$ from nanometer-sized region. Ultramicroscopy 41:211–223

Chapter 2
Electron Waves and Wave Propagation

Electron diffraction and imaging rely on the wave properties of electrons. A basic understanding of wave and wave properties is thus required for the interpretation of electron diffraction and electron imaging. Wave theory is also required for understanding electron probe formation using electron lenses. This chapter introduces waves, wave-related properties, and wave equations. While the basic concepts described here can be found in introductory text to electron microscopy, the discussions on wave propagation and wave coherence are conducted at a level equivalent to graduate courses in physics. For these sections, the readers are referred to books on quantum mechanics by Griffiths (2004) and on physical optics by Born and Wolf (1999) and Goodman (2004).

2.1 Wave Functions and the Wave Equation

There are two distinct, universal, properties associated with waves; one is the propagation of waves, and the other is local disturbance. If we think of a floating bouy anchored to the ocean floor, its vertical motion as waves pass under it maps out the wave amplitude as a function of time at one point. By contrast, a snapshot photograph of the ocean surface provides a map of the wave amplitude at one time as a function of space. Consider now a one-dimensional wave which propagates at a speed of v along the x direction. An observer moving with the wave, in the wave coordinate x', only sees a local disturbance, and the wave function is then simply given by $\phi = f(x')$. For a stationary observer, the origin of the x' coordinate moves by a distance of vt at time t. The relation between the two coordinates is given by $x' = x - vt$; thus, the 1D wave function seen by the stationary observer has the general form of

$$\phi = f(x - vt) \tag{2.1}$$

© Springer Science+Business Media New York 2017
J.M. Zuo and J.C.H. Spence, *Advanced Transmission Electron Microscopy*,
DOI 10.1007/978-1-4939-6607-3_2

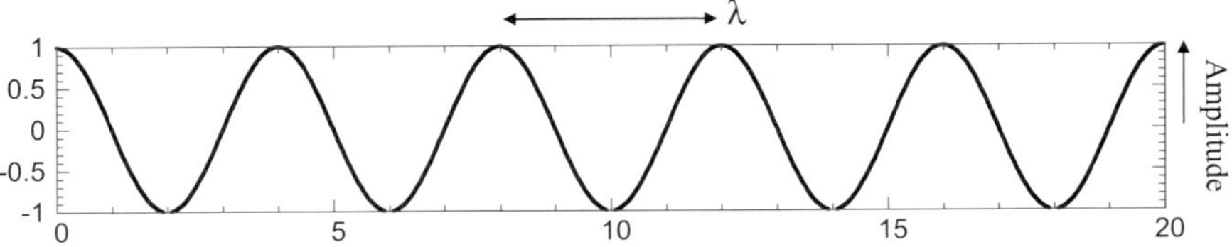

Fig. 2.1 Wave function of Eq. (2.3) plotted as function of x for $t = 0$ and $A = 1$

Double differentiating Eq. (2.1) on both sides by x and t gives the homogeneous wave equation of

$$\frac{\partial^2}{\partial x^2}\phi = \frac{1}{v^2}\frac{\partial^2}{\partial t^2}\phi. \tag{2.2}$$

The simplest wave is a sinusoidal wave in the form of

$$\phi = A\cos\left[\frac{2\pi x}{\lambda} - \omega t\right], \tag{2.3}$$

or

$$\phi = A\sin\left[\frac{2\pi x}{\lambda} - \omega t\right]. \tag{2.4}$$

Here, λ is the wavelength, ω is the frequency, and A is the wave amplitude. The wave velocity is then given by (Fig. 2.1)

$$v = \omega\lambda/2\pi. \tag{2.5}$$

Both the sine and cosine forms of wave function are the solutions of Eq. (2.2), so a general solution is a combination of these two, each with its own amplitude. By choosing an appropriate phase, the combination of sine and cosine functions can be expressed using a single cosine function so that

$$\phi = A\cos\left[\frac{2\pi x}{\lambda} - \omega t + \delta\right], \tag{2.6}$$

Here, δ is called the initial phase of the wave. The solution in Eq. (2.6) can be alternatively written as follows:

$$\phi = \mathrm{Re}\left\langle A\exp\left[i\left(\frac{2\pi x}{\lambda} - wt + \delta\right)\right]\right\rangle$$

Fig. 2.2 Plane wave in 3D

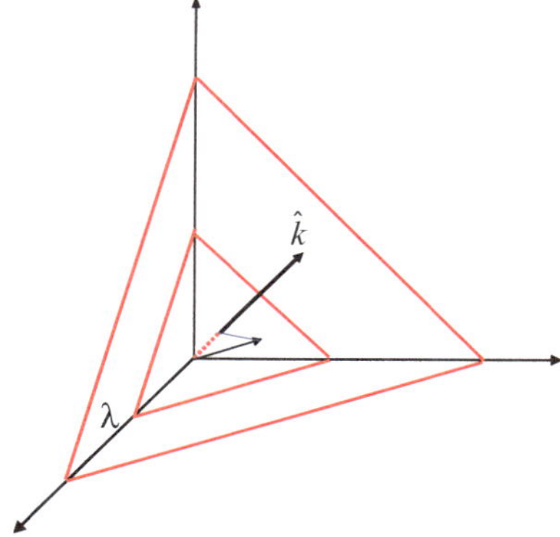

or

$$\phi = \phi_{\mathrm{o}}\exp\left[i\left(\frac{2\pi x}{\lambda} - wt\right)\right] = \phi_{\mathrm{o}}\exp(2\pi ikx)\exp(-i\omega t) \qquad (2.7)$$

by taking the real part afterward. The wave amplitude ϕ_{o} is complex in general.

So far, we have only discussed the 1D sinusoidal wave that is propagating along the x direction. In 3D, the sinusoidal wave can travel in any direction. Its direction is specified by a unit vector $\hat{k}$ (Fig. 2.2). An important point is that a 3D sinusoidal wave is no different from the 1D sinusoidal wave; the difference is that we instead choose to look at it at a different angle. A property of the wave of Eq. (2.6) is that the phase only changes in the direction of wave propagation (with x only). For this reason, a sinusoidal wave is also called plane wave because of its constant phase in the 2D plane normal to the wave propagation direction. The 3D equivalent is that the phase only depends on the distance along $\hat{k}$, which mathematically is given by $\hat{k} \cdot \vec{r}$. Putting this into Eq. (2.7), we have a general description of a sinusoidal wave in 3D of the form

$$\phi(\vec{r},t) = \phi_{\mathrm{o}}\exp\left[2\pi i\left(\frac{\hat{k}\cdot\vec{r}}{\lambda}\right)\right]\exp(-i\omega t) = \phi_{\mathrm{o}}\exp\left[2\pi i\vec{k}\cdot\vec{r}\right]\exp(-i\omega t). \qquad (2.8)$$

We call the vector $\vec{k} = \hat{k}/\lambda$ the wave vector and $\left|\vec{k}\right| = \frac{1}{\lambda} = \sqrt{k_x^2 + k_y^2 + k_z^2}$ the wave number. Correspondingly, the homogeneous wave equation in 3D is given by

$$\left(\frac{\partial^2}{\partial x^2} + \frac{\partial^2}{\partial y^2} + \frac{\partial^2}{\partial z^2}\right)\phi = \nabla^2\phi = \frac{1}{v^2}\frac{\partial^2}{\partial t^2}\phi \qquad (2.9)$$

The plane wave described in Eq. (2.9) is only one of many solutions to the three-dimensional wave equation. The other special solution is the spherical wave, which is emitted from a point source. The wave function of a spherical sinusoidal wave has the form

$$\phi = \frac{\phi_{\mathrm{o}}}{r} \exp(2\pi i k r) \exp(-i\omega t) \tag{2.10}$$

It can be shown that this is a solution of the homogeneous wave equation by using the spherical representation of Eq. (2.9)

$$\nabla^2 \phi = \frac{1}{r^2}\frac{\partial}{\partial r}\left(r^2 \frac{\partial}{\partial r}\right)\phi = \frac{1}{r}\frac{\partial^2}{\partial^2 r}(r\phi) = \frac{1}{v^2 r}\frac{\partial^2}{\partial t^2}(r\phi) \tag{2.11}$$

The spherical sinusoidal wave of Eq. (2.10) represents a diverging wave from the point of origin. By replacing k with $-k$, another solution representing a converging spherical wave is obtained. For plane waves, changing the sign of wave vector reverses the wave propagation direction.

By summing waves of different wave vectors together, three distinct important types of wave can be formed—the running waves treated above, the standing wave, and the pulse or wave packet. It is also possible to sum over a range of wave vectors and frequencies—in that case, the relationship between wave vector and frequency is called a dispersion relation, and this depends on the properties (the dielectric function in optics, or the electromagnetic potential for electron diffraction) of the medium in which the wave is traveling. We discuss all these effects in the following sections. The wave packet will turn out to be particularly important for our later treatment of STEM or electron nanodiffraction using a coherent probe.

2.2 Quantum Mechanical Wave of Electrons and Schrődinger Equation

Waves due to mechanical vibrations (such as the vibration of a guitar string) or electromagnetic oscillations (such as low-frequency radio waves observed directly on an oscilloscope) can be directly observed; e.g., we can measure both the amplitude and phase of these waves in principle. The quantum mechanical wave, including that of an electron, cannot be measured the same way. The mere fact that electrons are diffracted by crystals attests to the wavelike behavior of electrons. We also see the particle-like behavior of electrons inside electron microscopes; electrons are bent by the presence of electric or magnetic fields in a trajectory predicted by the theory of classical mechanics. In fact, we prefer to describe electrons as particles moving along optical paths (trajectories) on most of their trajectory through the electron microscope. Electrons are also detected by collision with another electron or atom. For example, when a photographic film is exposed to electrons, electrons are "captured" as speckles on the film after chemical processing—we can say that electrons travel as waves, but arrive as particles. (The likelihood of the electron arriving at a particular point is then proportional to the square of the electron wave function, described below.) Anyone who underexposes a lattice image or records low-dose electron images can testify to this, since the electron image is seen to be built-up from many individual electron arrivals, like raindrops. A discernible image or pattern only emerges when there are a sufficient number of these electrons. Thus, whenever we try to measure electrons, we only see particles. Summarizing this situation in the language of quantum mechanics, we have following properties, which are not limited to the electrons:

(1)   An electron can be wavelike or particle-like;
(2)   Its property is described by the electron wave function. Wave function itself is not measurable. What is measurable is the distribution of electrons, given by the square intensity of the wave:

$$I = |\phi|^2 = \phi \cdot \phi^*. \tag{2.12}$$

Here, ϕ^* is the complex conjugate of the wave function ϕ. The wave function must be normalized so the overall probability of finding the electron is 1, which is achieved by following integral

$$\int_{-\infty}^{\infty} \phi(\vec{r})\phi^*(\vec{r})d^3\vec{r} = 1. \tag{2.13}$$

(3)   The distribution can be any one of the measurable properties of the electrons. Each property is obtained by applying an appropriate operator ($\hat{A}$) to the wave function. For example, the operators of $\hat{A} = x$ and $\hat{A} = -ih\partial/\partial x$ give the position and momentum along the x direction, while $\hat{A} = t$ and $ih\partial/\partial t$ give the time and energy. What is measured is the expectation value, according to

$$\langle \hat{A} \rangle = \langle \phi^* | \hat{A} | \phi \rangle = \int \phi^*(\vec{r})\hat{A}\phi(\vec{r})d^3\vec{r}. \tag{2.14}$$

For electrons inside a potential field, $V(\vec{r})$, the wave function satisfies the Schrödinger equation in the form

$$-\frac{h^2}{8\pi^2 m}\nabla^2\phi - eV(\vec{r})\phi = i\frac{h}{2\pi}\frac{\partial}{\partial t}\phi. \tag{2.15}$$

The time period of electron wave oscillation, $2\pi/\omega$ (4×10^{-20} s for electrons of 100 keV), is however too short to detect for high-energy electrons—it is the differences in this frequency which appear as energy losses when multiplied by Planck's constant; the fundamental frequency is not believed by most physicists to be an observable. In a typical experiment performed inside an electron microscope, the measurement is over a much longer time period than this frequency. The time period is also much shorter than that of atomic vibrations in the order of 10^{-12} s (picoseconds). Thus, the potential can be assumed to be time independent. Then, the complex wave function can be separated into two parts, the time dependent and time-independent parts, such as $\phi(\vec{r}, t) = \phi(\vec{r}) \exp(-i\omega t)$. The time-dependent part is same to all electrons of the same energy, which can be taken out of the equation. For the rest of this book, we therefore use the so-called time-independent Schrödinger equation:

$$-\frac{h^2}{8\pi^2 m} \nabla^2 \phi - eV(\vec{r})\phi = E\phi, \tag{2.16}$$

where $E = \hbar\omega$ is the electron energy. For an electron traveling in vacuum, the potential is zero. Then, Eq. (2.16) reduces to

$$-\frac{1}{4\pi^2} \nabla^2 \phi = \frac{2mE}{h^2}\phi = k^2 \phi, \tag{2.17}$$

where $k = \sqrt{2mE/h^2}$. Equation (2.17) is then the same wave equation as in Eq. (2.9) for a single frequency. This equivalence enables us to apply our accumulated knowledge about waves from other fields, such as optics, to understand electron imaging and diffraction, while the specific quantum-mechanical wave properties of electrons are discussed in reference to electron-specimen interaction.

2.3 The Principle of Wave Superposition

The wave equation, in both the classical and quantum mechanical forms (Eqs. 2.9 and 2.15), is linear. By this, we mean that the equation only involves ϕ, and there are no higher order terms including products of ϕ in the wave equation. A general property of linear equations is that the sum of any two solutions is also a solution. Thus, if two waves of ϕ_1 and ϕ_2 are solutions of the wave equation, $\phi = \phi_1 + \phi_2$ is also a solution of the wave equation. Obviously, this extends to any number of such waves, and in general,

$$\phi = \sum_{i=1}^{N} \phi_i \tag{2.18}$$

is a solution, as are the individual solutions.

The property stated in Eq. (2.18) is referred as the principle of superposition. This enables us to write any wave function as a sum of solutions, such as the sinusoidal waves and the spherical waves from individual point sources.

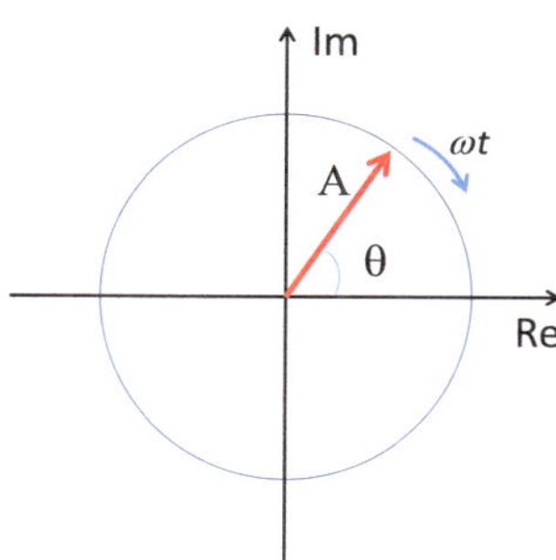

Fig. 2.3 Amplitude and phase diagram of a complex wave function

2.4 Amplitude and Phase Diagrams

The complex wave of Eq. (2.8) can be separated into three components: the amplitude, a position-dependent phase, and a time-dependent phase:

$$\phi = A\exp(i\theta)\exp(-i\omega t) \tag{2.19}$$

where $\theta = 2\pi\vec{k}\cdot\vec{r}+\delta$, and δ is the initial phase. The complex number of Eq. (2.19) can be plotted in the so-called complex plane, with the x-axis for the real and the y-axis for the imaginary parts of the number (see Fig. 2.3). The length of the vector is the amplitude A, and its angle with the x-axis is θ. The vector rotates as time progresses with frequency ω.

In the case that we have two waves with the same ω:

$$\varphi = [A_1\exp(i\theta_1)+A_2\exp(i\theta_2)]\exp(-i\omega t) \tag{2.20}$$

The resultant wave is a linear sum in the way that these two are added together as the addition of two vectors.

2.5 Coherence and Interference

The superposition of two waves of the same frequency results a new wave of the same frequency. The superposition of two waves of different frequencies, however, varies with time with the amplitude from $|A_1 + A_2|$ to $|A_1 - A_2|$. Since in a typical experiment, the electron intensity is measured over a certain limited period of time T, the question of practical importance is how the two types of superposition affect the intensity of waves, and the question arises as to whether waves of different frequency can interfere (the answer, as we now show, is "briefly").

Certain properties of wave superposition give rise to the concept of wave coherence. To examine the results of wave superposition, we start with Eq. (2.18), and for simplification, we look at the superposition of two one-dimensional waves:

$$\begin{aligned}
\phi(x,t) &= \phi_1(x,t) + \phi_2(x,t) \\
&= A_1 \exp[2\pi i(k_1 x - v_1 t) + i\delta_1] + A_2 \exp[2\pi i(k_2 x - v_2 t) + i\delta_2]
\end{aligned} \tag{2.21}$$

The intensity of the superimposed wave is given by:

$$\begin{aligned}
I(x,t) &= \phi(x,t)\phi^*(x,t) = A_1^2 + A_2^2 \\
&+ 2A_1 A_2 \cos[2\pi(k_1 - k_2)x - (\omega_1 - \omega_2)t + \delta_1 - \delta_2]
\end{aligned} \tag{2.22}$$

For an experiment carried out over an extended time T (a typical exposure time T in electron microscopes is in the order of seconds), we observe the average intensity

$$I_{\text{obs}}(x) = A_1^2 + A_2^2 + 2A_1 A_2 \langle \cos[2\pi(k_1 - k_2)x - (\omega_1 - \omega_2)t + \delta_1 - \delta_2] \rangle_T \tag{2.23}$$

Here, the observed intensity has three terms, the intensities of waves 1 and 2, respectively, plus an interference term in the form of cosine function. This term comes from the interference between the two waves. For this last term, we have several possibilities:

Case 1: $\omega_1 \neq \omega_2$ and $T \gg 2\pi/|\omega_1 - \omega_2|$, where we might call $2\pi/|\omega_1 - \omega_2|$ the "beat period," this being the wavering period we hear, for example, from a slightly out-of-tune piano whose two strings (for the same note) differ in frequency by this amount. (It is interesting to see how this condition, when multiplied by Planck's constant, takes on the form of the energy and time uncertainty principle, as further discussed below.) In this case, the positive and negative contributions of the cosine function cancel each other out, and the overall result is given by

$$I_{\text{obs}}(x) = A_1^2 + A_2^2 \tag{2.20}$$

Case 2: $\omega_1 = \omega_2$ and both δ_1 and δ_2 are constant; in this case, we have

$$I_{\text{obs}}(x) = A_1^2 + A_2^2 + 2A_1 A_2 \cos[2\pi(k_1 - k_2)x + \delta_1 - \delta_2] \tag{2.21}$$

and the intensity varies with x; the period of variation is determined by the difference between k_1 and k_2.

Case 3: $\omega_1 = \omega_2$ and both δ_1 and δ_2 vary with time randomly; in this case, the positive and negative contributions of the cosine function again cancel each other out, which gives rise to the following result

$$I_{\text{obs}}(x) = A_1^2 + A_2^2$$

Case 4: $\omega_1 \neq \omega_2$ but $T \ll 2\pi/|\omega_1 - \omega_2|$; in this case, the detection time is shorter than the "beat period" within which waves of different frequency can interfere, so we get the same results as cases 2 or 3. (This case is the basis for experiments in which laser light from two different, independent lasers, operating at

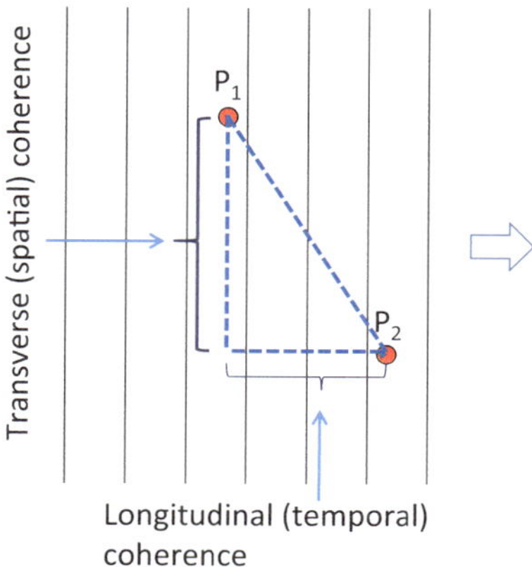

Fig. 2.4 Spatial coherence and temporal coherence between two wave-front points. The temporal coherence and spatial coherence are measured along and normal to the wave propagating direction, respectively

slightly different frequencies, can be shown to interfere if the detection time is short enough.)

Among the 4 cases above, cases 1 and 3, where the intensity of the total wave is simply the sum of the individual waves, are said to be incoherent. Case 2 is said to be coherent. Here, the intensity of the sum of two waves is the sum of individual wave intensities plus their interference effect, which depends on the relative phase difference between the two waves (An example of electron interference is shown in Fig. 2.11.) This interference effect is missing in the incoherent cases.

The above discussion was concerned with two opposite cases, one fully incoherent and the other fully coherent. In practice, wave interference often lies in between these two, which we call partially coherent. The degree of interference can vary continuously, and thus, it is useful to define a measure for the degree of partial coherence. To do so, we consider the general case of interference between two points on a propagating wave front, P_1 and P_2 as shown in Fig. 2.4, which are separated by a vector $\vec{r}$. The waves travel between these two source points and the detector through two different paths. Assuming the first wave arrives at time t and the second one arrives at $t + \tau$, the intensity averaged over a period of time is given by

$$
\begin{aligned}
\langle I \rangle &= \left\langle |\phi_1(t) + \phi_2(t+\tau)|^2 \right\rangle \\
&= \left\langle |\phi_1(t)|^2 \right\rangle + \left\langle |\phi_2(t+\tau)|^2 \right\rangle + 2\mathrm{Re}\{\langle \phi_1(t)\phi_2^*(t+\tau)\rangle\} \\
&= I_1 + I_2 + 2\mathrm{Re}\{\Gamma_{12}(\tau)\}
\end{aligned} \tag{2.24}
$$

where

$$
\Gamma_{12}(\tau) = \langle \phi_1(t)\phi_2^*(t+\tau)\rangle, \tag{2.25}
$$

which is known as the correlation function of the two waves. A correlation function is simply a function which measures the degree of similarity between two functions. When the two waves are the same, for example, $\Gamma_{11}(\tau) = \langle \phi_1(t)\phi_1^*(t+\tau)\rangle$, the function then defines the self-correlation (also known as autocorrelation function). Using these definitions, we rewrite Eq. (2.24) as

$$\langle I \rangle = I_1 + I_2 + 2\sqrt{I_1 I_2}\,\mathrm{Re}\{\gamma_{12}(\tau)\} \tag{2.26}$$

where

$$\gamma_{12}(\tau) = \frac{\Gamma_{12}(\tau)}{\sqrt{\Gamma_{11}(0)}\sqrt{\Gamma_{22}(0)}}. \tag{2.27}$$

The degree of coherence between the two waves (i.e., the extent to which the waves can interfere) can be measured from the maximum intensity and minimum intensity recorded in an interference pattern (such as Young's pinhole experiment), formed when these two waves are allowed to interfere. This is called the "visibility"

$$v = \frac{I_{\max} - I_{\min}}{I_{\max} + I_{\min}}.$$

The intensity of minimum ($I_{\min}$) and maximum ($I_{\max}$) is obtained with $\gamma_{12}(\tau) = -|\gamma_{12}(\tau)|$ and $\gamma_{12}(\tau) = |\gamma_{12}(\tau)|$, respectively. In case $I_1 = I_2$, then $v = |\gamma_{12}(\tau)|$. For this reason, $\gamma_{12}(\tau)$ is called the complex degree of partial coherence. The normalization against the autocorrelation function in the definition of $\gamma_{12}(\tau)$ ensures that its maximum will not exceed 1 (Born and Wolf 1999). The fully incoherent and coherent cases correspond to $|\gamma_{12}(\tau)| = 0$ and 1, respectively.

Experimentally, the longitudinal and transverse components of $\gamma_{12}(\tau)$ are measured as shown in Fig. 2.4 using interferometry. For the longitudinal component, the source points are selected to be on the same axis along the propagation direction using a Michelson interferometry (Steel 1985). They are selected by the time delay set by the path difference in the interferometer. The value of $\gamma_{12}(\tau)$ obtained in this way gives a measurement of longitudinal coherence, which is also known as temporal coherence. The transverse coherence can be measured by the Young's slit experiment, where the wave arrives at the detector from the two slits simultaneously with $\tau = 0$. This will be further discussed in Sect. 2.9.

2.6   Wave Packets and the Uncertainty Principle

So far, we have only discussed sinusoidal waves, which are continuous, infinite, and monochromatic with a single frequency. The sinusoidal wave can be only an approximation of the actual wave function of the free electrons for two reasons. First, it cannot be normalized since for the sinusoidal wave, we have

$$\int\limits_{-\infty}^{\infty} \phi(\vec{r})\phi^*(\vec{r})d^3\vec{r} = A^2 \int\limits_{-\infty}^{\infty} d^3\vec{r} = A^2(\infty).$$

Secondly, the emitted electrons have a finite energy spread (ΔE) as well as a finite angular distribution. Thus, instead of a single momentum k, there is a range of electron momenta (or $\vec{k}$). Since the sinusoidal wave of Eq. (2.8) is a solution of the wave equation, the wave function of a free electron can be expressed as a superposition of sinusoidal waves according to the principle of wave superposition

$$\phi(\vec{r},t) = \int \phi(\vec{k}) \exp\left[2\pi i \vec{k} \cdot \vec{r}\right] \exp\left(-\pi i \frac{hk^2}{m}t\right) d^3\vec{k} \qquad (2.28)$$

Here, $\phi(\vec{k})$ is the complex amplitude of the sinusoidal wave of wave vector $\vec{k}$. The resulting wave function is called a wave packet. This wave function can be normalized in the following way (the mathematics employed here is described in Appendix E)

$$\begin{aligned}
\int \phi(\vec{r},0)\phi^*(\vec{r},0)d^3\vec{r} &= \int\int \phi(\vec{k})\phi^*(\vec{k}')d^3\vec{k}d^3\vec{k}' \int \exp\left[2\pi i(\vec{k}-\vec{k}')\cdot\vec{r}\right]d^3\vec{r} \\
&= \int \phi(\vec{k})\phi^*(\vec{k}')\delta(\vec{k}-\vec{k}')d^3\vec{k}d^3\vec{k}' \\
&= \int \phi(\vec{k})\phi^*(\vec{k})d^3\vec{k} = 1
\end{aligned}$$

$$(2.29)$$

Both $\phi(\vec{k})$ and $\phi(\vec{r},t)$ describe the same electron wave function. The only difference is the representation, one in the real space and one in momentum or k space.

At $t = 0$, we have the wave function in integral form, known as a Fourier transform:

$$\phi(\vec{r},0) = \int \phi(\vec{k}) \exp\left[2\pi i \vec{k} \cdot \vec{r}\right]d^3\vec{k}. \qquad (2.30)$$

Inside an electron microscope, the emitted electrons emerge from a source, which can be real or virtual. (For example, the source is virtual when it is placed closer to a convex lens than the focal distance.) If we set $t = 0$ at the source position, then by applying an inverse Fourier transform (Appendix E), we obtain

$$\phi\left(\vec{k}\right) = \int \phi(\vec{r},0)\exp\left[-2\pi i\vec{k}\cdot\vec{r}\right]d^3\vec{r} \tag{2.31}$$

The emitted electrons in the TEM has a small divergence angle of few to tens of milliradians (mrad) and an energy spread of ~ 0.3 to ~ 2 eV. A useful approximation is as follows:

$$\phi\left(\vec{k}\right) = A\left(k_x,k_y\right)\frac{1}{\sigma\sqrt{2\pi}}e^{-(k_z-k_o)^2/2\sigma_k^2}, \tag{2.32}$$

where $A\left(k_x,k_y\right) = 1/\pi k_{\max}^2$ for $\sqrt{k_x^2+k_y^2} \leq k_{\max}$ and 0 otherwise. This model assumes that momentum is uniformly distributed within a disk along the x direction and y direction and has a Gaussian distribution along the z direction around the mean value of $k_o = 1/\lambda$.

The Fourier transform in the integral of Eq. (2.31) has the property that a broadly distributed function gives rise to a narrowly distributed Fourier spectrum or vice versa in a reciprocal relationship. For example, consider the Fourier transform of a Gaussian function in the form

$$\mathrm{FT}\left(e^{-ax^2}\right) = \int_{-\infty}^{\infty} e^{-ax^2}e^{-2\pi ikx}dx = \sqrt{\frac{\pi}{a}}e^{-\pi^2k^2/a}. \tag{2.33}$$

The standard deviation of the Gaussian function of e^{-ax^2} is $\sigma_x = 1/\sqrt{2a}$. Thus, $\sigma_k = \sqrt{a}/\sqrt{2}\pi$ and $\sigma_x\sigma_k = 1/2\pi$. In general, a wave packet of short duration in size has a broad range of electron momenta, while a sinusoidal wave extending over all space has only a single value of momentum. Such reciprocal relationships are broadly defined in the Heisenberg's uncertainty principles of quantum mechanics, which state that for two incompatible observables, the product of the measurement uncertainties (σ) is greater than the Planck's constant:

$$\begin{aligned}\sigma_x\sigma_p &\geq \hbar \\ \sigma_t\sigma_E &\geq \hbar\end{aligned}. \tag{2.34}$$

The same relationship also applies to position and momentum along y and z directions, as well as the angles and angular momentums. We note that Eq. (2.34) in the so-called modern representation of the uncertainty principle has $\hbar$ divided by 2. Heisenberg's original formulation had $\hbar$ instead of $\hbar/2$ on the right side of the inequality. Since the uncertainty principle only sets a lower limit, the two formulations only differ in the estimate of this limit. For a Gaussian function, we have $\sigma_x\sigma_k = 1/2\pi$. Since $p = hk$, $\sigma_x\sigma_p = \hbar$ for the Gaussian wave function.

2.7   The Gaussian Wave Packet and Its Propagation

It is instructive to examine the propagation of a Gaussian wave packet as described in Eq. (2.32). At $t = 0$, by converting the 3D integral in k space to cylindrical form, we have

$$
\begin{aligned}
\phi(\vec{r}, 0) &= \frac{1}{\sigma_k \sqrt{2\pi}} \int A\left(k_x, k_y\right) e^{-(k_z - k_o)^2/2\sigma_k^2} \exp\left[2\pi i \vec{k} \cdot \vec{r}\right] d^3\vec{k} \\
&= \frac{1}{\sigma_k \sqrt{2\pi}} \frac{1}{\pi k_{\max}^2} \int\limits_0^{k_{\max}} \int\limits_0^{2\pi} e^{2\pi i k_\parallel \rho \cos\theta} k_\parallel \, dk_\parallel \, d\theta \left\{ e^{2\pi i k_o z} \int\limits_{-\infty}^{\infty} e^{-k^2/2\sigma_k^2} e^{2\pi i k_z z} dk_z \right\}
\end{aligned}
\tag{2.35}
$$

where $\rho = \sqrt{x^2 + y^2}$ and θ is the angle between $\vec{k}$ and $\vec{\rho}$. Using the result in Eq. (2.33), we have for the integral over k_z

$$
\frac{1}{\sigma_k \sqrt{2\pi}} \int\limits_{-\infty}^{\infty} e^{-k^2/2\sigma_k^2} e^{2\pi i k_z z} dk_z = e^{-2(\pi\sigma_k z)^2}
$$

The integral over $k_\parallel$ is carried out first by integrating over θ, which gives the well-known zero-order Bessel function

$$
\begin{aligned}
\int\limits_0^{k_{\max}} \int\limits_0^{2\pi} e^{2\pi i k_\parallel \rho \cos\theta} k_\parallel \, dk_\parallel \, d\theta &= \int\limits_0^{k_{\max}} k_\parallel \, dk_\parallel \left[2\pi J_o\left(2\pi k_\parallel \rho\right)\right] \\
&= \frac{1}{2\pi\rho^2} \int\limits_0^{2\pi k_{\max} \rho} x J_o(x) dx
\end{aligned}
$$

where $x = 2\pi k_\parallel \rho$. Using the integral identity $\int_0^{x_o} x J_o(x) dx = x_o J_1(x_o)$ and putting the above results together, we obtain the following wave function at $t = 0$

$$
\phi(\vec{r}, 0) = e^{2\pi i k_o z} e^{-2(\pi\sigma_k z)^2} 2 \left[\frac{J_1\left(2\pi k_{\max} \rho\right)}{2\pi k_{\max} \rho}\right]
\tag{2.36}
$$

Thus, the wave function amplitude falls off away from $z = 0$ according to a Gaussian distribution, while normal to z, the function gives rise to the well-known Airy disk function $[2J_1(x)/x]^2$, which has a maximum at $\rho = 0$ and falls to its first zero at $x = 3.8317$ or $\rho = 0.6/k_{\max}$ and oscillates as ρ increases.

At $t \neq 0$, the electron wave function $\phi(\vec{r}, t)$ can be obtained using the propagation of individual plane waves that make up the wave packet, according to the integral in (2.28). This integral can also be separated into two parts similar to

Eq. (2.35). Here, we focus on the propagation along the z direction, for which we have

$$\frac{1}{\sigma_k\sqrt{2\pi}} \int_{-\infty}^{\infty} e^{-k_z^2/2\sigma_k^2} e^{2\pi i k_z z} \exp\left(-\pi i \frac{h(k_z+k_o)^2}{m}t\right) dk_z$$

$$= \frac{1}{\sigma_k\sqrt{2\pi}} \exp\left(-\pi i \frac{hk_o^2}{m}t\right) \int_{-\infty}^{\infty} e^{-ak_z^2} e^{2\pi i k_z z'} dk_z$$

$$= \frac{1}{\sigma_k\sqrt{2\pi}} \exp\left(-\pi i \frac{hk_o^2}{m}t\right) \sqrt{\frac{\pi}{a}} e^{-\pi^2 z'^2/a}$$

where

$$z' = z - \frac{hk_o}{m}t$$

and

$$a = \frac{1}{2\sigma_k^2}\left(1 + i\frac{2\pi\sigma_k^2 ht}{m}\right) = \frac{1}{2\sigma_k^2}\left(\frac{1+\Gamma^2 t^2}{1-i\Gamma t}\right) \tag{2.37}$$

With $\Gamma = \frac{2\pi\sigma_k^2 h}{m}$.

Putting the above results together, we have along the z-axis for the Gaussian wave packet

$$\phi(z,t) = C\sqrt{\frac{1-i\Gamma t}{1+\Gamma^2 t^2}} \exp\left[-2\pi^2\sigma_k^2\left(z - \frac{hk_o}{m}t\right)^2 / \left(1+\Gamma^2 t^2\right)\right] e^{2\pi i k_o z} \tag{2.38}$$

Here, C contains two phase terms, one is simply the phase oscillation with time according to the average frequency, and the other comes from the complex term in Eq. (2.37).

The wave packet is thus again described by a Gaussian function, and its center moves with the so-called group velocity hk_o/m. Its width increases with time according to $\sqrt{2}\pi\sigma_k/\sqrt{1+\Gamma^2 t^2}$. The rate of wave packet broadening is given by Γ.

As we will see in later chapters, the relationship between the frequency and wave vector $\omega\left(\vec{k}\right)$ is modified by electron interaction with the potential, and it becomes a general function of $\vec{k}$ in a complex form. In such cases, as long as the frequency spectrum is narrow, $\omega\left(\vec{k}\right)$ can be approximately expanded around the mean in a Taylor series of

$$\omega(k) = \omega(k_o) + (k - k_o)\omega'(k_o) + \frac{1}{2}(k - k_o)^2 \omega''(k_o) + \cdots$$

$$= 2\pi \left[k_o v_p + (k - k_o)v_g + \frac{1}{2}(k - k_o)^2 \Gamma + \cdots \right]$$

where $v_p = \omega(k)/2\pi k$ is the phase velocity, $v_g = (d\omega(k)/dk)/2\pi$ is the group velocity, and $\Gamma = (d^2\omega(k)/dk^2)/2\pi$ gives the dispersion.

In the case of a free electron, $v_p = E/hk = hk/2m$ and $v_g = (dE/dk)/h = hk/m$. Thus, the quantum mechanical phase velocity is half the speed of classical particles, while the group velocity is the same as the classical speed.

2.8   Temporal Coherence

A direct consequence of having an electron wave packet of finite length is the limited temporal coherence $\gamma_{12}(\tau)$, as defined in Eq. (2.27). To calculate $\gamma_{12}(\tau)$ for an electron wave packet, we approximate its wave function using the model of the quasi-monochromatic wave

$$\phi(z,t) = \phi_o \exp[2\pi ikz]\exp(-i\omega t)\exp(-i\delta(t)) \tag{2.41}$$

Here, the phase $\delta(t)$ is taken as constant within a coherence time τ_o, e.g., $\delta(t) = \Delta_n$ for the time period of $n\tau_o \leq t < (n+1)\tau_o$. However, from one coherent period to next, the value of Δ_n fluctuates randomly in the so-called random phase approximation.

The quasi-monochromatic wave applies to the experimental case when the electron momenta spread along the x and y directions are very small, and the emitted electrons are far from each other in time. Thus, the probability of having two or more electrons emitted within the coherence time is very small.

Assuming the amplitudes of two waves are the same, we obtain from the above wave function

$$\gamma_{12}(\tau) = \frac{\Gamma_{12}(\tau)}{\sqrt{\Gamma_{11}(0)}\sqrt{\Gamma_{22}(0)}} = \langle \exp(-i\omega\tau)\exp(i[\delta(t) - \delta(t+\tau)])\rangle$$

$$= \exp(-i\omega\tau)\lim_{T\to\infty}\frac{1}{T}\int_0^T \exp(i[\delta(t) - \delta(t+\tau)])dt \tag{2.39}$$

To evaluate the integral in Eq. (2.39), we first consider the case of $\tau > \tau_o$; e.g., the time delay exceeds the coherence time. The phase difference is then given by the difference between two random phases, which gives another random phase, so that $\delta(t) - \delta(t+\tau) = \Delta$. The integral over random phases averages to zero, and thus, $\gamma_{12}(\tau) = 0$ for $\tau > \tau_o$.

For $0 < \tau < \tau_0$, we have the following two scenarios

$$\delta(t) - \delta(t+\tau) = \begin{cases} 0, & 0 < t < \tau_0 - \tau \\ \Delta_n - \Delta_{n+1}, & \tau_0 - \tau < t < \tau_0 \end{cases}$$

This applies to every coherent period in the quasi-monochromatic wave. By summing up all coherent periods, we obtain the integral of Eq. (2.39) in the form

$$\gamma_{12}(\tau) = \exp(-i\omega\tau) \lim_{N \to \infty} \frac{1}{N\tau_0} \left\{ \sum_{n=0}^{N} \int_0^{\tau_0 - \tau} dt + \int_{\tau_0 - \tau}^{\tau_0} \exp[i(\Delta_n - \Delta_{n+1})] dt \right\}$$

$$= \exp(-i\omega\tau) \frac{1}{\tau_0} \int_0^{\tau_0 - \tau} dt$$

$$= \left(1 - \frac{\tau}{\tau_0}\right) \exp(-i\omega\tau) \tag{2.40}$$

for $0 < \tau < \tau_0$. Here, the second sum over the random phases averages to zero.

Together, the above results show that the visibility of interference fringes $|\gamma_{12}(\tau)|$ decreases linearly with the delay time and disappears beyond the coherence time. The path difference between the two waves thus must be smaller than $L = v\tau_o$ in order to observe the interference between the two waves.

ForThe longitudinal coherence between two waves of slightly different wavelengths ($\lambda_1 = \lambda$ and $\lambda_2 = \lambda - \Delta\lambda$) is defined as the length over which the two waves become completely out of phase with each other (e.g., by 180°) as shown in Fig. 2.5. According to this definition

$$2L_C = N\lambda = (N+1)(\lambda - \Delta\lambda)$$

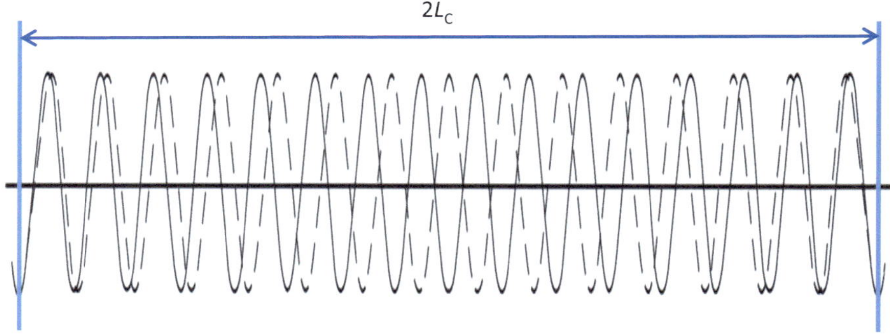

Fig. 2.5 Longitudinal coherence between two waves of different wavelengths (After D. Attwood, University of California, Berkeley)

where N is the number of periods where the two waves become in-phase again. From this, we obtain

$$N + 1 \approx N = \frac{\lambda}{\Delta\lambda}$$

and

$$L_C = \frac{\lambda^2}{2\Delta\lambda} \tag{2.41}$$

For the electron wave packet, it can be shown the same form is obtained by taking

$$L_C = \frac{1}{2}v\Delta t = \frac{\lambda^2}{2\Delta\lambda} = \frac{\lambda E}{2\Delta E}$$

where Δt is the uncertainty time obtained from the uncertainty principle. Thus, if we consider the coherence time is same as the uncertainty time ($\tau_o = \Delta t$), at L_C, we expect 50 % contrast in the interference. This expression for the temporal coherence of electron beams was first investigated by Mollenstedt and Ducker (1955).

2.9   Spatial Coherence

In defining the spatial, or transverse, coherence length (or width), we first consider two waves of the same wavelength, originating from two separate source points in space as shown in Fig. 2.6. Each gives rise to a set of wave fronts. Using the same criterion for

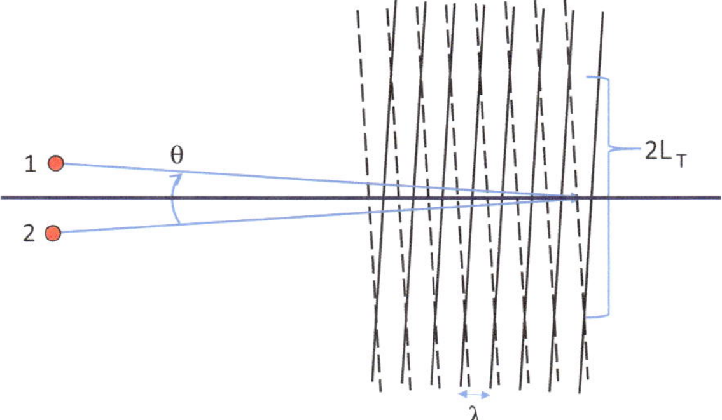

Fig. 2.6 Transverse coherence

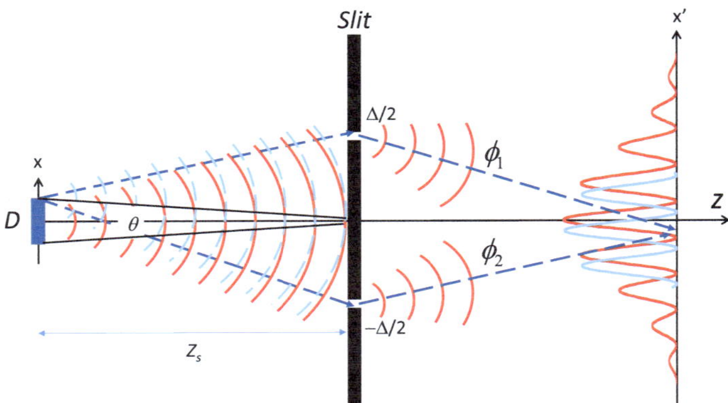

Fig. 2.7 Young's double-slit interference experiment for measurement of transverse coherence length

the definition of temporal coherence length, the transverse coherence length is defined as the lateral distance along a wave front over which there is a complete dephasing between the two waves. For small θ, according to Fig. 2.6, we have

$$2L_T \approx \frac{\lambda}{\theta}$$

Thus, the transverse coherence length is inversely proportional to the angle sustained by the two source points.

To generalize the above discussion for two discrete source points to a source of finite area, we consider the Young's double-slit interference experiment for the measurement of degree of transverse coherence. For simplicity, we will consider the two-dimensional case first, as shown in Fig. 2.7. Two narrow slits of the same width separated by a distance Δ are placed symmetrically relative to the source. The slits are illuminated by a one-dimensional source of finite width with intensity distribution $I(x)$. Interference is observed at the detector between the two waves selected by the two slits. Considering that electrons are typically emitted from areas about the size of an atom, and the source dimension is much larger than an atom, most electron sources can be considered to be ideally incoherent, that is, to consist of a statistically independent close-packed array of emitters. For such an incoherent extended source, each atomic point-source point generates an independent interference pattern at the detector. What is recorded then is the sum of the intensities of the interference patterns generated by each independent source point. To put this in mathematical form, we take a source point at x, so that the two waves at the slits can be written in the form

$$\phi_1 = A(x)\exp(2\pi ikz)\exp(-2\pi ik_x x')\exp(-i\delta_1)$$
$$\phi_2 = A(x)\exp(2\pi ikz)\exp(2\pi ik_x x')\exp(-i\delta_2).$$

$$(2.42)$$

where δ_1 and δ_2 are the phases of the two waves at the slits, and their difference is determined by the path difference from the source point to the two slits, which is simply $\delta_2 - \delta_1 = (2\pi/\lambda)x\Delta/Z_S$. For what follows, we will consider only the interference recorded at the center of the detector ($x' = 0$). In this case, the arrival time to the detector point is same from each slit, and thus, $\tau = 0$. According to Eq. (2.26), the intensity contribution from the source point at x is given by

$$\langle I(x)\rangle \mathrm{d}x = 2I(x)\big[1 + \mathrm{Re}\{\gamma'_{12}(x)\}\big]\mathrm{d}x$$

where

$$\gamma'_{12}(x) = \exp(2\pi ix\Delta/\lambda Z_S).$$

The overall complex degree of partial coherence is obtained by integrating over all source points, in the form

$$\gamma_{12}(\Delta, 0) = \int\limits_{-\infty}^{\infty} I(x)\exp(2\pi ix\Delta/\lambda Z_S)\mathrm{d}x$$

$$(2.43)$$

Extending this to a two-dimensional source, we have

$$\gamma_{12}\big(\Delta_x, \Delta_y, 0\big) = \frac{1}{I_0}\int\limits_{-\infty}^{\infty} I(x,y)\exp\big[2\pi i\big(x\Delta_x + y\Delta_y\big)/\lambda Z_S\big]\mathrm{d}x\mathrm{d}y$$

$$(2.44)$$

where $I_0 = \int_{-\infty}^{\infty} I(x,y)\mathrm{d}x\mathrm{d}y$ is the integrated source intensity. The above result is known as Van Cittert–Zernike theorem in optics (Born and Wolf 1999). According to this theory, the wave front from a small incoherent source will appear mostly coherent at a large source distance (Z_s). An intuitive explanation of this phenomenon can be provided based on the uncertainty principle. Considering first the range of electron momenta being measured, since the electrons are confined within the source before their emission, their directions are thus determined by the source angle, sustained by the source over the detector point. The range of electron momenta is proportional to the source angle, and thus, it is large for a detector placed close to the source. Intensity recorded by the detector is dominated by the contribution coming from the closest source point, which can be determined with a high degree of accuracy according to the uncertainty principle. For a detector placed far from the source, the momentum distribution is small, and our measurement will no longer be able to distinguish the contributions from specific source points. Consequently, all source points appear to be same, and they contribute almost equally at a large source distance.

For a one-dimensional source of size D, $I(x) = 1$ for $|x| \leq D/2$ and zero elsewhere. The Fourier transform of Eq. (2.43) gives the following result

$$\gamma_{12}(\Delta, 0) = \frac{1}{D} \int\limits_{-D/2}^{D/2} \exp(2\pi i x \Delta / \lambda Z_S) \mathrm{d}x = \frac{\sin(\pi D\Delta / \lambda Z_S)}{\pi D\Delta / \lambda Z_S},$$

and the first zero of $\gamma_{12}(\Delta, 0)$ occurs at $D\Delta / \lambda Z_S = 1$ or

$$\Delta = \lambda Z_S / D = \lambda / (D/Z_S) = \lambda / \theta$$

At the length of $L_T = \lambda / 2\theta$, we have

$$\gamma_{12}(\Delta/2, 0) = \frac{\sin(\pi/2)}{\pi/2} = 0.64.$$

To summarize, the coherence width at the TEM sample (the largest distance between points across the beam at which waves will interfere) is about $\lambda/2\theta$. Highly coherent conditions therefore require that the (ideally incoherent) source subtends a small angle θ at the sample (i.e., a well-collimated beam). This can be achieved by placing a large source (such as a star) at a very large distance, with the result that the detected intensity, although coherent, becomes weak. The coherence width from sunlight is in fact about a tenth of a millimeter. This trade-off between spatial coherence and intensity is of great importance in designing radiation sources, from synchrotrons to electron microscopes and neutron facilities, and is quantified by the quantity of emittance, which is the product of source area and the solid angle subtended by the source at the sample. Low emittance, coupled with high brightness, is a major goal for accelerator physicists and electron-optical designers.

2.10 Electron Refraction and the Refractive Index

Another wave property is refraction. Although we will see in later chapters that the effects of electron multiple scattering by a crystal can lead to multiple electron wave vectors inside the crystal, here we will ignore these effects and consider only the mean "refractive index" effect of the average electrostatic potential of the sample. The electron speeds up as it enters the sample, being attracted to the positive atomic nuclei, thereby gaining kinetic energy and a longer wave vector. We define V_o as the mean sample potential, in volts. Then, the magnitude K of the mean wave vector inside the sample is given by

$$K^2 = k_o^2 + 2meV_o/h^2 = k_o^2 + U_o$$

so that

$$K \approx k_o + U_o/2k_o \tag{2.45}$$

Here, U_o has the dimension of length^{-2}. Values of V_o are tabulated for various crystals [Appendix F and Okeeffe and Spence (1994); Kim et al. (1998); Kruse et al. (2006)].

The change in wavelength as the wave travels across the interface gives the well-known refraction effect, similar to the application of Snell's law in optics. By analogy with optics, an electron refractive index n can be defined using the approximation of Eq. (2.45) and Eqs. A.6 and A.7 in Appendix A:

$$n = K/k_o \approx 1 + U_o/2k_o^2 = 1 + U_o\lambda^2/2 = 1 + \frac{V_o}{\Phi}\frac{1 + e\Phi/m_e c^2}{1 + e\Phi/2m_e c^2}, \tag{2.46}$$

For $\Phi = 200$ kV and $V_o = 20$ V, $n = 1.0000997$. Thus, for high-energy electrons with small wavelength, the refractive index is close to 1. Since the electron refractive index is larger than unity (as for light, but unlike X-rays), an electron beam entering a surface at a low angle is bent toward the normal. The refraction effect is most noticeable when the electron beam is at a glancing angle to the sample surface. In that case, the diffraction condition depends strongly on U_o, but for most transmission microscopy on untilted slab samples, we will see that it is the components of the wave vectors K and k_o normal to sample surface which determine the diffraction condition, and these are approximately equal.

2.11   Wave Propagation

2.11.1   Huygens–Fresnel Principle

We start by asking how electron waves propagate from one plane to another in space. Or, if we know the wave amplitude and phase in plane A, how to calculate the wave in the plane A', which is at a distance of Z downstream, as shown in Fig. 2.8, illustrates.

Mathematically, the above problem can be solved using either a Green's function in quantum mechanics, or Kirchhoff's formulation of the solution of the wave equation. We will not repeat this approach here, but instead will follow the more intuitive approach based on Huygens' principle of wave propagation. Huygens proposed that propagation of waves in space involves the generation of spherical waves at every point on the wave front; these secondary waves propagate in the forward direction and form an envelope that becomes a new wave front. This

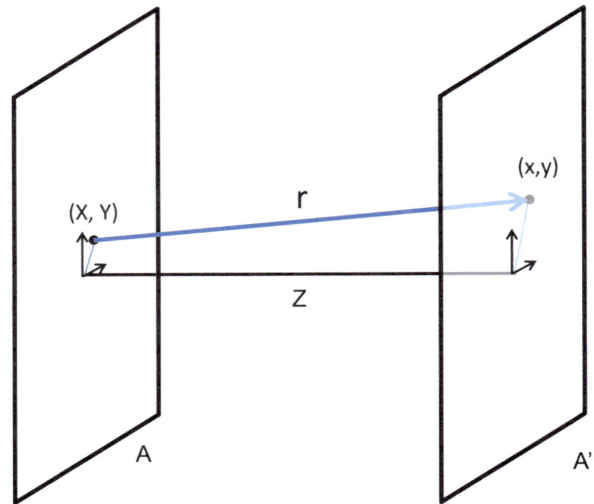

Fig. 2.8 Wave propagation from plane A to plane A', separated by distance Z. The (X, Y) and (x, y) are coordinates on A and A', respectively

principle was further developed by Fresnel. By combining Huygens' principle with the principle of interference, Fresnel was able to explain both the rectilinear propagation of light and also diffraction effects. To achieve this, Fresnel made a number of assumptions about the wave amplitude and phase. Together, the Huygens–Fresnel principle states:

(1) the wave front can be divided into small, finite sized, elements with each element acts the source of a secondary spherical wave;
(2) The secondary waves interfere with each other at point P of a distance Z away according to the principle of superposition;
(3) Each element contributes an amount of wave proportional to its wave amplitude ϕ_o and area dS;
(4) The contribution at point P is inversely proportional to the distance times the wavelength;
(5) There is an obliquity factor of $(1 + \cos \theta)/2$ to the contribution, which is 1 in the forward direction ($\theta = 0$) and zero in reverse direction ($\theta = \pi$); and
(6) The secondary waves vibrate at a quarter of the wavelength behind the primary disturbance.

Putting these points together mathematically, an area dS on the wave front contributes to the wave function at P by the amount

$$d\phi_P = -i\frac{\phi_o}{r\lambda}\left(\frac{1 + \cos \theta}{2}\right)e^{2\pi i k r}dS \tag{2.47}$$

and the total contribution at point P is an integration over the surface area of the wave front:

$$\phi_P = -i \iint_S \frac{\phi_o}{r\lambda} \left(\frac{1 + \cos\theta}{2}\right) e^{2\pi i k r} \, \mathrm{d}S \qquad (2.48)$$

The assumptions that lead to Eq. (2.48) emerge automatically in Kirchhoff's diffraction formula, in the form of an integral solution to the wave equation. A formal derivation of Eq. (2.48) based on Kirchhoff's integral can be found in Longhurst (1986) and Cowley (1995). For a comprehensive treatment and comparison of these results with the first Born approximation and reconciliation of their superficially different dependence on wavelength and the ninety degree scattering phase (factor i in Eq. 2.51), see the 7th edition of Born and Wolf (1999). For electron waves propagating in the forward direction inside an electron microscope, the small-angle approximation, as used in the paraxial equation of Chap. 6, is assumed, and the obliquity factor is taken to be approximately unity.

2.11.2   Propagation of Plane Wave and Fresnel Zones

We now apply the Huygens–Fresnel principle to the propagation of plane waves. The secondary waves are expected to produce another planar wavefront; thus, the result is known. By going through this exercise, we will introduce the concept of Fresnel zones in wave propagation. This concept is often employed in the calculation of electron diffraction from crystals containing defects. It is fundamental for X-ray focusing using zone plates.

A plane wave is represented by

$$\phi = e^{2\pi i \vec{k} \cdot \vec{r}} \qquad (2.49)$$

Here, the wave vector $\vec{k}$ is taken to be along Z as shown in Fig. 2.9. According to (2.47), each point in the planar wave front acts as a secondary source. The contribution from this point depends on the distance of r and $\cos\theta$; both are rotationally symmetrical around the Z-axis. Thus, the integral of (2.48) can be carried out in a circular area of

$$\mathrm{d}\sigma = 2\pi\rho\,\mathrm{d}\rho$$

For reasons that will later become clear, we divide the planar wave front into circular zones numbered from 1 to infinity. Inside each zone, the change in the angle is small. Then, the contribution from this zone is approximately given by

$$\phi_n \approx -i\left(\frac{1 + \cos\theta_n}{2}\right) \int_{r_{n-1}}^{r_n} \frac{1}{r\lambda} e^{2\pi i k r} 2\pi\rho\,\mathrm{d}\rho \qquad (2.50)$$

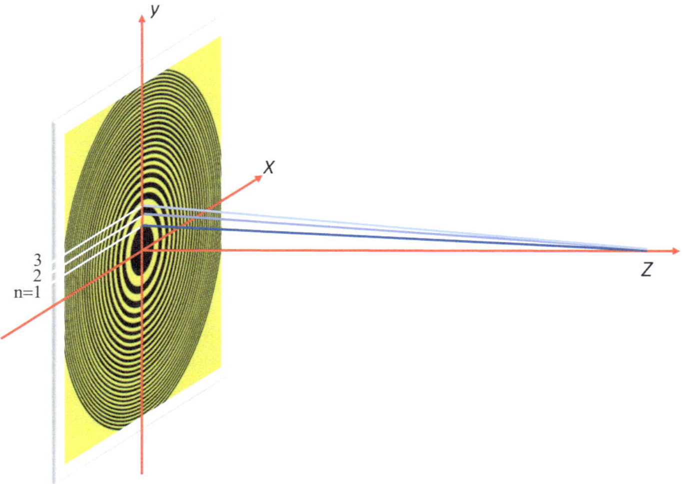

Fig. 2.9 The definition of Fresnel zones

To evaluate Eq. (2.50), we use the relationship

$$r^2 = z^2 + \rho^2$$

and

$$2r\mathrm{d}r = 2\rho\mathrm{d}\rho \tag{2.51}$$

Substituting (2.51) into (2.50), we obtain

$$\phi_n = -\left(\frac{1 + \cos\theta_n}{2}\right)\left(\mathrm{e}^{2\pi i k r_n} - \mathrm{e}^{2\pi i k r_{n-1}}\right) \tag{2.52}$$

Next, we define the zone radius in such a way that the radius of two neighboring zones differs only by half a wavelength such that (also see Fig. 2.9)

$$r_n = z + n\lambda/2 \tag{2.53}$$

The zones defined this way are called Fresnel zones. Putting the above radius into (2.52), we have

$$\phi_n = (1 + \cos\theta_n)(-1)^{n+1}\mathrm{e}^{2\pi i k z}$$

The contribution from the nth zone, ϕ_n, is positive or negative depending on whether n is odd or even. The total wave obtained at the point p from N Fresnel zones is the sum of contributions from all included zones:

$$\phi_P = |\phi_1| + (|\phi_3| - |\phi_2|) + \cdots + (|\phi_{N-1}| - |\phi_{N-2}|) - |\phi_N|$$

Here, N is an even number. When $Z \gg \lambda$, to a very good approximation, the terms inside each bracket are approximately equal and they cancel each other. Then, one may write

$$\phi_P = |\phi_1| - |\phi_N|$$

When N is taken to be a very large number, the angle θ approaches $90°$, and we have $\phi_1 = 2e^{2\pi ikz}$, $\phi_N = -e^{2\pi ikz}$, and

$$\phi_P = \frac{\phi_1}{2} = e^{2\pi ikz} \tag{2.54}$$

Thus, using Huygens–Fresnel principle we have successfully demonstrated that the secondary waves in a plane wave front give rise to another plane wave further away, as we expected. Equation (2.54) also shows that the wave function at the point P is half of the contribution from the first Fresnel zone, while contributions from the rest of zones cancel the other half.

2.11.3   Fresnel Diffraction—The Near-Field Small-Angle Approximation

For wavelengths much smaller than the size of the object (L), the range of diffraction angles, which can be estimated by the uncertainty principle, $\Delta k / k \sim 1/kL$, is small. Thus, the distance between the source point (X, Y) on plane A and the detection point (x, y) on plane A' can be approximated by

$$r = \sqrt{z^2 + (X - x)^2 + (Y - y)^2} \approx z + \frac{(X - x)^2 + (Y - y)^2}{2z}$$

If we assume that the wave distribution on plane A is $\phi_e(X, Y)$, the wave function at a distance z away is then approximately given by:

$$\phi(x, y) = -i \frac{e^{2\pi ikz}}{z\lambda} \iint \phi_e(X, Y) e^{\frac{\pi i}{\lambda z}[(x-X)^2 + (y-Y)^2]} dXdY, \tag{2.55}$$

which is called the Fresnel propagation equation. This equation can be used to explain a class of electron diffraction effects since the electron wavelengths employed in TEM are smaller than an atom. Fresnel diffraction can be observed directly in a TEM around the edge of a sample or an aperture in imaging mode, while Fraunhofer diffraction, to be discussed later, is observed at large distances or in the back focal plane of the objective lens of a TEM.

Next, we examine Fresnel diffraction from the edge of an aperture and the resulting Fresnel integral. These fringes are often used to correct for astigmatism in the electron microscope. To proceed, we consider the case of an opaque aperture with a straight edge. Suppose that this aperture covers half of the space in the x direction and the incident wave is a plane wave propagating along the z direction with $\phi_e(X, Y) = 1$ for $X > 0$, where the aperture is absent. Substituting this into (2.55), we obtain the following wave function:

$$\phi(x, y) = -i\frac{e^{2\pi ikz}}{z\lambda} \int\limits_{0}^{\infty} \int\limits_{-\infty}^{\infty} e^{\frac{\pi i}{\lambda z}\left[(x-X)^2 + (y-Y)^2\right]} dXdY \tag{2.56}$$

The integral in (2.56) is known as the Fresnel integral, which has the general form of (Fig. 2.10)

$$\int\limits_{0}^{s} e^{\frac{\pi i}{2}s^2} ds = \int\limits_{0}^{s} \cos\left(\frac{\pi}{2}s^2\right) ds + i\int\limits_{0}^{s} \sin\left(\frac{\pi}{2}s^2\right) ds = \overline{X} + i\overline{Y}$$

The complex value of the Fresnel integral can be plotted in 2D with the y-axis representing the imaginary part and the x-axis for the real part, and the result defines a curve known as Cornu's spiral. A plot of the Cornu's spiral is shown in Fig. 2.10. The s value is marked on the top part of the curve. The two spirals have an inversion symmetry with the center of two spirals at the s limit of positive and negative infinite. At these two limits, both $\overline{X}$ and $\overline{Y}$ approach the value of $\pm\frac{1}{2}$. Using this for (2.56), we have

Fig. 2.10 Cornu's spiral. The two spirals converge to $\pm(1/2, 1/2)$ as s moves from 0 to $\pm\infty$

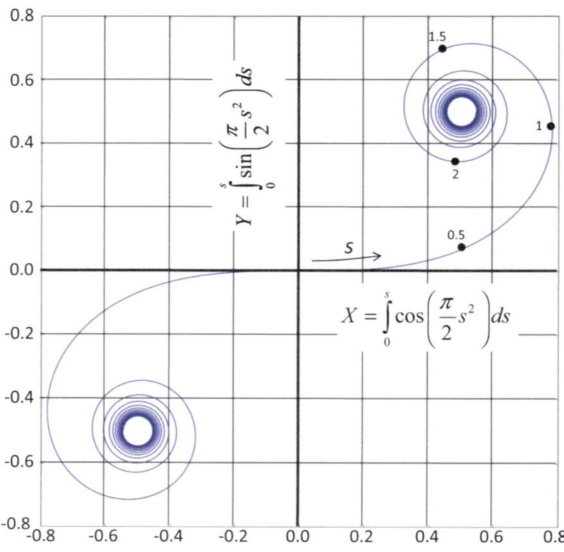

$$\phi(x,y) = c \int_0^\infty e^{\frac{\pi i}{\lambda z}(x-X)^2} \mathrm{d}X = c \int_{s_1}^\infty e^{\frac{\pi i}{2}s^2} \mathrm{d}s = c\left(\int_0^\infty e^{\frac{\pi i}{2}s^2} \mathrm{d}s - \int_0^{s_1} e^{\frac{\pi i}{2}s^2} \mathrm{d}s \right)$$

$$= c\left[\left(\frac{1}{2} + i\frac{1}{2} \right) - (\bar{X} + i\bar{Y}) \right]$$

(2.57)

Here, $s = \sqrt{2(x-X)^2/\lambda z}$ and $s_1 = \sqrt{2/\lambda z}x$. For $x > 0$, s_1 travels along the bottom left part of the spiral. Since the diffracted wave intensity $I(x,y) = |\phi(x,y)|^2$ is proportional to the square of the vector length from s_1 to (1/2, 1/2), the intensity oscillates as s_1 goes around the spiral and approaches—∞. On the other side, for $x < 0$, s_1 travels along the top right part of the spiral and the length decreases continuously. Figure 2.11 shows the intensity calculated from (2.57), and the oscillation expected for $x > 0$ and the monotonic decrease expected for $x < 0$ are both reflected in this plot.

Fresnel fringes are observed in an out of focus electron image. Figure 2.12 shows an example. It is an out of focus image of a sharp W tip (Beleggia et al. 2014). The tip is placed near a negatively biased electrode (-90 V) at a distance away. Electron wave propagation around the W tip creates the Fresnel fringes. They are deflected by the tip electric fields, giving rise to the shape of deflected wave, and the interference of the deflected waves is also shown in Fig. 2.12. The number of Fresnel fringes observed in an out of focus electron image is ultimately determined by the coherence of the electron source. A 300 kV FEI TEM equipped with a field emission gun was used to record the electron image here. The coherence properties of different electron sources and electron illumination are discussed in Chaps. 8 and 10.

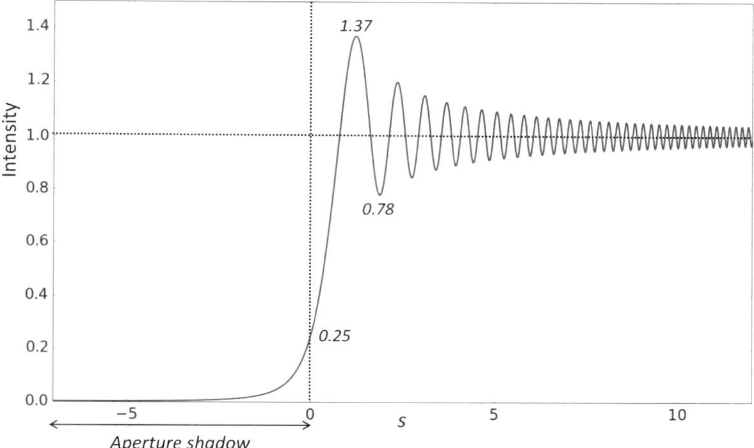

Fig. 2.11 Fresnel diffraction intensity from an opaque aperture with a straight edge at $s = 0$

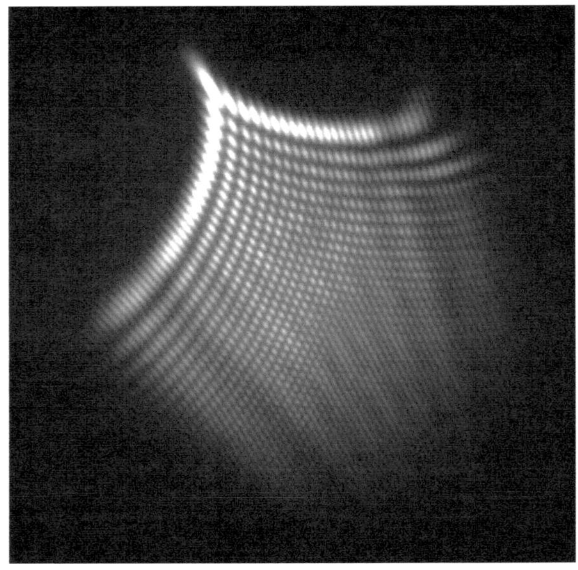

Fig. 2.12 Experimental image of Fresnel fringes and their interference formed by the propagation of coherent electron waves through the electric fields created by a sharp W tip and a negatively biased electrode (Beleggia et al. 2014). The image is approximately 6 mm away from the tip and electrode with a field view of 1 μm. The pattern of interference fringes results from overlap of waves from either side of the tip. (Image provided by Rafal Dunin–Borkowski, Ernst Ruska-Centre for Microscopy and Spectroscopy, Jülich, Germany)

2.11.4 Fraunhofer Diffraction—Far-Field Forward Diffraction

With the electron detector placed in the far field, and the detector size much larger than the extent of the object, to a good approximation for electron diffraction, we have

$$r = \sqrt{z^2 + (X - x)^2 + (Y - y)^2} \approx \sqrt{z^2 + x^2 + y^2} = R$$

and

$$kr = \sqrt{z^2 + (X - x)^2 + (Y - y)^2} \approx kr - \frac{x}{\lambda R}X - \frac{y}{\lambda R}Y$$

Most of electron diffraction occurs in the forward direction with $\cos\theta \approx 1$. Then, the integral of (2.48) can be simplified into the so-called Fraunhofer diffraction equation:

$$\phi_P = \frac{-ie^{2\pi ikr}}{R\lambda} \iint_S \phi_o(X, Y)e^{-2\pi i\left(k_x X + k_y Y\right)}\mathrm{d}X\mathrm{d}Y \tag{2.58}$$

With $k_x = \frac{x}{\lambda R}$ and $k_y = \frac{y}{\lambda R}$.

The Fraunhofer diffraction equation can be compared with the kinematical theory of diffraction in Chap. 4. Both involve the same type of integral known as a

Fourier transform. The difference is that in the kinematic theory of electron diffraction, an assumption is made about the nature of the electron scattering (that there is no multiple scattering), whereas the Fraunhofer diffraction equation simply relates the wave function at the far field to the exit-face wave function across the downstream face of the sample.

References

Beleggia M, Kasama T, Larson DJ, Kelly TF, Dunin-Borkowski RE, Pozzi G (2014) Towards quantitative off-axis electron holographic mapping of the electric field around the tip of a sharp biased metallic needle. J Appl Phys 116:024305

Born M, Wolf E (1999) Principles of optics: electromagnetic theory of propagation, interference and diffraction of light, 7th edn. Cambridge University Press, Cambridge

Cowley JM (1995) Diffaction physics, 3rd edn. Elsevier Science, Amsterdam

Goodman J (2004) Introduction to fourier optics, 3rd edn. Roberts and Company Publishers, Englewood

Griffiths DJ (2004) Introduction to quantum mechanics, 2nd edn. Pearson Prentice Hall, Upper Saddle River

Kim MY, Zuo JM et al (1998) Ab-initio LDA calculations of the mean Coulomb potential V_o in slabs of crystalline Si, Ge and MgO. Phys Status Solidi A 166:445–451

Kruse P, Schowalter M, Lamoen D, Rosenauer A, Gerthsen D (2006) Determination of the mean inner potential in III-V semiconductors, Si and Ge by density functional theory and electron holography. Ultramicroscopy 106:105–113

Longhurst RS (1986) Geometrical and physical optics, 3rd edn. Orient BlackSwan

Mollenstedt G, Ducker H (1955) Fresnelscher Interferenzversuch mit einem Bi-prisma fur Electronenwellen. Naturwissenschaften 42:41

O'Keeffe M, Spence (1994) On the average coulomb potential and constraints on the electron density in crystals. Acta Cryst A50:33–45

Steel WH (1985) Interferometry, 2nd edn. Cambridge University Press, Cambridge

Chapter 3
The Geometry of Electron Diffraction Patterns

An object illuminated by an incident wave field produces a characteristic diffraction pattern, measured as a function of scattering angle, at a large distance from the object. The analysis of such diffraction patterns obtained using X-rays, electrons, or neutrons, all with wavelengths of Ångstrom dimensions or less, forms the core of crystallographic techniques in materials research. This chapter describes the electron diffraction geometry for crystalline samples. We focus on transmission electron diffraction and discuss the generation of electron diffraction patterns, starting from a discussion of Bragg diffraction and followed by an introduction to the real and reciprocal lattices of crystals. Using these concepts, we then develop a full-fledged theory of the electron diffraction geometry covering both point diffraction patterns and CBED. Using this, we introduce the method used to assign a beam direction to each point in the diffraction pattern relative to the crystal coordinates. A thorough grasp of this concept is absolutely essential for any quantitative understanding of electron diffraction from thin crystals.

3.1 Bragg's Law

Our starting point is to consider diffraction by a set of lattice planes. Bragg himself obtained his law by applying the existing law for reflection of light from a thin oil film, which appears colored in reflected sunlight. Because it is thin, it reflects each component wavelength into a different direction. Bragg thought that the clearly visible facets of a good mineral crystal would act as mirrors for X-rays; by treating the crystal planes parallel to a facet as a stack of half-silvered mirrors, he could relate his law to the known Miller indices of the crystal planes, and, by assuming that there was considerable depth of penetration by the X-rays, so that millions of planes contributed, he came to understand that the reflections would be sharp in angle (for one wavelength only). This led to the idea that if polychromatic radiation is used (as Laue and coworkers has previously used, for their discovery of X-ray

© Springer Science+Business Media New York 2017

J.M. Zuo and J.C.H. Spence, *Advanced Transmission Electron Microscopy*,
DOI 10.1007/978-1-4939-6607-3_3

diffraction from crystals), then, unlike a thin oil film illuminated by sunlight, a thicker crystal would act as a monochromator. At the time when Bragg first derived his law, it was not possible to demonstrate or fully understand these thickness effects (the dynamical theory had not been developed); in his classic textbook (Kittel 1976), C. Kittel thus remarks that "Bragg's law can only be justified by (an equation equivalent to our) Eq. 3.14 (below)."

An incident electron wave is partially scattered by atoms, which make up the lattice plane. The nature of scattering is through Columbic interaction, which will be discussed in detail in Chap. 4. Based on atomic scattering, each lattice plane contributes a part to the diffracted wave observed at a far distance. For N number of successive lattice planes, assuming the electron wave is coherent, the sum of scattered waves gives (Fig. 3.1):

$$\phi = \phi_1 + \phi_2 + \phi_3 + \cdots + \phi_N$$

In the limit of weak atomic scattering, only a small part of the incident wave is scattered, thus each lattice plane contributes equally but different in phase. The contributions from two successive lattice planes are related by $\phi_i = \phi_{i-1} e^{i\alpha}$, where α is the phase difference between the two. The phase difference comes from the extra distances traveled by the incident and diffracted wave according to (also see Fig. 3.1)

$$\alpha = 2\pi \left(\frac{SQ}{\lambda} + \frac{QT}{\lambda} \right). \tag{3.1}$$

Strong diffraction is observed when all scattered waves are in phase with $\alpha = 2n\pi$ with n as an integer. For infinite lattice, e.g., the lattice dimension is much larger the incident beam size, by the law of reflection $\theta = \theta'$ and $SQ = QT = d \sin \theta$ with d the lattice plan spacing. The strong diffraction condition then reduces to the well-known Bragg's law

$$2d \sin \theta_B = n\lambda \tag{3.2}$$

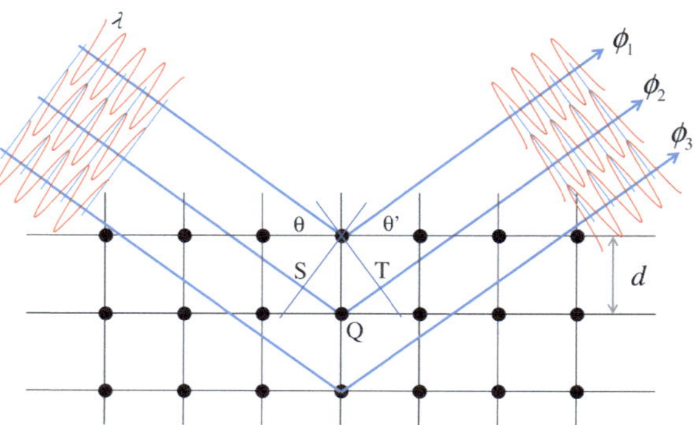

Fig. 3.1 Bragg diffraction by a set of lattice planes illustrated for an incident plane wave of wavelength λ at angle θ to the plane and interplanar distance d

Here θ_{B} is the Bragg angle. With the short electron wavelength, θ_{B} is small and

$$2d \sin \theta_{\mathrm{B}} \approx 2d\theta_{\mathrm{B}} = n\lambda \qquad (3.3)$$

Thus, the total scattering angle between direct and first-order scattered beam (twice the Bragg angle) is just equal to the wavelength divided by d. Later in Eq. 3.6, we will see that this holds for any reflection if d is replaced by the crystallographic d-spacing d_{hkl}. At 200 kV and $d = 1$ Å, θ_{B} is 12.5 mrad.

The weak scattering approximation, on which the Bragg's law is based, often breaks down for high-energy electrons when there are more than a few lattice planes involved in scattering. In such cases, the amplitude and phase of the scattered waves change among successive lattice planes, leading to the breakdown of Bragg's law. This topic will be further discussed in Chap. 5 on dynamic diffraction.

In transmission electron diffraction through a thin crystalline slab, because of the small Bragg angle, the diffracting lattice planes are closely parallel to the incident beam, and the crystal thickness is comparable to the electron beam size. Then this law of reflection does not hold. Rather, strong diffraction condition is obtained with

$$\theta + \theta' \approx 2\theta_{\mathrm{B}}$$

using the small-angle approximation. Thus, a strong diffraction spot can be observed in transmission electron diffraction even though the incident beam is not at the Bragg angle.

In both the Bragg and off-Bragg diffraction cases, the diffracted beam is at, or very close to, twice the Bragg angle to the incident beam, as shown in Fig. 3.2.

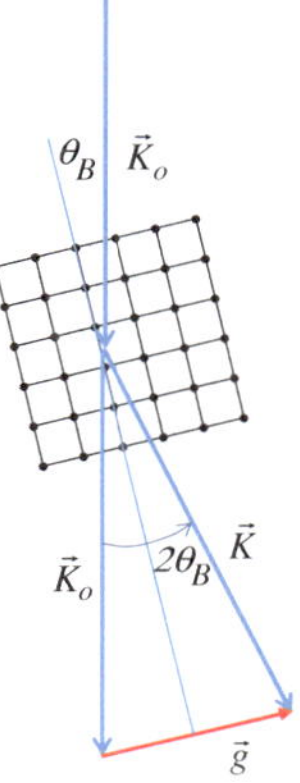

Fig. 3.2 Bragg diffraction in transmission geometry, where the incident and diffracted plane waves are represented by their wave vectors, $\vec{K}_o$ and $\vec{K}$, respectively

3.2   Laue Diffraction Condition

Next, we express the Bragg's law in vector form. The use of vectors then leads to a general mathematical framework for the reciprocal lattice for describing the diffraction geometry and the diffraction patterns.

Consider the diffraction geometry in Fig. 3.2, the incident and diffracted beams are defined by their respective wave vectors inside the crystal, $\vec{K}_o$ and $\vec{K}$. At the Bragg condition, we have

$$\vec{S} = \vec{K} - \vec{K}_o = \vec{g} \tag{3.4}$$

where $\vec{g}$ is a vector. Its direction is taken along the lattice plane normal direction and its length equals the inverse d spacing, e.g., $|\vec{g}| = 1/d$. The difference between incident and diffracted wave vectors, $\vec{S}$, is called the scattering vector. Since $|\vec{K}_o| = |\vec{K}| = 1/\lambda$, $|\vec{K} - \vec{K}_o| = 2\sin\theta_B/\lambda = |\vec{g}| = 1/d$ and thus the above equation is equivalent to Bragg's law. It is known as the Laue diffraction condition after Max von Laue (1879–1960).

3.3   Lattice d-Spacing and Crystal, Real, and Reciprocal Lattices

The crystal periodicity is defined by the crystal unit cell vectors, $\vec{a}$, $\vec{b}$, and $\vec{c}$, or according to their length, a, b, and c and angle $\alpha = \vec{b}^{\wedge}\vec{c}$, $\beta = \vec{c}^{\wedge}\vec{a}$, and $\gamma = \vec{a}^{\wedge}\vec{b}$. The three vectors together can be used to represent any vectors in the crystal in the so-called crystallographic coordinate. Each unit cell vector represents one of the coordinate axes. In this coordinate, the lattice plane (Fig. 3.4 for example) closest to the origin intercepts the three axes at x, y, and z. The Miller indices, (h, k, l), used to index the lattice plane, are obtained by following rules (Fig. 3.3):

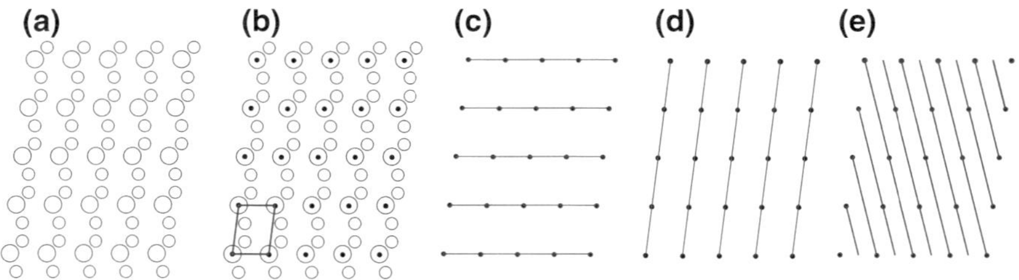

Fig. 3.3 Concepts of crystal lattice and lattice planes. **a** a hypothetical 2D crystal, **b** the identification of crystal lattice and the crystal unit cell, **c–e** three different lattice planes shown as lines in 2D

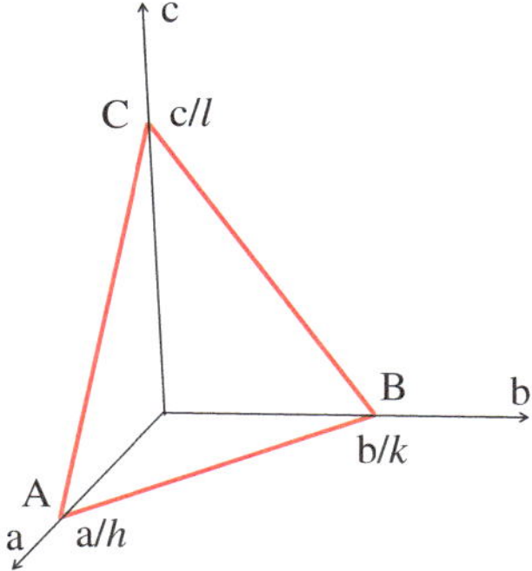

Fig. 3.4 Lattice plan and its Miller indices. The Miller indices are determined from the intercepts of the plane closest to origin

(1)   Determine the intercepts of the plane closest to the origin along the three crystallographic axes;

(2)   Record the intercepts in fractions of the unit cell dimensions;

(3)   Take the reciprocals of the intercepts and reduce them to the lowest integers.

For example, the lattice plane (3,1,1) has intercepts (1/3,1,1).

Appendix C describes the relationship between d_{hkl} and the unit cell dimensions for the various crystal systems. In case of a cubic crystal, the *d*-spacing can be simply calculated using

$$d_{hkl} = a/\sqrt{h^2 + k^2 + l^2}. \tag{3.5}$$

The $\vec{g}$ vectors of a crystal constitute the so-called reciprocal lattice. To introduce this concept, we will define the reciprocal lattice using three scattering vectors with Miller indices (100), (010), and (001), according to

$$\vec{a}^* = \vec{g}_{100}, \quad \vec{b}^* = \vec{g}_{010}, \quad \vec{c}^* = \vec{g}_{001}.$$

The $\vec{g}$ vectors are normal to at least two real space lattice vectors in the lattice plane. For example, $\vec{a}^*$ is normal to $\vec{b}$ and $\vec{c}$ in the (100) plane. On the other hand, $\vec{a}^* \cdot \vec{a} = 1$ because the projection of $\vec{a}$ along the (100) plane normal equals to $1/|\vec{g}_{100}|$. Using the same reasoning, the following relationships can be established between the real and reciprocal lattices:

$$\vec{a}^* \cdot \vec{a} = 1, \vec{a}^* \cdot \vec{b} = 0, \vec{a}^* \cdot \vec{c} = 0$$
$$\vec{b}^* \cdot \vec{a} = 0, \vec{b}^* \cdot \vec{b} = 1, \vec{b}^* \cdot \vec{c} = 0 \tag{3.6}$$
$$\vec{c}^* \cdot \vec{a} = 0, \vec{c}^* \cdot \vec{b} = 0, \vec{c}^* \cdot \vec{c} = 1$$

It can be shown that these relationships are satisfied by the following reciprocal lattice basis vectors, defined by the cross-products of the real-space lattice vectors, in the form

$$\vec{a}^* = \vec{b} \times \vec{c}/V_c, \quad \vec{b}^* = \vec{c} \times \vec{a}/V_c, \quad \vec{c}^* = \vec{a} \times \vec{b}/V_c. \tag{3.7}$$

where $V_c = \vec{a} \cdot (\vec{b} \times \vec{c})$ is the crystal unit cell volume.

In the reciprocal lattice, a vector from the lattice origin to a lattice point is given by

$$\vec{r}^*_{hkl} = h\vec{a}^* + k\vec{b}^* + l\vec{c}^*, \tag{3.8}$$

where h, k, and l are integers. Considering the lattice plane in Fig. 3.4, $\vec{r}^*_{hkl}$ is normal to the (hkl) plane because it is perpendicular to two vectors in the plane, $\overrightarrow{AC} = \vec{c}/l - \vec{a}/h$ and $\overrightarrow{AB} = \vec{b}/k - \vec{a}/h$. Further, $\vec{r}^*_{hkl} \cdot \vec{a}/h = 1$, so the lattice plane distance from the origin to the plane is the inverse of the length of $\vec{r}^*_{hkl}$. The length and direction are exactly the same as $\vec{g}_{hkl}$, thus we have

$$\vec{g}_{hkl} = \vec{r}^*_{hkl} = h\vec{a}^* + k\vec{b}^* + l\vec{c}^*. \tag{3.9}$$

In summary, the direction of a reciprocal lattice vector corresponds to the normal of corresponding (hkl) lattice planes. The length of the reciprocal lattice vector is equal to the reciprocal of the interplanar spacing for those (hkl) lattice planes. Since scattering vectors belong to the reciprocal lattice, we can now construct our scattering diagram using the reciprocal lattice for the crystal.

3.4   Transmission Electron Diffraction Patterns

A transmission electron diffraction pattern is formed when a collimated electron beam traverses a thin sample, giving rise to the diffraction spots recorded on a distant detector. Figure 3.5 shows the construction for such a diffraction pattern. The incident wave is represented by its wave vector inside the crystal $\vec{K}_o$. Following the convention established by Paul Peter Ewald (1888–1985), $\vec{K}_o$ is drawn toward the origin; its length defines the radius of the Ewald sphere, and its direction defines the incident beam orientation in the crystal reciprocal lattice. The radius can be taken as $K_o \approx 1/\lambda$. All elastically diffracted waves must have the same wavelength because of energy conservation, and thus fall onto this sphere. Reciprocal lattice points intercepting the Ewald sphere satisfy the Laue diffraction condition, and thus Bragg's law. The incident and diffracted beams make the same angle θ with the lattice plane at the Bragg condition ($\theta = \theta_B$).

In the so-called zone-axis orientation, the intersection of two sets of lattice planes defines the zone axis, which is a real space lattice vector, $[u, v, w]$. (We use square brackets to denote a direction, and curved brackets to denote planes). For a given zone axis, a reflection of (h, k, l) that is allowed in the diffraction pattern must satisfy:

Fig. 3.5 Ewald sphere
construction for transmission
electron diffraction

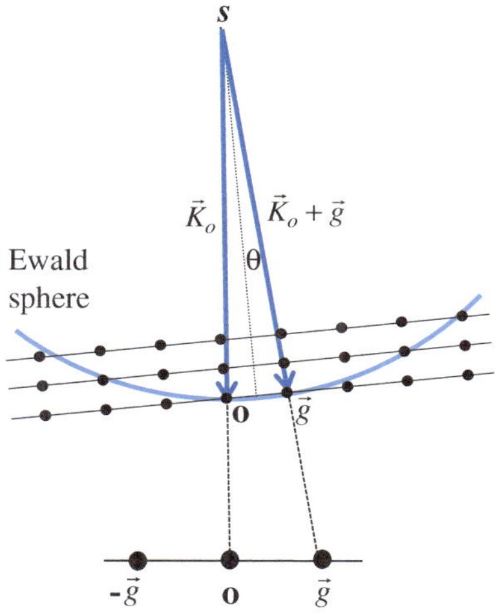

$$hu + kv + lw = n$$

Here n is an integer. Each n denotes a planar section of the reciprocal space, which is perpendicular to the zone axis. In diffraction, these planar sections are called the Laue zones. Zero-order Laue zone (ZOLZ) refers to $n = 0$, while others are called high-order Laue zones (HOLZ). The ZOLZ passes through the reciprocal lattice origin, normal to the beam. A HOLZ is any other reciprocal lattice plane parallel to this, not passing through the origin. (Other definitions are used in the crystallographic literature, however the preceding conforms to common usage in electron crystallography.)

An example of a single-crystal electron diffraction pattern is shown in Fig. 3.6. The lattice geometry of the ZOLZ is described by two, nonparallel, reflections $\vec{g}_1$ and $\vec{g}_2$, which are selected as the closest ones to the transmitted beam. The projected lattice of the HOLZs is the same as the ZOLZ, however, the lattice is shifted relative to the ZOLZ in general. The shift of HOLZs is described by the projection of an additional reflection $\vec{g}_3$ onto the ZOLZ, which is independent of $\vec{g}_1$ and $\vec{g}_2$, and it is also selected as the one closest to the origin. Once these three reflections are identified, all reflections can be indexed in integers using these three vectors as the basis:

$$\vec{g} = n\vec{g}_1 + m\vec{g}_2 + l\vec{g}_3 \qquad (3.10)$$

Experimentally, for a crystal of known structure, any two reflections in the ZOLZ can be used to identify the zone axis and the approximate crystal orientation by using the following relationship

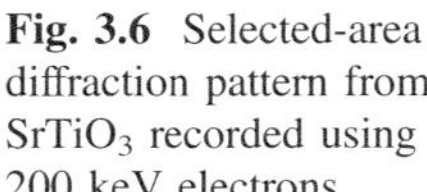

Fig. 3.6 Selected-area diffraction pattern from SrTiO$_3$ recorded using 200 keV electrons

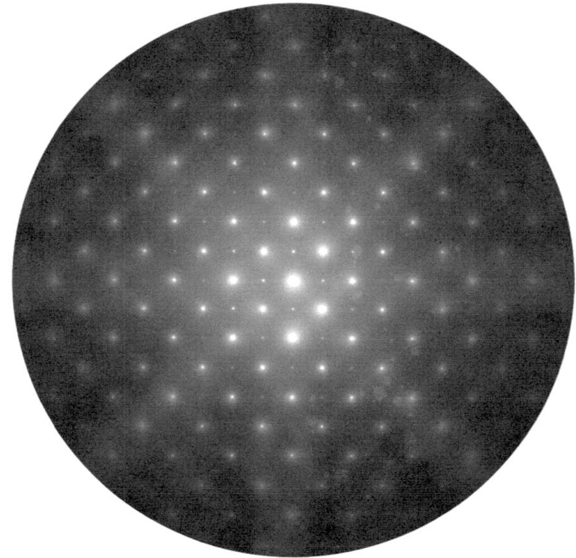

$$\vec{Z} = \vec{g}_1 \times \vec{g}_2 = \begin{pmatrix} \vec{a} & \vec{b} & \vec{c} \\ h_1 & k_1 & l_1 \\ h_2 & k_2 & l_2 \end{pmatrix} = u\vec{a} + v\vec{b} + w\vec{c}. \tag{3.11}$$

where u, v, and w are also integers which define a lattice vector. The crystallography convention is to reduce it to the shortest lattice vector by dividing the three numbers by a common integer.

3.5   Excitation Error

When constructing electron diffraction patterns following the above methods, it should be kept in mind that not all reciprocal lattice points intersecting the Ewald sphere give rise to diffracted beams. Only these close to the incident beam within a small angle of $\sim$100 mrad do so, for high-energy electrons (80–300 keV). Also, a reciprocal lattice point close to the Ewald sphere can give rise to a diffracted beam in the diffraction pattern. For the reasons discussed in the next two chapters, we will learn that diffraction intensity is strongly influenced by deviations from the Bragg condition as well as by the diffraction angle.

In order to determine which reflections are at the Bragg condition, which are not and by how much, we need a quantitative way of specifying the incident beam direction and its relationship to the lattice plane, as well the direction of diffracted beams for each reciprocal lattice point close to the Ewald sphere, for later use in the development of the kinematic and dynamic theory of diffraction intensities, as well as for use in computer programs. We consider here only thin crystals in the form of a parallel-sided slab, whose surface normal is approximately antiparallel to the beam direction. Samples of similar shape, but inclined to the beam, are considered in later chapters.

To start, we first examine the Ewald sphere construction in detail for a specific diffracted beam, as illustrated in Fig. 3.7. Again, by the requirement of elastic scattering, since the diffracted wave vector must fall on the Ewald sphere defined by the incident beam $\vec{K}_\mathrm{o}$ and satisfy the Bragg condition for $\vec{g}$, the diffracted beam is simply given by $\vec{K}_\mathrm{o} + \vec{g}$. For an incident beam to the right $(\vec{K}_\mathrm{o}')$, the diffracted beam $(\vec{K}_\mathrm{o}' + \vec{g})$ also tilts to the right (recall that the strong diffraction condition). The vector of $\vec{K}_\mathrm{o}' + \vec{g}$ may not then fall on the Ewald sphere, however. The distance of $\vec{K}_\mathrm{o}' + \vec{g}$ away from the Ewald sphere defines the so-called excitation error S_g. For a parallel-sided crystal, the direction of S_g must lie along the surface normal direction of the sample surface in order to keep the electron momentum continuous at the crystal surface. The sign, as well as its amplitude, of S_g can change.

Thus, the direction of a diffracted beam not exactly at the Bragg condition is given by $\vec{K}_\mathrm{o} + \vec{g} + \vec{S}_g$. It must satisfy the elastic scattering condition with

$$\left| \vec{K}_\mathrm{o} + \vec{g} + \vec{S}_g \right|^2 = \left| \vec{K}_\mathrm{o} \right|^2. \qquad (3.12)$$

If we take the surface normal to be perpendicular to $\vec{g}$ and take the component of $\vec{K}_\mathrm{o}$ along the surface normal direction to be same as K_o due to the small Bragg angle for electron diffraction, then

$$\left| \vec{K}_\mathrm{o} + \vec{g} + \vec{S}_g \right|^2 \approx \left| \vec{K}_\mathrm{o} + \vec{g} \right|^2 + 2K_\mathrm{o} S_g \qquad (3.13)$$

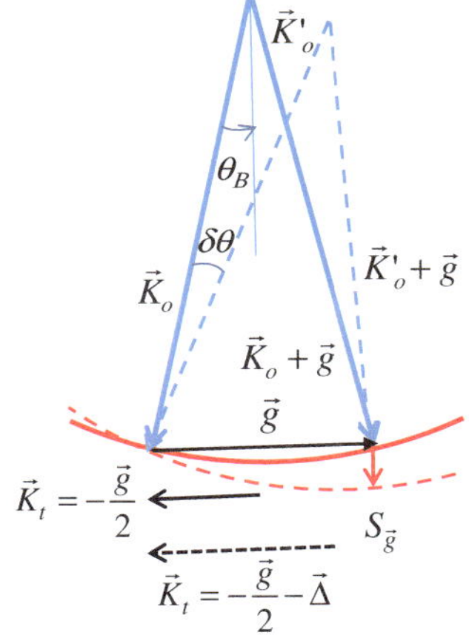

Fig. 3.7 Definition of excitation error (shown positive) for an incident beam of the Bragg condition, and its diffracted beam (represented by the *dash lines*)

Combining Eqs. (3.12) and (3.13), we thus obtain

$$S_g \approx \left(K_o^2 - \left|\vec{K}_o + \vec{g}\right|^2\right)/2K_o. \tag{3.14}$$

Equation (3.14) with $S_g = 0$ corresponds to Ewald condition (Eq. 3.4).

To calculate S_g using Eq. (3.14), we need to specify the direction of the incident beam $\vec{K}_o$ in the same reciprocal lattice coordinate as $\vec{g}$. For a crystal of known structure, the incident beam direction can be experimentally obtained by first determining the initial orientation of the crystal, and then using the degree of tilt (or rotation) applied to the crystal in the sample holder to determine the incident beam direction. Details of this procedure will be discussed in Chap. 10. For the discussion of electron diffraction, it is simpler to define the incident beam direction in a zone-axis coordinate and use the tangential component of the incident wave vector $\vec{K}_o$ in the zero order Laue zone, $\vec{K}_t$, to define the incident beam. Because the zone axis can be identified from the recorded experimental diffraction pattern, this method does not require a knowledge of the crystal initial orientation and its rotation.

Taking the zone-axis direction as $\hat{z}$, then in the zone-axis coordinate we have

$$\vec{K}_o = K_z\hat{z} + \vec{K}_t \tag{3.15}$$

Substituting this into Eq. (3.14) for a reflection belonging to ZOLZ, we obtain

$$S_g = \left(K_t^2 - \left|\vec{K}_t + \vec{g}\right|^2\right)/2K_o \tag{3.16}$$

The S_g is positive when the length of $\left|\vec{K}_o + \vec{g}\right|$ is shorter than K_o, thus its direction is into the sample at the beam entrant surface, corresponding to the case illustrated in Fig. 3.7. S_g is positive if the Ewald sphere enclosed the lattice point, otherwise negative.

To see how the excitation error changes as we deviate from the Bragg condition, we compare the two incident beam directions shown in Fig. 3.7. For $\vec{K}_o$, take $K_t = -g/2$ and $K_o = 1/\lambda$, we have

$$S_g = \lambda\left(\left(-\frac{g}{2}\right)^2 - \left(\frac{g}{2}\right)^2\right)/2 = 0$$

For the incident beam of $\vec{K}_o'$ to the right of $\vec{K}_o$, its component along the reciprocal lattice vector $\vec{g}$ is longer in length by an amount of Δ and

$$K_t' = -g/2 - \Delta,$$

where Δ is positive. The deviation from the Bragg angle is simply $\delta\theta \approx \Delta/(1/\lambda)$, which is positive corresponding to a deviation towards $\vec{g}$. Substituting K'_t into Eq. (3.14), we have

$$S_g = \lambda\left(\left(-\frac{g}{2}-\Delta\right)^2 - \left(\frac{g}{2}-\Delta\right)^2\right)/2 = g\Delta/(1/\lambda) = g\delta\theta. \tag{3.17}$$

Thus S_g and $\vec{K}_t$ are two related quantities commonly used to specify the beam direction in electron diffraction—the component $\vec{K}_t$ and the excitation error S_g. The intersection of the Ewald sphere with the ZOLZ forms a circle of lattice points (the Laue circle)—we will see in Sect. 3.9 that in three dimensions $\vec{K}_t$ is a vector which runs from the origin to the center of this circle. The deviation angle $\delta\theta$ can be separately determined with the help of Kikuchi lines in the case of point diffraction patterns similar to Fig. 3.6 or in case of CBED as discussed below.

The excitation error S_g has a simple interpretation in terms of the uncertainty principle. For components in the $\vec{z}$ direction, this becomes $\Delta K_z \Delta_z = 1$ (since $|K| = 1/\lambda$). The elastic scattering event is known to occur within a distance $\Delta z = t$ (the thickness of the sample), so the spread in z components of the distribution of elastically scattered wave vectors must be $\Delta K_z = 1/t = \Delta S_g$, which is just equal to the width of the kinematic rocking curve. In the two-beam theory of Chap. 5, an effective excitation error is introduced with a minimum value of the reciprocal of the extinction distance.

3.6   Kikuchi Lines and Their Geometry (Kinematic)

Kikuchi lines arise from the elastic scattering of inelastically scattered electrons. A discussion of the origin of Kikuchi lines can be found in Hirsch et al. (1977). The direction of inelastically scattered electrons is defined by the momentum transfer, which can be in any direction. Inelastic scattering gives rise to the intensity between Bragg peaks in the recorded point diffraction pattern. It has an angular distribution which is peaked around the electron propagation direction. The extent of angular distribution depends on the types and localization of inelastic scattering (Egerton (2011). Intuitively this can be understood based on the uncertainty principle. The volume impacted by inelastic scattering, or localization, can be determined experimentally in principle, for example by electron energy loss spectroscopy. For the localization of Δx, we have $\Delta k \approx 1/\Delta x$. Inelastic scattering involving optical phonons tends to be more localized than acoustic phonons, while excitation of core electrons is more localized than excitation of valence electrons. Inelastic scattering occurs throughout the volume of the sample under the electron beam illumination. For electron diffraction, we may think of the inelastically scattered electrons as originating from an electron source with a large convergence angle. An X-ray pattern with a very large convergence angle is known as a Kossel pattern. If we are only concerned with the geometry of the lines, we may treat Kikuchi patterns as a kind of "thickness averaged" electron Kossel pattern, since the inelastic electrons are generated continuously throughout the sample. It is also customary to neglect the small change in electron energy due to the energy loss which results from inelastic scattering.

Kikuchi patterns are observed for reasonably thick crystals in electron diffraction. The pattern is consists of pairs of lines (Fig. 3.8 is an example). The distance between the pair is same as the length of a reciprocal lattice vector. Some of the pairs appear as dark (deficient) and bright (excess) lines within the diffuse background. The deficient line is close to the direct beam where the inelastic scattering background is strong. It is deficient because the net redistribution of intensity by Bragg scattering is higher from low to high angles. Kikuchi patterns are important for electron diffraction for following reasons:

(1)   The pattern provides a sense of tilt direction because Kikuchi lines are fixed to the crystals; it moves as crystal rotates;
(2)   The pattern allows an accurate determination of crystal orientation and the incident beam direction;
(3)   Kikuchi lines, mapped over an area of the stereogram, provide a navigation map for the reciprocal space;
(4)   The pattern can be used to determine the sign and magnitude of excitation error or deviation from the Bragg condition $(\delta\theta)$;
(5)   The pattern can be used to determine the crystal symmetry.

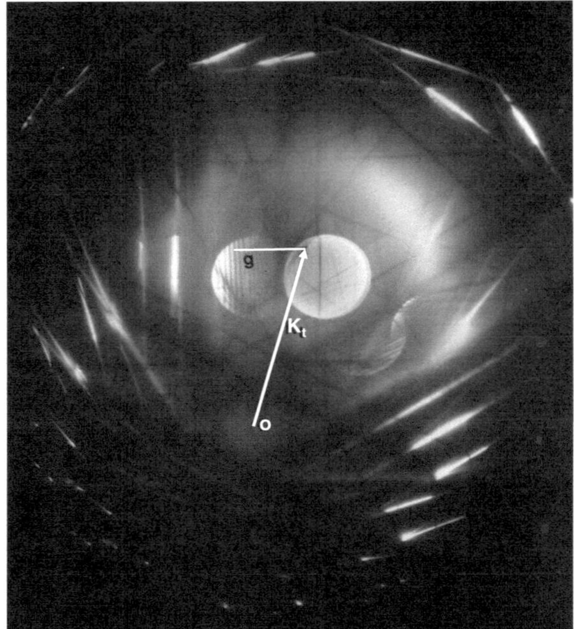

Fig. 3.8 A two-dimensional, off-zone axis, electron diffraction pattern recorded near the [111] zone axis in silicon at 120 kV using a convergent beam for CBED. The Kikuchi lines are visible in the background. The zone axis center is marked by *O*. (Courtesy of C. Deininger and J. Meyer, unpublished work)

The geometry of Kikuchi patterns and their dependence on the crystal structure and orientation can be understood based on the angular distribution of inelastically scattered electrons and Bragg diffraction of these electrons, as illustrated in Fig. 3.9. In the figure, the directional distribution of inelastically scattered electrons is represented by the inelastic illumination cone. Two Kossel cones are drawn on the two sides of the lattice plane for the lattice plane of $\vec{g}$. The cone has a tip to bottom edge distance of $1/\lambda$ and angle of θ_B to the lattice plane. Thus, all incident and diffracted beams satisfying the Bragg condition of $\vec{g}$ all reside on the cone. A Kikuchi line pair is given by the projection of these cones onto the observation plane. In an experiment, the observation plane is approximately parallel to the ZOLZ for the nearest zone axis. We therefore assume this ZOLZ as the observation plane throughout this book. In high-energy electron diffraction, the length of the side of the cone is very large compared to the length of a typical $\vec{g}$ vector, so that only a very small portion of the cone is observed, which is the case in transmission electron diffraction using a medium or large camera length ($L \sim 200$ mm for a 20 mm sized detector gives a detection angle of 100 mrad or 5.7°). Thus the projection of the cone may be approximated by a straight line, and this straight line is perpendicular to the $\vec{g}$ vector, as shown in Fig. 3.9. The pair of Kikuchi lines is formed by diffraction from both sides of the lattice plane. As illustrated, diffraction of incident beams on the line AB in the illumination cone gives the Kikuchi line C'D', while the line DC gives the line B'A' in the diffraction pattern. The intensity of Kikuchi line is determined by relative intensity of the two lines in the inelastic illumination cone and their transmission. Because the Kossel cones are fixed at the Bragg angle to the lattice plane, it rotates with the plane and consequently the Kikuchi lines move as crystal rotates.

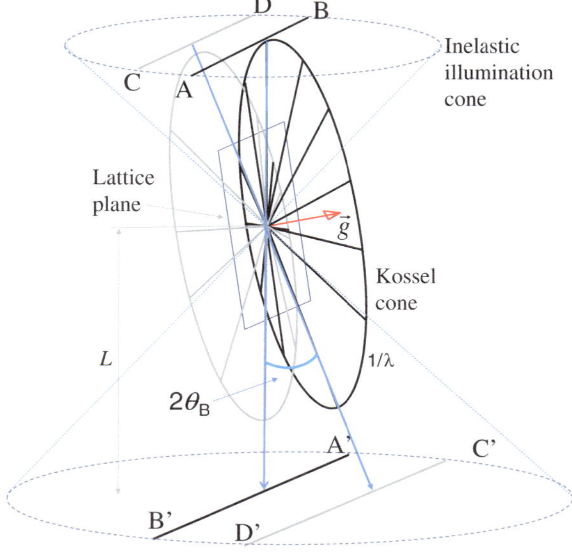

Fig. 3.9 Kikuchi line diffraction geometry. The inelastic illumination cone represents the range of inelastically scattered electrons in their directions. The Kossel cones are drawn on both sides of the lattice plane, and intersections of the Kossel cones with the detector give rise to a pair of Kikuchi lines

Next, we show that the center of two pairs of Kikuchi lines defines the zone-axis center in the diffraction pattern and show how they are constructed. To start, we consider two reflections from two sets of lattice planes, $\vec{g}$ and $\vec{h}$. The intersection of the two lattice planes in general defines the zone-axis direction. The lattice plane divides two Kossel cones, the plane normal (direction of $\vec{g}$ or $\vec{h}$) is perpendicular to the ZOLZ. Its intersection with ZOLZ forms a line, which we will call the center line. Two such center lines belonging to $\vec{g}$ and $\vec{h}$ are shown Fig. 3.10, they are labelled as LP$_\mathrm{g}$ and LP$_\mathrm{h}$. Their intersection marks the zone-axis direction (marked as ZC in Fig. 3.10). The intersection of the Kossel cones occurs at half the distance (θ_B) on both sides of the center line. The distance between a pair of Kikuchi lines corresponds to $2\theta_\mathrm{B}$. The whole ZOLZ pattern can be constructed following this method.

The interpretation of electron diffraction patterns is performed using simulations of Kikuchi patterns. For such a purpose, we need to develop a quantitative framework. First, we consider the coordinate system to be used in more detail. We take the $\vec{z}$ axis in the direction of the ZOLZ zone axis and the $\vec{x}$ axis in the direction of one of the reciprocal lattice vectors. The $\vec{y}$ axis runs in direction $\vec{z} \times \vec{x}$. The origin is fixed at the zone-axis center (also the center of the Laue circle as described in the last section) as shown in Fig. 3.11.

With the above coordinate system in place, we now return to Eq. (3.16) in order to obtain a description of Kikuchi lines. The Kossel cone is defined by $S_g = 0$. On expanding Eq. (3.16) in terms of components along x, y, and z directions (z is along the zone axis), we find that the geometric (kinematic) trajectory of the Kikuchi line $\vec{g}$ is given by (Zuo 1992)

$$K_y = -\frac{g_x}{g_y}K_x + \frac{g_z}{g_y}K_z - \frac{g^2}{2g_y} \tag{3.18}$$

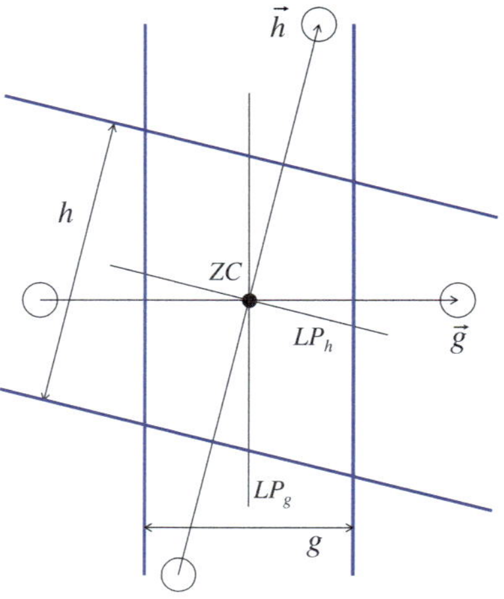

Fig. 3.10 Zone axis center (ZC) located by two pairs of Kikuchi lines for reflections $\vec{g}$ and $\vec{h}$

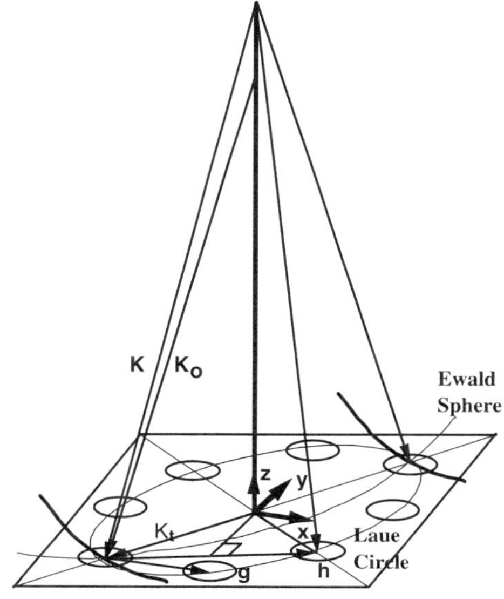

Fig. 3.11 Zone axis coordinate. The plane represents the ZOLZ. Intersection of the Ewald sphere defined by the incident beam (K inside the crystal and Ko outside the crystal) gives rise to the Laue circle. The incident beam direction is defined by the tangential component of K in the ZOLZ (Kt) drawn from the zone center to the incident beam

with respect to an origin at the zone center. Here, the z component of K is given to a good approximation by

$$K_z \approx \sqrt{K^2 - K_{xc}^2 - K_{yc}^2} \tag{3.19}$$

where K_{xc} and K_{yc} are the x and y components of a vector drawn from the zone center to the center of diffraction pattern. This approximation is equivalent to approximating a section of a cone by a straight line. It is also used in the HOLZ simulation program listed in Appendix 5 of Spence and Zuo (1992).

To demonstrate how Eq. (3.18) can be related to the intuitive approach we discussed before, we use the systematic case in as an example. The x-axis is taken at $90°$ from the systematic row direction, thus for the $+g$ reflection, we have $g_x = g_z = 0$, $g_y = g$ and

$$K_y = -\frac{g}{2}$$

For $-g$, $g_y = -g$ and

$$K_y = \frac{g}{2}$$

The distance between the two lines simply is g. It should be noted that line equation of (3.18) marks the deficiency line corresponding to the transmitted beam that satisfies the Bragg condition, e.g., $K_y = -g/2$ gives the deficiency line for reflection g.

3.7 Diffraction Pattern Indexing

Diffraction pattern indexing is performed by comparing the measured d-spacings referring to the powder diffraction files or a list of d-spacings generated using a crystallographic utility program. The angle between $\vec{g}_1$ and $\vec{g}_2$ is useful for indexing single crystal diffraction patterns, especially in case of a high index zone axis or a crystal with low symmetry. For crystals with unknown structure, the three dimensional unit cells can be reconstructed using a minimum of two zone-axis diffraction patterns with a known rotation angle, or from a single diffraction pattern with HOLZ reflections. Recent developments in diffraction tomography allow reconstruction of three-dimensional reciprocal space as well as diffraction intensities (Kolb et al. 2011). A reduced unit cell can be identified from these diffractions which then can be converted to the Bravais lattice (Zuo 1993).

The distance between the direct and diffracted beam is determined by the intersection of these beams with a detector placed at the distance L (camera length) away according to:

$$D = L \tan 2\theta_B \tag{3.20}$$

For small Bragg angles, one can use the approximation of $\sin\theta \approx \tan\theta \approx \theta$. This gives the relationship of

$$d \approx L\lambda/D \tag{3.21}$$

The d-spacing thus can be obtained by measuring the length of D in an experimental diffraction pattern using the above equation. Experimentally, the camera length L can be determined using a sample with known d-spacing, while the electron wavelength or acceleration voltage can be calibrated using high-order Laue zone (HOLZ) lines in CBED patterns.

During crystal rotation with a fixed incident beam, the diffracted beams are expected to stay at their positions because of the constant diffraction angles. Deviation from this arises in case of imperfect crystals due to strain or small crystals from the rotation of shape factor (to be discussed in the next Chapter). The intensity of diffracted beams changes as the diffracted beam goes in and out of the Bragg condition. The diffraction pattern geometry changes when the crystal is rotated from one zone axis to another. We can imagine that the Kikuchi lines are firmly connected to the crystal (and so rotate with it), while the Bragg spots are fixed to the screen, but fade as excitation errors increase.

Since in general it is difficult to align the crystal so the incident beam is exactly along the zone axis, a zone-axis diffraction pattern is loosely referred to the diffraction patterns recorded with the incident beam close to zone axis. The procedure for indexing a diffraction pattern is described in several texts (Hirsch et al. 1977; Lorretto 1994; Williams and Carter 2009). Appendix D of this book shows the indexed reciprocal lattices for many of the commonly encountered Bravais lattices and orientations. Appendix F gives the structure and cell constants for many commonly encountered crystals. These information are useful for indexing diffraction patterns as well as for students to get familiar with electron diffraction patterns.

3.8   One-Dimensional (Systematics) CBED

CBED is obtained by focusing the electron beam into a probe onto the sample. Instead of a parallel beam, the incident beam is now convergent. The convergence angle is determined by the aperture and the electron optics, which will be discussed in Chap. 10. Because of the convergent beam, the diffraction spots broaden into discs as illustrated in Fig. 3.12.

Figure 3.12 shows a simplified ray diagram for a CBED pattern in the systematics or "one-dimensional" case. Here, by a choice of orientation, the electron beam predominantly "see" only a single family of parallel crystal planes, so that the point diffraction pattern would consist of a single line of bright spots. The electron source is focused to a small probe on the surface of a thin crystalline sample. Then, the point P within the aperture in Fig. 3.12 represents an incident electron beam,

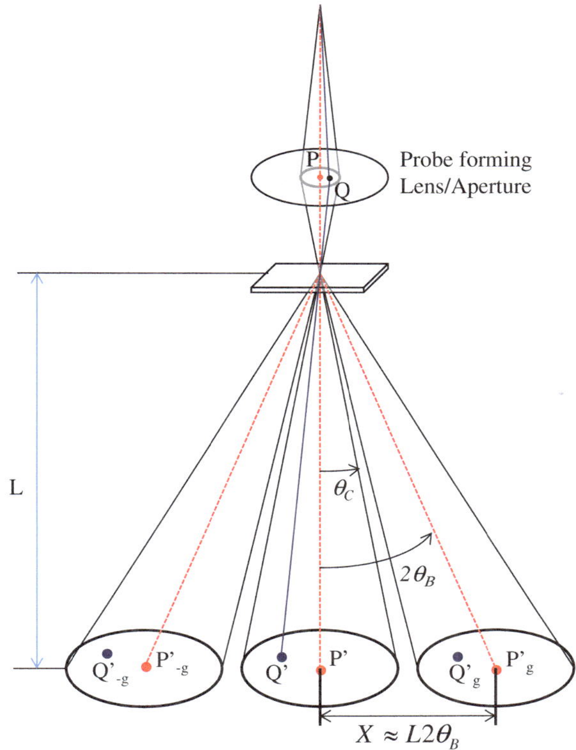

Fig. 3.12 Schematic ray diagram of CBED

which defines the direction of a plane wave at the sample. If we further assume that the crystal is a parallel-sided slab of perfectly crystalline material in which no inelastic scattering or defect scattering occur, then each such incident plane wave can only be scattered by multiples of twice the Bragg angle. Thus a beam such as P gives rise to a set of diffracted waves, which reach the detector at a family of points such as P'. A different source point Q similarly results in a different family of diffracted beams Q'. Since the angles are small, the distances X between points P' on the film are proportional to reciprocal lattice vectors g according to

$$X \approx L2\theta_{\mathrm{B}} \approx Lg\lambda \tag{3.22}$$

where L is the camera length, which is equivalent to the distance where the detector would be placed away from the sample in order to record the same diffraction pattern as in the TEM without the use of lens.

We note that the choice of a new source point Q in the illumination aperture slightly to the right of P actually produces a point of intensity to the left at Q' as shown. This is due to the inversion which occurs as rays pass through the sample. The point Q' may now be taken as the new origin of a point diffraction pattern, whose conjugate points Q'_{-g} and Q'_{g} in the $-g$ and g CBED disks.

Since the distances between points P' are fixed by the crystal structure for given experimental conditions, we see from the figure that the identification of a point in the central CBED disk defines a complete point diffraction pattern, with one point taken from each CBED disk. In two dimensions, we will see that these conjugate points lie on the two-dimensional reciprocal lattice. A CBED pattern may thus be thought of as a set of point diffraction patterns laid side by side. The set of points P' defines one such point pattern, while Q' defines another.

As each set of point diffraction patterns in CBED is associated with a specific and distinct incident beam direction, not all are at Bragg conditions. The change in the excitation error, as well as the direction of change, across the CBED disk is thus important for understanding the rich diffraction intensity patterns often observed in CBED. To see how the excitation error changes within a CBED disk for a particular reflection, consider Fig. 3.13, where we have taken the incident beam defined by the intersection of PQ and RS in the incident beam aperture as at Bragg condition ($Sg = 0$), a beam moving away along the line PQ and toward Q is then characterized by

$$S_g \approx g\delta\theta \tag{3.23}$$

Here, S_g is positive for a beam moving toward Q in Fig. 3.13 as $\delta\theta$ is positive. Further to the right toward Q inside the incident beam aperture leads to an increasingly positive excitation error. Correspondingly, a beam, moving to left toward P away from Bragg condition gives an increasingly negative excitation error. Negative Sg corresponds to "inside" the Bragg condition (with $\theta < \theta_{\mathrm{B}}$) and $|K_t| < g/2$. In the CBED pattern, the direction is reversed as the line PQ becomes $Q'P'$ in

Fig. 3.13. Thus, inside the diffracted CBED disk of $\vec{g}$, S_g is positive on the left side of the Bragg condition and negative on the right side. To a good approximation, the excitation error changes linearly across the CBED disk and along the direction of $\vec{g}$ according to Eq. (3.23). The slope of the change is minus the length of $\vec{g}$. The range of excitation errors within each disk is proportional to g and the convergence angle. For a line perpendicular to $\vec{g}$, such as RS in Fig. 3.13, since the component of $\vec{K}_t$ along $\vec{g}$ is unchanged, and S_g remains constant according to Eq. (3.16).

Now, the variation of intensity for a particular diffracted beam with the direction of the incident beam is known as a *rocking curve*. Thus, we may say that the CBED method displays a rocking curve simultaneously in every diffracted order. Figure 3.14 shows an experimental systematics CBED pattern obtained from the (111) planes of MgO at 120 kV. The broad vertical bright/dark bands of intensity result from diffraction by all planes (*hhh*) type (written {*hhh*}). Across these bands along the line *AA'*, intensity varies rapidly as the excitation error changes, while parallel to these bands the excitation error is constant for the {*hhh*} reflections. The oblique fine lines ("HOLZ lines") show the trajectories of points along which the Bragg condition is satisfied for a higher order Laue zone (HOLZ) reflection with different (nonsystematics) indices. These reflections are discussed in more detail in later chapters.

To calibrate the pattern, we must first measure the distance X in Fig. 3.14 corresponding to the first-order (111) reciprocal lattice vector. This may be done by measuring the distance between the edges of the disks as shown in the figure. We find (on the original print) $X = 5.6$ cm. This is the transverse distance on the print

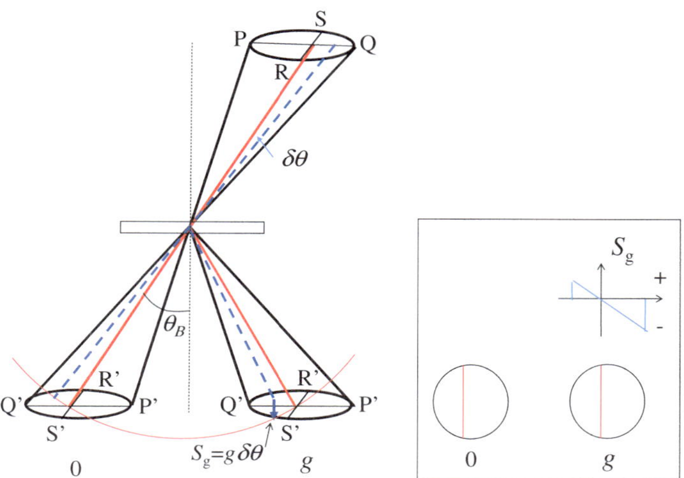

Fig. 3.13 This figure demonstrates the variation of excitation errors at different positions of the CBED disk. The beam marked by the full line (at intersection of PQ and RS) is at Bragg condition, while beam marked by the *dashed line* is associated with a positive excitation error (*Sg*)

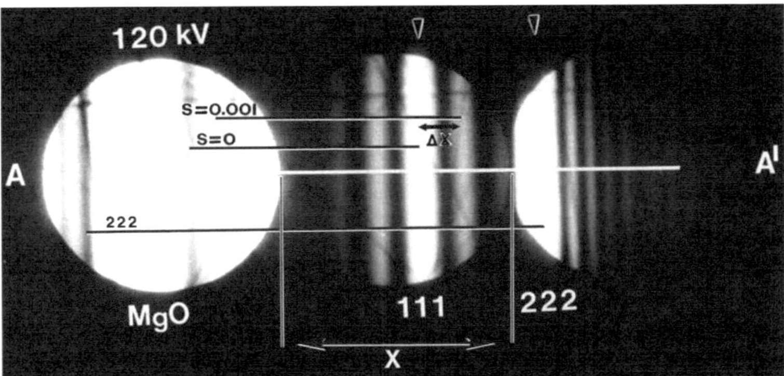

Fig. 3.14 Experimental CBED recorded from MgO in the (111) systematics orientation at 120 kV. The distance corresponding to the (222) lattice vector is shown. The line labelled $S = 0$ connects an incident beam direction [in the (000) disk] with the corresponding point in the (111) disk which is at the Bragg condition. The second line connects similar points for which the excitation error is 0.001 1/Å, as marked. Distances ΔX and X needed for assigning these excitation errors to the pattern are indicated. The arrowheads above indicate the (111) and (222) Bragg conditions

which corresponds to scattering through twice the Bragg angle, and it can be used to scale other measurements since it fixes L, the camera length, in Eq. (3.21).

The center of the vertical band of maximum intensity in the (111) disk corresponds to the (111) Bragg condition. (We will see in Chap. 5 that this may also be a band of minimum intensity at certain thicknesses.) The line (of length X) labelled $S = 0$ on Fig. 3.14 can thus be drawn. The end of this line in the (111) disk shows the point where the diffracted beam intensity is at the Bragg condition, while the start of the line in the (000) disk indicates the position of the corresponding (000) (plane-wave) beam. We note a darkening at this point, since most energy is diffracted into the first-order reflection at this orientation. Using $\lambda = 0.033491$ Å at 120 kV and $d_{111} = 2.42487$ Å (from Appendices 3 and 6, where the cell constant of 0.42 nm is given for MgO) Eq. (3.2) then gives $2\theta_B = 13.81$ milliradians (mrad) as the total scattering angle for this Bragg condition, where $S_g = \delta\theta = 0$.

We now consider another pair of points (marked $S = 0.001$), corresponding to a different incident beam direction. Equation (3.22) and Fig. 3.13 show that, in general, distance ΔX measured on the enlarged print of the CBED pattern is proportional to the corresponding change $\delta\theta$ in scattering angle. Thus, by proportion,

$$\Delta X/X \approx \delta\theta/2\theta_B$$

Using Eq. (3.23), we obtain

$$S_g \approx (\Delta X / X) g^2 / \lambda$$

Measurement from the print as shown in Fig. 3.14 gives $\Delta X = 1$ cm. With $g = (d_{111})^{-1}$ for MgO, we find $Sg = S = 0.001$ Å^{-1}, as indicated. Thus, the line marked $S = 0.001$ corresponds to an orientation "outside" the Bragg condition, where the total scattering angle is greater than the Bragg condition and the excitation error is negative. It would be necessary to reduce the exposure for the central beam in order to observe the corresponding reduction in intensity in the central disk at this first subsidiary maximum.

In a similar way, an excitation error may be assigned to every point across the first-order disk in Fig. 3.14. (This will be needed for comparisons with computed patterns.) Excitation errors may also be assigned to the higher order disks. For example, it is important to understand that the bright band in the second-order (222) disk marked with an arrowhead above it corresponds to a point at which the (222) Bragg condition is satisfied. The corresponding direct (000) beam is indicated at the left end of the line labelled g (222). At the right-hand end of this line, $S_{222} = 0$. It is important to appreciate that in this pattern, the Bragg condition has therefore been satisfied at *two* points, corresponding to the first- and second-order reflections in the systematics row.

3.9   Two-Dimensional CBED

In this section, we will extend the discussion of the Bragg condition to two-dimensional diffraction patterns, and further establish the zone-axis coordinate system that will be employed in the remainder of the book. In the pattern shown in Fig. 3.14, the Bragg condition was satisfied along two lines for two different reflections. If we now allow the incident beam to excite many reflections in the ZOLZ, it becomes possible to satisfy many Bragg conditions simultaneously. Figure 3.11 shows this situation. The incident beams are away from a major crystal zone axis. If we take the incident beam at the center of CBED disk, the Ewald sphere (ES) defined by this beam now intersects reciprocal lattice plane of the ZOLZ on a circle LC known as the Laue circle.

Consider the family of points in the CBED disks. Treat these as a scaled replica of the reciprocal lattice construction for the ZOLZ (see Appendix D). Then the two-dimensional vector $\vec{K}_t$ is the component of the incident wave vector in this plane. The origin is taken at the center of the Laue circle [not at the center of the (000) disk]. This origin point can sometimes be identified in experimental CBED patterns from the symmetrical pattern of Kikuchi lines which are seen to center at this point, as shown in Fig. 3.8. (These lines may be thought of as being attached to the crystal, see further discussions below.) Since they do not move as the beam direction varies across the central disk, they provide a fixed origin. We will also show that Laue circles are concentric on experimental CBED patterns. It is convenient to normalize the length of these vectors to unity when conducting measurements from a recorded pattern, namely, we call the distance X in Fig. 3.14 in the x direction one unit.

With this fixed choice of origin, the variation of incident beam directions across the (000) disk can now be specified by variations in the length and direction of $\vec{K}_t$, as shown in Fig. 3.15. The conjugate points for this beam direction are indicated by the reciprocal lattice. Three are shown in Fig. 3.15, one at the center of the CBED disks and two associated with the incident beams labelled as P and P'. Physically, as we explore different points in the central disk, the incident (plane-wave) beam direction varies, the directions of the corresponding diffracted beams vary, but both the direction of the crystal zone axis (indicated by inelastic scattering) and its origin remains fixed.

The Laue circle is simply defined by its radius, with its center fixed. Laue circles of different radii are concentric, with the minimum and maximum radii ($\mathrm{LC_{min}}$ and $\mathrm{LC_{max}}$ in Fig. 3.15) given by the shortest and longest length of $\vec{K}_t$. All Laue circles, on which the incident beams reside, fall in between these two.

For a ZOLZ reciprocal lattice vector $\vec{h}$ to be at the Bragg condition for the incident beam of wave vector $\vec{K}$, we require that its component $\vec{K}_t$ in the ZOLZ, when projected in direction $\vec{h}$, be equal to $-\vec{h}/2$, e.g., $\vec{K}_t \cdot \vec{h}/\left|\vec{h}\right| = -h/2$. To locate

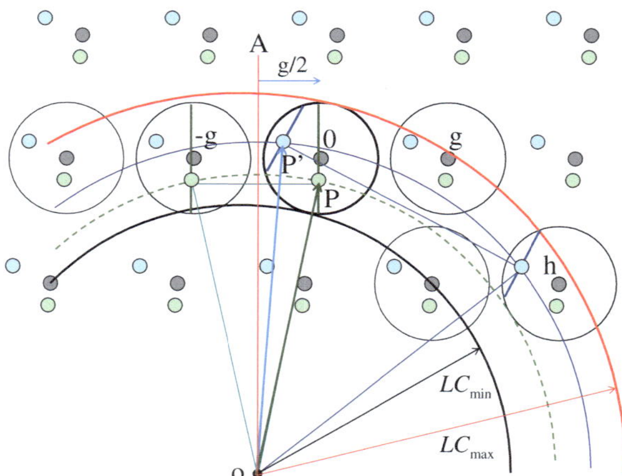

Fig. 3.15 Ewald sphere construction in CBED. The origin is fixed at the zone axis center. Laue circles of all incident beams with the 0 disk fall between the two circles marked as $\mathrm{LC_{min}}$ and $\mathrm{LC_{max}}$. The CBED disks are shown for selected reflections close to the Laue circles. Three separated point diffraction patterns are shown, one for the center of the CBED and two are associated with incident beams at P and P'. Both are at the Bragg condition, one for $-g$ and one for h reflection as labelled. The lines through P and P' mark all incident beams at Bragg condition within the incident beam disk for these two reflections, respectively

all incident beams within the central disk satisfying this condition, as example, we draw the line OA in Fig. 3.15 through the center of the Laue circles. This line is perpendicular to $-\vec{g}$. Then, the incident beams at the Bragg condition for $-\vec{g}$ can be simply located by drawing a line parallel to OA, at distance $g/2$ in the opposite direction of $-\vec{g}$. We will call this line the Bragg line of $-\vec{g}$. Since the incident beams must fall within the central disk, whether a particular reflection satisfies the Bragg condition in a CBED pattern depends on whether the central disk intersects the Bragg line of this reflection. In Fig. 3.15, two reflections at Bragg condition are identified, they are labelled as $-\vec{g}$ and $\vec{h}$. The incident beams labelled as P and P' satisfy the Bragg condition for $-\vec{g}$ and $\vec{h}$, respectively, with the intersecting Laue circles as shown in the figure.

When two Bragg lines intersect each other and fall within the central disk, we have the so-called three beam diffraction condition (0, g, and h). The four beam diffraction case will have three intersecting Bragg lines, and so on. The form of the intensity distributions in the CBED disks for special three-beam and four-beam cases will be discussed in Chap. 5.

3.10 High-Order Laue Zone (HOLZ) Lines

A HOLZ line is the locus of the Bragg condition for a HOLZ reflection $\vec{g}$. The lines therefore occur in pairs, a maximum of intensity in the outer HOLZ ring (excess line) and a corresponding minimum of intensity in the incident-beam disk (deficiency line). The geometry of HOLZ lines is the geometry of the Bragg condition projected onto the plane of observation. This construction also applies to high-index ZOLZ reflections.

Hence both Kikuchi lines and HOLZ lines arise from the same elastic Bragg-scattering mechanism. The difference between them lies in the source of electrons in each case—wide cones of inelastically scattered electrons inside the crystal for Kikuchi lines, and a smaller cone, generated by an external source, for HOLZ lines. In the following, we will discuss specifics related to the generation of HOLZ lines, but the results apply equally well to Kikuchi lines of reflections belonging to HOLZ.

The higher order reflections themselves may be visible as a bright outer ring of reflections if a small camera length is used, as shown in Fig. 3.16. These reflections, and their complements in the zero-order disk, are extremely useful in CBED, as we shall see later. For example, they may be used to confirm an orientation determination in non-centrosymmetric crystals, to check for stacking disorder in the beam direction, to observe tetragonal distortions in otherwise cubic crystals, to measure strains in crystals, and for the measurement of structure factors.

The radius R of this HOLZ ring is given from the Ewald sphere construction (see Fig. 3.16) as

$$R \approx \left(K_t^2 + 2H/\lambda\right)^{1/2} \qquad (3.24)$$

in reciprocal length units, where H is the height of the HOLZ layer considered and K_t is the component of incident-beam wave vector in the ZOLZ plane. (In zone-axis orientations, $K_t = 0$.) The approximation in Eq. (3.24) assumes that the wavelength $1/\lambda$ is much larger than both K_t and H. The height H may thus be obtained from a

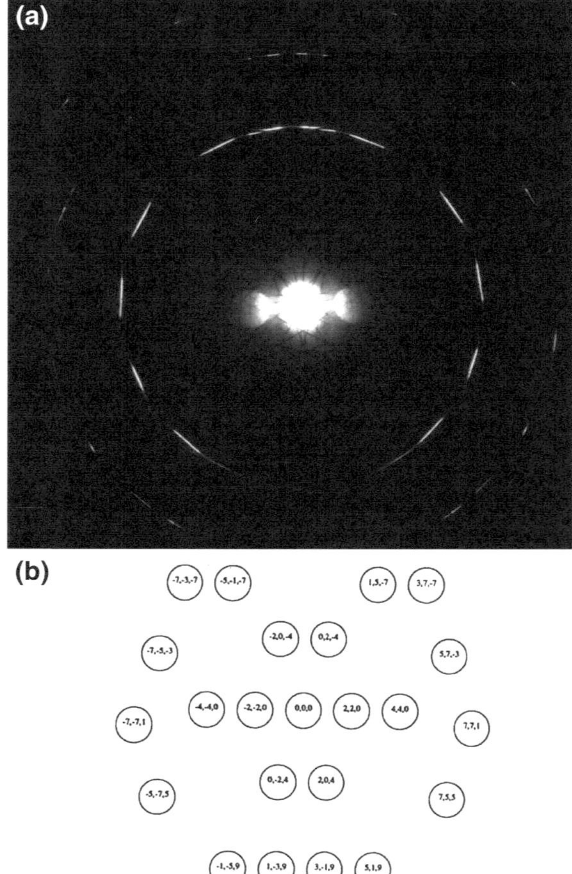

Fig. 3.16 a Experimental Si (−221) CBED pattern recorded at 100 kV showing three HOLZ rings. **b** computer-simulated pattern corresponding to the above experimental pattern with the ZOLZ and FOLZ reflections are shown

measurement of R if λ is known. In the absence of extinctions due to screw and glide symmetry elements, H may be taken as a measure of the spacing between reciprocal lattice planes in the beam direction. For the $[u, v, w]$ zone axis of a cubic crystal,

$$H_{uvw} = \frac{n}{a\sqrt{u^2 + v^2 + w^2}}$$

For fcc crystals, as a result of the centering of the lattice, $n = 1$ if $(u + v + w)$ is odd, and $n = 2$ if $(u + v + w)$ is even. For a bcc crystal, $n = 2$ if h, k, and l are all odd integers; otherwise, $n = 1$. Appendix C contains expressions for the other crystal classes; however, the centering of these lattices must also be considered.

Figure 3.17 shows the geometric construction of a HOLZ line. The position of the HOLZ line is defined by the angle α, the angle between incident beam and the zone axis, and is given by

$$\alpha = 90° - (90° - \theta_{\mathrm{B}}) - \beta = \theta_{\mathrm{B}} - \beta = \sin^{-1}(g\lambda/2) - \sin^{-1}(g_z/g) \qquad (3.25)$$

If this line is translated by the vector $\mathbf{g}$, we obtain the corresponding line in the outer HOLZ ring or "dark-field disk." For computer simulation, the position of the line can be also obtained directly by Eq. (3.18).

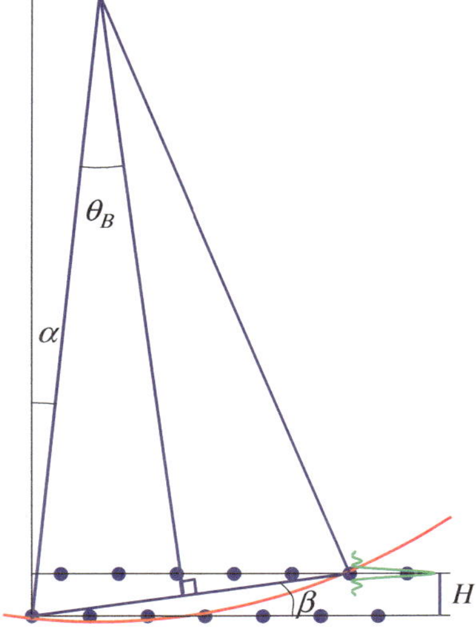

Fig. 3.17 HOLZ line construction. For details, see text

Fig. 3.18 Enlarged central
disk of Fig. 3.16. Many fine
HOLZ lines can be seen
across the disk. The lines
appear to extend outside with
a subtle difference in contrast.
The outside lines are Kikuchi
lines formed by Bragg
diffraction of inelastically
scattered electrons.
Intersection of HOLZ lines
allow a precise determination
of crystal orientation

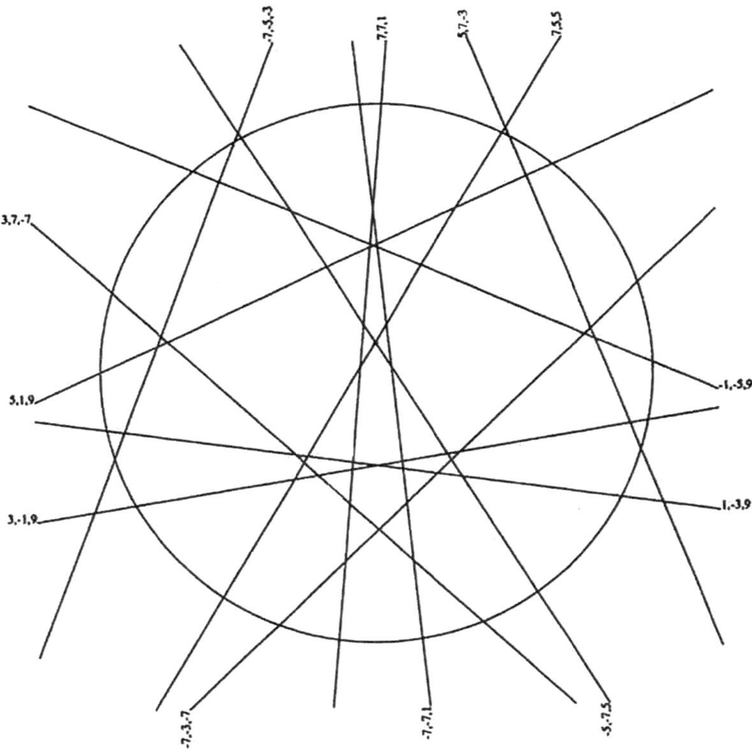

Fig. 3.19 Computer-simulated HOLZ lines corresponding to Fig. 3.18. The lines are indexed

The position of HOLZ lines depends on the microscope accelerating voltage and on the lattice constants through the Bragg law. The distance between the line intersections therefore provides a sensitive parameter for strain measurement, as discussed in Chap. 16. Unlike X-ray or neutron diffraction, it is difficult to measure the Bragg angle very accurately by electron diffraction using spot patterns, because of the electron-optical distortions in the projector lenses. The relative movement of three or more HOLZ lines in the zero disk avoids this difficulty and is therefore the best method for the measurement of high-voltage, lattice constants, and local strains.

HOLZ lines are narrow for the following reasons. As indicated on Fig. 3.17, Bragg scattering occurs strongly only when the Ewald sphere passes near the central maximum of the rocking curve. For HOLZ reflections, the sphere cuts this distribution at a steeper angle than in the ZOLZ, and intensity is therefore observed over a smaller angular range.

The indexing of HOLZ lines is performed in two stages. The first step is to index the diffraction pattern and identify possible zone axes. The second step is to simulate and index the HOLZ lines. On the figures in Appendix D, sufficient information is given to allow indexing of the entire three-dimensional lattice and hence the HOLZ lines. Fournier et al. (1989) also described a systematic technique for indexing diffraction patterns. A worked example of indexing is given in Steeds and Evans (1980). Figures 3.18 and 3.19 show experimental and computer-simulated patterns for the [−211] zone axis of silicon at 100 kV.

References

Egerton RF (2011) Electron energy-loss spectroscopy in the electron microscope, 2nd edn. Springer, New York

Fournier D, L′Esperance G, Saint-Jacques G (1989) Systematic procedure for indexing HOLZ lines in convergent beam electron diffraction patterns of cubic crystal, J Electr Mic Techn 13:123–149

Hirsch P, Howie A, Nicolson RB, Pashley DW, Whelan MJ (1977) Electron microscopy of thin crystals. Robert E. Krieger Publishing Company, Malaba

Kittel C (1976) Introduction to solid state physics. Wiley, New York

Kolb U, Mugnaioli E, Gorelik TE (2011) Automated electron diffraction tomography—a new tool for nano crystal structure analysis. Cryst Res Technol 46:542–554

Lorretto MH (1994) Electron beam analysis of materials, 2nd edn. Chapman and Hall, London

Spence JCH (1992) Electron channelling. In: Cowley JM (ed) Techniques of electron diffraction, vol 1. Oxford University Press, Oxford

Steeds JW, Evans NS (1980) In: Bailey G (ed) Proc. 38th Annual EMSA meeting. Claitors, Baton Rouge. p.188

Williams DB, Carter BC (2009) Transmission electron microscopy, a textbook for materials science, 2nd edn. Springer, New York

Zuo JM (1992) Automated lattice-parameter measurement from HOLZ lines and their use for the measurement of oxygen-content in $YBa_2Cu_3O_{7-\Delta}$ from nanometer-sized region. Ultramicroscopy 41:211–223

Zuo JM (1993) New method of Bravais lattice determination. Ultramicroscopy 52:459–464

Chapter 4
Kinematical Theory of Electron Diffraction

In this chapter, we develop the theory of transmission electron diffraction based on the assumption of single scattering or the so-called kinematical approximation. This approximation allows us to define some basic quantities in diffraction, such as atomic scattering and crystal structure factors and their relationships. It also provides a framework for further treatment of electron diffraction using dynamical theory in the next chapter.

While the similarly formulated kinematic theory for X-rays and neutrons form the core of X-ray and neutron diffraction analysis, historically, kinematic theory has found only limited use in electron diffraction, since the usual experimental conditions involve strong multiple scattering. However, recent developments in new electron diffraction techniques, and the popularity of monolayer substrates such as graphene, have renewed interests in kinematic theory. One of these developments is the introduction of precession electron diffraction (PED), which averages the electron diffraction intensity over a cone of incident beam directions. This technique, originally developed by Vincent and Midgley (1994), improves the agreement between single crystal electron diffraction intensities and crystal structure factors for structural analysis. Another development is electron diffraction from very small, nanometer-sized crystals. For electron powder diffraction of nanoparticles, the combination of orientation averaging and small crystal size makes electron powder data amenable to analysis based on kinematical theory.

We start by introducing kinematic theory based on the first-order Born approximation and the weak-phase-object approximation and then construct from this simple approximation the theory of atomic scattering, small crystals, and the effect of thermal vibrations. In the second part of the chapter, we define electron structure factors and relate them to the corresponding X-ray structure factors. These results will be needed in Chap. 5.

© Springer Science+Business Media New York 2017
J.M. Zuo and J.C.H. Spence, *Advanced Transmission Electron Microscopy*,
DOI 10.1007/978-1-4939-6607-3_4

The sign convention used in the following takes the incident plane wave to have the form of $\exp\left(2\pi i \vec{k} \cdot \vec{r}\right)$, which is similar to the quantum mechanical standard, except we have $\exp\left(i\vec{k}' \cdot \vec{r}\right)$ with $\vec{k}' = 2\pi\vec{k}$ in quantum mechanics. For a discussion of the various sign conventions used in electron diffraction, see Saxton et al. (1984) and Spence (2013), and the list of symbols at the beginning of this book. We have used the values for the fundamental physical constants, which are listed in Appendix B.

4.1 First-Order Born Approximation

Electrons diffract by interacting with an object through its electrostatic potential $V(\vec{r})$, measured in volts. Electron diffraction from a general, three-dimensional, potential is a complex problem. Rigorous solutions only exist for crystals with well-defined geometry, such as a semi-infinite crystal with a flat surface or a crystal slab. Accurate solutions, suitable for transmission electron diffraction through a parallel-sided crystal slab, will be given in the next chapter.

Here, we introduce the kinematical theory of electron diffraction, based on the first-order Born approximation. This approximation retains the first-order term in the so-called Born series, which describes the scattered wave in increasing orders corresponding to the number of scattering events. The Born series in principle provides a general solution to the electron diffraction problem, but in practice, it converges too slowly to be of practical use, except the first-order Born approximation (discussed here) and the distorted wave Born approximation (DWBA) (to be discussed in Chap. 13).

We start with the time-independent Schrödinger equation (Eq. 2.13) in the form of

$$\frac{1}{4\pi^2}\nabla^2\phi + k^2\phi = -\frac{2me}{h^2}V(\vec{r})\phi = -U(\vec{r})\phi \qquad (4.1)$$

where $k^2 = 2mE/h^2 = 2me\Phi/h^2$, with Φ standing for the electron acceleration voltage. The interaction potential $U(\vec{r}) = 2meV(\vec{r})/h^2$ is in the unit of Å^{-2}. It depends on the acceleration voltage, through the relativistic electron mass (see Chap. 2). Equation (4.1) can be transformed into a special form of the Lippmann–Schwinger integral equation:

$$\phi(\vec{r}) = \phi_\circ(\vec{r}) - \int d^3\vec{r}'\, G(\vec{r} - \vec{r}')U(\vec{r}')\phi(\vec{r}') \qquad (4.2)$$

Here, $\phi_o(\vec{r})$ is a solution of the homogeneous wave equation (in free space) $\frac{1}{4\pi^2}\nabla^2\phi_\circ + k^2\phi_\circ = 0$ and $G(\vec{r})$ is the solution of

$$\frac{1}{4\pi^2}\nabla^2 G(\vec{r}) + k^2 G(\vec{r}) = \delta(\vec{r}),$$

in the form of

$$G(\vec{r}) = -\pi\frac{e^{\pm 2\pi i k r}}{r}. \tag{4.3}$$

This solution can be verified by using the relationship of $\nabla^2(1/r) = -4\pi\delta(r)$, where $\delta(r)$ is the delta function defined in Appendix E. For electron scattering, we take $\phi_o(\vec{r}) = e^{2\pi i \vec{k}_o\cdot\vec{r}}$ for the incident wave and $G(\vec{r}) = -\pi\frac{e^{2\pi i k r}}{r}$ for the scattered wave, together they give rise to

$$\phi(\vec{r}) = e^{2\pi i \vec{k}_o\cdot\vec{r}} + \pi\int d^3\vec{r}'\,\frac{e^{2\pi i k|\vec{r}-\vec{r}'|}}{|\vec{r}-\vec{r}'|}U(\vec{r}')\phi(\vec{r}') \tag{4.4}$$

In the so-called first-order Born approximation, the potential scattering is assumed to be weak [thus, the second term is much smaller than the first term in Eq. (4.4)]. The wave function inside the integral, e.g., at the local potential, is taken approximately as the incident wave, $\phi(\vec{r})\approx e^{2\pi i \vec{k}_o\cdot\vec{r}}$. By making this approximation, we have implicitly assumed that only direct scattering from the incident wave contributes to the scattered wave, and thus, multiple scattering, involving other scattered waves, is excluded. This approximation is also called the single scattering, or kinematical, approximation. Higher order terms can be obtained by substituting the approximate solution back into Eq. (4.4). For example, the second-order solution is obtained from the first-order solution and so forth. The series obtained gives the Born series for multiple scattering. For the first-order approximation, we have the following simplified solution

$$\phi(\vec{r}) \approx e^{2\pi i \vec{k}_o\cdot\vec{r}} + \pi\int d^3\vec{r}'\,\frac{e^{2\pi i k|\vec{r}-\vec{r}'|}}{|\vec{r}-\vec{r}'|}U(\vec{r}')e^{2\pi i \vec{k}_o\cdot\vec{r}'} \tag{4.5}$$

This equation can be qualitatively understood as illustrated in Fig. 4.1, where an incident wave is incident on a potential field. A small volume of the potential field

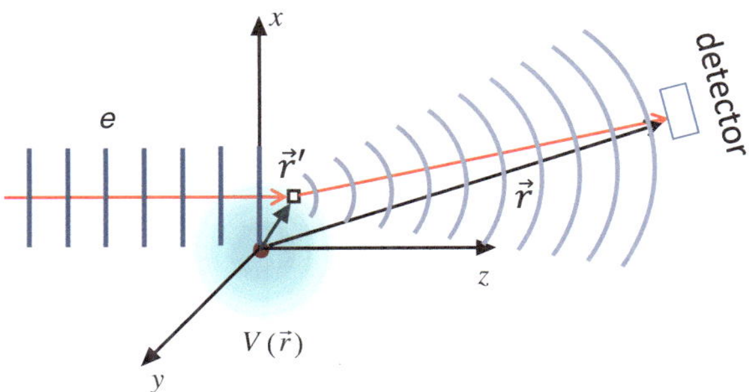

Fig. 4.1 Electron scattering geometry and coordinate

gives rise to a spherical wave whose amplitude is proportional to the potential and its phase is determined by the propagation of the incident wave.

Further simplifications can be made by having the detector placed far from the object as well as having a small object (Fig. 4.1). Under these conditions, we have $|\vec{r}| \gg |\vec{r}'|$. By replacing $|\vec{r} - \vec{r}'|$ with $|\vec{r}|$ in the denominator and taking

$$|\vec{r} - \vec{r}'| \approx r - \frac{\vec{r}' \cdot \vec{r}}{r}$$

In Eq. (4.5), we thus obtain

$$\begin{aligned}
\phi(\vec{r}) &\approx e^{2\pi i \vec{k}_o \cdot \vec{r}} + \pi \frac{e^{2\pi i k r}}{r} \int d^3 \vec{r}' e^{-2\pi i k \frac{\vec{r}' \cdot \vec{r}}{r}} U(\vec{r}') e^{2\pi i \vec{k}_o \cdot \vec{r}'} \\
&\approx e^{2\pi i \vec{k}_o \cdot \vec{r}} + \pi \frac{e^{2\pi i k r}}{r} \int d^3 \vec{r}' e^{-2\pi i (\vec{k} - \vec{k}_o) \cdot \vec{r}'} U(\vec{r}') \qquad (4.6) \\
&\approx e^{2\pi i \vec{k}_o \cdot \vec{r}} + f\left(\vec{k}, \vec{k}_o\right) \frac{e^{2\pi i k r}}{r}
\end{aligned}$$

where

$$f\left(\vec{k}, \vec{k}_o\right) = \pi \int d^3 \vec{r}' e^{-2\pi i (\vec{k} - \vec{k}_o) \cdot \vec{r}'} U(\vec{r}') \qquad (4.7)$$

Here, $f\left(\vec{k}, \vec{k}_o\right)$ defines the scattering amplitude for the incident and scattered wave vectors $\vec{k}_o$ and $\vec{k}$. The direction of the scattered wave vector $\vec{k}$ is determined by the detector position (along $\vec{r}$).

The integral in Eq. (4.7) is same as Fourier transform (see Appendix E). Thus, the scattering amplitude in the kinematical approximation is simply given by the Fourier transform (*FT*) of the interaction potential.

4.2   Weak-Phase-Object Approximation

An alternative approach to kinematical diffraction theory is to start with the so-called weak-phase-object (WPO) approximation. Since the WPO approximation is extensively used in electron microscopy, it is appropriate to introduce it here and to show how this approximation relates to the first-order Born approximation.

We examine the case where the object potential is approximated by a constant potential with small, weak, modulations. When the extent of such a potential is limited, an approximate solution to Eq. 2.13 known as the WKB, or Moliere "high-energy" approximation, in quantum mechanics can be made (see Wu and Ohmura (2011) for the relationship of this approach to "partial wave" treatments) regardless of the shape of potential, provided that the energy of the electrons is high

($E \gg V$). In that case, we have the following electron wave function in the near field of the sample, after the electron exits the potential field:(Wu and Ohmura 2011).

$$\phi_e(x, y, z) \approx \exp\left[i2\pi \int_0^t k(x, y, z)\mathrm{d}z\right] \qquad (4.8)$$

Here, the effect of potential is to introduce a position-dependent wave vector $k(x, y, z)$ described by:

$$k(x, y, z) = \sqrt{\frac{2me}{h^2}[\Phi + V(x, y, z)]} = \sqrt{k_o^2 + U(x, y, z)} \approx k_o + \frac{U(x, y, z)}{2k_o}. \qquad (4.9)$$

The second term accounts for the interaction potential. Thus, in the limit of a weak object potential, the effect of a scattering potential is approximately described by a phase change in the electron wave function.

Substituting (4.8) into (4.9) and omitting the constant phase, we have

$$\phi_e(x, y, z) \approx \exp\left[i\pi\lambda \int_0^t U(x, y, z)\mathrm{d}z\right] = \exp\left[i\pi\lambda\overline{U}(x, y)t\right] \qquad (4.10)$$

Here, $\overline{U}(x, y)$ is the potential averaged over the thickness (t) of the object. For a very thin crystal, the phase variation is small, we expand (4.10) to the first order, which gives:

$$\phi_e(x, y, z) \approx 1 + i\pi\lambda\overline{U}(x, y)t \qquad (4.11)$$

For $\overline{U}(x, y) \sim 5 \times 10^{-2}$ Å^{-2}, $\lambda \sim 0.02$ Å, Eq. (4.11) is a reasonable approximation for $t < 100$ Å.

We take a small volume of the potential. Using the weak-phase-object approximation, the electron wave at the "exit face" of this small volume is given by:

$$\phi_e(\vec{r}) = [1 + i\pi\lambda U(\vec{r})\mathrm{d}z]\phi_o(\vec{r}) \qquad (4.12)$$

Here, we take $U(\vec{r})$ as constant within the small volume. Under the weak scattering approximation where the effect of scattering on the incident wave is small, $\phi_o(\vec{r})$ can be taken as same as the incident wave with $\phi_o(\vec{r}) = \exp\left(2\pi i\vec{k}_o \cdot \vec{r}\right)$. Treating the exit wave from the small volume as a point source for spherical waves, its contribution to the wave at the distance r from this small volume has the form (details about wave propagation from near field to far field are provided in Chap. 2):

$$d\phi_s \approx -i \frac{[i\pi\lambda U(\vec{r}')]}{|\vec{r}-\vec{r}'|\lambda} \left(\frac{1+\cos\theta}{2}\right) e^{2\pi i k|\vec{r}-\vec{r}'|} e^{2\pi i \vec{k}_o \cdot \vec{r}'} d^3\vec{r}$$
$$\approx \frac{\pi U(\vec{r}')}{|\vec{r}-\vec{r}'|} e^{2\pi i k|\vec{r}-\vec{r}'|} e^{2\pi i \vec{k}_o \cdot \vec{r}'} d^3\vec{r}' \tag{4.13}$$

The approximation is for forward scattering in small angles, which is generally valid for electron diffraction. From Eq. (4.13), we thus obtain

$$\phi_s \approx \pi \int \frac{U(\vec{r}')}{|\vec{r}-\vec{r}'|} e^{2\pi i k|\vec{r}-\vec{r}'|} e^{2\pi i \vec{k}_o \cdot \vec{r}'} d^3\vec{r}'$$

which gives the same result as the first-order Born approximation (Eq. 4.6).

What's different here from, compared to the first-order Born approximation, is that we have treated electron scattering in a two-step processes: obtaining the near field wave function using the WPO approximation and then calculating the wave function at the detector (far field). By making a distinction between the far field and near-field wave functions, this approach becomes very useful when we discuss electron image formation.

4.3   Electron Atomic Scattering

As a starting point for developing electron diffraction theory, consider electron scattering by an atom of atomic number Z. The atomic potential has two contributions, one from the positive charged nucleus and the other from the atomic electrons. The relationship between the potential and charge is given by Poisson's equation (in SI units):

$$\nabla^2 V(\vec{r}) = -\frac{e[Z\delta(\vec{r}) - \rho(\vec{r})]}{\varepsilon_o} \tag{4.14}$$

According to Eq. (4.6), the total scattered wave is obtained by summing all scattered wave over the volume of the atom in the form of an integral of:

$$\phi_s \approx \frac{2\pi m e}{h^2} \frac{e^{2\pi i k r}}{r} \int V(\vec{r}') e^{-2\pi i (\vec{k}-\vec{k}_o)\cdot\vec{r}'} d^3\vec{r}'. \tag{4.15}$$

Here, $\vec{k}$ is the scattered wave vector and its direction is taken as same as $\vec{r}$. The scattered wave of an atom is proportional to the *FT* of the atomic potential.

The X-ray atomic scattering factor $f^x(s)$ is the Fourier transform of atomic charge density, where $\vec{s} = \vec{S}/2$ is half the scattering vector. The dimensionless scattering factor $f^x(s)$ has the units of number of electrons and is given by

$$f^x(s) = \int \rho(\vec{r}) e^{-4\pi i \vec{s} \cdot \vec{r}} d\vec{r} \tag{4.16}$$

The atomic potential seen by electron diffraction, according to (4.15), is related to charge density by Eq. (4.14). Hence, by analogy with the definition of the X-ray atomic scattering factor, we have in SI units

$$f^e(s) = \int V(\vec{r}) e^{-4\pi i \vec{s} \cdot \vec{r}} d\vec{r} \tag{4.17}$$

To relate the electron scattering factor (SF) to the X-ray SF, we use the relationship

$$\nabla^2 \int V(\vec{r}) e^{-4\pi i \vec{s} \cdot \vec{r}} d\vec{r} = \int \nabla^2 V(\vec{r}) e^{-4\pi i \vec{s} \cdot \vec{r}} d\vec{r} + \int V(\vec{r}) \nabla^2 e^{-4\pi i \vec{s} \cdot \vec{r}} d\vec{r} = 0.$$

And thus

$$\begin{aligned}
f^e(s) &= -\frac{1}{16\pi s^2} \int \nabla^2 V(\vec{r}) e^{-4\pi i \vec{s} \cdot \vec{r}} d\vec{r} \\
&= \frac{|e|}{16\pi^2 \varepsilon_o s^2} \int [Z\delta(\vec{r}) - \rho(\vec{r})] e^{-4\pi i \vec{s} \cdot \vec{r}} d\vec{r} \\
&= \frac{|e|}{16\pi^2 \varepsilon_o} \frac{(Z - f^x)}{s^2}
\end{aligned} \tag{4.18}$$

where we have taken $V(\vec{r}) = 0$ for large r. The definition of the electron atomic scattering factor here adopts the convention in the new version of the International Tables for Crystallography. The traditional definition is based on the first-order Born approximation of (4.15) and takes:

$$f^B(s) = \frac{2\pi m_e |e|}{h^2} \int V(\vec{r}) e^{-4\pi i \vec{s} \cdot \vec{r}} d\vec{r} = \frac{m_e e^2}{8\pi \varepsilon_o h^2} \frac{(Z - f^x(s))}{s^2} \tag{4.19}$$

where m_e is the electron rest mass. Using the unit of 1/Å for s, we have

$$f^B(s) = 0.023934 \frac{(Z - f^x(s))}{s^2} \, (\text{Å}) \tag{4.20}$$

The atomic scattering factor of $f^B(s)$ defined here can be related to the scattered wave intensity detected in the far field. At the electron detector, electrons can be counted in units of the number of electrons detected per second. The scattered intensity recorded depends on the area of the detector pixel (dA), the distance to the sample (r), the incident beam intensity (I_o), and the cross-sectional area of the beam (A_o). Taken together, we have

$$\frac{I_{sc}}{I_o} = \frac{|\phi_s|^2 dA}{|\phi_o|^2 A_o} = \frac{|\phi_s|^2 r^2 d\Omega}{|\phi_o|^2 A_o} \tag{4.21}$$

where $d\Omega = dA/r^2$ is the detector pixel solid angle in units of steradians. The standard practice in a scattering experiment is to normalize the measured intensity using the incident flux (I_o/A_o) and the detector solid angle. This leads to the definition of partial scattering cross section for elastic scattering:

$$\left(\frac{d\sigma}{d\Omega}\right) = \frac{I_{sc}}{(I_o/A_o)d\Omega} = \gamma^2 \left|f^B(s)\right|^2 \tag{4.22}$$

where γ is the relativistic constant. Thus, the square of $\gamma f^B(s)$ with units of area gives the scattering cross section. The X-ray scattering factor defined in the same way is given by

$$f(s) = \left(\frac{e^2}{4\pi\varepsilon_o m_e c^2}\right) f^x = 2.82 \times 10^{-5} f^x \,(\text{Å}) \tag{4.23}$$

For typical values of $s \sim 0.2$ 1/Å, the ratio $f^B/f \sim 10^4$. The electron scattering cross section in general is about $\sim 10^6$ to 10^8 times larger than X-ray dependent on scattering angle, beam energy, and type of atom. (A comparison electron scattering factors of selected elements is shown in Fig. 4.2.) For soft X-rays, the difference is least, and the strong interaction of soft X-rays may also lead to multiple scattering in samples more than a micron or so thick.

Fig. 4.2 Electron scattering factors of selected elements

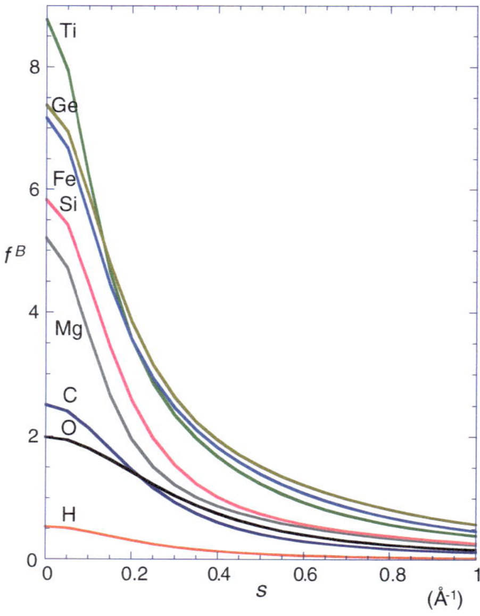

In most electron diffraction applications, we approximate atoms inside the material by spherical, free atoms or ions. The atomic electron density and its Fourier transform (the atomic scattering factor) can be obtained using several different theoretical approaches. Values for different atoms are tabulated in the International Tables for Crystallography, based on the so-called Dirac-Fock method (Doyle and Turner 1968). Electron atomic scattering factors can be calculated from the tabulated X-ray scattering factors using the Mott formula in Eq. (4.19). The International Tables for Crystallography also lists electron scattering factors in separate tables (Prince 2004). The X-ray atomic scattering factors can be approximately fitted using Gaussian functions. They are convenient for use in computer algorithms. The set most commonly used provided by Doyle and Turner uses five Gaussian functions plus a constant with a total of 9 parameters in the form of:

$$f^x(s) = \sum_{i=1}^{4} a_i e^{-b_i s^2} + c \tag{4.24}$$

At $s = 0$, from Eq. (4.16), we have the following result for a neutral atom:

$$f^x(0) = \int \rho(\vec{r}) \mathrm{d}\vec{r} = Z.$$

Thus, for the Gaussian fitting formula, we expect the following constraint on the fitting coefficients in the form of:

$$\sum_{i=1}^{4} a_i + c = Z.$$

This constraint should be checked in calculating electron scattering factors since a small deviation from this constraint can cause a large error at small scattering angles because of the $1/s^2$ weighting factor.

Parameterization of electron atomic scattering factor for all neutral atoms with s up to 6 Å^{-1} using five Gaussian functions is described and tabulated by Peng et al. (1996b).

Electrons belonging to different atomic states contribute to the X-ray scattering factor at different scattering angles. As a rule of thumb, core electrons with its distribution close to the nucleus dominate the high angle scattering, while valence electrons distributed further away from the nucleus mostly impact low angle scattering. Figure 4.3 shows the contribution of 3d and 4s Cu electrons to the scattering factor, normalized to unity. For 4s electrons, Fourier transform of the broad distribution gives a much narrow distribution in scattering factor.

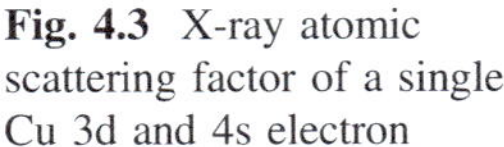

Fig. 4.3 X-ray atomic scattering factor of a single Cu 3d and 4s electron

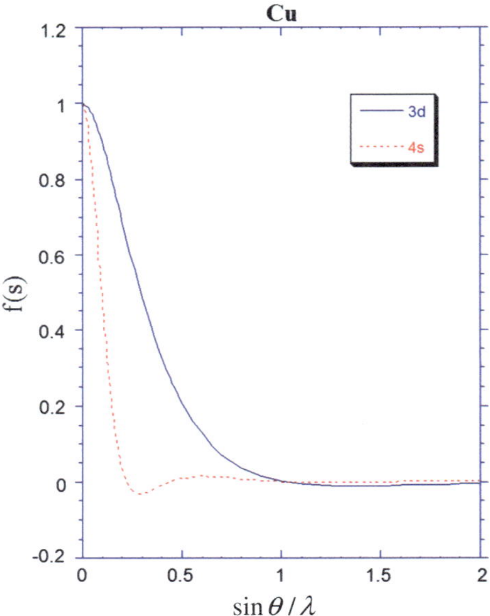

4.4   Kinematical Electron Scattering from a Monoatomic Small Crystal

Here, we extend our treatment of kinematical electron scattering from a single atom to a monoatomic small crystal. In the process, we will reinforce the concept of the reciprocal lattice and its relationship to Bragg diffraction. The starting point is Eq. (4.11). To evaluate this, we first provide a description of the crystal potential based on the convolution of the atomic potential with the lattice represented by δ-functions.

Inside a small crystal, each atom contributes to the potential at a point $\vec{r}$. The total potential is obtained by summing up the potential of each atom:

$$V\left(\vec{r}\right) = \sum_{n=0}^{N_1-1}\sum_{m=0}^{N_2-1}\sum_{l=0}^{N_3-1} V_A\left(\vec{r} - \vec{R}_{nml}\right) = \sum_{n=0}^{N_1-1}\sum_{m=0}^{N_2-1}\sum_{l=0}^{N_3-1} V_A\left(\vec{r} - n\vec{a} - m\vec{b} - l\vec{c}\right)$$

$$(4.25)$$

where N_1, N_2, N_3 are the number of atoms along a, b, and c, respectively, for the small crystal. In the atomic potential, subtraction of the lattice vector $\vec{R}_{nml}$ is equivalent to placing the atom on the lattice, which mathematically can be described by convolution between the atomic potential and the lattice L:

$$V\left(\vec{r}\right) = V_A(\vec{r}) \otimes L(\vec{r}) = V_A(\vec{r}) \otimes \sum_{n=0}^{N_1-1} \sum_{m=0}^{N_2-1} \sum_{l=0}^{N_3-1} \delta\left(\vec{r} - n\vec{a} - m\vec{b} - l\vec{c}\right)$$

The scattered wave is proportional to the Fourier transform of the potential. The FT of two convoluted functions is the product of the FT of each function. This gives:

$$\mathrm{FT}[V(\vec{r})] = \mathrm{FT}[V_A(\vec{r})] \cdot \mathrm{FT}[L(\vec{r})] \tag{4.26}$$

The Fourier transform of the atomic potential gives the atomic scattering factor and the Fourier transform of the lattice gives

$$\mathrm{FT}[L(\vec{r})] = \sum_{n=0}^{N_1-1} e^{-2\pi i \left(\vec{k}-\vec{k}_0\right)\cdot\vec{a}n} \sum_{m=0}^{N_2-1} e^{-2\pi i \left(\vec{k}-\vec{k}_0\right)\cdot\vec{b}m} \sum_{l=0}^{N_3-1} e^{-2\pi i \left(\vec{k}-\vec{k}_0\right)\cdot\vec{b}l}$$

$$= \frac{\sin\left[\pi\left(\vec{k}-\vec{k}_0\right)\cdot\vec{a}N_1\right]}{\sin\left[\pi\left(\vec{k}-\vec{k}_0\right)\cdot\vec{a}\right]} \frac{\sin\left[\pi\left(\vec{k}-\vec{k}_0\right)\cdot\vec{b}N_2\right]}{\sin\left[\pi\left(\vec{k}-\vec{k}_0\right)\cdot\vec{b}\right]} \frac{\sin\left[\pi\left(\vec{k}-\vec{k}_0\right)\cdot\vec{c}N_3\right]}{\sin\left[\pi\left(\vec{k}-\vec{k}_0\right)\cdot\vec{c}\right]} \tag{4.27}$$

Equation (4.19) is the product of three $\sin(Nx)/\sin(x)$ functions. To examine its overall properties, it is useful to examine the first function, which has an infinite number of maxima at the condition:

$$\left(\vec{k}-\vec{k}_0\right)\cdot\vec{a} = h \tag{4.28}$$

Here, h is integer from $-\infty$ to ∞. The maxima has the amplitude of N, which becomes progressively more pronounced with increasing N (see Fig. 4.4). For sufficiently large N, (4) reduces to a periodic array of delta functions with the spacing of $1/a$.

The condition of (4.28) restricts the scattered beam to a set of two-dimensional planes perpendicular to x. The planes are separated by a distance $1/a$, and each plane is indexed by the integer h. At a point on the plane, the distance to the origin is given by $2s = \sqrt{(h/a)^2 + \Delta k_y^2 + \Delta k_z^2}$. The second and third functions in Eq. (4.19) place similar conditions as Eq. (4.20) for observing the intensity maxima. Together, they constitute following diffraction conditions:

$$\left(\vec{k}-\vec{k}_0\right)\cdot\vec{a} = h$$

$$\left(\vec{k}-\vec{k}_0\right)\cdot\vec{b} = k \tag{4.29}$$

$$\left(\vec{k}-\vec{k}_0\right)\cdot\vec{c} = l$$

Fig. 4.4 sin(Nx)/sin
(x) function plotted for
$N = 10$ and 20 with $a = 1$

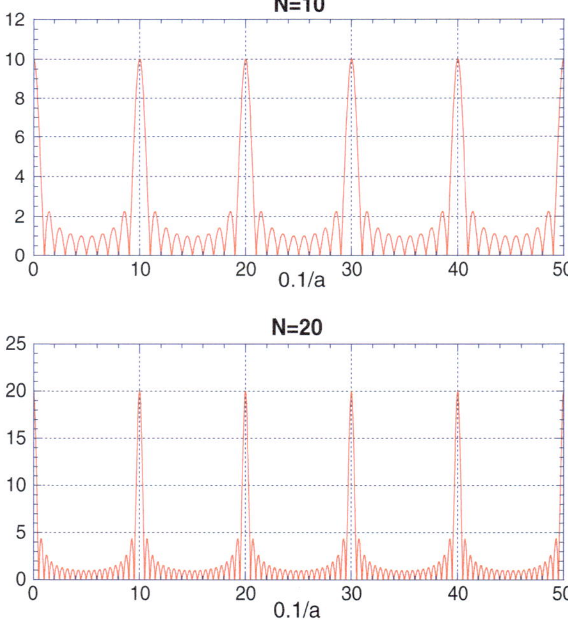

The above conditions can be met by taking

$$\vec{S} = \left(\vec{k} - \vec{k}_{\mathrm{o}}\right) = h\vec{a}^* + k\vec{b}^* + l\vec{c}^*$$

Which is the same as the Laue diffraction relation that we developed earlier, based on Bragg's law in Chap. 3. Thus, for a small crystal, diffraction peaks are observed at the Bragg condition according to kinematical theory.

4.5   Electron Crystal Structure Factors and the Diffracted Intensity from a Small Crystal

To extend the above treatment to the general case of crystals having more than a single atom inside the unit cell, we modify Eq. (4.17) to include a number of atoms and types of atoms in the unit cell:

$$V\left(\vec{r}\right) = V_{\mathrm{uc}}(\vec{r}) \otimes L(\vec{r}). \tag{4.30}$$

Here, the first term $V_{uc}(\vec{r})$ describes the potential in the crystal unit cell resulting from the contribution of each atom in the cell. Following the example of 4.18, the FT of the crystal potential can be written as follows:

$$V(\vec{S}) = V_{uc}(\vec{S}) \cdot L(\vec{S})$$

Furthermore, the unit cell potential can be written as a sum of atomic potentials over n atoms in the unit cell:

$$V_{uc}(\vec{r}) = \sum_{i=1}^{n} V_i(\vec{r}) \otimes \delta(\vec{r} - \vec{r}_i)$$

From this, we obtain in reciprocal space:

$$V_{uc}(\vec{S}) = \sum_{i=1}^{n} V_i(\vec{S}) e^{-2\pi i \vec{S} \cdot \vec{r}_i}$$

The lattice function of a general crystal has the same properties of the lattice in the monoatomic crystal. According to the last section, diffraction peaks are obtained when $\vec{S} = \vec{g} = h\vec{a}^* + k\vec{b}^* + l\vec{c}^*$. Fourier transform of the unit cell potential at the peak position gives the Fourier coefficients of the periodic potential:

$$V_g = \frac{1}{V_c} \sum_{i=1}^{n} V_i(\vec{g}) e^{-2\pi i \vec{g} \cdot \vec{r}_i} = \frac{1}{V_c} \sum_{i=1}^{n} V_i(\vec{g}) e^{-2\pi i (hx_i + ky_i + lz_i)} \tag{4.31}$$

Here, (x, y, z) are components of the atomic position vector, V_c for the unit cell volume, and $V_i(\vec{g})$ is the Fourier coefficient of the potential of ith atom.

Having defined the Fourier transform of the crystal potential and the lattice, we are now in a position to derive the scattered intensity, which can be used to calculate the kinematical diffraction intensity from a small crystal. The incident wave is assumed to be uniform on the crystal in the form of a plane wave, corresponding to a coherent, parallel (collimated), electron beam. Using the results in Eqs. (4.15), (4.21), (4.27), and (4.31), we have for a small crystal:

$$I_{sc}\left(\frac{\text{electrons}}{\text{sec}}\right) = \frac{I_o}{A_o}\left(\frac{\text{electrons}}{\text{sec} \times \text{area}}\right)\left(\frac{d\sigma}{d\Omega}\right)\Delta\Omega = \frac{I_o}{A_o}\left|F_g^B\right|^2 \left|\Lambda(\vec{S}_g)\right|^2 \Delta\Omega \tag{4.32}$$

where I_o/A_o is the incident flux in units of the number of electrons per second and per unit area, $\vec{S}_g$ is the deviation of the scattering vector from the Bragg condition as defined by

$$\vec{S}_g = \vec{k} - \vec{k}_o - \vec{g} \tag{4.33}$$

The electron structure factor F_g^B is related to the Fourier coefficients of the periodic crystal potential:

$$F_g^B = \frac{2\pi m_e e}{h^2} V_c V_g, \tag{4.34}$$

and Λ is the so-called lattice shape function, given by the Fourier transform of the lattice function in (4.30). This function has two contributions, one from the periodic lattice and second from the termination of the lattice due to the shape of the small crystal. For a small parallelepiped crystal with N_1, N_2, and N_3 unit cells along each crystallographic axes, this function as we saw before has the form of

$$\Lambda\left(\vec{S}_g\right) = \frac{\sin\left[\pi\vec{S}_g \cdot N_1\vec{a}\right]}{\sin\left[\pi\vec{S}_g \cdot \vec{a}\right]} \frac{\sin\left[\pi\vec{S}_g \cdot N_2\vec{b}\right]}{\sin\left[\pi\vec{S}_g \cdot \vec{b}\right]} \frac{\sin\left[\pi\vec{S}_g \cdot N_3\vec{c}\right]}{\sin\left[\pi\vec{S}_g \cdot \vec{c}\right]} \tag{4.35}$$

Further discussion on small crystal diffraction and shape functions can be found in Chap. 17.

4.6   Integrated Diffraction Intensity of a Rotating Crystal

The deviation from the Bragg condition of (hkl) reflection is expressed by the excitation error as the deviation from the Bragg condition $\vec{k} - \vec{k}_o = \vec{g} + \vec{S}_g$. When $\vec{S}_g = 0$, we have

$$\Lambda\left(\vec{S}_g = 0\right) = N_1 N_2 N_3,$$

and

$$I_{sc}\left(\vec{S}_g = 0\right) = \frac{I_o}{A_o}\left|F_g^B\right|^2\left|\Lambda\left(\vec{S}_g = 0\right)\right|^2 \Delta\Omega = \frac{I_o}{A_o}\left|F_g^B\right|^2\left(\frac{V_{\text{sample}}}{V_c}\right)^2 \Delta\Omega, \tag{4.36}$$

Which gives the kinematic diffraction intensity at the Bragg peak.

Experimentally, it is often not possible to measure diffraction intensity at the exact Bragg angle (the exception is CBED, which is discussed in later Chapters). An example is powder diffraction, where the recorded diffraction intensity is integrated over the orientations of powder crystals. A whole range of values is obtained for S_{hkl} from large negative to large positive as long as the crystallites are small enough and they are randomly oriented. Another case is precession electron

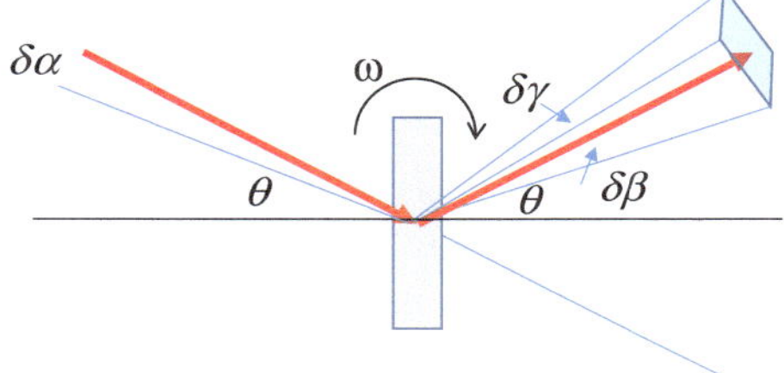

Fig. 4.5 Electron Bragg diffraction of a rotating crystal recorded by an area detector

diffraction (see Chap. 10 for details), where the incident beam is intentionally rotated in relatively large angles, and the effect on S_{hkl} is similar to powder diffraction.

We next consider diffraction from a rotating crystal as illustrated in Fig. 4.5 with the incident electron beam fixed, which can be taken as equivalent to precession electron diffraction with the rotating beam and a fixed crystal. There are three angles here to consider, α, β, and γ, with the latter two defined by the two-dimensional detector. Change in α is introduced by crystal rotation, for time δt, we have $\delta \alpha = \omega \delta t$. Thus, the integrated intensity is given by

$$E = \int I_{\text{sc}} d\Omega dt = \frac{I_{\text{o}}}{A_{\text{o}} \omega} \left| F_g^B \right|^2 \int \left| \Lambda \left(\vec{S}_g \right) \right|^2 d\beta d\gamma d\alpha. \tag{4.37}$$

The change in crystal orientation, or detector angles, leads to a change in the excitation error. It can be shown (Warren 1990)

$$\frac{\sin 2\theta}{\lambda^3} d\beta d\gamma d\alpha = \frac{1}{V_c} dS_x dS_y dS_z, \tag{4.38}$$

where S_x, S_y and S_z are along $\vec{a}^*$, $\vec{b}^*$ and $\vec{c}^*$ directions, respectively. Substituting Eq. (4.38) into Eq. (4.37) leads to three integrals of the following type

$$\int \left(\frac{\sin[\pi SNa]}{\sin[\pi Sa]} \right)^2 dS$$

When the integral is carried in the neighborhood of the Bragg peak ($\theta \approx \theta_g$) with S small enough so that only one peak is included, we then have approximately

$$\int \left(\frac{\sin[\pi SNa]}{\sin[\pi Sa]} \right)^2 dS \approx \int \frac{\sin^2[\pi SNa]}{(\pi Sa)^2} dS = N. \tag{4.39}$$

Using this result and Eq. (4.38), we obtain

$$E = \int I_{\mathrm{sc}} \mathrm{d}\Omega \mathrm{d}t = \frac{I_{\mathrm{o}}}{A_{\mathrm{o}}\omega} \left| F_g^B \right|^2 \frac{\lambda^3 V_{\mathrm{sample}}}{V_c^2 \sin 2\theta_g} \tag{4.40}$$

Thus, integration over three directions leads to a dependence on sample volume, instead of the square of sample volume as predicted for the peak Bragg intensity in kinematic approximation.

4.7 Atomic Thermal Vibrations and Effect on Electron Scattering

So far we have treated the crystal as static and rigid, with atoms fixed in their positions. In a real crystal, atoms are far from being static for two reasons. One is due to quantum mechanics, where the uncertainty principle dictates that atoms must deviate from its equilibrium position in order to have a finite kinetic energy. Such atomic motions are called zero-point motions and are independent of temperature. The second reason is due to thermal excitation. The thermal energy is stored in atomic motion. Atoms are bond together through attractive and repulsive forces. Atomic vibration propagates in a crystal in the so-called phonon modes, compatible with crystal symmetry. A proper treatment of crystal diffraction must take account of the effects of atomic vibrations on elastic scattering, as well as inelastic scattering. For this section, we will focus on elastic scattering and take account of the effects of atomic vibration on the observed diffraction intensity. Inelastic thermal scattering will be discussed in Chap. 13. For simplicity, we will consider scattering from a monoatomic crystal including atomic vibration. The treatment method, as well as conclusions, obtained here also applies to the general case of crystals with multiple atoms. The monoatomic crystal case allows an expression of the theory in the simplest mathematic form.

Consider the structure of a monoatomic crystal at time t. Because of atomic vibrations, the structure is no longer periodic and the position of nth atom is given by

$$\vec{r}_n(t) = \vec{R}_n + \vec{u}_n(t),$$

where $\vec{u}_n(t)$ is the time-dependent displacement from the lattice site. The diffraction intensity recorded at time t is simply a sum of scattering from all atoms:

$$I(t) = \left| \sum_n f(q) \mathrm{e}^{-i\vec{q}\cdot\vec{r}_n(t)} \right|^2 = f^2(q) \sum_{n,m} \mathrm{e}^{i\vec{q}\cdot\left(\vec{R}_n - \vec{R}_m\right)} \mathrm{e}^{i\vec{q}\cdot[\vec{u}_n(t) - \vec{u}_m(t)]}$$

Here, $\vec{q} = 2\pi\left(\vec{k} - \vec{k}_{\mathrm{o}}\right)$. The experimental diffraction intensity is recorded over a period of time, typically much longer than the period of atomic vibrations. Thus, experimentally, we have:

$$I = \langle I(t)\rangle = f^2(q) \sum_{n,m} e^{i\vec{q}\cdot\left(\vec{R}_n - \vec{R}_m\right)}\left\langle e^{i\vec{q}\cdot[\vec{u}_n(t) - \vec{u}_m(t)]}\right\rangle$$

To simplify this further, let $u_{qn} = \vec{q}\cdot\vec{u}_n(t)$, then

$$I = \langle I(t)\rangle = f^2(q) \sum_{n,m} e^{i\vec{q}\cdot\left(\vec{R}_n - \vec{R}_m\right)}\left\langle e^{iq(u_{qn} - u_{qm})}\right\rangle \tag{4.41}$$

The average in (4.41) can be expanded in the case of small atomic displacements using the relationship:

$$e^{iqx} \approx 1 + iqx - \frac{1}{2}(qx)^2 + i\frac{1}{6}(qx)^3 + \cdots$$

and

$$\langle e^{iqx}\rangle \approx 1 - \frac{1}{2}\left\langle(qx)^2\right\rangle + \cdots \approx e^{-q^2\langle x\rangle^2/2}$$

Applying this to Eq. (4.41), we obtain the following equation:

$$I = f^2(q) \sum_{n,m} e^{i\vec{q}\cdot\left(\vec{R}_n - \vec{R}_m\right)} e^{-\left(qu_{qn}\right)^2/2} e^{-\left(qu_{qm}\right)^2/2} e^{-q^2\langle u_{qn}u_{qm}\rangle} \tag{4.42}$$

The last term describes the effect of correlated atomic vibration; this term can be expanded based on the assumption of small atomic vibrations according to:

$$\exp\left(-q^2\langle u_{qn}u_{qm}\rangle\right) \approx 1 - q^2\langle u_{qn}u_{qm}\rangle + \cdots.$$

The first term gives the zero order diffracted intensity, while the second term leads to an expression for thermal diffuse scattering, which will be discussed in Chap. 13. The zero order term from Eq. (4.42) gives

$$I_{\mathrm{o}} = f^2(q)e^{-2M} \sum_{n,m} e^{i\vec{q}\cdot\left(\vec{R}_n - \vec{R}_m\right)} = \left[f(q)e^{-M}\right]^2 \sum_{n,m} e^{i\vec{q}\cdot\left(\vec{R}_n - \vec{R}_m\right)}$$

With

$$M = q^2\langle u_q\rangle^2/2 = B(\sin\theta/\lambda)^2$$

For isotropic atomic vibration, the average atomic displacement is independent of direction and thus $\langle u_q\rangle^2 = \langle u\rangle^2$, which is the mean-square vibrational amplitude of the atom. Here, B is the Debye–Waller factor defined by

$$B = 8\pi^2\langle u\rangle^2 \tag{4.43}$$

In the Debye model for lattice vibrations, a linear dispersion relationship is assumed and the excitation spectrum parameterized using a Debye temperature Θ_D. Then, the mean-square vibrational amplitude is given by

$$\langle u\rangle^2 = \frac{3h^2}{4\pi^2 M_A k_B \Theta_D}\left[\frac{1}{4} + \frac{T^2}{\Theta_D^2}\int_0^{\Theta_D/T}\frac{x\,dx}{\exp(x)-1}\right]$$

where M_A is the atomic mass. Values of B and the integral above are tabulated in the International Tables for Crystallography, and a tabulation for elemental crystals and comparison with measurements can be found in Sears and Shelley (1991), Peng et al. (1996a) and Gao and Peng (1999). The later publication also included 17 compound crystals with the zincblende structure (see also Reid 1983). Figure 4.6 plots the values for several elemental crystals in the temperature range of 80–900 K. The first temperature-independent term in the parentheses results from zero-point motion, and this produces appreciable effects in electron diffraction which cannot be neglected or removed by cooling (Humphreys and Hirsch 1968).

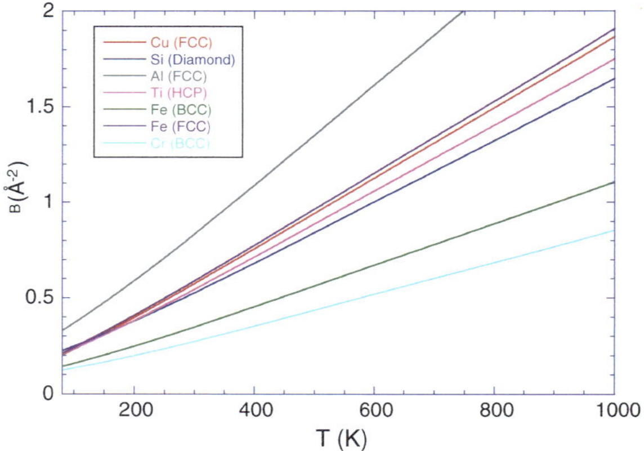

Fig. 4.6 Debye–Waller factors of selected elemental crystals from temperatures of 80–1000 K. (Data from Gao and Peng 1999)

4.8 Electron Structure Factors

Now we are in position to define the FT of atomic potential in Eq. (4.31) by combining electron atomic scattering and the effect of atomic thermal vibrations. If B, s, and V_c are instead given in angstrom units, then V_g is given in volts as follows:

$$
\begin{aligned}
V_g &= \frac{1.145887}{V_c} \sum_{i=1}^{n} \frac{\left[Z_i - f_i^x(s)\right]}{s^2} \exp\left(-B_i s^2\right) e^{-2\pi i (h x_i + k y_i + l z_i)} \\
&= \frac{47.877647}{V_c} F_{hkl}^B
\end{aligned}
\tag{4.44}
$$

Equation (4.44) allows electron "structure factors" V_g and F_g^B to be evaluated from tabulations of X-ray atomic scattering factors $f_i^x(s)$ if B_i is known. Two other quantities commonly used in the literature are U_g and ξ_g, given by

$$
U_g = \frac{\gamma}{\pi V_c} F_g^B = \frac{2m|e|}{h^2} V_g
\tag{4.45}
$$

and

$$
\xi_g = \frac{1}{\lambda |U_g|} = \frac{\pi}{\sigma |V_g|} = \frac{\pi V_c}{\gamma \lambda \left| F_g^B \right|}
\tag{4.46}
$$

where γ is the relativistic constant. Many of these useful relationships are collected together in Appendix A. The definition of extinction distance ξ_g is based on the two-beam intensity expression at the Bragg condition as we will see in next Chapter; it depends on the amplitude of the structure factors.

Since these quantities depend on the details of the scattering experiment (accelerating voltage), they are not true "structure factors," unlike V_g and F_g, which are properties of the crystal alone, as defined above. For V_g in volts and U_g in Å^{-2}, we then have

$$
U_g = 0.006648403 \left(1 + 1.956951 \times 10^{-6} \Phi\right) V_{\vec{g}}
\tag{4.47}
$$

If the X-ray structure factor is defined by

$$
F_g^x = \sum_{i=1}^{n} f_i^x(s) \exp\left(-B_i s^2\right) e^{-2\pi i (h x_i + k y_i + l z_i)}
\tag{4.48}
$$

with the electronic charge density (in electrons per cell) as follows:

$$\rho(\vec{r}) = \frac{1}{V_c} \sum_g F_g^x e^{-2\pi i \vec{g} \cdot \vec{r}} \tag{4.49}$$

then Eqs. (4.48), (4.45), and (4.44) give the following expression for the retrieval of an X-ray structure factor F_g^x from electron diffraction data:

$$
\begin{aligned}
F_g^x &= \sum_{i=1}^n Z_i \exp\left(-B_i s^2\right) e^{-2\pi i \vec{g} \cdot \vec{r}_i} - \left(\frac{8\pi^2 \varepsilon_0 h^2 V_c s^2}{\gamma m_e e^2}\right) U_g \\
&= \sum_{i=1}^n Z_i \exp\left(-B_i s^2\right) e^{-2\pi i \vec{g} \cdot \vec{r}_i} - \left(\frac{C V_c s^2}{\gamma}\right) U_g
\end{aligned}
\tag{4.50}
$$

Here, the numerical constant $C = 131.2625$ if s, V_c, and U_g are given in angstrom units.

The role of the temperature factor in structure-factors calculation and conversions between X-ray and electron structure factors is important. The atomic vibrational amplitude u is appreciable even at 0 K, where in many materials, it falls to only about half its room-temperature value. Thus, the observable crystal potential is a temperature-dependent quantity—the static potential computed from band-structure calculations is not an experimental observable. From Eqs. (4.44) and (4.50), we may draw the following conclusions:

1. Debye–Waller factors must be known in order to take account of atomic vibrations on atomic scattering factors.
2. Debye–Waller factors must be known in order to compare the results of measurements taken at different temperatures. Since the neutron diffraction studies or lattice dynamical calculations required to determine $\langle u_q \rangle^2$ have been completed for relatively few crystals, this has created considerable difficulty in the past for comparisons of structure-factor measurements reported by different groups, who may work at different temperatures.
3. A knowledge of the Debye–Waller factor B_i is essential in order to convert a structure factor U_g measured by electron diffraction at temperature T into the corresponding X-ray structure factor F_g^x, at temperature T. A knowledge of B_i is similarly required to convert measured X-ray scattering factors into electron scattering factors for the same temperature.
4. From Eq. (4.50) we see that, at small scattering angles, small changes in F_g^x result in large changes in U_g. Thus, if U_g is known to a particular percent error, F_g^x may be deduced to a greater accuracy.

The asymptotic behavior of the scattering factors for large and small values of s must be considered (Peng and Cowley 1988). To obtain the desired asymptotic behavior in which the electron scattering factor converges to the mean potential, it is convenient to use Eq. (4.44) written in the form

$$
V_0 = \lim_{s->0} \frac{1.145887}{V_c} \sum_{i=1}^{n} \frac{\left[\sum_{j=1}^{4} a_i \left(1 - e^{-b_i s^2} \right) \right]}{s^2} \exp\left(-B_i s^2 \right)
$$
$$
= \frac{1.145896}{V_c} \sum_{i=1}^{4} a_i b_i.
$$

(4.51)

4.9   Electron-Optical Potential

An additional imaginary potential may be added to the electron interaction potential $U(\vec{r})$ in order to describe the depletion of the elastic wave field by inelastic scattering (absorption). The total potential is then known as an optical potential, given by

$$
U(\vec{r}) = U^C(\vec{r}) + iU'(\vec{r})
$$

(4.52)

An additional correction to the real potential representing virtual inelastic scattering has been neglected, since this is a very small effect (Rez 1978). $U^C(\vec{r})$ is the real crystal potential (describing the interaction of the incident electron with the crystal electrons and the nuclei) and $U'(\vec{r})$ a second real potential which accounts for depletion of the elastic wave field by inelastic scattering. The term "absorption" is used to describe this depletion, since the probability that the electron is scattered back into the original state is very small. The use of such a phenomenological absorption potential in HEED has been justified theoretically (Yoshioka 1957). For high-energy electrons, there are three important inelastic scattering mechanisms: (1) inelastic scattering resulting from the excitation of crystal electrons, (2) excitation of plasmons, and (3) excitation of phonons. The contributions of plasmon scattering are confined to very low scattering angles, and they decrease rapidly as the scattering angle increases. To a good approximation, their effect can be taken into account by considering the mean absorption (U'_0) only. The contribution to $U'_{\vec{g}}$ ($\vec{g} \neq 0$) comes mostly from phonon scattering (Yoshioka and Kainuma 1962; Hall and Hirsch (1965); Radi (1970)). The mean absorption describes an overall attenuation of the incident electrons. Figure 4.7 shows calculated phonon absorption potential for copper. The total absorption is seen to deviate only slightly from the phonon scattering at very low angles.

Calculations for total absorption coefficients, including plasmon excitation, single electron excitation, and phonon scattering, are given in Radi (1970) and Humphreys and Hirsch (1968) for a range of crystal structures. Otherwise measurements and calculations for particular crystals are scattered throughout the literature (see Reimer and Kohl (2008) for a summary). A comparison of measured

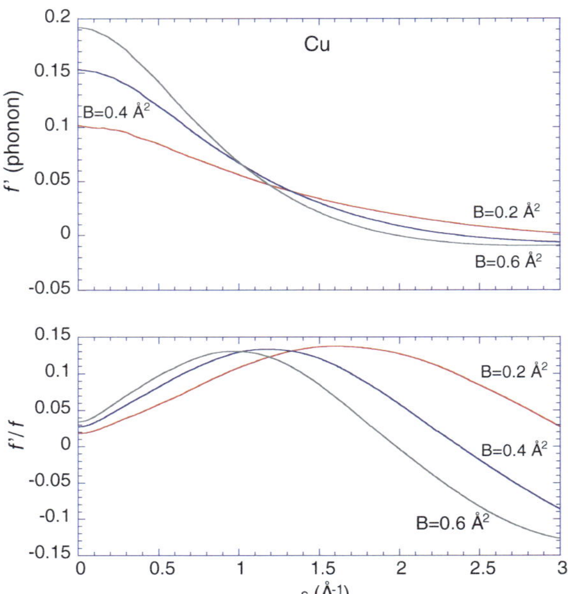

Fig. 4.7 (*Top*) The phonon absorption potential for copper at three values of Debye–Waller factor as a function of *s* for 100 kV electrons. (*Bottom*) ratio of absorption potential versus real potential

and experimental values for Al, Cu, Au, Si, Ge, MgO, and NaCl is given by Weickenmeier and Kohl (1991), in which a Fortran program is offered on request.

Absorption has two effects on all crystals in electron diffraction. The first is the average absorption, which gives rise to a mean complex wave vector

$$k_z = k_{oz} + U_o/2k_{oz} + iU_o'/2k_{oz} \tag{4.53}$$

The imaginary part yields a damping term

$$\exp\left(-4\pi U_o' t/2k_{oz}\right) = \exp(-t/\Lambda) \tag{4.54}$$

which multiples all the diffracted beam intensities, where

$$\Lambda = k_{oz}/\left(2\pi U_o'\right) \tag{4.55}$$

is the equivalent elastic penetration depth of the incident electron.

A second effect of absorption has become known as the anomalous transmission effect, by analogy with the Borrmann effect in X-ray diffraction. This is due to the terms V_g'. This will be further discussed in next Chapter.

The absorption coefficients for phonon scattering may be calculated analytically most simply using an Einstein model. This model assumes that the atoms in the crystal vibrate independently of each other, in contrast to more accurate models based on discrete vibrational modes. Thus, the thermal diffuse scattering calculated using the Einstein model cannot predict the thermal diffuse streaks which are often observed in experimental patterns and which can be related to the phonon dispersion (Kitamura 1966). A more realistic model is the Debye model. However, it has also been shown (Hall and Hirsch 1965) that for gold, at least, the Debye model and Einstein model give similar absorption coefficients. It is not clear whether the same conclusion holds for more complex crystals; however, the incorporation of lattice vibration models greatly complicates the theory. The mean atomic vibration amplitude $\langle u^2 \rangle$ is related to the Debye–Waller factor $\exp(-Bs^2)$ by Eq. (4.43).

A Fortran program for computing the phonon scattering contribution to the absorption coefficients is also described in Bird and King (1990), together with tabulated values for Al, Cu, Ag, Au, C, and Ga. Other tabulations are provided by Weickenmeier and Kohl (1991) and Peng et al. (1996a). It is commonly assumed that other electronic processes make a significant contribution only for $g = 0$. These programs can provide values of the absorption coefficient U'_g, and hence V'_g through Eq. (4.45) (with primes added to U_g and V_g). They require the Debye–Waller factor for the crystal of interest and are based on an Einstein model. In summary, calculations based on the Einstein model, using the Debye–Waller factor as input, appear to be a useful first approximation. The Debye–Waller factor can be found from experimental X-ray or neutron diffraction results, from theoretical calculations (see above discussions).

Equations (3.2b), (3.3), and (3.4) yield the Fourier coefficients of the elastic portion of the interaction potential for a crystal (in Å^{-2}):

$$U_g = \frac{\gamma}{\pi V_c} \sum_{i=1}^{n} f_i^B(s) \exp\left(-B_i s^2\right) e^{-2\pi i \vec{g} \cdot \vec{r}_i} \tag{4.56}$$

where B_i is the Debye–Waller factor as defined following Eq. (3.4). An atomic absorptive coefficient $f'_i(s)$ can be defined similarly by

$$U'_g = \frac{\gamma}{\pi V_c} \sum_{i=1}^{n} f'_i(s) \exp\left(-B_i s^2\right) e^{-2\pi i \vec{g} \cdot \vec{r}_i} \tag{4.57}$$

where U'_g is the imaginary part of the Fourier coefficient of the optical potential. The "atomic" absorption coefficient $f'_i(s)$ with $s = g/2$ is then given by Hall and Hirsch (1965):

$$f'(g) = \frac{1}{k} \int f^B(\vec{q}) f^B(\vec{q} - \vec{g}) \left[\exp\left(-Bg^2/4\right) - \exp\left(-B\left(q^2 - (\vec{q} - \vec{g})^2\right)/4\right) \right] d^2\vec{q} \tag{4.58}$$

The integral may be evaluated on a grid of B and g values (Bird and King 1990), or analytically if the atomic scattering factor is first expanded as a sum of Gaussians (Weickenmeier and Kohl 1991).

References

Bird DM, King QA (1990) Absorptive form-factors for high-energy electron-diffraction. Acta Cryst A46:202–208

Doyle PA, Turner PS (1968) Relativistic Hartree-Fock X-ray and electron scattering factors. Acta Crystallogr A 24:390

Gao HX, Peng LM (1999) Parameterization of the temperature dependence of the Debye-Waller factors. Acta Cryst A55:926–932

Hall CR, Hirsch PB (1965) Effects of thermal diffuse scattering on propagation of high energy electrons through crystals. Proc Roy Soc London Ser A 286:158

Humphreys CJ, Hirsch PB (1968) Absorption parameters in electron diffraction theory. Philos Mag 18:115

Kitamura N (1966) Temperature dependence of diffuse streaks in single crystal Si electron diffraction patterns. J Appl Phys 37:2187–2188

Peng LM, Cowley JM (1988) Errors arising from numerical use of the Mott formula in electron image simulation. Acta Cryst A44:1–5

Peng LM, Ren G, Dudarev SL, Whelan MJ (1996a) Robust parameterization of elastic and absorptive electron atomic scattering factors. Acta Cryst. A52:257–276

Peng LM, Ren G, Dudarev SL, Whelan MJ (1996b) Debye-Waller factors and absorptive scattering factors of elemental crystals. Acta Cryst A52:456–470

Prince E (ed) (2004) International tables for crystallography volume C: mathematical, physical and chemical tables. Kluwer, Academic Publishers, Boston

Radi G (1970) Complex lattice potentials in electron diffraction calculated for a number of crystals. Acta Crystallogr A 26:41

Reid JS (1983) Debye-Waller factors of zinc-blend-structure materials—a lattice dynamical comparison. Acta Cryst A39:1–13

Reimer L, Kohl H (2008) Transmission electron microscopy (4th). Springer, Berlin

Rez P (1978) The theory of inelastic scattering in electron microscopy of crystals, Thesis D.Phil., University of Oxford

Saxton WO, O'Keefe MA, Cockayne DJH, Wilkens M (1984) Sign conventions in electron-diffraction and imaging. Ultramicroscopy 12:75–78

Sears VF, Shelley SA (1991) Debye-Waller factor for elemental crystals. Acta Cryst. A47:441–446

Spence JCH (2013) High resolution electron microscopy, 4th edn. Oxford University Press, Oxford

Vincent R, Midgley PA (1994) Double conical beam-rocking system for measurement of integrated electron-diffraction intensities. Ultramicroscopy 53:271–282

Warren BE (1990) X-ray diffraction, Reprint edn. Dover Publications, Mineola

Weickenmeier A, Kohl H (1991) Computation of absorptive form-factors for high-energy electron-diffraction. Acta Cryst A47:590–597

Wu T-Y, Ohmura T (2011) Quantum theory of scattering. Dover Publications, Mineola

Yoshioka H (1957) Effect of inelastic waves on electron diffraction. J Phys Soc Jpn 12:618–628

Yoshioka H, Kainuma Y (1962) Effect of thermal vibrations on electron diffraction. J Phys Soc Jpn 17:134

Chapter 5
Dynamical Theory of Electron Diffraction for Perfect Crystals

In this chapter, we outline the dynamical theory of high-energy transmission electron diffraction. There have been three major, and related, approaches to the subject; each was developed for specific applications. The first is based on the study of few-beam solutions (and cases reducible to them) in arbitrary orientations, following Bethe's original work (Bethe 1928) aimed at explaining the first demonstration of electron diffraction by Davisson and Germer. This powerful Bloch wave approach was developed first in Europe and Japan for the study of reflection electron diffraction, for transmission diffraction from simple structures, for Kikuchi lines, and for HOLZ line analysis and phase measurement. It has been applied most successfully to small unit cell crystals. The second approach, developed in the UK in the late 1950s, was developed for nanometer-resolution ("diffraction-contrast") imaging of crystals with defects and treats electron propagation in nanometer-sized columns of crystal, by considering scattering among a small number of diffracted beams under the so-called column approximation. This approach approximates the atomic structure of defects by a thickness-dependent rigid and uniform displacement in the so-called Howie–Whelan equations (Howie and Whelan 1961). The multislice method as the third approach was developed in Australia. This method has been shown to be highly efficient for numerical simulations of large unit cell crystals where hundreds of beams may be involved and can also be used for disordered materials or small nanostructures through the use of a computational superlattice (Cowley and Moodie 1957). As a numerical method, it is less well suited for developing theoretical insights, but because of its flexibility it has produced the most widely used algorithm in electron microscopy for the interpretation of high-resolution electron microscopy (HREM) images, and more recently for simulating STEM images (see Spence (2013), Kirkland (2010) for a comparison of these theories and the computational algorithms based on them).

The theory described here is based on solving the Schrödinger equation by expanding the electron wave function inside the crystal using a set of Bloch waves as originally proposed by Bethe (1928). The same formulation can also be used to solve the Howie–Whelan equation for imperfect crystals. Thus, one can start out

© Springer Science+Business Media New York 2017

J.M. Zuo and J.C.H. Spence, *Advanced Transmission Electron Microscopy*,
DOI 10.1007/978-1-4939-6607-3_5

with the theory of dynamic diffraction involving many beams and solved it analytically in case of a few beam cases to build a knowledge base for dynamic diffraction, including the effect on HOLZ lines. Quantitative results are obtained by many-beam calculations, including dynamical diffraction involving defects.

Students new to the field looking for pedagogically sound reviews are particularly referred to Hirsch et al. (1977), Humphreys (1979), Metherall (1975), Peng et al. (2004).

5.1 Many-Beam Theory, Wave-Mechanical Approach

Reviews of the Bloch wave method for solving the problem of high-energy transmission electron diffraction in the ZOLZ approximation can be found in Humphreys (1979) and Metherall (1975). We now extend these treatments by the renormalized eigenvector method of Lewis et al. (1978) to include HOLZ effects, acentric crystals, absorption, and inclined boundary conditions. Our aim is to provide the theoretical basis for quantitative electron diffraction intensity analysis, from which specialized two- and three-beam cases may be extracted in subsequent sections to illustrate the effects of dynamic diffraction and introduce basic concepts.

We consider a collimated incident electron beam of the form $\phi_{\mathrm{o}} = \exp\left(2\pi i \vec{k}_{\mathrm{o}} \cdot \vec{r}\right)$. Then, the Schrödinger equation describing high-energy electron diffraction, after some rearrangements of Eq. (2.16) (Chap. 2), is

$$\frac{1}{4\pi^2} \nabla^2 \phi(\vec{r}) + U(\vec{r})\phi(\vec{r}) + k_{\mathrm{o}}^2 \phi(\vec{r}) = 0, \tag{5.1}$$

where $U(\vec{r})$ is the electron interaction potential described in Chap. 4. Relativistic correction is included by using the relativistic mass m for electrons. Justification for the use of Eq. (5.1) and implied lack of spin effects can be found in the reviews mentioned above.

The interaction potential $U(\vec{r})$ is periodic; that is, the value repeats itself after a lattice vector translation or $U(\vec{r}) = U\left(\vec{r} + \vec{R}_n\right)$. A solution of Eq. (5.1) for such a potential must have the following properties, according to the Bloch theorem:

$$\phi(\vec{r}) = e^{2\pi i \vec{k} \cdot \vec{r}} \varphi(\vec{r}) \tag{5.2}$$

where $\varphi(\vec{r}) = \varphi\left(\vec{r} + \vec{R}_n\right)$ has the same period as the interaction potential and $\vec{k}$ is the Bloch wave vector. A 3D periodic function can be expanded as a Fourier series, whose spatial frequency corresponds to reciprocal lattice vector $\vec{g}$. Thus, we have for the interaction potential

$$U(\vec{r}) = \sum_g U_g e^{2\pi i \vec{g}\cdot\vec{r}}$$

Here, U_g a Fourier coefficient of the interaction potential. Similarly,

$$\varphi(\vec{r}) = \sum_g C_g e^{2\pi i \vec{g}\cdot\vec{r}}$$

The number of Bloch wave solutions to Eq. (5.1) equals to the number of Fourier coefficients included in the Fourier series expansion of the interaction potential. The electron wave function inside the crystal is a sum of those Bloch waves:

$$\phi(\vec{r}) = \sum_i c_i e^{2\pi i \vec{k}_i\cdot\vec{r}} \sum_g C_g^i e^{2\pi i \vec{g}\cdot\vec{r}} \tag{5.3}$$

Here, i is the index of a Bloch wave and c_i is the coefficient representing the excitation of ith Bloch wave. The result of inserting these two equations into Eq. (5.1) and equating coefficients yields the standard dispersion equations of high-energy electron diffraction (HEED)

$$\left[K^2 - (\vec{k}+\vec{g})^2 \right] C_g + \sum_h U_{gh} C_h = 0 \tag{5.4}$$

Here, the subscript i in $\vec{k}$ is skipped, so it represents all Bloch waves, and

$$K^2 = k_o^2 + U_o \tag{5.5}$$

gives the incident wave number inside the crystal.

The complex electron interaction coefficients are as in Sect. 4.9

$$U_g = 2m|e|V_g/h^2 = U_g^C + U_g'' + iU_g'$$

with V_g a Fourier coefficient of the total crystal potential in volts. It has three components, a contribution from crystal potential U^c, from absorption U' and a correction due to modification of the crystal potential by virtual inelastic scattering U''. In noncentrosymmetric (acentric) crystals, both U_g^C and U_g' are the complex Fourier coefficients of real potentials with the period of the lattice. The most important contribution to U' and U'' for $g > 0$ comes from inelastic phonon scattering. Details of the evaluation of the absorption potential can be found in Sect. 4.9. The U'' term, including exchange and correlation between the beam electron and crystal electrons, is significantly smaller than U' and is neglected.

As in the theory of electronic band structure in crystals, our general approach is to force Eq. (5.4) into the form of an eigenvalue equation and then solve it to yield eigenvalues and eigenvectors. The constants c_i must be obtained from the boundary conditions and so depend on the shape of the crystal. In order to satisfy the boundary conditions, the tangential components of the incident and Bloch wave vectors inside the crystal must be matched at the crystal entrance surface. Thus, for a crystal slab, for all Bloch waves,

$$\vec{k}_t = \vec{K}_t = \vec{k}_{ot}.$$

We now let

$$\vec{k} = \vec{K} + \gamma\vec{n} \tag{5.6}$$

where $\vec{n}$ is a unit vector out of, and normal to, the slab (against the beam), and both $\vec{k}$ and γ may be complex, to allow for absorption. If we introduce the expansion

$$\begin{aligned} K^2 - (\vec{k} + \vec{g})^2 &= K^2 - (\vec{K} + \vec{g})^2 - 2(\vec{K} + \vec{g}) \cdot \vec{n}\gamma - \gamma^2 \\ &= 2KS_g - 2(\vec{K} + \vec{g}) \cdot \vec{n}\gamma - \gamma^2 \end{aligned}$$

then

$$\left[2KS_g - 2(\vec{K} + \vec{g}) \cdot \vec{n}\gamma - \gamma^2\right]C_g + \sum_h U_{gh}C_g = 0 \tag{5.7}$$

is obtained with excitation errors S_g defined in Eq. (3.14). Thus,

$$2KS_g = K^2 - (\vec{K} + \vec{g})^2 \tag{5.8}$$

in general, and

$$2KS_g = -2\vec{K}_t \cdot \vec{g} - g^2 \tag{5.9}$$

for ZOLZ reflections.

As discussed in Chap. 3, we may think of the CBED pattern as a function of $\vec{K}_t$, a vector which originates in the center of the zone axis and extends to a point of interest in the central disk. The reciprocal lattice links this point to points in all the other disks. Four diffraction conditions in particular are commonly encountered in CBED: the axial (systematics) orientation, the zone-axis orientation (for which $\vec{K}_t = 0$ at the center of the (000) disk), the two-beam Bragg condition ($\vec{K}_t = -\vec{g}/2$), and the three-beam condition (described in Sect. 5.6.1) which is important for phase determination.

Equation (5.7) includes both the forward-scattered waves of interest for CBED and the backscattered waves important for RHEED. We define $g_n = \vec{g} \cdot \vec{n}$ and $K_n = \vec{K} \cdot \vec{n}$. In transmission diffraction, K_n is large, γ is small ($\gamma \ll K_n$) for the forward-scattered waves and large ($\gamma \approx -2K_n$) for the backscattered waves. The excitation coefficients of the backscattered waves are very small and usually neglected in high-energy transmission electron diffraction. Neglect of the backscattered waves (thus, $\gamma \ll K_n$) leads to a very small γ^2, and dropping this term gives

$$2KS_g C_g + \sum_h U_{gh} C_h = 2K_n \left(1 + \frac{g_n}{K_n}\right) \gamma C_g. \qquad (5.10)$$

This, with Eq. (5.10) includes all HOLZ effects, boundary inclination effects, and absorption terms, and may be applied to acentric crystals. The most important approximation has been the neglect of backscattering. Equation (5.10) can be reduced to an eigenvalue equation by renormalizing the eigenvector:

$$B_g = \left(1 + \frac{g_n}{K_n}\right)^{1/2} C_g$$

If the surface normal is approximately antiparallel to the beam, so that $K_n \gg g_n$, then g_n/K_n is negligible, and we have

$$2KS_g C_g + \sum_h U_{gh} C_h = 2K_n \gamma C_g \qquad (5.11)$$

The latter equation may be written in matrix form

$$AC = 2K_n \gamma C \qquad (5.12)$$

where the off-diagonal entries of the "structure matrix" A are U_{gh}, while the diagonal entries are the excitation error terms $2KS_g$. Here, C is a column vector. For centrosymmetric crystals without absorption, A is real, symmetric, and Hermitian. For centrosymmetric crystals with absorption, A is complex and not Hermitian.

By neglecting g_n/K_n, Eq. (5.12) includes HOLZ effects in an approximate way, and the crystal tilts through the term in K_n. The eigenvalues γ obtained by solving Eq. (5.8) have a physical interpretation as the change in electron wave vector, in the direction of the surface normal. If n-beams are included, the structure matrix A is $n \times n$, and Eq. (5.12) gives n eigenvalues and n eigenvectors. Thus, there are N electron wave vectors excited inside the crystal, instead of a single wave vector corresponding to the incident beam in the kinematical theory. Each electron wave vector in dynamic theory gives rise to a set of plane waves, in total there are $N \times N$ different plane waves involved. These define the wave field inside the crystal according to Eq. (5.3). Alternatively, Eq. (5.3) may be regrouped according to the "Darwin representation" (Hirsch et al. 1977) of n plane waves propagating in the crystal, each in the direction of $\vec{K} + \vec{g}$, namely

$$\phi(\vec{r}) = \sum_g \phi_g e^{2\pi i (\vec{K} + \vec{g}) \cdot \vec{r}} \tag{5.13}$$

Then, the wave amplitude at crystal thickness t becomes

$$\phi_g(t) = \sum_i c_i C_g^i e^{2\pi i \gamma^i t} \tag{5.14}$$

The distance t is defined along the surface normal direction. The excitation coefficients c_i are determined by matching the incident waves to waves inside the crystal at the entrance surface. That is, we set $t = 0$ in Eq. (5.14) and then solve the resulting linear equation for c_i. This can be most elegantly expressed by writing Eq. (5.14) in matrix form

$$
\begin{pmatrix} \phi_o(t) \\ \phi_g(t) \\ \vdots \end{pmatrix}
=
\begin{pmatrix} C_o^1 & \cdots & C_o^N \\ C_g^1 & \cdots & C_g^N \\ \vdots & \ddots & \vdots \end{pmatrix}
\begin{pmatrix} e^{2\pi i \gamma^1 t} & \cdots & 0 \\ \vdots & \ddots & \vdots \\ 0 & \cdots & e^{2\pi i \gamma^N t} \end{pmatrix}
\begin{pmatrix} c_1 \\ \vdots \\ c_N \end{pmatrix}
$$

At the sample entrance surface $t = 0$, in which case we have

$$
\begin{pmatrix} \phi_o(0) \\ \phi_g(0) \\ \vdots \end{pmatrix}
= C
\begin{pmatrix} c_1 \\ \vdots \\ c_N \end{pmatrix}
\tag{5.15}
$$

The excitation coefficients c_i are found by premultiplying by C^{-1} on both sides of Eq. (5.15). Upon substituting c_i into Eq. (5.15), the wave field inside the crystal is found to be

$$
\begin{pmatrix} \phi_o(t) \\ \phi_g(t) \\ \vdots \end{pmatrix}
= C
\begin{pmatrix} e^{2\pi i \gamma^1 t} & \cdots & 0 \\ \vdots & \ddots & \vdots \\ 0 & \cdots & e^{2\pi i \gamma^N t} \end{pmatrix}
C^{-1}
\begin{pmatrix} \phi_o(0) \\ \phi_g(0) \\ \vdots \end{pmatrix}.
\tag{5.16}
$$

The matrix $S = C\left\{ e^{2\pi i \gamma^i t} \right\} C^{-1}$ is called the "scattering matrix" and relates the incident waves to the scattered waves at crystal thickness t.

If absorption is not included, the inverse of the eigenvector matrix is the transposed conjugate of the eigenvector matrix, i.e., $C^{-1} = C^{\dagger}$. If there is only one incident plane wave ($\phi_o(0) = 1$ and $\phi_g(0) = 0$), we find $c_i = C_o^{i*}$ without absorption, and $c_i = C_o^{-1i}$ with absorption. Quantities C_o^{-1i} are the elements of the first column of the inverse of the matrix whose elements are C_g^i (column i, row g). The intensity of a particular Bragg beam (for a given incident plane wave $\vec{K}$) is then found from

$$I_g(K_x, K_y) = \left|\phi_g(K_x, K_y)\right|^2 = \left|\sum_i c_i(K_x, K_y) C_g^i(K_x, K_y) \exp\left[2\pi i \gamma^i(K_x, K_y)t\right]\right|^2$$

(5.17)

where t is again the crystal thickness.

5.2 Howie–Whelan Equations

The above many-beam theory was developed by solving the Schrödinger equation for transmission electron diffraction from a periodic potential in the so-called wave-mechanical formulation (Hirsch et al. 1977). For electron diffraction from imperfect crystals, the periodic symmetry breaks down, and a new approach must be developed to deal with local structure and defects. The theory below was developed for wave propagations inside the crystal to provide a theoretical foundation for doing this. Here, we introduced it for crystals without defects. As can be seen below, the treatment ignores the HOLZ effects, and thus for perfect crystal, the theory is not as accurate as the above many-beam theory. For the treatment of crystals with defects, this theory has been found to be adequate. This section thus serves as the basis for the further developments in Chap. 15, on the subject of electron diffraction from imperfect crystals. It can be also used as an alternative approach for teaching dynamical theory. A full treatment can be found in Chap. 8 of the book by Hirsch et al. (1977).

We consider a thin slice in a column of crystal as shown in Fig. 5.1. Electron waves at the top and bottom of the slice are marked as ϕ_i and ϕ_e, respectively. The slice is considered thin enough, so that only the projected potential needs be considered. Using the weak phase object approximation in Chap. 4, the exit wave can be related to the incident wave using:

$$\phi_e = (1 + i\pi\lambda U(x, y)dz)\phi_i$$

(5.18)

Fig. 5.1 Diffraction by a column of crystal

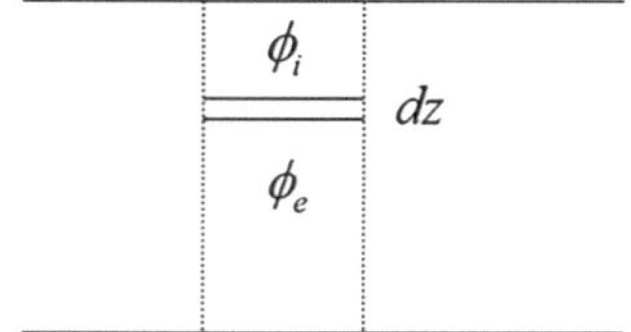

Both the projected potential and the incident and exit waves can be expanded in a Fourier series, based on the crystal periodicity, which leads to the following:

$$U(x, y) = \sum U_g \exp(2\pi i \vec{g} \cdot \vec{r})$$

and

$$\phi_e = \phi_o^e \exp\left(2\pi i \vec{k} \cdot \vec{r}\right) + \phi_g^e \exp\left(2\pi i \left(\vec{k} + \vec{g} + \vec{S}_g\right) \cdot \vec{r}\right) + \cdots$$

$$= \sum_g \phi_g^e \exp\left(2\pi i \left(\vec{k} + \vec{g} + \vec{S}_g\right) \cdot \vec{r}\right)$$

$$\phi_i = \sum_g \phi_g^i \exp\left(2\pi i \left(\vec{k} + \vec{g} + \vec{S}_g\right) \cdot \vec{r}\right)$$

Substituting these expressions into Eq. (5.18), we obtain

$$\sum_g \phi_g^e \exp\left(2\pi i \left(\vec{k} + \vec{g} + \vec{S}_g\right) \cdot \vec{r}\right)$$

$$= \left\{1 + i\pi\lambda \sum_{g'} U_{g'} \exp(2\pi i \vec{g}' \cdot \vec{r})\right\} \sum_h \phi_h^i \exp\left(2\pi i \left(\vec{k} + \vec{h} + \vec{S}_h\right) \cdot \vec{r}\right)$$

$$= \sum_g \phi_g^i \exp\left(2\pi i \left(\vec{k} + \vec{g} + \vec{S}_g\right) \cdot \vec{r}\right)$$

$$+ i\pi\lambda \sum_g \sum_h \phi_h^i U_{gh} \exp\left(2\pi i \left(\vec{k} + \vec{g} + \vec{S}_h\right) \cdot \vec{r}\right)$$

The sum is over diffracted waves inside the crystal. Since these waves vary with position, the equation is only valid when:

$$\phi_g^e = (1 + i\pi\lambda U_o dz)\phi_g^i + i\pi\lambda dz \sum_{h \neq g} \phi_h^i U_{gh} \exp\left(2\pi i \left(\vec{S}_h - \vec{S}_g\right) \cdot \vec{r}\right)$$

By writing $\phi_g^e = \phi_g^i + d\phi_g^i$, we have

$$\frac{d\phi_g}{dz} = i\pi\lambda U_o \phi_g + i\pi\lambda \sum_{h \neq g} \phi_h U_{gh} \exp\left(2\pi i \left(\vec{S}_h - \vec{S}_g\right) \cdot \vec{r}\right) \tag{5.19}$$

Thus, the general theory of electron diffraction can be also expressed as a system of linear differential equations, as first used by Darwin for X-ray diffraction and then derived by Howie and Whelan for electron diffraction.

To solve Eq. (5.19), we use the wave amplitude expression based on the Bloch waves in Eq. (5.14), by taking

$$\phi_g(z) = C_g \exp\left(2\pi i(\gamma - S_g)z\right)$$

Substitute this into Eq. (5.19), which then gives

$$2k_{\mathrm{o}}(\gamma - S_g)C_g = U_{\mathrm{o}}C_g + \sum_{h \neq g} U_{gh}C_h \tag{5.20}$$

Compared to Eq. (5.11), the mean interaction potential has now been separated out. The other terms are the same if the incident beam is taken to be normal to g.

5.3 Two-Beam Theory

In this section, we give the results of the two-beam theory of electron diffraction in the approximation that there is only one strong diffracted beam in the diffraction pattern. This beam, together with the direct beam, constitutes a two-beam solution to the above many-beam theory. Apart from the understanding it provides, the main uses of two-beam theory in CBED come from its ability to provide a quick estimate of sample thickness.

The two-beam theory may be obtained from Eq. (5.11) if only one Fourier coefficient U_g is retained. Figure 5.2 illustrates the diffraction geometry from a parallel-sided crystal slab. The diffraction lattice plane contains the surface normal direction with $\vec{n} \perp \vec{g}$. Further, because of the small Bragg angle for high-energy electrons, we employ the following approximations to simplify the mathematical

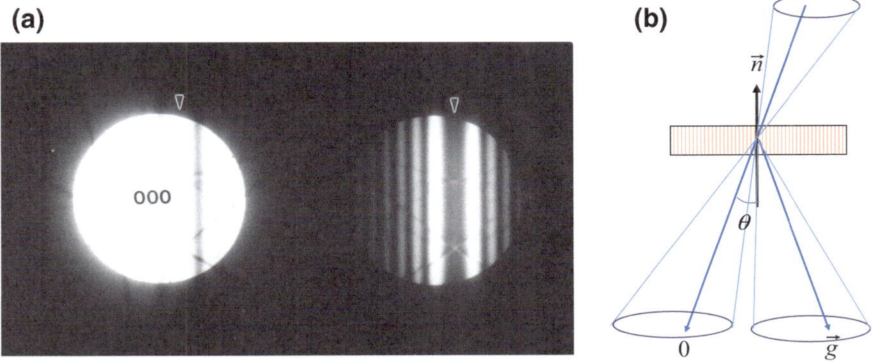

Fig. 5.2 a MgO (220) two-beam CBED pattern at 120 kV. The *arrows* indicate the Bragg conditions. **b** Two-beam diffraction geometry

derivations: $\cos\theta \approx 1$, $U_\mathrm{o} \approx 0$, and $K_n \approx K \approx k_\mathrm{o} = 1/\lambda$. If absorption is also neglected, Eq. (5.12) becomes

$$\begin{pmatrix} -2K\gamma & U_{-g} \\ U_g & 2KS_g - 2K\gamma \end{pmatrix}\begin{pmatrix} C_\mathrm{o} \\ C_g \end{pmatrix} = 0 \tag{5.21}$$

The values of γ are found by setting the determinant of the structure matrix equal to zero, i.e., $|A - 2K\gamma I| = 0$, where I is the identity matrix. This leads to the following quadratic equation:

$$\gamma^2 - \gamma S_g - \left|U_g/2K\right|^2 = 0$$

which gives

$$\gamma^{1,2} = \left[S_g \mp \sqrt{S_g^2 + \left|U_g/K\right|^2}\right]/2 \tag{5.22}$$

where the superscript 1 refers to $-$ and 2 to $+$. If we set $\omega = \cot\beta = S_g\xi_g$ where $\xi_g = K/\left|U_g\right|$ is called extinction distance, then Eq. (5.21) yields the ratio of C_o and C_g in the form

$$\frac{C_g}{C_\mathrm{o}} = \frac{\cos\beta \mp 1}{\sin\beta} = \begin{cases} -\sin(\beta/2)/\cos(\beta/2) \\ \cos(\beta/2)/\sin(\beta/2) \end{cases}$$

Substituting this back to the eigenequations leads to

$$\begin{pmatrix} C_\mathrm{o}^1 & C_\mathrm{o}^2 \\ C_g^1 & C_g^2 \end{pmatrix} = \begin{pmatrix} \cos(\beta/2)e^{-i\alpha_g/2} & \sin(\beta/2)e^{-i\alpha_g/2} \\ -\sin(\beta/2)e^{i\alpha_g/2} & \cos(\beta/2)e^{i\alpha_g/2} \end{pmatrix} \tag{5.23}$$

And

$$\begin{pmatrix} C_\mathrm{o}^1 & C_\mathrm{o}^2 \\ C_g^1 & C_g^2 \end{pmatrix}^{-1} = \begin{pmatrix} \cos(\beta/2)e^{i\alpha_g/2} & -\sin(\beta/2)e^{-i\alpha_g/2} \\ \sin(\beta/2)e^{i\alpha_g/2} & \cos(\beta/2)e^{-i\alpha_g/2} \end{pmatrix} \tag{5.24}$$

If the results of Eqs. (5.23), (5.24), and (5.22) are substituted in Eq. (5.14), we obtain for the diffracted wave:

$$\phi_g(t) = \sum_i c_i C_g^i \exp\left(2\pi i\left(\gamma^i - S_g\right)t\right)$$

$$= \sin(\beta)e^{-i\alpha_g}\exp\{-\pi i S_g t\}\sin\left\{\pi z\sqrt{S_g^2 + \left|U_g/K\right|^2}\right\}$$

$$= \frac{1}{\sqrt{1+\omega^2}}e^{-i\alpha_g}\exp\{-\pi i S_g t\}\sin\left\{\frac{\pi t}{\xi_g}\sqrt{1+\omega^2}\right\}$$

The intensity I_g is thus as a function of sample thickness t, structure factor U_g, accelerating voltage, and excitation error S_g:

$$I_g(t) = \left|\phi_g(t)\right|^2 = \frac{1}{1+\omega^2}\sin^2\left\{\frac{\pi t}{\xi_g}\sqrt{1+\omega^2}\right\} \tag{5.25}$$

with

$$I_0 = 1 - I_g \tag{5.26}$$

Figure 5.3 shows this function plotted as a function of both thickness and orientation. The dimensionless parameter $\omega = S_g\xi_g$ has been used, with $\xi_g = K/\left|U_g\right|$. The variation of the intensity with thickness is known as "Pendellosung," or thickness fringe oscillations. At a particular thickness, the intensity variation with excitation error (direction of the incident beam) gives a good impression of the intensity variation seen in an experimental systematics CBED disk around the Bragg condition, as shown in Fig. 5.2 (Fig. 3.14 is another example). The fine HOLZ lines seen crossing the (220) disk in Fig. 5.2, however, cannot be

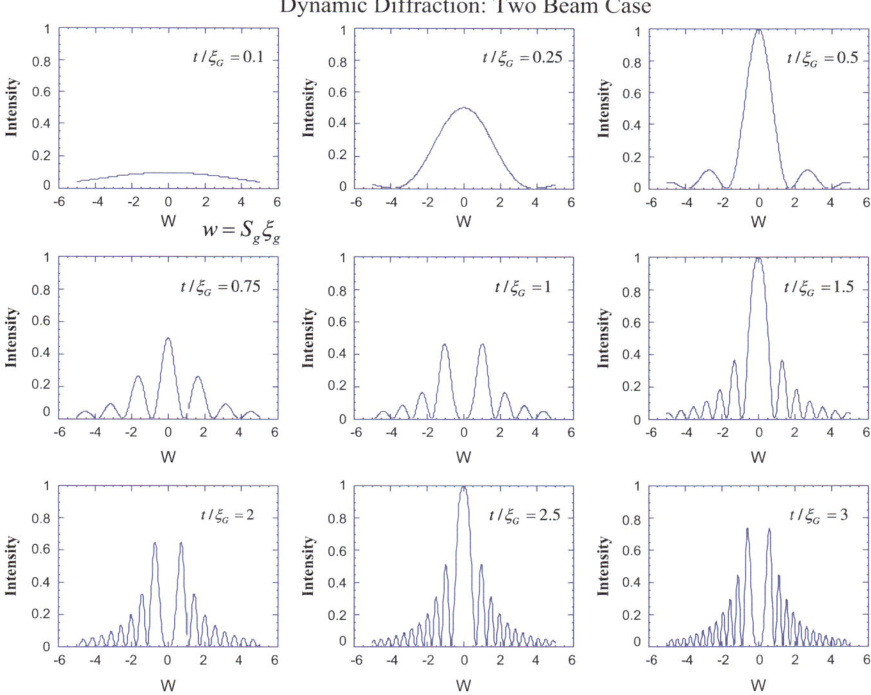

Fig. 5.3 The variation in the intensity of diffracted beam with incident beam direction (represented by ω), plotted for different crystal thicknesses as defined by the ratio of t/ξ_g. Two-beam theory

reproduced in a two-beam theory based solely on U_{111}. The calibration procedure for S_g given in Chap. 3 could enable data taken from Fig. 3.14 along the line AA' to be compared directly with Eq. (5.17), and values of U_g and t found to produce the best fit. This is the basis of the techniques described in Chap. 12.

The intensity minima of Eq. (5.25) occur at S_g values given by

$$S_g^2 = \frac{n^2}{t^2 (K/K_n)^2} - \frac{|U_g|^2}{K^2},\tag{5.27}$$

which may be related to the incident beam angle θ through Eq. (3.23). Here, n is an index for the intensity minima, with the first minima away from $\omega = 0$ marked as 1. The factor K/K_n in Eq. (5.27) is a correction term when the surface normal direction is taken into account. This term is often neglected in the literature. If so, then the thickness measured is the effective crystal thickness, $t^{\mathrm{eff}} = t(K/K_n) = t/\cos\theta$, where θ is the angle between the beam direction and the surface normal. The intensity at the intensity maxima is given by

$$I_g(t) = \left[\frac{\pi t |U_g|}{K_n}\right] / (1 + x^2),\tag{5.28}$$

where x is given by the solutions of $x = \tan(x)$.

At the Bragg condition $S_g = 0$, the intensity becomes

$$I_g(t) = \left|\phi_g(t)\right|^2 = \sin^2\left\{\frac{\pi t^{\mathrm{eff}} |U_g|}{K}\right\}\tag{5.29}$$

The periodicity of the intensity with thickness when $S_g \neq 0$ is $L = \xi_g / \sqrt{1 + \omega^2}$.

5.4 The Concept of the Dispersion Surface

In this section, we introduce the dispersion surface construction. The dispersion surfaces are formed by the locus of allowed wave vectors for a given total beam electron energy, as determined by the accelerating voltage. All wave vectors inside the crystal are restricted to lie on the dispersion surfaces. Mathematically, the dispersion surfaces are described by Eq. (5.12). The two-beam results of the last section will be used as an example to illustrate the dispersion surface construction. The concept of dispersion surfaces in electron diffraction is similar to the concept of energy bands in solid-state physics. They differ, however, in the physical quantities represented. The energy bands of band theory plot the allowed total energy E of a crystal electron for a given crystal momentum $\hbar\vec{k}^j$. (Here, $\vec{k}^j$ labels the Bloch wave state j.) The dispersion surfaces of high-energy electron diffraction plot the opposite; that is, they plot the locus of allowed momentum $\hbar\vec{k}^j$ for an incident beam electron of given total energy E_o. For elastic electron diffraction, the beam electron has a constant total energy E_o. As a result of the interaction of the incident electron with the crystal potential, the allowed momentum (and corresponding kinetic energy) of an incident electron varies between Bloch wave states $\vec{k}^j$. The difference between the kinetic and total energy of a Bloch wave state is taken up by potential energy.

As an example, the two-beam dispersion surfaces shown in Fig. 5.4 were constructed according to the following procedure:

1. Approximate dispersion surfaces are plotted, using the "empty lattice" approximation. Here, all the interaction parameters are set to zero: $U_{gh} = 0$. From Eq. (5.4), this gives

$$\left(\vec{k}^j + \vec{g}\right)^2 = K^2 \tag{5.30}$$

Hence, in the empty lattice approximation, the dispersion surfaces are a set of spheres of radius K centered on each reciprocal lattice point g. These are shown as thin lines in Fig. 5.4. (The aspect ratio of the figure has been exaggerated for clarity—in high-energy electron diffraction, the ratio g/K in particular is much smaller than shown in the figure.)

2. A vector is drawn in the direction normal to the entrance surface of the crystal, and intersecting the vector $\vec{K}$, as shown by the arrow in Fig. 5.4.

3. The dynamical dispersion constants γ^i are calculated from Eq. (5.12), now using nonzero U_{gh} values. In the two-beam case, γ may be obtained from Eq. (5.22). According to Eq. (5.6), the values of γ^i are measured along the surface normal direction, starting from a point on that K sphere which is centered on the origin

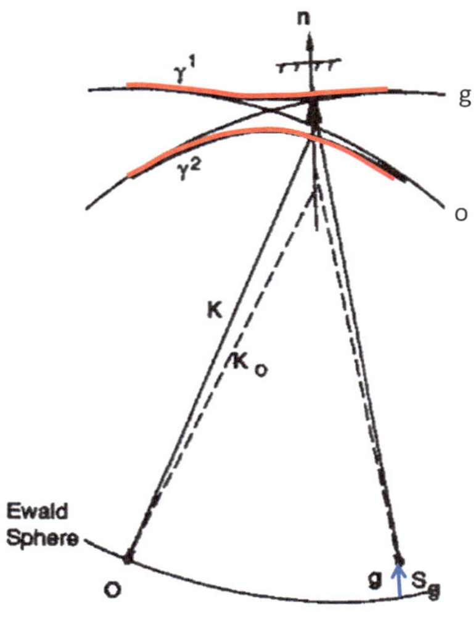

Fig. 5.4 Dispersion surfaces in the two-beam approximation. Here, γ and S_g are measured in the direction of the surface normal n. Surfaces are numbered from top down. 0 and g are spheres of radius K (not to scale) about 0 and g

of the reciprocal lattice. Thus, the incident beam direction $\vec{K}$ must be drawn first. Then points on the dispersion surfaces are drawn at distances γ^i measured from the end point of $\vec{K}$ along the surface normal direction. The complete dispersion surfaces are obtained by repeating this for each possible beam direction, and these are shown as bold lines in Fig. 5.4.

Two methods have been used in the literature to present calculated or measured dispersion surfaces. The first is similar to Fig. 5.4—here, the momenta $\vec{K}$ are plotted against the x coordinate of the incident beam direction $\vec{K}_t$ (as defined in Chap. 2). The plotting can be made easier by assuming that the surface normal is along the z-axis, with the x-axis as $\vec{K}_t$. Figure 5.5a shows the two-beam dispersion surfaces plotted in this fashion. The alternative method, mostly used by the Norwegian groups, uses γ as the y-axis, assuming that the surface normal is antiparallel to the beam direction and the incident beam's K sphere is taken as the x-axis. Only a plane section of the dispersion surfaces is plotted; thus, the K spheres are shown as circles. In such a diagram, the K spheres appear flattened. The departure from flatness is actually very small in high-energy electron diffraction within the angular range typically shown (about 0.1 rad). Figure 5.5b shows the two-beam dispersion surfaces plotted in this way. The bold lines are the dispersion surfaces, the thin line is the K sphere drawn about reflection g. The x-axis is the incident beam K sphere. Each point on the incident beam K sphere corresponds to a different incident beam direction.

For the two-beam dispersion surfaces, a gap opens near the Bragg condition for reflection g, where the K spheres about o and g intersect. The gap has width (measured vertically) $|U_g|/K\,|$ at the Bragg condition. The gap between the dispersion surfaces and the K spheres decreases as one moves away from the Bragg condition.

The momenta of the incident beam electron states inside the crystal, as represented by the dispersion surfaces, are also physical observables, as are energies in crystal band theory. One way to observe these momenta is to use a wedge-shaped crystal, since this has a different boundary condition at its entrance and exit surfaces (Lehmpfuhl and Reissland 1968). The principle is demonstrated in Fig. 5.6, where n_i and n_e are the entrance and exit surface normal directions, respectively. The Bloch wave vectors $\vec{k}^1$ and $\vec{k}^2$ are determined by the entrance surface boundary condition. However, the transmitted wave vectors $\vec{K}_o^1$ and $\vec{K}_o^2$ are determined by the boundary condition at the exit surface. The difference between the entrance and exit surface boundary condition and the difference in the Bloch wave vectors causes the angular splitting in the transmitted beam. This splitting is proportional to the gap between the dispersion surfaces. The same splitting is expected for the diffracted beams. For simplicity, we do not show the diffracted beams in Fig. 5.6. An experiment has been performed by recording a series of point diffraction patterns while rotating a thin crystal around the axis normal to a plane section of the dispersion surface of interest. The component of the incident beam direction along the rotation axis was held constant. Then, a plane section (normal to the rotation

Fig. 5.5 a Calculated two-beam dispersion surfaces. The abscissa is dimensionless and the ordinate is given as $(K - k_z)$ in reciprocal angstroms. **b** Similar to **a** but plotted in such way that the asymptotic sphere about the origin is taken as a straight line (an excellent approximation at high energy). Two-beam theory

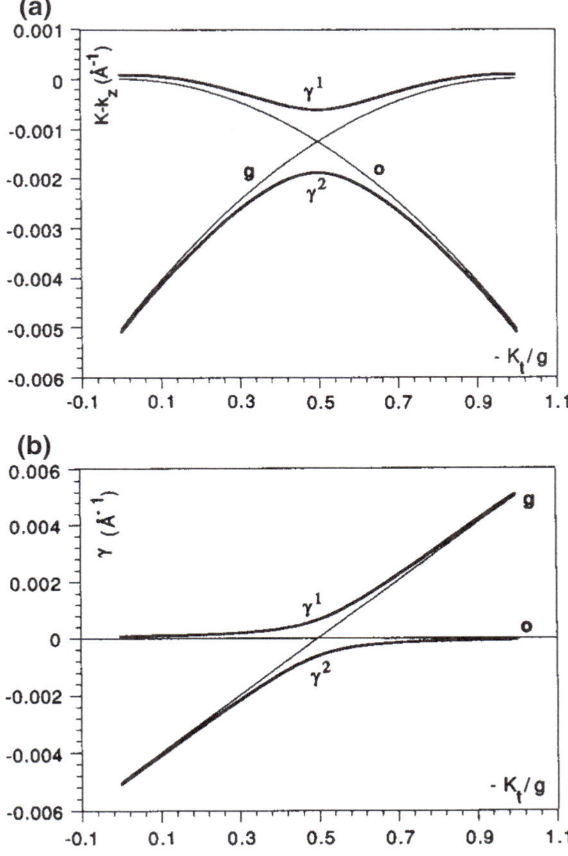

Fig. 5.6 Dispersion surfaces for a wedge-shaped sample. The entrance surface normal is n_i and the exit-face normal is n_e. Each diffracted beam becomes split—here in two-beam theory, the direct beam is split into K_o^1 and K_o^2 as shown

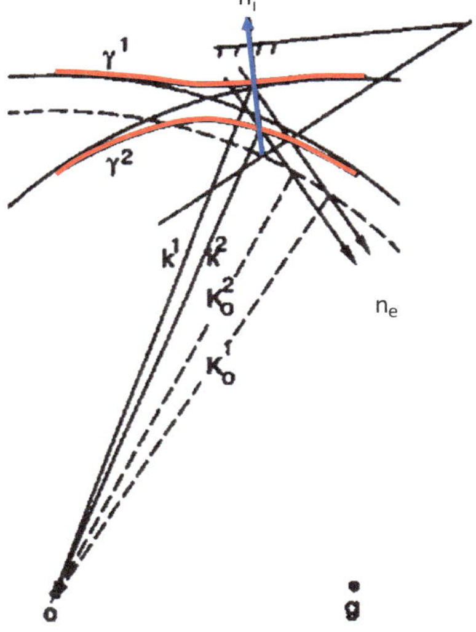

axis) of the dispersion surfaces can be mapped out experimentally. The results shown in Fig. 5.7 demonstrate that these dispersion surfaces can be observed experimentally.

Fig. 5.7 Fine structure in the (220) reflection from an MgO wedge crystal is shown in **a** as it is rotated under the beam (Lehmpfuhl and Reissland 1968). A point pattern is formed, and the film is moved as the crystal is rotated about the (220) direction. **b** Shows higher resolution (slower rotation) while **c** shows calculated dispersion surfaces for this case, in good agreement. The width of the point is proportional to the strength of the Bloch wave

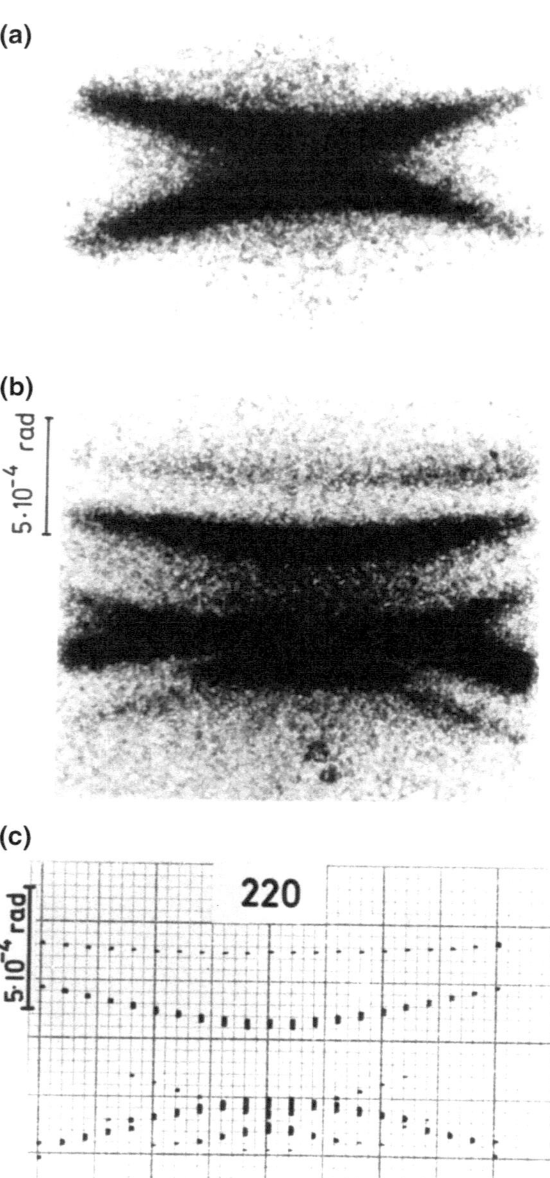

Parts of the dispersion surfaces can also be observed in conventional point diffraction patterns, within the fine structure of HOLZ lines, or in Kikuchi line patterns. These effects are further discussed in Sect. 5.6.4 and demonstrated experimentally in Fig. 5.14.

5.5   Absorption and Its Effects in a First-Order Approximation

Absorption has two effects on all crystals in electron diffraction. The first is the average or mean absorption, whose effect has been described in Chap. 4, Sect. 4.8. A second effect of absorption has become known as the anomalous transmission effect, by analogy with the Borrmann effect in X-ray diffraction. This is due to the terms U_g' described in Chap. 4, Sect. 4.8. The effect in electron diffraction was first observed by Honjo and Mihama (1954), who used a wedge-shaped crystal to produce spot spitting (doublets) in diffraction patterns [similar to Lehmpfuhl and Reissland (1968)], so that each spot of the doublet can be related to one of the dispersion surfaces. They then detected a difference in intensity between the two spots. In CBED, this anomalous transmission effect is revealed by the intensity asymmetry around the Bragg condition in the transmitted (zero-order) disk, instead of the symmetry which we expect from the two-beam intensity expression of Eq. (5.25). The two-beam theory has been applied to electron diffraction with absorption, and comparisons made with experiment in the work of Hashimoto et al. (1962). In following, we shall apply the perturbation theory to the problem of absorption in the two-beam case.

In high-energy electron diffraction, the absorption potential is much smaller than the crystal potential, typically less than one-tenth of the crystal potential. Thus, absorption may be treated as a perturbation. A general treatment of perturbation theory in electron diffraction can be found in the ref. Zuo (1991). Here, we will give a brief description of the nondegenerate case, which will be sufficient for the use of the rest of this section, the degenerate case will be treated in Sect. 5.6.3. Starting with Eq. (5.11), which can be written in the matrix form:

$$AC = C\Upsilon$$

Let

$$\Upsilon = \begin{pmatrix} 2K_n\gamma^1 & \cdots & 0 \\ \vdots & \ddots & 0 \\ 0 & 0 & 2K_n\gamma^N \end{pmatrix}$$

Let $A = A_0 + \delta A$, where δA is a small perturbation resulting from a small change in the potential U_{gh} or the beam direction S_g or both, then

$$A_{\mathrm{o}}C + \delta AC = C\Upsilon \tag{5.31}$$

We assume that the eigenvalue and eigenvector matrices Υ_{o} and C_{o} corresponding to A_{o} are known. The eigenvector matrix of A may be written as a combination of the eigenvectors of A_{o} and $C = C_{\mathrm{o}}(1 + \varepsilon)$, where ε is the coefficient matrix. Multiplying both sides of Eq. (5.31) by C_{o}^{-1} gives

$$\left\{\Upsilon_{\mathrm{o}} + C_{\mathrm{o}}^{-1}\delta AC\right\}\varepsilon = \varepsilon\Upsilon$$

If $A' = \Upsilon_{\mathrm{o}} + C_{\mathrm{o}}^{-1}\delta AC$, then we can write a transformed equation in the basis of the states of A_{o},

$$A'\varepsilon = \varepsilon\Upsilon \tag{5.32}$$

where the matrix A' is defined by

$$A'_{ii} = \Upsilon_{ii} + \delta A'_{ii} \tag{5.33}$$

And

$$A'_{ij} = \delta A'_{ij} = \sum_{km} C^{-1}_{ik}\delta A'_{km}C_{mj} \tag{5.34}$$

We seek changes in the ith Bloch wave and assume that this Bloch wave is nondegenerate. This means that no other Bloch wave has an eigenvalue which is close to that of the ith Bloch wave. This condition may be written as follows:

$$\left|\Upsilon_{ii} - \Upsilon_{jj}\right| \gg \left|A'_{\mathrm{max}}\right| \tag{5.35}$$

where $\left|A'_{\mathrm{max}}\right|$ is the largest off-diagonal element of the transformed matrix A'. Then, the ith Bloch wave is said to be nondegenerate. In this case, the coefficient matrix is approximately obtained by Zuo (1991)

$$\varepsilon_{ij} \approx \delta A'_{ij}/\left(\Upsilon_{\mathrm{o}ii} - \Upsilon_{\mathrm{o}ij}\right) \text{ and } \varepsilon_{ii} \approx 1$$

The change in the eigenvector is just

$$\Upsilon_{ii} = \Upsilon_{\mathrm{o}ii} + \delta A'_{ii} + \sum_{j}\left|\delta A'_{ij}\right|^{2}/\left(\Upsilon_{\mathrm{o}ii} - \Upsilon_{\mathrm{o}ij}\right)$$

Next, we will use this result under the condition (5.35) to explain the effects of absorption in two-beam case. Therefore, from Sect. 5.3,

$$A_{\mathrm{o}} = \begin{pmatrix} 0 & U_{-g} \\ U_g & 2KS_g \end{pmatrix}, \quad \delta A = \begin{pmatrix} 0 & iU'_{-g} \\ iU'_g & 0 \end{pmatrix}$$

The solutions to A_{o} are given in Eqs. (5.22) and (5.23). Equation (5.34) yields

$$\delta A'_{11} = ie^{i\alpha_g}\left[U'_{-g}\cos(\beta/2)\sin(\beta/2) + \sin(\beta/2)U'_g\cos(\beta/2) \right]$$

$$= i\left|U'_g\right|\sin(\beta)e^{i\left(\alpha_g - \alpha'_g\right)}$$

and similarly

$$\delta A'_{22} = -i\left|U'_g\right|\sin(\beta)e^{i\left(\alpha_g - \alpha'_g\right)}$$

and

$$\delta A'_{12} - \delta A'^{*}_{21} = i\left|U'_g\right|\left\{\cos^2(\beta/2)e^{i\left(\alpha_g - \alpha'_g\right)} - \sin^2(\beta/2)e^{i\left(\alpha_g - \alpha'_g\right)}\right\}e^{i\alpha_g}$$

Here, α_g and α'_g are the phases of U_g and U'_g. For centrosymmetric crystals, $\alpha_g = \alpha'_g = 0$ or π; thus,

$$\delta A'_{11} = -\delta A'_{22} = i\left|U'_g\right|\sin(\beta) = \frac{i\left|U'_g\right|}{\sqrt{1+\omega^2}} \tag{5.36}$$

And

$$\delta A'_{12} = -\delta A'^{*}_{21} = iU'_g\cos(\beta) = \frac{iU'_g\omega}{\sqrt{1+\omega^2}} \tag{5.37}$$

The terms $\delta A'_{11}$ and $\delta A'_{22}$ are thus the first-order corrections to the eigenvalues 1 and 2 of Eq. (5.22), while $\delta A'_{12}$ and $\delta A'_{21}$ contribute to the corrections to the eigenvectors resulting from the introduction of an absorption potential. Near the Bragg condition ω is small, so that the corrections to the eigenvectors may be neglected. Then, the amplitudes for two-beam diffraction with absorption are given by

$$\phi_{\mathrm{o}}(t) = \cos^2(\beta/2)e^{-iXt} + \sin^2(\beta/2)e^{iXt} \tag{5.38}$$

$$\phi_g(t) = -\cos(\beta/2)\sin(\beta/2)\left\{e^{-iXt} - e^{iXt}\right\}$$

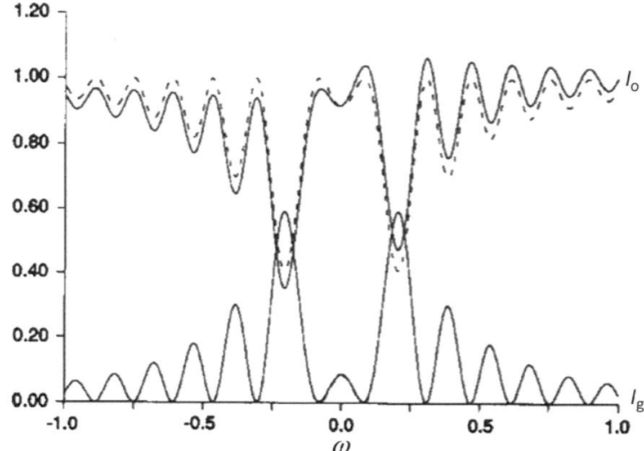

Fig. 5.8 Intensity of beams I_o and I_g according to two-beam theory with absorption, as function of deviation parameter ω around Bragg condition $\omega = 0$. The thickness is 1.9 ξ_g. The full lines show $U'_g/U_g = 0.1$ and $U_o = 0$, while dashed lines show $U'_g/U_g = 0$ and $U_o = 0$

where

$$X = \frac{\pi|U_g|}{K}\sqrt{1+\omega^2} + i\frac{\pi|U'_g|}{K}\frac{1}{\sqrt{1+\omega^2}} \tag{5.39}$$

The mean absorption effect has been neglected here. Rocking curves for o and g based on Eq. (3.95) with $\left|U'_g\right|/|U_g| = 0$ and 0.1 are plotted in Fig. 5.8. It is seen from this figure that the absorption potential causes an intensity asymmetry around the Bragg angle in the rocking curve of the (000) reflection, the intensity at $-S_g$ being lower than the intensity at $+S_g$. This asymmetry arises because Bloch wave 1 is strongly excited on the side where S_g is less than zero, while Bloch wave 2 is strongly excited on the side where $+S_g$ is larger than zero. According to Eq. (5.36) Bloch wave 1 is strongly absorbed, while Bloch wave 2 is less absorbed. As discussed in many texts [e.g., Hirsch et al. (1977)], this is attributed to the fact that, in real space, Bloch wave 1 has maxima of intensity located on the atomic sites (resulting in a higher probability of inelastic scattering by localized processes), while Bloch wave 2 has maxima which fall between the atoms. (We are here making the independent Bloch wave approximation. The total wave field in the crystal actually contains contributions from both Bloch waves.)

For noncentrosymmetric crystals, $\varphi_g \neq \phi'_g$ in general, and the corrections to the eigenvectors cannot be neglected. In the kinematic approximation,

$$I_g \propto \left|U_g + iU'_g\right| = |U_g|^2 + \left|U'_g\right|^2 + 2|U_g|\left|U'_g\right|\sin\left(\alpha_g - \alpha'_g\right) \tag{5.40}$$

Because of the sine function, there is a difference between the intensities of the g and $-g$ reflections; this difference is

$$\Delta I_g = I_g - I_{-g} \propto 4 \left| U_g \right| \left| U_g' \right| \sin\left(\alpha_g - \alpha_g' \right)$$

In X-ray diffraction, this effect is called anomalous dispersion (Karle 1989). Its effects on electron diffraction have been discussed by Bird (1990) and Tafto (1987). In two-beam theory, by including the correction to the eigenvalues for an acentric crystal, it can be shown that the inclusion of absorption potential leads to a small correction to the diffractive wave at the Bragg condition

$$\Delta\phi_g\left(t, S_g = 0\right) = i \frac{\left| U_g' \right|}{\left| U_g \right|} \sin\left(\alpha_g - \alpha_g' \right) e^{i\alpha_g} \sin\left(\frac{\pi \left| U_g \right| t}{K} \right) \tag{5.41}$$

If the effect of absorption in the wave amplitude ϕ_g is neglected, then

$$\phi_g\left(t, S_g = 0\right) = i e^{i\alpha_g} \sin\left(\frac{\pi \left| U_g \right| t}{K} \right)$$

Thus, the intensity at the Bragg condition is approximately

$$I_g(t, S_g = 0) = \left\{ 1 + 2 \frac{\left| U_g' \right|}{\left| U_g \right|} \sin\left(\alpha_g - \alpha_g' \right) \right\} \sin^2\left(\frac{\pi \left| U_g \right| t}{K} \right) \tag{5.42}$$

From this equation, we would expect the same intensity asymmetry between I_g and I_{-g} at the Bragg condition as predicted by the kinematic approximation. The intensity away from the Bragg condition is given by rather complicated expressions.

Thus, in the two-beam case, this absorption potential causes two anomalous dispersion effects. One is the difference between the intensities of the central beam at orientations $+ S_g$ and $-S_g$. The other is an intensity difference between the g and $-g$ reflections at their respective Bragg conditions in acentric crystals.

Absorption coefficients may be measured using the intensity distribution in the zero-order disk of a CBED pattern at the Bragg condition. The phases of the Fourier coefficients of the absorption potential may be measured by comparing the intensities of the g and $-g$ reflections near the Bragg condition.

5.6 Many-Beam Effects

In this section, we describe many-beam effects in electron diffraction using three-beam theory specialized to some particular solutions for centrosymmetric crystals, two-beam theory with Bethe potential, and perturbation theory for non-centrosymmetric crystal and HOLZ lines. The section is intended for advanced readers and students who want to obtain in-depth knowledge of dynamical electron diffraction, its effects and theory for describing these effects.

5.6.1   Three-Beam Theory and Particular Solutions for Centrosymmetric Crystals

The three-beam case is the simplest case of many-beam diffraction. While a general three-beam theory for dynamic diffraction has yet to be developed, specialized solutions for centrosymmetric crystals under some particular diffraction symmetry are available. The three-beam solution contains within it all the essential features needed to explain such phenomena as the critical voltage effect, dynamical shifts of HOLZ lines, and the principles of structure-factor phase measurement using multiple scattering. None of these effects can be understood using two-beam theory. Since they are among the most interesting and useful applications of electron dynamic diffraction, an understanding of three-beam theory is important. Here, we describe the solutions of three-beam cases for centrosymmetric crystals. A general discussion of three-beam theory for noncentrosymmetric (acentric) crystals is deferred to Sect. 5.6.2. Eigenvectors and solutions for many other solvable cases related by symmetry (up to twelve beams) can be found in Fukuhara (1966).

There are six possible three-beam geometries. These are shown in Fig. 5.9. In each case, the scattering is dominated by the interactions between the three beams o, g, and h. The first three cases (a, b and c) are known as systematic three-beam cases. The rest are nonsystematic cases. If these three beams only are retained and we take $\vec{n}$ antiparallel to $\vec{K}$, Eq. (5.12) becomes

$$\begin{pmatrix} 0 & U_{-g} & U_{-h} \\ U_g & 2KS_g & U_{gh} \\ U_h & U_{hg} & 2KS_h \end{pmatrix} \begin{pmatrix} C_o \\ C_g \\ C_h \end{pmatrix} = 2K\gamma \begin{pmatrix} C_o \\ C_g \\ C_h \end{pmatrix} \tag{5.43}$$

This equation was first studied in detail in the important early papers of Fues (1949) and Kambe (1957). The three eigenvalues γ are given by the roots of the secular equation

$$|2K\gamma I - A| = 0 \tag{5.44}$$

This gives a cubic equation, whose closed-form solution can be found in mathematical texts but which is too lengthy and complicated to be of much use. Instead, Eq. (5.43) is usually solved using approximations, or for special cases involving symmetry.

In centrosymmetric crystals, the structure factors are real, and $U_g = U_{-g}$ if the origin of the reciprocal lattice is taken at the center of the symmetry. This simplifies Eq. (5.43), but a transparent solution is still not possible. In the following,

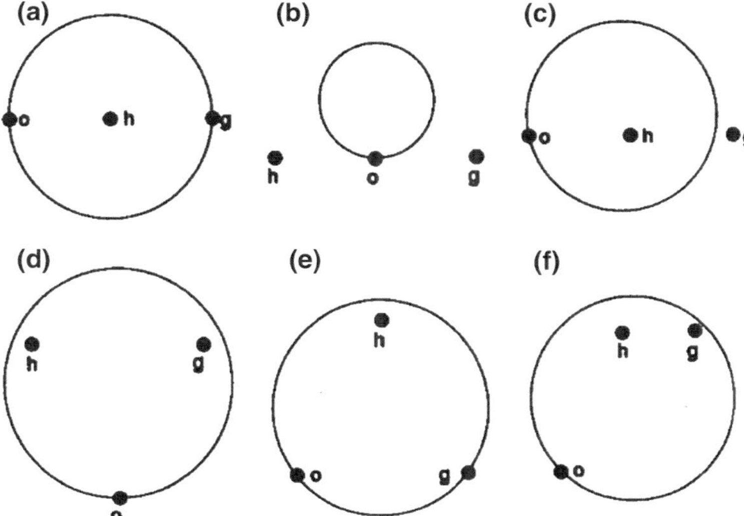

Fig. 5.9 All possible three-beam geometries. The Laue *circle* and position of reciprocal lattice points are shown. Symmetry reduction makes many cases easily soluble along certain lines in three-beam CBED patterns

we describe two particular solutions to Eq. (5.43) for centrosymmetric crystals. One involves degeneracy in eigenvalues, and the another requires some symmetry.

It was found by Gjonnes and Hoier (1971) that for centrosymmetric crystals, two of the three eigenvalues of Eq. (3.51) become degenerate at a special point on the dispersion surfaces. This occurs when

$$2KS_g = \frac{U_g\left(U_{gh}^2 - U_h^2\right)}{U_h U_{gh}}$$
$$2KS_h = \frac{U_h\left(U_{gh}^2 - U_g^2\right)}{U_g U_{gh}} \tag{5.45}$$

The degenerate eigenvalue is then given by

$$2K\gamma = -U_g U_h / U_{gh} \tag{5.46}$$

In the diffraction pattern, this degeneracy causes an intensity minimum. The position of this minimum, as defined by the excitation errors of Eq. (5.45), depends on the values of the three structure factors involved, in particular on the sign of the product $U_{-g}U_h U_{gh}$. This sign dependence is easily observed in experimental CBED patterns. Thus, the sign of the product $U_{-g}U_h U_{gh}$ is determined by the sign of S_h at the degeneracy point (S_h and S_g have the same sign). The sign of the excitation error can easily be determined, as discussed in Chap. 3. Thus by noting where the intensity minimum occurs, we may determine the sign of the triplet $U_{-g}U_hU_{gh}$. Since all structure factors U_g have phases of $0°$ or $180°$ in centric crystals (corresponding to a sign of plus or minus), this procedure is equivalent to a determination of the sum of the phases of the structure factors. The phase sum is known as a "three-phase structure invariant". [An n-phase structure invariant may similarly be defined from the determinantal Eq. (5.44).] This topic is discussed in more detail in the next section on noncentrosymmetric crystals, for which centrosymmetric crystals form a special case.

In the three-beam geometry of Fig. 5.9a, e, there is a mirror symmetry for centrosymmetric crystals. In this case, Eq. (5.43) becomes

$$
\begin{pmatrix} 0 & U_g & U_h \\ U_g & 0 & U_h \\ U_h & U_h & 2KS_h \end{pmatrix} \begin{pmatrix} C_o \\ C_g \\ C_h \end{pmatrix} = 2K\gamma \begin{pmatrix} C_o \\ C_g \\ C_h \end{pmatrix}
\tag{5.47}
$$

using the fact that $U_g = U_{-g}$ and $U_{gh} = U_h$. Because of the symmetry in this equation, there are only two possibilities: (1) $C_o = C_g$, or (2) $C_o = -C_g$ and $C_h = 0$. In the first case, Eq. (3.54) is reduced to a 2×2 matrix, and to a 1×1 matrix in the second case. These matrices are

$$
\begin{pmatrix} U_g & U_h \\ 2U_h & 2KS_h \end{pmatrix} \begin{pmatrix} C_o \\ C_h \end{pmatrix} = 2K\gamma \begin{pmatrix} C_o \\ C_h \end{pmatrix}
\tag{5.48}
$$

and

$$
-U_gC_o = 2K\gamma C_o
\tag{5.49}
$$

The two Bloch waves given in Eq. (5.48) are symmetric because $C_o = C_g$ while the Bloch wave in Eq. (5.49) is antisymmetric. These results may be deduced more elegantly and generally using group theoretical arguments (Kogiso and Takahashi 1977; Spence 1988). Equations (5.48) and (5.49) are easily solved and give

$$
2K\gamma^1 = -U_g \quad \text{and} \quad 2K\gamma^{2,3} = \frac{1}{2}\left(2KS_h + U_g \pm \sqrt{\left(2KS_h - U_g\right)^2 + 8U_h^2} \right)
\tag{5.50}
$$

The degeneracy in this symmetric three-beam case occurs when $\gamma^1 = \gamma^3$, at an excitation error

$$
2KS_h = \left(U_h^2 - U_g^2 \right)/U_g
$$

This agrees with the result of Eq. (5.45) if we use the fact that $U_{gh} = U_h$, as assumed in Eq. (5.47).

5.6.2 Two-Beam Theory with Weak-Beam Effects

In Sect. 5.3, we assumed an ideal two-beam case, in which all other beams have zero intensity. In reality, there are always weak beams present. One way to include these weak-beam effects on two-beam diffraction is to use the perturbation method of Bethe. The criteria for classifying a beam h as a weak beam is that $2KS_h \gg |U_g|$. Thus, $|S_g|$ must be comparable with $|U_g|/K$ to justify treating g as a strong beam.

Applying these conditions to Eq. (5.12), we have

$$C_h = -\frac{\sum\limits_{h'} U_{hh'} C_{h'}}{2KS_h - 2K_n\gamma} \approx -\frac{U_h C_{\mathrm{o}}}{2KS_h} - \frac{U_{hg} C_g}{2KS_h}$$

with second-order terms neglected. Substitution of this equation into Eq. (5.12) yields a modified two-beam equation

$$\begin{pmatrix} 2KS_{\mathrm{o}}^{\mathrm{eff}} - 2K\gamma & U_{-g}^{\mathrm{eff}} \\ U_g^{\mathrm{eff}} & 2KS_g^{\mathrm{eff}} - 2K\gamma \end{pmatrix} \begin{pmatrix} C_{\mathrm{o}} \\ C_g \end{pmatrix} = 0$$

where

$$2KS_{\mathrm{o}}^{\mathrm{eff}} = -\sum_h \frac{|U_h|^2}{2KS_h}, \quad 2KS_g^{\mathrm{eff}} = 2KS_g - \sum_h \frac{|U_{gh}|^2}{2KS_h} \tag{5.51}$$

and

$$U_g^{\mathrm{eff}} = U_g - \sum_h \frac{U_h U_{gh}}{2KS_h} \tag{5.52}$$

This effective structure factor, first introduced by Bethe, incorporates weak-beam effects within the two-beam approximation. It is therefore known as the Bethe potential. The two-beam intensity, now using the Bethe potential, is given by

$$I_g(t) = \frac{\left|U_g^{\mathrm{eff}}\right|^2}{K^2\left(S_g^{\mathrm{eff}} - S_{\mathrm{o}}^{\mathrm{eff}}\right)^2 + \left|U_g^{\mathrm{eff}}\right|^2} \sin^2\left\{\frac{\pi t}{K}\sqrt{K^2\left(S_g^{\mathrm{eff}} - S_{\mathrm{o}}^{\mathrm{eff}}\right)^2 + \left|U_g^{\mathrm{eff}}\right|^2}\right\} \tag{5.53}$$

Two new effects are seen to result from the inclusion of additional weak beams. First, the Bragg law no longer strictly applies, as we see from the appearance of the effective excitation error in Eq. (5.53). This is the origin of the displaced Kikuchi lines and HOLZ lines, which will be discussed in Sects. 3.6 and 5.4. The precise position of the intensity maximum near the Bragg angle thus depends on the values of the other structure factors. Second, the extinction distance now also depends on orientation. Other effects, including those involving structure-factor phases, will be discussed later. The effective potential is sometimes called the second Bethe approximation in the literature. It is the simplest approximation for many-beam effects and has been widely used to explain, for example, the critical voltage effect [see Sect. 5.2 in Spence and Zuo (1992)].

5.6.3 *Three-Beam Theory—Noncentrosymmetric Crystals and the Phase Problem*

Our understanding of the noncentrosymmetric case is relatively recent. For acentric crystals, the Fourier coefficients F_g^x of $\rho(r)$ are complex, and values of the structure-factor phases F_g^x are required to enable the crystal charge density to be synthesized. These phases depend on the choice of origin in the crystal. Therefore, researchers in both X-ray and electron crystallography have for many years sought to find practical solutions to this famous "phase problem". Rather than simply reading-in an entire CBED pattern into a computer and attempting to match it to a set of phases, it is useful to determine first which portions of a CBED pattern are most sensitive to changes in a given structure-factor phase. Our aim in the following is to use the results of three-beam dynamical theory for noncentrosymmetric crystals to find the sensitive region, and to indicate the general form of the intensity pattern. For accurate phase measurement in practice, many-beam calculations will be required.

It is well known that the single-scattering or kinematic theory of diffraction does not allow phases to be measured, nor, since Eq. (5.25) contains only $\left| U_g \right|^2$, does the simple two-beam dynamical theory. If absorption is included exactly (or by using perturbation theory), small differences appear between the g and $-g$ rocking curves in two-beam theory due to the difference between the phases of the structure factors of the real potential and those of the absorption potential. These are responsible for the observed asymmetry in Kikuchi lines (Bird and Wright 1989) and inner-shell energy-loss spectra (Tafto 1987). We note here in passing that the phase of the electron structure factor U_g is not equal to that of the X-ray structure factor F_g^x. This can be seen from Eq. (4.50).

For reviews of phase measurement by X-ray diffraction using many-beam effects in acentric crystals, see Chang (2004) and Hummer and Billy (1986). (This has followed somewhat similar lines to the electron diffraction work, with a number of additional complications arising from the vector nature of the electromagnetic wave field. The error in phase measurement for low-order reflections in acentric crystals by X-ray diffraction is about $45°$. Using electrons, it is less than $1°$.) For electron diffraction, Høier and Marthinsen, building on earlier work by Kambe and Gjønnes, have provided the basic theory for three-beam diffraction and channeling in acentric crystals in a series of papers (Hoier and Marthinsen 1983; Marthinsen and Hoier 1986, 1988; Hoier et al. 1988).

The simplest way to expose the phase dependence of three-beam intensities is to use the Bethe potential described in Sect. 5.6.2, if one of the three beams, h, is weaker than the other two. We write the structure factors as $U_g \exp(i\alpha_g)$. In the three-beam case for an acentric crystal, we then have

$$\left| U_g^{\text{eff}} \right|^2 = \left| U_g - \frac{U_h U_{gh}}{2KS_h} \right|^2$$
$$= \left| U_g \right|^2 \left[\left(1 - \frac{|U_h||U_{gh}|}{2KS_h|U_g|} \cos \Psi \right)^2 + \left(\frac{|U_h||U_{gh}|}{2KS_h|U_g|} \sin \Psi \right)^2 \right] \tag{5.54}$$

where

$$\Psi = \alpha_h + \alpha_{-g} + \alpha_{g-h} \tag{5.55}$$

is the three-phase invariant, or sum of the phases of the corresponding structure factors. This three-phase invariant is independent of the choice of origin in the crystal, as we expect from the intensities of electron diffraction, since the vectors h, $-g$, and $g - h$ form a closed triangle. Thus, the intensities in three-beam diffraction depend on both the amplitudes of the three structure factors and the three-phase invariant involved. The influence of the phases is strongest if

$$|U_h||U_{gh}|/2KS_h|U_g| = 1 \tag{5.56}$$

The Bethe approximation is the best approximation for the systematic three-beam cases of Fig. 5.9a, c. In these cases, $g = 2h$ and $2KS_h \approx h^2$ near the Bragg condition for g. The condition (5.56) may be satisfied by varying the electron high voltage, using the dependence of the structure factor U_g on accelerating voltage [see Eqs. (4.55) and (4.47)], and this is possible in certain favorable cases. Then, the systematics CBED three-beam intensity distribution may be used to measure the three-phase invariant to an accuracy of about 1° (Zuo et al. 1989b; Jiang et al. 2010).

The most general three-beam case, however, is the nonsystematics three-beam case shown in Fig. 5.9f. Because of the freedom here in choosing reflections g and h, the excitation error $2KS_h$ may now be varied in two dimensions over the CBED intensity distribution, and the voltage may also be varied to satisfy condition (5.56). In this way, the greatest sensitivity to phases may be obtained. Thus, the nonsystematics three-beam CBED method is the most versatile and general method of phase determination in acentric crystals. Although Eq. (5.54) applies to this case also under certain conditions, the nonsystematic three-beam case is best described by the "Kambe" approximation, as used by Kambe in his classic paper (Kambe 1957). In the following, we describe the Kambe approximation, and the nonsystematic three-beam case in the language of perturbation theory.

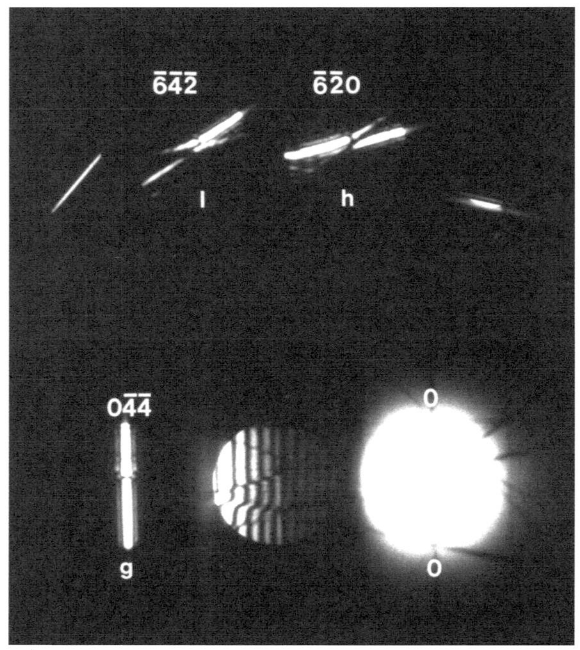

Fig. 5.10 Nonsystematic three-beam CBED pattern from silicon recorded at 100 kV. The two beam at the Bragg condition $l = (-6, -4, -2)$ and $h = (-6, -2, 0)$ differ by the $(0, -2, -2)$ reflection containing vertical bands. The gap in intensity along the Bragg hyperbolas in reflection l and h is shown

For the nonsystematic three-beam diffraction of Fig. 5.10 and illustrated in Fig. 5.11, when the coupling between g and h is far greater than the coupling between o and g and h, that is, if

$$\left|U_{gh}\right| \gg \left|U_g\right| \ \text{.or.} \ \left|U_h\right| \tag{5.57}$$

then the structure matrix of the three-beam Eq. (5.43) may be separated into a primary structure matrix and a perturbation matrix as follows:

$$A_o = \begin{pmatrix} 0 & 0 & 0 \\ 0 & 2KS_g & U_{gh} \\ 0 & U_{hg} & 2KS_h \end{pmatrix}, \quad \delta A = \begin{pmatrix} 0 & U_{-g} & U_{-h} \\ U_g & 0 & 0 \\ U_h & 0 & 0 \end{pmatrix} \tag{5.58}$$

The eigenvalues and eigenvectors of A are readily solved using the two-beam solution given in Sect. 5.3. They are

$$\tau_1 = \gamma_o^1 = 0; \quad \tau_{2,3} = \gamma_o^{2,3} = \left[S_g + S_h \mp \sqrt{(S_g - S_h)^2 + \left|U_{gh}/K\right|^2}\right]/2 \tag{5.59}$$

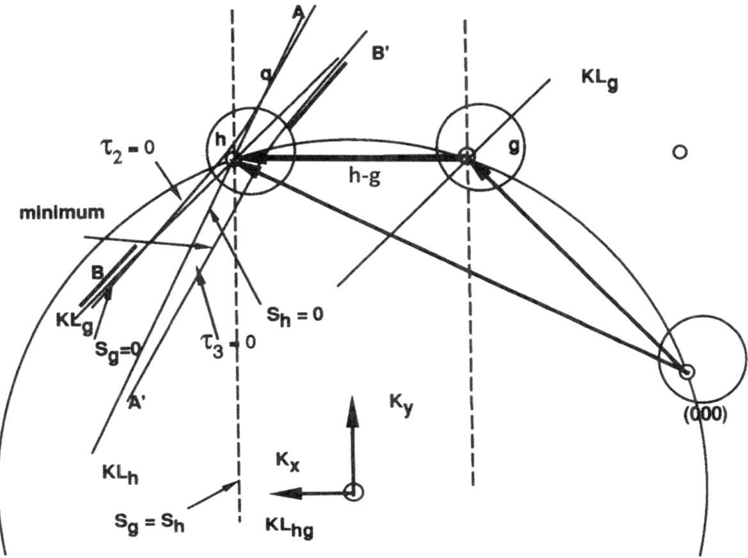

Fig. 5.11 General form of a three-beam CBED pattern. Kikuchi lines KL_h and KL_g (along which S_g and S_h are zero) are shown. The intensity distribution has extrema on the hyperbola where $\tau_2 = 0$ and $\tau_3 = 0$, as shown. The intensity approaches two-beam form at A and A' and fades toward B and B'. For the CdS, example discussed in the text, $g = (4, 1, -2)$ and $h - g = (0, 0, 2)$. The position of the minimum is shown for $\psi = 0$

and

$$C_{\mathrm{o}} = \begin{pmatrix} 1 & 0 & 0 \\ 0 & \cos(\beta/2)e^{i\alpha_{gh}} & -\sin(\beta/2) \\ 0 & \sin(\beta/2) & \cos(\beta/2)e^{-i\alpha_{gh}} \end{pmatrix} \tag{5.60}$$

where β is defined by $\cot \beta = \left(S_g - S_h\right)/\left|U_{gh}\right|$. Using these results, we obtain the three-beam equation, with the states of A_{o} as the basis (Zuo 1991):

$$\begin{pmatrix} 2K\gamma_{\mathrm{o}}^1 & A'_{12} & A'_{13} \\ A'_{21} & 2K\gamma_{\mathrm{o}}^2 & 0 \\ A'_{31} & 0 & 2K\gamma_{\mathrm{o}}^3 \end{pmatrix} \begin{pmatrix} \varepsilon_1 \\ \varepsilon_2 \\ \varepsilon_3 \end{pmatrix} = 2K\gamma \begin{pmatrix} \varepsilon_1 \\ \varepsilon_2 \\ \varepsilon_3 \end{pmatrix} \tag{5.61}$$

Here,

$$A'_{12} = A'^{*}_{21} = U_{-g}\cos(\beta/2)e^{i\alpha_{gh}} + U_{-h}\sin(\beta/2)$$

$$A'_{13} = A'^{*}_{31} = -U_{-g}\sin(\beta/2) + U_{-h}\cos(\beta/2)e^{-i\alpha_{gh}}$$

In the region $\gamma_{\mathrm{o}}^2 \approx 0$, $\left|2K\gamma_{\mathrm{o}}^3\right| \gg \left|A'_{12}\right|$ or $\left|A'_{13}\right|$, to first order $\varepsilon_3 \approx 0$. Then Eq. (5.61) are reduced to the twofold degenerate perturbation equation

$$\begin{pmatrix} 2K\gamma_{\mathrm{o}}^1 & A'_{12} \\ A'_{21} & 2K\gamma_{\mathrm{o}}^2 \end{pmatrix} \begin{pmatrix} \varepsilon_1 \\ \varepsilon_2 \end{pmatrix} = 2K\gamma \begin{pmatrix} \varepsilon_1 \\ \varepsilon_2 \end{pmatrix} \tag{5.62}$$

The solutions of this type of equation are similar to the two-beam case of Sect. 5.6.2. Similar results may be obtained in the region $\gamma_{\mathrm{o}}^3 \approx 0$. Combining these results with Eq. (5.17), we obtain the intensity expression in the regions defined by $\gamma_{\mathrm{o}}^2 \approx 0$ and $\gamma_{\mathrm{o}}^3 \approx 0$ (Zuo et al. 1989a):

$$I_h(S_g) = \frac{(2KS_g)^2}{(2KS_g)^2 + \left|U_{hg}\right|^2} \sin^2\left\{\frac{\pi t}{K}\sqrt{\left|U_h^{\mathrm{eff}}\right|^2}\right\} \tag{5.63}$$

where

$$\left|U_h^{\mathrm{eff}}\right|^2 = \left|U_h\right|^2 \frac{(2KS_g)^2}{(2KS_g)^2 + \left|U_{hg}\right|^2}\left[\left(1 - \frac{\left|U_g\right|\left|U_{hg}\right|}{2KS_g|U_h|}\cos\Psi\right)^2 + \left(\frac{\left|U_g\right|\left|U_{hg}\right|}{2KS_g|U_h|}\sin\Psi\right)^2\right] \tag{5.64}$$

The expression for the intensity of reflection g is obtained by interchanging g and h in Eqs. (5.63) and (5.64). The effective potential in Eq. (5.64) is the same as the Bethe potential in Eq. (5.54) if $2KS_g \gg \left|U_{hg}\right|$, which is the condition for the Bethe approximation. Figure 5.12 shows how the ratio $\left|U_h^{\mathrm{eff}}\right|/|U_h|$ depends on the phase invariant. The effective structure factor is most sensitive to the phase invariant near the minimum shown.

These results may be used to draw some qualitative conclusions about the main features of nonsystematic three-beam CBED patterns. These features are shown in Fig. 5.11, which shows three CBED disks in a general nonsystematic pattern in which the Bragg condition is satisfied at the center of each disk g and h. Kikuchi lines are shown at KL_g and KL_h. These lines run normal to the respective g vectors. We note the line KL_g in disk h, which arises due to multiple scattering. Lines KL_{h-g} and KL_{g-h} are the Kikuchi bands belonging to reflection $g - h$. Along these lines $S_g = S_h$. The Kikuchi line of maximum intensity outside the disks (due to inelastic scattering) continues inside the disks (due mainly to elastic scattering) as the locus of the Bragg condition, but is severely perturbed near the center of the disks. Instead of following the geometric locus defined by the Bragg condition ($S_g = 0$), the lines separate, as shown in Fig. 5.11. The resulting gap between the lines at the Bragg condition has been studied for many years since the early work on Kikuchi line patterns of Shinohara (1932). In X-ray diffraction, it is known as the Renninger effect (James 1950). It is this gap which is measured in the intersecting Kikuchi line

Fig. 5.12 Variation of the effective potential U_h^{eff} with excitation error for equal increments in the phase invariant between 0 and 90. This quantity is related to the intensity in three-beam CBED patterns along AB ($S_g < 0$) and $B'A'$ ($S_g > 0$) (see Fig. 5.11). Asymptotes are $|U_h|$

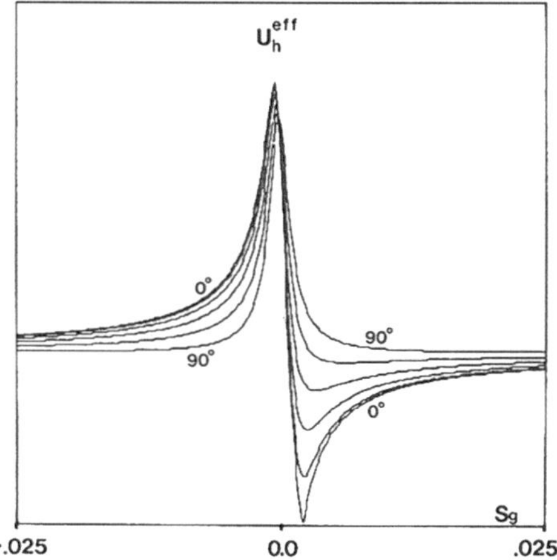

(IKL) and intersecting HOLZ line (IHL) methods (Høier 1972; Tafto and Gjonnes 1985). Figure 5.11 shows the hyperbola $\tau_2 = 0$ and $\tau_3 = 0$ (or $\gamma_o^2 = 0$ and $\gamma_o^3 = 0$). According to Eq. (5.63), the locus of maximum intensity follows these two hyperbolas. On the hyperbolas, the incident beam is constrained to move such that

$$2KS_g = \frac{|U_{gh}|^2}{2KS_h} \tag{5.65}$$

We call $\gamma_o^2 \approx 0$ the upper hyperbola and $\gamma_o^3 \approx 0$ the lower. On the upper hyperbola, both S_g and S_h are negative. They are positive on the lower hyperbola; thus, Eq. (5.65) gives both hyperbolas. The change in excitation error across the gap shown is found from the roots of Eq. (5.59) to be

$$|\Delta S_h| = |U_{gh}|/K \tag{5.66}$$

along the line $S_g = S_h$, that is, KL_{h-g}, in Fig. 5.11. This equation is the basis of the IKL and IHL methods, since it offers a simple method of measuring U_{gh}.

The intensities along the hyperbola also vary. The intensity of beam h decreases as the hyperbola approaches the line $S_g = 0$, due to the leading term in Eq. (5.63). Similar conclusions may be drawn about the intensity of beam g. A maximum or minimum effective potential (5.64) results in an intensity maxima or minima. According to Eq. (5.64), the maxima and minima occur at the coordinates (Zuo et al. 1989a, b)

$$S_h = \frac{|U_{gh}|}{4K\cos\Psi}\left\{\frac{|U_h|}{|U_g|} - \frac{|U_g|}{|U_h|} \pm \sqrt{\left(\frac{|U_h|}{|U_g|} - \frac{|U_g|}{|U_h|}\right)^2 + 4\cos^2\Psi}\right\} \tag{5.67}$$

with S_g given by Eq. (5.65). The plus sign in Eq. (5.67) gives the maximum, the minus sign gives the minimum. For centrosymmetric crystals, $\Psi = 0$ or π. Then, the minima occur at

$$S_g = \pm \frac{|U_{gh}|}{2K} \frac{|U_g|}{|U_h|} \quad \text{and} \quad S_h = \pm \frac{|U_{gh}|}{2K} \frac{|U_h|}{|U_g|} \qquad (5.68)$$

These results were derived under the Kambe "strong coupling" approximation where $|U_{gh}| \gg |U_g|$ or $|U_h|$. Under this condition, Eq. (5.68) agrees with (5.45), the exact solutions of Gjønnes and Høier for centrosymmetric crystals. The position of this minimum immediately tells us whether the three-phase invariant Ψ equals 0 or π in a centrosymmetric crystal. In Fig. 5.11, the minimum occurs on the lower hyperbola $A'B'$ if $\Psi = 0$, and on the upper hyperbola AB if $\Psi = \pi$. This is also illustrated in Fig. 5.10, where the $(-6, -2, 0)$ and $(-6, -4, -2)$ reflections in silicon are strongly excited at 100 kV. These reflections are coupled (differ) by the (022) reciprocal lattice vector. The arrow indicates the minimum of intensity, which occurs for a positive excitation error on the lower hyperbola, showing that the sum of the phases Ψ is zero for these reflections. This is consistent with the fact that the phases of the $(-6, -2, 0)$, $(-6, -4, -2)$, and (022) in silicon are all zero.

For noncentrosymmetric crystals, the measurement of the excitation error at the minimum position may give the values of the phase invariant from Eq. (5.67). The dependence of the nonsystematic three-beam intensity on the phase invariant and the flexibility one has in choosing different reflections g and h (and thus different three-phase invariants) make it an ideal method for measuring phase invariants. It has been used to study the phase of the (002) reflection in CdS by Zuo et al. (1989), based on a comparison of experimental and simulated many-beam intensities. The automation of this process (by defining a chi-square goodness-of-fit index) is described in Chap. 11.

The preceding theory applies to elastic Bragg scattering. Yet similar gaps are seen in Kikuchi line patterns. The assumption is made that the inelastic scattering responsible for Kikuchi lines is generated continuously throughout the crystal and so can be described by integrating Eq. (5.63) over thickness. This affects the last term in Eq. (5.63), but does not alter the expression for the gap.

In summary, for a centric crystal, it may be possible to determine whether the three-phase invariant is 0 or π by direct inspection of the position of the minimum in nonsystematic three-beam CBED patterns (Hurley and Moodie 1980). For acentric crystals, the value of a phase invariant may be estimated from the position of the intensity minimum along the Bragg lines or may be measured more accurately by comparing computer-simulated CBED patterns with experimental patterns. The experimental feasibility of this approach was demonstrated by Nakashima et al. (2013) for α-Al_2O_3. In their work, individual structure-factor phases were obtained by inspecting features within three-beam CBED patterns. Using nine structure-factor phases, Nakashima et al. determined the structure of α-Al_2O_3 to within a few picometers per atom from the known crystal structure.

5.6.4 Dynamic HOLZ Intensities and Positions. Dispersion Surfaces for HOLZ Lines. How the Bragg Law Depends on Local Composition

Owing to their potential usefulness for the measurement of strains, accelerating voltage, and composition, we now extend the discussion of Sect. 3.10 on HOLZ lines to the dynamical case. The study in Sect. 3.10 was based on the simple Bragg law, but in Sect. 5.6.2, we have already seen that this may not be accurate if more than two beams are excited. In this section, we quantify this effect and also provide a geometrical account of the dynamical displacement of HOLZ lines showing how, in principle, the HOLZ Bragg condition depends on structure factors for other excited beams and therefore on local composition. This effect is suggested schematically in Fig. 5.13b. The experimental evidence for these effects and their applications are discussed in Chap. 16, where suggestions are also made for procedures which minimize these errors due to multiple scattering.

The problem of "dynamical HOLZ line shifts" has been studied by many authors. The earliest relevant work was actually devoted to the problem of the anomalies which were observed in the position and intensities of Kikuchi lines (Pfister 1953; Menzel-Kopp and Menzel 1962; Gjonnes and Watanabe 1966). These had been analyzed using three-beam considerations as long ago as in 1932 (Shinohara 1932). A complete analysis using explicit three-beam expressions was given by the Norwegian group (Gjonnes and Hoier 1969; Høier 1972, 1973), and these results can also be applied directly to the problem of HOLZ line shifts. Buxton (1976) analyzed the problem by treating the HOLZ reflections as a weak perturbation of the ZOLZ reflections in zone-axis orientations, which was followed by Jones et al. (1977). Other works have concentrated on the problem of correction schemes for strain measurement (Bithell and Stobbs 1989; Lin et al. 1989; Zuo 1992; Armigliato et al. 2000; Houdellier et al. 2006, also see Chap. 16). For a three-beam analysis of HOLZ lines, see Britton and Stobbs (1987).

We expect the HOLZ intensity for beam h to follow the trajectory defined by Zuo (1991):

$$S_h - \gamma^i = 0 \tag{5.69}$$

This may simply be derived by using the condition for the *dynamical* Bragg condition for the strongest Bloch wave,

$$K^2 - (\vec{k} + \vec{h})^2 \approx K^2 - (\vec{K} + \vec{h})^2 - 2K\gamma = 2K(S_h - \gamma) = 0 \tag{5.70}$$

where $\vec{k}$ is the Bloch wave vector AO in the ZOLZ in Fig. 5.13b. (The kinematic expression for the HOLZ line trajectory, Eq. (3.18), was based on the condition $\gamma = 0$ and should be compared with the above.) We now discuss the geometric interpretation which these results give to dynamical HOLZ line shifts.

Figure 5.13b shows the scattering geometry for a HOLZ line h crossing the (000) ZOLZ disk at the Bragg condition for h. Because of the large number of wave vectors, we depart temporarily from the definitions of wave vectors given at the beginning of the book. The first-order reflection g in the ZOLZ is slightly off the Bragg condition. (The HOLZ line h and the first-order Bragg condition **g** would therefore lie at the extreme left-hand edges of the central and first-order CBED disks if the center of the axial disk corresponded to the zone axis in an experimental pattern.) The scale of the diagram has been distorted for clarity—on a diagram drawn to scale, it is not possible to show all relevant detail. (Typical dimensions at 100 kV might actually be $K \approx 27$, $|g| \approx 0.5$, the distances $AD \approx 0.005$ and $DG \approx 0.01$, all in reciprocal angstroms, and scattering angles all less than $10°$.) We shall use the expression "K-sphere" to mean a sphere of radius K, and "dynamical Bragg condition" to mean the dominant minimum in an experimental HOLZ line pattern. The figure shows several sets of wave vectors. The boldest lines show the true dynamical wave vectors $\vec{K}_o$ and $\vec{K}_o + \vec{h}$ outside the crystal at the dynamical Bragg condition. Here, $\vec{K}_o$ makes an angle θ with the zone axis, which can be measured. This dynamical Bragg condition is defined by the intersection of a K-sphere about $\vec{h}$ with dispersion surface 1, which we assume to be dominant.

Continuous lines show Bloch wave vectors excited inside the crystal. These originate on the dispersion surfaces, and (in two dimensions) their components $\vec{k}^{(i)}_{\mathrm{perp}}$ in the direction normal to the surface normal $\vec{n}$ must equal that of $\vec{K}_o$. A second Bloch wave vector $\vec{k}^{(2)}$ (not shown) might therefore originate at E. Although the various ZOLZ Bloch wave vectors $\vec{k}^{(i)}$ travel in different directions inside the crystal, since they have the same component $\vec{k}^{(i)}_{\mathrm{perp}}$ normal to $\vec{n}$ they combine to form a single beam on leaving the crystal. For the HOLZ reflections, each beam remains split into its Bloch wave components after leaving the crystal.

Short-dashed lines indicate kinematic wave vectors $\vec{K}'$, based on the true accelerating voltage. We assume initially that this true accelerating voltage is known, so that $\vec{K}'$ (inside the crystal) and $\vec{K}'_o$ (outside the crystal) can be determined using Eqs. (1.2) and (2.45). These wave vectors are then used solely as construction lines to provide a first approximation to the shape of the dynamical dispersion surfaces AB and EF, as described in Sect. 5.4. No such wave vectors are actually excited inside the crystal in these directions at the dynamical Bragg condition. Exploring the intensity along a line in the central disk (i.e., varying θ) corresponds to moving the wave vectors which meet at A (the dynamical Bragg condition) to C and F, where a secondary minimum might occur due to branch 2 of the dispersion surfaces. The Ewald sphere, of radius K', drawn about C, therefore passes through $\vec{h}$ at the Bragg condition. Then, the projection of $\vec{K}'$ onto $\vec{h}$ must equal half the length of $\vec{h}$, as required by the simple Bragg law [Eq. (3.14)].

In order to find the true diffracted wave vectors $\vec{K}_o + \vec{h}$ leaving the crystal, we need first to find the allowed dynamical wave vectors $\vec{k}^{(i)}$ inside the crystal. This requires both a differential equation (the Schrödinger equation) and a boundary

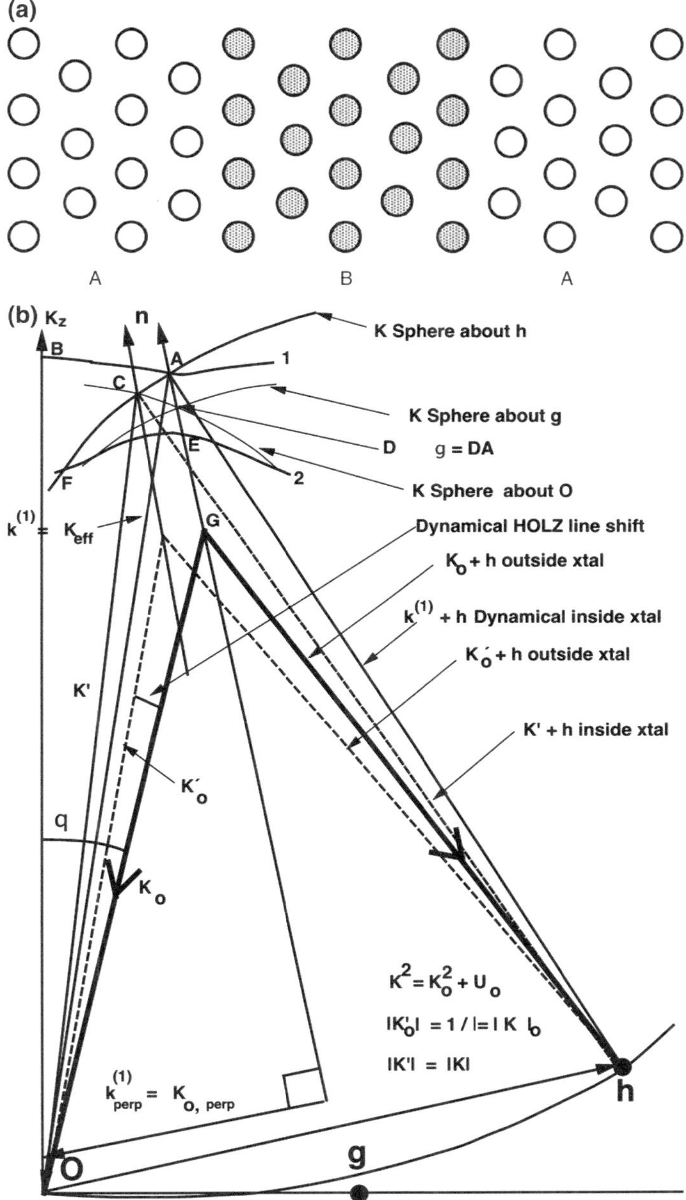

Fig. 5.13 **a** A hypothetical crystal containing different types of atoms on the same lattice. There is a small difference in the directions into which HOLZ beams would be diffracted between region A and region B. **b** The origin of dynamical shifts on HOLZ lines. Dashed rays (such as K'_o) are kinematical "construction lines," defining a kinematic Bragg condition at C. Actual HOLZ line intensity for h is a minimum in direction K_o, in the central disk (at θ to the zone axis), because ZOLZ potential opens a gap DA, shifting Bragg condition from C to A. Medium lines are dynamical waves inside the crystal. Thick lines are dynamical beams outside the crystal. The surface normal is n. The angle between K_o and K'_o is the dynamical HOLZ line shift. K_{eff} is the wave vector which would be derived, using simple Bragg law, from experimental measurement of θ, given the length of h

condition. Equation (5.11) provides solutions to the differential equation, giving the allowed Bloch wave labeling wave vectors $\vec{k}^{(i)}$ inside the crystal. The boundary condition requires that the components of the incident and diffracted wave vectors in the plane of the surface $\vec{K}_{o,\mathrm{perp}}$ and $\vec{k}^{(i)}_{\mathrm{perp}}$ be equal. Figure 5.13b is an attempt to summarize these boundary conditions geometrically, while also imposing the energy and momentum conservation conditions. The figure tells us little about the intensities of the Bragg beams—we assume here that there is a minimum of intensity in the central disk whenever the K-sphere about h crosses a dispersion surface. This then defines an angle θ and hence a point in our one-dimensional central disk. (In fact, the branch of the dispersion surface which is nearest to the K-sphere drawn about the $\vec{g}$ vector of interest contributes most to the intensity in beam g. That is, the scattering kinematics deviate as little as possible from that allowed in "vacuum.")

The effect of "switching on" the lattice potential is therefore to open up a gap DA near D, so that the true dispersion surfaces (on which all wave vectors inside the crystal must commence) becomes curved, as shown at AB and EF. If the HOLZ line is treated as a weak perturbation, then the Bragg point of interest moves from C to A. Weaker lines (fine structure) may also be seen in the outer ring of HOLZ disks at an incident beam direction at which the K-sphere intersects branch 2 at F. This fine structure is shown experimentally in Fig. 5.14a. In this sense, the fine structure in HOLZ disks can give a map of the ZOLZ dispersion surface. (It should be distinguished from the "splitting" fine structure on HOLZ lines which occurs due to strain in crystals. This occurs on a finer scale and is discussed in Chap. 16.)

The length of the diffracted wave vector $\vec{K}_o$ is fixed by Eq. (5.5); its direction is therefore also now fixed (in two dimensions) by the requirement that it originates on the surface normal as shown. There is therefore a small difference between the directions of the wave vectors $\vec{K}_o$ and $\vec{K}'_o$, which is the dynamical HOLZ line shift. Changes in composition, causing changes in structure factors and hence in γ, will alter the shape of branch 1 and the distance DA. In two-beam theory, this distance is just $1/\xi_g$. If the gap DA widens, for example, the figure shows that A moves to the left, and the angle between $\vec{K}_o$ and $\vec{K}'_o$ will increase. Thus, the direction Θ in which the minimum of intensity for HOLZ line h occurs in the central disk depends on structure factor g for the ZOLZ. Use of the simple Bragg law [Eq. (3.3)] corresponds to the use of wave vectors $\vec{K}'_o$ rather than $\vec{K}_o$, but this, if based on the true accelerating voltage, would not agree with the measured angle at which the HOLZ minimum occurs.

Equation (5.70) gives a dynamically corrected HOLZ line equation

$$K_y = -\frac{g_x}{g_y}K_x + \frac{g_z}{g_y}\left(K_z - \frac{K\gamma}{g_z}\right)K_z - \frac{g^2}{2g_y} \tag{5.71}$$

using the same notation as in Eq. (3.18). This dynamical expression differs from the kinematic expression [Eq. (3.18)] only by the term in γ. Over a small region of the

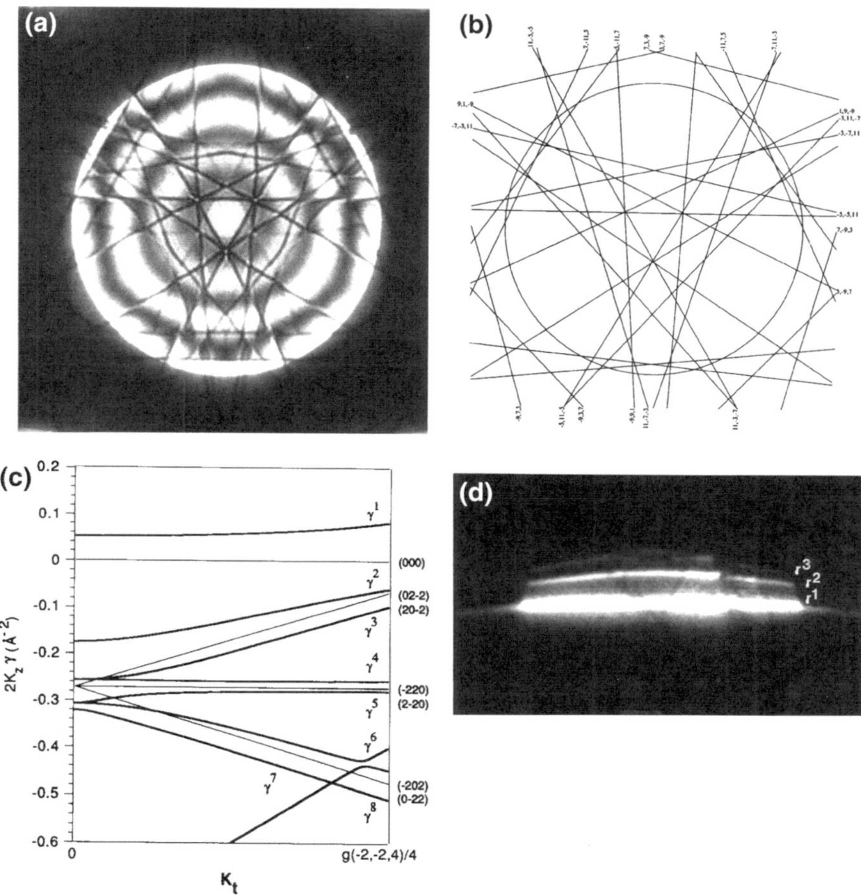

Fig. 5.14 a HOLZ lines in the central disk of silicon at 100 kV. [111] orientation at 183 °C.
b Kinematic calculation of the pattern in **a** for the voltage which gives the best fit (98.5 kV). The
true voltage is 100.0 kV. **c** A section of dispersion surfaces for Si [111] at 100 kV from the zone
center to $K_t = (-2, -2, 4)/4$. **d** HOLZ line in outer ring (*dark* field) corresponding to **a** for the (9,
−9, 1) reflection, showing the fine structure. The lines are labeled to correspond with the ZOLZ
dispersion surfaces shown in **c**. (Fig. 5.13b illustrates this effect)

dispersion surface, we might assume that the dispersion γ is approximately constant.
Then, the effects of dynamical dispersion may be thought of as a correction to the
accelerating voltage and accommodated by a change in the term K_z in Eq. (5.71).
However, this correction to the high voltage differs from zone to zone because of the
weighting g_z.

The strongest experimental evidence for HOLZ line shifts comes from the work of
Lin et al. (1989), who found variations of several kilovolts when using Eq. (2.4) to
determine the microscope accelerating voltage from indexed HOLZ lines taken from
different zone axes of the same silicon crystal. For example, they found $E_o = 195$.

6 kV at the [112] pole (based on IHL), but $E_o = 198.4$ kV at the [356] pole. Using complete many-beam computations to match the line positions, the calculated values become 198.1 and 198.5 kV for the same axes. (The additional correction for U_o has not been included. This is about 20 V for silicon.) The reason for these changes is that, in the zone-axis center, γ is positive; thus the effective K_z is lower, and this can be simulated by lowering the high voltage. The value of γ decreases as one moves away from the center of a high-symmetry zone axis.

In summary, the following conclusions may be drawn:

1. Dynamical effects will always result in the accelerating voltage being underestimated near a center of a zone axis, if Eq. (3.18) is used.
2. The dynamical correction is least at high-index zone axes or may be greatly reduced by avoiding zone axes altogether, where γ is small (Zuo 1992).
3. The dynamical correction increases with atomic number (for a similar projected density of atoms).
4. The correction will be least (other things being equal) in smaller unit cell crystals in which the reciprocal lattice is sparse.
5. Since branch 1 is relatively flat at the zone center, it has frequently been assumed that the correction is orientation-independent over the central disk.
6. The correction is least from the highest HOLZ layer. For example, it was found (Lin et al. 1989) that, at the [113] pole of silicon, the voltage used to match the FOLZ was 0.6 kV lower that required to match the TOLZ, using Eq. (3.18).
7. Errors of several kilovolts are likely in measurements of accelerating voltage based on Eq. (3.18). If the perturbation correction given below is used, this error is reduced to perhaps 200 V. Automated refinement, using many profiles across the lines matched to dynamical calculations, can reduce the error to as little as 14 V (Zuo 1992).

An expression for the dynamical correction to the accelerating voltage has been given (Lin et al. 1989; Zuo 1992) based on Bethe's perturbation method described in Sect. 5.6.2. This gives the correction (increase) to the "kinematic" voltage $\mathcal{E}'_o$ obtained by applying Eq. (3.18) to a HOLZ pattern with known lattice spacings as

$$\Delta\mathcal{E}_o = \frac{300K^2\gamma^{(1)}}{(1 + 1.956 \times 10^{-6}\mathcal{E}_o)nH} \tag{5.72}$$

where nH is the height of the HOLZ layer in reciprocal angstroms, and γ and K are measured in similar units. The true microscope voltage is approximately $\mathcal{E}'_o + \Delta\mathcal{E}_o$. We may now use the results of Sect. 3.3 on the Bethe potential to approximate the value of γ which is closest to the incident beam K-sphere. Away from the Bragg condition, where S_g is large, we may expand Eq. (5.51) to first order, so that

$$\gamma^{(1)} \approx S_o^{\text{eff}} = -\frac{1}{2K}\sum_{g\neq 0}\frac{|U_g|^2}{2KS_g} \tag{5.73}$$

Near the zone center, we have $2KS_g = -g^2$, and thus,

$$\gamma^{(1)} \approx \frac{1}{2K} \sum_{g \neq 0} \frac{|U_g|^2}{g^2} \tag{5.74}$$

where the sum is over structure factors U_g in the ZOLZ.

Figure 5.14a shows a silicon [111] zone-axis HOLZ pattern at 100 kV, recorded at a temperature of -183 °C. Figure 5.14b shows a simulated HOLZ pattern based on the kinematic approximation of Eq. (3.18). The high voltage has been varied to match the experimental HOLZ pattern of Fig. 3.11a near the center of the disk. This kinematic matching gives a high voltage of 98.5 kV. Figure 5.14c shows the calculated dispersion surfaces of the Si [111] zone axis from $\vec{K}_t = 0$ to $\vec{K}_t = (-2, -2, 4)/4$, obtained using the Bloch wave method described in Sect. 5.1. The bold lines are the dispersion surfaces, and the thin lines are the kinematical K spheres. Figure 5.14d shows the fine structure in the CBED disk for the $(9, -9, 1)$ HOLZ reflection. This HOLZ reflection has a very weak interaction with the other HOLZ reflections; thus, the fine structure shown is produced by the ZOLZ dispersion surfaces, in the manner discussed above. The fine structure in Fig. 5.14d is thus a direct image of the various branches of the dispersion surfaces shown in Fig. 5.14c, as indicated by the labeling on the figure.

From Fig. 5.14b, a measurement of the high voltage using the kinematic approximation yields the value 98.5 kV. This is 1.5 kV less than the actual value of 100 kV, obtained by an independent method. From Fig. 5.14c, we see that $2K\gamma$ is about 0.05 Å^{-2} near the [111] zone-axis center and, for the Si [111] zone axis, g_z is about 1/9.4 Å^{-1} for the FOLZ. At 100 kV, $K = 1/0.037$ Å^{-1}. From Eq. (5.71), the effective wavelength is therefore

$$\lambda^{\text{eff}} = 1 \Big/ \left(K_z - \frac{K\gamma}{g_z} \right) = 0.037325 \text{ Å}$$

The wavelength at 98.5 kV is 0.03732 Å. This is therefore in good agreement with the predicted value of the effective wavelength.

A different type of fine structure has also been observed on the satellite reflections produced by semiconductor multilayers (Gong and Schapink 1991).

References

Armigliato A, Balboni R et al (2000) TEM/CBED determination of strain in silicon-based submicrometric electronic devices. Micron 31:203–209
Bethe H (1928) Theory on the diffraction of electrons in crystals. Ann Phys 87:55–129
Bird DM (1990) Absorption in high-energy electron-diffraction from noncentrosymmetric crystals. Acta Cryst A46:208–214

Bird DM, Wright AG (1989) Phase dependence of Kikuchi patterns. I. Theory. Acta Cryst A45:104–109

Bithell EG, Stobbs WM (1989) The simulation of HOLZ line positions in electron-diffraction patterns—a 1st order dynamical correction. J Microsc 153:39–49

Britton EG, Stobbs WM (1987) The analysis and application of dynamic effects in HOLZ patterns. Ultramicroscopy 21(1):1–11

Buxton BF (1976) Bloch waves and higher-order Laue zone effects in high-energy electron-diffraction. Proc R Soc Lond A 350:335–361

Chang S-L (2004) X-ray multiple-wave diffraction: theory and application. Springer, New York

Cowley JM, Moodie AF (1957) The scattering of electrons by atoms and crystals. I. A new theoretical approach. Acta Crystallogr 10:609–619

Fues E (1949) Zur Deutung der Kossel=Möllenstedtschen Elektroneninterferenzen konvergenter Bündel an dünnen Plättchen II. Z Phys 125:531–538

Fukuhara A (1966) Many-ray approximation in the dynamical theory of electron diffraction. J Phys Soc Jpn 21:2645–2662

Gjonnes J, Hoier R (1969) Multiple-beam dynamic effects in Kikuchi patterns from natural spinel. Acta Crystallogr A 25:595

Gjonnes J, Hoier R (1971) Application of non-systematic many-beam dynamic effects to structure-factor determination. Acta Crystallogr A 27:313

Gjonnes J, Watanabe D (1966) Dynamical diffuse scattering from magnesium oxide single crystals. Acta Crystallogr 21:297–302

Gong H, Schapink FW (1991) Fine details in satellite HOLZ reflection disks of CBED from a GaAs/AlAs multilayer. Ultramicroscopy 35(3–4):171–184

Hashimoto H, Howie A, Whelan MJ (1962) Anomalous electron absorption effects in metal foils: Theory and comparison with experiment. Proc R Soc Lond A 269:80–103

Hirsch P, Howie A, Nicolson RB, Pashley DW, Whelan MJ (1977) Electron microscopy of thin crystals. Robert E. Krieger Publishing Company, Malaba, Florida

Høier R (1972) Displaced lines in Kikuchi patterns. Phys Status Solidi A 11:597–610

Hoier R (1973) Multiple-scattering and dynamical effects in diffuse electron-scattering. Acta Cryst A 29:663–672

Hoier R, Marthinsen K (1983) Effective structure factors in many-beam X-ray-diffraction—use of the 2nd Bethe approximation. Acta Crystallogr A 39:854–860

Hoier R, Zuo JM, Marthinsen K, Spence JCH (1988) Determination of structure factor phase invariants from non-systematic many-beam effects in convergent-beam patterns. Ultramicroscopy 26:25–30

Honjo G, Mihama K (1954) Fine structure due to refraction effect in electron diffraction pattern of powder sample part II. Multiple structures due to double refraction given by randomly oriented smoke particles of magnesium and cadmium oxide. J Phys Soc Jpn 9:184–198

Houdellier F, Roucau C, Clement L, Rouviere JL, Casanove MJ (2006) Quantitative analysis of HOLZ line splitting in CBED patterns of epitaxially strained layers. Ultramicroscopy 106: 951–959

Howie A, Whelan MJ (1961) Diffraction contrast of electron microscope images of crystal lattice defects. II. Development of a dynamical theory. Proc R Soc Lond Ser A 263:217

Hummer K, Billy H (1986) Experimental-determination of triplet phases and enantiomorphs of noncentrosymmetric structures. I. Theoretical considerations. Acta Crystallogr A 42:127–133

Humphreys CJ (1979) Scattering of fast electrons by crystals. Rep Prog Phys 42:1825–1887

Hurley AC, Moodie AF (1980) The inversion of the three-beam intensities for scalar scattering by a general centrosymmetric crystal. Acta Cryst A 36:737–738

James RW (1950) The optical principles of the diffraction of X-rays. G. Bell and Sons, London

Jiang B, Zuo JM et al (2010) Combined structure-factor phase measurement and theoretical calculations for mapping of chemical bonds in GaN. Acta Cryst A66:446–450

Jones PM, Rackham GM, Steeds JW (1977) Higher-order Laue zone effects in electron-diffraction and their use in lattice-parameter determination. Proc R Soc Lond A 354:197

Kambe K (1957) Study of simultaneous reflexion in electron diffraction by crystals I. Theoretical treatment. J Phys Soc Jpn 12:13–25

Karle J (1989) Linear algebraic analyses of structures with one predominant type of anomalous scatterer. Acta Crystallogr A 45:303–307

Kirkland EJ (2010) Advanced computing in electron microscopy, 2nd edn. Springer, New York

Kogiso M, Takahashi H (1977) Group-theoretical method in many-beam theory of electron-diffraction. J Phys Soc Jpn 42:223–229

Lehmpfuhl G, Reissland A (1968) Photographical record of the dispersion surface in rotating crystal electron diffraction pattern. Z Naturforsch 23A:544–549

Lewis AL, Villagrana RE, Metherell AJF (1978) Description of electron-diffraction from higher-order Laue zones. Acta Cryst A34:138–139

Lin YP, Bird DM, Vincent R (1989) Errors and correction term for HOLZ line simulations. Ultramicroscopy 27:233–240

Marthinsen K, Hoier R (1986) Many-beam effects and phase information in electron channeling patterns. Acta Crystallogr A 42:484–492

Marthinsen K, Hoier R (1988) Determination of structure-factor phase invariants and effective structure factors in non-centrosymmetric crystals. Acta Crystallogr A 44:558–562

Menzel-Kopp C, Menzel E (1962) J Phys Soc Jpn 17 (Suppl. BII) 80

Metherall AJF (1975) Diffraction of electrons by perfect crystals. In: Valdre U, Ruedl E (eds) Electron microscopy in materials science. The Commission of the European Communities, Luxembourg

Nakashima PNH, Moodie AF et al (2013) Direct atomic structure determination by the inspection of structural phase. Proc Natl Acad Sci USA 110:14144–14149

Peng LM, Dudarev SL, Whelan MJ (2004) High energy electron diffraction and microscopy. Oxford University Press

Pfister H (1953) Ann Phys 11:239

Shinohara K (1932) Diffraction of cathode rays by single crystals, Part 111.-Simultaneous reflection. Sci Pap Inst Phys Chem Res Tokyo 20:39–51

Spence JCH (1988) Experimental high-resolution electron microscopy. Oxford University Press, New York

Spence JCH (2013) High resolution electron microscopy, 4th edn. Oxford University Press, Oxford, UK

Spence JCH, Zuo JM (1992) Electron microdiffraction. Plenum, New York

Tafto J (1987) Reciprocity in electron energy-loss spectra from noncentrosymmetric crystals. Acta Crystallogr A 43:208–211

Tafto J, Gjonnes J (1985) The intersecting Kikuchi line technique—critical voltage at any voltage. Ultramicroscopy 17:329–334

Zuo JM (1991) Perturbation-theory in high-energy transmission electron-diffraction. Acta Cryst A47:87–95

Zuo JM (1992) Automated lattice-parameter measurement from HOLZ lines and their use for the measurement of oxygen-content in $YBa_2Cu_3O_{7-\Delta}$ from nanometer-sized region. Ultramicroscopy 41:211–223

Zuo JM, Hoier R, Spence JCH (1989a) 3-beam and many-beam theory in electron-diffraction and its use for structure-factor phase determination in non-centrosymmetric crystal-structures. Acta Cryst A45:839–851

Zuo JM, Spence JCH et al (1989b) Accurate structure-factor phase determination by electron-diffraction in noncentrosymmetric crystals. Phys Rev Lett 62:547–550

Chapter 6
Electron Optics

In this and next two chapters, we introduce the construction and properties of magnetic lenses, aberration correctors, and electron sources. They are critical components of modern electron microscopes; understanding the physics as well as their functions is important for both basic and advanced applications of electron diffraction and microscopy in general. Introductory accounts of magnetic lenses and their optics can be found in the books by Hall (1966), Hawkes (1972), and especially Grivet (1972). An introduction to aberration correction from a user point of view is provided by Erni (2010), while the theory behind aberration correction is described in the advanced book by Rose (2013). Modern lens design uses computed solutions of the Laplace equation and subsequent numerical solution of the ray equation; such materials can be found in special treaties of electron optics (Mulvey and Wallington 1973; Carey 1987; Orloff 2008; Tsuno 2008), and they are not covered here.

The intuitive trajectory-based approach to electron optics is employed here. First, construction of round electron magnetic lenses is described, followed by a derivation of the paraxial equation and its solutions for electron paths inside the magnetic lens. Results are then used to derive the properties of magnetic lenses, such as image rotation, cardinal points, focal length, and aberrations. The lens characteristics are obtained for simple models, showing the effects of lens excitation and geometry.

The writing of this chapter and next two chapters has benefited from following references: (1) the lectures given by Prof. Rose at University of Illinois, Urbana-Champaign, in 2012, and the lecture notes that Prof. Rose has kindly prepared, (2) the book by Grivet, (3) the book chapters by Krivanek et al. (2008), Swanson and Schwind (2008) and Inada et al. (2009), and other references cited in the chapter.

© Springer Science+Business Media New York 2017
J.M. Zuo and J.C.H. Spence, *Advanced Transmission Electron Microscopy*,
DOI 10.1007/978-1-4939-6607-3_6

6.1 Magnetic Lenses

Electrons are focused by the force exerted by the electric and/or magnetic fields in an electron lens. Inside a magnetic lens (without the external electric field), the electron motion is described by:

$$\vec{F} = m\frac{\mathrm{d}^2\vec{r}}{\mathrm{d}t^2} = -e\vec{V} \times \vec{B}. \tag{6.1}$$

Because electrons are negatively charged, the force is opposite to the cross product of $\vec{V} \times \vec{B}$. To focus electrons in a short distance, the magnetic field must be strong, the field must also be confined, so there is no additional interaction when electrons are away from the lens. In order to have a uniform focus, the field should be rotationally symmetric as much as possible.

A basic lens design that meets the above requirements is consisted of a round magnetic loop, which is made out of the yoke, pole, and pole piece, and the windings for passing electric current (see Fig. 5.1 for a schematic illustration). The lens ideally has the cylindrical symmetry with its center axis serving as the optical axis. The yoke, pole, and pole piece are made of typical materials of soft iron with a high permeability. High permeability keeps the magnetic field generated by the electric current within the magnetic loop, except at the pole-piece gap. The magnetic field in the gap, as characterized by its distribution and strength, is determined by the gap geometry, and the lens excitation strength, which is determined by the number of windings times the current, NI. The gap is measured by the gap distance and the bore diameter as illustrated in Fig. 6.1. For a magnetic lens with a short focal distance, such as the objective lens, the gap distance is designed to be about a few millimeters. The magnetic field is confined mostly in the gap region, and this field acts as the lens for the electrons.

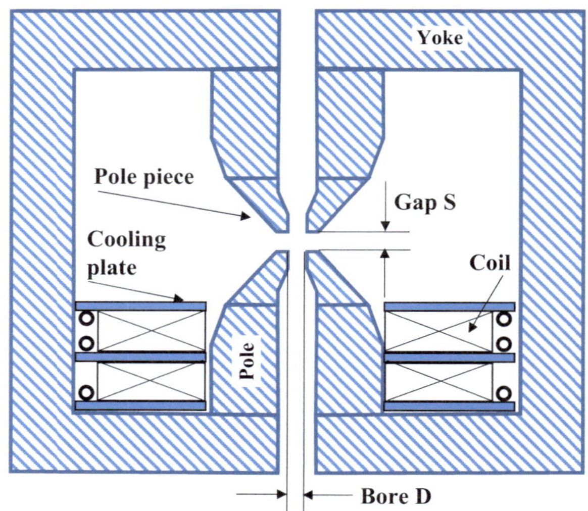

Fig. 6.1 Schematic diagram showing the cross-sectional view of the construction of a round magnetic lens with cylindrical symmetry (after Tsuno 2008)

Fig. 6.2 A dodecapole, the central aperture for the beam path is surrounded by twelve coils. (From CEOS GmbH, with permission)

Another type of electron lenses is the electric and/or magnetic multipoles made up of pairs of dipoles. Inside a magnetic multipole, the pole pieces are arranged symmetrically with each pole piece surrounded by its own coil (an example is shown in Fig. 6.2). Fields are generated by passing currents in each coil. The amount of current in each coil is kept same within the experimental accuracy, but the directions of currents are opposite for any two neighboring coils. Inside an electrostatic multipole, fields are formed by applying a voltage of alternating polarity to the electrodes. The multipole is named based on the number of poles ($2m$, m for multiplicity). For example, the quadrupole has $2m = 4$, and a hexapole (also known as sextupole) has $2m = 6$. Inside a conventional TEM, weakly excited magnetic quadrupoles are used for correcting defects in the magnetic lens focus. They are also used in energy-loss spectrometers for focusing the electron spectrum. The hexapoles, or the octupoles, also serve as major components in aberration correctors. The multipole lenses in general have following characteristics:

(1) They possess a finite rotation symmetry, but lack the continuous rotation symmetry of a round lens;
(2) Multipoles have large, primary, aberrations that are absent in an ideal round lens;
(3) The largest field is perpendicular to the optical axis. Because of this, a quadrupole can provide a short focus using far less power than a round lens.

For the reason 3, multipole lenses have found exclusive use in focusing high-energy charged particles, for accelerators such as synchrotrons. In what follows, we will first focus on the round magnetic lenses. The properties of electron multipoles will be discussed in the next chapter on aberration correction.

The motion of an electron inside a magnetic lens is described by its position and velocity according to classical mechanics. If we define the optical axis as the z-axis, then by rewriting Eq. (6.1) using the electron position in the Cartesian coordinate of $\vec{r} = (x, y, z)$, we obtain

$$m\frac{d^2\vec{r}}{dt^2} = -e\frac{d\vec{r}}{dt} \times \vec{B} \tag{6.2}$$

or

$$m\frac{d^2x}{dt^2} = e\left(V_z B_y - V_y B_z\right)$$
$$m\frac{d^2y}{dt^2} = e\left(V_x B_z - V_z B_x\right) \tag{6.3}$$

The absolute value of the electron velocity remains constant inside the magnetic field, thus

$$\left|\frac{d\vec{r}}{dt}\right| = \frac{ds}{dt} = \frac{dz}{dt}\sqrt{1 + \left(\frac{dx}{dz}\right)^2 + \left(\frac{dy}{dz}\right)^2} = V_\text{o}. \tag{6.4}$$

If we take the initial electron velocity as along the optical axis, then the initial force on the electrons comes from the x and y components of the fringe magnetic field (in combination, they contribute to a radial component, B_r, because of the cylindrical symmetry). The initial force generates an angular velocity (V_θ) that makes the electrons rotate. This annular velocity in turn interacts with the axial fringe magnetic field (B_z) and generates a force that bends the electrons toward the optical axis.

The magnetic field inside the gap can be described based on the scalar magnetic potential $A(\vec{r})$. According to Maxwell equations, we have $\nabla \times \vec{B} = 0$ in a space with no electric field and current. In analogy to the electrostatic potential, the magnetic field in this space is given by the gradient of the scalar magnetic potential with $\vec{B} = \nabla A$ and

$$\nabla \cdot \vec{B} = \Delta A = \left(\frac{\partial^2}{\partial x^2} + \frac{\partial^2}{\partial y^2} + \frac{\partial^2}{\partial z^2}\right)A = 0 \tag{6.5}$$

Inside the magnetic lens, the magnetic pole-piece surface has a constant potential, $A(\vec{r})|_{\vec{r}\in S} = A_\text{o}$. Thus, the magnetic field inside the pole-piece gap can be calculated by solving Eq. (6.5) using the boundary condition of a constant potential at the pole-piece surfaces. For a cylindrical lens, the potential only depends on the in-plane radius r, and z with $A = A(r, z)$, and Eq. (6.5) can be rewritten as

$$\Delta A = \left(\frac{1}{r}\frac{\partial}{\partial r}r\frac{\partial}{\partial r} + \frac{\partial^2}{\partial z^2}\right)A = 0 \tag{6.6}$$

The electron beam diameter inside the TEM is small on the order of μm compared to the bore diameter of the lens on the order of millimeters. The angle to the optical axis is also small, on the order of tens of mrad. Thus, electrons see mostly the magnetic potential in the vicinity of the optical axis. At a very small distance to the optical axis, the magnetic potential of a round lens can be expanded in a Taylor expansion series according to

$$A(r,z) \approx \sum_{n=0}^{\infty} a_n(z) r^{2n}. \tag{6.7}$$

Here $a_o(z) = A(0,z)$ is the axial potential. Substitute (6.7) into (6.6), we readily derive the following recurrence formulae

$$4a_{n+1}(z)(n+1)^2 = -a_n''. \tag{6.8}$$

Using this, we can show that the magnetic potential near the optical axis is completely determined by the axial potential in the form (Hawkes 2008)

$$A(r,z) \approx a_o(z) - a_o''(z)r^2/4 + a_o''''(z)r^4/64 - \cdots \tag{6.9}$$

The magnetic field is obtained by taking derivative of the potential in Eq. (6.9), which gives:

$$
\begin{aligned}
B_r &\approx -B'(z)r/2 + B''''(z)r^3/16 - \cdots \\
B_z &\approx B(z) - B''(z)r^2/4 + \cdots
\end{aligned}
\tag{6.10}
$$

Here, the prime ($'$) marks the partial differentiation against z. Equation (6.10) shows that the magnetic field employed for electron focusing is determined entirely by the field along the optical axis ($B(z)$) and its derivatives.

We are now going to derive the so-called electron paraxial ray equation, also known as path equation. First we note that the electron velocity, which gives rise to the force, is related to the electron trajectory, more specifically the slope of the trajectory as illustrated in Fig. 6.3 according to

$$
\begin{aligned}
\frac{V_x}{V} &= \frac{dx/dt}{ds/dt} = \frac{x'}{\sqrt{1+x'^2+y'^2}} \\
\frac{V_y}{V} &= \frac{dy/dt}{ds/dt} = \frac{y'}{\sqrt{1+x'^2+y'^2}} \\
\frac{V_z}{V} &= \frac{dz/dt}{ds/dt} = \frac{1}{\sqrt{1+x'^2+y'^2}}
\end{aligned}
\tag{6.11}
$$

where s is the path length, $x' = dx/dz$ and $y' = dx/dz$. For electrons traveling close to the optical axis, the curvature is small. Under the so-called paraxial

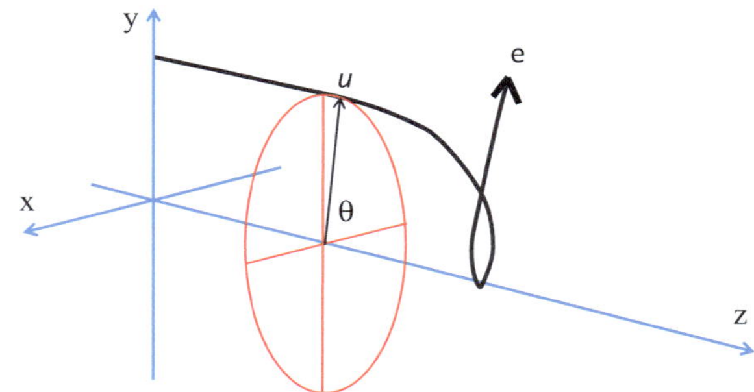

Fig. 6.3 Rotational coordinate for electron path with u for distance to the z-axis (in a lens with a straight optical axis) and θ for rotational angle. The position of the electron is specified by $\vec{r} = (u, \theta, z)$

approximation, $V_x \approx x'V$, $V_y \approx y'V$ and $V_z \approx V$, the electron equations of motion, as described in Eq. (6.3), can be reduced to:

$$m\frac{\mathrm{d}^2 x}{\mathrm{d}t^2} \approx mV^2 x'' = e\left(V_z B_y - V_y B_z\right) \approx -eV(y'B + yB'/2)$$
$$m\frac{\mathrm{d}^2 y}{\mathrm{d}t^2} \approx mV^2 y'' = e\left(V_x B_z - V_z B_x\right) \approx eV(x'B + xB'/2)$$

$$(6.12)$$

This allows a description of the electron path in the x and y coordinate as it goes through the lens. Alternatively, the path can be separated according to the distance to the optical axis in term of radius and the azimuthal angle, which describes the electron rotation. In describing the focusing properties of a magnetic lens, we use the distance to the optical axis. For this purpose, it is convenient to introduce the following complex notation for the electron position (Rose 2013, Sect. 3.1.1):

$$\omega(z) = x(z) + iy(z) = u(z)\mathrm{e}^{i\theta(z)}$$

$$(6.13)$$

where $u(z)$ is the distance to the axis and $\theta(z)$ is the azimuthal angle in the xy plane. By multiplying the second equation in Eq. (6.12) by the imaginary phase and adding the two equations together, we obtain

$$\omega'' - i\frac{eB}{mV}\omega' - i\frac{eB'}{2mV}\omega = 0$$

$$(6.14)$$

Next, we substitute Eq. (6.13) into Eq. (6.14) and obtain after some mathematical manipulations (for details, refer to Sect. 4.1 of Rose 2013):

$$u'' + 2i\theta'u' + i\theta''u - \theta'^2 u - i\frac{eB}{mV}\left(u' + i\theta'u\right) - i\frac{eB'}{2mV}u = 0$$

$$(6.15)$$

This complex equation contains two variables and two equations, one for the real part and one for the imaginary part. It can be simplified considerably by choosing:

$$\theta' = \frac{\mathrm{d}\theta}{\mathrm{d}z} = \frac{eB}{2mV} = \sqrt{\frac{e}{8m\Phi}}B(z) \tag{6.16}$$

Here, Φ is the electron accelerating voltage and m is the electron relativistic mass. Using Eq. (6.16) in Eq. (6.15), we have cancellation of several terms and a simplified equation in the form

$$u'' + \frac{eB^2}{8m\Phi}u = \frac{\mathrm{d}^2 u(z)}{\mathrm{d}z^2} + \frac{e}{8m\Phi}B(z)^2 u(z) = 0 \tag{6.17}$$

This equation, together Eq. (6.16), is known as the paraxial ray equation.

In summary, the electron path in a magnetic lens is completely determined by the magnetic field along the optical axis. The distance to the axis is determined by Eq. (6.17), while integration of Eq. (6.16) over z gives the electron rotation angle. The approximation we used to obtain the above results is the paraxial approximation.

The design of a round magnetic lens thus comes down to what magnetic axial field will provide the desired electron path. The field along the optical axis can be only influenced by the magnetic potential of the pole piece and its geometry. Thus, there are no separate "knobs" to dial to obtain a field distribution as desired. This makes the design of magnetic lens akin to an inverse problem, e.g., what boundary conditions will give the desired axial potential and electron path? The usual practice is to start with a good lens design based on experience and then optimizes its performance numerically using an iterative procedure (Tsuno 2008).

6.2 Fundamental Rays and Conjugate Planes

Here, we examine the imaging property of an ideal magnetic lens as described by the paraxial ray equation. The small-angle approximation used to derive the paraxial equation was first introduced by Gauss in his study of optical lenses. The properties of various optical elements summarized using simple quantities such focal lengths and position of principle planes derived under this approximation are known as Gaussian optics. As shown in this section, the paraxial equation for the magnetic lens gives the same properties as in Gaussian optics.

First, we note that Eq. (6.17) is a second-order linear differential equation, which has two independent solutions. These two solutions define two independent electron paths, which we will call as fundamental rays. All other electron paths can be expressed as a linear combination of fundamental rays.

Figure 6.4 illustrates how the two fundamental rays are defined from the object plane to the image plane: one is the principle ray and the other is the axial ray. These two paths are described by the normalized, unitless distance $u_\pi(z)$ and $u_\alpha(z)$, respectively. The actual distance can be obtained after multiplying it with an

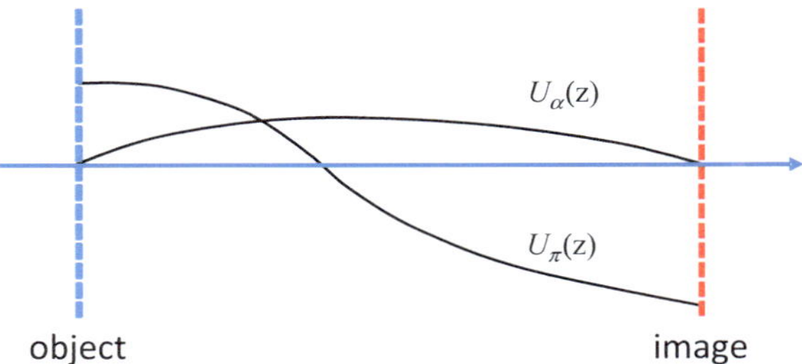

Fig. 6.4 Principle and axial rays as solutions of the paraxial equation. The *dashed lines* marked as object, and image indicates 2D planes. They are conjugate planes according the HL relationship

appropriate length. At the object plane at $z = z_o$, the principle ray is at the distance of $u_\pi(z_o) = 1$ to the axis and parallel to the optical axis with $u'_\pi(z_o) = 0$. The axial ray is defined by $u_\alpha(z_o) = 0$ and $u'_\alpha(z_o) = 1$. An arbitrary ray path can be obtained by linear combination

$$u(z) = u_o u_\pi(z) + \alpha_o u_\alpha(z). \tag{6.18}$$

Here, u_o and α_o are constants. The image plane is defined by the intersection of the axial ray with the optical axis, which occurs at $z = z_i$.

The lens described by the paraxial equation has the imaging property of an ideal lens. This can be shown using the two fundamental rays that we just introduced. Take the distance to the optical axis for an point in the object plane as d_o, all rays passing through this point as defined in Eq. (6.18) will have

$$u(z_o) = u_o u_\pi(z_o) + \alpha_o u_\alpha(z_o) = u_o = d_o.$$

At the image plane,

$$u(z_i) = d_o u_\pi(z_i) + \alpha_o u_\alpha(z_i) = d_o u_\pi(z_i)$$

Thus, all rays passing through the same point on the object plane converge to the same point on the image plane. In 3D, u_o is a complex number with its phase corresponding to the azimuthal angle in the xy plane for the incident beam. A focused beam thus has the shape of a pencil. The constant α_o is also a complex number, corresponding to different rays on the pencil.

The image magnification is defined by the ratio of image distance to the object distance:

$$M = u(z_i)/u(z_o) = u_\pi(z_i). \tag{6.19}$$

It is entirely determined by the intersection of the axial ray with the optical axis (which defines the image plane position).

The two fundamental rays that we have defined are distinct in the way of one representing parallel rays specified entirely by their lateral positions, and one representing pencil rays on the optical axis specified by their slopes at the object plane.

The lateral positions and slopes of any two trajectories are connected with each other in the so-called Helmholtz–Lagrange (HL) relation. To show this, take any two solutions of the paraxial equation, $u_\gamma(z)$ and $u_\chi(z)$, they satisfy following equations:

$$u_\gamma'' + Tu_\gamma = 0$$
$$u_\chi'' + Tu_\chi = 0$$

where $T = eB(z)^2/(8m\Phi)$. We multiply the two equations with u_χ and u_γ, respectively, and subtract the products from each other, which gives

$$u_\chi u_\gamma'' - u_\gamma u_\chi'' = \frac{\mathrm{d}}{\mathrm{d}z}\left(u_\chi u_\gamma' - u_\gamma u_\chi'\right) = 0 \tag{6.20}$$

and

$$u_\chi u_\gamma' - u_\gamma u_\chi' = \text{const.} = 1 \tag{6.21}$$

The Helmholtz–Lagrange relation applies to any two rays as long as they are the solutions of the paraxial equation. Let us take $\chi = \alpha$ for the axial ray and $\gamma = \pi$ for the principle ray; combining Eqs. (6.20) and (6.19) then gives the following relationship:

$$M = u(z_i)/u(z_\mathrm{o}) = u_\pi(z_i) = \frac{1}{u_\alpha'(z_i)} = \frac{u_\alpha'(z_\mathrm{o})}{u_\alpha'(z_i)} \tag{6.22}$$

Thus, the magnification is determined by the slope of the axial ray at the image plane, while intersection of axial ray with the optical axis gives the image plane. The above equation also shows that the angular magnification, $u_\alpha'(z_i)/u_\alpha'(z_\mathrm{o})$, is the inverse of image magnification. The angle of the pencil rays is reduced by $1/M$; this effectively makes the beam angle much smaller inside TEM for all lenses after the objective lens.

6.3 Thin Lens

The magnetic field inside a pole piece for electron focusing extends only a short distance along the optical axis. For an object placed far outside the gap field, the electron beam emanated from the object is close to being parallel at a distance to the optical axis, electron deflection occurs within the short lens field, which results in very little change in the axial distance in case of a weak lens. Deflection thus can be approximated as taking place at a single plane. Under the above conditions, the magnetic lens may be considered as a thin lens as in optics.

Consider the principle ray u_π coming from the left side of the thin lens at $u_\pi(z_o) = 1$ and being parallel to the optical axis $\left(u'_\pi(z_o) = 0\right)$. We perform integration of the paraxial equation starting from the object plane z_o to a plane z. Assuming u_π remains the same under the above thin lens approximation with $u_\pi(z_o) = 1$; integration of u_π then gives

$$u'_\pi(z) - u'_\pi(z_o) = u'_\pi(z) = -\int_{z_o}^{z} T(\zeta)\mathrm{d}\zeta \tag{6.23}$$

Here, $T(z) = |e|B(z)^2/8m\Phi$. The z can be taken to the right side of the thin lens and just outside the lens field. Then for the practical purpose, the integral in Eq. (6.23) can be taken into infinite. The principle ray of u_π emerges from the lens with an asymptotic slope of

$$u'_\pi(\infty) = -\int_{-\infty}^{\infty} T(\zeta)\mathrm{d}\zeta.$$

This asymptotic ray intersects with the optical axis and gives the lens back focal point f. Further to the left at the center of the thin lens, the asymptotic rays intersect with u_π, giving rise to

$$0 = 1 + u'_\pi(\infty)f$$

From this, we thus obtain for a thin lens,

$$f = -1/u'_\pi(\infty) = 1/\int_{-\infty}^{\infty} T(\zeta)\mathrm{d}\zeta \tag{6.24}$$

The above results, and the calculations of magnetic lens properties in general, require the accurate knowledge of the field distribution $B(z)$. Obtaining such knowledge often requires elaborated experimental or theoretical efforts, which are not easy to carry out. Durandeau (1957) suggested that for a magnetic lens with the bore diameter of D and gaps distance S, its axial field can be treated as equivalent to that of a solenoid of diameter of 2D/3 and diagonal length of $L = \sqrt{S^2 + 0.45D^2}$. This ingenious solution works for a remarkable range of pole-piece geometries. Comparison with the calculated field distributions revealed excellent agreements between the solenoid model and the calculated fields for $S/D \sim 1$. Reasonable agreement was also obtained for $S/D = 0.5$. The agreement breaks down seriously when $S/D = 0.2$ (see Mulvey and Wallington 1973). The agreement also falls off at high lens excitation because the strength of magnetic fields in a magnetic lens is ultimately limited by the nonlinear magnetic susceptibility of ferromagnetic

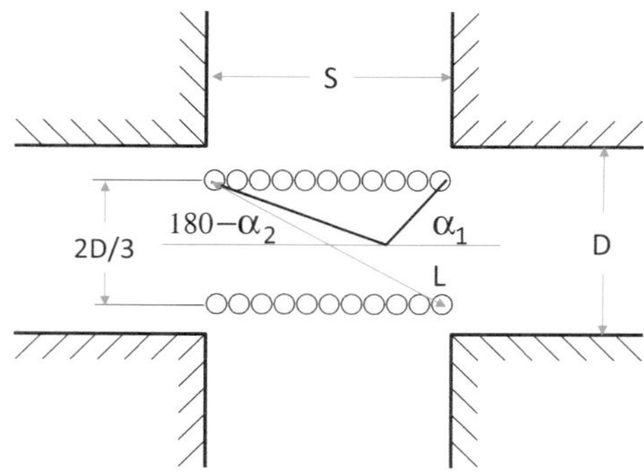

Fig. 6.5 Illustration of the solenoid model for a magnetic lens pole piece of bore D and gap S (after Mulvey and Wallington 1973)

materials, while such limit does not exist for solenoid lenses, except the maximum current that can be carried out in the coil. The magnetic field of a solenoid has the simple solution of

$$B(z) = \frac{\mu_o NI}{2S}(\cos \alpha_1 - \cos \alpha_2) = 2\pi \times 10^{-7}\left(\frac{NI}{S}\right)(\cos \alpha_1 - \cos \alpha_2) \qquad (6.25)$$

where the angles of α_1 and α_2 are defined in Fig. 6.5, N is the number of windings, and I is the current supplied to the lens windings in Amperes. B_p is in Tesla and S is in meters.

Insights into the focal properties of magnetic lenses can also be obtained by using analytical models. A popular model, known as Glaser's bell-shaped model, is that $B(z)$ increases and decreases symmetrically and smoothly across the gap with its maximum value B_{max} at the center of the lens according to the Lorentzian function:

$$B(z) = \frac{B_{\text{max}}}{1 + z^2/a^2} \qquad (6.26)$$

At $z = a$, $B(z)$ decreases by half. Thus, the distribution has a full width at half maximum (FWHM) of $2a$. According to the solenoid model, at the center, we have

$$B_{\text{max}} = \frac{\mu_o NI}{L} = \frac{\mu_o NI}{\sqrt{S^2 + 0.45D^2}}$$

And furthermore

$$2a = 0.96L.$$

Thus, B_{max} is proportional to the lens excitation (NI) for an unsaturated magnetic lens with a fixed geometry.

Integration of Eq. (6.24) using the above field gives the focusing power as

$$1/f = \frac{\pi}{16} \frac{e}{m\Phi} aB_{\mathrm{max}}^2 \tag{6.27}$$

An example given by Egerton (2005, p. 42) for $B_{\mathrm{max}} = 0.3T$ and $a = 3$ mm gives the focusing power $1/f = 93$ m^{-1} and focal length $f = 11$ mm without the relativistic correction for the electron mass. This focal length obtained with the thin lens approximation underestimates the exact solution for the same conditions by 14 %.

From the above analytical approximation, we can draw following conclusions about the focusing power of a thin magnetic lens (also see Sect. 2.4, Egerton 2005):

(1) $1/f$ increases with the strength and spatial extent of the magnetic field;
(2) $1/f$ decreases as the acceleration voltage increases. Thus, faster electrons are deflected less in the same magnetic field;
(3) The focal length is changed by lens excitation. This is simply done by adjusting the lens current. Consequently, the lens current must be highly stabilized to maintain a constant focus. The stability required amounts typically to better than a few parts per million;
(4) In addition to the focus action, electrons rotate azimuthally inside the magnetic field dependent on $B(z)$. Reversing the magnetic field direction results in a change in rotation direction, but not the focusing power;
(5) The focusing power is always positive for a round magnetic lens because it depends only on the power of $B(z)$. Thus, there is no the round electron magnetic lens equivalent of a diverging (concave) optical lens with a negative focal length.

6.4   Thick Lenses

6.4.1   Glaser's Bell-Shaped Model

Electron magnetic lenses capable of providing high focusing power and short focal distances are strong lens with a large peak magnetic field strength and extended field distribution (relative to the object or image position). Deflection of the electron trajectory inside the strong lens occurs over a distance comparable to the focal distance. Thus, the approximation for thin lens no longer applies. They must be treated as thick lenses. Mathematical solutions for the electron trajectory in a thick lens are only available for a few analytical models. Numerical solutions are obtained in practice. In what follows, we will use the solutions obtained for the Glaser's bell-shaped model to illustrate how the electron path can be used to determine the cardinal points of a thick lens, and how these cardinal points provide a description that can be used to determine image formation in an ideal thick lens.

The Glaser's bell-shaped model provides the closed-form analytical solutions to all optical properties, which makes it popular for studying the properties of magnetic lenses. However, the Lorentzian distribution imbedded in this model falls off much too slowly to make it useful for designing real lenses.

Substitution of the bell-shaped magnetic field model of Eq. (6.26) into Eq. (6.17) yields the paraxial equation in the form of

$$\frac{d^2u}{dz^2} + \frac{k^2/a^2}{1+z^2/a^2}u = 0 \tag{6.28}$$

where

$$k^2 = eB_{max}^2 a^2/(8m_e\Phi^*), \tag{6.29}$$

and $\Phi^* = \gamma\Phi$ for the relativistic corrected acceleration voltage. To transform Eq. (6.28) into a simpler form, we introduce two new variables y and φ so that

$$z/a = \cot\varphi$$

and

$$u/a = y(\varphi)/\sin\varphi$$

The variable φ ranges from π to 0 with $z = -\infty$ to $z = \infty$. At $z = 0$, $\varphi = \pi/2$. Substituting y and φ into Eq. (6.28) gives the following transformed equation:

$$\frac{d^2y}{d\varphi^2} + \omega^2 y = 0 \tag{6.30}$$

with $\omega^2 = 1 + k^2$, which is a dimensionless, characteristic, parameter of the lens. Equation (6.30) has simple solutions in the general form of

$$y(\varphi) = u\sin\varphi/a = c_1\cos(\omega\varphi) + c_2\sin(\omega\varphi)$$

The coefficients c_1 and c_2 are determined by the initial conditions for the electron trajectory. For the principle ray u_π coming from the left side of the thin lens at $u_\pi(-\infty) = 1$ and parallel to the optical axis $\left(u'_\pi(-\infty) = 0\right), y'(\pi) = 0$ and $y(\pi) = 1$. This gives $c_1 = 0$ and (Hawkes and Kasper 1996, p. 696):

$$u_\pi = -\frac{\sin\omega(\varphi - \pi)}{\omega\sin\varphi}. \tag{6.31}$$

(Figure 6.6 plots u_π for several ω values.) Similarly, for the principle ray $u_{\bar{\pi}}$ coming from the left side of the lens, parallel to the optical axis, with $u_{\bar{\pi}}(\infty) = 1$, we have:

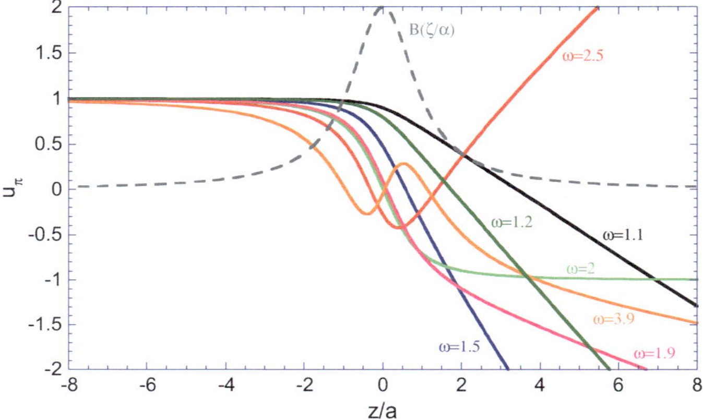

Fig. 6.6 The principle ray of u_π plotted for different ω according to Glaser's bell-shaped lens model. All rays are parallel and at distance of 1 to the optical axis on the object side, and they cross the optical axis at the same point. The shape of $B(z)$ is shown in the *dashed line*

$$u_{\overline{\pi}} = \frac{\sin \omega\varphi}{\omega \sin \varphi} \tag{6.32}$$

The slope of this ray path, at any point, is given by taking the derivative against z. For $u_{\overline{\pi}}$, $u'_{\overline{\pi}}(z) = \mathrm{d}u'_{\overline{\pi}}(\varphi)/\mathrm{d}\varphi \cdot \mathrm{d}\varphi/\mathrm{d}z$ and it can be shown

$$u'_{\overline{\pi}}(-\infty) = -\frac{1}{a\omega}\sin \omega\pi. \tag{6.33}$$

The focus occurs when the path intersects with the optical axis at positions where $\sin \omega\varphi = 0$, that is where

$$\varphi = n\pi/\omega \text{ with } \quad 0 < \varphi < \pi. \tag{6.34}$$

Since the value of ω is determined by the lens excitation k^2. At high lens excitation with $k^2 > 3$ and $\omega > 2$, we can have several real foci. The more common practice is to excite lens such as $\omega \leq 2$, then there is only a single focus.

6.4.2   Cardinal Points and Planes

While the ray paths in a thick magnetic lens can be highly complex as the examples in Fig. 6.6 show, a simpler description of the lens can be provided in analogy to an optical lens by defining its cardinal points on the optical axis. These cardinal points include the following:

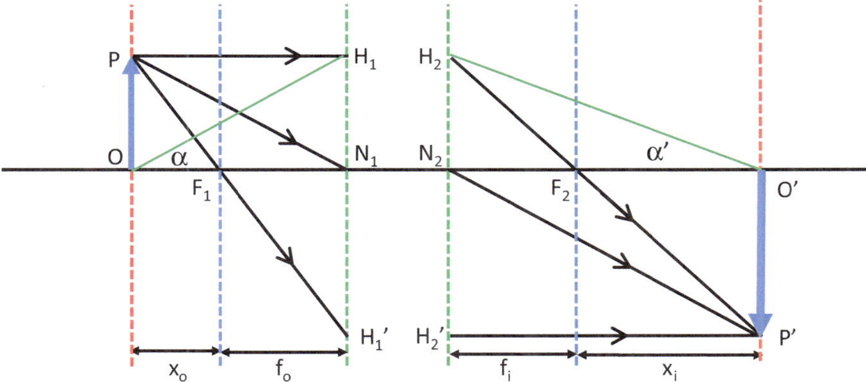

Fig. 6.7 Image formation in a thick lens and its principle points and planes

(1) Front focal point at z_{F_1} and back focal point at z_{F_2};
(2) Front principle point at z_{H_1} and back principle point at z_{H_2};
(3) Front nodal point at z_{N_1} and back principle point at z_{N_2};

The imaging process as illustrated in Fig. 6.7 is helped by further introducing:

(1) Front and back focal planes, which are two planes perpendicular to the optical
     axis that pass through the front and rear focal points;
(2) Front and back principle planes that are defined similarly as focal planes but
     passing through the front and rear principle points.

A ray emanated from an object point O parallel to the optical axis is deflected at
the back principle plane; it intersects with the optical axis at the back focal point
and is imaged at the image point M. Similarly, a ray emanated from O and passing
through the front focal point is deflected at the front principle plane and imaged at
the point M.

The principle planes and focal planes are special planes. According the HL
relationship, the principle planes are two special conjugated planes for which the
magnification is 1 ($M = 1$). The front focal plane is the conjugated plane of $+\infty$ and
the back focal plane is the conjugated plane of $-\infty$. The distance between the
principle plane and focal plane defines the focal distance. There are two focal
distances for the front and back, respectively.

The front and back nodal points are special points having the property that a
pencil ray aimed at one of the two nodal points appears to have come out from the
other nodal point with the same angle. Thus, the nodal points preserve angles in the
way of what the principal planes do for off-axis distance. For an electron magnetic
lens with no electric fields on both sides of the lens, the front and rear nodal points
coincide with the front and rear principal points, respectively.

6.4.3   Lens Equation

The image position, size, and orientation are completely determined by the focal distance as Fig. 6.7 illustrates. The procedures for locating the image of an object of an ideal lens are the following:

(1)   Draw from P to H_1 parallel to the axis; take H_1 to H_2;
(2)   Draw a ray from H_2 to F_2 and extend beyond F_2;
(3)   Draw from P to F_1 and extend to $H_1{}'$; take $H_1{}'$ to $H_2{}'$;
(4)   Draw a parallel ray from $H_2{}'$; the intersection of 2 and 4 gives the image point P', and the plane containing P' is the image plane.

$O'P'$ is an image of OP. The triangle defined by three points $H_2F_2N_2$ is similar to $F_2O'P'$, and $F_1N_1H_1{}'$ is similar to F_1PO, while H_2N_2 equals to H_1N_1 and $H_2{}'N_2$ equals to $H_1{}'N_1$. From these relationships, we have

$$\frac{O'P'}{f_{\mathrm o}} = \frac{OP}{x_{\mathrm o}} \quad \text{or} \quad \frac{O'P'}{OP} = \frac{f_{\mathrm o}}{x_{\mathrm o}}$$

Similarly

$$\frac{OP}{f_i} = \frac{O'P'}{x_i} \quad \text{or} \quad \frac{OP}{O'P'} = \frac{f_i}{x_i}$$

Combining these two, we have the well-known Newton's equation:

$$\frac{O'P'}{OP} = \frac{f_{\mathrm o}}{x_{\mathrm o}} = \frac{x_i}{f_i} \quad \text{or } x_i x_{\mathrm o} = f_{\mathrm o} f_i. \tag{6.35}$$

This equation allows a determination of image position from the focal distance. For the thin lens described in Sect. 6.3, $f_i = f_{\mathrm o} = f$. Then, if we take the distance from the object plane to the principle plane of H_1 as U and the distance from the image plane to the principle plane of H_2 as V, we have

$$U = x_{\mathrm o} + f \text{ and } V = x_i + f.$$

It can be easily shown that

$$UV - (U + V)f = 0$$

or in the familiar form of Gaussian form of lens equation:

$$\frac{1}{U} + \frac{1}{V} = \frac{1}{f} \tag{6.36}$$

From this equation, three possible cases emerge: (1) $U < f$; image is virtual, erect, and magnified, (2) $f < U < 2f$; image is real, inverted, and magnified, and (3) $U > 2f$; image is real, inverted, and reduced.

The image magnification can be obtained directly without involving the lens equation. We note that the triangles defined by three points of OPN_1 and $N_2O'P'$ are similar; from this we have the following equation:

$$M = \frac{O'P'}{OP} = \frac{V}{U}.$$

(6.37)

For the lens acceptance angle, under the small-angle approximation,

$$\tan \alpha \approx \alpha = \frac{N_1 H_1}{U} \, , \, \tan \alpha' \approx \alpha' = \frac{N_2 H_2}{V}$$

And thus

$$\frac{\alpha'}{\alpha} = \frac{U}{V} = 1/M$$

(6.38)

This is same as Eq. (6.22) obtained from the HL relationship for the electron path. Together, the above discussions demonstrate a three-step process in electron optics:

(1)   Determination of axial magnetic field of the lens,
(2)   Solving the paraxial equation for electron paths in the form of the principle and axial rays, and
(3)   Determination of asymptotic cardinal points from the electron path.

The parameters obtained thus allow a complete description of image formation in an ideal lens.

6.4.4   Determination of Cardinal Points from the Electron Path

To define the position of cardinal points of a magnetic lens, it is helpful to examine the asymptotic rays of the paraxial ray equation. Figure 6.8 illustrates the method for doing this for a solution to the paraxial equation: $u_\pi(z)$. It describes a ray, parallel to the optical axis, coming from the right hand side of the lens and emerging from the lens on the left hand side with the following asymptotic behaviors

$$u_\pi(z) \xrightarrow{z \to \infty} 1 \text{ and } u'_\pi(z) \xrightarrow{z \to \infty} 0$$

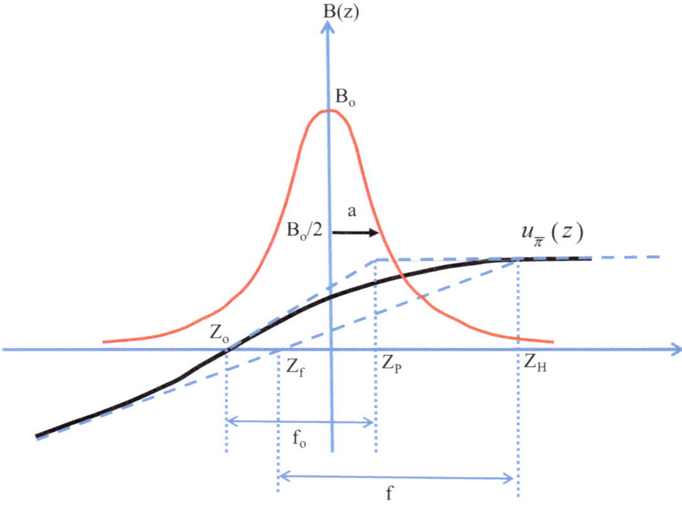

Fig. 6.8 Definition of cardinal points and planes from the electron path of $u_{\bar{\pi}}(z)$ (after Grivet 1972, Fig. 93a)

For the principle ray $u_{\bar{\pi}}(z)$ obtained from the Glazer's bell model, we have

$$\lim_{z \to -\infty} u_{\bar{\pi}}(z) = z u_{\bar{\pi}}'(-\infty) + \lim_{z \to -\infty} \left[u_{\bar{\pi}}(z) - z u_{\bar{\pi}}'(z) \right]$$

Using Eq. (6.33) for the slope, we have

$$\lim_{z \to -\infty} \left[u_{\bar{\pi}}(z) - z u_{\bar{\pi}}'(z) \right] = \frac{1}{\omega} \lim_{\varphi \to \pi} \left[\frac{\sin \omega \varphi}{\sin \varphi} + \omega \cos \varphi \cos \omega \varphi - \frac{\sin \omega \varphi}{\sin \varphi} \cos^2 \varphi \right]$$

$$= \frac{1}{\omega} \lim_{\varphi \to \pi} \left[\sin \varphi \sin \omega \varphi + \omega \cos \varphi \cos \omega \varphi \right]$$

$$= -\cos \omega \pi$$

Combining these results, we have the asymptotic ray in the following form

$$\lim_{z \to -\infty} u_{\bar{\pi}}(z) = z \left(\frac{1}{a\omega} \sin \omega \pi \right) - \cos \omega \pi. \tag{6.39}$$

This ray intersects the optical axis at

$$z_F = a\omega \cot \omega \pi \tag{6.40}$$

And it intersects the incident asymptotic ray on the right at

$$z_H = a\omega(1 + \cos \omega\pi)/\sin \omega\pi. \tag{6.41}$$

The distance between the focal and principle points gives the focal distance in the form of

$$f = z_H - z_F = a\omega/\sin \omega\pi. \tag{6.42}$$

The focus f may fall inside or outside the lens field. When f falls within the lens field, the image formed can nevertheless be found by the constructions of geometric optics if a virtual object is used. The use of asymptotic elements arises when a projector lens is used to form a real image at high magnification using the image formed by a preceding lens as object. The projector lenses in the TEM include the condenser, intermediate, and project lenses. They are distinguished from the object or gun lens, which use a physical specimen as object.

Liebmann and Grad (1951) reported a detailed study of the variation in f with lens excitations, which is measured by the quantity of

$$\kappa = \frac{NI}{\sqrt{\Phi^*}}.$$

The coefficient $k^2 = eB_{\max}^2 a^2/(8m_e\Phi^*)$ in the paraxial equation is proportional to κ^2 (see Eq. 6.26). The focal length f first decreases with κ, then it reaches a minimum f_m. Further increase leads to an increase in f. Liebmann and Grad (1951) and Durandeau (1956, 1957) found that for all practical purposes the variations in f with κ for different lens gap geometries scale to a single, universal curve when they are plotted as f/f_m versus $NI/(NI)_o$, where $(NI)_o$ is the current-turns at the minimum focal distance. Further, in the useful range of the gap and bore ratio of $0.5 < D/S < 5$, we have

$$(NI)_o \approx 13.5\sqrt{\Phi^*}$$

and

$$f_m \approx 0.5L = 0.5\sqrt{S^2 + 0.45D^2}.$$

6.5   The Objective Lens

The objective lens, which immediately follows the specimen, is the most important lens of the microscope. It is designed with a short focal distance and a large magnification. Since angular magnification is inversely proportional to lateral magnification (Eq. 6.22), the magnification provided by this lens ensures that rays travel at very small angles to the optical axis in all subsequent lenses. We shall see that lens aberrations increase sharply with angle, so that it is the objective lens, in which rays make the largest angle with the optical axis and determines the final quality of the image or the smallest probe in STEM.

Glaser was the first to show, using his bell-shaped magnetic field model, the condition for obtaining the minimum focal length, and the highest magnification of a magnetic lens is to put the specimen at the middle of the gap of a pair of symmetrical pole pieces (Cosslett 1991). This is known as an immersion lens. The field maintained on the illuminating side of the object is called as the prefield, which acts as another condenser lens and plays no part in the image formation. The extent of the remaining field available for image formation depends on the object position. Thus, the position of lens cardinal points, as well as aberration coefficients, to be discussed later, depends on the object position Z_o. This position becomes an important electron optical parameter for the objective lens. By convention for an immersion lens, Z_o is specified to produce an image at infinity (infinite magnification), and this corresponds to have the specimen at the exact focus.

According to (6.34), in the Glaser's bell-shaped lens model, $Z_o = \pi/\omega$ for $n = 1$. The slope of $u_{\bar{\pi}}(z)$ at this point is $\sin(\pi/\omega)/a$. Following the same procedures in Sect. 6.4, we find the cardinal points for the immersion lens (as defined in Fig. 6.8):

$$f_o = a/\sin(\pi/\omega),\, Z_o = a\cot(\pi/\omega),\, Z_P = -a\cot(\pi/2\omega) \qquad (6.43)$$

The minimum is obtained when $\omega = 2$ with $f_{o\,\mathrm{min}} = a$, and the immersion foci coincide at the center of the lens. From $\omega^2 = 1 + k^2$ and Eq. (6.29), we thus obtain

$$f_{o\,\mathrm{min}} = a = \sqrt{24m_e/e}\sqrt{\Phi^*}/B_{\mathrm{max}} = 1.17 \times 10^{-5}\sqrt{\Phi^*}/B_{\mathrm{max}}\,(m)$$

With Φ^* in volts and B in Tesla.

6.6 The Objective Prefield

For an objective lens of short focal distance, the specimen will be placed well within its lens field. The field on the illuminating side of the specimen is known as the prefield and has the effect of a strong condenser lens placed before the specimen. The modern trend has been toward the use of increasingly symmetrical "condenser-objective" lenses. This configuration allows the convenient switching from the HREM to the probe-forming mode for CBED or STEM on the same specimen region. Figure 6.9 shows a ray entering a "bell-shaped-field" objective parallel to the axis with the lens excitation at $k^2 = 3$, and the object focus is right at the center of the lens. According to Grivet (1972), for high-energy electrons with $\Phi > 100\,\mathrm{kV}$, the lens is highly saturated but the field distribution, $B(z)$, only broadens slowly so that the focal distance and the related spherical and chromatic aberration coefficients remain relatively constant with

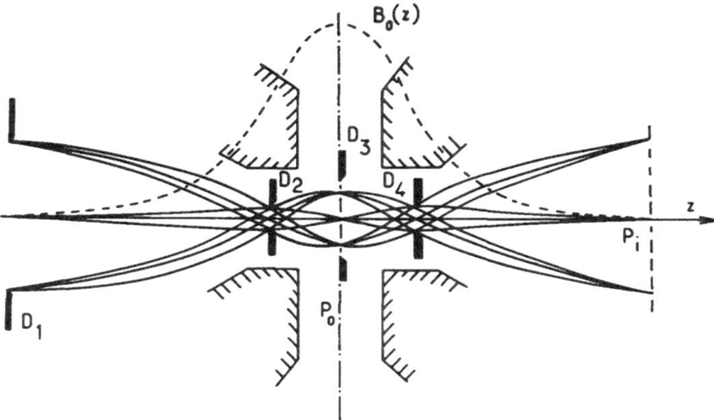

Fig. 6.9 Electrons trajectories in a symmetrical condenser-objective lens. Here, D_1 and D_4 are the condenser and the objective aperture, respectively. D_3 marks the specimen position, D_2 marks the condenser-objective aperture (from Grivet, Electron optics, 1972)

$$a(\text{mm}) = 1.27 + 0.057 \times I(\text{kA}) \approx f(\text{mm}). \tag{6.44}$$

Further discussion on the prefield focal length, focal distance, and the demagnification of the incident beam by the prefield can be found in paper by Mulvey and Wallington (1973). The significance of this for electron diffraction is such that

(1)   The specimen is located at the position where the axial field is strongest;
(2)   A broad beam is formed on the specimen using a focused beam at the front focal plane, and a focused beam is obtained using a broad beam on the object side;
(3)   The incident-beam convergence, and hence the resolution in reciprocal space, is strongly affected by the strength and aberrations of the prefield lens;
(4)   Aberrations of the prefield lens change the focus of off-axis beams, including tilted beams as in precession electron diffraction;
(5)   High-quality condenser lenses are thus required to work with the strong objective prefield lens.

References

Carey DC (1987) The optics of charged particle beams. Harwood Academic Publishers, London
Cosslett VE (1991) Fifty years of instrumental development of the electron microscope. In: Barer R, Cosslett VE (eds) Advances in optical and electron microscopy. Academic Press, London, pp 215–267
Durandeau P (1956) Construction des lentilles electroniques magnetiques. J De Physique Et Le Radium 17:A18–A25

Durandeau P (1957) A study of magnetic electron lenses, abstract number: A1960-10956. Annales de la Faculte des Sciences de l'Universite de Toulouse 21:1–88

Egerton RF (2005) Physical principles of electron microscopy: an introduction to TEM, SEM, and AEM. Springer, New York

Erni R (2010) Aberration-corrected imaging in transmission electron microscopy: an introduction. Imperial College Press, London

Grivet P (1972) Electron optics: translated by PW Hawkes, revised by A Septier (2nd English edn). Oxford Pergamon Press, Oxford

Hall CE (1966) Introduction to electron microscopy, 2nd edn. Mcgraw Hill Publishing Company, New York

Hawkes PW (1972) Electron optics and electron microscopy. Taylor & Francis Ltd., London

Hawkes PW (2008) Aberrations. In: Orloff J (ed) Handbook of charged particle optics, 2nd edn. CRC Press, Boca Raton

Hawkes PW, Kasper E (1996) Principles of electron optics: applied geometrical optics. Academic Press, San Diego

Inada H, Kakibayashi H, Isakozawa S, Hashimoto T, Yaguchi T Nakamura K (2009) Hitachi's development of cold-field emission scanning transmission electron microscopes. In: Advances in imaging and electron physics. Elsevier, Amsterdam

Krivanek OL, Dellby N, Murfitt MF (2008) Aberration correction in electron microscopy. In: Orloff J (ed) Handbook of charged particle optics, 2nd edn. CRC Press, Boca Raton

Liebmann G, Grad EM (1951) Imaging properties of a series of magnetic electron lenses. Proc Phys Soc B 64:956

Mulvey T, Wallington MJ (1973) Electron lenses. Rep Prog Phys 36:347

Orloff J (ed) (2008) Handbook of charged particle optics, 2nd edn. CRC Press, Boca Raton

Rose HH (2013) Geometrical charged-particle optics, 2nd edn. Springer, Berlin

Swanson LW, Schwind GA (2008) Review of ZrO/W Schottky cathode. In: Orloff J (ed) Handbook of charged particle optics, 2nd edn. CRC Press, Boca Raton

Tsuno K (2008) Magnetic lenses for electron microscopy. In: Orloff J (ed) Handbook of charged particle optics, 2nd edn. CRC Press, Boca Raton

Chapter 7
Lens Aberrations and Aberration Correction

Aberrations in a lens refer to departure from the point-to-point imaging prescribed by the paraxial equation introduced in Chap. 6. This equation does not include diffraction effects, whereby a point-like object forms a blurred disk, because of the limited acceptance angle of the lens or lens aperture. Aberrations are categorized into two types: chromatic and geometric aberrations. Chromatic aberration is caused by wavelength dispersion, due to changes in the electron velocity (and electron wavelength), and resulting change in the electron path in a magnetic lens. This type of aberration disappears if all electrons have the same wavelength or energy (monochromatic). Geometric aberrations, on the other hand, are present even for electrons of the same energy. The largest geometric aberration arises from defects in the magnetic lens or from the use of magnetic fields with low rotational symmetry, such as a multipole. In an ideal round magnetic lens with continuous rotational symmetry, geometric aberrations are caused by the higher order terms, beyond the first-order approximation included in the paraxial equation (6.17). These terms become significant when the electron travels at significant distances and/or angles from the optical axis.

This chapter treats the complex subject of aberrations in sufficient details to provide an introduction to the basic concepts, as well as the means for their calculation, using the trajectory method. This is followed by a description of the effects of aberrations of different order. The last part of this chapter introduces the concepts of aberration correction.

7.1 Lens Aberrations

The starting point for examining the aberrations of magnetic lenses arises when we include higher order terms in the paraxial lens equation. This leads to the following nonlinear equation (Rose 2011; also see Hawkes 2008 for a general treatment of aberrations):

© Springer Science+Business Media New York 2017
J.M. Zuo and J.C.H. Spence, *Advanced Transmission Electron Microscopy*,
DOI 10.1007/978-1-4939-6607-3_7

$$u'' + \frac{eB^2}{8m\Phi} u = P(u, \bar{u}, u', \bar{u}', z) \tag{7.1}$$

where P is a perturbation function comprising all nonlinear terms involving the distance and slope of the electron path described by u and u' and their complex conjugates: $\bar{u}$ and $\bar{u}'$. For the subsequent discussion, u is taken as a complex number with its amplitude representing distance and its phase for angle in the xy plane normal to the optical axis. In the absence of higher order terms and with $P = 0$, the electron rotation is taken out from u according to Eq. (6.16). Thus, the phase of u is constant for paraxial solutions. A solution is then sought for Eq. (7.1) with a small departure from the paraxial equation, with small P, and with $P = 0$ to a first-order approximation. Because P itself is dependent on the solution, Eq. (7.1) is then solved iteratively starting from the first order. This approach is based on the mathematical method of variation of parameters, also known as variation of constants, which is a general method for solving inhomogeneous linear ordinary or partial differential equations.

A solution to the inhomogeneous Eq. (7.1) is proposed based on a combination of the solutions of the homogeneous paraxial equation (the principle and axial rays of $u_\pi(z)$ and $u_\alpha(z)$, Sect. 6.2):

$$u(z) = C_\pi(z)u_\pi(z) + C_\alpha(z)u_\alpha(z) \tag{7.2}$$

where $C_\pi(z)$ and $C_\alpha(z)$ are complex, differentiable functions. They are complex because P is a complex function. Consequently, $C_\pi(z)$ and $C_\alpha(z)$ represent four independent variables. Since there are only two equations, they must be constrained to reduce the number of variables to two. The condition we will impose is that:

$$C_\pi'(z)u_\pi(z) + C_\alpha'(z)u_\alpha(z) = 0. \tag{7.3}$$

Using this, we obtain the second-order derivative of u in the form:

$$u''(z) = C_\pi'(z)u_\pi'(z) + C_\alpha'(z)u_\alpha'(z) + C_\pi(z)u_\pi''(z) + C_\alpha(z)u_\alpha''(z) \tag{7.4}$$

By combining Eqs. (7.2) and (7.4) and taking into account that $u_\pi(z)$ and $u_\alpha(z)$ are the solutions of the homogeneous paraxial ray equation, we obtain the following equation:

$$C_\pi'(z)u_\pi'(z) + C_\alpha'(z)u_\alpha'(z) = P \tag{7.5}$$

Next, we multiply both sides of Eq. (7.5) by $u_\pi(z)$ and $u_\alpha(z)$ and combine the results with the help of the Helmholtz–Lagrange relation. The results are the following solutions:

$$C_\alpha'(z) = Pu_\pi(z), \ C_\pi'(z) = -Pu_\alpha(z) \tag{7.6}$$

Thus, the coefficients of $C_\pi(z)$ and $C_\alpha(z)$ can be obtained simply by integrating Eq. (7.6) starting from a position z_o, which gives the following expression for the electron ray path:

$$u(z) = u^{(1)}(z) + u_\alpha(z) \int_{z_o}^{z} P u_\pi(z)\mathrm{d}z - u_\pi(z) \int_{z_o}^{z} P u_\alpha(z)\mathrm{d}z \qquad (7.7)$$

Selection for the starting position (z_o) for the integration is made at the point where the two solutions of the inhomogeneous and homogenous equations meet. For the homogenous, paraxial equation, $u(z_o) = u_o u_\pi(z_o) + u'_o u_\alpha(z_o)$ at z_o with u_o and u'_o taken as constants. This path is taken as the first-order solution in Eq. (7.7):

$$u^{(1)}(z) = u_o u_\pi(z) + u'_o u_\alpha(z). \qquad (7.8)$$

Overall, Eq. (7.7) gives a solution that can be expressed as an expansion of polynomials involving successively higher order terms:

$$u(z) = \sum_{n=1}^{\infty} u^{(n)}(z) \qquad (7.9)$$

To obtain the higher order terms, the integration in Eq. (7.7) is carried out in steps or iterations; in each iteration the previous solution for u and u' is used to calculate P and the coefficients of $C_\pi(z)$ and $C_\alpha(z)$. Thus, each step yields successively higher order terms in Eq. (7.9). Since the constants u_o and u'_o are small, since electrons travel close to the optical axis in a magnetic lens, a few iterations often suffice to obtain a sufficiently accurate approximation for the actual electron path. The number of required steps in the calculation is lower than the order of the term to be calculated.

The primary aberrations of a cylindrical magnetic lens consist of the third-order spherical aberration and the first-order chromatic aberration. These aberrations can be obtained by substituting $u(z) = u^{(1)}(z)$ in Eq. (7.7) and by examining the axial ray at the image plane at z_i as defined by $u_o = 0$, $u'_o = \omega$, $u_\alpha(z_i) = 0$ (Fig. 7.1).

Fig. 7.1 Illustration of ideal and aberrated electron paths for the axial ray

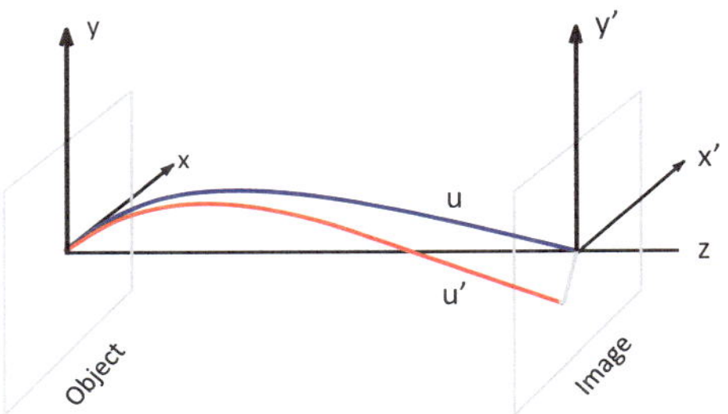

To calculate the chromatic aberration for the axial ray using Eq. (7.10), we take $u^{(1)}(z) = \omega u_\alpha(z)$. Substituting this into Eq. (7.7) gives

$$\Delta u(z_i) = u^{(1)}(z_i) - u(z_i) = -u_\pi(z_i) \int_{z_o}^{z_i} P u_\alpha(z)\mathrm{d}z = -M \int_{z_o}^{z_i} P u_\alpha(z)\mathrm{d}z. \qquad (7.10)$$

The change in electron beam energy gives a first-order perturbation term of the form

$$P = -\frac{\Delta\Phi}{\Phi}\frac{eB^2}{8m\Phi}u \qquad (7.11)$$

The combination of Eqs. (7.10) and (7.11) then gives the axial chromatic aberration coefficient C_c of a round magnetic lens:

$$\Delta u(z_i) = -M\frac{\Delta\Phi}{\Phi}C_c\omega, \; C_c = \int_{z_o}^{z_i} \frac{eB^2}{8m\Phi}u_\alpha^2\mathrm{d}z \qquad (7.12)$$

Since all terms in the above integral are positive, the C_c value of a round magnetic lens is also positive.

The third-order spherical aberration (C_s) is a geometrical aberration that arises from the third-order terms (u^3, u^2u', and uu'^2) in the perturbation function. Expressions for C_s for a round magnetic lens were first given by Scherzer, Glaser, and others. According to Scherzer (1936),

$$C_s = \frac{1}{16}\int_{z_o}^{z_i} \left\{ \frac{\eta^4 B^4}{\Phi^{*2}}u_\alpha^4 + 2\left(u_\alpha B' + u_\alpha' B\right)^2 \frac{\eta^2 u_\alpha^2}{\Phi^*} + 2\frac{\eta^2 B^2}{\Phi^*}u_\alpha^2 u_\alpha'^2 \right\}\mathrm{d}z \qquad (7.13)$$

where $\eta = \sqrt{e/2m_e}$ and u_α is the *axial* ray as in (7.12). Another expression for C_s was found by Glaser (1956) which does not involve derivatives of the electron path and has the form

$$C_s = \frac{1}{48}\int_{z_o}^{z_i} \left\{ 4\frac{\eta^4 B^4}{\Phi^{*2}} - \frac{\eta^2 BB''}{\Phi^*} + 5\frac{\eta^2 B'^2}{\Phi^*} \right\}u_\alpha^4\mathrm{d}z \qquad (7.14)$$

Other expressions for C_s are listed in the book of Hawkes and Kasper (1996a, b), Eqs. (24.59)–(24.70). It is thus obvious from the Scherzer's formulation that for an object forming real images (u_α is real), *the spherical aberration C_s of a round magnetic lens is always positive and unavoidable, as is C_c.*

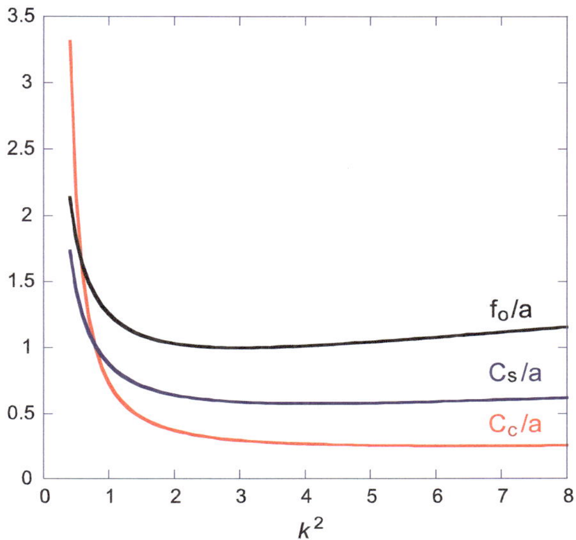

Fig. 7.2 Dependence of the objective focal distance, spherical aberration, and chromatic aberration coefficients as function of lens excitation k^2 for Glaser's bell-shaped model

The spherical and chromatic aberration of Glaser's bell-shaped field model was obtained analytically by Glaser and Lammel (1943). For an immersion lens with the object at the focal position and $M = \infty$,

$$\frac{C_s}{a} = \left(\frac{\pi k^2}{4\omega^3} - \frac{1}{8}\frac{4k^2 - 3}{4k^2 + 3}\sin\frac{2\pi}{\omega} \right) \frac{1}{\sin^4(\pi/\omega)} \tag{7.15}$$

and

$$\frac{C_c}{a} = \frac{\pi k^2}{2\omega^3}\frac{1}{\sin^2(\pi/\omega)} \tag{7.16}$$

where we recall from Sect. 6.5.1, $\omega^2 = 1 + k^2 = 1 + eB_{\max}^2 a^2/(8m_e\Phi^*)$. Figure 7.2 plots C_s and C_c, together with the objective focal distance, as a function of the lens excitation k^2. Both decrease initially as k^2 increases, before reaching a minimum. The minimum of C_c occurs near $k^2 = 4$ with a value of

$$\left(\frac{C_c}{a}\right)_{\min} \approx 0.58.$$

The minimum of C_s occurs at very high excitation beyond the practical range of excitations used for the condenser-objective lens. At $k^2 = 3$, $C_s/a = 0.3$.

In the design of magnetic lenses, the ratio of C_s to the focal length f (C_s/f) is used as a figure of merit in defining the quality of a magnetic lens. Spherical aberration data for various types of lenses can be found in Septier (1967), El-Kareh and El-Kareh (1970), Riecke (1982), Szilagyi (1988), and Hawkes and Kaspar (1996a, b).

7.2   Aberration Coefficients

It is helpful to describe aberrations in terms of their effect on image formation. In order to do this, we define first the aberration-free image formed by an ideal lens as described by the paraxial equation. The aberration-free image will then be used as a reference for the description of aberrations. For an ideal image, summarizing the results given in Sect. 6.3, we have the following rules:

(1)   All the rays emanated from a point A in the object plane, after traveling through the electron magnetic lens, converge toward the same conjugate point A' in the image plane (Fig. 7.3);
(2)   All object points on a plane normal to the optical axis (P) have their image in another plane (P'), which is also normal to the optical axis. The plane P' is said to be conjugate to P; e.g., for each point in P, there is an equivalent point in P';
(3)   The magnification of the lens (M) is defined by the ratio of image and object distances. It is constant and does not depend on the object position in P, nor on the image position in P';

In rule 1, the points A and A' are conjugate across the lens system. Thus, an electron traveling along all the paths from starting from A to A' takes exactly the same time, and every path is a possible trajectory for the electron. The lens system is then said to be stigmatic for points A and A'. This rule is only approximately obeyed in a real lens, however, for several reasons. First, electrons emanating from a single object point may arrive anywhere in the vicinity of the ideal image point; the electron intensity distribution gives rise to a disk-like image, which is called the disk of confusion. The disk of confusion can be caused either by lens aberrations, or by a focussing error (defocus effect). In addition to aberrations or defocus, diffraction caused by the limited acceptance angle of the lens, or any aperture

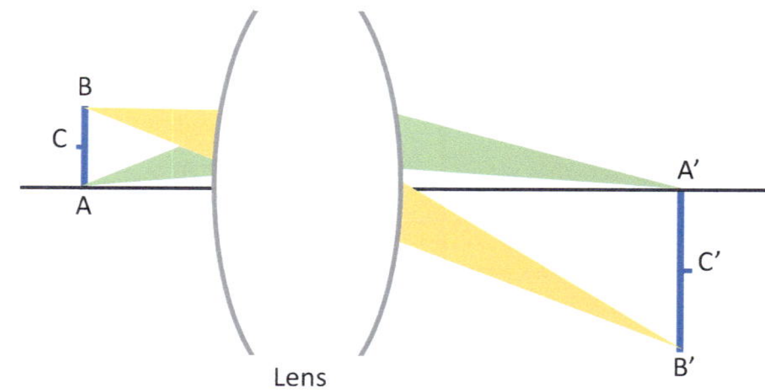

Fig. 7.3 Aberration-free image formation by an ideal lens, where $A'C'B'$ is the planar image of the planar object ACB and $AC/AB = A'C'/A'B'$

placed after the lens, also contributes to the disk of confusion. The image disk at A' due to diffraction is known as an Airy pattern. The intensity of the Airy pattern is peaked at the ideal image point.

The property described by rule 2 is known as "isoplanatic" and describes a large field of view where all image points are at the same focus. Under this condition, aberrations in a real lens depend only on the scattering angles, not on the object coordinates. Aberrations having this property are known as aperture or axial aberrations. Aberrations without this property are known as off-axis aberrations.

Rule 3 defines the geometrical transformation between object and image points. Consider a group of object points that form a pattern with a set of distances between neighboring points, such as ACB in Fig. 7.3. These distances are magnified or demagnified by the lens depending on the object position. The ratios among image distances, however, stay the same, as long as the magnification for different points is the same. Thus, while the image itself may be rotated or inverted after the lens, the pattern formed by conjugate points in the image should be similar to the object points. The transformation from object points to image points given by an ideal lens is thus linear with constant magnification across the entire image. In a real lens, while the image may appear similar to the object, close inspection, however, may often reveal some form of distortion. Distortion can be directly observed in a diffraction pattern when the diffraction spots of a regular pattern appear deformed in a distorted pattern. The presence of image distortion is equivalent to a variation of magnification factor with lateral object position.

The aberration clearly depends on the electron path. When all rays emanating from an object point and passing through the lens aperture plane (pupil or back focal plane), the rays are identified by their field position (x, y) on the object plane and their pupil position $(\theta_{px}, \theta_{py})$. The pupil position is related to the ray angle (θ_x, θ_y) to the optical axis at the object plane. In general, the lens aberrations are described by the four variables $(x, y, \theta_x, \theta_y)$. Among different types of aberrations, the aberrations can be further separated for these that are intrinsic to the symmetry of the lens and other aberrations coming from a breakdown of the lens symmetry due to lens defects or misalignment. These extrinsic aberrations are called parasitic aberrations. For a round lens, the third-order intrinsic aberrations are known in optics as five Seidel aberrations, which are (1) spherical aberration (proportional to θ^3), (2) off-axial coma aberration (proportional to $r\theta^2$, $r = \sqrt{x^2 + y^2}$ the distance to the optical axis), (3) off-axial astigmatism (proportional to $r^2\theta$), (4) curvature of image field (proportional to $r^2\theta$), and (5) distortion (proportional to r^3). Among these five aberrations, the spherical aberration is most important for the objective lens, which is followed by the off-axial coma aberration in terms of theoretical importance. The other three, namely off-axial astigmatism, curvature of image field, and distortion, are more important for the lenses after the objective lens, since the object (or the magnified object image) for the intermediate and projector lenses is no longer small as for the objective lens. The aberrations produced by the beams passing through the lens at distances away from the optical axis become significant.

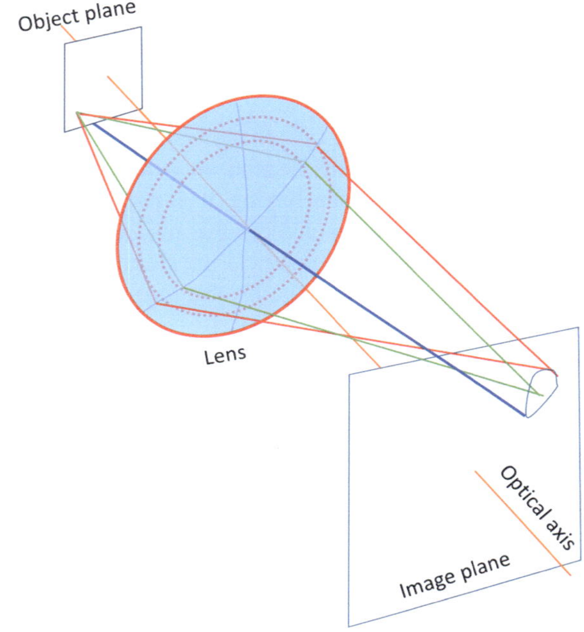

Fig. 7.4 Illustration of the off-axial coma aberration

In what follows, we will be mainly concerned with axial aberrations, e.g., the aberrations that are independent of the field position. This is justified for the objective lens where the object field is small. For the off-axial coma aberration as illustrated in Fig. 7.4, the effect of coma is that rays emanating from an off-axis point in the object plane are focused further away from the optical axis. This creates a trailing, comet-like, blur directed away from the optical axis. Inside a conventional TEM, a magnetic lens with considerable coma is capable of forming sharp images near the optical axis, if the lens is aligned with the electron beam along the optical axis (the coma-free axis), but the image becomes increasingly blurred away from center, thus limiting the field of view.

So far, we have described imaging in terms of geometrical optics. Since a change in the electron path leads to a phase shift in the electron wave, a complete description of image formation in a real lens can be obtained by considering waves and how these waves are transformed by a lens. In this way, the amount of aberration can be simply measured based on the size of the phase shifts. For example, in imaging, a useful rule of thumb is that the system can be regarded as diffraction limited if any phase shift from other sources corresponds to a path difference of less than one-quarter of the wave length. This is the so-called Rayleigh limit. Similarly, an aberration is considered as significant if it causes a shift in phase of more than $\pi/2$.

Figure 7.5 illustrates wave propagation before and after a lens for a point object placed on the optical axis. Following rule 1 for imaging by an ideal lens, the action of an ideal lens produces a spherical wave on the right that converges to the image point at O' for a point object at O. The lens converts a diverging spherical wave from a

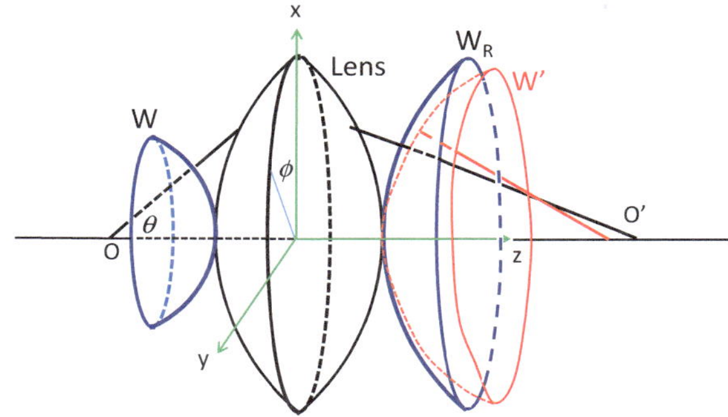

Fig. 7.5 Transformation of object wave front W emitted by the point O in a lens. W_R is from the ideal lens, while W' represents an aberrated wavefront

point object to a spherical wave converging to a point in the image. Note that a lens of infinite width would be needed to form an ideal converging spherical wave. The actual electron wave front inside the lens can be complex. However, the above statement holds for the electron wave seen at a distance away from the lens, which we will call the asymptotic wave with its surface normal corresponding to the asymptotic ray prescribed by the electron path. For a point object on the optical axis, its asymptotic exit wave after the lens is centered at the imaging point with radius of curvature as R; then, the spherical wave front is given in the xyz coordinate by

$$x^2 + y^2 + (z - R)^2 = R^2 \qquad (7.17)$$

For electrons traveling at a small angle to the optical axis, z is small compared to the image distance R. Equation (7.17) can then be approximated by a parabolic surface

$$z_R(x, y) \approx \frac{x^2 + y^2}{2R}.$$

Lens aberrations produce a distorted wave front $z_A(x, y)$, which we will call as wave aberrations. The pattern of wave aberrations, as well as the magnitude of the aberrations, can be related to the ray aberration(s) present in the lens that we have discussed. Using the wave front from the ideal lens as described in Eq. (7.17) as a reference, we describe the position-dependent wave-front aberrations as:

$$\Delta z(x, y) \approx z_A(x, y) - \frac{x^2 + y^2}{2R} \qquad (7.18)$$

The wave-front aberration function of Eq. (7.18) is expected to depend also on the object position. By neglecting the object position in Eq. (7.18), we have assumed the object is very small and lies very close to the optical axis. This approximation is justified for the objective lens since it operates with the object placed close to the focal distance $z_o \approx f$ for high magnification with a field of view limited to tens or hundreds of nanometers. The field of view in STEM is even smaller, as probe formation in STEM can be considered as the reciprocal case of high-resolution imaging. In both cases, we can limit ourselves to the consideration of so-called axial aberrations. For this purpose, it is more convenient to express the wave-front aberration function in terms of the angles:

$$\theta_x = x/f = \theta \cos\phi \text{ and } \theta_y = y/f = \theta \sin\phi \qquad (7.19)$$

where θ and ϕ are defined in Fig. 7.5. Using this, we rewrite Eq. (7.18) as

$$\Delta z\left(\theta_x, \theta_y\right) \approx z_A\left(\theta_x, \theta_y\right) - \frac{f^2\theta^2}{2R} = z_A\left(\theta_x, \theta_y\right) - \frac{f\theta^2}{2M}. \qquad (7.20)$$

Here we have to use the relation $R = Mf$ with M for lens magnification.

The wave-front aberration function can be expanded in a power series involving θ and ϕ of different orders (Uhlemann and Haider 1998):

$$\begin{aligned}
\chi(\omega, E) &= \frac{2\pi}{\lambda}\Delta z(\omega, E) \\
&= \mathrm{Re}\left\{\frac{1}{2}\bar{\omega}^2 A_1 + \frac{1}{2}\omega\bar{\omega}C_1 + \frac{1}{3}\bar{\omega}^3 A_2 + \omega^2\bar{\omega}B_2 \right. \\
&\quad + \frac{1}{4}\bar{\omega}^4 A_3 + \frac{1}{4}(\omega\bar{\omega})^2 C_3 + \omega^3\bar{\omega}S_3 \\
&\quad + \frac{1}{5}\bar{\omega}^5 A_4 + \omega^3\bar{\omega}^2 B_4 + \omega^4\bar{\omega}D_4 \\
&\quad \left. + \frac{1}{6}\bar{\omega}^6 A_5 + \frac{1}{6}(\omega\bar{\omega})^3 C_5 + \cdots \right\}
\end{aligned} \qquad (7.21)$$

Here $\omega = \bar{\omega}* = \theta_x + i\theta_y$. The symbols A, B, C, D, and S mark different types of aberrations, while the subscript denotes the order of aberration. These coefficients

Table 7.1 A list of axial wave aberrations and their description. Their contributions to wave-front aberration are illustrated in Fig. 7.6

Symbol	Name	Complex or real
A_1	Twofold astigmatism	Complex
C_1 or Δf	Defocus	Real
B_2	Axial coma	Complex
A_2	Threefold astigmatism	Complex
A_3	Fourfold astigmatism	Complex
C_3 or C_s	Third-order spherical aberration	Real
S_3	Twofold star aberration	Complex
A_4	Fivefold astigmatism	Complex
B_4	Axial coma	Complex
D_4	Three-lobe aberration	Complex
A_5	Sixfold astigmatism	Complex
C_5	Fifth-order spherical aberration	Real

are complex, except for the aberration coefficients named C. Table 7.1 gives a description of each aberration and indicates whether it is complex or real. The phase angle ϕ is given in the lens coordinate. The power series does not include the zero-order coefficient since the wave front is centered on the optical axis where two wave fronts coincide.

An alternative wave-front expansion has been given by Krivanek et al. (1999, 2008) using a different notation for the aberration coefficients:

$$
\begin{aligned}
\chi(\theta, \phi) &= \frac{2\pi}{\lambda} \Delta z\left(\theta_x, \theta_y\right) \\
&\approx \sum_n \sum_m \left\{ \left[c_{n,m,a}\theta^{n+1} \cos(m\phi) + c_{n,m,b}\theta^{n+1} \sin(m\phi) \right] / (n+1) \right\}
\end{aligned}
\tag{7.22}
$$

In Krivanek's notation, the first sum is over integer n, starting from 0 and ending with the highest order of aberration included. For each order n, the sum over m is taken either from 0 or from 1 up to $n + 1$, under the condition that $m + n$ is odd. The $n = 0$ term gives a linear term for the tilt of the wave front, which is not included in Eq. (7.21). A comparison between the two notations in Eqs. (7.21) and (7.22) and the earlier notation by Saxton for aberrations up to third-order can be found in the review by Hawkes (2008).

The first-order aberration, C_1, is actually not an aberration as it is equivalent to a change in focus with

$$
\frac{1}{2} C_1 \omega\bar{\omega} = \frac{1}{2} C_1 \left(\theta_x + i\theta_y\right)\left(\theta_x - i\theta_y\right) = \frac{1}{2} C_1 \theta^2
\tag{7.23}
$$

Comparison with Eq. (7.20) gives $C_1 = \Delta f / M$. The aberration A_1 describes twofold astigmatism. Its effect can be examined by reformulating the complex expression by considering $A_1 = A_{1a} + iA_{1b} = |A_1| \exp(i\phi_o)$, which gives

$$
\begin{aligned}
\mathrm{Re}\left(\frac{1}{2} A_1 \bar{\omega}^2\right) &= \mathrm{Re}\left(\frac{1}{2}(A_{1a} + iA_{1b})\left(\theta_x - i\theta_y\right)\left(\theta_x - i\theta_y\right)\right) \\
&= \frac{1}{2} A_{1a}\left(\theta_x^2 - \theta_y^2\right) + A_{1b}\theta_x\theta_y \\
&= \frac{1}{2}\theta^2(A_{1a}\cos 2\phi + A_{1b}\sin 2\phi) \\
&= \frac{1}{2}\theta^2 A_1 \cos(2\phi - \phi_o)
\end{aligned}
$$

Thus, twofold astigmatism gives a rotation-angle-dependent change in focus with twofold symmetry. The effect is that the lens focus is strong in one direction and weak in the orthogonal direction. This particular parasitic aberration is common and routinely corrected in TEM using a stigmator, which produces a weak quadrupole magnetic field. The axial coma (B_2) introduces a complex wave-front distortion. To examine its effect, we simplify the B_2 term in Eq. (7.21) by

$$\mathrm{Re}\left(B_2\omega^2\bar{\omega}\right) = |B_2|\theta^3\,\cos(\phi - \phi_o) = |B_2|\theta^2\theta_{x'}$$

where $\theta_{x'} = \theta\cos(\phi - \phi_o)$ with x' defines the coma axis. The aberration is quadratic, antisymmetric and increases with distance along a particular direction.

The third-order spherical aberration (C_3 or C_s) is a geometrical aberration, originating from the fourth-order terms in the inhomogeneous electron path equation (Eq. 7.13). It is positive and produced by all round magnetic lenses as we have discussed in Sect. 7.1. The effect of C_s gives rise to a positive wave-front distortion proportional to θ^4:

$$\frac{1}{4}C_s(\omega\bar{\omega})^2 = \frac{1}{4}C_s\theta^4.$$

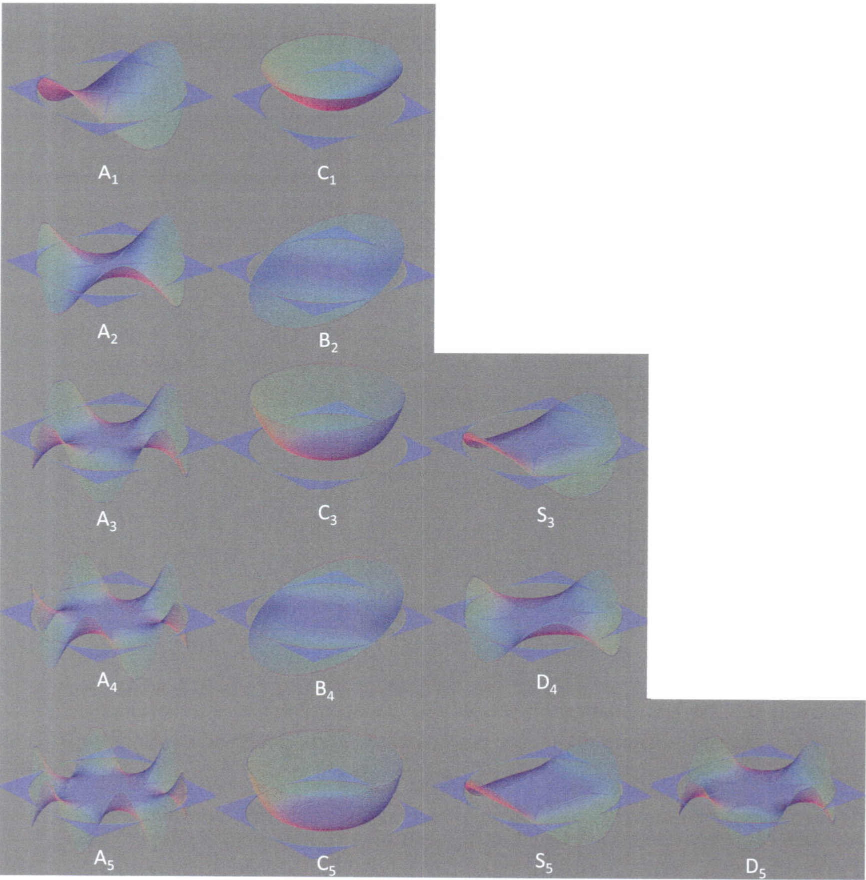

Fig. 7.6 Wave-front aberration from the contribution of each aberration coefficient in Eq. (7.21)

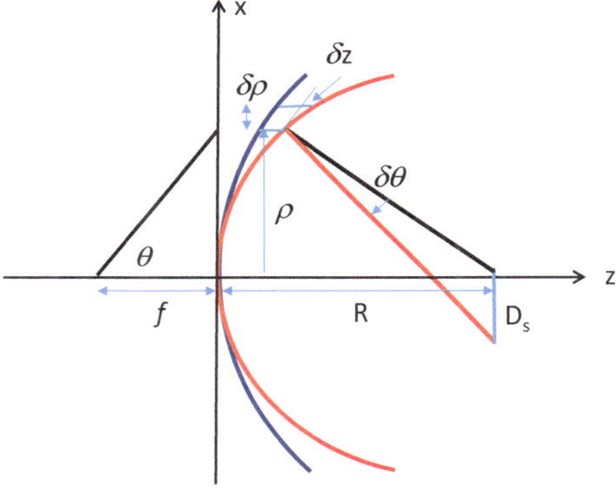

Fig. 7.7 Calculation of D_s from the wave-front distortion resulting from third-order spherical aberration

This comes about because the axial rays at a larger angle to the optical axis travel further away from the optical axis and interact more with the stronger off-axis magnetic field, consequently they experience a stronger focusing effect and are brought to a premature focus, beyond which they broaden out and create a disk of confusion further downstream at the image plane, intersecting it as shown in Fig. 7.6 at a lateral distance (D_s) from the ideal image point on the optical axis. The resulted disk of confusion is rotationally symmetric. Its size can be related to the wave-front distortion by considering Fig. 7.6, where (Fig. 7.7)

$$\delta\rho \approx \frac{D_s}{R} \approx \frac{\delta z}{\delta\rho} = \frac{1}{f}\frac{d}{d\theta}\left(\frac{1}{4}C_s\theta^4\right) = \frac{C_s\theta^3}{f}$$

Taking into account $R = Mf$, we have

$$D_s = MC_s\theta^3$$

The radius D_s increases rapidly as the cube of the angle. In order to reduce the effect of spherical aberration, a small aperture is used to reduce the acceptance angle. However, the smallest aperture that can be used is limited by diffraction effect, which acts to increase the size of the disk of confusion in the same manner described by Rayleigh's criterion for the resolution of a telescope imaging a star. Since C_s is always positive for a magnetic lens, and it is large even at the condition for optimal lens design and excitation (see Chap. 6), for many decades it was the most important factor in determining electron microscope resolution, prior to the development of electron aberration correctors, i.e., aberration correction.

In Sect. 7.1, we stated that C_s is positive-definite, under the conditions for a rotationally symmetric magnetic lens, forming a real image of a real object. This very important result, which had been the major limitation on TEM resolution for high-resolution electron microscopy over many years, was first shown by Scherzer (1936). Other implicit assumptions behind this result are that the lens is not a mirror, and that the lens has no space-charge or conductors on the axis which could give rise to discontinuities in the electrostatic or magnetic potential. Further the focusing fields are static rather than time-dependent. Under the above conditions, C_s could not be eliminated. Equation (7.14) and similar expressions for C_s were used in the unsuccessful search for round-lens fields with $C_s = 0$. In the search for reduced spherical aberration, relaxation of the above conditions was analyzed by Scherzer in a famous 1947 paper (Scherzer 1947). The simplest approach is to relax the requirement for rotational symmetry. This approach, after considerable earlier and unsuccessful efforts by a number of outstanding research groups, led to the successful development of aberration correctors half a century later. A comprehensive historical review of aberration correction can be found in Hawkes (2009), Krivanek et al. (2008), and Rose (2008).

7.3 Multipole Fields and Quadrupole Focal Properties

The simplest form of aberration corrector is a single multipole, correcting for lens astigmatism (the "stigmator" used in all electron microscopes). For example, a quadrupole is used to correct twofold (A_1) astigmatism or a hexapole for threefold (A_2) astigmatism. Correction of spherical aberration uses a combination of multipoles or hexapoles and round lenses. It relies on both the primary and the secondary aberrations associated with the multipole fields and their combined effects.

To examine the multipole fields, we start by assuming that the thickness of the multipole (l) is much larger than the distance between two opposite poles ($2a$). Then, the potential field can be approximated as planar, independent of z. This field satisfies the 2D Laplace equation of

$$\Delta\Psi(r, \theta) = \left(\frac{1}{r}\frac{\partial}{\partial r}r\frac{\partial}{\partial r} + \frac{1}{r^2}\frac{\partial^2}{\partial\theta^2} + \frac{\partial^2}{\partial z^2}\right)\Psi(r, \theta) = 0, \qquad (7.24)$$

where $\Psi = A$ or V for the magnetic or electrostatic multipoles, respectively. For a multipole having $2m$ dipoles, a general solution of Eq. (7.24) that meets the symmetry requirement is given by

$$\Psi(r, \theta) = \Psi_m \left(\frac{r}{a}\right)^m \cos[m(\theta - \theta_o)] \qquad (7.25)$$

where the angle θ_o defines the orientation of the multipoles in the polar coordinate system. If we assume the potential on the pole surface is uniform, $|\Psi_m| = \mu_o nI$ with nI standing for the ampere-turns in each pole, in the case of a magnetic multipole with infinite permeability $\mu \to \infty$ and assume the loss of ampere-turns in the pole is small.

Next, we consider a multipole of finite thickness. Its field is modeled by having Φ_m dependent on z and by including higher order terms, in order to describe the fringing field near the pole tips. Together, we then have the field in the form of (Rose 2011)

$$\Psi(r,\theta,z) = \sum_{\lambda=0}^{\infty} (-1)^{\lambda} \frac{m!}{\lambda!(m+\lambda)!} \left(\frac{r}{4a}\right)^{2\lambda} \left(\frac{\partial^{2\lambda}\Psi_m(z)}{\partial z^{2\lambda}}\right) \left(\frac{r}{a}\right)^m \cos[m(\theta-\theta_o)].$$

$$(7.26)$$

The $\Psi_m(z)$ in the above expansion series can be written as $\Psi_m(z) = \Psi_o(0)h(z)$, where $h(z)$ is called the characteristic function of the multipole lens. The distribution of $h(z)$ can be measured experimentally in case of a magnetic multipole or calculated by using numerical methods. This function is approximately constant in the middle section of the multipole. Thus, a common model in treating multipole lenses is to approximate $h(z)$ using a box-shaped function with an effective width of L, whose integrated area is the same as for $\int h(z)\mathrm{d}z$. This approximation, known as SCOFF for sharply cut-off fringe field, is a very useful tool for theoretical investigation, capable of describing the principle behavior of a corrector system by providing analytical relations. For practical purposes, we can take $L \approx l + 1.1a$ according to (Grivet 1972).

For a quadrupole with $m = 1$, Eq. (7.26) gives

$$\frac{\Psi(x,y,z)}{\Psi_1(0)} = 2\frac{h(z)}{a^2}xy - \frac{xy(x^2+y^2)}{6a^4}h''(z) + \cdots$$

And to first-order approximation:

$$B_x \approx -2\Psi_1(0)\frac{h(z)}{a^2}y$$

$$B_y \approx -2\Psi_1(0)\frac{h(z)}{a^2}x \qquad\qquad (7.27)$$

$$B_z \approx -2\Psi_1(0)\frac{1}{a^2}xyh'(z)$$

Substituting Eq. (7.27) into Eq. (6.3) using $V_z \approx V$, and $V_x \approx V_y \approx 0$ (there is no rotation in a multipole lens), we obtain the following equations of motion for a quadrupole

$$\frac{\mathrm{d}^2x}{\mathrm{d}z^2} + \beta^2 h(z)x \approx 0$$

$$\frac{\mathrm{d}^2y}{\mathrm{d}z^2} - \beta^2 h(z)y \approx 0 \qquad\qquad (7.28)$$

where

$$\beta^2 = \frac{2\Psi_1(0)}{a^2}\sqrt{\frac{e}{2m_e\Phi^*}} = \frac{\mu_o nI}{a^2}\sqrt{\frac{2e}{m_e\Phi^*}}$$

Using the SCOFF approximation, Eq. (7.28) has simple solutions of the general form

$$x(z) = c_1 \cos(\beta z) + c_2 \sin(\beta z)$$
$$y(z) = c_1 \cosh(\beta z) + c_2 \sinh(\beta z)$$

The coefficients c_1 and c_2 are determined by the initial conditions for the electron trajectory following the same procedures as we used in Sect. 6.4.1. For the principle ray u_π coming from the left side of a thin lens at $u_\pi(-L/2) = 1$ and parallel to the optical axis $u'_\pi(-L/2) = 0$. This gives the principle ray u_π in the x and y sections as

$$x(z) = \cos(\beta(z+L/2))$$
$$y(z) = \cosh(\beta(z+L/2)). \tag{7.29}$$

The asymptotic rays for the above principle paths in the SCOFF approximation are determined by their position and slope at $z = L/2$. They intersect the optical axis and give the image focal points in the x and y planes at

$$z_{Fx} = \frac{L}{2} + \frac{\cos \beta L}{\beta \sin \beta L}, \quad z_{Fy} = \frac{L}{2} - \frac{\cosh \beta L}{\beta \sinh \beta L} \tag{7.30}$$

And they intersect the incident asymptotic ray on the left at

$$z_{Hx} = \frac{L}{2} + \frac{\cos \beta L - 1}{\beta \sin \beta L}, \quad z_{Hy} = \frac{L}{2} - \frac{\cosh \beta L - 1}{\beta \sinh \beta L}$$

The distance between the focal and principle points gives the focal distance of

$$f_x = \frac{1}{\beta \sin \beta L}, \quad f_y = -\frac{1}{\beta \sinh \beta L}.$$

Figure 7.8 illustrates the cardinal points for the quadrupole. Along x, the quadrupole behaves as a convex focusing lens, while along y, the lens is divergent as for a concave lens. Acting together, these produce a line focus, corresponding to twofold astigmatism (A_1). At small excitations of the quadrupole lens with $\beta L < 0.2$, $f_x \approx -f_y \approx 1/\beta^2 L$ and $Z_{Hx} \approx Z_{Hy} \approx 0$. When $\beta L > 0.2$, the ratio of $|f_x/f_y|$ increases rapidly with excitation and the principle planes also move away from the center.

7.4   Aberrations of Hexapole Fields

For multipoles with multiplicity $m > 2$, their first-order focal lengths are infinite because the paraxial equation only takes into account constant and linear fields. In a hexapole, for example, with $m = 3$, the fields depend on u^2. Hexapoles, when used

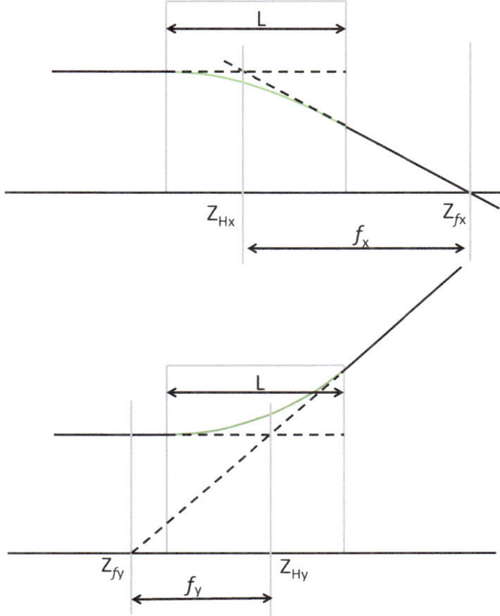

Fig. 7.8 Line focus of a quadrupole lens. Plotted for $\beta L = 1$

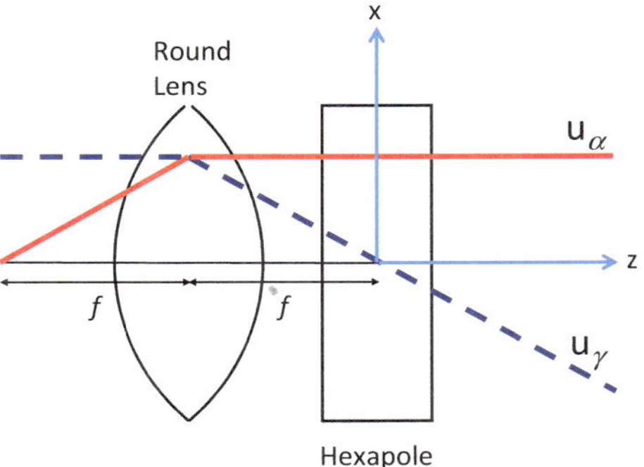

Fig. 7.9 Compound lens system of a round lens and a hexapole

alone, introduce threefold astigmatism (A_2). This property has been employed in TEM for A_2 correction. In general, multipoles with $m > 2$ enable adjustment of a variety of aberrations without significantly altering the electron paths. For these reasons, they play a central role in aberration correction.

Here we consider the combination of a round lens and a hexapole in the special configuration of Fig. 7.9. The hexapole is placed at the aperture plane of the round lens at $z = z_a$. According to Eq. (7.26), the hexapole fields give rise to a third-order term and its derivative yields the second-order perturbation term

$$P_2 = 3i\eta \Psi_3(z)\bar{u}^2 / \sqrt{\Phi^*} = H\bar{u}^2 \tag{7.31}$$

Here $\Psi_3(z)$ is complex with $\Psi_3(z) = \Psi_{3c}(z) + i\Psi_{3s}(z)$ dependent on the azimuthal orientation of the hexapole. Substituting Eq. (7.31) into (7.1) leads to following equation of motion for the compound lens of Fig. 7.9:

$$u'' + \frac{\eta^2}{\Phi^*} u = H\bar{u}^2 \tag{7.32}$$

This equation can be solved by using the iterative perturbation method of Sect. 7.1. The starting point concerns the fundamental rays of the paraxial equation ($H = 0$). They are unaltered by the hexapole field because it has no focusing effect.

A single hexapole has second-order, axial astigmatism as its primary aberration. To show this, we consider the axial ray of $u^{(1)}(z) = \omega u_\alpha(z)$. The second-order path deviation is obtained according to Eq. (7.10)

$$u^{(2)}(z_i) = u^{(1)}(z_i) - u(z_i) = -M \int_{z_0}^{z_i} H\bar{u}^2 u_\alpha(z)\mathrm{d}z$$

$$\approx -M\bar{\omega}^2 \int_{z_0}^{z_i} Hu_\alpha^3(z)\mathrm{d}z = -M\bar{\omega}^2 A_2$$

Here

$$A_2 = \int_{z_0}^{z_i} Hu_\alpha^3(z)\mathrm{d}z$$

is the second-order, axial astigmatism coefficient. According to Fig. 7.6, this aberration forms a threefold figure. The azimuthal orientation of the threefold aberration figure is determined by the imaginary part of the complex coefficient.

Other primary aberrations arise for incident rays that have components projected onto the principle and axial rays. To examine this further, it is helpful to consider u_α together with u_γ with following asymptotic properties:

$$u_\alpha(z_0) = 0, \quad u_\alpha'(z_0) = 1;$$
$$u_\gamma(z_0) = 1, \quad u_\gamma'(z_0) = m_\gamma \text{ and } u_\gamma(z_a) = 0$$

Thus selecting u_γ (instead of u_π) as one of two fundamental rays allows us to consider the slope of u_γ as a free parameter. Now, take the first-order solution of Eq. (7.32) as

$$u^{(1)} = \alpha u_\alpha + \gamma u_\gamma \tag{7.33}$$

Here both u_α and u_γ are real quantities, representing the distance to the optical axis in the rotational coordinates of the paraxial equation, while α and γ in general are complex to represent rays of different azimuthal angles. Taking this into account, we have to first order

$$\bar{u}^2 \approx \left(\overline{\alpha u_\alpha + \gamma u_\gamma} \right)^2 = \bar{\alpha}^2 u_\alpha^2 + \overline{\alpha\gamma} u_\alpha u_\gamma + \bar{\gamma}^2 u_\alpha^2$$

Substituting this into (7.31), we obtain the second-order path deviation

$$u^{(2)} = \bar{\alpha}^2 u_{\alpha\alpha} + 2\overline{\alpha\gamma} u_{\alpha\gamma} + \bar{\gamma}^2 u_{\gamma\gamma} \tag{7.34}$$

and

$$u_{\alpha\alpha} = u_\alpha \int_{z_0}^{z} H u_\alpha^2 u_\gamma \, \mathrm{d}z - u_\gamma \int_{z_0}^{z} H u_\alpha^3 \, \mathrm{d}z$$

$$u_{\alpha\gamma} = 2u_\alpha \int_{z_0}^{z} H u_\gamma^2 u_\alpha \, \mathrm{d}z - 2u_\gamma \int_{z_0}^{z} H u_\alpha^2 u_\gamma \, \mathrm{d}z \tag{7.35}$$

$$u_{\gamma\gamma} = u_\alpha \int_{z_0}^{z} H u_\gamma^3 \, \mathrm{d}z - u_\gamma \int_{z_0}^{z} H u_\gamma^2 u_\alpha \, \mathrm{d}z$$

By fixing the orientation of the hexapole,

$$\Psi_3(z) = i\Psi_{3s}(z)$$

Then all the aberration coefficients of Eq. (7.35) become real.

To calculate the third-order aberrations of the hexapole field, we now proceed to the next step of iteration by substituting u with $u \approx u^{(1)} + u^{(2)}$. From Eqs. (7.33) and (7.34), we have

$$\left(\overline{u^{(1)} + u^{(2)}} \right)^2 = \left(\bar{\alpha} u_\alpha + \bar{\gamma} u_\gamma + \alpha^2 u_{\overline{\alpha\alpha}} + 2\alpha\gamma u_{\overline{\alpha\gamma}} + \gamma^2 u_{\overline{\gamma\gamma}} \right)^2$$

Following the same procedures and notations used to obtain the second-order aberrations, we find the third-order terms

$$u^{(3)} = \alpha^2 \bar{\alpha} u_{\chi\overline{\alpha\alpha}} + \alpha^2 \bar{\gamma} u_{\gamma\overline{\alpha\alpha}} + \alpha\bar{\alpha}\gamma u_{\alpha\overline{\alpha\gamma}} + \alpha\bar{\gamma}\gamma u_{\gamma\overline{\alpha\gamma}} + \bar{\alpha}\gamma^2 u_{\alpha\overline{\gamma\gamma}} + \gamma^2 \bar{\gamma} u_{\gamma\overline{\gamma\gamma}} \tag{7.36}$$

Similarly to Eq. (7.35), the third-order aberration rays are linear combinations of u_α and u_γ with integral coefficients which depend on z. This statement is applicable to any aberrated ray with the index combination of X (for example, $X == \alpha\bar{\alpha}\bar{\alpha}$). It is therefore useful to write it in a general form as (Haider et al. 2008):

$$u_X(z) = u_\alpha(z)c_X(z) + u_\gamma(z)C_X(z)$$

$$u'_X(z) = u'_\alpha(z)c_X(z) + u'_\gamma(z)C_X(z)$$

At the image plane $z = z_i$,

$$C_X(z_i) = \frac{u_X(z_i)}{u_\gamma(z_i)}$$

Giving the image aberration

$$C_X(z_o) = \frac{u_X(z_i)}{M}$$

This is the equivalent aberration in the object plane. The other coefficient c_X determines the aberration in the diffraction image at $z = z_a$, where $u_\gamma = 0$. Because of these properties, C_X and c_X are known as image and slope coefficients.

The aberrated rays must have a symmetry that is consistent with the rotational symmetry of the lens. For a round lens, under the rotational transformation $\alpha = \alpha e^{i\theta}$ and $\gamma = \gamma e^{i\theta}$, the aberrated ray must be unchanged or invariant except for a simple rotation. Thus, for an aberrated ray with the prefactor of $\alpha^{N_\alpha}\gamma^{N_\gamma}\bar{\alpha}^{N_{\bar{\alpha}}}\bar{\gamma}^{N_{\bar{\gamma}}}$,

$$\alpha^{N_\alpha}\gamma^{N_\gamma}\bar{\alpha}^{N_{\bar{\alpha}}}\bar{\gamma}^{N_{\bar{\gamma}}} \longrightarrow \alpha^{N_\alpha}\gamma^{N_\gamma}\bar{\alpha}^{N_{\bar{\alpha}}}\bar{\gamma}^{N_{\bar{\gamma}}} e^{i\left(N_\alpha - N_{\bar{\alpha}} + N_\gamma - N_{\bar{\gamma}}\right)\theta} = \alpha^{N_\alpha}\gamma^{N_\gamma}\bar{\alpha}^{N_{\bar{\alpha}}}\bar{\gamma}^{N_{\bar{\gamma}}} e^{i\theta}$$

Thus, we have the constraint of $N_\alpha - N_{\bar{\alpha}} + N_\gamma - N_{\bar{\gamma}} = 1$, for a round lens. For the third order, in terms of hexapoles in Eq. (7.36), $N_\alpha - N_{\bar{\alpha}} + N_\gamma - N_{\bar{\gamma}} = 1$, the same as the round lens. Thus, we have reached a remarkable and important conclusion that the secondary aberrations of the hexapole field possess the same symmetry as the round lens. This result provided the basis for intensive discussion on aberration correction (Crewe and Kopf 1980).

Using the coordinates illustrated in Fig. 7.9, the fundamental rays at the hexapole are then specified by

$$u_\alpha = f, \text{ and } u_\gamma = -\frac{z}{f}$$

Here we have used the condition $u'_\alpha(-2f) = 1$, and $u_\gamma(0) = 0$ for these two rays. To evaluate the second order, primary aberrations of a hexapole field in Eq. (7.35), we use the SCOFF approximation and take $H(z) = 0$ for $|z| > L/2$ and $H(z) = H_o$ otherwise. Thus, we have following results for $z > L/2$:

$$u_{\alpha\alpha} = f \int_{-L/2}^{L/2} H_{\mathrm{o}} f^2 \left(\frac{z}{f}\right) \mathrm{d}z + \frac{z}{f} \int_{-L/2}^{L/2} H_{\mathrm{o}} f^3 \mathrm{d}z = H_{\mathrm{o}} f^2 Lz$$

$$u_{\alpha\gamma} = 2f^2 \int_{-L/2}^{L/2} H_{\mathrm{o}} z^2 \mathrm{d}z = \frac{1}{6} f^2 H_{\mathrm{o}} L^3$$

$$u_{\gamma\gamma} = z \int_{-L/2}^{L/2} H_{\mathrm{o}} \left(\frac{z}{f}\right)^2 \mathrm{d}z = \frac{1}{12} \frac{H_{\mathrm{o}}}{f^2} L^3 z$$

For $z < -L/2$, $u_{\alpha\alpha} = u_{\alpha\gamma} = u_{\gamma\gamma} = 0$. For $L/2 < z < -L/2$, we will only consider $u_{\alpha\alpha}$. It can be shown that in this region,

$$u_{\alpha\alpha} = -\frac{H_{\mathrm{o}}}{2} f^2 \left(z + \frac{L}{2}\right)^2$$

At the image plane, the primary aberration of threefold astigmatism is given by

$$A_2 = C_{\alpha\alpha}(z_i) = \frac{u_{\alpha\alpha}(z_i)}{u_{\gamma}(z_i)} = -H_{\mathrm{o}} f^3 L = 3\eta \Psi_{3s} f^3 L / \sqrt{\Phi^*}$$

The first term in the second-order aberration of Eq. (7.36) corresponds to spherical aberration C_s. At the image plane,

$$C_S = C_{\alpha\bar{\alpha}\bar{\alpha}}(z_i) = \frac{u_{\alpha\bar{\alpha}\bar{\alpha}}(z_i)}{u_{\gamma}(z_i)} = -2 \int_{z_o}^{z_i} H u_{\alpha}^2 u_{\bar{\alpha}\bar{\alpha}} \mathrm{d}z = -H_{\mathrm{o}}^2 f^4 \int_{-L/2}^{L/2} \left(z + \frac{L}{2}\right)^2 \mathrm{d}z \tag{7.37}$$

$$= -\frac{1}{3} H_{\mathrm{o}}^2 f^4 L^3 = -3|\eta \Psi_{3s}|^2 f^4 L^3 < 0$$

Thus, we have the following conclusions about the spherical aberration of a hexapole field:

(1) It is negative and opposite to the sign of C_s in a round lens;
(2) It is a secondary aberration, derived from the primary aberration of threefold astigmatism of the hexapole field;
(3) The magnitude of C_s is proportional to the cubic power of the effective thickness of the hexapole. Thus, it is only significant for a thick hexapole;
(4) The magnitude of C_s is proportional to the square of the hexapole excitation, $|\Psi_3|^2 = (\mu_{\mathrm{o}} nI)^2$;
(5) It is inversely proportional to the electron accelerating voltage.

7.5   C$_s$ Correctors

The first successful TEM C_s corrector was constructed in Germany by Haider et al. (1995) and was used to demonstrate atomic-resolution imaging in 1998 (Haider et al. 1998). The corrector used the design of two hexapoles and two round lenses (Rose 1981, 1990). This design has since become the most widely adopted design for TEM instruments. For dedicated STEM instruments, work at the Cavendish in 1997 resulted in a successful first-generation *Cs*-corrected instrument using a combination of quadrupoles and octupoles (Krivanek et al. 1997). The large off-axis aberrations of the octupole–quadrupole arrangement made it more suitable for scanning probe instruments (where electrons are confined to a small region around the optical axis), while the hexapole corrector is better suited to the off-axis TEM mode with its isoplanatic requirement, as well as in STEM.

The use of a hexapole field for C_s correction takes advantage of its third-order aberrations when it is combined with a round lens, which is equivalent to these of a round lens. The C_s of the hexapole field is also negative. For these reasons, a hexapole field may be used to correct the third-, and higher, order aberrations of the round lenses. However, the hexapole field comes with the primary, second-order aberrations. They must be canceled first by another field in order to realize the benefits of higher order aberration coefficient correction. Cancelation of second-order aberrations is achieved in a system consisting of two identical hexapoles and a telescopic round-lens doublet, as illustrated in Fig. 7.10.

To examine the aberrations of the double hexapole corrector, we take the coordinate at the center; hexapoles 1 and 2 are located at $z = -2f$ and $2f$, respectively, and

$$u_\alpha = -f, \text{ and } u_\gamma = \frac{z + 2f}{f} \quad \text{for } z < -f$$

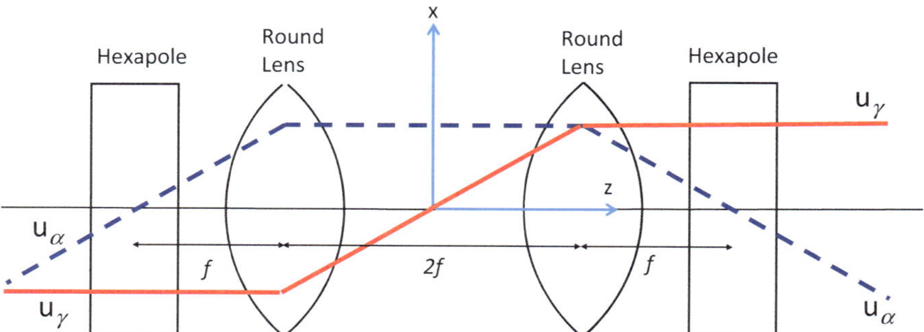

Fig. 7.10 Design of a hexapole corrector with two hexapoles and two round lens in a mirror symmetrical arrangement, where the mirror plane and two hexapoles are in the aperture planes of the two round lenses

$$u_\alpha = f, \text{ and } u_\gamma = -\frac{z - 2f}{f} \quad \text{for } z > f$$

Then, $u_\alpha(z)$ is antisymmetric (odd) with $u_\alpha(z) = -u_\alpha(-z)$, and both $u_\gamma(z)$ and $H(z)$ are symmetric (even) with $u_\gamma(z) = u_\gamma(-z)$ and $H(z) = H(-z)$. The term $H(z)$ is symmetrical if the direction of the currents of the two round lenses are opposite to each other. This requirement is necessary since P_2 is complex (Eq. 7.31) and the rotation angle of the round lens must be included. The rotation cancels for the round lens when their currents are opposite.

The second-order path deviations involve the following four integrals:

$$I_1 = \int_{z_o}^{z_i} Hu_z^2 u_\gamma dz, I_2 = \int_{z_o}^{z_i} Hu_\alpha^3 dz, I_3 = \int_{z_o}^{z_i} Hu_\gamma^2 u_\alpha dz, \text{ and } I_4 = \int_{z_o}^{z_i} Hu_\gamma^3 dz$$

The terms I_2 and I_3 are integrals of odd functions, which results in even functions and thus zero integrant when the region of integration covers both hexapoles. The other two integrals, I_1 and I_4, can be broken down into two separate integrals for the two hexapoles, respectively. Within the field of each hexapole, the function is antisymmetric so the resulted integrant is also zero. Thus, the symmetric design of the double corrector shown in Fig. 7.10 in theory will cancel all secondary aberrations. In practice, misalignment will lead to parasitic second-order aberrations. Minimization of these aberrations requires flexibility and accuracy in the beam alignment as well as real-time diagnosis procedures, which are critical to the success of aberration correction.

The spherical aberration of the two round lenses and hexapoles simply adds up in the corrector, with the total spherical aberration given by

$$C_3 = C_{3r} - \frac{2}{3}H_o^2 f^4 L^3$$

Here C_{3r} is the spherical aberration of the round lenses, including the objective lens that the corrector is designed to correct for. The C_{3r} contributions from the round transfer lenses are not included here because the reduction of acceptance angle by the magnification (M) of the objective lens reduces their effects by a factor of M^3, close to 10^4 for $M = 20$. The cancelation of the system's spherical aberration is achieved by setting

$$|H_o| = \frac{1}{f^2} \sqrt{\frac{3C_{3r}}{2L^3}}$$

The off-axial coma limits the field of view in the object plane in an electron-optical system. In optics, an optical system which achieves the same resolution for all imaging points is called an aplanat. Inside the corrector, the off-axial third-order coma must be also eliminated as well so that all points are imaged at the

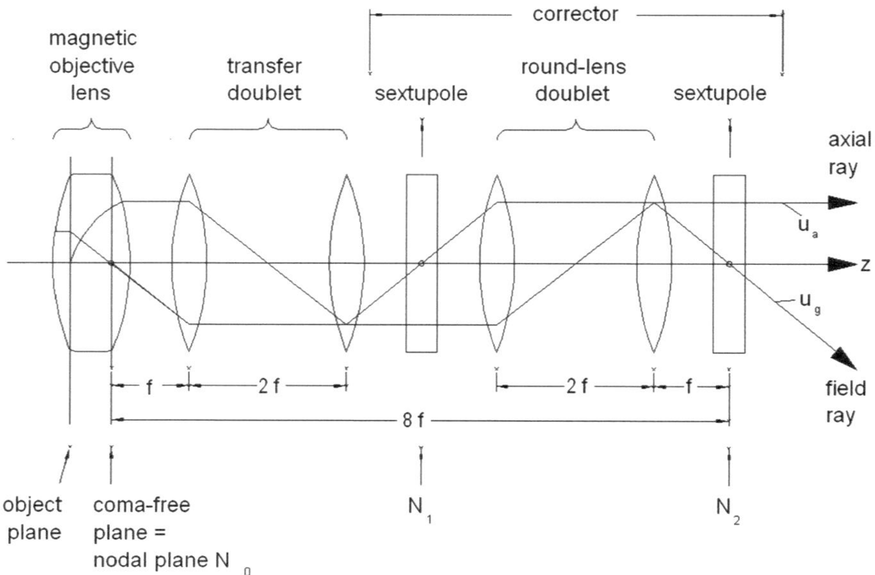

Fig. 7.11 Schematic optical diagram showing the integration of the double hexapole corrector after the objective lens using a transfer lens to achieve the electron-optical aplanat. The paths of the fundamental paraxial rays are also illustrated in the diagram (from H. Rose)

same resolution, within an extended object area that is sufficient to meet the experimental requirements.

The optical element has a so-called coma-free point on the optical axis such that no coma is introduced when the center ray of an off-axial ray bundle intersects this coma-free point. The coma-free point of the objective lens is located within its field region, while the coma-free point of the corrector is at the center of its first hexapole. To obtain an aplanat, the field ray u_γ must intersect both points. For the spatially separated hexapole and objective lens, this can achieved by introducing another telescopic round-lens doublet, where the coma-free point of the doublet is its front and back focal points. Thus by incorporating the doublet, we can image the coma-free plane of the objective lens into the coma-free plane of the corrector without introducing any coma. This design is the basis of all correctors built for commercial TEMs by the CEOS Company (see Fig. 7.11).

Some other optical elements in the toolbox of electron optics, beside the hexapoles we have just discussed, are other multipoles, especially quadrupoles and octupoles, and magnetic/electrostatic sectors. Together, they produce a rich variety of optical effects that can be used for aberration correction, as well as correction of C_c.

The complex optical effects of an electron-optical system for aberration correction, and also for other purposes, derive from two basic phenomena: combination aberrations and misalignment aberrations. These aberrations are absent when an element is operated alone. For example, the spherical aberration of the hexapole is a combination aberration produced in a thick hexapole, which is absent in an ideal infinite thin 2D field ($C_s = 0$ for $L = 0$). To further illustrate combination aberrations, let us consider a quadrupole placed in front of a octupole, the quadrupole produces an elliptical or line beam (see Sect. 7.3), which then passes through the octupole, and the shape of the beam coming out of the octupole depends strongly on the extent of elliptical distortion, its relative orientation to the octupole, and the octupole excitation. For any two electron-optical elements, Krivanek observed, in his notation for aberrations, under the condition of small-beam distortions, the combining $C_{n,m}$ aberration of the first element with the $C_{u,v}$ of the second element gives rise to one or two aberrations of $C_{n+u-1,|m-v|}$, or together with $C_{n+u-1,m+v}$, when $m + v$ is less or equal to the highest multiplicity allowed for a given order. For example, $C_{2,3}$ and $C_{3,4}$ only give rise to $C_{4,1}$. When the beam distortion produced by element 1 is no longer small, all aberrations of the $n + u - 1$ order are excited according to Krivanek et al. (2008).

Misalignment aberrations are produced when the beam is shifted or tilted relative to the optical axis of an optical element. In a multipole, lower orders of aberrations than the multipole's primary aberrations are produced when a round beam passes at a distance away from the optical axis. Take an ideal octupole for example, and a round and parallel beam traveling through along its optical axis produces only $C_{3,4a}$ as its primary aberration. When the beam is shifted in the octupole and it arrives at the sample position tilted by an angle $\tau = (\alpha, \beta)$, corresponding to placing the octupole in front of a round lens and the sample at the back focal plane of the round lens, the threefold astigmatism, the twofold astigmatism, and the probe shift caused by the octupole change as shown by Krivanek et al. (2008):

$$
\begin{aligned}
C_{2,3a} &= 3\alpha C_{3,4a} & C_{3,2b} &= -3\beta C_{3,4a} \\
C_{1,2a} &= 3\left(\alpha^2 - \beta^2\right)C_{3,4a} & C_{1,2b} &= -6\alpha\beta C_{3,4a} \\
C_{0,1a} &= \left(\alpha^3 - 3\alpha\beta^2\right)C_{3,4a} & C_{0,1b} &= \left(\beta^3 - 3\alpha^2\beta\right)C_{3,4a}
\end{aligned}
\tag{7.38}
$$

For small α and β, the second-order aberrations are the largest among all of these produced. In general, a small beam shift in a multipole of multiplicity m produces a primary aberration of $C_{m-2,m-1}$. Thus, shifting the beam in an octupole is used to adjust $C_{2,3}$ in the quadrupole–octupole correctors.

Figure 7.12 shows the arrangement of quadrupoles and octupoles in the second-generation C3/C5 correctors constructed by NION (Seattle, WA). Quadrupole 1 (Q1) converts the round beam into an elliptical beam in octupole 1 (O1), Q2 changes it back into a round beam in O2, and Q3 changes it into an elliptical beam in O3, while Q4 changes it back to a round beam. Now the third-order octupole aberration contribution $C_{3,4}$ has the same scattering angle dependence along two orthogonal directions normal to the axis as does the isotropic spherical aberration C_s we wish to cancel, but with the opposite sign. The idea then is to align the long axis of the ellipse in the first octupole O1 with the direction which produces cancelation, then convert the beam to an ellipse whose long axis is

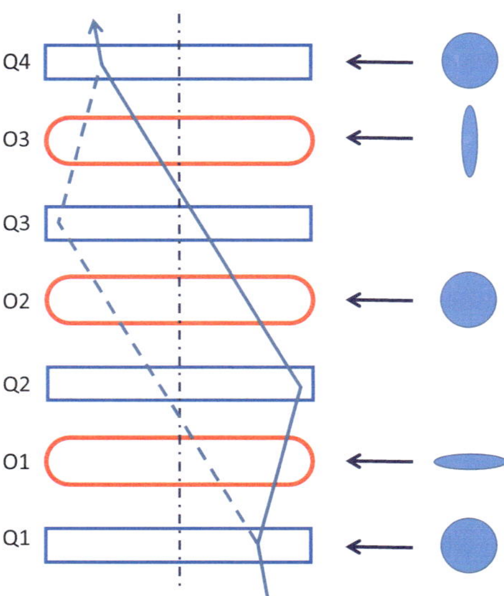

Fig. 7.12 Illustration of first-order electron trajectories (full line in *x-z* and *dashed line* in *y-z* planes) and beam shapes at the indicated positions in the second-generation NION (Seattle, WA) quadrupole–octupole corrector (after Krivanek et al. 2008)

orthogonal to the first ellipse, and pass it though another octupole O3 in order to cancel the aberration in the remaining direction.

The overall C_c in the objective lens and corrector system increases as additional fields are introduced according to Eq. (7.12). The hexapole has no first-order focal properties, and thus, it does not contribute to the increased C_c. By minimizing the excitation of the round lenses inside the correctors (with $f \sim 3$ cm), their contribution can be kept small compared to that of the objective lens. Overall, in a hexapole corrector system, C_c increases by ~ 15–25 %. The correction of chromatic aberration has proceeded more slowly and is more difficult, but was foreseen in Scherzer's (1947) paper. A detailed design was given by Rose (1971) for an instrument which corrects chromatic, spherical, and off-axial coma aberrations—finally an *achroplanatic* microscope. This complex instrument has been constructed, based on quadrupole and octupole elements, and first results were reported by Kabius et al. (2009). Chromatic aberration (including the significant amount introduced by the corrector itself) is corrected using electric–magnetic quadrupole fields.

A chromatically corrected lens with $C_c = 0$ brings all rays to the same focus at all energies. The energy spread in the beam both due to the source and due to energy losses in the sample must be considered. For a STEM without post-specimen lenses, the second effect is absent. Chromatic aberration correctors are expected to be important in lower voltage machines (where the beam energy spread is otherwise a larger fraction of the beam energy). In thicker samples where electron energy losses dominate, the advantages of having all the electrons which pass through the sample contribute to the in-focus image are clear, since image background and radiation dose would be reduced.

References

Crewe AV, Kopf D (1980) A sextupole system for the correction of spherical aberration. Optik 55:1–10

El-Kareh AB, El-Kareh JCJ (1970) Electron beams, lenses, and optics. Orlando Academic Press, Orlando

Glaser W (1956) Elektronen- und Ionenoptik. Handbuch der physik 33:123–395

Glaser W, Lammel E (1943) Arch Elektrotech 37:347–356

Grivet P (1972) Electron optics: translated by PW Hawkes, revised by A Septier (2nd English edn). Oxford Pergamon Press, Oxford

Haider M, Rose H, Uhlemann S, Schwan E, Kabius B, Urban K (1998) A spherical-aberration-corrected 200 kV transmission electron microscope. Ultramicroscopy 75:53–60

Haider M, Muller H, Uhlemann S (2008) Present and future hexapole aberration correctors for high-resolution electron microscopy. In: Advances in imaging and electron physics, vol 153. Elsevier Academic Press Inc, San Diego

Haider M, Braunshausen G, Schwan E (1995) Correction of the spherical-aberration of a 200-KV TEM by means of a hexapole-corrector. Optik 99(4):167–179.

Hawkes PW (2008) Aberrations. In: Orloff J (ed) Handbook of charged particle optics, 2nd edn. CRC Press, Boca Raton

Hawkes PW (2009) Aberration correction past and present. Philos T Roy Soc A 367:3637–3664

Hawkes PW, Kasper E (1996a) Principles of electron optics: applied geometrical optics. Academic Press, San Diego

Hawkes PW, Kasper E (1996b) Principles of electron optics: basic geometrical optics. Academic Press, San Diego

Kabius B, Hartel P, Haider M, Muller H, Uhlemann S, Loebau U, Zach J, Rose H (2009) First application of C_c-corrected imaging for high-resolution and energy-filtered TEM. J Electron Microsc 58:147–155

Krivanek OL, Dellby N, Spence AJ, Camps RA, Brown LM (1997) In electron microscopy and analysis. Phys C Ser, Inst Phys, Bristol 153:35–40

Krivanek OL, Dellby N, Lupini AR (1999) Towards sub-angstrom electron beams. Ultramicroscopy 78:1–11

Krivanek OL, Dellby N, Murfitt MF (2008) Aberration correction in electron microscopy. In: Orloff J (ed) Handbook of charged particle optics, 2nd edn. CRC Press, Boca Raton

Riecke WD (1982) Practical lens design. In: Hawkes PW (ed) Magnetic electron lenses. Springer, Berlin

Rose H (1971) Aplanatic electron-lenses. Optik 34:285

Rose H (1981) Correction of aperture aberrations in magnetic systems with threefold symmetry. Nucl Instr Meth 187:187–199

Rose H (1990) Outline of a spherically corrected semiaplanatic medium-voltage transmission electron-microscope. Optik 85:19–24

Rose H (2008) History of direct aberration correction. In: Peter WH (ed) Advances in imaging and electron physics. Elsevier, Amsterdam

Rose H (2011) Electron optics. University of Illinois, Urbana-Champaign. http://cbed.matse.illinois.edu/download/Rose_optics_of_magnetic_lenses.pdf

Scherzer O (1936) Über einige fehler von elektronenlinsen. Z. Physik 101:593–603

Scherzer O (1947) Sphärische und chromatische korrektur von elektronen-linsen. Optik 2:114–132

Septier A (1967) Focusing of charged particles. Academic Press, New York

Szilagyi M (1988) Electron and ion optics. Springer, New York

Uhlemann S, Haider M (1998) Residual wave aberrations in the first spherical aberration corrected transmission electron microscope. Ultramicroscopy 72:109–119

Chapter 8
Electron Sources

Electrons are emitted from solids by overcoming the electron potential barrier through one of four mechanisms illustrated in Fig. 8.1. The potential barrier arises from the electron interaction with the positively charged atomic nuclei and other negative charged electrons. To avoid charging during electron emission, the solid must be highly conductive. For this reason, only metals and metallic solids are used as electron emitters. In a metal, electrons are filled in energy states up to the Fermi level at zero K temperature, the difference between the vacuum level and the Fermi level in the absence of external applied electric fields is called the work function, which is also the height of the potential barrier for electrons inside a metal. The mean inner potential is the sum of the Fermi energy and the work function. Electron guns in conventional TEMs are constructed using one of three types of electron emitters: thermionic sources using the hairpin tungsten, lanthanum hexaboride filaments, and Schottky or field emission sources. Their workings and properties are described in this chapter. Photocathodes based on photoemission driven by a pulsed laser beam are used for time-resolved electron diffraction and imaging.

8.1 Source Properties

Electron microscopes are designed and constructed based on one specific type of electron source. In selecting a TEM for electron diffraction and imaging, the following electron source properties are considered:

1. Source brightness. The brightness β_s of a source is defined as the emission current density J per unit solid angle. For small beam angles α, this is given by

$$\beta_s = J/\pi\alpha^2 \tag{8.1}$$

© Springer Science+Business Media New York 2017

J.M. Zuo and J.C.H. Spence, *Advanced Transmission Electron Microscopy*,
DOI 10.1007/978-1-4939-6607-3_8

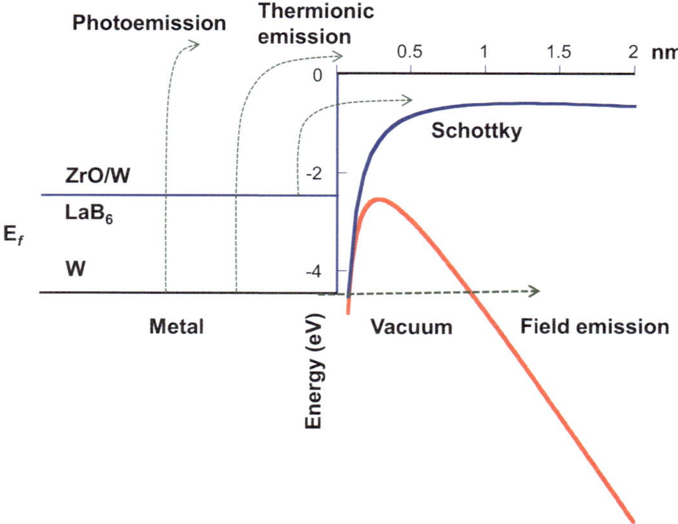

Fig. 8.1 Schematic illustration of thermionic, Schottky, field and photo emission. The potential difference between the vacuum and the Fermi level (E_f) defines the work function and the potential barrier for electron emission. The barrier is reduced by applying a biasing voltage for Schottky and field emission

The brightness is constant at all points along the optical axis, from the source tip to the detector, even if lens aberrations are permitted. This follows from the electron-optical analog of the Helmholtz–Lagrange theorem which, in simplified form, gives the product of area, energy, and local solid angle as constant along the beam path. Thus, the effect of a demagnifying lens, for example, is to increase current density J (proportional to M^{-2}), and increase the solid angle $\pi * \alpha^2$ (proportional to M^{-2}), leaving brightness constant. The source brightness is important whenever a small probe is required. Since the beam convergence angle is fixed by an aperture as required by experiment, high brightness ensures a sufficient probe current density for electron image or diffraction pattern recording. The product of α and the emitting diameter is called the emittance. Accelerator physicists seek small emittance and large brightness to obtain a highly coherent but intense source.

2. Source size, which is the effective object size seen by the first condenser lens. Reducing the source size improves the lateral coherence of the illuminated condenser aperture and consequently the quality of electron illumination.

3. Source energy spread. Electron energy spread is determined by two factors in an electron gun: (a) the distribution of the emitted electrons' energy and (b) Coulomb interaction between emitted electrons. The contribution from the

second factor measured along the beam is called the Boersch effect (Boersch 1954). For thermionic emission at temperature of 2500 K, the half-width energy distribution ΔE is about 0.5 eV. The measured values of ΔE range from 1.5 to 3 eV. The contribution due to the Boersch effect is about 1–2 eV or more. A large Boersch effect is observed when the electron beam forms a crossover inside the electron gun when the electron speed is relatively low. In field emitters, by avoiding the initial crossover, the energy spread can be kept low. The electron energy spread affects the temporal coherence of the electron beam and information transfer in high resolution electron microscopy.

4. Emission stability, as defined by the variance of intensity over a period of time. Any fluctuations in the electron source brightness result in noise in the electron diffraction signals acquired over time. For example, extended acquisition time is required for 3D electron imaging or diffraction.

In addition, the following factors must be considered for time-resolved electron diffraction:

1. Electron pulse duration. Electrons are emitted in the form of electron pulses using photocathodes and a pulsed laser of suitable wavelength, each pulse is characterized by a pulse duration or width. Pulse width determines the experimental time resolution.
2. Number of electrons per pulse. For diffraction patterns recorded using a single electron pulse ("single shot"), the number of electrons in the pulse determines the amount of diffracted electrons.
3. Repetition rate. The time between two sequential pulses is determined by the repeat frequency of the pulsed laser.
4. Coulomb interactions between electrons acting across the beam causes unwanted additional beam divergence, which degrades spatial coherence, just at the Boersch effect due to electron repulsion along the beam direction degrades temporal coherence, producing unwanted additional energy spread in the beam. As the electron beam velocity approaches the speed of light, it can be shown that the magnetic and electric fields which cause these repulsion forces exactly cancel, allowing large numbers of electrons to be packed into extremely brief pulses, perhaps sufficient to form a snapshot image in a few picoseconds. It is for this reason, to avoid loss of coherence in femtosecond electron pulses, that many modern fast electron diffraction cameras operate at MeV energies.

In what follows, we describe the thermionic, Schottky and cold-field emission sources and conclude with a summary of their properties. For the principles and applications of pulsed electron sources using photocathodes, see Mourou and Williamson (1982), Williamson and Zewail (1993), Aeschlimann et al. (1995), Elsayed-Ali (1995), Helliwell and Rentzepis (1997), King et al. (2005), Zewail (2006).

8.2 Thermionic Emission Source

For thermionic emission, electrons are emitted from a filament by heating. An emitted electron must overcome the work function (ϕ_W, see Fig. 8.1), which is a characteristic of the solid on the order a few eV for metals. Among metals, cesium has the lowest work function at 2.14 eV, while the work function of tungsten ranges from 4.32 to 5.22 eV depending on which surface facet the electrons are emitted from. Thus, for electrons to carry sufficient energy to escape into vacuum in thermionic emission, the filament must be heated to high temperature. Only a few metals, or metallic solids, are capable of doing this. Among these, tungsten and lanthanum hexaboride (LaB_6) with their melting temperatures of 3650 and 2483 K, respectively, are the two most widely used. According to quantum mechanics, the Fermi distribution of electrons broadens as temperature increases. At sufficiently high temperatures, the number of electrons has energy greater than ϕ_W is proportional to $\exp[-E/k_BT](E \geq \phi_W \gg k_BT)$. Not all of electrons in the high energy states escape the solid. For those that do, some return to the solid and do not contribute to electron emission. By approximating the local emitting surface as a uniform metal surface, the maximum current density j_T emitted by at temperature T can be obtained from the Richardson–Dushman equation

$$j_T = AT^2 \exp(-\phi_W/k_BT) \tag{8.2}$$

where A is a material-dependent constant ($A \approx 120$ Å K^{-2} cm^{-2} for tungsten), Boltzman's constant $k_B = 1.38 \times 10^{-23}$ J K^{-1}, and ϕ_W is the work function of the emitter (see Table 8.1). Tungsten is used for thermal emitters because of its stability at high temperatures. Modern TEMs use LaB_6 tips as thermionic sources because their lower work function leads to higher brightness as well as reduced operating temperature (Table 8.1). They require somewhat better vacuum conditions than tungsten thermal emitters (to reduce damage from positive-ion collisions).

Figure 8.2 shows the construction of an electron gun fitted with the thermionic filament as the cathode. The filament, Wehnelt and anode form three electrodes, hence the name of the triode gun. A tungsten wire in a hairpin shape is used to mount and heat the filament. This wire is connected via a variable bias resistor to the Wehnelt, which is in turn connected to the negative high-voltage terminal. The variable resistor forms the gun bias (V) or beam-current control. This bias also allows the triode guns to run in the "autobias" mode. The Wehnelt is negatively biased at potential $\Phi = -U$ with $1 < U < 50$ kV, while the anode is grounded, with $\Phi = 0$. The circuit can then be analyzed by analogy with a triode vacuum tube. Any temporary increase in beam current causes an increase in the negative Wehnelt bias, which tends to decrease the beam current. Additional accelerating anodes are then used to bring the beam up to the full kinetic energy required.

An important feature of the triode gun is that electrons are emitted only within the circular zero equipotential on the filament tip. The size of this area grows smaller with increasing bias, vanishing to a point at the filament tip at the cut-off

Table 8.1 Electron sources compared at 100 kV

	Thermionic emission			Field emission		
	Hairpin	Point	LaB6	Thermal	Cold	ZrO/W[a]
Brightness ($\mathrm{\mathring{A}}/$ cm^2 str.)	5×10^5 at 100 kV	1×1^6 at 100 kV	5×10^6 at 100 kV	10^8 at 100 kV	10^8–10^9 at 100 kV	5×10^8
Crossover diameter	$\sim 20\ \mu$	$\sim 5\ \mu$	$\sim 10\ \mu$	$\sim 200\ \mathrm{\mathring{A}}$	50–100 $\mathrm{\mathring{A}}$	150 $\mathrm{\mathring{A}}$
Life (h)	~ 30	~ 10	~ 500	500–1000	$\sim 10^3$	5000
Operating T (K)	2800	2900	1800–1900	1000–1800	R.T.	1800
Vacuum (torr)	$\sim 10^{-5}$	$\sim 10^{-5}$	$\sim 10^{-6}$	10^{-8}–10^{-9}	10^{-9}–10^{-10}	$\sim 10^{-8}$
Emission current (μA)	~ 100	~ 10	~ 50	50–100	5–20	
Current density (A/cm^2)	~ 3	~ 6	20–30	10^5–10^6	10^4–10^6	
Current density (μA/str.)	$\sim 10^5$	$\sim 10^5$	$10^5 \sim$	$\sim 10^2$	~ 10	0.1–1
Current stability (short term)	Good	Good	Good	A little worse	A little worse	<1 % RMS
Current stabil. (long term)	Good	Good	Good	Good	A little worse	<1 %
Current efficiency (%)	100	100	100	100	~ 5	0.1–0.5
Work function (eV)	4.5 (mean val.)	4.5 (mean val.)	2.7 (mean val.)	4.7 for (100)	4.35 for (310)	
Energy spread (eV)	1–2	0.5–2	2–3	~ 0.5	~ 0.2	0.3–1.0
Melting point (° C)	3370	3370	2200	3370	3370	3370

[a]Approximate values provided by L. Swanson for 25 kV

bias. This explains the changing appearance of the focused illumination spot as the filament current is increased, increasing both the filament temperature and gun bias (see Fig. 5.5 in Williams and Carter 2009) or Fig. 4.7 in Reimer and Kohl (2008). The spot contracts as electrons are drawn from a progressively smaller region of the filament tip. A full treatment of self-biased guns can be found in Haine and Einstein (1952).

Since any axially symmetric system of electrodes forms a lens, the Wehnelt-anode system images the tip onto the plane D. [For a comprehensive treatment of electron optics, see Part IX, "Electron Guns" in Hawkes and Kasper (1996)] A "crossover" is therefore formed as shown, and this forms the minimum cross section of the beam. It is a demagnified image of this crossover which is used to form the probe in electron diffraction experiments.

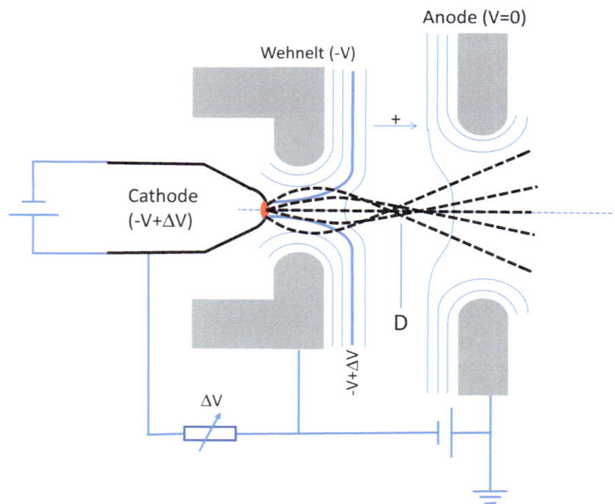

Fig. 8.2 Construction of a thermionic gun

8.3   Schottky Emission Source

Schottky (Walter Hermann Schottky 1886–1976) pointed out that the energy barrier to thermal emission could be reduced by the application of an external field, resulting in increased emission. The lowered potential barrier has a maximum due to the attractive potential between an emitted electron and its image charge inside the sharp tip. Together the total potential experienced by an emitted electron is given by

$$\phi(z) \approx \phi_{\mathrm{W}} - \frac{e^2}{16\pi\varepsilon_0 z} - eEz, \tag{8.3}$$

where we have assumed that the tip is mounted along the z axis with the end of the tip at $z = 0$. The peak potential is obtained at

$$z_{\mathrm{max}} = \sqrt{\frac{e}{16\pi\varepsilon_0 E}}.$$

At the peak, the work function is lowered to

$$\phi_{\mathrm{W}}^{\mathrm{eff}} \approx \phi_{\mathrm{W}} - e\sqrt{\frac{eE}{4\pi\varepsilon_0}} \tag{8.4}$$

In the construction of a Schottky FEG (see Fig. 8.3), the cathode typically consists of a ZrO coated, and (100) faceted, tungsten tip with a radius ranging from 0.1 to 1 μm. ZrO coating reduces the work function of $W(100)$ from 4.55 to 2.7 eV. The cathode protrudes beyond the Wehnelt, which acts as the suppressor here, by

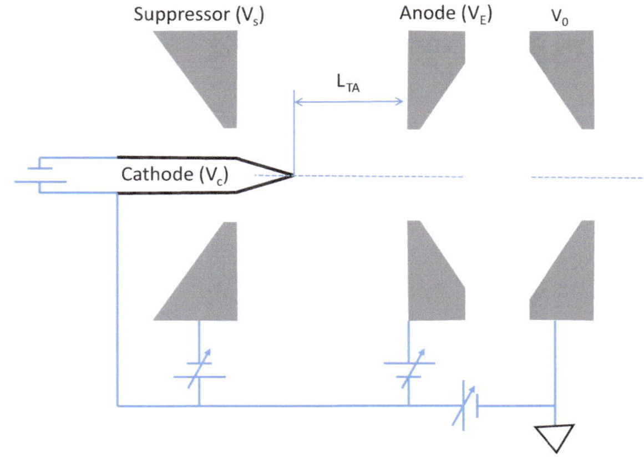

Fig. 8.3 Emitter geometry for the ZrO/W emission

hundreds of microns. Two anodes, in Butler's original design (Butler 1966), are placed in front of the cathode. High external fields are obtained by biasing the sharp tip. This tip field is given very approximately by the potential (V) between the tip and the first extraction anode

$$E = \beta(r)V, \tag{8.5}$$

modified by the shape factor $\beta(r)$ with r the radius of the tip. The shape factor $\beta(r)$ also changes with the distance from the emitter tip to the first anode, L_{TA}. In the range of $L_{\mathrm{TA}} = 550\text{–}1400$ μm and $r = 300\text{–}1400$ nm (Swanson and Bell 1973),

$$\beta(r) \approx 3.5 \times 10^9 / L_{\mathrm{TA}}^{0.63} r^{0.96} \left(\mathrm{m}^{-1}\right) \tag{8.6}$$

with L_{TA} in μm and r in nm. The extraction voltage V ranges from 4 to 8 kV.

A small aperture of hundreds of μm is placed on the first anode. This small aperture excludes electron emission from the shank of the tip emitter. Additionally, the two anodes also act together as an electron lens; they are weakly excited with a long focal distance to avoid any large lens aberrations due to irregularities on the aperture. Change of focus is achieved by varying the ratio V_E/V_o.

The lowering of the work function amounts to ~ 0.4 eV for a Schottky emitter of tip radius 1 μm and a field strength of $E = 10^8$ V/m. Under these conditions, the barrier width is too large for field emission at room temperature. To escape the barrier, the electrons have to be raised to higher energy levels, where the barrier width is smaller because of the approximately triangular shape of the potential field (see Fig. 8.1). In a thermally assisted process, e.g., by heating the cathode to high temperatures, a portion of the electrons in the tail of the Fermi distribution now have the kinetic energy needed for thermally assisted field emission. Because the distribution is broad, some also have the kinetic energy needed to overcome the now lowered work function $\phi_{\mathrm{W}}^{\mathrm{eff}}$, leading to thermionic emission. The operation of these two processes can be distinguished experimentally by plotting $\log I$ versus the

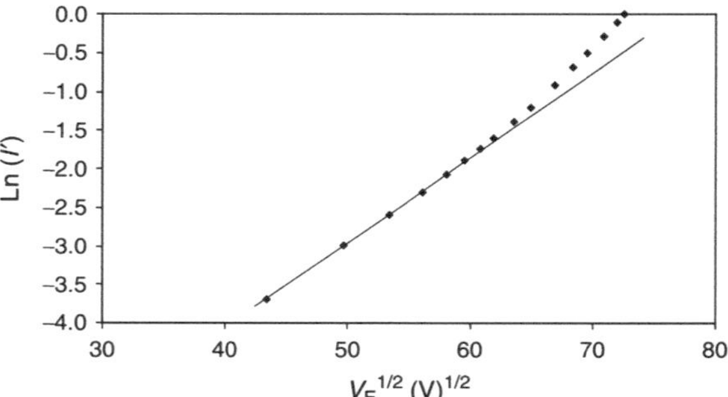

Fig. 8.4 Schottky plot of emission current from a ZrO/W emitter at the fixed cathode temperature of $T = 1800$ K with a tip radius of $r = 320$ nm (from Swanson and Schwind 2009, reproduced with permission)

square root of the extraction voltage in a so-called Schottky plot (Fig. 8.4). According to Eq. (8.2) with the substituted effective work function, there is a linear relationship between $\log I$ and the extraction voltage V_E, if the cathode temperature is kept constant. By increasing the extraction voltage, which is experimentally accessible, I is measured directly using a detector. Figure 8.4 shows that the emission initially increases with increasing $\sqrt{V_E}$ as for thermionic emission. At higher extraction fields, above $|E| = 4 \times 10^8$ V/m, we see a stronger increase of emission, which is especially noticeable at lower cathode temperatures. This increase indicates the onset of field emission.

The amount of tunneling current in Schottky emission is measured by the dimensionless parameter q (Swanson and Bell 1973)

$$q = \frac{h(4\pi\varepsilon_o e)^{1/4}E^{3/4}}{2\pi^2 m_e^{1/2} k_B T} = 1.656 \times 10^{-4}\frac{E^{3/4}}{T} \tag{8.7}$$

with E in V/m. Here q is the fraction of current from tunneling. For example, at $q = 0.3$, 30 % current can be attributed to tunneling. For small q values ($q < 0.3$), electron emission is adequately described by the Schottky equation:

$$j_S = \frac{4\pi m_e (k_B T)^2}{h^3}\exp\left[\left(\sqrt{\frac{e^3 E}{4\pi\varepsilon_o}} - \phi_W\right)/k_B T\right]. \tag{8.8}$$

The region between $q = 0.3$–0.7 is called the extended Schottky regime, where it has been found empirically (Swanson and Schwind 2008)

$$j_{ES} = j_S\frac{\pi q}{\sin \pi q} \tag{8.9}$$

The Schottky guns employed in a TEM typically operate in the extended Schottky regime.

The effect of the electric field between the tip and the anode can be accounted for by a virtual lens, as illustrated in Fig. 8.5. A virtual source is formed behind the

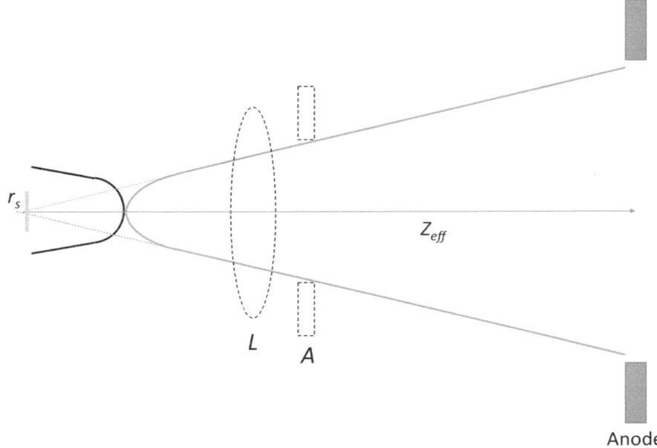

Fig. 8.5 Virtual lens L and the virtual aperture A in front a field emitting tip

tip. The angular width of the electron beam is limited by the virtual aperture. Aberrations are also introduced by the virtual lens, so that careful optical design is required in order to minimize the source size. The aberrations of planar and spherical electrodes may be calculated (Liebl 1989; Scheinfein et al. 1993).

The electron current is measured after the electrodes. The measured current is related to the current density J_s and the emitter radius by

$$I_m = J_s \left(\frac{r_s}{m}\right)^2,\tag{8.10}$$

where m is the angular magnification between the virtual source and the tip surface. Then m is approximately related to the tip radius by

$$m = 8.713 \times 10^{-5}(\beta r)^{0.42}.$$

The radius of the virtual image r_s gives the effective source in the calculations of the spatial coherence (Sect. 2.9). At the anode, the normalized coherence function is approximately described by Scheinfein et al. (1993)

$$\gamma(r) = \exp\left(-\pi^2 r_s^2 / \lambda^2 z_{eff}^2\right)\tag{8.11}$$

where z_{eff} is the distance from the virtual source to the anode. Then, the coherence width on the anode L_T, where $\gamma(L_T) = 1/e$ is

$$L_T = \frac{\lambda z_{eff}}{\pi r_s} = \frac{\lambda}{\pi \alpha_s}.\tag{8.12}$$

Both the Zr coverage and the Zr/O ratio on the emitter tip are critical in maintaining a low work function and the stability of electron emission. To keep the current drift to a minimum, the partial pressures of O_2 and H_2O must be kept $<1 \times 10^{-8}$ torr. At the high operating temperature at ~ 1800 K, the emitter tip with

Zr/O reshapes with the assistance of surface diffusion, and it becomes faceted with preferred (100) or (110) and (112) facets. The (100) facet is stabilized under a high extraction field in a process called emitter dulling. This process is driven by the balance between the surface tension and the applied field stress, which must be larger than a critical level for this process to occur. Once the (100) facet develops, the emitter dulling ceases and a stable emitter tip shape is formed. The shape is stable at a high operating temperature, with any atomic displacements caused by positive-ion bombardment annealed away. This makes the Schottky emitter less sensitive to the environment, compared to the cold-field emitter, which will be discussed next.

8.4 Cold-Field Emission Source

At still higher fields, achieved by sharply reducing the tip radius, electron tunneling through the potential barrier becomes possible at room temperature in a process called cold-field emission, as first explained by Fowler and Nordheim (1928). [See Shimoyama and Maruse (1984) for a comparison of the three modes of electron emission]. It was Albert Crew who developed the first successful scanning electron microscope equipped with a cold FEG (Crewe 1964; Crewe et al. 1968).

Field emission guns are constructed using a pointed cathode tip and at least two anodes. The tip is made from single-crystal tungsten and sharpened using electrochemical etching. It is not heated (except for cleaning and sharpening by "flashing"), and electron emission commences abruptly once the electric field at the tip exceeds a certain value E_C. Since $E_C \approx 0.5$ V per angstrom, emission occurs at voltages between a few hundred and several thousand volts, depending on tip sharpness and crystallographic orientation. [Different crystal faces have different work functions—(310) and (111) oriented W tips are often used because of their low work functions.] Since tip sharpness is so critical, ultrahigh vacuum conditions (better than 10^{-10} torr) are required to prevent positively charged ions, focused along the field lines and running toward the tip, from destroying the tip. Emission is initially very high after flashing, but it is not stable and settles to a lower and stable value after gas molecules adsorbed on the surface alter the work function.

From the uncertainty principle, we might expect field emission to commence once the uncertainty in axial position coordinate Δz for an electron at the tip surface exceeds the width of the barrier potential. If the extraction field is E, then the width of a triangular barrier is approximately $\Delta z = \phi/eE$, where ϕ is the work function in electron volts. The positional uncertainty is $\Delta z = h/4\pi\Delta p$, with Δp the momentum uncertainty and h Planck's constant. Using plane-wave states at the Fermi level, $\Delta p = (2m\phi)^{1/2}$. Thus, the onset of field emission is expected at

$$E = E_C = 4\pi(2m)^{1/2}\phi^{3/2}/he \tag{8.13}$$

This agrees approximately with the results of the more accurate calculation, first given by Fowler and Nordheim (1928) (see also Everhart 1967). They obtained an expression (the elemental Fowler–Nordheim equation, as reviewed by Forbes (1999)) for the emission current density as

$$J_{\mathrm{F}} = A\frac{a}{\phi_{\mathrm{w}}}E^2 \exp\left[-b\phi_{\mathrm{w}}^{3/2}/E\right] \tag{8.14}$$

where A is the area of emission, E is the applied electric field given approximately by Eq. (8.5) in V/m, while the ϕ is the local work function of the emitting surface given in electron volts. Both a and b are universal constants, their values are

$$a = e^3/8\pi h = 1.5414 \times 10^{-6}\,\mathrm{eV\,V}^{-2}$$

$$b = \frac{4}{3}\sqrt{2m_e}/eh = 6.8309 \times 10^{9}\,\mathrm{eV}^{-3/2}\,\mathrm{Vm}^{-1} \tag{8.15}$$

The elemental Fowler–Nordheim equation assumes a triangular potential barrier. Other assumptions made in deriving this equation are as follows: (1) metal band structure has been approximated by a free-electron energy band, (2) electrons are in thermodynamic equilibrium at zero temperature, (3) a smooth emitting surface with a uniform local work function and applied field, and (4) a classical electron–electron interaction represented by an image potential. Later treatments (see Murphy and Good 1956 and review by Forbes) used a more realistic potential barrier rounded by the image force. The generalized Fowler–Nordheim equation includes two correction terms, one for the prefactor and one for the exponential, both are functions of $\sqrt{E}/\phi_{\mathrm{w}}$.

Plots of $\ln(J_F/E^2)$ versus $1/E$ are found to be linear and known as Fowler–Nordheim plot. An example of early FEG design can be found in Crewe et al. (1968). The design considerations involved in combining magnetic lenses with a field emission gun are described in Venables and Cox (1987). Such a magnetic gun lens was incorporated into the Vacuum Generators HB501S STEM.

We now consider the ultimate quantum limits on source brightness and discuss the relations between source brightness, degeneracy, and coherence. Field emission electron sources are currently the brightest particle sources available, considerably brighter than either synchrotron/wiggler systems or the sun. The brightness β_{s} is defined as the current density J per unit solid angle Ω. Thus, the number of particles received in a time interval δt within a solid angle $\delta\Omega$, through a detecting surface δA can be written as

$$\delta n = \frac{\beta_{\mathrm{s}}}{e}\delta A \delta t \delta \Omega \tag{8.16}$$

The beam degeneracy η is defined as the mean number of particles per cell of phase space. In a flux of quantum mechanical particles with momentum p and speed v, the phase space volume element is defined by $\delta^3 x \delta^3 p$ and the number of particles within the element is given by

$$\begin{aligned}
\delta n &= \text{(number of particles/cell of phase space)}\\
&\quad \times \text{(number of occupied cells)}\\
&= \eta\left(2\delta^3 x\delta^3 p/h^3\right)\\
&= 2\eta(vt\delta A)\left(p^2\delta p\delta\Omega\right)/h^3.
\end{aligned} \tag{8.17}$$

The maximum theoretical brightness is obtained with $\eta = 1$ as (Spence et al. 1994)

$$\beta_{\mathrm{s,max}} = 2evp^2\delta p/h^3 = 2ep^2\delta E/h^3 = 2e\delta E/h\lambda^2. \tag{8.18}$$

At this limit, there are two electrons of opposite spin in a phase space cell of volume h^3. Thus, from Eq. (8.17), the degeneracy can be regarded as the ratio of actual to maximum brightness (Spence et al. 1994)

$$\eta = \frac{\beta_{\mathrm{s}}}{\beta_{\mathrm{s,max}}}. \tag{8.19}$$

The volume of a cell in phase space may be also expressed in terms of the longitudinal and transverse coherence length, L_{c} and L_{T}, according to

$$\left(p^2\delta p\delta\Omega\right)\left(L_{\mathrm{c}}L_{\mathrm{T}}^2\right) = h^3. \tag{8.20}$$

Using Eq. (8.20) and

$$L_{\mathrm{c}} = vT_{\mathrm{c}}\,(T_{\mathrm{c}}\ \text{coherence time}) \tag{8.21}$$

leads to an alternative, but equivalent, expression for the degeneracy

$$\eta = \frac{J}{e}L_{\mathrm{T}}^2 T_{\mathrm{c}}. \tag{8.22}$$

In other words, the degeneracy is the mean number of electrons per coherence time traversing a coherence area (normal to the beam). This is unity if $J = e/L_{\mathrm{T}}^2 T_{\mathrm{c}}$, so that one electron crosses the coherence area per coherence time. From the above relations, it can be shown that the experimental brightness can be determined from the wavelength, the current density J, and the coherence area at the image plane

$$\beta_{\mathrm{s}} = 2JL_{\mathrm{T}}^2/\lambda^2. \tag{8.23}$$

Since λ^2 is inversely proportional to the electron energy, the maximum brightness is proportional to the relativistically corrected accelerating voltage.

From Eq. (8.22), we see that the degeneracy, which is unity if there are two electrons of opposite spin in every cell in phase space, can be measured experimentally if we can estimate the beam current, the coherence area, and the coherence time. A rough estimate of the coherence width is given by the width of all the

Fresnel fringes seen at a sharp edge (or from the width of all fringes produced by an electron biprism). The coherence time can be obtained from Eq. (8.21) if the energy spread in the beam is known using an energy-loss spectrometer. The result for a modern field emitter will give degeneracy in the neighborhood of 1E-6 and might be compared with a Helium–Neon laser, where, for photons, any number of bosons can occupy a state, the degeneracies are much larger than unity. It is clear that there is much scope for improvement in electron sources before the quantum limit of brightness, corresponding to a degeneracy or unity, is reached.

The greatest challenge in adopting the cold-field emitter technology for electron microscopy was to improve field emission stability and its operation. This was greatly helped by advances in UHV technology, and understanding the processes of surface "build-up" and "dulling," in which the tip is heated and shaped under an intense field, causing surface tungsten atoms to migrate. The interested reader is referred to the field ion microscopy literature (Tsong 1990). For the operation of the field emitter, improvements were also made on automatic V_0/V_E ratio control for the Butler lens, noise reduction for the electronics, and the incorporation of beam monitors for user operation (Inada et al. 2009).

Achieving an acceleration voltage higher than 80 kV with the Butler lens electron gun has proven to be difficult because of discharge between the emitter and electrodes. A multistage electrode gun was developed for higher acceleration voltages. The distance between the emitter and ground electrode in the multistage electrode gun was approximately 10 times that in the Butler lens gun (Tonomura 1984).

Table 8.1 summarizes the properties of the electron sources currently in use. The entries for the zirconiated tungsten Schottky source (provided by L. Swanson) are given for 25 kV, since these sources have so far only been used in SEM instruments at lower voltages; the values at 100 kV will be slightly higher. More details of these sources, which tolerate poorer vacuum conditions than cold-field emission tips, can be found in Samoto et al. (1985), McGinn et al. (1991), and Speidel and Brauchle (1987).

References

Aeschlimann M, Hull E et al (1995) A picosecond electron-gun for surface-analysis. Rev Sci Instrum 66:1000–1009

Boersch H (1954) Experimentelle bestimmung der energieverteilung in thermisch ausgelösten elektronenstrahlen. Z Phys 139:115–146

Butler JW (1966) 6th International Congress Electron Microscopy (Kyoto):191

Crewe AV (1964) Scanning techniques for high voltage microscope. In: Proceedings of the AMU-ANL, high voltage electron microscope mtg (Argonne National Laboratory), pp 68–81

Crewe AV, Eggenberger DN et al (1968) Electron gun using a field emission source. Rev Sci Instrum 39:576–583

Elsayed-Ali HE (1995) Time-resolved reflection high-energy electron diffraction of metal surfaces. Proc SPIE 2521:92–102

Everhart TE (1967) Simplified analysis of point-cathode electron sources. J Appl Phys 38 (13):4944–4957

Forbes RG (1999) Refining the application of Fowler-Nordheim theory. Ultramicroscopy 79:11–23

Fowler RH, Nordheim L (1928) Electron emission in intense electric fields. Proc R Soc Lond A 119:173–181

Haine ME, Einstein PA (1952) Characteristics of the hot cathode electron microscope gun. Br J Appl Phys 3:40

Hawkes PW, Kasper E (1996) Principles of electron optics: applied geometrical optics. Academic Press, San Diego

Helliwell JR, Rentzepis PM (eds) (1997) Time resolved diffraction. Oxford Science Publications, Oxford

Inada H, Kakibayashi H, Isakozawa S, Hashimoto T, Yaguchi T, Nakamura K (2009) Hitachi's development of cold-field emission scanning transmission electron microscopes. In: Advances in imaging and electron physics. Elsevier, Amsterdam

King WE, Campbell GH, Frank A, Reed B, Schmerge JF, Siwick BJ, Stuart BC, Weber PM (2005) Ultrafast electron microscopy in materials science, biology, and chemistry. J Appl Phys 97:111101

Liebl H (1989) The image aberration caused by the acceleration field between concentric spherical electodes. Optik 83:129–135

McGinn JB, Swanson LW, Martin NA, Gesley MA, McCord MA, Viswanathan R, Hohn FJ, Wilson AD, Naumann R, Utlaut M (1991) 100 KV Schottky electron-gun. J Vac Sci Technol B 9(6):2925–2928

Mourou G, Williamson S (1982) Picosecond electron-diffraction. Appl Phys Lett 41:44–45

Murphy EL, Good RH (1956) Thermionic emission, field emission, and the transition region. Phys Rev 102:1464–1473

Reimer L, Kohl H (2008) Transmission electron microscopy, 4th edn. Springer, Berlin

Samoto N, Shimizu R, Hashimoto H, Tamura N, Gamo K, Namba S (1985) A stable high-brightness electron gun with Zr/W-tip for nanometer lithography. I. Emission properties in Schottky- and thermal field-emission regions. Jap J Appl Phys 24:766–771

Scheinfein MR, Qian W, Spence JCH (1993) Aberrations of emission cathodes: nanometer diameter field-emission electron sources. J Appl Phys 73(5):2057–2068

Shimoyama H, Maruse S (1984) Theoretical considerations on electron optical brightness for thermionic, field and T-F emissions. Ultramicroscopy 15:239–254

Speidel R, Brauchle P (1987) Electron-beam test system with field emission gun. Optik 77:46–54

Spence JCH, Qian W, Silverman MP (1994) Electron source brightness and degeneracy from fresnel fringes in field emission point projection microscopy. J Vac Sci Technol A 12:542–547

Swanson LW, Bell AE (1973) Recent advances in field electron microscopy of metals. Adv Electron Electron Phys 32:193–309

Swanson LW, Schwind GA (2008) Review of ZrO/W Schottky Cathode. In: Orloff J (ed) Handbook of charged particle optics, 2nd edn. CRC Press, Boca Raton

Swanson LW, Schwind GA (2009) A review of the cold-field electron cathode. In: Advances in imaging and electron physics, vol 159: cold field emission and the scanning transmission electron microscope. Academic Press, Cambridge

Tonomura A (1984) Applications of electron holography using a field emission electron microscope. J Electron Microsc 33(2):101–115

Tsong TT (1990) Atom-probe field ion microscopy. Cambridge University Press, Cambridge

Venables JA, Cox G (1987) Computer modelling of field emission gun scanning electron microscope columns. Ultramicroscopy 21:33–45

Williams DB, Carter BC (2009) Transmission electron microscopy, a textbook for materials science, 2nd edn. Springer, New York

Williamson JC, Zewail AH (1993) Ultrafast electron-diffraction—velocity mismatch and temporal resolution in crossed-beam experiments. Chem Phys Lett 209:10–16

Zewail AH (2006) 4d ultrafast electron diffraction, crystallography, and microscopy. Annu Rev Phys Chem 57:65–103

Chapter 9
Electron Detectors

High-energy electrons can be detected by a direct measurement of the electron current or by converting the electron energy into optical, electrical, or chemical signals, which then can be analyzed and visualized. A simple setup, called the Faraday cage, is used to measure the beam current. The cage is made by covering a hole drilled into a metal stub with a small diaphragm. Electrons passing through the diaphragm produce an electric current, while the small diaphragm prevents the escape of backscattered and secondary electrons generated inside the hole. The typical beam current employed in electron imaging or diffraction is of the order $I = 10^{-13}$–10^{-10} A; thus, the measurement requires a low-current-level electrometer, which is commercially available. This setup is used mostly for instrument calibration.

Unlike X-ray diffraction which was largely developed using point detectors, TEM from the beginning was based on area detectors, and more recently array detectors, for image and diffraction pattern recording. Point detectors are used exclusively in SEM and STEM. Electron detectors are constructed based on the principles of electron energy conversion and signal processing. The detectors are made to meet various electron detection needs in the format of point (or serial), area or array (1D or 2D) detectors. In what follows, we will first introduce the scintillator–photomultiplier detectors and then discuss electron area and array detectors.

9.1　Scintillator–Photomultiplier Detectors

A widely used electron point detector is the Everhart–Thornley detector (Everhart and Thornley 1960), designed to detect secondary electrons in a SEM. The energy of secondary electrons is low (<50 eV). To detect these electrons, a positively biased collection grid is placed between the sample and the scintillator, which is further biased at 10 kV. Electron detection is made by converting some of the ~10 keV electron energy into photons using a scintillator, then by collecting

© Springer Science+Business Media New York 2017

J.M. Zuo and J.C.H. Spence, *Advanced Transmission Electron Microscopy*, DOI 10.1007/978-1-4939-6607-3_9

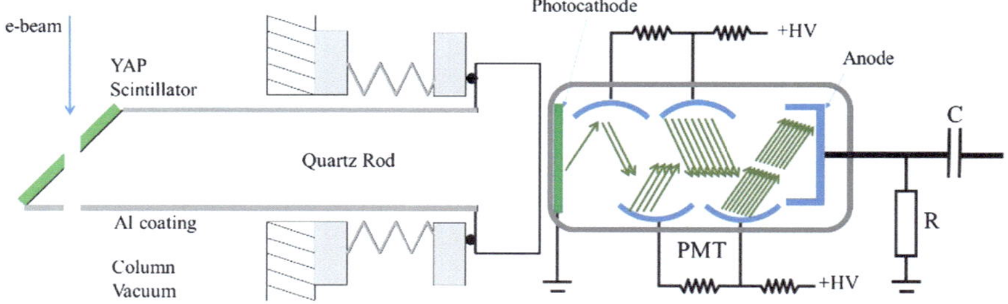

Fig. 9.1 Schematic design of an ADF detector based on the scintillator–photomultiplier combination for electron detection. The performance of the ADF detector is critical to the performance of Z-contrast STEM imaging

these photons using a light pipe and detecting them using a photomultiplier tube (PMT).

The annular dark-field (ADF) detector used for STEM uses the same detection principle as the Everhart–Thornley detector (see Fig. 9.1) (Kirkland and Thomas 1996), except that high-energy electrons can be converted directly into photons without the need of acceleration. This is done by using a scintillator placed directly inside the TEM column. In the ADF detector, a quartz or glass rod of ~ 1 in. in diameter acting as the light pipe is used to collect photons. It is coated with a thin film of Al, except at the two ends of the rod. This configuration provides the maximum light transmission efficiency almost without optical loss. Inside the column, the rod has a beveled end with a 45° angle and is used to mount the scintillator. On the other end, the rod is fabricated into a vacuum flange. With the use of an O-ring, the rod provides an efficient vacuum seal compatible with high vacuum. The PMT is placed outside the TEM column.

Scintillators are materials capable of producing a flash of light when struck by an energetic particle in a process called scintillation. Inside an electron scintillator, electrons are converted to photons by cathodoluminescence (CL). High-energy electrons impinging on the scintillator excite valence band electrons to the conduction band and defect levels (created by dopant atoms or impurity); some of these excited electrons relax to the ground state by emitting a photon of light. The amount of photons (n) given out decreases with time exponentially, and thus $n/n_o = \exp(-t/T)$, where T stands for the characteristic decay time. Only a small number of atoms in a scintillator material are capable of giving out light and acting as luminescence centers. They are created by adding a trace amount of dopants and impurities, which are selected based on their emission characteristics of wavelength, decay time, and emission efficiency.

There are a number of electron scintillators available for different applications in electron microscopy. These include cerium-doped single-crystal yttrium aluminum garnet (YAG, $Y_3Al_5O_{12}$) and perovskite (YAP, $YAlO_3$) (Autrata et al. 1978), P-46 (YAG, crystalline powders), P-47 (Y_2SiO_5: Ce), P-43 (GOS, Gd_2O_2S: Tb), CaF_2(Eu), P-20 ((Zn, Ca)S doped with silver, P22A of (Zn, Cd)S: Cu, Al or ZnS:

Table 9.1 List of electron scintillators and their characteristics

Material (dopant)	type	λ (nm)$^{\$}$	T$^{\$}$	η (%,)^a	Lifetime (h/50 nA)*
P43 (Tb)	Powder	545	1.5 ms	12–20 %&	
P46 (Ce)	Powders	550	70 ns	3	>10^4
P47 (Ce)	Powders	400	120 ns	7	10^2
P20 (Ag)	Powders	530	80 µs	20	
CaF$_2$ (Eu)	Single crystal	435	1 µs	2	10^4
ZnS(Ag)	Powders	455	80 µs	12	
YAG(Ce)	Single crystal	550	70 ns	4–5	>10^4
YAP(Ce)	Single crystal	370	25 ns	7	>10^4

Here, λ refers to peak emission wavelength, T for decay to 10 %, η for the estimated energy conversion efficiency, and lifetime for damage lifetime measured in hours under a 50-nA electron beam
[a](2011), Table 2.2, Faruqi and Subramaniam (2000), Kirkland and Thomas (1996)
$^{\$}$ Decay time from: http://www-group.slac.stanford.edu/tf/nlcta/PEP-LER/AST%20-%20Phosphors.htm

Cu, Al, Au, and ZnS scintillators. Among these listed, P20 and P20A are used for fluorescent screens; both have high-energy conversion efficiency, but their decay times are on the order of tens of microseconds. Table 9.1 summarizes the characteristics of electron scintillators in terms of peak emission wavelength, time of decay to 10 % intensity level, energy conversion efficiency, and damage lifetime measured in terms of hours for a beam current of 50 nA. The actual florescence spectrum has a distribution of wavelengths. For simplicity, we have only listed the peak emission here. Similarly, the florescence decay has a fast and slow component; we avoid this complication by using the time of decay to 10 %.

The energy conversion efficiency of a scintillator depends on its thickness and the electron energy. Thus, the values quoted in Table 9.1 represent only an estimate of the best efficiency that can be achieved. For the P20 scintillator, Fan and Ellisman (1996) performed a systematic measurement of luminescence brightness under the 80–400 kV electron beam for scintillators prepared at 12–58 µm thick. Their study shows that the maximum brightness is obtained at 12 µm thickness for 80 kV electrons and 58 µm for 250 kV electrons. At 120 kV, luminescence brightness increases as the scintillator thickness increases starting from 12 µm and reaches a maximum at thickness between 22 and 32 µm and then falls off as the thickness further increases. The maximum is observed because of two underlying processes in electron energy conversion. One is electron scattering accompanied by energy absorption, which increases with thickness up to the limit of electron penetration depth. The other is light transmission, which decreases with thickness due to light scattering and absorption within the scintillator.

Other considerations in selecting a scintillator are scintillator uniformity and radiation resistance. Single-crystal YAG has more uniform light emission and high resistance to beam radiation and thus is a favorite for electron detection in analytical electron microscopy.

Photomultiplier tubes are versatile photodetectors with outstanding detection characteristics. Photons are detected using a photocathode in PMT with a low work function. When a photon strikes the photocathode, the emitted photoelectron is then focused using a focusing electrode onto a series of dynodes biased at successively higher positive voltages. The dynodes are made of metal coated with a material of high secondary electron yield, such as BeO, MgO, and GaAsP. They serve as electron multipliers as each accelerated electron generates multiple (~ 2) secondary electrons that are accelerated again to the next dynode. In this way, a low current emitted by the photocathode can be amplified by a factor of 10–10^8 depending on the number of dynodes (up to 19 according to Hammatsu, Japan). The amplified electron current is then measured using an anode made of fine mesh electrodes and output as an analog or digital signal. Photon detection can be made very fast in a PMT. The time response is determined by the electron transit time, e.g., the time it takes for the electrons to travel from the photocathode to the anode, which is on the order of a few ns. Since electrons only travel inside the vacuum, the whole device is sealed inside an evacuated glass tube.

The photocathode inside a PMT uses alkali metals with very low work functions, such as the bialkali (Sb-Rb-Cs, Sb-K-Cs) photocathodes that respond to a broad spectrum of photons in the ultraviolet to visible range. Its conversion efficiency (photocathode sensitivity) varies with the photon wavelength. For a bialkali photomultiplier tube, the peak conversion efficiency is achieved near the 400 nm wavelength.

ADF detectors use a scintillator fabricated out of single-crystal YAP or P47 with a hole in the center and an elliptical shape, so it is circular in projection. Although the luminous efficiency of YAP or P47 (about 7 %) is less than that of P20 powder phosphor (about 20 %), they offer a much shorter decay time, which is important in STEM since the acquisition rate is about few µs. Single-crystal YAP also offers uniform emission and radiation resistance similar to YAG but at a higher luminous efficiency. The collected photons in the ADF detector are detected using a head-on PMT (an example is Burle type 8575 of 52 mm in diameter with 12 stages and a bialkali photocathode). It has a semitransparent photocathode coated on the inner surface of the entrance window. The peak emission of YAP at 370 nm matches closely to the peak efficiency of the PMT at 400 nm.

9.2 Characteristics of Point Detectors

A point electron detector like the ADF detector gives a signal when it is activated. The output signal from an ideal point detector is proportional to the detector integration time, the area of the detector, and the electron beam current density.

This linear relationship may, or may not, exist in a real detector. To quantify this, we measure a detector by its response in terms of the detector gain, dynamic range, linearity, and detective quantum efficiency (DQE) based on the following definitions:

- Average gain g, it is the ratio between the average output signal (minus the background) and the average number of incident electrons. Averaging is carried out over a number of repeated measurements under the same electron detection conditions;
- Linearity, a detector's response is said to be linear if the difference in output signals is proportional to the difference in incident current densities. Such response appears as a straight line in the signal versus electron intensity plot;
- Dynamic range (DR), it is defined by the ratio of the largest and smallest signals recorded by a detector. As the smallest signal that can be measured is often determined by the detector's readout noise (σ_R), the dynamic range can be measured by

$$DR = 20\, \log_{10}(I_{\max}/\sigma_R)(\text{DB}). \tag{9.1}$$

where DB stands for decibels. The linear dynamic range is obtained by taking $I_{\max}$ as the largest linear signal of the detector;
- Detector quantum efficiency (DQE). DQE measures the detector's noise performance by taking the squared ratio of output signal-to-noise ratio (SNR) to input SNR:

$$\text{DQE} = (\text{SNR}_{\text{out}}/\text{SNR}_{\text{in}})^2. \tag{9.2}$$

A DQE of unity is only possible if every beam electron is detected, with no other noise added to the shot noise. Noise is introduced in a real detector from three unwanted sources—Poisson-distributed dark current noise, conversion noise, and Gaussian-distributed readout noise, in addition to the irreducible quantum noise of the electron beam itself (shot noise). Because these noise sources cannot be avoided, DQE is always smaller than 1 in a real detector.

9.3   Characteristics of ADF Detectors

In an ADF detector, secondary electrons are generated and measured in a cascading chain of detection events as shown in Fig. 9.2. The number of secondary electrons generated after the multiplier per incident electron can be estimated according to

$$N_{SE} = G_{\text{PMT}} n_{pe} n_e = G_D n_e \tag{9.3}$$

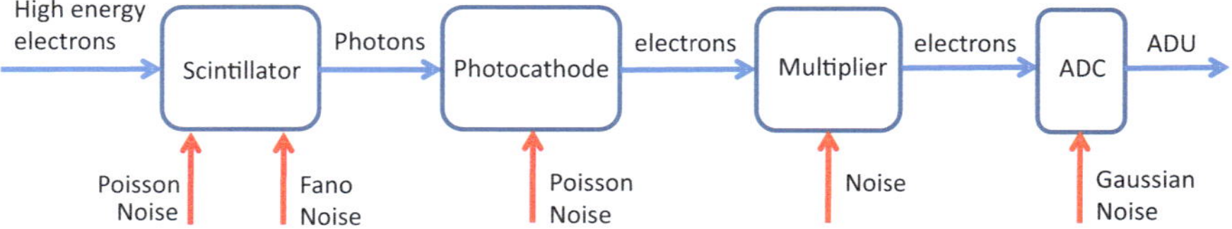

Fig. 9.2 Electron detection chain of events in a scintillator–photomultiplier detector and the noise introduced in each step

where n_e is the number of incident electrons on the detector, n_{pe} number of photoelectrons generated by the PMT photocathode per incident electron, and G_D the secondary electron gain factor. The number of photoelectrons generated per incident electrons depends on (1) E_o, the incident electron energy, (2) η_C energy conversion efficiency of the scintillator, (3) E_{ph} average energy of emitted photons, (4) ε_T scintillator photon transmission coefficient, (5) ε_{QR} quartz rod photon collection efficiency, (6) ε_S surface transmission coefficient, n number of optical surfaces, (7) G_{PMT} PMT gain factor, and (8) q_{PMT} quantum efficiency of PMT. Together, they yield

$$n_{pe} = q_{PMT} (\varepsilon_S)^n \varepsilon_{QR} \varepsilon_T \frac{\eta_C E_o}{E_{ph}} n_e = \varepsilon_A n_e \tag{9.4}$$

where ε_A stands for the overall conversion efficiency. Using values of ε_T, ε_{QR}, ε_S at 0.7, 0.5, and 0.95 and $n = 3$ (Kirkland and Thomas 1996), $\eta_C = 0.07$ (Table 9.1), for 100 kV electrons and $E_{ph} = 3.35$ eV, $q_{PMT} = 0.25$ (Burle, type 8575 PMT), we have $\varepsilon_A \approx 150$.

In the ADF detector, the PMT is operated in the so-called continuous current mode so that the secondary electron current is integrated over time T, amplified, and converted to a digital signal using an ADC (analog digital converter). In addition to photoelectrons generated by incident electrons, there is a background level of photoelectrons (n_b) from accidental electrons or particles hitting the scintillator in the absence of a direct electron beam exposure, and spontaneous emission in the scintillator and the photocathode. The output signal from the detector is obtained from

$$S = G_S n_e - S_b \tag{9.5}$$

With G_S standing for system gain and S_b for the background signal (which depends on the PMT offset current I_o as well as G_{PMT}); both can be adjusted during STEM operation. For the quantitative analysis of ADF images, we must convert the measured signal to the number of detected electrons using

$$n_e = (S + S_b)/G_S$$

Both the background signal level and the system gain depend on the electron beam energy and the PMT gain factor (see Ishikawa, Lupini et al. (2014)).

The PMT gain G_{PMT} and offset I_o are controlled by the so-called contrast and brightness values set during STEM image acquisition. The rule of thumb is that the offset decreases as brightness increases, while the gain increases exponentially as the contrast value increases. A detailed calibration for the JEOL ADF detector was reported by He and Li (2014). The image background level also changes with the contrast level because S_b is changed by the PMT gain factor. Characterization by Ishikawa, Lupini et al. (2014) in a controlled experiment without specimen shows that the background increases with increasing gain and the background also increases with the electron beam on.

The smallest signal that can be detected by the ADF detector is a single electron hitting the detector during the acquisition time T. The secondary electron current generated by a single incident electron is $G_D n_e e/T$. A PMT can detect a maximum secondary electron current, I_{max}, which is defined by the current with 2 % deviation from being linear, with a background current level I_b. For the type 8575 PMT, $I_{max} = 0.15$ A, and $I_b = 1$ to 5 nA, which is very small compared to the secondary electron current produced by a single incident electron. Taking the single electron as the smallest signal, we can estimate the dynamic range of the ADF detector using

$$DR = 20 \log_{10}(I_{max}/(G_D n_e e/T)), \tag{9.6}$$

which gives $DR = 95$ DB for $G_{PMT} = 2^{20}$, and $q_{pmt} = 0.25$, $n_{ph} = 630$. The actual dynamic range in an ADF detector is set by the lower limit of the PMT and ADC.

Noise in the detected signal comes from the shot or Poisson noise of incident electrons, and noise in the energy conversion process called Fano noise. For high-energy electrons, a disproportionate part of the energy deposition occurs in the late stage of electron scattering when the electron is slowed by successive events of inelastic scattering. This together with the randomness of inelastic scattering lead to fluctuations in energy deposition and the Fano noise (Fano 1947; Zuo 1996; Meyer and Kirkland 1998), which introduces an additional factor of F to the shot noise

$$\mathrm{var}(n_e) = n_e(1 + F),$$

with var standing for variance. Other sources come from variance in the gain factor, and variance in the background signal. Using the error propagation rule, we have

$$\mathrm{var}(S) = \mathrm{var}(G_S)n_e^2 + G_S^2 \mathrm{var}(n_e) + \mathrm{var}(S_b), \tag{9.7}$$

Combining Eq. (9.7) with Eq. (9.2) gives

$$\mathrm{DQE} = \frac{S^2 / \left[G_S^2 \mathrm{var}(n_e) + \mathrm{var}(G_S) n_e^2 + \mathrm{var}(S_b) \right]}{n_e}$$

$$= \left(1 - \frac{S_b}{G_S n_e} \right)^2 \left[1 + F + n_e \frac{\mathrm{var}(G_S)}{G_S^2} + \frac{1}{n_e} \frac{\mathrm{var}(S_b)}{G_S^2} \right]^{-1} < 1. \tag{9.8}$$

When $G_S \gg S_b$, or with a large number of incident electrons, eqn. can be simplified as

$$\mathrm{DQE} \approx \frac{1}{1 + F + n_e \frac{\mathrm{var}(G_S)}{G_S^2}} \tag{9.9}$$

Thus, the DQE of the scintillator–photomultiplier-based ADF detector, operated with a low background level, is largely limited by the Fano noise and variations in the gain. The variance in the system gain G_S is dominated by randomness in secondary electron multiplication and detector nonuniformity.

Figure 9.3 shows an example of single-electron detection using an ADF detector and measured noise levels. The STEM image was acquired without a specimen in a JEOL2200FS TEM using a JEOL ADF detector with contrast at 2400 and brightness at 4000 (maximum of 4096 for both), and a small condenser aperture. What is plotted in Fig. 9.3a is the histogram of counts recorded in the image. Four separate peaks at equal distances are observed corresponding to $n = 0$, 1, 2 and 3 electrons. These are accidental electrons that hit the ADF detector in the absence of scattering in a specimen. The measured variance scales linearly with the number of electrons. For this particular detector, the slope is measured at 0.046, corresponding to $\sqrt{\mathrm{var}(G_S)}/G_S \approx 21\%$. This large value comes from the nonuniformity of the detector. For single-electron detection, if we take an estimate of the Fano factor

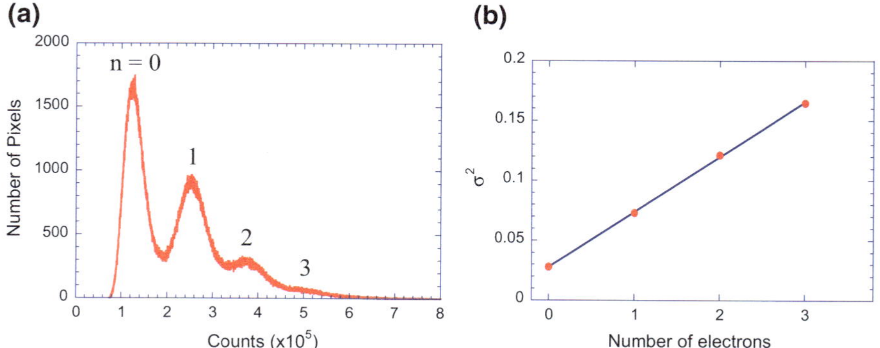

Fig. 9.3 **a** Histogram of measured signals in an ADF image acquired without the specimen. Four separate peaks are observed corresponding to the background peak ($n = 0$), single-, double-, and triple-electron peaks. **b** The gain-normalized Gaussian peak variance plotted as function of number of electrons

from the best-scenario experimental data as 0.1 (Grob et al. 2013), the DQE of the ADF detector is then estimated at 0.87 with the PMT gain set between 2^{14} and 2^{20}. The largest contributions to the reduced DQE come from the Fano factor and any nonuniformity in the detector. The DQE initially decreases with increasing number of incident electrons because variance in the gain contributes to linear noise in an ADF detector. As the number of electrons increases and the detector becomes uniformly illuminated, nonuniformity no longer contributes to noise and the DQE is expected to increase up to the limit set by Fano noise.

9.4   CCD Cameras

CCD cameras, first introduced to electron microscopy in the late 1980s (Spence and Zuo 1988), have largely overtaken film as the method of choice for electron image and diffraction pattern recording in materials science. The most suitable of these are based on cooled, slow-scan charge-coupled devices (CCD) coupled to YAG single-crystal scintillator (Autrata et al. 1978) screens as shown in Fig. 9.4a. A typical CCD camera might contain 1024×1024 or 2048×2048 square pixels, each 14 μm on edge, with 14- or 16-bit readout capacity. Electron images are recorded in these cameras in three stages: (1) converting electrons into photons in an electron scintillator, (2) transporting photons to the CCD array via fiber optical or lens couplings, and (3) converting photons to well electrons and reading out well electrons in the CCD. A CCD can be also used to detect electrons directly, however, because damage to the CCD by direct electron exposure, direct detection using a CCD has a very short lifetime. This damage is avoided by converting the electron energy to photons using a radiation resistant scintillator. The downside of this detection process is the reduced spatial resolution (beam broadening) due to electron and photon propagation within the scintillator and photon propagation in the optical coupling.

CCD cameras combine good electron detection characteristics (details are discussed in the next section) with the capacity for immediate real-time display of electron images. Online image processing, including the characterization of the transfer function of the microscope and automated electron-optical alignment, has made the CCD system the most popular device for electron imaging of inorganic samples, whereas the Image Plate to be discussed later has advantages for electron diffraction (more pixels, better linearity and higher dynamic range) and for the recording of diffuse scattering in diffraction patterns (less blooming near Bragg peaks).

A CCD as illustrated in Fig. 9.4b consists of an array of isolated metal-oxide-semiconductor (MOS) capacitors (Janesick et al. 1987). By biasing the gate of a MOS capacitor, a potential well is formed underneath. An incident photon excites an electron–hole pair and the electron is collected in the potential well. Only a limited number of electrons can be hold inside the potential well of a CCD pixel. The maximum number is called the well depth. When this is exceeded, electrons

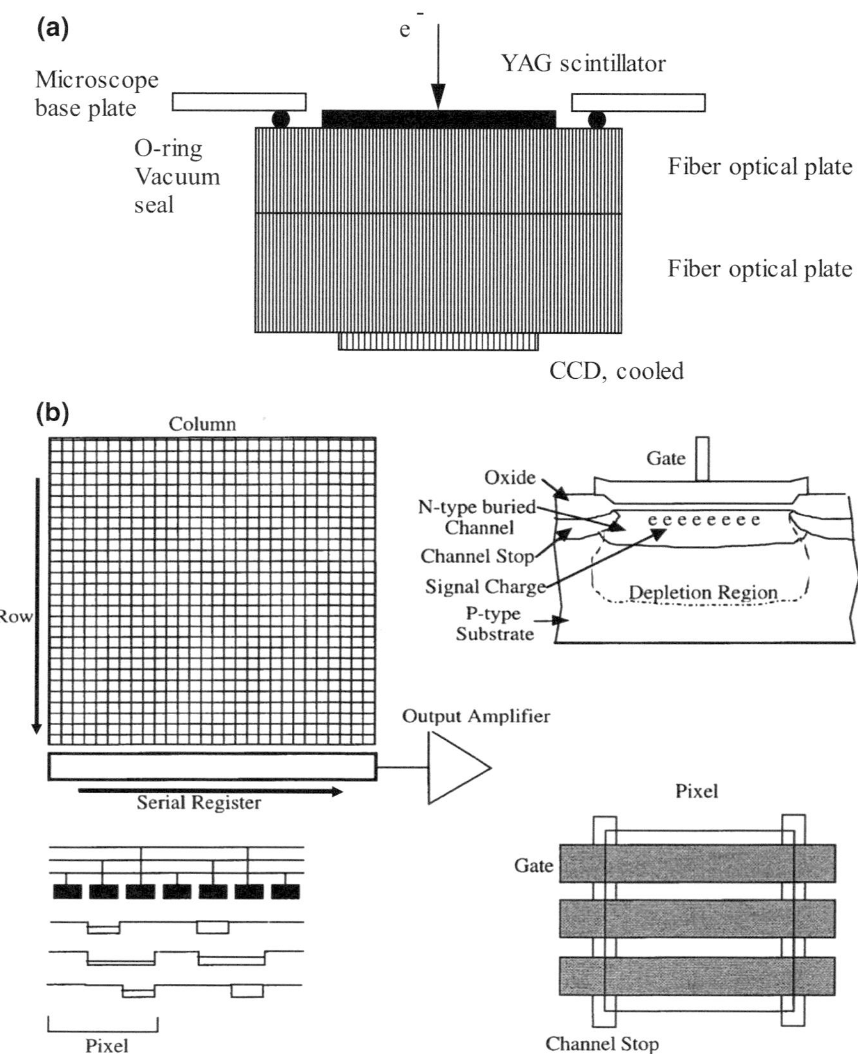

Fig. 9.4 a Schematic diagram of a CCD camera using a YAG scintillator and fiber-optics coupling. **b** Clockwise, a CCD array with a serial register, each square represents a CCD pixel. Within a pixel, each gate is a MOS capacitor and the figure shows a potential well-being formed in the depleted region after biasing. Each pixel has three gates. The *bottom left* figure shows how the bias of each gate is changed for charge accumulation and transfer. From Zuo (2000)

spill into neighboring wells and create the so-called blooming effect. CCD uses the three-phase structure with three gates per pixel. By controlling the bias of each gate, charges are accumulated and transferred. A channel stop made of heavily *p*-doped polycrystalline silicon is used to separate neighboring pixels in a row. In a full-frame (slow scan) CCD, the entire image area is active and the charge is transferred row by row to a serial register placed next to the CCD array, then transferred pixel by pixel and measured by an A/D converter along the directions indicated by arrows in Fig. 9.4b. In this way, the well electrons are measured and converted into analog to digital units (ADU), which are output as a number by the CCD controller. The readout speed of a full-frame CCD can reach ~ 10 frames per second. The CCD is usually cooled by liquid nitrogen or an electronic Peltier cooler to reduce the dark current. Dark current comes from thermally generated electrons in the CCD. The dark current can be subtracted, but the noise associated with it remains, which can be taken as the square root of the dark current.

Faster frame transfer rates up to video rate can be obtained using a interline CCD, which has its own electronic shutter (operation of a full-frame CCD requires an external shutter). Only part (~ 25 %) of a pixel in an interline CCD is active for recording light, while the rest is masked to prevent light from striking the CCD. The active and masked CCD pixels form alternative columns for imaging and charge storage. When taking an image, charge from the CCD is cleared first with a trigger electronic pulse. Another pulse is sent to the CCD at the end of the exposure, and this causes the charge in the active columns to be shifted simultaneously into the adjacent storage column, which takes less than 1 ms. Image readout is performed on the storage columns in the same way as a full-frame CCD. Since the storage columns are masked off, readout can be performed at a slower rate while the imaging columns continue to accumulate charge for the next image. Alternatively, the imaging columns can be rendered effectively light-insensitive by discarding the accumulated charge using the electronic shutter. By draining excess well electrons, images with high brightness or electron diffraction patterns with a large dynamic range can be recorded using an interline CCD without blooming. Interline CCDs also allow video-rate readout and recording of electron diffraction patterns; this is very useful for in situ electron diffraction experiments. However, they are not suited for electron imaging of radiation-sensitive materials because of their poor sensitivity and DQE due to the fact that approximately 75 % of the CCD surface is masked off and inactive for photodetection.

For a 2D detector, resolution is an important consideration. The resolution limit of a scintillator is defined as the full width at half maximum height of the distribution of scattered electrons in the scintillator. Resolution improves with a thinner scintillator because there is less electron and light scattering with a smaller thickness. However, a decrease in electron scattering also results in a loss of light due to the reduced electron energy absorption. The optimal scintillator thickness is therefore a compromise between these two demands.

The as-recorded CCD signal contains many artifacts, including readout noise, dark current and nonuniform gain variations across the image field. Possible imperfections in the fiber-coupled systems include Newton interference fringes at fiber-optics interfaces, "chicken-wire" patterns in the fiber optics due to sub-bundle formation, and Moire patterns. Most image processing software will therefore treat CCD images initially as follows. A dark current image I_{Dark} is recorded with the shutter closed for the same time as the exposure of interest I_{Raw}. A flat-field or gain image I_{Gain} is also recorded with the shutter open using uniform illumination (no sample), and an exposure time which gives about the same average count as that expected in the final image. This image records variations in gain across the image field, due, for example, to "chicken-wire" patterns in the fiber optics, or spatial variations in the doping levels in the YAG. A so-called gain-normalized and dark current-subtracted image

$$I = C \frac{I_{\text{Raw}} - I_{\text{Dark}}}{I_{\text{Gain}} - I_{\text{Dark}}} \tag{9.10}$$

is then formed, where C equals the average intensity of $I_{\text{Gain}} - I_{\text{Dark}}$. This corrected image is then used for subsequent analysis, such as deconvolution and comparison with image simulations.

9.5   Detector Characteristics of CCD Cameras

A CCD camera can be considered as an array of point detectors, and thus, its response is characterized by the same characteristics of a point detector (Sect. 9.2) plus the characteristics specific to imaging, namely the camera's resolution and uniformity. Characterization of CCD cameras involves measurements of the gain g, resolution (modulation transfer function or MTF) and detective quantum efficiency (DQE) of the camera. Experimental measurements have shown that the signal recorded by a CCD camera is linear in electron intensity up to near its saturation level (e.g., see Fig. 9.5). Uniformity is achieved through gain normalization Eq. (9.10).

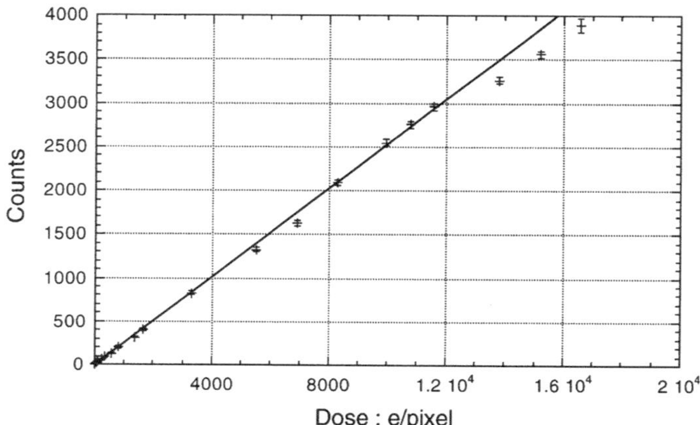

Fig. 9.5 The measured CCD output (in counts) versus electron dose for an earlier model of Gatan slow-scan CCD camera (model 679) with an anti-reflection YAG. The CCD of this camera uses a 12-bit A/D converter and the maximum count is 4096. From Zuo (2000)

The camera gain, g, is defined as the ratio of the average number of CCD output I units (ADU, analog to digital unit) to the electron dose per pixel N_e.

$$g = \frac{\bar{I}}{N_e} \qquad (9.11)$$

Its measurement therefore requires an absolute measurement of beam current, using, for example, a Faraday cup or the known response of an Image plate, which must be converted to the number of beam electrons per pixel, to obtain N_e. $\bar{I}$ is given as a digital number by the output of the software used to control the CCD camera. Gain can be optimized by the choice of scintillator and scintillator thickness at the expense of spatial resolution. A high gain value may be obtained if fine-grained phosphor such as P43 is used instead of YAG. However, a high gain is not necessarily advantageous, since it results in a loss of the dynamic range. The gain for CCD cameras is typically set at several ADU per beam electron for a 14-bit ADC.

The resolution of the slow-scan CCD camera is described by its point-spread function (PSF), defined as the image of an object which is much smaller than one detector pixel. Unlike film, the YAG/CCD combination spreads the light generated by a single-beam electron over several CCD pixels, so that the PSF has an extended tail, and this effect must be removed by image processing. (This can be done because this pixel "cross talk" affects mainly low, rather than high, spatial frequencies.) The MTF (modulation transfer function) is defined as the modulus of the Fourier transform of the PSF and would be constant (equal to c below) for all spatial frequencies for an ideal delta function PSF. The MTF function has been modeled for an early Gatan CCD camera by the function (Zuo 1996):

$$M(\omega) = \frac{a}{1 + \alpha\omega^2} + \frac{b}{1 + \beta\omega^2} + c \qquad (9.12)$$

where ω is spatial frequency in units of 1/pixel (with a maximum value of 0.5/pixel), and the remaining quantities are fitting constants (e.g., $a = 0.40$, $b = 0.46$, $c = 0.15$, $\alpha = 3328$, and $\beta = 13.51$). The first two terms model the head and tail (due to reflected photons in the YAG and electron backscattering) of the PSF, while c represents an ideal response.

There are three main methods for measuring MTF; each has its advantages and disadvantages. The so-called noise method uses the noise power spectrum obtained from a uniformly illuminated image to measure MTF. Since there is no other setup involved other than having a uniform illumination, this method is simple and can be applied quickly. The image must be recorded at medium and high dose in order to minimize the contributions from readout noise and dark current in the noise power spectrum (Zuo 1996). Indeed, because of this, the noise method tends to overestimate the MTF at high spatial frequencies. Figure 9.6 shows as an example the measured MTF of a CCD with a YAG single-crystal scintillator for three different incident electron voltages using the noise method. The other method for measuring

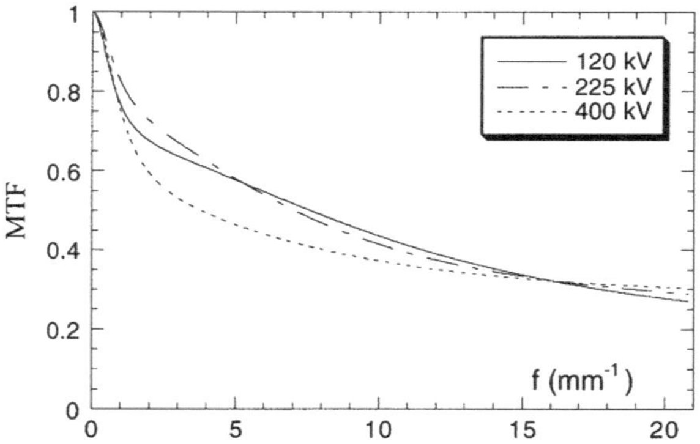

Fig. 9.6 Experimentally measured MTF at 120, 225, and 400 kV for an earlier model of Gatan SSC cameras (model 679) with a YAG scintillator. From Zuo (2000)

the MTF is to use a slanted edge. By taking derivatives of the recorded edge image, the point-spread function (inverse Fourier transform of MTF) can be obtained directly (Tatian 1965; Samei et al. 1998). To avoid edge scattering as well as X-rays generated by incident electrons, the edge should be placed as close as possible to the detector. Secondly, the edge must be sharp with a roughness much smaller than the CCD pixel size in the recorded image. In practice, a pointer above the observation screen is often used to record the edge image. Because of the edge roughness and edge scattering (both are magnified by having the edge far above the detector), this method underestimate the MTF at high frequencies. The difficulties of the noise and edge methods can be avoided by using holographic fringes. By setting up a cosine-modulated image, this method can measure MTF at low and medium frequencies. A comparison between the three methods can be found in McLeod and Malac (2013).

Having obtained a measurement of the MTF, this must then be used to remove the spreading effect of the PSF from HREM images recorded using a CCD camera. A number of methods have been evaluated, including the simple Fourier deconvolution routine. For the particular type of noise present in CCD images, best results have been obtained using the Richardson–Lucy maximum-likelihood algorithm, as used for the analysis of the Hubble telescope images. Details are given in Zuo (2000). An alternative approach to deconvolution, with its attendant noise amplification problems, is to convolute simulated images with the measured PSF before comparing them with experimental images.

The DQE of a CCD camera is defined as either the averaged DQE of individual pixels or the collective performance of all pixels for a spatial frequency recorded in the image. In electron diffraction, the measurement of noise in the intensity of a diffracted beam requires the DQE of individual pixels. This is not the case for imaging, where the collective performance of all pixels is more important.

The DQE of a single pixel is defined by the square of the ratio of its output signal-to-noise ratio to input signal-to-noise ratio according to Eq. (9.2). However, because of the cross talk among neighboring pixels, the variance of output signal is dependent on contributions from electrons impinging on neighboring pixels. Let us consider the case of an image recorded with a uniform illumination of N_e average number of electrons per pixel. The variance recorded in a pixel is as follows:

$$\text{var}(n_e) = \sum_{l,k} h_{l,k}^2 \text{var}(N_e) = N_e \sum_{l,k} h_{l,k}^2 = mN_e, \tag{9.13}$$

which contains a mixing factor $m = \sum_{l,k} h_{l,k}^2$, where $h_{l,m}$ stands for the camera's point-spread function. The value of $h_{l,m}$ specifies the contribution from the neighboring pixel at pixel distances of (l,m). By definition, $\sum_{l,m} h_{l,m} = 1$. Combining Eq. (9.13) and Eq. (9.2), the DQE of a single pixel can be shown to be given by the following:

$$\text{DQE}(n_e) = \left(\frac{\text{SNR}_{\text{out}}}{\text{SNR}_{\text{in}}}\right)^2 = \frac{\bar{I}/\text{var}(I)}{N_e/\text{var}(n_e)} = \frac{mg\bar{I}}{\text{var}(I)} \tag{9.14}$$

where m can be obtained from the area under the MTF, g is the gain, $\bar{I}$ is the average number of CCD counts per pixel, and var(I) is the variance in that number. These quantities are obtained by recording a uniformly illuminated field at various illumination conditions. By applying the rules for the manipulation of variance to the processes of scintillation and subsequent electron–hole pair production in the CCD, an expression can be obtained for the DQE as a function of dose, including contributions from the linear noise (Δ), background noise (B), and noise in the electron–photon–electron/hole conversion process (C) (Zuo (2000)):

$$\text{DQE}(n_e) = \frac{mg\bar{I}}{\text{var}(I)} = \left[1 + C + n_e\Delta + \frac{B}{n_e}\right]^{-1} \tag{9.15}$$

Figure 9.7 shows the measured DQE for a YAG/CCD camera at several electron accelerating voltages, which are compared with the DQE model without the linear noise term.

The frequency-dependent DQE is an important quantity in electron imaging or holography. Ideally, autocorrelation as measured by the power spectrum (amplitude square of Fourier transform) of the recorded image should be the same as that of the electron intensity. Because of MTF and detector noise, both the magnitude and noise of the Fourier spectrum are changed. A more appropriate measure can be obtained from the frequency-dependent DQE

$$\text{DQE}(u, v) = \left(\frac{\text{SNR}_{\text{out}}(u, v)}{\text{SNR}_{\text{in}}(u, v)}\right)^2 \tag{9.16}$$

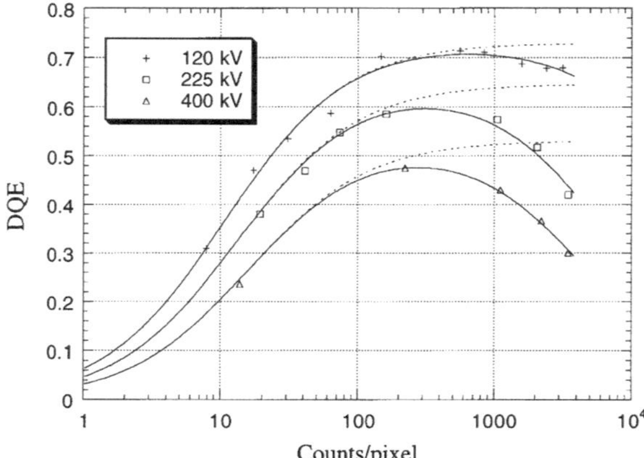

Fig. 9.7 DQE measured at 120, 225 and 400 kV as function of counts per pixel for the Gatan SSC camera of model 679 with a YAG scintillator. The *dashed line* is the projected DQE without linear noise. From Zuo (2000)

where u and v stand for the spatial frequency in the x and y directions. $\mathrm{DQE}(u, v)$ can be measured directly using the holographic fringe method. Take the incident beam intensity obtained from a biprism as follows:

$$I_{\mathrm{in}}(x) = n_e[1 + 2\varepsilon \cos(2\pi u x)]$$

We have

$$\mathrm{SNR}_{\mathrm{in}}(u, 0) = \sqrt{n_e}\varepsilon \text{ for } u \neq 0; \tag{9.17}$$

and

$$\mathrm{SNR}_{\mathrm{out}}(u, 0) = \frac{n_e \varepsilon g(u, 0) \otimes MTF(u, 0)}{\sqrt{\mathrm{var}[I_{\mathrm{out}}(u, 0)]}} \approx \frac{n_e \varepsilon MTF(u, 0)}{\sqrt{\mathrm{var}[I_{\mathrm{out}}(u, 0)]}} \tag{9.18}$$

The approximation is made for a flat gain function with $g(u, 0) \approx g\delta(u, 0)$. By combining Eqs. (9.16), (9.17), and (9.18) together and extending them to 2D coordinates, we obtain

$$\mathrm{DQE}(u, v) \approx \frac{g n_e \varepsilon MTF^2(u, v)}{\mathrm{var}[I_{\mathrm{out}}(u, v)]} \tag{9.19}$$

The measurement of $\mathrm{DQE}(u, v)$ using holographic fringes is time-consuming and can be performed only in certain types of TEMs (e.g., most TEMs used for cryo-electron microscopy are not equipped for electron holography). Alternatively, a uniformly illuminated image can be used for the analysis. Since the shot noise behaves almost as white noise, its power spectrum is flat for all frequencies. Thus, it has been found convenient to measure $\mathrm{DQE}(u, v)$ using Eq. (9.19) by measuring the MTF using the edge method and $\mathrm{var}[I_{\mathrm{out}}(u, v)]$ from noise power spectrum and taking $\varepsilon = 1$ (Mooney 2007).

9.6 Direct Detection Cameras

A recent development is the use of back-thinned CMOS monolithic active pixel sensors (MAPS) for direct electron detection (Turchetta 2003; Evans et al. 2004; Faruqi et al. 2005; Milazzo et al. 2005; Deptuch et al. 2007; Battaglia et al. 2009). These detectors which are in essence "optical static random-access memory" offer small pixel size and fast readout. This technology was first explored for applications in charged particle tracking (Prydderch et al. 2003). The same features that enhance particle tracking can be used to improve TEM detector resolution. CMOS detectors are available in large format (≥ 4 k $\times$ 4 k). The CMOS construction, based on a thin active region, can be back-thinned and radiation-harden for direct electron detection. Initial application of back-thinned CMOS pixel sensors for cryo-electron microscopy demonstrated significantly improved spatial resolution in the form of high-frequency information transfer, single-electron detection, and improved DQE over conventional CCD cameras at higher incident electron energies important for biological applications (~ 300 kV) (Milazzo et al. 2011). For in situ electron microscopy, current-generation CMOS active sensors for direct electron detection offer readout rates of about 400 full frames per second (which is more than ten times the readout speed of CCD cameras technology). The fast readout and single-electron detection sensitivity makes these new CMOS active sensors attractive for in situ electron microscopy.

The way these solid-state detectors operate can be compared to a rolling shutter, which continuously adds a few hundred low-dose images per second into computer memory. What is recorded is the charge due to electrons or holes generated as the beam electron traverses a thin, biased, silicon membrane. This detection process eliminates photons as the intermediate products, and thus, the resolution loss and pixel cross talk due to light scattering in the scintillator and fiber-optic interfaces in the scintillator-fiber-optic CCD-based cameras, as well as the additional noise coming from the photon conversion processes. However, radiation damage and consequent limitations on the lifetime of the device becomes a concern, depending on the cost of the active element. Radiation hard devices have therefore been developed, and tested in TEMs. Two main types have been developed, the Medipix device and the monolithic active pixel device (MAPS) based on CMOS technology (Faruqi and McMullan 2011; Contarato et al. 2007). For the Medipix chip, each pixel has its own readout electronics (amplifier, discriminator, and counter) capable of accepting count rates of up to 1 MHz. Thus, the pixels in the Medipix count independently, whereas the MAPS data must be processed frame by frame. This detector shows noiseless readout, high dynamic range, and a DQE at zero spatial frequency and low dose rates approaching unity for beam energies below about 120 kV, but falling off beyond that energy (McMullan et al. 2007).

Figure 9.8 illustrates a MAPS pixel for electron detection formed from reverse bias diodes (labeled n+). Inside each pixel are two control diodes forming an n-MOS transistor and a sensor diode acting as a capacitor. The diodes are formed by introducing an n+ diffusion region in the p epitaxial layer. The capacitor

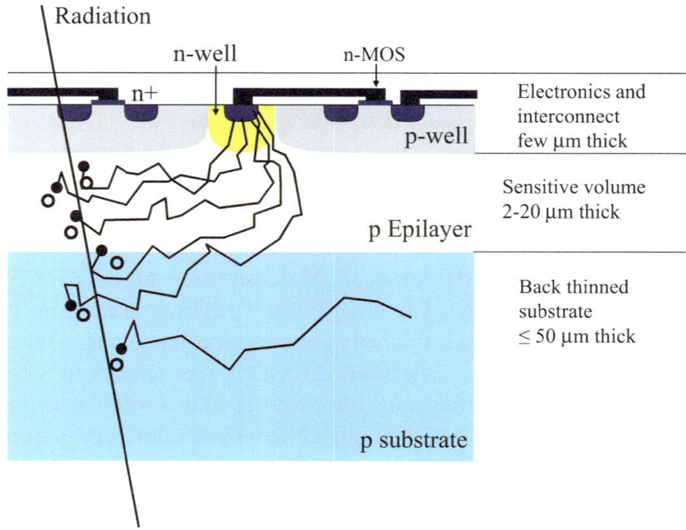

Fig. 9.8 Cross section of a MAPS pixel for direct electron detection. Electron in an electron–hole pair is attracted to the positively biased pool of electrons (called n-well). Most collection occurs in the epilayer, whose use assures 100 % fill-factor. The collected electrons are converted to the voltage and amplified within the pixel itself

voltages and readout are controlled by the n-MOS transistor (shown with wires attached). Underneath the n-MOS and sensor diodes, p-wells, beside the n-well diode, are formed by heavy doping; they act by confining the motion of electrons. According to Milazzo et al. (2011), an incident electron of 200–300 keV will generate about 1000 electrons in the epitaxial silicon layer, which is much higher than the background level (~ 50 electrons). These generated electrons are collected by one or more sensor diodes, as they diffuse randomly. Only a limited number of electrons can be held in the sensor diode, the limited dynamic range of a single recording with these detectors is compensated for by rapid "continuous" readout and accumulation of frames in a computer. For each frame, the voltages of the sensor array are read out, digitized, and stored using the electronics built on the top of the pixels.

The rapid readout of CMOS also allows for drift correction by subframe averaging (sample movement during the exposure) within the limits of additional noise due to multiple readouts. Dynamic range increases as separately readout images are summed. If we consider a sum of M exposures, with a small fixed background count N_b per exposure per pixel, and N_s signal counts per exposure per pixel, the signal-to-noise ratio S/N, assuming Poisson fluctuations (shot noise) on the beam electron arrivals, is as follows:

$$S/N \sim MN_s/\sqrt{MN_s + MN_b} \sim \sqrt{MN_s}(1 - N_b/2N_s).$$

Hence, S/N increases as the square root of the number of exposures, a law of diminishing returns. Taking $S = 5$, and given N_b, and N_s, this then gives the number of frames M needed for statistical significance of the accumulated image. We see that if readout and electronic noise become large, a limit is set to the maximum dynamic range possible, hence the importance of getting low noise per exposure. These detectors are the detectors of choice for the study of radiation-sensitive samples in materials science and cryo-electron microscopy.

9.7   Film and Image Plates

Film and, more recently, image plates are two area detectors that have been used for electron image recording. Both use a continuous film instead of a discrete array of detectors like CCD to capture and store electron energy. A digitizer is used to transfer the recorded image into a digital one.

Film for electron microscopes consists of a photographic emulsion layer on a transparent thick (7 mil or 0.18 mm) plastic support base. Photographic emulsions suitable for electron microscopy are between 10 and 30 μm thick. The emulsion is made of gelatin with suspended fine particles of silver halides. The average size of particles is about 1 μm depending on the film. The proportion of gelatin and silver halides is about 1:1. When the emulsion is exposed to high-energy electrons, the energy transferred from the beam electron converts the silver halide to silver and forms a latent image. After the exposure, film is developed in a chemical solution which converts crystals with latent image much more rapidly to metallic silver. The undeveloped silver halides are dissolved in the chemical solution of fixer. The energy required for the conversion of silver halide to silver is about 7 eV; thus, a single high-energy electron is capable of rendering many particles of silver halide crystal developable. This makes film exposure to high-energy electron almost a single-hit process compared to response of film to light that requires the cooperative action of many photons to render the micron-sized grains of film developable.

Quantitative measurement of film is done by measuring the optical density of film using a film scanner. The optical density is defined as $D = \log_{10}(1/T)$, and this is related to the original electron dose E by

$$D = D_s(1 - \exp(-\alpha E)) + D_0 \approx \beta E + D_0 \tag{9.20}$$

where D_s is the saturation density (= 7.8), D_0 (= 0.1) the background (fog), and $\beta = \alpha D_s$ is the speed of the film (0.41). (Figures in parenthesis are for Kodak SO-163. Ilford Ortho Plus film is also used). Speed is partly determined by the development conditions—faster film has larger grain size. Thus, film is linear up to an optical density of about $D_s/4$ and is used in cryo-em around an optical density of about unity, at a magnification about 50,000 and incident electron dose of perhaps 20 electrons per square Angstrom. The dynamic range may be estimated as 30 dB for 6 μm pixels, or 45 dB for the 24 μm pixels used by CCDs Zuo (2000).

Film has much smaller pixels than either CCD or IP, resulting in a total number of pixels several times greater than even the IP. The DQE of film is limited by granular noise (Zeitler 1992) and rises from 0.4 to about 0.75 at a density of about unity (Kodak SO-163) with a gradual falloff at higher density. It was preferred until recently for single-particle cryo-EM applications because these require the largest possible number of pixels but modest dynamic range for the weak-contrast bright-field images. As a result, CCD detectors were less successful in this field.

Image plates (IP) are flexible, reusable electronic film sheets sized to fit into the regular film cassettes of an electron microscope. They consist of a layer of photostimulable phosphors of BaFX (X = Cl, Br) doped with Eu^{2+} and an elastic protective layer above and the supporting material below. For the Fuji 25 μm IP, the phosphor layer is about 110 μm thick (Mori et al. 1990). The mechanism of photostimulable phosphors is described by Takahashi et al. (1984) for X-ray and UV excitations. The process is similar for electrons. Briefly, by electron excitation, Eu^{2+} is ionized to Eu^{3+}. Some of the electrons excited to the conduction band are trapped at F^+-centers of the crystal, which are about 2 eV below the conduction band. In this way, part of the incoming electron energy is stored. The trapped electrons return to Eu^{3+} ions and convert them into excited Eu^{2+} upon illumination by visible light during readout, this process is called photostimulation. Luminescence is emitted when the excited Eu^{2+} returns to its ground state, and the wavelength of luminescence is 390 nm. At 120 keV, each beam electron creates about 2000 charged F-centers, which during later readout, generates about 650 detected photons.

Unlike film, IPs are highly linear, reusable and may be exposed to room light before electron exposure. Their pixel size (about 19 μm) is determined by the laser-scanning readout device, such as the Fuji FDL5000. The plates are linear up to a saturation of about 10^5 beam e^-/pixel, with a gain of about 0.85 at 120 keV (Zuo et al. 1996).

The PSF of IPs is determined both by the plates and spot size of the laser reader and changes little at beam energies up to 400 keV. Unlike CCD cameras, noise is dominated by quantum noise in the beam, and the PSF lacks the undesirable tail due to light scattering in the YAG. It is therefore ideal for the measurement of weak diffuse scattering beside a strong Bragg peak in diffraction patterns. The reduced "blooming" of the IP compared to the CCD means that HREM images recorded on IPs have a granular appearance, due to fluctuations in ionization along different electron paths and the large pixel size compared to film. The DQE of the IP system is seen to be much higher than that of a CCD camera at low dose. CBED patterns and diffuse scattering recorded on IPs lack the blooming effect of CCDs, showing greatly improved fine detail, resulting in a better match with calculations (Zuo 2000). These properties mean that IP is also well suited for image recording with radiation-sensitive materials or where large numbers of pixels are needed (e.g., electron holography).

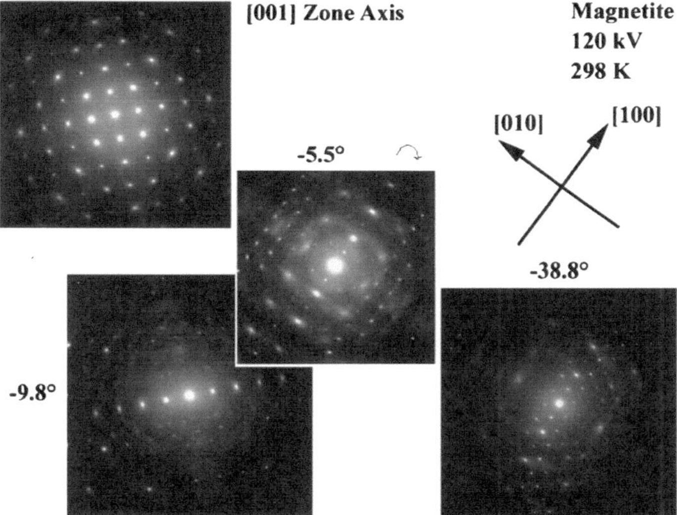

Fig. 9.9 As-recorded electron diffraction patterns from magnetite (Fe3O4) at 298 K for four different orientations using a 15-eV energy-filtering window and imaging plates

Since the image plates replace the film in a (modified) microscope film camera, the image plate is compatible with the CCD and ELS systems and might be best used for the study of weak diffuse scattering and HOLZ features in conjunction with an imaging filter. An example of the use of the image plate in electron diffraction is given by Shindo et al. (1990), who studied the dynamical diffuse electron scattering from Cu3Pd alloys as a function of temperature. Figure 9.9 shows another example of diffuse scattering recorded using image plates (Zuo et al. 2000).

A comparison of the strengths and weaknesses of film, CCD and Image Plate (IP) detectors can be found in Zuo (2000), where the gain, linearity, resolution, dynamic range, and detective quantum efficiency (DQE) of these systems are compared. In summary, the IP system offers more pixels, similar or better dynamic range, excellent linearity, and better DQE at low dose than the CCD camera but lacks the capacity for online image analysis. IPs also avoid the undesirable "blooming" feature associated with the CCD, which must be removed by image deconvolution. The CCD and IP have similar gain. A major advantage of film is the resolution. With high resolution and a large area, film gives the largest field of view, about 4 times of that of the IP. For cryo-electron microscopy, an image detector is required with a very large number of pixels (to ensure a sufficient number of pixels per particle in a many-particle image) with modest dynamic range (since the images have low contrast). Until the advent of direct electron detectors, film was best suited to these requirements and was also an ideal archival storage medium, offering protection against the loss of digital information, the cost of powered digital storage, and changes in digital media and formats.

A review of the use of film in electron microscopy can be found in Valentine (Valentine (1966)) and Spence (Spence 2013). The detective quantum efficiency (DQE) of film is limited to about 0.7 by background fogging at low exposures, by shot noise statistics at medium exposures, and by nonlinearities and saturation at higher exposures. The number of gray levels obtainable with film is less than 256 (8 bits).

References

Autrata R, Schauer P, Kuapil J (1978) A single crystal of YAG-new fast scintillator in SEM. J Phys E: Sci Instrum 11(7):707–708

Battaglia M, Contarato D et al (2009) A rad-hard CMOS active pixel sensor for electron microscopy. Nucl Instrum Meth A 598:642–649

Contarato D, Bussat JM, Denes P, Greiner L, Kim T, Stezelberger T, Wieman H, Battaglia M, Hooberman B, Tompkins L (2007) CMOS monolithic pixel sensors research and development at LBNL. Pramana 69(6):963–967

Deptuch G, Besson A et al (2007) Direct electron imaging in electron microscopy with monolithic active pixel sensors. Ultramicroscopy 107:674–684

Egerton RF (2011) Electron energy-loss spectroscopy in the electron microscope, 2nd edn. Springer, New York

Evans DA, Allport PP, Casse G, Faruqi AR, Gallop B, Henderson R, Prydderch M, Turchetta R, Tyndel M, Velthuis J, Villani G, Waltham N (2004) CMOS active pixel sensors for ionising radiation. Paper presented at 4th international workshop on radiation imaging detectors, University of Glasgow, Glasgow

Everhart TE, Thornley RFM (1960) Wide-band detector for micro-microampere low-energy electron currents. J Sci Instrum 37:246

Fan GY, Ellisman MH (1996) Optimization of thin-foil based phosphor screens for CCD imaging in tem in the voltage range of 80–400 kV. Ultramicroscopy 66:11–19

Fano U (1947) Ionization yield of radiations. II. The fluctuations of the number of ions. Phys Rev 72:26–29

Faruqi AR, McMullan G (2011). Electronic detectors for electron microscopy. Q Rev Biophys 44(3):357–390

Faruqi AR, Subramaniam S (2000) CCD detectors in high-resolution biological electron microscopy. Q Rev Biophys 33:1–27

Faruqi AR, Henderson R et al (2005) Direct single electron detection with a CMOS detector for electron microscopy. Nucl Instrum Meth A 546:170–175

Grob P, Bean D, Typke D, Li X, Nogales E, Glaeser RM (2013) Ranking TEM cameras by their response to electron shot noise. Ultramicroscopy 133:1–7

He DS, Li ZY (2014) A practical approach to quantify the ADF detector in STEM. EMAG Conf 2013:522

Ishikawa R, Lupini AR, Findlay SD, Pennycook SJ (2014) Quantitative annular dark field electron microscopy using single electron signals. Microsc Microanal 20:99–110

Janesick JR, Elliott T, Collins S, Blouke MM, Freeman J (1987) Scientific charge-coupled-devices. Opt Eng 26:692–714

Kirkland EJ, Thomas MG (1996) A high efficiency annular dark field detector for STEM. Ultramicroscopy 62:79–88

McLeod RA, Malac M (2013) Characterization of detector modulation-transfer function with noise, edge, and holographic methods. Ultramicroscopy 129:42–52

McMullan G, Cattermole DM, Chen S, Henderson R, Llopart X, Summerfield C, Tlustos L, Faruqi AR (2007) Electron imaging with Medipix2 hybrid pixel detector. Ultramicroscopy 107 (4–5):401–413

Meyer RR, Kirkland A (1998) The effects of electron and photon scattering on signal and noise transfer properties of scintillators in CCD cameras used for electron detection. Ultramicroscopy 75:23–33

Milazzo AC, Leblanc P et al (2005) Active pixel sensor array as a detector for electron microscopy. Ultramicroscopy 104:152–159

Milazzo AC, Cheng AC et al (2011) Initial evaluation of a direct detection device detector for single particle cryo-electron microscopy. J Struct Biol 176:404–408

Mooney P (2007) Optimization of image collection for cellular electron microscopy. In: Methods in cell biology. Academic Press, New York

Mori N, Oikawa T, Harada Y, Miyahara J (1990) Development of the imaging plate for the transmission electron-microscope and its characteristics. J Electron Microsc 39(6):433–436

Prydderch ML, Waltham NJ, Turchetta R, French MJ, Holt R, Marshall A, Burt D, Bell R, Pool P, Eyles I, Mapson-Menard H (2003) A 512 × 512 CMOS monolithic active pixel sensor with integrated ADCS for space science. Nucl Instrum Meth A 512:358–367

Samei E, Flynn MJ et al (1998) A method for measuring the presampled MTF of digital radiographic systems using an edge test device. Med Phys 25:102–113

Shindo D, Hiraga K, Oikawa T, Mori N (1990) Quantification of electron diffraction with imaging plate. J Electron Microsc 39(6):449–453

Spence JCH (2013) High resolution electron microscopy, 4th edn. Oxford University Press, Oxford

Spence JCH, Zuo JM (1988) Large dynamic-range, parallel detection system for electron-diffraction and imaging. Rev Sci Instrum 59:2102–2105

Takahashi K, Kohda K, Miyahara J, Kanemitsu Y, Amitani K, Shionoya S (1984) Mechanism of photostimulated luminescence in BaFCl-Eu^{2+}, BaFBr-Eu^{2+} phosphors. J Lumin 31:266–268

Tatian B (1965) Method for obtaining transfer function from edge response function. J Opt Soc Am 55:1014

Turchetta R (2003) CMOS monolithic active pixel sensors (MAPS) for scientific applications. Presented at 9th workshop on electronics for LHC experiments, Amsterdam

Valentine RC (1966) The response of photographic emulsions to electrons. In: Barer R, Zeitler E (eds) Advances in optical and electron microscopy. Academic Press, New York

Zeitler E (1992) The photographic-emulsion as analog recorder for electrons. Ultramicroscopy 46:405–416

Zuo JM (1996) Electron detection characteristics of slow-scan CCD camera. Ultramicroscopy 66:21–33

Zuo JM (2000) Electron detection characteristics of a slow-scan CCD camera, imaging plates and film, and electron image restoration. Microsc Res Tech 49:245–268

Zuo JM, McCartney MR et al (1996) Performance of imaging plates for electron recording. Ultramicroscopy 66:35–47

Zuo JM, Pacaud J, Hoier R, Spence JCH (2000) Experimental measurement of electron diffuse scattering in magnetite using energy-filter and imaging plates. Micron 31:527–532

Chapter 10
Instrumentation and Experimental Techniques

Many of the instrumental requirements for electron diffraction, particularly the needs for small electron probes, will be found to be similar to those for analytical electron microscopy. In fact, modern analytical TEMs provide special CBED or nanobeam modes, or both, together with the modes for low- and high-magnification TEM, diffraction, and EDX. This development resulted partly from the fact that the requirements for EDX (large tilt, small probe, low contamination) exactly match those for nanodiffraction. Analytical TEMs designed for STEM also feature an improved vacuum system and instrument stability required for electron nanodiffraction. Most importantly, there is a clear scientific merit here since microanalysis complements electron diffraction in analytical TEM by providing both chemical and crystallographic information.

This chapter focuses on instrumentation and experimental techniques. We will discuss electron probe formation and examine the working of several illumination systems, including electron beam deflectors. We then move on to electron diffraction techniques, specimen rotation, energy filtering, and radiation effects. A brief introduction to analytical TEM was provided in Chap. 1. Further introductory material on TEM operation can be found in the textbook by Williams and Carter (2009). For additional reading, in addition to the cited references, we refer readers to the books by Egerton (2011) and Reimer (1995) on energy filters and the instrument manuals for the illumination optics in specific microscopes.

Specimen preparation is a special topic deserving separate attention that is beyond the scope of this book. We refer readers to the texts and review articles by Goodhew (1972), Hirsch et al. (1977), and Özdöl et al. (2012). An important development in TEM sample preparation is the use of focused ion beams of Ga^+ ions for cross sectioning of various samples. The ion column can be integrated together with an electron column in an SEM in the so-called dual-beam configuration. Image can then be formed using either column. This allows precise control over the position and thickness of the cross section, which has opened up new

© Springer Science+Business Media New York 2017
J.M. Zuo and J.C.H. Spence, *Advanced Transmission Electron Microscopy*,
DOI 10.1007/978-1-4939-6607-3_10

avenues for electron imaging and diffraction and failure analysis in general, including the characterization of semiconductor devices. Further details about ion beam techniques can be found in Giannuzzi and Stevie (2005) and Mayer et al. (2007).

10.1 Electron Beam Illumination

Condenser lenses work with the objective lens to deliver the electrons from the gun crossover to the specimen. Because of this, they have special roles in electron diffraction and imaging. As the major parts of the illumination system, they are designed to meet the following imaging and small probe analysis needs:

(1) The image quality is directly related to illumination uniformity, as well as intensity. A good illumination system thus must be able to provide uniform illumination for electron imaging at low, medium, or high magnification and a successively smaller electron beam of increasing intensity at larger magnifications. The electron beam should be kept as parallel (well collimated) as possible in order to maintain the electron coherence required for high-resolution TEM imaging and diffraction;
(2) High intensity small probes are required for nanodiffraction and small beam analysis using EDX or EELS in analytical TEMs. Thus, the same condenser lenses used for TEM illumination must also be able to form small probes.
(3) In STEM, the condenser lenses are configured to provide the large convergence angles required for high-resolution STEM and STEM/EDX/EELS; and
(4) Similarly, the condenser lenses used to provide various convergence angles for CBED may also be used to form small parallel beams.

10.1.1 Illumination Using Two Condenser Lenses

A basic TEM illumination system consists of two condenser lenses (C1 and C2). Figure 10.1 shows how they work with the objective prefield lens to provide illumination for TEM imaging, focused probe for CBED, and STEM. The C1 lens images the source crossover onto the object plane of the C2 lens. It is used as a strongly demagnifying lens. Using this, the C1 lens determines the effective electron source size for subsequent electron lenses. The C2 focal distance is varied, together with the condenser aperture, to control the electron beam. In forming the TEM illumination, the C2 focal distance is set to transfer the image of the C1 lens to the front focal plane of the objective prefield lens for a broad beam on the specimen. The C2 lens is only weakly excited in the STEM configuration of Fig. 10.1. Here,

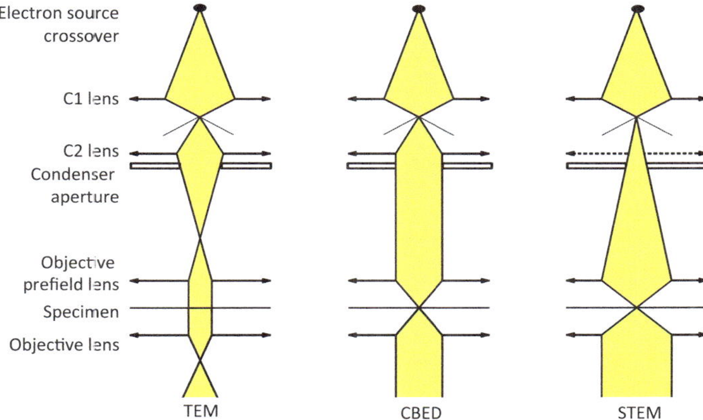

Fig. 10.1 Two condenser lens illumination system working with a single field, symmetrical, condenser–objective lens for TEM and focused probes for CBED and STEM

the objective prefield works together with the C1 lens to provide a two-stage demagnification for the electron crossover and thus a large convergence angle. A focused probe with a smaller convergence angle for electron nanodiffraction can be produced by using the C2 lens to form a nearly parallel beam, with the specimen placed at the back focal plane of the objective prefield.

This two condenser lens setup works with the symmetrical condenser–objective lens design of Riecke and Ruska (1966). The changeover from the TEM illumination to STEM is made by reducing the C2 excitation and adjusting the objective lens to image the C1 crossover onto the specimen.

The two condenser lens setup, however, does have a number of drawbacks. The primary drawback is there is very little flexibility in controlling the TEM illumination. In order to have a parallel beam on the specimen, the C2 lens is set to bring electrons into a crossover in the front focal plane of the objective prefield. Thus, the only way to change the size of the illuminated area is by changing the CA size or by having a convergent or divergent beam, for a smaller or larger beam on the specimen, respectively. Since having a parallel beam is critical for the performance of TEM imaging and electron diffraction, this becomes a major limitation. This is also evident in the illumination for CBED as shown in Fig. 10.1. The strong objective prefield here remains constant in both TEM and CBED modes, which helps to maintain the optical alignment and stability of the objective lens. Once the beam is focused by the objective prefield onto the specimen to form a small probe, its convergence angle becomes fixed by the size of the beam, which is determined by the CA size. Since there are only a few apertures available in a TEM, this places a limit on their selection and on the ability to meet the different needs of TEM and STEM imaging.

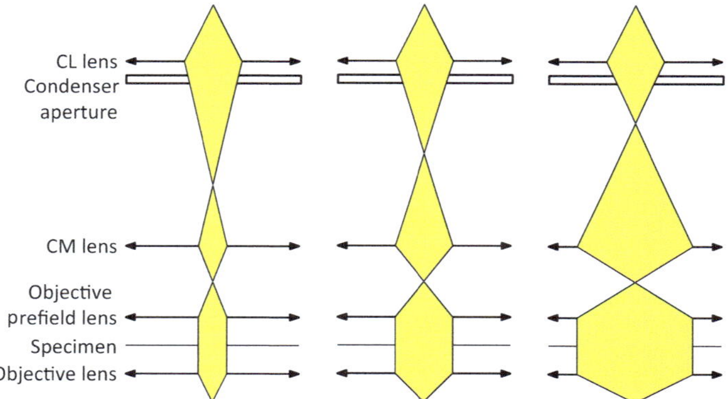

Fig. 10.2 Use of the condenser minilens (CM) at three different setting for small, medium, and large TEM illumination field

10.1.2   The Use of Condenser Minilens

A solution to the limitations of the two condenser lens setup is to add an auxiliary condenser lens, called a condenser minilens or CM, placed immediately above the condenser–objective lens (Van der Mast et al. (1980)). Figure 10.2 shows the use of a CM at three different settings for small, medium, and large parallel beams, to be used for electron imaging at high ($M > 200$ K), medium ($M > 50$ K), and low magnifications. The CM lens takes the crossover formed by the last condenser lens (CL, C2 in the two condenser lenses setup) and images it onto the front focal plane of the objective prefield lens, which then forms a parallel beam on the specimen. Adjustment to the parallel beam can be made by increasing the CL lens excitation, which reduces the CL focus length and moves the beam crossover closer to the CM lens. A focused beam is formed with the CM lens, and the objective prefield lens works together as a telescopic lens as shown in Fig. 10.3.

The size of the parallel beam formed with the help of the CM lens can be related to the angle subtended by the condenser aperture and the magnifications of the CL and CM lenses by

$$d = \frac{\alpha}{M_{\mathrm{CL}} M_{\mathrm{CM}}} f_{\mathrm{obj}}. \tag{10.1}$$

The values of M_{CL} and M_{CM} are related since the object distance of the CM lens is fixed by its focal length, which in turn fixes the focal length of the CL lens. To increase the beam size, M_{CL} is first reduced, which further reduces M_{CM}. Conversely, a smaller beam is obtained by increasing M_{CL} to the limit when M_{CM} becomes equal to unity. Further reduction on the beam size can be obtained by switching off the CM lens (Fig. 10.3a). Then, the beam size is given by

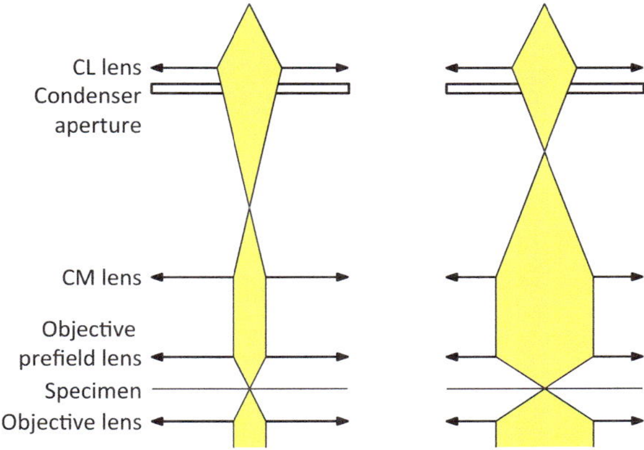

Fig. 10.3 Focused electron probes with a small and large convergence angle obtained using different settings for the CL and CM lenses

$$d = \frac{\alpha}{M_{\mathrm{CL}}} f_{\mathrm{obj}}, \qquad (10.2)$$

with M_{CL} fixed by the distances to the front focal plane of the objective prefield and the crossover position in front of the CL lens.

In forming a focused probe, the crossover in front of the CL lens is imaged onto the specimen. The convergence angle of a focused probe can be changed by using the different settings (focal lengths) of the CM lens, according to

$$\theta = \frac{\alpha}{M_{\mathrm{CL}} M_{\mathrm{CM/OP}}}, \qquad (10.3)$$

where $M_{\mathrm{CM/OP}}$ is the combined magnification of the CM lens and the object prefield lens.

10.1.3   A Third Condenser Lens and Kohler Illumination

The C2 lens acts as the last condenser lens (CL) in the two condenser illumination system. Above the C2 lens, we have the crossover formed by the C1 lens as the object, which moves when a different setting (spot size) is used. The crossover also moves when a change is made in the electron gun lens focal length by varying the extraction voltage in a field emission gun or the gun bias in a thermionic gun. Below the lens, its image position is fixed in order to work with the CM lens (or the objective prefield with the CM switched off). Thus, in a two condenser lens system, the C2 lens has to be readjusted each time when a change is made to the C1 lens or the electron gun. In applications such as STEM or electron nanodiffraction, a different setting of the C1 lens or the electron gun is often used from the one used

Fig. 10.4 Use of three
condenser lenses with the C1
and C2 lenses configured as
the zoom lens

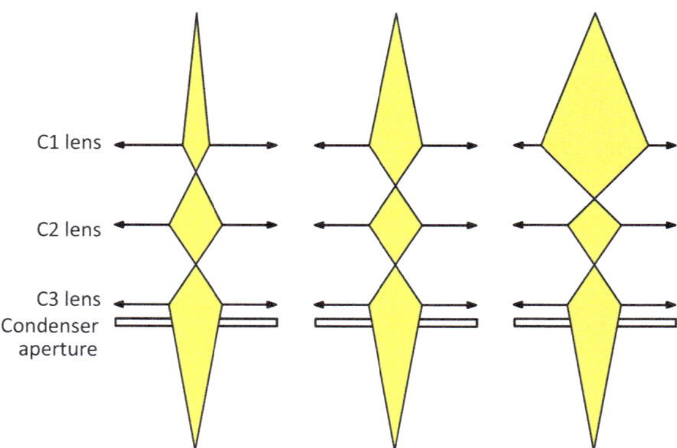

for the microscope alignment and specimen examination. Thus, a fixed object
position for the CL lens is desired to improve the instrument usability.

A solution to the above-stated problem is to introduce a third condenser lens, C3,
as shown in Fig. 10.4. (Benner et al. 1990) Fig. 10.4 shows how the C1 and C2
lenses work in a three condenser lens setup as a zoom lens and how they can be
used to provide various demagnifications of the electron gun crossover, while
keeping the object and image distances the same. Similarly, the C1 and C2 lenses
can be configured to work with different positions of the gun crossover, while
keeping the image distance of the C2 lens fixed.

The three condenser lenses setup described above can be used to realize Köhler
illumination for electron imaging and diffraction (Benner and Probst (1994)). The
Kohler illuminating system is characterized by two conjugate planes: (1) The
(electron) source is imaged into the front focal plane of the objective prefield lens
by the condenser system and (2) the condenser aperture is imaged into the specimen
plane by the objective prefield lens. The first point provides a uniform illuminating
field on the specimen, which is important for achieving high-quality images. The
size of illuminated specimen area is defined by the condenser aperture. Both are
critical for electron nanodiffraction.

10.1.4   Beam Current

The brightness β_s of a source is conserved and constant even if aberrations and
aperture stops are permitted (Chap. 8). In the previously discussed electron illumination systems, the source is demagnified by various magnifications using the C1
lens or by the zoom lens consisting of the C1 and C2 lenses. The net effect is to
increase current density J and increase the source angle α (proportional to $1/M$) for
the effective source seen by the CL lens, leaving brightness constant. The condenser
aperture, which is placed at an intermediate position before the objective lens and at
or after the CL lens, only lets through a part of the illuminating cone. Its diaphragm

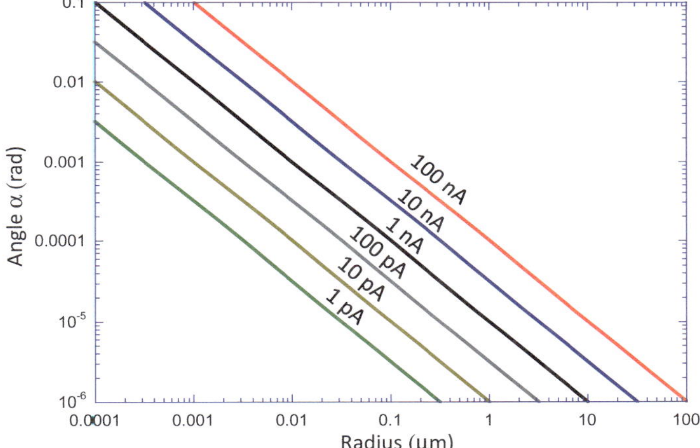

Fig. 10.5 The probe radius plotted against the probe convergence angle for brightness 10^{12} A/m^2/sr

size determines the size of the electron cone and the numerical aperture produced by the CL lens. Thus, the beam current after the condenser aperture is given by

$$I_b = \beta_s \pi^2 \alpha_c^2 (d_s/2)^2 \tag{10.4}$$

where α_c is the condenser aperture half angle. The beam current can be increased by enlarging the illumination aperture α_c (provided that it remains filled with electrons), or by increasing the effective source diameter d_s using a smaller source demagnification, or by increasing the intrinsic source brightness.

At the specimen level, and a fixed beam current, the beam size is determined by

$$d = \sqrt{\frac{4I_b}{\beta_s \pi^2} \frac{1}{\alpha_b}} . \tag{10.5}$$

Figure 10.5 shows the calculated probe radius plotted against the probe convergence angle for a source brightness of 10^{12} A/m^2/sr and a range of beam currents from 1 pA to 100 nA. The typical current employed in STEM is between 10 pA to hundreds of pA, which gives a convergence angle on the order of tens of mrad for an Angstrom-sized probe.

10.1.5   Coherence and Coherent Current

The smallest probes can only be obtained using fully coherent illumination. For the STEM or CBED modes using a focused probe, the electron beam coherence is

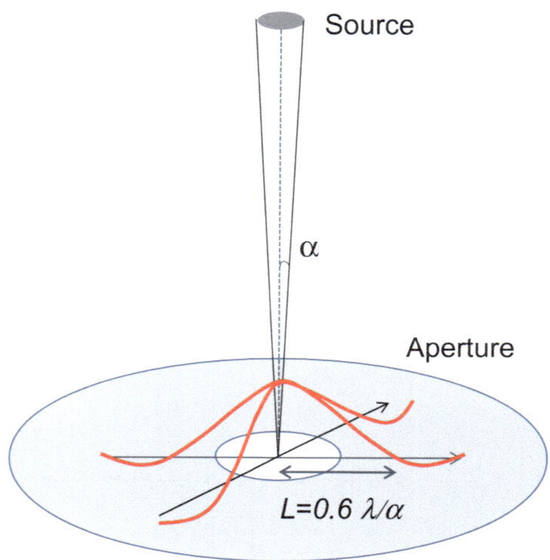

Fig. 10.6 Relationship between a finite source and the lateral coherence length L at the condenser aperture. The source size is defined by the angle (α) shown subtended at the aperture

defined by the coherence width L at the aperture which illuminates the sample. According to the Zernike–Van Cittert theorem (Eq. 2.47, Chap. 2), the degree of coherence between the electron wave functions at two different points in an aperture plane far away from the electron source is given by the Fourier transform of the source intensity distribution. The source seen by the condenser aperture is the source image formed before the CL lens in Fig. 10.2. If we assume the source has a uniform and ideally incoherent intensity distribution within a circular disk, the coherence function at the aperture is then given by

$$\gamma_{12}(r,0) = \lambda J_1(2\pi\alpha r/\lambda)/2\alpha r. \tag{10.6}$$

with J_1—a first-order Bessel function, r—the radial distance in the aperture plane, and α—the angle subtended by the electron source (see Fig. 10.6). The lateral coherence length L, which is used in the literature, is defined by the distance r to the first zero of J_1, which has the value of

$$L = 0.6 \ \lambda/\alpha. \tag{10.7}$$

This provides an estimate of the distance between points in the plane of CA at which the wave field is capable of producing strong interference fringes in a Young's slit experiment with slit spacing L. If the diameter of the CA is $2R_a$, we then have following possibilities:

(1) If $2L \ll 2R_a$, the illumination aperture can be considered to be incoherently filled and so treated as an ideally incoherent effective source. This is the situation for conventional TEM systems using a tungsten or LaB_6 source under most operating conditions. Then, CA can be treated as an ideally incoherent source, within which each point acts as a statistically independent emitter of electrons. (A useful exercise is to calculate L for a LaB_6 source

operating at the smallest probe size.) The probe formed further downstream by this incoherent source, filling the illumination aperture, will then be partially coherent.

(2)   If $2L > 2R_a$, the illumination aperture is coherently filled, and the radiation can be considered to originate from a point source. Then, the entire optical system beyond CA is filled with perfectly coherent radiation, and the probe may be treated as perfectly coherent. This is often a good approximation for field emission gun (FEG) instruments. With a focused probe, the sample is then illuminated by an aberrated, converging spherical wave.

We see that coherence is increased by decreasing d_s, that is, by increased demagnification of the source, by using lower accelerating voltage, and by decreasing the size of CA. For a Schottky emission source, the emission diameter is between 20 and 30 nm according to Botton (2007). At a distance of 10 cm away from the electron source image, a factor of 10 source demagnification provides a coherence length from 100 to 150 μm. However, the actual coherence length is smaller because of the gun lens aberrations (see Chap. 8).

Note that, in forming a focused probe, the source is imaged onto the sample by the probe-forming lens, while the illumination aperture CA (called the "objective aperture" on STEM instruments) is imaged onto the detector in STEM by the objective lens. CA subtends a semi-angle θ_c, at the sample, while the geometrical electron source image (of diameter d_s) subtends a semi-convergence angle α at CA. Thus, a measurement of the coherence can be made by examining interference in overlapping CBED disks obtained with a crystal (see Chap. 14). In this book, we shall refer loosely to "coherent CBED" as the case $2L > 2R_a$ and "incoherent CBED" as the case $2L \ll 2R_a$, noting that these labels refer to the coherence conditions in the illumination aperture and that in the second case, the probe itself is partially coherent.

A figure of merit for imaging or diffraction using coherent electrons is the coherent current, defined by the electron current available at a certain degree of coherence (Lichte and Lehmann 2008). To define this, we go beyond the above flat disk approximation for the electron source by assuming a rotational symmetric Gaussian function for the source intensity distribution

$$I(r) = \frac{I_o}{\sqrt{\pi}r_s}\exp\left[-\left(\frac{r}{r_s}\right)^2\right] \tag{10.8}$$

According to Eq. 2.47, the coherence function of this source is given by

$$\gamma(\theta,0) = \exp\left[-(\pi\theta r_s/\lambda)^2\right] \tag{10.9}$$

where θ is the semi-acceptance angle of the CA. For a coherently illuminated CA, the coherent current I_{coh} is given by

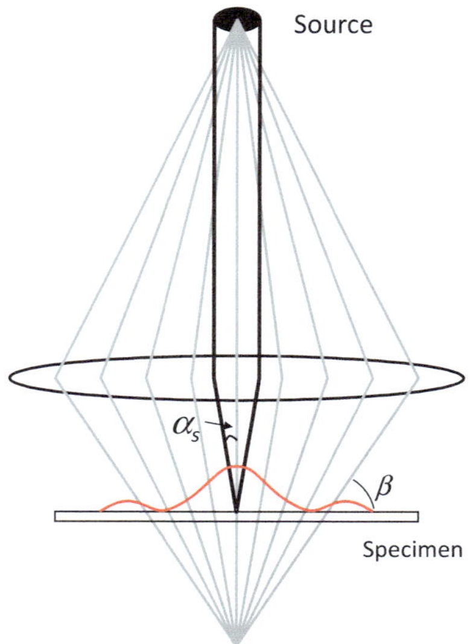

Fig. 10.7 Specimen illumination from a finite source in a symmetrical condenser–objective lens with the specimen placed at the focal distance of the objective prefield

$$I_{\mathrm{coh}}(\theta) = \pi^2 \theta^2 r_s^2 \beta_s = -\ln[\gamma(\theta,0)]\beta_s \lambda^2, \qquad (10.10)$$

which shows that the available coherent current is proportional to log of the degree of coherence and the so-called "reduced brightness" $\beta_s \lambda^2$ of the electron source. The electron source brightness increases linearly with accelerating voltage Φ, while the electron wavelength is inversely proportional to the square root of the accelerating voltage. Thus, it can be concluded that both the reduced brightness and the coherent current are a constant property of the emitter, independent of the electron wavelength or accelerating voltage.

Next, we examine the coherence at the specimen level. Figure 10.7 considers the general case of a specimen illuminated in a symmetrical condenser–objective lens. An effective electron source is formed above the objective prefield lens; its distance to the lens depends on the application, which can be at the front focal plane of the prefield lens as in the TEM illumination or far from the prefield lens as in STEM. In either case, we have individual plane waves coming from the effective source in the far-field approximation. The coherence length of those plane waves is defined by the source convergence angle α_s according to Eq. (10.7)

$$L_s = 0.6\lambda/\alpha_s = 1.2\lambda f/d_s \qquad (10.11)$$

The extent of coherence at the specimen in a TEM illumination can be obtained from the width of the band of Fresnel fringes seen in images of the edge of a sample, or from electron holography experiments using an electron biprism. In electron holography, the biprism is placed at an intermediate position between the diffraction pattern at the back focal plane and the image (see Chap. 16 for further details). The electron source crossover is imaged into the diffraction pattern, and its size defines the lateral coherence and the fringe visibility in the electron hologram. Similarly, in selected area electron diffraction, the specimen coherence length determines the number of lattice planes which interfere and thus the sharpness of diffraction peaks, as long as the selected area is larger than the coherence length.

In an analytic TEM using a strong objective prefield lens, the illumination formed at the specimen level is not perfectly parallel, nor perpendicular, to the specimen. This fact was pointed out by Christenson and Eades in their paper "Skew thoughts on parallelism" (Christenson and Eades 1988). The angular range in the illumination has two contributions in addition to the source convergence angle that we have just discussed. The "nonparallelism," as defined by Christenson and Eades, refers to the point to point variation in the β angle inside the illuminated area as shown in Fig. 10.7. Then, we can take the range of β in the illumination field to define the degree of nonparallelism. The nonparallelism is intended for a TEM illumination, and it should not be confused with the convergence angle that we have referred to for a focused probe, which is the range of incident angles at a single point, even though a point exists strictly only in geometric optics. In Fig. 10.2, the electron source crossover is projected onto the front focal plane of the objective prefield lens, and the range of incident angle is defined by its aberrations. A larger variation in incident angles over the illuminated area is obtained by defocusing the final condenser lens. All of those effects are well known. A less well-known fact is the second variation in the incident angles over the illuminated area, which is caused by the spiral path of the electrons in the magnetic field of the objective lens (ϕ), which gives rise to an angular spread

$$\phi = 0.404 \, rB_z \text{ mrad,} \tag{10.12}$$

This spread, introduced by having a wide electron beam, is dependent on the beam radius (r in μm) and axial magnetic field B_z, in tesla. The spread becomes a serious issue when a large parallel illumination is needed, for example, when imaging defects from a large area of crystals (Eades 2006).

10.2   Probe Formation

Small electron probes are employed in electron nanodiffraction, microanalysis, or STEM. To form a focused probe, the crossover from the last condenser lens is imaged by the objective prefield lens onto the specimen plane. The difference between the objective lens focal length and the focal length required to image the crossover onto the sample is defined as the defocus. Under "incoherent" conditions (i.e., $2L \ll 2R_a$), the total probe diameter d_o is given approximately at Gaussian focus by adding in quadrature the various contributions to d_o. Thus,

$$d_{\mathrm{o}}^2 = d_{\mathrm{s}}^2 + d_{\mathrm{d}}^2 + d_{\mathrm{sa}}^2 + d_{\mathrm{c}}^2 + d_{\mathrm{f}}^2 \tag{10.13}$$

where d_{s} is the geometrical source image diameter, d_{d} is the diffraction broadening equal to 0.6 $\lambda/\theta_{\mathrm{c}}$ with θ_{c} for the convergence angle, d_{sa} is the contribution from lens aberrations (in a TEM without a probe corrector, it is equal to 0.5 $C_{\mathrm{s}}\theta_{\mathrm{c}}^3$ in the plane of least confusion, not the Gaussian image plane), and d_{c} is the contribution from chromatic aberration, given by $(\Delta E_{\mathrm{o}}/E_{\mathrm{o}})\,C_{\mathrm{c}}\theta_{\mathrm{c}}$, with ΔE_{o} the energy spread in the electron beam. This last term can be neglected as a first approximation. In Eq. (10.13), d_{f} is the contribution $2\theta_{\mathrm{c}}\Delta f$ from a small focusing error Δf. For a typical modern TEM instrument with $C_{\mathrm{s}} = 2$ mm at 100 kV, the contributions of diffraction d_{d} and spherical aberration d_{sa} are equal at an angle of about 7 mrad. Equation (10.13) cannot strictly be used for coherent conditions, with the widths of intensity distributions added in quadrature.

The smallest probe is obtained by minimizing all the quantities in Eq. (10.13). The d_{s} can be made smaller than d_{d} and d_{sa} by combining a small physical source with large demagnification. Then, the probe formation becomes "diffraction limited" and the illumination necessarily coherent ($2L > 2R_{\mathrm{a}}$). Then, detailed computations are required for the probe shape for particular values of the lens aberrations, the defocus Δf, λ, and θ_{c}.

For a convergent beam of electrons, the lens aberrations introduce an angle-dependent phase, $\chi(k_x, k_y)$, with x and y standing for the coordinates perpendicular to the optical axis of the electron lens. The phase $\chi(k_x, k_y)$ from the objective lens aberrations is described in Chap. 7. For electron nanodiffraction, we must also consider the electron source wave function $\phi_S(x, y)$ formed by the last condenser lens and its contribution to the electron probe. The electron probe on the sample is an image of $\phi_S(x, y)$ magnified by the lens magnification M. According to the image formation theory that will be discussed in Chap. 11, the actual image is a convolution of $\phi_S(x, y)$ with the objective lens resolution function $T(x, y)$

$$
\begin{aligned}
\phi_{\mathrm{P}}(x, y) &= \phi_S(-x/M, -y/M) \otimes T(x, y) \\
&= \int_{-\infty}^{\infty} \phi_S\!\left(-M\vec{k}_{\mathrm{t}}\right) A\!\left(\vec{k}_{\mathrm{t}}\right) \exp\!\left[i\chi\!\left(\vec{k}_{\mathrm{t}}\right)\right] \exp\!\left(2\pi i \vec{k}_{\mathrm{t}} \cdot \vec{r}\right) \mathrm{d}\vec{k}_{\mathrm{t}} \\
&= FT\!\left(\phi_S\!\left(-M\vec{k}_{\mathrm{t}}\right) A\!\left(\vec{k}_{\mathrm{t}}\right) \exp\!\left[i\chi\!\left(\vec{k}_{\mathrm{t}}\right)\right]\right)
\end{aligned}
\tag{10.14}
$$

Here, we have used $\vec{k}_{\mathrm{t}} = k_x\vec{x} + k_y\vec{y}$ and FT as a shorthand for Fourier transform, and $A\!\left(\vec{k}_t\right)$ is the aperture function with a value of 1 for $\left|\vec{k}_{\mathrm{t}}\right| < \theta_c/\lambda$ and 0 beyond. The electron beam energy spread and the chromatic aberration are neglected in Eq. (10.14). The equation also assumes that the illuminating electron wave is perfectly coherent across the condenser aperture.

A focused electron probe on the sample is formed by placing the electron beam crossover far away from the front focal plane of the objective lens as shown in

Fig. 10.3. This gives a demagnified, sharp electron source image on the sample with magnification $M \ll 1$. The size of the electron probe, in this case, is largely determined by the objective lens resolution function $T(x, y)$. In reciprocal space, the demagnified electron source has a broad, spherical wavelike spectrum of wave vectors. For a conventional TEM (aberration-corrected probes will be discussed in Chap. 14), those calculations have been published by Mory et al. (1987). The intensity distribution of the probe at the sample is given by

$$I_{\mathrm{p}} = \left| \phi_{\mathrm{p}}(x, y) \right|^2 = \phi_{\mathrm{p}}(x, y)\phi_{\mathrm{p}*}(x, y).$$

For a conventional TEM without the probe C_{s} corrector, the wave-front aberration function χ is given by

$$\chi\left(\vec{k}\right) = \pi\left(\Delta f \lambda k_{\mathrm{t}}^2 + \frac{1}{2} C_s \lambda^3 k_{\mathrm{t}}^4 + \vec{k}_{\mathrm{t}} \cdot \vec{r}_{\mathrm{p}} \right) \tag{10.15}$$

Here, $\vec{r}_{\mathrm{p}}$ is the probe coordinate. The required focus setting Δf can be defined as which minimizes the radius of the probe area, which contains, say, 70 % of the beam intensity Mory et al. (1987). Calculations based on Eq. (10.15) then show that the following values must be used to obtain this "most compact" probe:

$$\theta_c = 1.27 \, C_s^{-1/4} \lambda^{1/4} \tag{10.16}$$

$$\Delta f = -0.75 \, C_s^{1/2} \lambda^{1/2} \tag{10.17}$$

This gives the minimum probe diameter (containing 70 % of the intensity) as follows:

$$d(70 \text{ %}) = 0.66 \, \lambda^{3/4} C_s^{1/4} \tag{10.18}$$

For experimental reasons, it may be easier to measure the probe displacement for which the intensity at a sharp edge in a STEM image falls from 80 to 20 %. As judged by this criterion, the smallest probe is obtained with the constant 0.66 above replaced by 0.4. Experimental measurements of coherent probe widths, in rough agreement with the above theoretical estimates, can be found in Berger et al. (1987).

Each plane-wave component of the coherent, subnanometer probe illuminates the specimen to the extent of the illumination coherence length defined in Sect. 10.1.5. The Fourier synthesis of all components concentrates the energy into a localized region. The amount of energy remaining in the tails of the distribution depends on the focus setting. As an example, for an analytical pole piece ($C_{\mathrm{s}} = 1.2$ mm), we find d(70 %) = 0.244 nm at 200 kV. In practice, extremely stable conditions are required to achieve this performance and the effects of tip vibration, sample movement, stray fields, and electronic instabilities in the lens, and accelerating voltage must also be considered.

To form a small parallel beam, the electron beam crossover is placed close to, or at, the front focal plane. The electron source in this case is magnified ($M \gg 1$), which is used to reduce the electron beam convergence angle for this "parallel-beam" diffraction. Meanwhile, the sample remains at the back focal plane of the objective prefield lens, far away from the electron source image. Thus, the electron beam seen at the specimen level is a defocused image of the source. This large under-focus must be included as a part of the lens aberration function in Eq. (10.14) (Zuo et al. (2004)). To demonstrate this, we assume a Gaussian distribution for the magnified electron source after the objective prefield lens (with a magnification of M):

$$\phi_S(r/M) = A \, \exp\left(-a^2 r^2 / M^2\right) \tag{10.19}$$

where a gives the crossover half width at the amplitude of A/e. The Fourier transform of this Gaussian probe is as follows:

$$\phi_S(k_t) = \frac{A\sqrt{\pi}}{a} \exp\left[-k_t^2 / (a/M\pi)^2\right]. \tag{10.20}$$

Substituting Eq. (10.20) into Eq. (10.14) gives

$$\phi_P(x,y) = FT\left(\frac{A\sqrt{\pi}}{a} \exp\left[-k_t^2 / (a/M\pi)^2\right] A\left(\vec{k}_t\right) \exp\left[i\chi\left(\vec{k}_t\right)\right]\right) \tag{10.21}$$

Thus, the width of the beam in the reciprocal space is reduced by a factor of $1/M$. The Gaussian half width of the defocused electron beam is ~ 0.05 mrad in the JEOL2010F TEM formed using a 10-μm condenser aperture Zuo et al. (2003). The real space probe observed at the specimen level is a convolution of the magnified source with

$$T(x,y) = FT\left\{A\left(\vec{k}_t\right) \exp\left[i\chi\left(\vec{k}_t\right)\right]\right\}. \tag{10.22}$$

The dominant probe features come from $T(x,y)$ as shown in Fig. 10.8 for a comparison between an experimental probe and simulation based on $T(x,y)$ alone Zuo et al. (2004).

In electron nanodiffraction using a focused probe, a small condenser aperture is used so that the diffraction pattern recorded consists of small diffraction disks. The small disks help with the determination of diffraction peak positions for lattice analysis, for example, local strain measurements (Cowley 2004). For a small convergent angle at ~ 1 mrad, the size of the focused probe is diffraction limited, with its intensity distribution given by

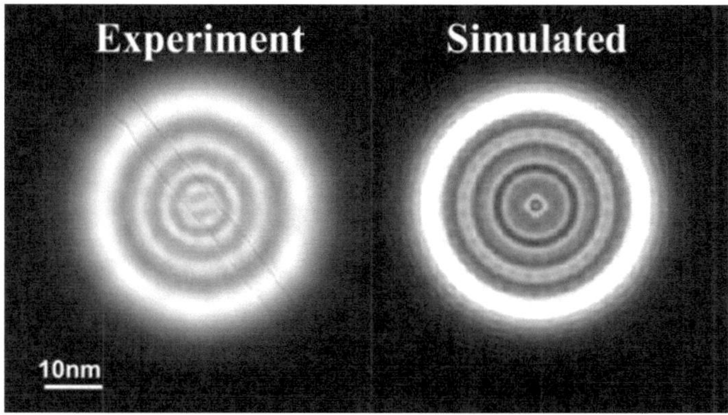

Fig. 10.8 Experimental and simulated electron nanobeam used in nanoarea electron diffraction (NAED). The simulation used Cs = 1 mm and $\Delta f = -360$ nm

$$I(r) \propto \left[\frac{J_1(2\pi r\, \sin\theta/\lambda)}{\pi r\, \sin\theta/\lambda} \right]^2 \tag{10.23}$$

where θ is the beam's half-convergence angle, and J_1 is the first-order Bessel function. The first zero of $J_1(x)$ occurs at $x = 3.832$, which gives the so-called Rayleigh criterion for resolution:

$$r_{\mathrm{o}} = 0.61\frac{\lambda}{\theta} \tag{10.24}$$

The intensity distribution in Eq. (10.24) can be fitted approximately by a Gaussian function with a full width at half maximum (FWHM) of

$$d_{\mathrm{FWHM}} = 0.52\frac{\lambda}{\theta} \tag{10.25}$$

The diffraction-limited probe size increases as the convergence angle decreases. Since the probe size defines the spatial resolution for a beam with a small convergence angle, improvements in angular resolution in diffraction pattern are thus obtained at the expense of spatial resolution.

Figure 10.9 shows an example of a small focused probe formed inside a probe C_s-corrected FEI Titan microscope operated at 300 kV using a condenser aperture of 50 μm in diameter. The microscope was operated in the so-called μProbe TEM scan mode, in which the probe corrector was operated like an additional condenser lens. The probe recorded has a FWHM of 2.7 nm. Diffraction patterns recorded using this probe consist of small diffraction disks with a convergence semi-angle of 0.37 mrad, according to Eq. (10.25).

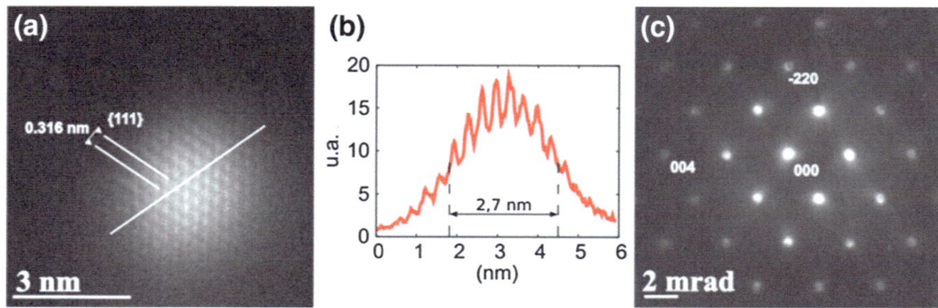

Fig. 10.9 **a** Electron probe formed inside a FEI Titan TEM. The image was recorded as the probe passing through a (110) Si crystal with the lattice fringes clearly seen within the probe. **b** electron nanodiffraction pattern recorded with the probe in (a). From Beche et al. (2009)

Since the STEM instruments under consideration are also capable of forming atomic-resolution images for direct examination of materials structure, we consider in Chap. 14 different probes and detectors which produce the most faithful image of a crystal in STEM.

10.3   Beam Deflectors and Scanning

A set of two deflectors (double deflection coils) are placed below the CL lens and above the CM lens to deflect the beam. A TEM has at least two other sets of double deflection coils, one placed below the electron high-tension accelerator and one below the objective lens for image or diffraction shift. The beam deflectors are used to provide beam shift, bright-field beam tilt, and dark-field beam tilt. When driven by an external scan generator, they are used to scan the probe in a raster over the specimen and to form STEM images by coupling the scan together with a detector. In electron diffraction mode, they can be configured in a number of ways for beam rocking, conical scan, as used in precession (Vincent and Midgley 1994), and scanning electron nanodiffraction (Zuo and Tao 2011). While the electron beam can be deflected by either electric or magnetic fields, modern TEMs use magnetic coils for beam deflection for reasons explained below.

Each deflector has two pairs of coils on either side of the electron beam (Fig. 10.10 shows one such pair). They are extended over a large arc in the so-called saddle yoke configuration (the other configuration is toroidal yoke, see Munro (1975)) and used to generate a homogeneous magnetic field. The angle of deflection in a combined, uniform, electric, and magnetic fields is given by $\theta = p_x/p_z$, where Reimer (1998)

Fig. 10.10 Saddle yoke magnetic coils for electron beam deflection. A horizontal magnetic field is produced by a pair of coils of N turns with current I flowing in opposite directions. Electrons traveling vertically experience a force F as shown

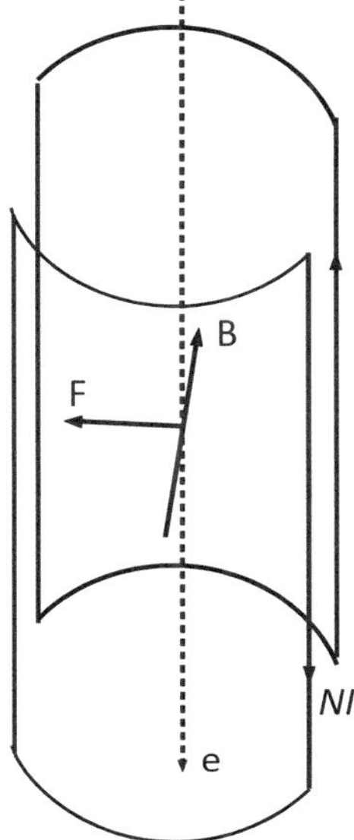

$$p_x = \int_0^T F_x \mathrm{d}t = -e \int_0^T \left(E_x + VB_y\right)\mathrm{d}t = -\frac{e}{V}\int_0^h \left(E_x + VB_y\right)\mathrm{d}z = -eh\left(E_x/V + B_y\right)$$

$$(10.26)$$

Here, h is the height of the field, which can be taken approximately as the height of the coil in a magnetic deflector. Thus, in the absence of the electric field, we have

$$\theta = \frac{eh}{mV}B_y = \frac{ehB_y}{\sqrt{2m_oE(1 + E/2E_o)}} \qquad (10.27)$$

For example, a 100 keV beam of electrons is deflected by 5.1° in a distance of $h = 1$ cm by a magnetic field of $B = 0.01$ T. To generate a similar amount of deflection using an electric field would need a field of about 18 kV/cm for 100 keV electrons. The required field is proportional to the electron beam energy. Thus, electrostatic deflection requires high tension in order to provide fields strong enough to deflect the high-energy electron beam, whereas magnetic fields can be used with a relatively low current. A deflector using magnetic fields is therefore simpler to design. Electrostatic deflection is used primarily for the purpose of fast electron beam blanking, up to gigahertz frequencies. When this is done near the electron source, only a small deflection angle is needed.

When a pair of deflection coils are arranged perpendicular to each other, they apply uniform forces on the beam electrons along the horizontal (x and y) directions. Together, they can be used to shift or tilt the beam along any direction in the

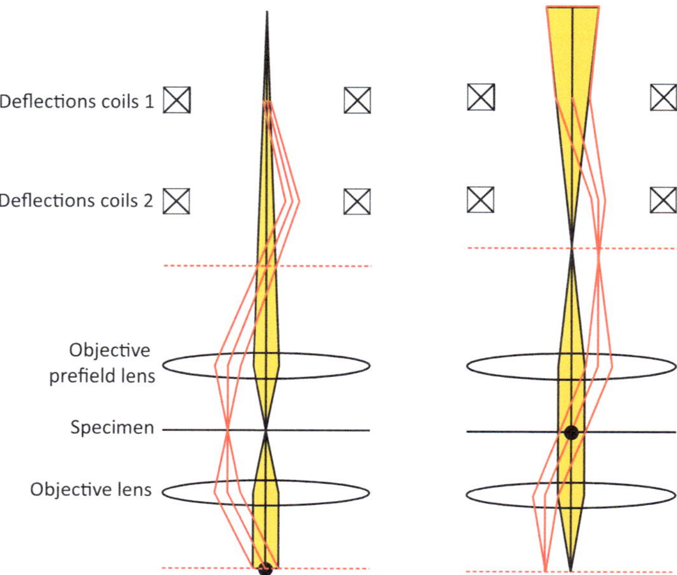

Fig. 10.11 Beam deflection coils used for beam shift (*left*) and beam tilt (*right*). The dark disk marks the pivot point, and dash lines mark the front and back focal planes of the prefield and objective lenses, respectively

x–y plane. Two deflectors working in opposite senses are used to shift or tilt the beam; the individual deflector excitations are different for these two operations. To shift the beam in a TEM, the current in the deflector coils is varied by two illumination-shift controls. Beam shift is used to center the illumination on the TEM screen as it corrects for any horizontal drift of the electron gun or small misalignments of the condenser lenses. Separate controls are provided for beam tilt, one for bright-field tilt, used to compensate slight misalignment of the objective lens, and one for dark-field tilt, which is used to center a diffracted beam along the optical axis for diffraction contrast imaging.

Figure 10.11 compares the beam shift with beam tilt using the double deflection coils in a TEM with a condenser–objective lens. For simplicity, the CM and the objective prefield lenses are shown as a single lens above the specimen. Consider a ray along the optical axis. To shift this ray at the specimen, it must be first deflected away from, and then toward, the optical axis by the first and second deflectors successively. Finally, the beam must intersect the optical axis at the front focal plane of the lens above the specimen, which then brings it to the specimen running parallel to the optical axis. To shift the beam, we actually tilt the beam. For other rays in the beam, because of the small convergence angle, the same tilt is achieved so they all converge to the same point on the specimen. The amount of beam shift is proportional to the tilt angle. To tilt the beam, it is first deflected away from the axis and then back toward the optical axis in such a way that all rays in the beam converge to the same point on the front focal plane as undeflected rays, but now shifted laterally.

It is important to separate beam shift from beam tilt; e.g., when we shift the beam laterally across the specimen, we do not tilt the beam. A beam tilt, for example, will require a realignment of the objective lens, as well as crystal rotation, in high-resolution TEM imaging. In STEM, a tilt introduced by the probe shifts will severely limit the area that can be resolved at atomic resolution. In precession electron diffraction, the beam shift accompanying the beam tilt limits the smallest electron beam area that can be obtained. In TEM dark-field imaging, when a beam shift is accompanied by an additional beam tilt, the change in the incident beam direction affects the diffraction conditions and thus the image contrast. Theoretically, a complete separation of beam tilt and shift (tilt-shift purity) is only achievable with an ideal lens. The tilt-shift purity is very hard to obtain in real lenses away from the optical axis because of the lens aberrations. In practice, the tilt-shift purity is optimized by a process called pivot point alignment. A controlled oscillation, called a wobbler, is applied to the double deflection coils, to rock the beam about a "pivot point" or "rocking point," which is a point on the specimen in the case of beam tilt or a point in the diffraction pattern in the case of beam shift. The ratio of the two deflector excitations is adjusted so that the beam stays at the pivot point, while the wobbler is on. There are two adjustable directions—a main one (X direction) and the perpendicular correction (Y direction). Only a small correction along the Y direction is needed, in order to compensate for a small rotation between the upper and the lower deflectors. In case of beam tilt, because it depends on specimen height, the pivot point must be readjusted whenever the specimen height is changed.

The controllers for the deflection coils in a conventional TEM are only capable of running at relatively low frequencies. An external scan generator is used to drive the deflection coils for STEM using two separate sawtooth signals, one to drive the x-scan and one for the y-scan. The x-scan is relatively fast changing at frequency f_x. The y-scan runs at a much slower frequency $f_y = f_x/n$ (see Fig. 10.12), where n is an integer. In a digital scan generator, the sawtooth signal consists of discrete steps,

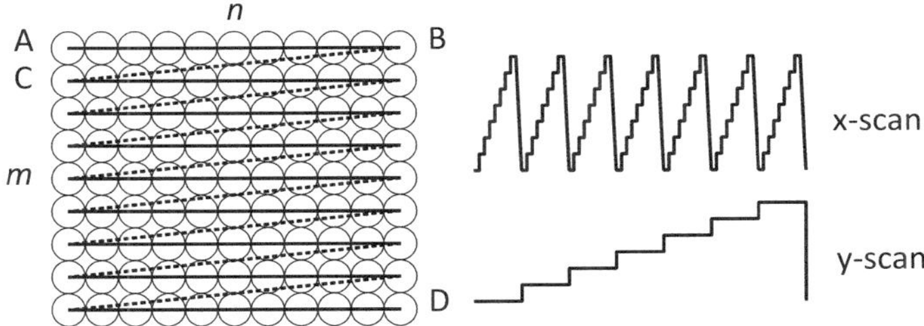

Fig. 10.12 A single frame raster scan using digital sawtooth waveform illustrated below for driving the x- and y-scan. A and D mark the start and end of frame scan; AB marks the first-line scan. As the x-scan reaches the end of the line, the probe flies back and starts a new scan, and the flyback time corresponds to the drop time for the sawtooth signal after Egerton (2005)

and the integer n then corresponds to the number of steps in the x-scan sawtooth signal. In a single period of the x-scan, the y-scan signal stays constant, and thus, the electron probe moves in a straight line, for example, from A to B in Fig. 10.12. Once the beam reaches B, the beam is deflected back along the x-axis as quickly as permitted by the flyback part of the x-scan signal. At the same time, the y-scan signal has stepped up so that the beam is displaced in the y direction, instead of returning to A, and moved to point C. By repeating this, the beam is sequentially deflected to cover a rectangular area on the specimen. Once the probe reaches the last point Z, it is quickly returned to A, and the scan repeats itself for the next frame. In analog TV or video technology using a CRT for display, continuous signals are used to drive the deflection coils and the acquisition process is coupled to the display scan.

10.4   Electron Diffraction Techniques

10.4.1   Selected Area Electron Diffraction (SAED)

Selected area electron diffraction is formed using the TEM illumination (Fig. 10.2), which is spread out over a large area of the specimen. The diffraction pattern is recorded by placing an aperture at the image plane of the objective lens as shown in Fig. 10.13. This plane is conjugate to the sample. Only electron beams passing through this aperture contribute to the recorded diffraction pattern. The diffraction pattern comes from the specimen area defined by the image of the selected area aperture for an ideal lens. With the aperture centered on the optical axis, a small area of the specimen is selected. This area is much smaller than the size of the aperture image because of the objective lens magnification. A TEM equipped with an imaging (not probe) aberration corrector comes close to providing an ideal objective lens. Without the corrector, rays belonging to a diffracted beam are at an angle to the optical axis and are displaced away from the center because of the spherical aberration of the objective lens (C_s). The displacement is proportional to $C_s\alpha^3$, where α is twice the Bragg angle (Fig. 10.13). The smallest area that can be selected in SAED is thus limited by the objective lens aberrations. This limitation is removed by using a TEM aberration corrector (see Chap. 7).

SAED is the most popular diffraction technique in TEM. The technique can be applied to study both crystalline and noncrystalline structures. The large area illumination is useful for recording diffraction patterns from polycrystalline samples or for averaging over a large volume (e.g., a large number of nanoparticles). SAED can also be used for low-dose electron diffraction, which is required for studying radiation-sensitive materials, such as organic molecules. For small area analysis, the nanoarea electron diffraction technique described next is more appropriate.

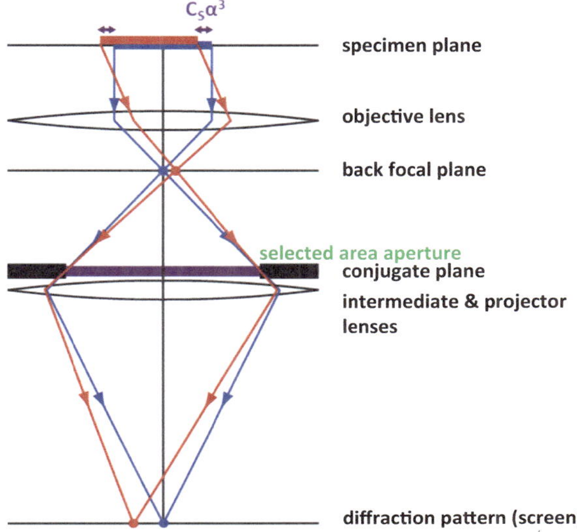

Fig. 10.13 Schematic illustration of selected area electron diffraction in conventional TEM. (Provided by Jun Yamasaki of Nagoya University, Japan)

Alternatively, an aberration-corrected TEM coupled with a small aperture can be used. For example, Yamasaki et al. demonstrated that coherent diffraction can be achieved from areas as small as ~ 10 nm using this technique (Morishita et al. 2008).

10.4.2   Nanoarea Electron Diffraction (NAED) and Nanobeam Diffraction (NBD)

NAED uses a small (nanometer-sized) parallel beam, with the condenser/objective setup shown in Fig. 10.2 (Zuo et al. 2004), or by using Kohler illumination mode together with the use of a small condenser aperture. A CM lens is required for the formation of a nanometer-sized parallel beam, as discussed in Sect. 10.2. For a condenser aperture of 10 microns in diameter, the probe diameter is ~ 50 nm with an overall magnification factor of 1/200 in the JEOL 2010 electron microscopes (JEOL, USA). The beam size is much smaller than can be achieved using a selected area aperture. Diffraction patterns recorded in this mode are similar to SAED patterns. For crystals, the diffraction pattern consists of sharp diffraction spots. The major difference is that the diffraction volume is defined directly by the electron probe in NAED since all the electrons illuminating the sample are recorded in the diffraction pattern. NAED in a FEG microscope also provides higher beam intensity than SAED (the probe current intensity using a 10-micron condenser II aperture in JEOL 2010F is $\sim 10^5$ e/s-nm^2) Zuo et al. (2004).

The small probe size allows the selection of an individual nanostructure and reduction of the background in the electron diffraction pattern from the surrounding

materials. Application examples are given in Chap. 17 to the structural characterization of individual nanoparticles and carbon nanotubes.

A focused probe can be formed by weakening the CL lens and placing the
crossover at the front focal plane of the CM lens. This results in a focused probe on
the specimen, which is placed at the focal plane of the objective prefield lens. When
using a small condenser aperture with a small convergence angle, the probe size
becomes diffraction limited in a FEG TEM, as discussed in Sect. 10.2. The
diffraction patterns recorded in this case consist of small disks (see Fig. 10.9 for an
example). This coherent CBED or nanodiffraction technique is useful for probing
local structures (Cowley 2004).

10.4.3 Convergent-Beam Electron Diffraction (CBED)

CBED is recorded using a focused electron probe at the specimen. Compared to the
diffraction techniques that we have discussed so far (Fig. 10.14), CBED differs
from STEM or NBD in terms of the convergence angle as well as the electron probe
size. In addition, in the conventional CBED mode, the incident plane-wave components of the illumination are considered to be incoherently related. The aperture
CA is also conjugate to the diffraction pattern in CBED mode. Using the additional
minilens placed above or in the objective prefield, it is possible to vary the convergence angle by changing the strength of the minilens for CBED. The resulting
convergence angle is several times larger than used in NBD, but still significantly
smaller than the convergence angle used in an aberration-corrected STEM.

As we have discussed in Chap. 3, CBED patterns are formed from transmitted
and diffracted disks (see Fig. 10.15 for an example); the size of the disk determines
the range of excitation for each reflection. Thus, the convergence angle is a very
important parameter in CBED. Its choice depends on the application. Along a zone
axis, the ideal CBED disk size is twice the Bragg angle of the lowest order ZOLZ

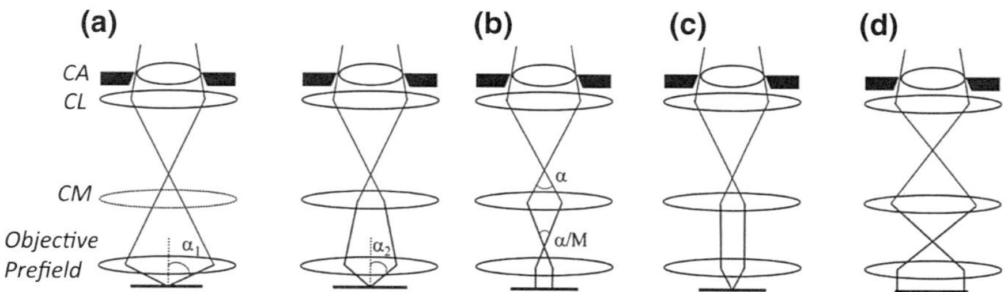

Fig. 10.14 Comparison between CBED, NAED, NBD, and TEM illumination for SAED. The
sample is located at the lower end of the diagrams

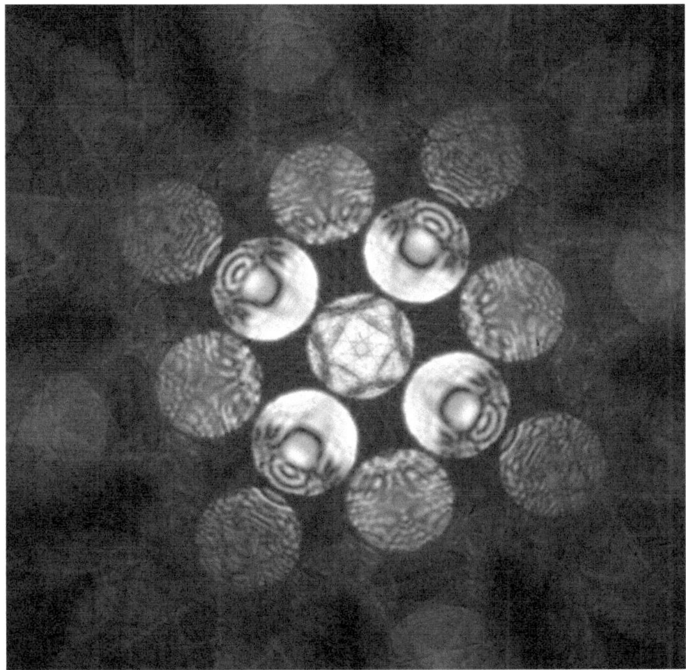

Fig. 10.15 CBED pattern recorded from spinel ($MgAl_2O_4$) at 120 kV, energy filtered, using LEO 912 TEM by Syo Matsumura, Yoshitsugu Tomokiyo, and Jian Min Zuo

reflection, in order to fill the diffraction space as nearly as possible with scattered rays. In an off-zone-axis orientation, a large CBED disk can be used to extend the number of HOLZ lines recorded in the transmitted disk. As the desired convergence angle changes from one crystal to another or one application to another, a TEM designed for CBED provides a range of excitations of the CM lens so it can be used to vary the convergence angle as shown in Fig. 10.14. The size of the CBED disk for a fixed CM lens excitation is determined by the condenser aperture size and the focal length of the probe-forming lens Eq. (10.3). Experimentally, by having several condenser apertures from a few microns to several tens of microns, it is possible to cover a range of convergence angles for many materials science applications.

So far, we have assumed a focus setting which will produce the most compact probe for CBED work. However, CBED patterns can be recorded using an under-focused or overfocused probe, provided that the sample is a parallel-sided slab of defect-free crystal. Theoretically, if inelastic scattering is ignored, it is easily shown that, for small focus changes, CBED patterns are independent of focus setting Δf. (We assume that the diffraction disks do not overlap, i.e., $\theta_c < \theta_B$.) Then, variable convergence angle is obtained by changing the CL focal length. In practice, there are two major effects to this. The first comes with an enlarged probe. As different beams now go through different parts of the sample, defocus effectively creates a projected sample image in the recorded CBED pattern. Since most of prepared samples have a small wedge angle, or are bent, the enlarged probe thus introduces a position dependence on the sample thickness and angle. (The fact that CBED patterns do change in practice when the probe is moved is mainly due to variations in thickness and crystal orientation under the probe.) This added information can be useful in the study of crystals with a large unit cell, as demonstrated by Tafto et al. (1998). In most applications, the complexity of such patterns can be avoided by using a focused probe. The second effect comes in when a highly coherent probe is used for CBED; defocus then leads to wave propagation, interference, and a complex intensity distribution. In a perfect crystal, the interference is avoided when the CBED disks do not overlap. However, in a real sample, because of surface or other imperfections, their scattering leads to interference and contrast that is highly dependent on the sample and the sample position in the under- or overfocused conditions.

If coherent CBED disks are allowed to overlap, it then becomes possible to form a scanning transmission (STEM) lattice image. By observing this STEM lattice image, it thus becomes possible (in thin crystals) to stop the probe on the region at which a CBED pattern is required. We shall see that it is quite possible by this method to obtain CBED patterns from different regions within a single unit cell and that these show different site symmetries, or alternatively, by averaging over one or several unit cells, to obtain their average symmetry. In order to obtain sufficient intensity from a probe of subnanometer dimensions, an instrument fitted with a field emission gun is needed for this type of work. For the analysis of perfect crystals, the most important benefit of a field emission gun is the improved plane-wave coherence at the specimen level. This also makes it sensitive to the contributions from defects in a real crystal. However, because of the small focused probe, the pattern has reduced contributions from thickness variations and bending under the probe.

For very thin crystals, the resulting patterns may be interpreted as electron holograms. Coherent CBED patterns formed with a very large illumination aperture have a special name, called ronchigrams. The interpretation of ronchigrams is discussed in Chap. 14 on STEM, since these provide the simplest and most accurate method of aligning the instrument and of measuring the optical constants of the probe-forming lens.

For nonoverlapping CBED disks, the question is whether these CBED patterns from very thin crystals can be interpreted quantitatively and so used to solve crystal structures? Here, we are not interested in analyzing the angular variation of intensity within the CBED disks (as for space-group determination as discussed in the next chapter), but rather using the average disk intensity to measure structure-factor magnitudes based on the kinematic (single-scattering) theory. The angular width of the rocking curve is inversely proportional to sample thickness, so that we might expect the intensity to be constant within each disk for a sufficiently thin crystal (certainly for a monolayer). Such a kinematic convergent-beam

(KCBED) (or "blank disk") method has been investigated (McKeown and Spence 2009) and found to have the following advantages:

(i)   It allows use of the smallest electron beam diameter for solving true nanocrystal structures.

(ii) Since the beam energy is spread out throughout the disks, the (000) disk intensity may be measured without saturating the detector, so that "absolute" intensity measurements can be made, comparing the intensity of the zero-order beam with the Bragg intensities.

(iii) One has a test, which is independent of the (unknown) crystal structure, for the presence of unwanted multiple scattering, if the structure is known to be noncentrosymmetric. In that case, these CBED patterns will only be centrosymmetric (in accordance with Friedel's law) if the scattering is kinematic. (Friedel's law is violated in the presence of multiple scattering.)

Experimentally, one needs to obtain good quality CBED patterns from all the major zone axes of the crystal, which may be difficult for a very small nanocrystal, depending on the degree of symmetry and radiation damage limitations. Full details of the method, as used to solve the structure of a spinel crystal with about 0.03-nm resolution, are given in McKeown and Spence (2009). Here, a three-dimensional map of the crystal potential was obtained, including the positions of the oxygen atoms. To solve the phase problem, the remarkable "charge-flipping" algorithm was used.

10.4.4   Large-Angle Methods

Various instrumental techniques have been developed to obtain an angular view of a diffracted order which is greater than the Bragg angle. These methods have been reviewed by Vincent (1989). Such an angular expansion is required for space-group determination of crystals with a large unit cell, in which overlap of low orders may occur at such a small illumination angle θ_c that little or no rocking curve structure can be seen within the orders. It has also been discovered that many narrow high-order reflections may be observed simultaneously using large-angle techniques. The intensity of these reflections appears to agree with kinematic structure factors and so can sometimes be used to assist in solving structures or to measure a static lattice modulation, for example, which modulates structure-factor intensities.

Closely related to the large-angle methods are ronchigrams and shadow images described in the chapter on STEM; however, they differ according to the angular range over which the illumination is coherent. In this section, we deal only with "incoherent" conditions and the application of techniques used to prevent the overlap of orders.

Several groups have experimented with external control of the incident beam tilt currents, which control the direction of the illumination onto the sample (Krakow and Howland 1976; Eades 1980; Tanaka et al. 1980). The essential idea is to operate the microscope in the selected area mode, so that a "point" diffraction pattern is formed. The incident beam is then rocked over a certain angular range, while ensuring that the same selected area of the sample contributes to the diffraction pattern. For every incident beam direction in this angular scan, the intensity of the transmitted beam is displayed on a video monitor. This gives a wide-angle view of the (000) CBED disk. In practice, deflection coils both before and after the sample are required, since the beam must be both "rocked" and "unrocked", so that it remains fixed on the stationary detector (see Fig. 10.18). A single scan generator drives both the deflection coils and the display. Alignment of the system (so that the beam remains on the same specimen point during the scan) and compensation for image rotation are both important. The angular resolution of the system is about 0.1 mrad, and the area of the crystal which can be studied typically has a diameter of about 250 nm, limited by spherical aberration. The angular range of the scan is typically about $3°$. We note that other diffracted beams in the "point" pattern can be excluded by using a small on-axis detector, so that an angular view which is larger than the Bragg angle can be obtained without any overlap of orders. It is also possible to display other diffracted beams ("dark-field" disks) by deflecting the pattern at the detector to bring the spot of interest onto the axis. Using multiple detectors, several orders can also be displayed simultaneously over a large angular range.

The major advantage of this "double-rocking" technique (Eades 1980) is the great flexibility of the scan patterns which can be generated (if these are under software control). "Hollow cone" illumination conditions (which result from the use of an annular illumination aperture) can be simulated readily (Tanaka and Terauchi 1985c). In particular, if the beam is tilted until one of the HOLZ lines in the outer ring is centered on the axial detector, and the beam then scanned on the surface of a cone about the axis, a wide-angle view of HOLZ structure is obtained (Kondo et al. 1984). Disadvantages of double-rocking methods include the need for new electronics, the inefficient serial detection, tedious alignment procedures, and the large specimen area from which the pattern is obtained.

The second method (known as the "Tanaka" or LACBED method, see Tanaka et al. (1980)) allows parallel detection of the entire wide-angle pattern and requires no instrumental modifications. The pattern is again, however, obtained from a rather large area of sample. A clear description of the method is given in Eades (1984). Figure 10.16 shows the principle of the method, while Fig. 10.17 shows a pattern from (111) silicon taken at 120 kV by this method. In Fig. 10.16, the CBED probe has been focused on the object plane of the objective lens, while the sample is moved up by a distance dS, forming an image of the electron source in the plane of the selected area aperture. A source image is formed in every diffracted order, as shown. The aperture can then be used to isolate one source image and so prevent other diffracted beams from contributing to the image. Because the source images are small at the crossover, the illumination cone can be opened up to a semi-angle which is larger than the Bragg angle. The price to be paid for this is the large area of sample illuminated by the out-of-focus probe at the sample. In addition, different

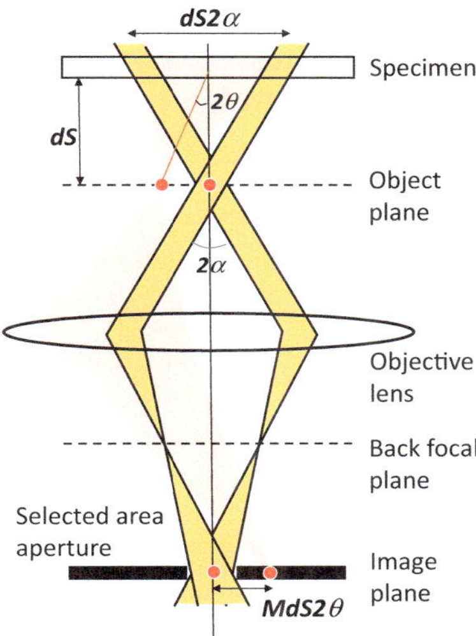

Fig. 10.16 Schematic illustration of Tanaka's LACBED method. Here, α and θ denote the half-convergence and Bragg angles, respectively, and M is the objective lens magnification (after Vincent 1989)

Fig. 10.17 Large-angle CBED pattern recorded from Si [111] at 120 kV. (Provided by John Steeds, Bristol University)

regions of the sample contribute to different parts of the diffraction pattern. Patterns may be obtained with the probe focused either above or below the sample—best results seem to be obtained with it below the sample (for TEM instruments).

The simplest procedure for obtaining LACBED patterns (in principle) is to form an in-focus image of the electron probe in the plane of the selected area aperture and then to adjust the sample height slightly to take the sample out of focus using the specimen height control on an TEM, but an alternative (more advantageous) procedure is based on changing the objective lens focus, as follows:

1. Set up the microscope in the selected area mode using a medium-sized condenser aperture. Set the eucentric height and focus the image.
2. Over focus the objective lens somewhat (excess lens current). The viewing screen will then be conjugate to a plane just below the sample, and the image of the sample will appear out of focus.
3. Adjust the condenser lens until the probe comes into focus on this same plane, just below the sample.
4. Introduce the selected area aperture and center it about the central probe image (spot).
5. Switch to diffraction mode and replace the condenser aperture with the largest size available. This aperture controls the contribution of spherical aberration to the probe size.

An important finding is that the use of the smallest selected area aperture together with the largest permissible defocus minimizes the contribution of inelastic scattering to the pattern. This effect has been studied in detail by Jordan et al. (1991) (also see Sect. 10.6). A similar technique is used to image other diffracted orders. Here, the order of interest is brought onto the optical axis using the dark-field tilt controls.

If the geometric probe size and the effects of spherical aberration are both small [see Eq. (10.13)], the diameter D of the region from which the pattern is obtained is given approximately by

$$D = 2\alpha\, d\mathrm{S} \tag{10.28}$$

where 2α is the beam convergence angle. The smallest dS which allows separation of the diffraction orders should be used to minimize D. A small geometric source image (consistent with sufficient intensity on the viewing screen) also facilitates the separation of orders—this is controlled by the demagnification settings of the condenser lenses. Under the large-angle conditions at which these patterns are formed, the effects of spherical aberration (proportional to α^3) dominate those due to diffraction. By assuming that the geometric source size is negligible, the size D of the sample region contributing to the Tanaka pattern may be calculated for various illumination semi-angles α by setting the size of the axial aberrated probe image equal to the distance between the centers of these images. Thus,

$$0.5 C_s \alpha^3 = 2\theta_B dS = \theta_B D / \alpha \qquad (10.29)$$

Here, it is assumed that the objective lens is focused on the plane of least confusion for the probe-forming lens and that the probe images just touch in the plane of the selected area aperture; θ_B is the first-order Bragg angle. We see that D increases as the fourth power of the illumination semi-angle θ_c. Tables of values for D and θ_c are given by Eades (1984). It is found that rather small selected area apertures are needed. Typically, for $\alpha = 40$ mrad, a selected area aperture of about 10 micron diameter is needed. For a first-order d_{hkl} spacing of 0.2 nm, this gives a wide-angle pattern from an area of $D = 1$ micron at a defocus of 13 microns at 100 kV.

Applications of the method can be found in Fung (1984). A variation of this method has also been demonstrated which makes it possible to record simultaneously on a single micrograph most of the CBED pattern, together with several diffracted orders at the Bragg condition (Terauchi and Tanaka 1985).

For defects, large-angle CBED technique can characterize individual dislocations, stacking faults, and interfaces (Morniroli and Gaillot 2000; Morniroli 2003; Morniroli et al. 2006). For applications to defect structures and structure without three-dimensional periodicity, parallel-beam illumination with a very small beam convergence is required.

10.4.5   Precession Electron Diffraction

Precession electron diffraction (PED) is a technique pioneered by Vincent and Midgley (1994). The principle of the method is shown in Fig. 10.18, in which for simplicity, we have omitted the CM and condenser–objective lenses above and below the specimen. In PED, the incident electron beam is made to rotate around the microscope optical axis, maintaining a constant angle—the "precession angle," by using the beam deflectors (Own et al. 2005). To compensate for motion of the diffracted beams as the incident beam rotates, the outgoing beams are deflected back using the deflectors below the specimen. The technique is similar to the double-rocking technique we discussed for the recording of LACBED patterns (Eades 1980), in which case the beam is made to scan over a rectangular area instead of precession around a circle. By recording electron diffraction patterns with the incident electron beam in precession, PED is able to provide the electron diffraction intensity integrated in angle across the Bragg condition for many reflections, provided that the recording time is much longer than the time it takes for one precession. Compared to CBED, which records the diffraction intensity for every incident beam directions, PED records one intensity integrated over the precession angle in a way similar to the rotational method in X-ray diffraction. It may be shown that this angular integration reduces the effects of multiple scattering, as first discussed by Blackman (1939) and tested experimentally by Horstmann and Meyer (1965). This can be understood whether we imagine that there is one

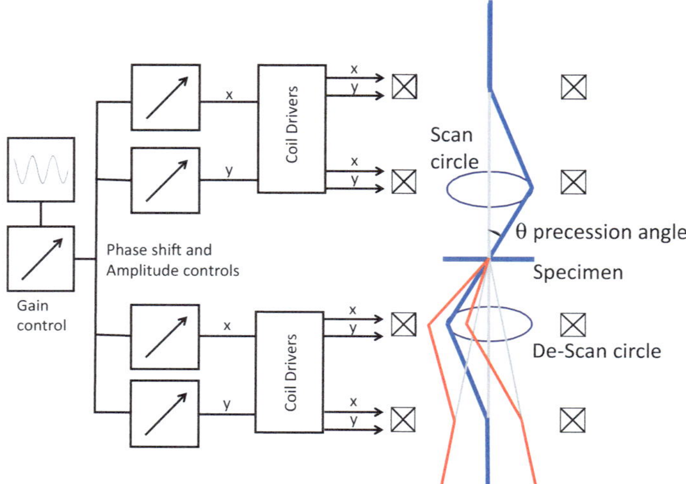

Fig. 10.18 Schematic illustration of precession electron diffraction setup and controls using the deflector coils above and below the specimen

extinction distance (in 2-beam theory) associated with every point (every excitation error) within a CED disk. By integrating over many such points, the precession signal averages over many extinction distances and so smooths out the oscillations with thickness due to the Pendellosung effect (Gjonnes 1997). Because of this unique feature, PED has found many applications in electron crystallography for solving crystal structures.

PED is implemented by driving the x and y deflection coils before and after the specimen synchronously using the oscillating sine wave obtained from a signal generator, which is the phase shifted and amplitude adjusted for the x- and y-scan drivers. The same waveforms are used to drive the coils below the specimen. This is schematically shown in Fig. 10.18. The result after careful adjustments is that at the lower part of the beam deflector coils, the incident beam scans sequentially around a circle, which is then brought back to the specimen ideally to a fixed point so the rotating incident beam form a cone of a constant angle. Thus, a focused beam should stay focused in PED, and sharp diffraction spots should stay similarly sharp. In practice, there are a number of complications in performing PED. First, because of the large precession angle, the objective prefield lens aberrations must be compensated in order to have a small beam, which becomes harder to do as the precession angle increases. Because of this, compromise is often made between the need for a small spot size and the need for large precession angle. Second, the deflection coils on both sides of the specimen are not symmetrically arranged in a TEM, since not all microscopes have deflection coils immediately after the specimen. Thus, in order to achieve good results, the excitation of the various coils must be accurately adjusted.

10.4.6   Selected Area Diffraction in STEM

The drawback of performing SAED in a conventional TEM, where the objective lens spherical aberration limits the selected area to about 100 nm or more, can be largely avoided by performing electron nanodiffraction in a STEM. There are several ways to perform selected area electron diffraction in a STEM. Sharp diffraction spots can be obtained by using the objective prefield lens to form a small parallel probe on the specimen. The diameter of the region of the specimen with near-parallel illumination depends on the diameter of the condenser aperture. Using a small aperture (10 μm or less), the illumination may be as small as a few tens of nanometers, and diffraction pattern spots are then as sharp as these obtained by a parallel beam in a TEM. For applications where having sharp diffraction spots are not so critical, such as phase identification or orientation mapping, a focused probe can be used with correspondingly higher spatial resolution.

The recording of SAED patterns can be made in conjunction with STEM imaging using an annular dark-field detector with a low-camera-length setting and a large inner cutoff angle. Because STEM imaging is performed in diffraction mode, no additional optical adjustment is needed between imaging and diffraction. Once the image is obtained, nanodiffraction patterns can be recorded in several ways:

(1)   By positioning the electron probe at specific specimen positions, selected based on the STEM image.
(2)   By applying a small, fast scan of the beam during the recording of the pattern (0.1 s exposure time or longer). Then, the area giving rise to the diffraction pattern can be increased significant beyond the diameter of the electron probe (Cowley 2004).
(3)   By recording scanning electron nanodiffraction patterns, which will be the subject of the next section.

Unlike SAED performed in TEM, the beam convergence angle is separately controlled from the selected area for electron diffraction in STEM. Because of this, some unique applications can be made. One is to acquire diffraction patterns over a small rectangular area defined by the STEM scan coils, which has special applications in atomic-resolution STEM. The condenser aperture is coherently illuminated, so that large overlapping CBED disks interfere. The interference pattern changes sensitively as the electron probe moves from one atomic column to another, contributing to the image contrast observed in bright-field STEM (Spence and Cowley 1978). Interpretation of coherent CBED (or "coherent nanodiffraction") patterns, however, is complicated because we need to know the exact probe position as well as the phase of electron waves, including the phase from lens aberrations and electron multiple scattering. In this sense, the interpretation of these patterns is exactly as complicated as the interpretation of HREM images. These patterns do, however, reveal the local point symmetry of the crystal as reckoned

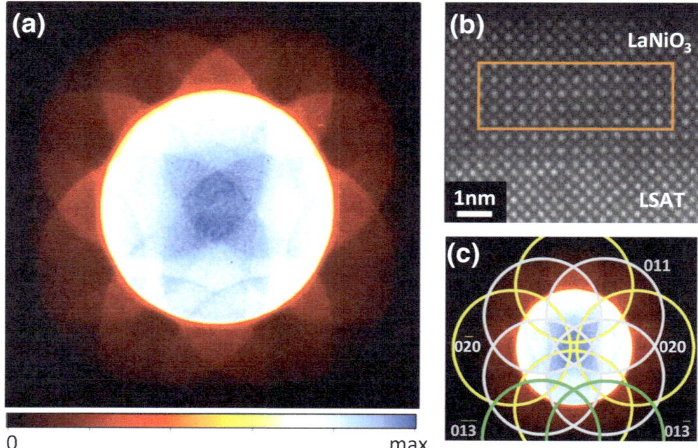

Fig. 10.19 Position-averaged CBED (PACBED). **a** Experimental PACBED pattern recroded by scanning the electron probe across the boxed area in (**b**). **b** HAADF-STEM image of a 5-nm-thick LaNiO$_3$ film on (LaSr)AlTiO$_3$. **c** Same pattern as in (**a**), with pseudocubic Miller indices. Produced from Hwang et al. (2012) with permission

about the center of the beam, and this effect has been used to locate the STEM probe on particular atoms, for collection of EELS spectra (Spence and Lynch 1982). This method has been used to determine the atomic structure and to classify the anti-phase domains which occur in alloys of CuAu (Zhu and Cowley 1983) and is reviewed in Cowley and Spence (1981).

By averaging over a region of specimen, the so-called position-averaged CBED (PACBED) removes all the interference between overlapping CBED disks (LeBeau et al. 2010a; Hwang et al. 2012). As the example in Fig. 10.19 shows, the patterns show a remarkable resemblance to CBED patterns recorded with an incoherent probe. In an aberration-corrected STEM, the electron probe can be smaller than 1 Å. The smallest specimen region that can be scanned in order to fully remove the coherence effect is a unit cell. The actual volume probed in a PACBED experiment depends on electron probe propagation. Since the electrons are no longer confined to a single atomic column as in a channeling situation, the actual volume is larger than the region scanned by the electron probe. Nonetheless, PACBED has the highest spatial resolution among all diffraction techniques for probing structure on the scale of the unit cell.

A major application of PACBED is the determination of crystal thickness for quantitative analysis of STEM image contrast. This technique when combined with quantitative techniques described in later chapters could be used to study local symmetry, polarization, and crystal stoichiometry.

10.4.7   Scanning Electron Nanodiffraction

Using the deflection coils, scanning electron nanodiffraction (SEND) patterns can be recorded from an area of the sample for every probe position, to provide spatially resolved structural information. This can be done either in a TEM or STEM. Diffraction patterns are recorded using a two-dimensional digital detector, for example, a CCD camera. Compared to PACBED, which records one diffraction pattern over many probe positions, SEND collects the full 4D data, in the form of two spatial coordinates, the (x, y) in the real space and the (k_x, k_y) in the reciprocal space. Digital recording and storage of large dataset are relatively new. Diffraction pattern recording had been a particular issue in the early dedicated STEMs, which did not have the camera system installed in the conventional TEMs. Early attempts by Cowley used a video recorder for scanning electron nanodiffraction (Cowley (1993)). However, the quality of diffraction patterns recorded this way was poor, and data analysis of taped videos was difficult. These issues are solved by using CCD cameras for diffraction pattern recording in SEND. (Characteristics of CCD cameras for electron image recording are described in Chap. 9.) The state-of-the-art CMOS cameras are also capable of fast readout. This, combined with the computing power of modern computers, has improved the acquisition and processing of electron diffraction patterns greatly.

Once the 4D dataset is collected using SEND, bright- and dark-field STEM images can be obtained simultaneously from SEND in the simplest form of analysis by integrating the diffraction intensities of the direct beam and diffracted beams, respectively. In this way, SEND works like STEM. A major distinction is that with the diffraction patterns recorded and stored, other information can be extracted off-line to form images, beyond the simple integrated intensities. For example, diffraction patterns can be indexed and analyzed for orientation and phase mapping. This analysis can be done at a very fine scale when the SEND patterns are acquired using an nanometer-sized probe and step size, which is unique to electron diffraction. This last option is simply not available using the fixed, STEM detectors. The trade-off here, of course, is that one will be dealing with a far more complex, and larger, dataset.

In SEND, the speed of the probe scan is limited by the diffraction pattern acquisition time. When an Å-sized electron probe is used in SEND, the electron dose density (the number of electrons per unit area) is very high. Only radiation hard materials can be studied in this way. The electron dose density can be lowered significantly by using a larger electron probe or by lowering the brightness of the electron gun. Because total electron elastic scattering cross sections are larger than inelastic for elements heavier than copper, electron diffraction patterns can be recorded at a lower dose level for heavier elements. Nevertheless, there are major benefits in reducing radiation damage by using low-dose SEND to study radiation-sensitive materials, even for lighter organic molecules. Downing and Glaeser (1986) showed that electron images recorded using illumination spots of 100–200 nm from thin paraffin crystals and purple membrane improve the image contrast by a factor of 3–5 compared to electron images taken with a large illumination spot of 3 μm. The improvement in image contrast was attributed to the reduced beam damage induced by specimen movement. In SEND, the beam damage is limited to only area of the specimen illuminated by the electron beam, and thus, each diffraction pattern is recorded under nearly identical specimen condition. The new "direct electron" detectors take advantage of this effect also, by summing many very brief exposures for which the effects of beam-induced motion are corrected during data merging.

Scanning electron diffraction can be carried out by first selecting an area of interest, dividing this area into a number of pixels, placing the electron probe at each of these pixels, and recording the diffraction patterns at each pixel (Zuo and Tao 2011). Data acquisition is automated using either dedicated hardware to synchronize the scan and diffraction pattern (NanoMegas SPRL, Brussels, Belgium) or computer control of the TEM and the electron camera. An implementation of SEND using the second approach is reported by Kim et al. (2015), which involves the automation of TEM deflection coils and diffraction pattern acquisition using a custom script written in the DigitalMicrograph® (DM, Gatan Inc, Pleasanton, CA) script language. The electron microscope is controlled using the script by communicating with the host processor built into the TEM. This technique does not require additional hardware other than the computer and the electron detector that are already installed on the TEM. The main drawback is that the speed of acquisition is limited by the camera readout speed or the speed of beam deflection inside the TEM, whichever is slower.

In the method reported by Kim et al. (2015), the electron beam scanning is performed in TEM mode and carried out using the deflection coils to shift the beam under computer control. Two types of access to the TEM are used for the scanning process: The first retrieves the values of the illumination deflection coils and stores the values as real number x and y, and the second shifts the electron beam by the amount x and y. The x and y values, however, only refer to the setting of the deflection coils, which need to be calibrated into distances in nanometers. For this purpose, two scanning vectors are established along the vertical and horizontal directions. The calibration is carried out under a standard magnification in TEM mode. The reference value of (x_1, y_1) is first obtained from the initial beam position. The electron beam is then horizontally shifted to position "2," and (x_2, y_2) are obtained. Using the calibrated magnification, the distance (d) between "1" and "2" can be set to a fixed value. Then, the horizontal scanning vector $(\vec{R}_x)$ can be calculated as follows:

$$\vec{R}_x = \frac{1}{d}(x_2 - x_1, y_2 - y_1) \ (\text{nm}) \tag{10.30}$$

Similarly, the vertical scanning vector $(\vec{R}_y)$ can be calculated from (x_1, y_1) and (x_3, y_3) as follows:

$$\vec{R}_y = \frac{1}{d}(x_3 - x_1, y_3 - y_1) \ \text{(nm)} \tag{10.31}$$

Once calibrated, the electron beam can be shifted to a specific position by a combination of the two vectors in Eqs. (10.30) and (10.31).

For recording electron diffraction patterns, a CCD camera mounted on axis under the microscope column can be used. The typical exposure time per diffraction pattern is 0.1 s.

Figure 10.20 shows an example of SEND applied to a nanostructured Au disk. The SEND patterns were acquired over the area of 210×210 nm^2 in 30×30 pixels, corresponding to a step size of 7 nm. The diffraction patterns were recorded over a period of 10 min. Figure 10.20b shows one of 900 diffraction patterns acquired from SEND. The diffracted beams appear as small disks corresponding to 4.2 mrad of full convergence angle. The electron probe was formed in a JEOL TEM with the LaB$_6$ gun at the low-dose condition (Kim et al. (2015)). To demonstrate the imaging capability of SEND, the diffraction intensity between two circles as marked in Fig. 10.20b was integrated from the diffraction patterns. The intensity sum for every single diffraction pattern was then mapped in the raster image, as shown in Fig. 10.20c. For the mapping, three regions of the diffraction pattern were

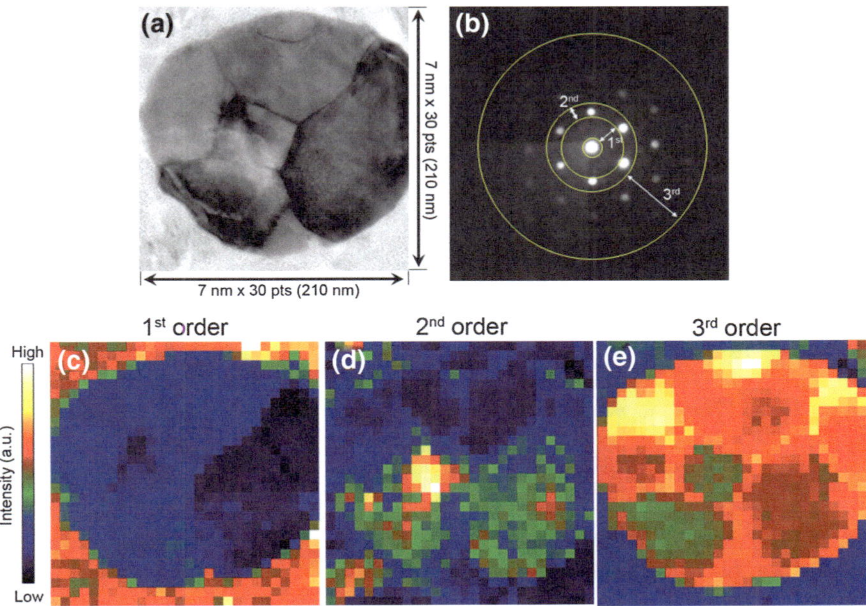

Fig. 10.20 SEND of a small Au disk (**a**) A BF image of nanostructured Au disk and (**b**) a selected diffraction pattern acquired from SEND. The diffraction intensity is integrated for the areas of 1, 2, and 3 represented in (**b**). The corresponding intensity maps are shown in (**c**), (**d**), and (**e**), respectively

selected as marked in Fig. 10.20c, d, and e: (1) an annular area between the direct beam and the first ring (marked as 1), (2) the second ring (marked as 2), and (3) the remaining area of the third ring, akin to the use of an annular dark-field (ADF) detector in STEM. For the first region (Fig. 10.20c), the amorphous region (C film) has high intensity, while the Au nanodisk shows low intensity. This is expected since the amorphous scattering is strong where there are no Bragg spots from the Au nanodisk. Figure 10.20d shows the variation in the integrated intensity over the grains of Au nanoparticles. This reflects the orientation change across the grains.

When acquiring SEND patterns, ideally the center of the (000) reflection in the diffraction pattern should stay at the same location on the detector, while the electron probe is scanned across the sample. Improper alignment of microscope, however, can lead to diffraction pattern shift during beam scanning, which affects the post-diffraction data analysis. To minimize this effect, first, the standard procedures for microscope alignment are followed for correcting the shift-tilt purity of the illumination deflection coils. The optical axis alignment of the objective lens is set using the bright-field tilt and high-voltage centering. This is followed by the adjustment of the intermediate lens focus, since defocus of this lens can have a large effect on diffraction pattern shift. The pattern movement can be minimized by changing the diffraction focus with the help of the beam shift function on the TEM, which applies a wobble to the deflection coils causing the electron beam to shift back and forth on the sample automatically without tilting the beam. The diffraction focus is adjusted until the direct beam does not move in the diffraction mode.

SEND can also be carried out in combination with precession for scanning PED. This technique will be further discussed in Chap. 16 on the measurement of strain.

10.5　Specimen Holders and Rotation

The popular double-tilt specimen holder used for electron crystallography has one rotation axis parallel to the axis of the holder with its rotation angle commonly referred as primary or x-tilt. The second rotation axis is perpendicular to the axis of the holder (Fig. 10.21). This rotation angle is called secondary or y-tilt. While the x-tilt axis is always perpendicular to the electron beam, the Y-axis remains in the plane of the specimen and its angle with the electron beam changes as the specimen is rotated about the x-tilt axis. The x-tilt axis is in the plane of the specimen only when the y-tilt is zero. In addition to the specimen tilt, the holder can be mechanically shifted along and perpendicular to the holder axis in three directions.

To determine the orientation of the electron beam relative to the specimen and the specimen position after rotation, it is helpful to define two coordinate systems: One is the microscope coordinate $(\vec{x}_m, \vec{y}_m, \vec{z}_m)$—this is fixed relative to the microscope as the name implies, and the other is the specimen coordinate $(\vec{x}_s, \vec{y}_s, \vec{z}_s)$ fixed to the sample. We take $\vec{x}_m$ to be along the double-tilt holder x-tilt axis, $\vec{z}_m$ along the microscope optical axis, with its direction opposite to the electron beam, and $\vec{y}_m$

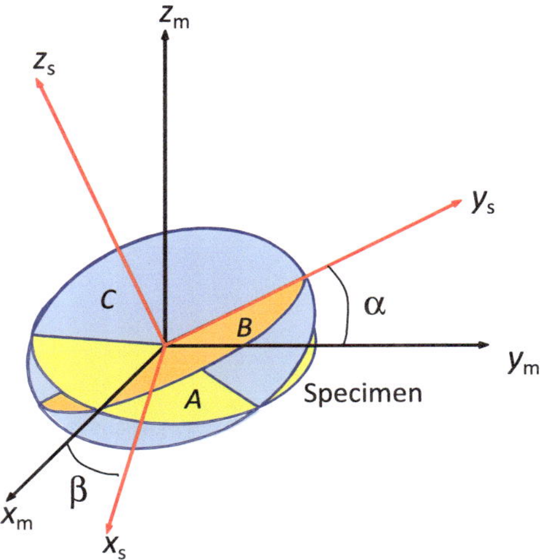

Fig. 10.21 The microscope coordinate and the specimen coordinate in a double-tilt specimen holder. A, B, and C marks the initial, intermediate, and final specimen planes and their relationship with the two tilt angles α and β

given by the vector product of $\vec{z}_m$ and $\vec{x}_m$. For the specimen coordinate system, we choose it so that $\vec{z}_s$ is the normal to the specimen, $\vec{y}_s$ is the direction in the plane of the specimen along the y-tilt axis, and $\vec{x}_s$ is the vector product of $\vec{y}_s$ and $\vec{z}_s$. The origins of the two coordinates are taken as the same at the intersection of the x- and y-tilt axes and with the z position, the so-called eucentric height position, which corresponds to the height of the specimen inside the TEM at which its image seen on the microscope screen no longer moves laterally as the specimen tilts. The two coordinates can be related by following the changes in the coordinates of the specimen coordinate axes in the microscope coordinate system. First, the x-tilt transforms the initial specimen coordinate $(\vec{x}_{s0}, \vec{y}_{s0}, \vec{z}_{s0})$ to $(\vec{x}_s', \vec{y}_s', \vec{z}_s')$ according to

$$\left(\vec{x}_s', \vec{y}_s', \vec{z}_s'\right) = (\vec{x}_{s0}, \vec{y}_{s0}, \vec{z}_{s0})\,T_\alpha \tag{10.32}$$

with

$$T_\alpha = \begin{pmatrix} 1 & 0 & 0 \\ 0 & \cos\alpha & -\sin\alpha \\ 0 & \sin\alpha & \cos\alpha \end{pmatrix}. \tag{10.33}$$

Similarly, the y-tilt transforms the intermediate specimen coordinate $\left(\vec{x}_s', \vec{y}_s', \vec{z}_s'\right)$ to the final coordinate $(\vec{x}_s, \vec{y}_s, \vec{z}_s)$ by

$$(\vec{x}_s, \vec{y}_s, \vec{z}_s) = \left(\vec{x}_s', \vec{y}_s', \vec{z}_s'\right)T_\beta \tag{10.34}$$

with

$$T_\beta = \begin{pmatrix} \cos\beta & 0 & \sin\beta \\ 0 & 1 & 0 \\ -\sin\beta & 0 & \cos\beta \end{pmatrix}. \tag{10.35}$$

Together, we have

$$(\vec{x}_s, \vec{y}_s, \vec{z}_s) = (\vec{x}_m, \vec{y}_m, \vec{z}_m) T_\alpha T_\beta = (\vec{x}_m, \vec{y}_m, \vec{z}_m) T \tag{10.36}$$

with

$$T = \begin{pmatrix} \cos\beta & 0 & \sin\beta \\ \sin\alpha\sin\beta & \cos\alpha & -\sin\alpha\cos\beta \\ -\sin\beta\cos\alpha & \sin\alpha & \cos\alpha\cos\beta \end{pmatrix}. \tag{10.37}$$

In the above equation, the order of the tilts, or how they are carried out, does not matter. For example, it is not essential to know whether the x-tilt is done first followed by the y-tilt or vice versa. It also makes no difference whether the amount of tilt is made in a single or multiple steps. Only the final tilt angles and their signs are needed relate the microscope and specimen coordinates.

From the above result, we have the following cosine relationships between the specimen coordinate and the microscope axis, $\cos(\vec{z}_m \wedge \vec{x}_s) = -\sin\beta\cos\alpha$, $\cos(\vec{z}_m \wedge \vec{y}_s) = \sin\alpha$, and $\cos(\vec{z}_m \wedge \vec{z}_s) = \cos\alpha\cos\beta$, as given in the paper by Kelly et al. (1994). These cosines then give the direction of the electron beam (represented by $\vec{z}_m$) relative to the specimen axes as follows:

$$\begin{aligned} (\cos(\vec{z}_m \wedge \vec{x}_s), \cos(\vec{z}_m \wedge \vec{y}_s), \cos(\vec{z}_m \wedge \vec{z}_s)) \\ = (-\cos\alpha\sin\beta, \sin\alpha, \cos\alpha\cos\beta) \end{aligned} \tag{10.38}$$

Electron diffraction is often used to determine the angle between two crystal grains. In such applications, it is helpful to obtain the angle between two electron beam directions in the specimen coordinates. The direction cosines of the electron beam relative to the specimen axes after the specimen is tilted to another orientation by angles α' and β' are obtained simply by substituting α and β with α' and β' in Eq. (10.38). The angle θ between two specimen orientations is given by the scalar product of the two directional vectors of the electron beam, given by

$$\cos\theta = \cos(\beta' - \beta)\cos\alpha\cos\alpha' + \sin\alpha\sin\alpha'. \tag{10.39}$$

A similar relationship can be obtained for the tilt-rotate holder, in which the second rotation axis is fixed along $\vec{z}_s$. Details are given in the paper by Kelly et al. (1994). For the case of a single-tilt holder, $\beta = 0$, then we have simply

$$\cos \theta = \cos \alpha \cos \alpha' + \sin \alpha \sin \alpha' = \cos(\alpha' - \alpha). \qquad (10.40)$$

The projected position of an area of interest on the specimen changes as the specimen is rotated using the double- or single-tilt holder. If we define the specimen position by its coordinates on the specimen axes as (x_s, y_s, z_s), then its microscope coordinate can be obtained using the coordinate transformation of Eq. (10.36)

$$\vec{r} = (\vec{x}_s, \vec{y}_s, \vec{z}_s) \begin{pmatrix} x_s \\ y_s \\ z_s \end{pmatrix} = (\vec{x}_m, \vec{y}_m, \vec{z}_m)T \begin{pmatrix} x_s \\ y_s \\ z_s \end{pmatrix} \qquad (10.41)$$

Thus,

$$\begin{pmatrix} x \\ y \\ z \end{pmatrix} = T \begin{pmatrix} x_s \\ y_s \\ z_s \end{pmatrix} \qquad (10.42)$$

This equation can be used to keep the specimen area of interest approximately at the center of the screen in an automated electron diffraction experiment using a computer-controlled goniometer and a double-tilt holder. It is approximate because the precision of a TEM specimen holder, while very good, is not sufficient at high magnification. Other factors influencing the specimen position and accuracy during rotation are mechanical backlash in the tilt or shift mechanisms. Kelly et al. (1994) found that an accuracy of better than $0.2°$ is achievable for the x-tilt. The y-tilt has a slightly bigger error.

10.6   Energy Filtering

Removal of the inelastic background is an option when using an electron microscope equipped with an electron energy filter. The purpose of the filter is to remove from diffraction patterns all those electrons which, on traversing the sample, loose more than a few electron volts in energy and contribute to the background intensity. The important inelastic processes are phonon or plasmon scattering and single-electron excitation. [For a review of energy-loss processes in electron microscopy, see Botton (2007) and Egerton (2011).] Phonon scattering involves relatively large inelastic scattering angles, but very small energy losses (perhaps 30 meV). These are not excluded by elastic filtering. Plasmon losses involve larger energies (about 15 eV) and small scattering angles. Plasmon excitations or higher energy losses can be filtered out by dispersing the electrons according their energies using the magnetic or electrostatic fields inside an electron energy filter and using a slit equivalent to a few eV in width around the elastic (zero-loss) electron beam. The advantages of zero-loss energy filtering are outlined in more detail below in the discussion of imaging filters, however, a glance at Fig. 10.15 will indicate the improvement in the quality of data to be expected. This improvement affects all the diffraction techniques discussed in this book, especially for quantitative CBED. Duval et al. (1970) was the first to demonstrate that by placing the diffraction pattern at the object plane of the imaging filter, most of the background intensities disappeared especially at small scattering angles. The other major application of energy filtering is electron spectroscopic imaging (ESI) used for composition mapping (Reimer 1995).

The thickness of the sample is important in determining the need for filtering. In very thin samples, there is little inelastic scattering and hence no requirement for elastic filtering. But such ideal samples only exist in 2D materials like graphene. For other materials, very thin samples are known to be strongly influenced by thin-film relaxation effects (Treacy and Gibson 1986). Thus, for the study of thicker material which is more representative of the bulk, elastic energy filtering combined with a detector system with large dynamic range is required for quantitative intensity analysis. For example, in order to obtain accuracy in structure-factor measurement or atomic position determination using dynamical scattering comparable to that achieved in X-ray crystallography, an energy filter, tuned to the elastic peak, is essential. An image of a wedge formed from elastically filtered scattering will be seen to reach a maximum intensity at some thickness, beyond which it becomes dark, since at larger thickness, virtually all electrons have been inelastically scattered. Other benefits of zero-loss energy filtering are as follows:

1. The dramatic reduction in background significantly improves the contrast of fine HOLZ lines and other features and thus enables the measurement of strain and determination of symmetry at greater accuracy.
2. It enables the study of diffuse scattering whether it is thermal or it comes from defects or from modulations of the crystal structure. In relatively thick samples, the diffuse scattering is buried by the inelastic background if elastic energy filtering is not used.
3. Filtering allows much thicker crystals to be examined without incurring the penalty of radiation damage, which would result if higher accelerating voltages were used.
4. The use of greater thickness (without background) for the study of defects. This is new information, which was not previously extractable due to the presence of the background.

There are several types of in-column energy filters that are named according to the shape of the electron path, such as the Ω, α, or γ energy filters. The other is the post-column Gatan imaging filter (GIF). The in-column filter is placed between the intermediate and the projector lenses of the TEM and can be used in combination with all forms of electron detectors. The GIF is integrated with an electron camera and placed below the camera chamber. Its use for electron diffraction typically requires switching the TEM to a special low-camera-length setting.

The optics of an imaging energy filter is shown in Fig. 10.22. While the details can differ, all energy filters have the optical elements of entrance image plane,

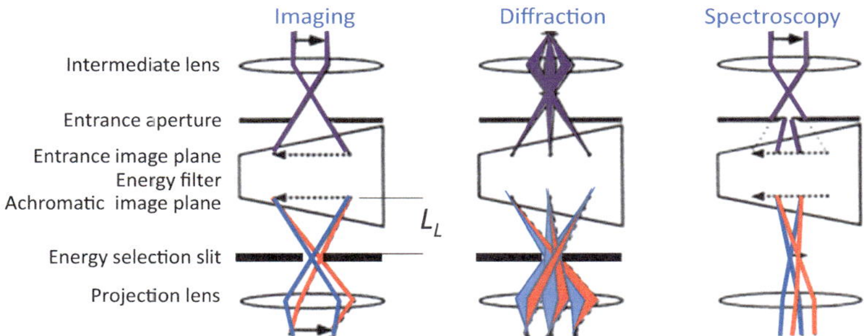

Fig. 10.22 The optics of energy filter for electron imaging, diffraction, and energy-loss spectroscopy (modified from Tanaka 1999, with permission)

achromatic image plane (where a focused image is formed without the separation of "color" or energy), and an energy dispersion plane, where the energy selection slit is placed. The intermediate and projector lenses here refer to the lenses immediately above and below the energy filter, which are the part of intermediate and projection lens systems in a TEM for the in-column energy filters. In GIF, they are replaced by focusing coils and additional multipoles (see Sect. 10.6.6 for details). The function of the intermediate lens is to transfer the image or diffraction pattern to the image entrance plane, while the projection lens looks at the achromatic image plane or the energy dispersion plane in the imaging and diffraction mode or spectroscopic mode, respectively. A diffraction pattern is formed at the energy dispersion plane when we have an image at the achromatic image plane, or, in reverse, an image is formed at the energy dispersion plane when we have a diffraction pattern at the achromatic image plane.

Figure 10.22 shows the striking improvement which results in the quality of CBED patterns from the use of a Ω-energy filter, even when film recording is used (Mayer et al. 2007). These patterns were recorded on the Zeiss Omega model 912 TEM-STEM. The exposure time used in Fig. 10.22 was 1 s for the unfiltered recording and 3 s for the filtered pattern. These times were arranged to produce an approximately equal optical density on the film, so that a valid comparison could be made. Before the imaging filters became widely available, energy filtering was performed by scanning the electron image or diffraction pattern over an EELS system with a point or array detector, by measuring the zero-loss peak intensity. The image recording time for a $10^3 \times 10^3$ pixel image using the Omega system is at least 1000 times less than that required by the scanned readout system, for the same dose. (This assumes a parallel detector for the EELS capable of 1D imaging.) The advantages of the Omega filter increase rapidly with the number of pixels. Since the experimental observation in Fig. 10.22 is that the background between the Bragg reflections is almost entirely removed by elastic filtering, we must conclude that this background is due to multiple, coupled phonon and plasmon scattering. The phonon scattering events provide the large angular change, and the associated plasmon losses then allow these electrons to be removed by elastic filtering. This interpretation is consistent with the relatively large thickness used ($t = 270$ nm).

For CBED, an optimum sample thickness exists with energy-filtered data. For very thin crystals, there is little inelastic scattering and so no requirement for filtering. However, the CBED disks show no useful contrast variation. At very large thickness, all scattering is inelastic, and no elastic signal can be recorded. In the simplest model (Hirsch et al. 1977), the thickness dependence of, for example, the plasmon-loss electrons is given by the product of the multiply scattered Bragg beam intensity with an appropriate term of the Poisson distribution. The zero-loss intensity will be given, as a function of thickness, by the expressions in Chap. 5.

For the large-angle shadow-imaging techniques described in Sect. 10.4.4, we have pointed out that the selected area aperture used in a plane conjugate to the electron source performs an energy-filtering function. If such an image is formed on an instrument fitted with an imaging energy filter (which normally has insufficient energy resolution to exclude the phonon-scattered background), this aperture may be used to exclude the phonon scattering, while the energy filter excludes larger energy-loss processes which scatter through smaller angles. By varying the height of the aperture, the cutoff angle may be varied. This technique has been analyzed in detail by Jordan et al. (1991). In this way, a very great improvement in the quality of LACBED patterns should be possible.

10.6.1   First-Order Focusing by Magnetic Sectors

Energy selection is made by bending the electron beam in an electric or magnetic field. As beam bending depends on the electron energy (see Eq. (10.26)), the beam disperses, producing a spectrum of energies. Within an electric field, electrons are deflected and accelerated along the direction of the electric field. On the other hand, electrons are bent normal to the velocity of the electrons, while electron energy is conserved in a magnetic field. For this and other practical reasons, most electron energy filters use the fields produced by magnetic sectors (see Fig. 10.23). In the literature, they are also called bending magnets or magnetic prisms. An electron energy filter uses at least one or more bending magnets.

For use with an area detector for electron energy filtering, the filter must have double-focusing capability to function also as an imaging lens. In an electron energy spectrometer used for EELS, a strong focus along the energy dispersion direction is also required in order to achieve high-energy resolution.

In what follows, we will show that a circular magnetic sector with a small gap (as shown in Fig. 10.23) functions like an imaging lens in addition to electron beam bending and energy dispersion. A full analytical treatment of the first-order focusing and aberrations of magnetic sector fields taking into account the higher order field terms has been given by Rose (1995). The discussion below is based on the more intuitive approach provided by Enge (1967) using simplified models of the magnetic fields. It is used to treat the first-order focusing and dispersion of magnetic sectors.

Fig. 10.23 Comparison of filtered and unfiltered CBED patterns, recorded on film (provided by Joachim Mayer, MPI)

We start with the following model. The magnetic sector has a uniform B field and the median radius of R, where $R = mV/eB$ is the cyclotron radius of the electron moving inside B. In a magnetic sector with a small gap, the fringe field can be made to fall off rapidly (Rose 1995)). Thus, to first order, we use the SCOFF (sharply cutoff fringe field) approximation introduced in Chap. 7 for the magnetic sector field. We use cylindrical coordinates (r, θ, y) with y in the out-of-plane direction in Fig. 10.23. Then, according to Newton's second law, we have the following equations of motion

$$m\frac{\mathrm{d}\dot{r}}{\mathrm{d}t} = mr\dot{\theta}^2 - er\dot{\theta}B$$
$$m\frac{\mathrm{d}\left(r^2\dot{\theta}\right)}{\mathrm{d}t} = er\dot{r}B \tag{10.43}$$

where $\dot{r} = dr/dt$, $\dot{\theta} = d\theta/dt$, and m is the relativistic mass of the electron. Next, we adopt curved axes for the moving electron as shown in Fig. 10.23. The z-axis is taken as the optical axis of the magnetic sector, which follows the path of the electrons entering and traveling along the center radius of the magnetic sector, while the x and y axes are normal to z in the horizontal and vertical directions, respectively. In this coordinate system, $r = R + x$ where x is small compared to R for a small diameter electron beam. Considering this and $\dot{\theta} = -\omega$ when $r = R$, to first order, we have from the second equation of Eq. (10.43),

$$R^2 \left(\dot{\theta} + \omega \right) \approx \omega R x \text{ and } \dot{\theta} \approx -\omega \left(1 - \frac{x}{R} \right), \tag{10.44}$$

where $\omega = eB_z/m$ is the cyclotron frequency. Substituting Eq. (10.44) into the first equation in Eq. (10.43), we have, to first order,

$$\frac{d^2}{dt^2} x + \omega^2 x = 0,$$

Which transform into the following equation using $z = R\omega t$:

$$\frac{d^2}{dz^2} x + \frac{1}{R^2} x = 0. \tag{10.45}$$

Along the vertical direction, since the field is uniformly vertical inside the magnetic sector, we have

$$\frac{d^2}{dz^2} y = 0. \tag{10.46}$$

Equations (10.45) and (10.46) have simple solutions of the form of

$$x(z) = a_1 \cos(z/R) + a_2 \sin(z/R)$$
$$y(z) = b_1 + b_2 z$$

The coefficients in the above equations are determined by the initial conditions for the electron trajectory following the same procedures we used in Chap. 6. For the principle ray u_π ($u = x$ or y), the ray comes from the left side of the magnetic sector at $u_\pi(0) = 1$ and parallel to the optical axis ($u'_\pi(0) = 0$). This gives the principle ray u_π in the x and y sections as

$$x_\pi(z) = \cos(z/R).$$
$$y_\pi(z) = 1 \tag{10.47}$$

The asymptotic rays for these principle paths in the SCOFF approximation are determined by their position and slope at $z = \phi R$. Focus only occurs in the x direction; it intersects the optical axis at

$$z_{Fx} = R\left(\phi + \frac{\cos \phi}{\sin \phi}\right). \tag{10.48}$$

It intersects the incident asymptotic ray on the left at

$$z_{Hx} = R\left[\phi + \frac{\cos \phi - 1}{\sin \phi}\right]. \tag{10.49}$$

The distance between the focal and principle points gives a focal length of

$$f_x = \frac{R}{\sin \phi}. \tag{10.50}$$

Using the same analysis, the focal length and the principle plane positions on the left can also be determined. They are the same when the sector is symmetric. Thus, the sector magnet acts as a thick lens with a focal length given by Eq. (10.50) in the horizontal (x) direction, with principal planes at a distance from the entrance or exit surfaces, respectively, given by Eq. (10.49). Since no focus occurs in the vertical direction (y direction), $f_y = \infty$.

The focal length of a magnetic sector can be adjusted by introducing an inclined entrance surface as shown in Fig. 10.24. The inclination is measured by the angle β, which can be either positive or negative. With positive β, the effect of inclination is to remove or add part of the field from the electrons traveling at greater or smaller distance from the sector median, respectively. This is equivalent to introducing two wedged sectors with opposite fields (see Fig. 10.24b). The strength of these fields is the same as in the main magnet sector, so that the electron radius of curvature equals R, as in the main magnet. A change in the slope is introduced for electrons entering at position x_0 with

$$x' = x_0 \tan \beta / R.$$

The effect of this is shown in Fig. 10.24, and it introduces a small divergence, equivalent to a thin divergent lens that increases the overall focal length. For negative β, the effect is the opposite with a reduced focal length. Thus, by using different inclination angles at the entrance and exit surfaces, a magnetic sector with different front and back focal lengths can be designed.

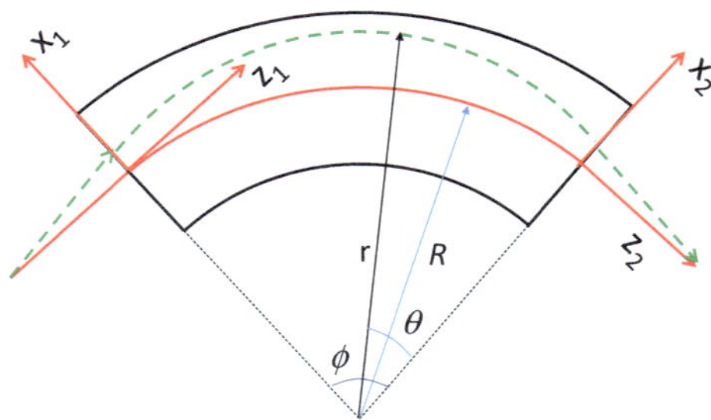

Fig. 10.24 Electron motion in the sector magnetic field

10.6.2   Energy Dispersion

So far, we have only considered incident electrons with the same velocity and thus the same energy. These electrons have the same orbital radius given by $p = mv = eBR$. To find the effect of an energy change ΔE, we note that to this, new energy corresponds a new momentum and a new equilibrium orbit with

$$p + \Delta p = p(1 + \delta) \approx p\left(1 + \frac{\Delta E}{2E}\right) = eB(R + \delta x)$$

and

$$\delta x = R\frac{\Delta E}{2E}. \tag{10.51}$$

The shift δx defines a new optical axis for the electrons with momentum $p + \Delta p$. The path of an electron entering at $x = 0$ with a slope of $x' = 0$ and momentum $p + \Delta p$ can be shown to have the following form

$$x(z) = \delta_x[1 - \cos(z/R)]. \tag{10.52}$$

The energy-dependent shift in the optical axis thus gives rise to an energy dispersion proportional to the change in the energy and the cyclotron radius R. For example, if we take $R = 10$ cm, which is obtained with $B = 0.011$ T for the electrons of 100 keV in energy, a change of 1 eV gives a dispersion of 0.5 μm for a 90° magnetic sector. A dispersion of 1 μm/eV for 200 keV electrons is typically targeted in an electron energy filter, which is achieved through the use of multiple magnetic sectors for in-column filters, or the use of additional quadrupole lenses for magnification in post-column filters.

An illustrative example is the recently developed high-energy resolution electron monochromator developed by Krivanek et al. (2009). A bending radius of 10 cm is used to produce an energy dispersion of 0.3 μm eV^{-1} for 200 keV electrons. Together with two quadrupole lenses placed in front of it, the first magnetic sector demagnifies the electron source image by 6 times to about 1 nm. Thus, the maximum energy resolution which can be obtained is 3.3 meV. To be able to use a larger energy selection slit, a system of four quadrupole lenses placed immediately after the first magnet prism is then used to provide a variable magnification of the spectrum from 0 μm eV^{-1} (used to cancel the sector dispersion) to 200 μm eV^{-1}. At 200 μm eV^{-1}, a 2-μm slit gives 10 meV energy resolution.

10.6.3   Vertical Focusing Using Fringing Fields

We now show how the fringe field, which has been neglected so far in the SCOFF approximation for a sector magnetic field, can be used to provide focus along the vertical direction, following the derivation given by Enge (1967). Figure 10.25 shows a schematic distribution of the fringe field of a magnetic sector. At the edge, the field decreases away from the magnet over a length comparable to the magnetic gap. For the vertical field, the SCOFF is a reasonable approximation with a small gap. Along the radial direction, the fringing field produces a field component away from the middle section for an inclined surface. This field generates a force on the electrons along the vertical direction, according to Newton's second law,

$$m\frac{d^2y}{dt^2} = -er\dot{\theta}B_r, \tag{10.53}$$

This can be transformed into the following equation using $z = vt$ and $v = r\dot{\theta}$

$$\frac{d^2y}{dz^2} = -\frac{1}{R}\frac{B_r}{B} \tag{10.54}$$

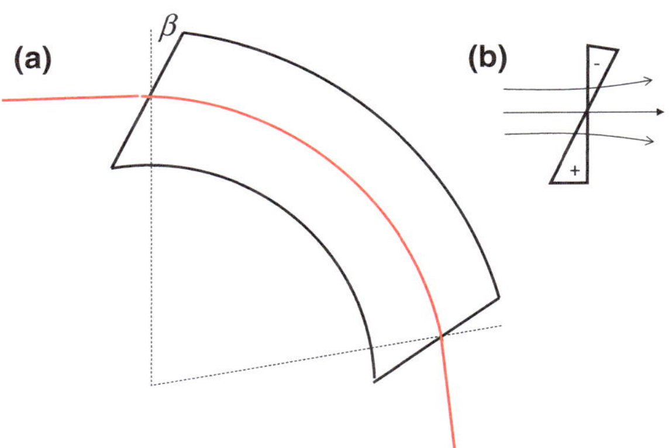

Fig. 10.25 Inclined entrance and exit surfaces

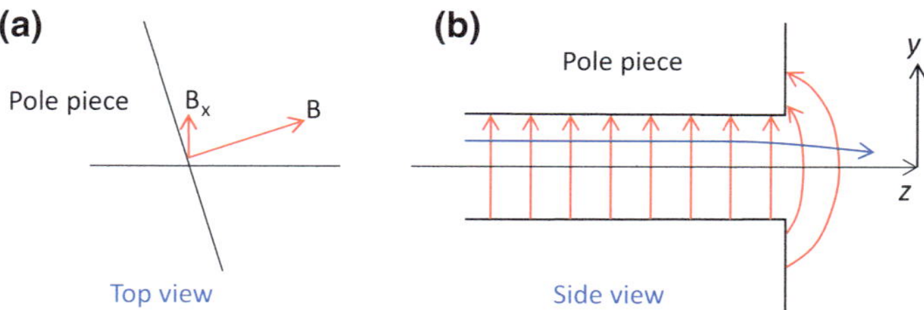

Fig. 10.26 Fringe field of magnetic sector

By integrating Eq. (10.54) once, we obtain

$$\Delta y' = -\frac{1}{RB} \int B_r \mathrm{d}z \tag{10.55}$$

The field B_r is related to B_y by $\nabla \times \vec{B} = 0$, which gives

$$\frac{\partial B_r}{\partial y} = \frac{\partial B_y}{\partial r}$$

And according to Fig. 10.26b and Enge (1967)

$$\frac{\partial}{\partial y} \int B_r \mathrm{d}z = \frac{\partial}{\partial r} \int B_y \mathrm{d}z = B \tan \beta.$$

For small y, B can be taken as approximately constant and $\int B_r \mathrm{d}z = By \tan \beta$. Substituting this result into Eq. (10.55), we obtain

$$\Delta y' = -\frac{y \tan \beta}{R}. \tag{10.56}$$

For the principle ray, with the zero slope initially, and entering at the position $y = 1$, the fringing field thus provides a focal length of

$$f_y = \frac{R}{\tan \beta}, \tag{10.57}$$

This is positive for a positive inclination angle β. At $\beta = 0$, the focal length is infinitely long, and thus, there is no vertical focusing. Double focusing is obtained in a magnetic sector with $\beta > 0$. This the basic principle behind the design of electron imaging energy filters.

Another method to provide vertical focusing is to use conical magnetic sectors, where the gap between the two poles increases linearly with the radius, compared to the uniform gap we have discussed so far. Inside the sector, the field scales with distance according to

$$B_y = B_{\mathrm o} \left(\frac{r}{R}\right)^{-p}.$$

This is accompanied by fields along the radial direction, which gives rise to vertical focusing as we saw in above discussions. Further details on focusing in a conical sector can be found in Enge (1967) and Rose (1995).

10.6.4   Sector Fields, Paraxial Equations, and Second-Order Aberrations

The lack of rotational symmetry in the sector fields introduces large second-order aberrations. To discuss these aberrations, it is necessary to go beyond the first-order treatment that we just outlined. Such treatments can be found in the theory given by Rose (1995). While the approach is similar to the theory of the round magnetic lenses described in Chap. 7, the optical axis is no longer straight. We will use the curved (curvilinear) coordinates (x, y, z) that we have introduced before, where z represents the curved optical axis. First, the magnetic field inside the pole-piece gap with middle section symmetry is described by the scalar magnetic potential $\Psi(\vec{r})$ expressed in the power series expansion

$$\Psi(\vec{r}) = \Psi_{1s}(z)y + \Psi_{2s}(z)2xy + \Psi_{3s}(z)\left(3x^2y - y^3\right)$$
$$+ \frac{1}{8}\left[2\Gamma(z)\Psi_{2s}(z) - \Psi''_{1s}(z)\right]\left(x^2y + y^3\right) + \cdots \tag{10.58}$$

Here, $\Psi_{1s}(z)$, $\Psi_{2s}(z)$, and $\Psi_{3s}(z)$ stand for the dipole, quadrupole, and hexapole field components, respectively, on the optical axis. The curvature $\Gamma(z)$ of the optical axis is given by

$$\Gamma(z) = -\eta\Psi_{1s} = -1/R.$$

and

$$\eta = \left[2m_e\Phi\left(1 + e\Phi/2m_eC^2\right)/e\right]^{1/2}$$

The magnetic field is given by the gradient of the scalar magnetic potential with $\vec{B} = \nabla\Psi$. Using this, we have for the dipole component

$$\Psi_{1s}(z) = \left.\frac{\partial}{\partial y}\Psi\right|_{\substack{x=0 \\ y=0}} = -B_y(0,0,z) = -B(z). \tag{10.59}$$

The quadrupole field component, as we discussed in the previous section, can be introduced by an inclined sector surface, or conical sector field, or both.

The paraxial ray equations of the magnetic sector including the quadrupole term are given by Rose (1978)

$$x''(z) + \left[\eta^2 \Psi_{1s}^2(z) + 2\eta \Psi_{2s}(z)\right] x(z) = \frac{1}{2}\eta \Psi_{1s}(z)\chi$$

$$y''(z) - 2\eta \Psi_{2s}(z)y(z) = 0 \tag{10.60}$$

where

$$\chi = \frac{1 + E_o/m_o c^2}{1 + E_o/2m_o c^2}\frac{\Delta E}{E_o}. \tag{10.61}$$

Thus, the right-hand side of Eq. (10.60) gives the dispersion along x, proportional to the fractional change of χ with beam energy and the dipole field.

Each of the paraxial ray in Eqs. (10.60) gives two independent solutions. A separate solution represents the energy dispersion. Energy filters in general take the diffraction pattern as the object in image filtering, and an intermediate image is formed after the diffraction plane as shown in Fig. 10.27. The fundamental rays are defined at these two distinct planes:

$$
\begin{aligned}
x_\alpha(z_i) &= 0, \; x'_\alpha(z_i) = 1, \; \chi = 0 \\
x_\gamma(z_d) &= 0, \; x'_\gamma(z_d) = 1, \; \chi = 0 \\
x_\chi(z_i) &= 0, \; x'_\chi(z_i) = 0, \; \chi = 1 \\
y_\beta(z_i) &= 0, \; y'_\beta(z_i) = 1 \\
y_\delta(z_d) &= 0, \; y'_\delta(z_d) = 1
\end{aligned}
\tag{10.62}
$$

Here, z_i and z_d denote the intermediate image and diffraction planes, respectively. The fundamental rays x_α and x_γ and the ray parameters α and γ are defined in Fig. 10.27.

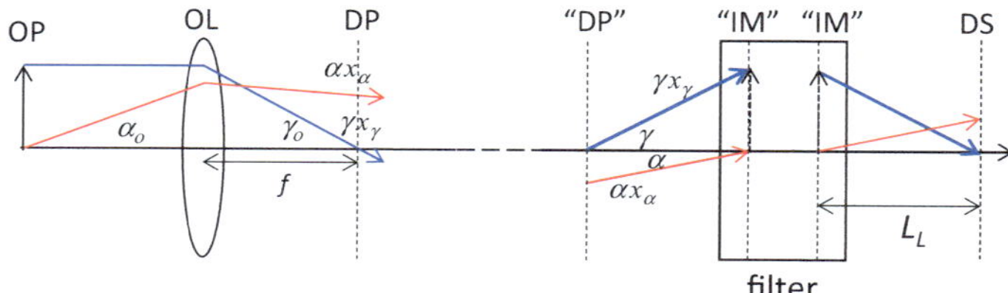

Fig. 10.27 Fundamental paraxial rays entering the filter and their relationship to fundamental rays of the objective lens (OL). Here, OP for object plane, DP for diffraction plane, and "DP" and "IM" for diffraction pattern and image in image filtering mode and image and diffraction pattern in diffraction filtering mode. DS stands for dispersion plane

An off-axis ray is described by a combination of fundamental rays according to

$$x = x^{(1)} = \alpha x_\alpha + \gamma x_\gamma + \chi x_\chi ,$$
$$y = y^{(1)} = \beta y_\beta + \delta y_\delta \tag{10.63}$$

where

$$\alpha = x_i', \ \beta = y_i', \ \gamma = y_d', \ \text{and} \ \delta = y_d'$$

are the slopes at the intermediate and diffraction planes, respectively.

The dominant aberrations of the sector fields are 2nd order, which are introduced by including higher order terms in Eq. (10.58) that were neglected in the paraxial equations of (10.60). They are calculated at the final image and dispersion planes, z_I and z_E, using the Eikonal method described by Rose (1995). Below is a summary of the results for the second-order aberrations:

$$x^{(2)}(z_I) = -\frac{x_{\gamma I}}{L}\frac{\partial H_I^{(3)}}{\partial \alpha}$$
$$y^{(2)}(z_I) = -\frac{y_{\delta I}}{L}\frac{\partial H_I^{(3)}}{\partial \beta} \tag{10.64}$$

and

$$x^{(2)}(z_E) = \frac{x_{\alpha E}}{L}\frac{\partial H_E^{(3)}}{\partial \gamma}$$
$$y^{(2)}(z_E) = -\frac{y_{\beta I}}{L}\frac{\partial H_E^{(3)}}{\partial \delta} \tag{10.65}$$

Here, L is the distance L_L between the intermediate image plane "IM" and the dispersion plane "DS" in Fig. 10.27, and $H^{(3)}$ is a third-order Eikonal in the form given by Rose (1995)

$$\begin{aligned}
H^{(3)} = {}& \left(A_{\alpha\alpha\alpha}\alpha^3 + 3A_{\alpha\alpha\gamma}\alpha^2\gamma + 3A_{\alpha\gamma\gamma}\alpha\gamma^2 + A_{\gamma\gamma\gamma}\gamma^3\right)/3 \\
& + \left(B_{\alpha\beta\beta}\alpha\beta^2 + 2B_{\alpha\beta\delta}\alpha\beta\delta + B_{\alpha\delta\delta}\alpha\delta^2\right)/2 + \left(B_{\gamma\beta\beta}\gamma\beta^2 + 2B_{\gamma\beta\delta}\alpha\beta\delta + B_{\gamma\delta\delta}\gamma\delta^2\right)/2 \\
& + \left(C_{\alpha\alpha\chi}\alpha^2\chi + 2C_{\alpha\gamma\chi}\alpha\gamma\chi + C_{\gamma\gamma\chi}\gamma^2\chi\right)/2 + \left(C_{\beta\beta\chi}\beta^2\chi + 2C_{\beta\delta\chi}\beta\delta\chi + C_{\delta\delta\chi}\delta^2\chi\right)/2 \\
& + C_{\alpha\chi\chi}\alpha\chi^2 + C_{\gamma\chi\chi}\gamma\chi^2
\end{aligned} \tag{10.66}$$

Here, $A_{\lambda\mu\upsilon}$, $B_{\lambda\mu\upsilon}$, and $C_{\lambda\mu\upsilon}$ are second-order aberration coefficients, and they are all functions of the coordinate z along the optical axis. These coefficients separate into the categories of geometrical aberrations ($A_{\lambda\mu\upsilon}$ and $B_{\lambda\mu\upsilon}$) and chromatic aberrations ($C_{\lambda\mu\upsilon}$). The geometrical aberrations are further classified, based on their effects, into aperture aberrations, which give a different focus for rays arriving at different distances from the optical axis, distortion, which affects the image distance but not the focus, and mixed aberrations. For an example of experimentally observed aberrations at the energy-dispersive plane, see Fig. 10.31.

10.6.5   In-Column Energy Filters

The first in-column energy filter, capable of giving high-quality energy-filtered images and diffraction patterns, is now known as the Castaing–Henry filter (Castaing et al. 1967). In the Castaing–Henry filter, a triangular magnetic sector acting as a double prism is combined with an electrostatic mirror to form a symmetric prism-mirror-prism system, which forms an inverted image of the object, like a thick lens for electrons of the primary energy, while electrons of lower energy are dispersed by the prism magnetic field. To reflect the electrons, an electrostatic mirror supplies the same high voltage as the acceleration voltage of the incident electrons, but with the opposite polarity, so the electrons are deaccelerated, reflected, and reaccelerated again. High-voltage stability issues prevent the Castaing–Henry filter from operating at the acceleration voltages higher than about 80 kV. To overcome this limitation, the electron imaging filters developed later have all used magnetic fields.

Figure 10.28 shows the Ω filter designed and constructed by JEOL Ltd, Japan, for 200 kV TEMs. The Ω filter has four bending magnets with an electron path shaped like the Greek letter Ω (and hence its name). The intermediate magnetic fields (BM2 and BM3 in Fig. 10.28) are opposite in polarity to these of the first and final bending magnets. Because of this arrangement, the net deflection angle

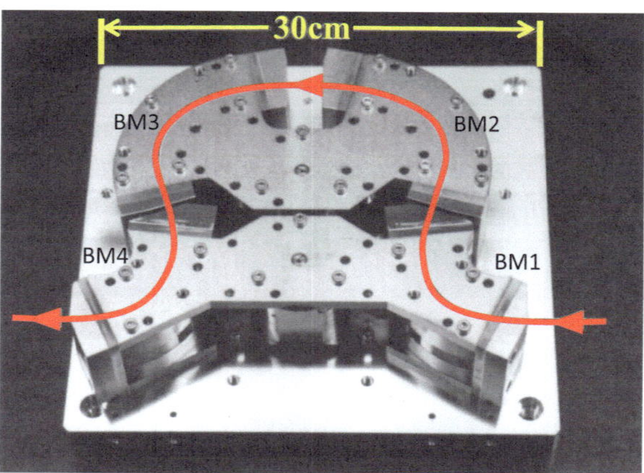

Fig. 10.28 The JEOL in-column energy Ω filter for 200 kV JEOL2200FS TEM (from Tanaka et al. 1999, reproduced with permission)

achieved by all four bending magnets is $0°$. Aberrations are the major design issues for electron energy filters as we have discussed in the last section. An important feature in the original design of the Ω energy filter is the central plane symmetry, which cancels out the distortion and the aperture aberrations on the final image. The Ω filter was developed first for the MeV microscope project in France (Jouffrey (2009)). It was further developed by Rose and Plies (1974), Lanio (1986), and Tsuno et al. (1997). The Ω energy filters available commercially differ in their focusing modes and by the number of crossovers on the x and y axes.

Another design is the α energy filter, which has recently been successfully developed into an electron monochromator by Krivanek et al. (2009). It has three bending magnets; the first and final magnets are the same, while two intermediate bending magnets are arranged symmetrically across the middle section of the filter. By having all three bending magnets excited with the same polarity, the electron path follows the shape of the letter α with a total deflection angle of $360°$. An analysis of the α energy filter performance for imaging is given by Perez et al. (1984) using two bending magnets.

The Ω energy filter is a part of the TEM projection system, as opposed to a post-column attachment like the Gatan imaging filter (GIF). Because of this, the filter must work like an intermediate lens. However, it is not like a round magnetic lens because its optical axis depends on the geometry of the bending magnets and their excitations. To work like an intermediate lens, the filter must be able to form (1) a stigmatic undistorted image of the object normal to the TEM optical axis and (2) a stigmatic image of the diffraction pattern at the energy selection slit plane, in order to achieve a good energy discrimination without altering the illumination uniformity significantly.

The process of image energy filtering can be compared to dark-field TEM imaging. In that case, a part of the diffraction pattern is selected in the back focal plane of the objective lens using an aperture and allowed to form an image some distance away. For energy filtering of an image, the diffraction pattern is first formed at the energy selection (slit) plane and selected using a slit. The propagation of the selected diffraction pattern again forms an image. Thus, in both cases, we rely on the separation in the diffraction plane based on the electron momentum. For energy filtering of a CBED pattern, the filter accepts the probe image formed by the objective and intermediate lenses. The filter forms a series of probe images at the slit plane, dispersed according to the beam energy. An energy selection slit is placed around the zero-loss image to obtain the energy-filtered CBED pattern.

10.6.6 *Post-Column Imaging Filters*

A post-column energy filter is placed below the TEM viewing screen (Krivanek et al. 1992). As shown in the previous sections, a single magnetic sector is capable of forming an energy-filtered image. By placing a slit in the diffraction plane after the exit surface of the magnetic sector, where the energy dispersion is made, an energy-filtered image can be obtained in the same way as for the in-column filters. Such an image, however, would be of little use because of the large image distortions and aberrations of the magnetic sector. The magnetic sector spectrometers are widely used as attachments to TEMs for EELS. Thus, there is a considerable interest in integrating image filtering and EELS capabilities in these systems.

The first imaging filter developed by Gatan (Pleasanton, CA), known as GIF200 or simply GIF, employed a single magnetic sector of 10 cm bending radius for an energy dispersion of 1.8 μm/eV at 100 keV primary energy. The sector has tilted and curved entrance and exit surfaces. A quadrupole and sextupole placed before the sector are used to correct the second-order aberrations in the spectrum. The post-sector electron-optical system consists of 6 quadrupoles and 5 sextupoles, which combine aberration and distortion correction with image projection. Two of the six quadrupoles are placed in front of the energy-selecting slit for spectrum focusing. They are used to magnify the dispersion of the magnetic sector by about $5 \times$ from 1.8 μm/eV at 100 keV primary energy to about 10 μm/eV. The four quadrupoles placed after the energy selection slit are configured to either project the spectrum image onto the CCD detector, with a variable magnification along the x and the y directions, or project a magnified image of the object. In the second case, the four quadrupoles are used to cancel energy dispersion in normal imaging (without energy filtering). The 5 sextupoles are used in various ways to correct the second-order aberrations in energy-filtered or nonfiltered image. For details, see Fig. 10.29 and Krivanek et al. (1992).

The object is magnified by the GIF by about 15x. Because of this, the optimum operation of a GIF requires that the TEM projector lens to be weakened by a factor of about 15 relative to its normal operation condition. This reduced TEM magnification makes the GIF sensitive to any small movements of the image after the

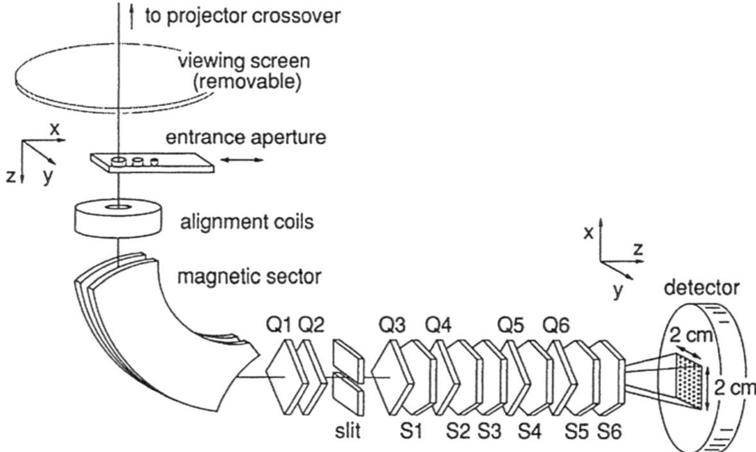

Fig. 10.29 Schematic diagram of the Gatan imaging filter. (From Gatan, reproduced with permission)

TEM projection lens, since they would cause a noticeable blurring of the final image. To reduce this effect, additional magnetic shielding is used in the TEM viewing chamber against stray electromagnetic fields.

The new generation of imaging filters from Gatan (GIF quantum, see Gubbens, et al. (2010)) uses eight dodecapole lenses instead of the individual quadrupoles and sextupoles used in earlier generations of the GIF. The poles of the dodecapole are individually excited so they can be configured electronically to generate any combinations of dipole field for beam deflection, quadrupole for focusing, and sextupole, octupole, decapole, and dodecapole for aberration correction. This flexibility enables a control of optical properties up to 6th order. Two of the eight dodecapole lenses are placed in front of the energy selection slit, one in front the magnetic sector. The rest are after the slit in a 2 + 1 + 2 configuration. Another change made in the GIF quantum is the reduction of the bending radius of the magnetic sector from 10 cm in the GIF 200 to 7.5 cm. This reduces the dispersion of the magnetic sector by 25 % to 1.35 μm/eV and helps the filter performance for lower energy electrons.

10.6.7 *Isochromaticity, Filter Acceptance, and Distortion*

Isochromaticity describes the angular acceptance and geometric distortions of imaging filters, which are important characteristics for energy filtering of electron diffraction patterns Rose (1995). Isochromaticity is defined by the range of electron energies for each detector position. Ideally, this should be the same across the whole detector area. However, aberrations at the slit plane prevent a complete separation between the zero-loss beam and the energy-loss beams. The degree of separation is expressed as the nonisochromaticity I_s (Tsuno et al. 1999),

$$I_s = S/D \text{ (eV)} \tag{10.67}$$

where S is the beam diameter on the slit plane in units of μm and D is the dispersion in units of μm/eV. In the presence of second-order aberration, S is limited by the so-called aperture aberrations, $A_{\gamma\gamma\gamma}$ and $B_{\gamma\delta\delta}$ in Eq. (10.66), in the form of

$$S = \left(A_{\gamma\gamma\gamma}/2 + B_{\gamma\delta\delta}/4\right)\Omega^2, \tag{10.68}$$

where the angle

$$\Omega = \sqrt{\delta^2 + \gamma^2} = M_i \rho_o/L, \tag{10.69}$$

This is proportional to the image radius ρ_o at the object plane, the image magnification M_i at the intermediate image plane in front of the filter, and inversely to L, the distance between the object and intermediate image. The nonisochromaticity I_s

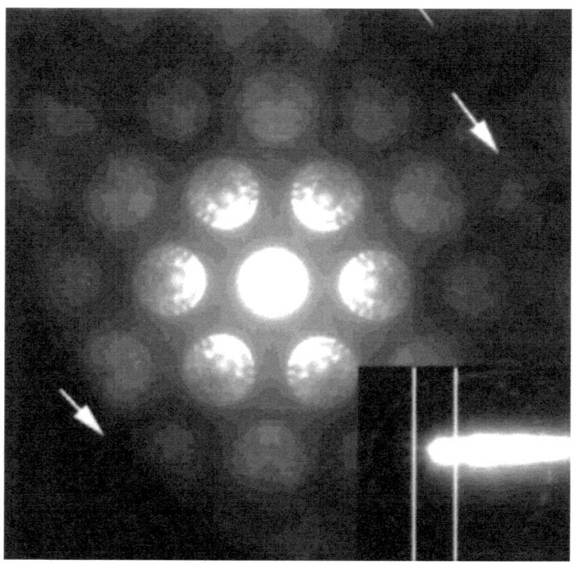

Fig. 10.30 Energy-filtered Si [111] zone-axis CBED pattern using the uncorrected Ω filter of LEO 912 electron microscope operated at 120 kV and with 15 eV energy slit. The energy spectrum is shown at *lower right* corner. The *white lines* mark the position of the energy slit. *Arrows* indicate the energy-filtering envelope

for the uncorrected Omega filter installed in LEO 912 is 15 eV, while the JEOL Omega filter has I_s of 1–2 eV (Fig. 10.28).

Figure 10.30 shows a zero-loss energy-filtered Si[111] CBED pattern and the corresponding spectrum (see inset) using the LEO 912 electron microscope with an energy slit of 15 eV wide. The envelope shown in Fig. 10.28 is due to the non-isochromaticity that we have discussed. The effect is to make a circle in the object plane into an ellipse, the center of which is displaced by a distance proportional to the square of the radius according to Eq. (10.68). A disk-shaped object forms a flattened cone-shaped aberration figure. The LEO 912 Ω filter uses the uncorrected design of Lanio (1986). A fully corrected filter, such as the Mandoline filter, has much smaller aberrations and thus enables a large acceptance angle (Uhlemann and Rose 1994).

Figure 10.31 shows the experimentally observed aberration figure for silicon using the LEO 912. For a large object, the aberration figures overlap on the energy dispersion plane with those for different energy losses. Using a small energy slit, the primary beam energy of the filtered pattern is selected at different radii from the optical axis because of the nonisochromaticity. By using a narrower energy slit, it is possible to reduce the contributions from electrons of different energies. However, this also results in a smaller field of view. For energy filtering on the zero-loss peak, the problem is most severe for metals with a large contribution from plasmon losses. In the CBED mode, the image is that of the focused probe and its diffraction (see inset in Fig. 10.30). They are sufficiently separated in the Leo 912 Ω filter for the use of energy filtering of low-order reflections (see Fig. 10.30). In fact, there is no significant difference between the zero-loss disk intensities obtained with the 8 and 16 eV energy slits.

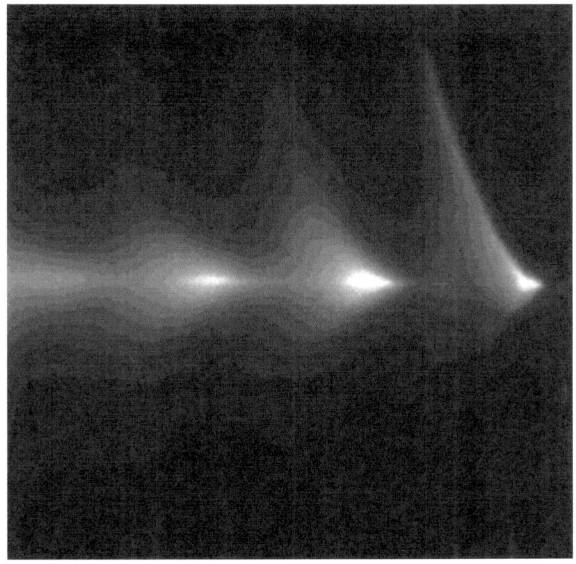

Fig. 10.31 Experimentally observed spectrum from silicon using the uncorrected Ωd Ω filter of LEO 912 electron microscope operated at 120 kV in image mode

We have seen in the above examples the limits to the performance of energy filtering imposed by second-order aperture aberrations in an uncorrected Ω filter. To measure this performance, we introduce the transmissivity T or acceptance of the filter, which is defined by

$$T = \left(\pi \rho_{\mathrm{o}} \alpha_{\mathrm{o}}\right)^2 \tag{10.70}$$

Here, ρ_o and α_o are the radius of the object and its semi-angle accepted by the filter. Since the product of $\rho_o \alpha_o$ is constant in an imaging system, $\sqrt{T}$ is also referred as transferred phase space. For an ideal filter without aberration, the transmissivity is optimized for a detector size of radius ρ_d in the form of Uhlemann and Rose (1994)

$$T = \left(\frac{\pi \rho_d D \cdot \Delta E}{2L \cdot M_f M_p}\right)^2, \tag{10.71}$$

Here, D denotes dispersion of the filter, ΔE is the energy slit width used for filtering, M_f is the magnification of the filter, and M_p is the magnification of the projector lens in case of an in-column filter or that of quadrupole lenses in a GIF. The transmissivity of a filter with second-order aberrations is limited by the largest uncorrected aperture aberration A_2, which gives (Uhlemann and Rose 1996)

$$T = \left(\frac{\pi D \cdot \Delta E}{2M_f}\right)^2 \frac{1}{A_2}. \tag{10.72}$$

The estimated transmissivity for the Castaing and Henry filter and for the uncorrected Ω filter is 0.04 and 0.3 10^{-3} μm^2, respectively, while correction of the second-order aperture aberration improves the transmissivity by a factor of more than 10 (Uhlemann and Rose 1996).

Geometrical distortions arise from the use of noncylindrical lenses inside the energy filter. Large distortions are caused by optical misalignment, which is an issue with the GIF in its low-camera-length setting. In a Ω filter, the second-order distortion, as measured in the object plane, has the following form

$$x_D^{(2)} = \left(\gamma^2 A_{\alpha\gamma\gamma} + \delta^2 B_{\alpha\delta\delta}/2\right)/M_i$$
$$y_D^{(2)} = \gamma\delta B_{\alpha\beta\delta}/M_i$$

For the image of a square grid aligned with the coordinate axes, the $\gamma^2 A_{\alpha\gamma\gamma}$ term deforms the squares into rectangles in the x direction and the distortion increases quadratically with γ. The term $\delta^2 B_{\alpha\delta\delta}$ deforms a straight grid line normal to the x-axis into parabolic lines, while the term $\gamma\delta B_{\alpha\beta\delta}$ deforms the square into a trapezoid. The amount of distortion can be measured using a standard calibration sample and corrected using numerical methods.

The so-called corrected filters employ separate magnet quadrupoles and hexapoles for aberration corrections. For example, the aberration-corrected Mandoline filter designed by Uhlemann and Rose (1994) has nine symmetrically arranged correction elements. Half of the second-order aberrations vanish due to the filter's symmetry. The remaining second-order aberrations in the achromatic image plane and energy selection plane are eliminated by symmetrically arranged hexapole fields at the nine correction elements. The correction of the third-order axial aberrations at the energy selection plane is performed by octupole fields at selected correction elements.

10.7 Radiation Effects and Low-Dose Techniques

It is well known that for most biological and organic materials, and for many inorganic materials, the structure of the specimen is rapidly degraded by electron beam irradiation during the TEM observation when using a large magnification, or in the STEM analysis mode (Cowley 1999). The damage occurs in some cases during a small fraction of a second. Other known beam effects include the following:

(1) A change from one crystalline phase to another. For example, in thin films of Al_2O_3 and AlF_3, reduction to metallic Al has been observed (Muray et al. 1985).

(2) Electron beam directed growth. Titanium dioxide nanostructures may grow under electron irradiation at the interfaces of small Au particles supported on

rutile surfaces in an environmental TEM under oxygen partial pressure and high temperature (Chee et al. 2011), and small crystals of Pd on MgO faces may be transformed to single crystals of PdO (Ou and Cowley 1988). Inside a liquid cell, focused electron beam-induced deposition of nanostructures can be made using a liquid precursor (Donev and Hastings 2009; Evans et al. 2011).

(3) Sculpting, atom by atom. Using a focused electron probe, it is possible to remove atoms individually for some compounds. The loss of materials, presumably by dispersion into the vacuum, leads to the use of the electron beam for the drilling of very small holes in thin films for the construction of devices, for DNA decoding, for example (Venkatesan et al. 2010). A review of the electronic processes involved in this hole-drilling can be found in Berger et al. (1987). The STEM probe is also capable of a kind of direct-write, inorganic lithography, since a focused field emission beam will "write" a line of silicon with subnanometer resolution when scanned across silicon dioxide, and many other beam-induced reactions have been studied and used for pattern formation (Jiang et al. 2003). The mechanism of beam damage for STEM imaging of oxides is now understood to be due to the electric field-induced migration of atoms, where the cylindrical field is generated by beam-induced ionization around the STEM probe (for a review of radiation damage due to electron beams in insulators (see Jiang 2016).

(4) Beam-induced atomic motion. Dislocation motion has been observed in CdTe at atomic resolution due to irradiation (Sinclair et al. 1982). Dramatic HREM images of the radiation-induced motion of atoms on the surfaces of small gold particles have been recorded in real-time (Iijima and Ichihashi 1986)—in these and similar experiments, fast video recording has revealed many new aspects of atomic processes, including direct imaging of moving dislocation kinks in silicon Kolar et al. (1996).

Application of STEM to radiation-sensitive materials largely depends on how well the electron beam-induced damage under the illuminating probe is managed during the diffraction pattern or image recording. For X-ray diffraction, because of its large elastic and inelastic scattering path lengths, most photons do not interact at all with the sample, pass through it, and hit the axial beam dump. To record the wanted diffraction pattern, a large volume of crystals must be illuminated so the small fraction of photons which do diffract produce sufficient intensity for recording. The large volume keeps most of the total number of photons that are annihilated in the production of photoelectrons at a minimum, per unit volume. The electron in a TEM, however, can lose any amount of energy on scattering and does so repeated as it moves through the sample.

There are two main types of processes associated with electron energy loss in insulators: One is ballistic knock-on damage, and the other is the ionization process and subsequent atomic rearrangement, known as radiolysis, a process extensively studied in the X-ray and UV literature. Beam heating may also be important, as discussed in detail in Reimer and Kohl (2008). General reviews of damage mechanisms in TEM can be found in Hobbs (1978, 1979), Urban (1980), Egerton et al. (2004), and Egerton et al. (2006) for inorganics and in Isaacson et al. (1973), Glaeser and Taylor (1978), and Henderson (1995) for organics. Discussion of radiation damage in X-ray diffraction can be found in the special issue devoted entirely to radiation damage (Journal of Electron Spectroscopy and Related Phenomena, volume 170 (2009)).

Electron beam damage in a CaF_2–Al_2O_3–SiO_2 glass has been investigated, for example, using time-dependent Ca $L_{2,3}$- and O K-edge electron energy-loss spectroscopy. It appears that there is a threshold dose rate, below which the damage involving the formation of O defects may not be detected, at any total dose. This suggests that this threshold dose rate effect may result from competition between the damage and recovery processes. The accumulation of damage can only occur when the damage rate is higher than the recovery rate. For surface sputtering process in TEM, the recovery rate is negligible. Therefore, there is no threshold dose rate for surface sputtering (Jiang and Spence (2012)).

For the ballistic knock-on process, the energy transferred in an elastic collision between the beam electron and the nucleus exceeds the displacement energy. This energy varies from tenths of an electron volt for Van der Waals bonding, to tens of electron volts for ionics. A Frenkel pair, consisting of an interstitial and a vacancy, is produced. The rate is proportional to the product of the local electron intensity and a cross section, and so it depends on local diffraction conditions. Thus, enhanced damage is seen (by TEM diffraction contrast) inside bend contours due to the concentration of electron flux along these lines (Fujimoto and Fujita 1972). Some typical values of the knock-on threshold for bulk polycrystalline material at room temperature are as follows: graphene 140 keV, silicon 145 keV, copper 400 keV, molybdenum 810 keV, and gold 1300 keV. In a crystal, in addition to channeling effects, the threshold energy depends on the direction of transferred momentum, being smallest for atoms knocked-on into close-packed directions. Second, the displacement threshold is reduced for atoms at surfaces and defects. Typical displacement cross sections are 5×10^{-23} cm^2, much smaller than ionization values.

Ionic materials, such as the alkali halides, are highly sensitive to ionization damage through radiolysis. The excitations whose decay processes may produce damage are as follows: (1) plasmons; (2) valence excitations; (3) inner-shell excitations; and (4) excitons. The Bethe theory (Egerton 2011) for single-electron excitation gives a cross section inversely proportional to the square of the electron velocity, so that ionization damage decreases at higher accelerating voltage. (This cross section is about 10^{-18} cm^2 at 100 kV for a carbon atom.) Electronic excitations can couple to the nuclear system if their lifetime exceeds vibrational periods (the Debye frequency is about 10^{14} Hz), if they can transfer sufficient energy to break bonds, and if they are localized. Plasmon excitations are delocalized, but the decay processes may involve localized processes. Thus, a rough guide in ionic materials has been that for damage to occur, the electron–hole recombination energy should exceed the lattice binding energy, a condition fulfilled, for example in CaF_2, but not in MgO, where radiolysis is not seen. Inner-shell excitation may also be important, since, although rarer, they transfer more energy to the system which appears in the form of Auger electrons and X-rays, which may cause damage. Valence excitations generate secondary electrons which can be damaging. Radiolysis may have a threshold in dose, a highly nonlinear temperature dependence, and in the alkali, halides may lead to the formation of halogen bubbles and metallic precipitates (Hobbs 1978, 1979).

For metallic and covalently bonded materials, because of screening and reversibility, ionization damage is normally not seen in TEM at energies below the beam energy threshold for ballistic "knock-on" damage, except perhaps at point defects in semiconductors. (This threshold is about 150 kV for aluminum, increasing with atomic number and with a weak temperature dependence.)

The main damage mechanisms in hydrocarbons are ionization, leading possibly to bond breaking, or heating. Loss of hydrogen (4 eV bond energy) and C–H bond breaking are common, and secondary processes such as cross-linking and the diffusion of hydrogen must be considered. Mass loss may occur and has been measured, leading finally to a carbon-rich deposit. Damage to organic crystals has traditionally been measured by the time taken for diffraction spots to fade. The angular dependence of these spots can be modeled using a Debye–Waller factor to give an average lattice displacement arising from the damage, which may then be measured as a function of dose, temperature, and accelerating voltage (Jeng and Chiu 1984). In this way, the critical dose which starts to produce damage may be defined. High-order reflections, corresponding to fine detail, fade first. Alternatively, the Patterson function may be studied as a function of dose. The effects of increasing dose from 1 to 14 e Å^{-2} on the periodically averaged electrostatic potential of a paraffin layer, reconstructed from transmission electron diffraction patterns, can be found in Dorset and Zemlin (1985)—the projected hydrocarbon chains lose fine detail and become more rounded. Other critical dose measurements are summarized in Reimer and Kohl (2008)—they vary at 100 kV from about e Å^{-2} for amino acids, to 8 e Å^{-2} for polymers and gelatin, to 30 e Å^{-2} for coronene, to 240 e Å^{-2} for Cu-phthalocyanine, to 360 e Å^{-2} for the nucleic acid guanine, and to 18 000 e Å^{-2} for CuCl_{16}-phthalocyanine. A dose of 1 e Å^{-2} is about 10^5 times greater than that needed to kill *E-coli* bacteria. Aromatics are found to be more resistant than aliphatics, since the delocalized π bonds in the benzene rings of the aromatics dissipate energy better.

Damage to organics may be reduced by several methods, including cooling, the use of higher voltages, embedding in a resin or ice, elemental substitution, the use of pulses so brief that they outrun the damage processes, or use of an environmental cell. X-ray lasers have demonstrated the ability to outrun damage, where pulses of 20 fs duration provide 0.2-nm resolution diffraction from nanocrystal. High-energy (MeV) electron beam diffraction cameras are also approaching this capability. The Bethe energy-loss formula shows that energy deposited per unit length is inversely proportional to the square of the beam velocity, so that higher beam energy reduces ionization energy deposited. A gain of two in spot-fading times is found from 100 to 200 kV and three at 1 MeV. However, similar ionization processes occur at the detector, so that detector sensitivity decreases with beam energy (see Chap. 9).

References

Beche A, Rouviere JL, Clement L, Hartmann JM (2009) Improved precision in strain measurement using nanobeam electron diffraction. Appl Phys Lett 95:123114

Benner G, Bihr J et al (1990) A new illumination system for an analytical transmission electron microscope using a condenser objective lens. In: Proceedings of XIIth ICEM Seattle. San Francisco Press, SanFrancisco, pp 138–139

Benner G, Probst W (1994) Köhler illumination in the tem: fundamentals and advantages. J Microsc 174(3):133–142

Berger SD, Salisbury IG et al (1987) Electron energy-loss spectroscopy studies of nanometer-scale structures in alumina produced by intense electron-beam irradiation. Philos Mag B 55:341–358

Blackman M (1939) On the intensities of electron diffraction rings. Proc Roy Soc Lond A 173:68–82

Botton G (ed) (2007) Analytical electron microscopy. Science of microscopy. Springer, New York

Castaing R, Hennequin JF et al (1967) The magnetic prism as an optical system. In: Septier (ed) Focussing of charged particles. Academic, New York, pp 265–293

Chee SW, Sivaramakrishnan S, Sharma R, Zuo JM (2011) Electron-beam-induced growth of Tio2 nanostructures. Microsc Microanal 17:274–278

Christenson KK, Eades JA (1988) Skew thoughts on parallelism. Ultramicroscopy 26:113–132

Cowley JM (1993) Configured detectors for STEM imaging of thin specimens. Ultramicroscopy 49:4–13

Cowley JM (1999) Electron nanodiffraction. Microsc Res Tech 46:75–97

Cowley JM (2004) Applications of electron nanodiffraction. Micron 35:345

Cowley JM, Spence JCH (1981) Convergent beam electron microdiffraction from small crystals. Ultramicroscopy 6:359–366

Donev EU, Hastings JT (2009) Electron-beam-induced deposition of platinum from a liquid precursor. Nano Lett 9:2715–2718

Dorset DL, Zemlin F (1985) Structural changes in electron-irradiated paraffin crystals at < 15 K and their relevance to lattice imaging experiments. Ultramicroscopy 17(3):229–235

Downing KH, Glaeser RM (1986) Improvement in high-resolution image quality of radiation-sensitive specimens achieved with reduced spot size of the electron-beam. Ultramicroscopy 20:269–278

Duval P, Hoan N et al (1970) Réalisation d'un dispositif de filtrage en énergie des images de microdiffraction électronique. Nouv Rev d'Optique Appliquée 1:221–228

Eades JA (1980) Zone-axis patterns formed by a new double-rocking technique. Ultramicroscopy 5:71–74

Eades JA (1984) Zone-axis diffraction patterns by the Tanaka method. J Electron Microsc Tech 1:279–284

Eades A (2006) Obtaining TEM images with a uniform deviation parameter. Ultramicroscopy 106:432–438

Egerton RF (2005) Physical principles of electron microscopy: an introduction to TEM, SEM, and AEM. Springer, New York

Egerton RF (2011) Electron energy-loss spectroscopy in the electron microscope, 2nd edn. Springer, New York

Egerton RF, Li P et al (2004) Radiation damage in the TEM and SEM. Micron 35:399–409

Egerton RF, Wang F et al (2006) Beam-induced damage to thin specimens in an intense electron probe. Microsc Microanal 12:65–71

Enge HA (1967) Deflecting magnets. In: Septier A (ed) Focusing of charged particles. Academic Press, New York

Evans JE, Jungjohann KL et al (2011) Controlled growth of nanoparticles from solution with in situ liquid transmission electron microscopy. Nano Lett 11:2809–2813

Fujimoto F, Fujita H (1972) Radiation damage induced by channeling of high energy electrons. Phys Status Solidi A 11:K103–K104

Fung KK (1984) Large-angle convergent-beam zone axis patterns. Ultramicroscopy 12:243–246

Giannuzzi LA, Stevie FA (eds) (2005) Introduction to focused ion beams: instrumentation, theory, techniques and practice. Springer, New York

Gjonnes K (1997) On the integration of electron diffraction intensities in the Vincent-Midgley precession technique. Ultramicroscopy 69:1–11

Glaeser RM, Taylor KA (1978) Radiation damage relative to transmission electron microscopy of biological specimens at low temperature: a review. J Microsc 112:127–138

Goodhew PJ (1972) Specimen preparation in materials science, North-Holland

Gubbens A, Barfels M, Trevor C, Twesten R, Mooney P, Thomas P, Menon N, Kraus B, Mao C, McGinn B (2010) The GIF quantum, a next generation post-column imaging energy filter. Ultramicroscopy 110:962–970

Henderson R (1995) The potential and limitations of neutrons, electrons and X-rays for atomic-resolution microscopy of unstained biological molecules. Q Rev Biophys 28:171–193

Hirsch P, Howie A, Nicolson RB, Pashley DW, Whelan MJ (1977) Electron microscopy of thin crystals. Robert E. Krieger Publishing Company, Malaba

Hobbs LW (1978) Radiation damage in electron microscopy of inorganic solids. Ultramicroscopy 3:381–386

Hobbs LW (1979) Radiation effects in analysis of inorganic specimens by TEM. In: Hren JJ, Goldstein JI, Joy DC (eds) Introduction to analytical electron microscopy. Springer, Boston

Horstmann M, Meyer G (1965) Zeit Fur Physik 182:380

Hwang J, Zhang JY, Son J, Stemmer S (2012) Nanoscale quantification of octahedral tilts in perovskite films. Appl Phys Lett 100:191909

Iijima S, Ichihashi T (1986) Structural instability of ultrafine particles of metals. Phys Rev Lett 56:616–619

Isaacson M, Johnson D, Crewe AV (1973) Electron beam excitation and damage of biological molecules; its implications for specimen damage in electron microscopy. Radiat Res 55:205–224

Jeng T-W, Chiu W (1984) Quantitative assessment of radiation damage in a thin protein crystal. J Microsc 136:35–44

Jiang N (2016) Beam damage by the induced electric field in transmission electron microscopy. Micron 83:79–92

Jiang N, Spence JCH (2012) On the dose-rate threshold of beam damage in TEM. Ultramicroscopy 113:77–82

Jiang N, Hembree GG, Spence JCH, Qiu J, de Abajo FJG, Silcox J (2003) Nanoring formation by direct-write inorganic electron-beam lithography. Appl Phys Lett 83:551–553

Jordan IK, Rossouw CJ, Vincent R (1991) Effects of energy filtering in lacbed patterns. Ultramicroscopy 35(3–4):237–243

Jouffrey B (2009) On the high-voltage stem project in Toulouse. In: Advances in imaging and electron physics: cold field emission and the scanning transmission electron microscope. Academic Press, Amsterdam

Kelly PM, Wauchope CJ et al (1994) Calculation of overall tilt angles for a double tilt holder in a TEM. Microsc Res Tech 28:448–451

Kim KH, Xing H, Zuo JM, Zhang P, Wang HF (2015) TEM based high resolution and low-dose scanning electron nanodiffraction technique for nanostructure imaging and analysis. Micron 71:39–45

Kolar HR, Spence JCH, Alexander H (1996) Observation of moving dislocation kinks and unpinning. Phys Rev Lett 77:4031–4034

Kondo Y, Ito T et al (1984) New electron diffraction techniques using electronic hollow-cone illumination. Jpn J Appl Phys 23:L178–L180

Krakow W, Howland LA (1976) A method for producing hollow cone illumination electronically in the conventional transmission microscope. Ultramicroscopy 2:53–67

Krivanek OL, Gubbens AJ, Dellby N, Meyer CE (1992) Design and 1st applications of a post column imaging filter. Microsc Microanal Microstruct 3:187–199

Krivanek OL, Ursin JP, Bacon NJ, Corbin GJ, Dellby N, Hrncirik P, Murfitt MF, Own CS, Szilagyi ZS (2009) High-energy-resolution monochromator for aberration-corrected scanning transmission electron microscopy/electron energy-loss spectroscopy. Philos T Roy Soc A367:3683–3697

Lanio S (1986) High-resolution imaging magnetic energy filters with simple structure. Optik 73:99–107

LeBeau JM, Findlay SD, Allen LJ, Stemmer S (2010) Position averaged convergent beam electron diffraction: theory and applications. Ultramicroscopy 110:118–125

Lichte H, Michael L (2008) Electron holography—basics and applications. Rep Prog Phys 71:016102

Mayer J, Giannuzzi LA, Kamino T, Michael J (2007) TEM sample preparation and FIB-induced damage. MRS Bull 32:400–407

McKeown J, Spence JCH (2009) The kinematic convergent beam method for solving nanocrystal structures. J Appl Phys 106:074309

Morishita S, Yamasaki J, Nakamura K, Kato T, Tanaka N (2008) Diffractive imaging of the dumbbell structure in silicon by spherical-aberration-corrected electron diffraction. Appl Phys Lett 93:183103

Morniroli JP (2003) CBED and LACBED analysis of stacking faults and antiphase boundaries. Mater Chem Phys 81:209–213

Morniroli JP, Gaillot F (2000) Trace analyses from LACBED patterns. Ultramicroscopy 83: 227–243

Morniroli JP, Marceau RKW et al (2006) LACBED characterization of dislocation loops. Philos Mag 86:4883–4900

Mory C, Colliex C, Cowley JM (1987) Optimum defocus for STEM imaging and microanalysis. Ultramicroscopy 21:171–177

Munro E (1975) Design and optimization of magnetic lenses and deflector systems for electron beams. J Vac Sci Technol 12:1146–1150

Muray A, Scheinfein M, Isaacson M, Adesida I (1985) Radiolysis and resolution limits of inorganic halide resists. J Vac Sci Technol B 3:367–372

Ou HJ, Cowley JM (1988) The surface-reaction of Pd/MgO studied by scanning reflection electron-microscopy. Phys Status Solidi A 107:719–729

Own CS, Marks LD, Sinkler W (2005) Electron precession: a guide for implementation. Rev Sci Instrum 76(3):033703

Özdöl VB, Srot V, van Aken PA (2012) Sample preparation techniques for transmission electron microscopy. In: Handbook of nanoscopy. Wiley-VCH Verlag GmbH & Co

Perez JP, Sirven J, Sequela A, Lacaze JC (1984) Etude, au premier ordre, d'un systeme dispersif, magnetique, symetrique, de type alpha. J Phys (Paris) 45(C2):171–174

Reimer L (ed) (1995) Energy-filtering transmission electron microscopy. Springer, New York

Reimer L (ed) (1998) Scanning electron microscopy. Springer, New York

Reimer L, Kohl H (2008) Transmission electron microscopy (4th). Springer, Berlin

Riecke WD, Ruska E (1966) A 100 kV transmission electron microscope with single-field condenser objective. In: 6th international congress for electron microscopy, Kyoto, Japan

Rose H (1978) Aberration correction of homogeneous magnetic deflection systems. Optik 51:15–38

Rose H (1995) In energy-filtering transmission electron microscopy, Edited by L. Reimer. Springer, Berlin

Rose H, Plies E (1974) Design of a magnetic energy analyzer with small aberrations. Optik 40 (3):336–341

Sinclair R, Ponce FA, Yamashita T, Smith DJ, Camps RA, Freeman LA, Erasmus SJ, Nixon WC, Smith KCA, Catto CJD (1982) Dynamic observation of defect annealing in cdte at lattice resolution. Nature 298:127–131

Spence JCH, Cowley JM (1978) Lattice imaging in STEM. Optik 50:129–142

Spence JCH, Lynch J (1982) STEM microanalysis by transmission electron-energy loss spectroscopy in crystals. Ultramicroscopy 9:267–276

Tafto J, Zhu YM, Wu LJ (1998) A new approach towards measuring structure factors and valence-electron distribution in crystals with large unit cells. Acta Cryst A54:532–542

Tanaka M, Terauchi M (1985) Whole pattern in convergent-beam electron diffraction using the hollow-cone beam method. J Electron Microsc 34:52–55

Tanaka M, Saito R, Ueno K, Harada Y (1980) Large-angle convergent-beam electron-diffraction. J Electron Microsc 29:408–412

Tanaka M, Tsuda K, Terauchi M, Tsuno K, Kaneyama T, Honda T, Ishida M (1999) A new 200 kV Ω-filter electron microscope. J Microsc 194:219–227

Terauchi M, Tanaka M (1985) Simultaneous observation of zone-axis pattern and $\pm$g-dark-field pattern in convergent-beam electron-diffraction. J Electron Microsc 34:347–356

Treacy MMJ, Gibson JM (1986) The effects of elastic relaxation on transmission electron-microscopy studies of thinned composition-modulated materials. J Vac Sci Technol B 4:1458–1466

Tsuno K, Kaneyama T, Honda T, Tsuda K, Terauchi M, Tanaka M (1997) Design and testing of omega mode imaging energy filters at 200 kV. J Electron Microsc 46:357–368

Tsuno K, Kaneyama T, Honda T, Ishida Y (1999) Design of omega mode imaging energy filters. Nucl Instrum Meth A 427:187–196

Uhlemann S, Rose H (1994) The mandoline filter—a new high-performance imaging filter for sub-eV EFTEM. Optik 96:163–178

Uhlemann S, Rose H (1996) Acceptance of imaging energy filters. Ultramicroscopy 63:161–167

Urban A (1980) Radiation damage in inorganic materials in the electron microscope. Electron Microsc 4:188

Van der Mast KD, Rakels CJ, Le Poole JB (1980) A high quality multipurpose objective lens. In: Proceedings of European congress electron microscope, vol 1. The Hague, pp 72–73

Venkatesan BM, Shah AB, Zuo JM, Bashir R (2010) DNA sensing using nanocrystalline surface-enhanced Al_2O_3 nanopore sensors. Adv Func Mater 20:1266–1275

Vincent R (1989) Techniques of convergent beam electron-diffraction. J Electron Microsc Tech 13:40–50

Vincent R, Midgley PA (1994) Double conical beam-rocking system for measurement of integrated electron-diffraction intensities. Ultramicroscopy 53:271–282

Williams DB, Carter BC (2009) Transmission electron microscopy, a textbook for materials science, 2nd edn. Springer, New York

Zhu J, Cowley JM (1983) Micro-diffraction from stacking-faults and twin boundaries in fcc crystals. J Appl Crystallogr 16:171–175

Zuo JM, Tao J (2011) Scanning electron nanodiffraction and diffraction imaging. In: Pennycook S, Nellist P (eds) Scanning transmission electron microscopy. Springer, New York

Zuo JM, Vartanyants I, Gao M, Zhang R, Nagahara LA (2003) Atomic resolution imaging of a carbon nanotube from diffraction intensities. Science 300:1419–1421

Zuo JM, Gao M, Tao J, Li BQ, Twesten R, Petrov I (2004) Coherent nano-area electron diffraction. Microsc Res Tech 64:347–355

Chapter 11
Crystal Symmetry

The symmetry of physical properties of any kinds must include the point symmetry of the crystal, according to Neumann's principle (Nye 1957). In ferroelectric and ferroelastic crystals, experimental determination of local symmetry provides critical experimental insight into the constitutive phases and their stability governed by free energy. In crystallography, symmetry determines the smallest structural unit (asymmetric unit), whose atomic structure is determined.

This chapter thus deals with the subject of crystal symmetry and its determination. We start with an introduction of symmetry and symmetry groups and then proceed to the concepts of point and translational symmetry in crystals and the space groups. The sections that follow deal with experimental determination of crystal symmetry and its applications using CBED and other diffraction techniques.

Introductory accounts of crystal symmetry, including an explanation of how to read the International Tables for Crystallography, can be found in McKie and McKie (1992), Sands (1994), Jackson (1991), and Stout and Jensen (1989). The abbreviated "teaching edition" of the International Tables for Crystallography and the book by Sands are particularly recommended. We shall describe crystals without a center of symmetry as "acentric" and those possessing a center of symmetry as "centric." The terms "vertical" and "horizontal" are used for convenience and refer to the orientation of a sample in the microscope, for which a preferred coordinate system is defined by the electron beam.

11.1 Symmetry Operations and Symmetry Groups

Some general concepts about symmetry operations and symmetry groups are needed before we get into details of crystal symmetry. A symmetry operation can be a (1) rotation, (2) mirror, (3) reflection, and (4) translation, or a combination of translation with (1) or (2). Symmetry operation is a geometric transformation and isometric since the distance (metrics) is preserved.

© Springer Science+Business Media New York 2017

297

J.M. Zuo and J.C.H. Spence, *Advanced Transmission Electron Microscopy*,
DOI 10.1007/978-1-4939-6607-3_11

Two points p and p' are symmetry related when the symmetry operation S takes point p to $S(p)$ and

(1) $p = p'$. An example is lattice translation. Since the lattice points are indistinguishable, translation leads to another lattice point, which is the same,
(2) $p = S(p')$ or $p = S(S(p'))$,..., or $p' = S(p)$ or $p' = S(S(p))$,.... The symmetry operation, or successive applications of the same operation, moves one point to the other.

The symmetry of an object is defined by its symmetry operations. This typically is discussed in reference to a property of the object (f). In diffraction, f represents the Fourier coefficients of the modified electron density for X-ray diffraction or the optical potential for electrons. An object is said to have a specific symmetry S, if

$$f(p) = f(S(p)), \tag{11.1}$$

e.g., the crystal property is invariant under the symmetry operation.

The assemblage of all symmetry operations belonging to an object gives rise to a symmetry group. Each operation is an element of the group. Two objects are said to have the same symmetry if their symmetry operations belong to the same symmetry group.

In mathematics, a group (G) is an algebraic structure of elements, whose operations satisfy the following four rules:

(1)   Closure: if S_a and S_b belong to G, then $S_c = S_a S_b$ is also in G,
(2)   Associativity: $(S_a S_b) S_c = S_a (S_b S_c)$ for all S_a, S_b, S_c in G,
(3)   Identity: there is an element $1 \in$ G such that $S_a 1 = 1 S_a = S_a$ for all $S_a \in$ G,
(4)   Inverse: if $S_a \in$ G, then there is an element a $S_a^{-1} \in$ G such that $S_a S_a^{-1} = S_a^{-1} S_a = 1$.

A symmetry group has all the characteristics of a mathematical group. The closure, for example, makes the symmetry group to be self-consistent, e.g., combinations of symmetry operations in a group produce another symmetry operation since the object must remain symmetric during successive applications of symmetry operations. For the same reason, the onefold rotation axis (discussed below) is included as a symmetry operation, representing the identity.

A subset (H) of group G forms a subgroup when H is also a group. A trivial subgroup is the identity operation, which is a group by itself. A set of generators is a subset of the group (not necessarily a subgroup) that the repeat applications of generators on themselves or each other can produce all the elements in the group. An example of generators is lattice translation, there are an infinite number of possible translations in an infinite crystal, but all of them can be obtained by a combination of lattice translations along a, b, and c.

11.2 Point Groups

Inversion, rotation, or mirror operations refer to a fixed point (center), line (axis), or plane, respectively. Because of this, they are called as point symmetry operations. Together they form point groups. In what follows, we shall examine the elemental point symmetry operations first, and then discuss their combinations.

Rotational Symmetry: If an object repeats itself every 360°/n of rotation, then it is said to have the n-fold rotational symmetry. The axis, along which the rotation is carried out, is referred to as the rotation axis, and the axis itself is a symmetry element. While n can be any integer number (e.g., $n = 5$) for a nonperiodic object, only following rotation axes are permissible in a periodic crystal (the quasicrystals are ordered but not periodic):

(1) Onefold rotation, an object returns to itself after 360° rotation.
(2) Twofold rotation axis $[p^{'} = S_2(P), p = S_2(p^{'})]$, where S_2 is a rotation of 180°. Two points related by a twofold axis have the rotation axis at the middle and normal to their connection line. In crystallography, a filled oval shape is used to represent the twofold axis at where it intersects the viewing page.
(3) Three, four, or sixfold rotation axis, where a rotation of 360°/n is carried out with $n = 3$, 4, or 6. The symmetry related points sit on the corners of an equal-sided triangle, square, or hexagon, respectively, while the rotation axis is at the center and normal to the plane containing these points. A filled triangle, square, or hexagon is used for their representation. (see Table 11.2 for a list of symmetry symbols.)

Mirror Symmetry: A mirror symmetry operation is performed by first taking an imaginary cut to separate the object into two halves, and then one of the halves is reflected as by a mirror. If the resulted mirror image reproduces the other half of the object, then the object is said to have the mirror symmetry, e.g., $f(p) = f(S_m(p))$ where m stands for the mirror. The mirror plane is a symmetry element, symbolized by the letter m.

Center of Symmetry: In this operation, $f(p) = f(S_I(p))$ with I standing for inversion through a point. If we take the coordinate's origin at the inversion center, then $S_I(p) = -p$. The object has a center of symmetry only if S_I applies to all points within the object.

Roto-inversion axes ($\bar{n}$): The symmetry operation consists of a rotation by 360°/n about an axis, followed by inversion through a point. The element of symmetry is the roto-inversion axis. A onefold roto-inversion axis is the same as inversion or the center of symmetry, and it is denoted by the symbol of $\bar{1}$. Other possible roto-inversion axes are $\bar{2}$, $\bar{3}$, $\bar{4}$, and $\bar{6}$. The operation of $\bar{2}$ starts first by rotating the object by 180°, and then inverting it through the inversion center. The combined operations are equivalent to having a mirror plane perpendicular to the $\bar{2}$ axis that passes through the center of inversion.

Figure 11.1 illustrates some of the symmetry elements present in the molecule of methane. The molecule has 4 threefold rotation axes along the C-H bond, 3 twofold rotation axes (one is shown in Fig. 11.1 at the intersection of H1-C-H2 and H3-C-H4 planes), 6 mirror planes (two are shown in Fig. 11.1 for the H1-C-H2 and H3-C-H4 mirror planes), and 3 fourfold roto-inversion axes. These symmetry operations belong to the point group of $\bar{4}3m$ (see below for the meaning of the point group symbols).

There are 32 point groups in three-dimensional crystals, with each group representing one possible combination of point symmetry operations. All crystals can be classified into one of 32 three-dimensional point groups (the crystal classes).

Before we get further into the details of the 32 crystal classes, we shall introduce the Hermann-Mauguin (H-M) symbols to describe the symmetry present in each crystal class. The H-M notation is designed for space groups in the International Tables for Crystallography. Simple rules allow a deduction of point groups from the H-M notation. The point group symbol has the general form of ijk, representing symmetry elements in primary (i), secondary (j), and tertiary (k) directions. A short hand form is sometimes adopted, where j or k or both is skipped, if the symmetry is 1 in the secondary or tertiary direction. The H-M notation is one of the standard ways to represent point groups, and the other popular method is to use the stereographic projection to represent symmetry operations (Fig. 11.2 shows an example). In the H-M symbols, the highest rotation axis is given first (primary), e.g., a point group with both fourfold and twofold rotation axes that starts with 4 for the

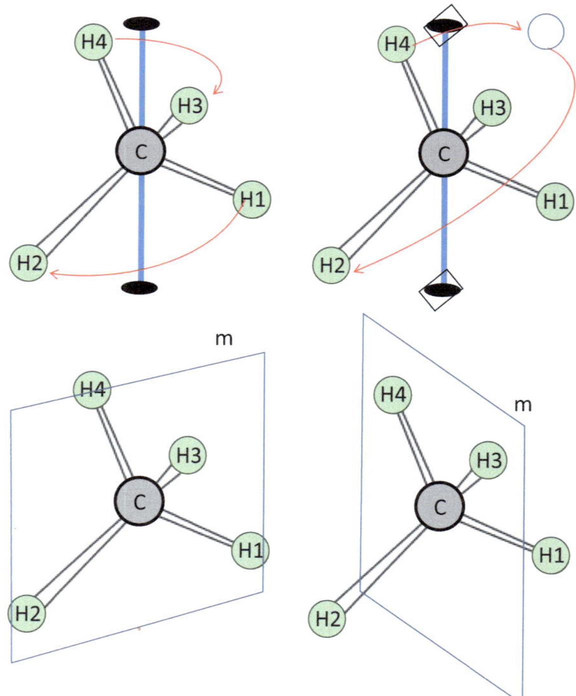

Fig. 11.1 Selected symmetry elements of methane (CH$_4$) molecule. The H atoms, which are labeled to guide the eye, are transformed by the action of a twofold rotation axis and by the action of the $\bar{4}$ roto-inversion axis (*upper*) and the action of mirrors (*lower*)

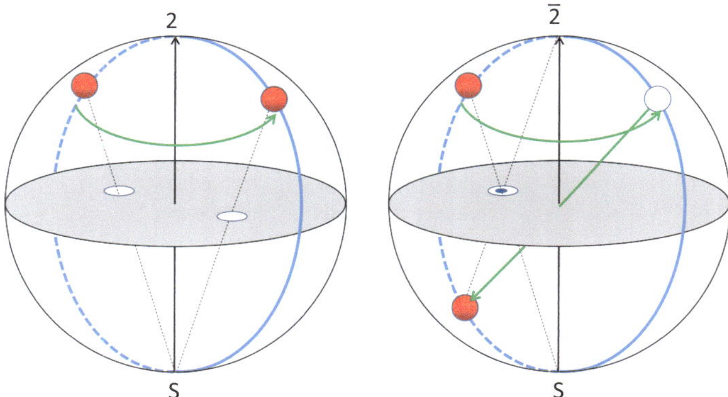

Fig. 11.2 Stereo projection and generation of 2 and $\bar{2}$(or *m*) symmetry. A three-dimensional point is represented by its projection on the horizontal plane with *open* and *filled circles* representing the opposite sides

fourfold axis. A mirror symmetry normal to the fourfold axis is noted as 4/m, where the slash indicates that the mirror plane is normal to the fourfold axis, while a parallel mirror is simply noted as 4m. The same convention applies to other combinations of rotation axes and mirrors.

The H-M symbols show which symmetry operations are unique (essential) to the point group. A symmetry operation is unique when it exists by itself and is not a product of other symmetry operations. The center of symmetry is an example with the exception of the point group $\bar{1}$; it is derived from other symmetry operations and thus not recorded in the H-M symbols. Take the point group 2/m, for example. The symbol tells that there is a twofold rotation axis and a mirror plane normal to the rotation axis. Because the combination of these two operations gives the inversion symmetry, $\bar{1}$ is also a symmetry operation in this point group, but it is not present in the H-M symbol. Altogether, the point group 2/*m* contains the following symmetry operations, 1 (identity), twofold rotation axis, horizontal mirror plane, and inversion.

Abbreviated form is used in the H-M symbols when the represented symmetry operations are sufficient to generate the rest in the point group. An example is the orthorhombic 2/*m* 2/*m* 2/*m*, which is written in the abbreviated form of *mmm*. Another example is cubic *m3m*, the full symbol of this point group is 4/*m* 3 2/*m*. The International Tables for Crystallography A lists both the abbreviated form and the full symbol for each space group.

The unique symmetry operations are taken along special directions in the different crystal classes. When there is more than one symmetry direction, the order of the H-M symbols carries specific information. An example is the orthorhombic system, in which there are three symmetry directions that coincide with the three crystallographic axes, a, b, and c. A point group with two unique mirrors and one twofold rotation axis is written as *mm2*, *2mm*, or *m2m*, where the twofold axis is

Table 11.1 Example of point groups in the H-M symbols and the direction of rotation axis for each 7 crystal systems

Crystal system Point group	Primary	Secondary	Tertiary	*Stereo projection Symbol*
Cubic $m\bar{3}m$ or $4/m\ \bar{3}\ 2/m$	$4/m$ parallel to a, b, and c	$\bar{3}$ roto-inversion axis parallel to 4 body diagonals	$2/m$ parallel to 6 edge diagonals	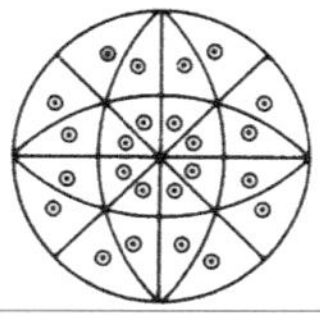
Tetragonal $4/mmm$ or $4/m\ 2/m\ 2/m$	$4/m$ parallel to c	$2/m$ parallel to a and b	$2/m$ parallel to a, b bisectors	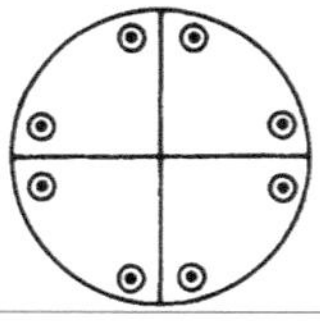
Hexagonal 62 or 622	sixfold rotation axis parallel to c	twofold rotation axis parallel to a, b and $(-a-b)$	twofold rotation axis parallel to bisectors of a, b and $(-a-b)$	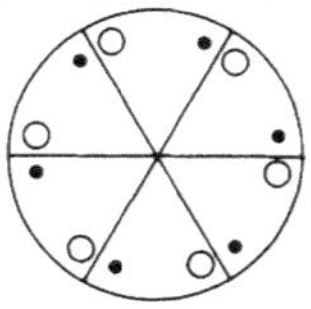
Hexagonal (Rhombohedral) $\bar{3}\ 2/m$ or $\bar{3}\ 2/m\ 1$	$\bar{3}$ parallel to c	$2/m$ parallel to a, b and $-a-b)$	onefold rotation axis parallel to bisectors of a, b and $(-a-b)$	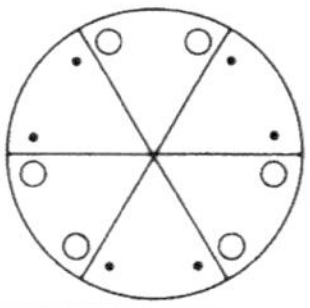
Orthorhombic $mm2$	m parallel to a	m parallel to b	twofold rotation axis parallel to c	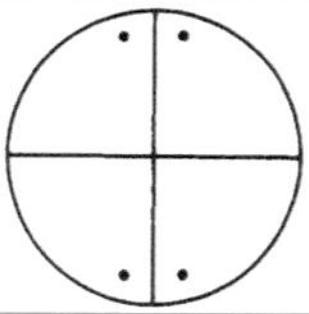
Monoclinic $2/m$	$2/m$ parallel to b			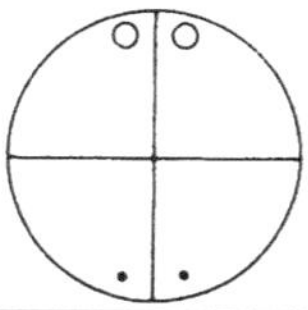
Triclinic $\bar{1}$	With inversion symmetry			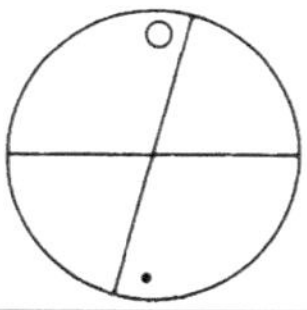

taken along c, a, and b, respectively. In such cases, there is a standard form listed in the International Tables for Crystallography. The standard for our example here is $mm2$. However, the crystallographic axes are sometimes selected for reasons other than symmetry, thus it helps to know what the H-M symbols tell us and how they are derived in order to work with nonstandard point groups. To further help understand what the symbols mean, we listed one example below for each 7 crystal systems in Table 11.1.

Now we are ready for a full list of 32 point groups, which are organized based on the 7 crystal systems together with some comments (point groups with the inversion symmetry are marked in **bold**, for example, **2/m**):

(1) *Triclinic*: 1, $\bar{1}$.
(2) *Monoclinic*: 2, m, **2/m**. The only symmetry direction is taken along b, m is normal to b.
(3) *Orthorhombic*: 222, $mm2$ (or $m2m$ or $2mm$), **mmm**. The three parts of the H-M symbol indicating the symmetry with respect to the a, b, and c axes, respectively. The m is denoted by mirror plane normal direction. Thus, mmm stands for three mirrors normal to a, b and, c axes.
(4) *Tetragonal*: 4, $\bar{4}$, **4/m**, 422, $4mm$, $\bar{4}2m$ (or $\bar{4}m2$), **4/mmm**. The first three have no additional symmetry with respect to the a and b axes. The fourfold rotation axis is always taken to be parallel to the c-axis. For those groups with symmetry along other the other axes, e.g., as in $4/mm$, see Table 11.1 for their directions.
(5) *Trigonal and Rhombohedral*: Symbol 3,$\bar{3}$, 321 (or 312), $3m1$ (or $31m$), **$\bar{3}m1$** (or **$\bar{3}1m$**). The symmetry is referred to hexagonal axes with the threefold symmetry taken to be parallel to the c-axis. Additional symmetry operations are taken along the a and b axes or their bisectors. A third axis is added along $-a$–b, which is equivalent to a and b by symmetry.
(6) *Hexagonal*: 6, $\bar{6}$, 6/m, 622, $6mm$, $\bar{6}2m$ (or $\bar{6}m2$), **6/mmm**. The symbols here are similar to that of the trigonal point group symbols except that we now have sixfold instead of threefold symmetries along the c-axis.
(7) *Cubic*: 23, **m3**, 432, $\bar{4}3m$, **m3m**. The symbols of the cubic point group symbols refer symmetry with respect to the cubic axes first and the body diagonals second, which is followed by symmetry operations with respect to the face diagonals if they are present (see Table 11.1 for an example). Among those point groups, $m3m$ has the highest symmetry.

11.3 Lattice and Space Groups

A crystal is comprised of unit cells stacked in parallel, giving rise to translational periodicity. The periodicity of translation is captured by the crystal lattice, consisted of discrete points. (The crystal lattice is thus a mathematical abstraction from the crystal structure, which is an arrangement of atoms in space. We note that the lattice is always centric.) The unit cell is defined by its dimensions (a, b, and c) and angles (α, β, and γ); for their definition see Fig. 11.3 The entire lattice can be reproduced by moving up and down from one lattice point, one unit cell length at a time, along one of the three unit cell axes.

There are only 14 distinct lattices (Bravais lattices) that can be constructed based on the constraint of periodicity in three dimensions, and they are illustrated in Fig. 11.4. The Bravais lattices build upon the seven crystal systems of different symmetry, which places following constraints on the unit cell dimensions and angles (for their minimum symmetry, see Table C.3 in Appendix C):

1. Triclinic, $a \neq b \neq c$ and $\alpha \neq \beta \neq \gamma \neq 90°$;
2. Monoclinic, $a \neq b \neq c$ and $\alpha = \gamma = 90°$, $\beta \neq 90°$;
3. Orthorhombic, $a \neq b \neq c$ and $\alpha = \beta = \gamma = 90°$;

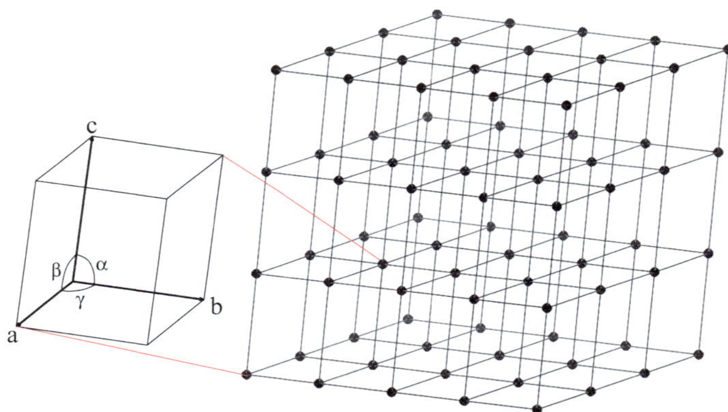

Fig. 11.3 A 3D lattice comprised of discrete points (lattice points) and its unit cell

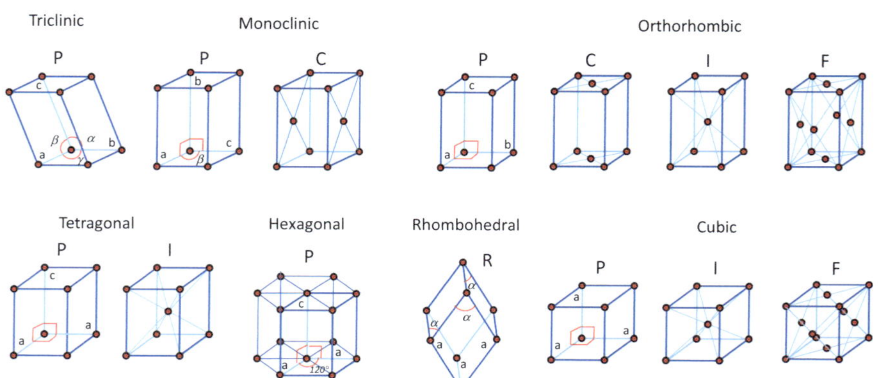

Fig. 11.4 The 14 Bravais lattice and 7 crystal systems

4. Tetragonal, $a = b \neq c$ and $\alpha = \beta = \gamma = 90°$;
5. Hexagonal, $a = b \neq c$ and $\alpha = \gamma = 90°$, $\beta = 120°$;
6. Trigonal, $a = b = c$ and $\alpha = \beta = \gamma \neq 90°$;
7. Cubic, $a = b = c$ and $\alpha = \beta = \gamma = 90°$.

Each gives a primitive (P) lattice with one lattice points per unit cell. Additional lattice points can be added on the faces or center of the parallelepiped unit cell, giving rise to different unit cells, labeled as I (body-centered), F (centered on every face), or C (centered on two faces). These have 2 (I and C) or 4 (F) lattice points per unit cell. The lattice centering must be consistent with the symmetry, for example, C centering is not allowed in the cubic crystal system, while the F centering is not compatible with the tetragonal symmetry and can be reduced to an I-centered lattice.

It should be noted that there are many ways to define a unit cell for a given lattice. For example, a rhombohedral primitive unit cell can be defined for the fcc unit cell by taking the three shortest lattice vectors as the three axes of the unit cell. The unit cells for the 14 Bravais lattices are thus chosen to reflect the full symmetry of each lattice.

The lattice translation involves an integer number of unit cell edges. Translation can be also incorporated with rotation and mirror. The introduction of translation element adds 2 new types of symmetry elements. Both are compatible with the translations of the Bravais lattice.

(1) *Screw Axes*: A screw axis combines a rotation axis with a translation along the axis. There are n-1 screw axes for a n-fold rotation, which is symbolized by N_m with $m = 1 \ldots n-1$, where $n = 2,3,4,6$. The combined symmetry operations are performed by a rotation of 360°/n degrees, followed by a translation along the rotation axis by mx/n, where x is the shortest lattice distance parallel to the screw axis. For a twofold screw axis 2_1 along the b axis in the monoclinic system, there is only one translation by $b/2$, two successive 2_1 operations result one lattice translation along b. The threefold rotation gives 2 possible screw axes 3_1 and 3_2. With the threefold axis taken along the c-axis in the trigonal system, we have of $c/3$ and $2c/3$ translation, respectively, for 3_1 and 3_2.

(2) *Glide Planes*: A glide plane is constructed by a mirror operation followed by a translation. The amount of translation is half of the shortest lattice distance along the translation direction except the diamond glides (d, discussed below.) The translation direction can be along the crystallographic axes for $a/2$, $b/2$, or $c/2$, or face diagonals for $(a + b)/2$, $(a + c)/2$, or $(b + c)/2$. (The face diagonal glides are all marked by n, while glides along the axes are named by their direction, e.g., a, b, or c.) In a centered unit cell (faced-centered F or body-centered I), the glide is one-half of the primitive cell vector, for example, $(a + b)/4$, $(a + c)/4$, or $(b + c)/4$ in a face-centered unit cell, $(a + b + c)/4$ in a body-centered unit cell. This former is called diamond glide since it is one of symmetry elements in the

face-centered diamond structure. Glide planes are very common in inorganic structures, and they are generally absent in protein structures.

In a 3D molecular crystal, an identical molecular orientation is obtained only after repeat applications of the symmetry. For example, in case of a 4_3 screw axis, the same orientation is obtained after 4 repeated applications and at 3 unit cells away ($4*3/4 = 3$). The same holds for glide planes.

In the H-M symbols, the glide planes and screw axes are reduced to point symmetries by simply removing the translation from the screw axes to obtain the rotation axes and replacing the glides planes by mirrors. For example, a 4_3 screw axis yields a fourfold rotation axis in the point group.

A rotation axis cannot be distinguished from the corresponding screw axis or a glide plane from the mirror plane by the external crystal faces. Because of this, symmetry elements involving translations are also referred to as internal symmetry elements, in contrast to the point groups or crystal classes, which can be determined by *examining the external shapes of crystals*. In principle, the internal and external symmetries can be distinguished if molecules or atoms can be resolved on the surfaces. For example, every other molecule would stick out of a lattice plane (crystal face) by $1/2$ of the lattice plane distance in case of a 2_1 screw axis. This requires high-resolution microscopy in order to see the molecules.

The space groups combine the symmetry elements of the point groups with the translational symmetries. If the symmetry operations of a space group are simply the point group operations added on top of the Bravais lattice points, we have a symmorphic space group. Since all translations belong to the lattice, a symmorphic space group does not have any nonprimitive translations. There are 73 such space groups. Introducing screw axes and glide planes extends the number to 230 space groups, which are listed in the International Tables for Crystallography A. All crystals can be classified into one of these groups. The H-M symbol of a space group starts with the type of Bravais lattice whether it is primitive or centered and the type of centering. This is then followed by the symmetry of the characteristic directions in the same order as in the symbol of the corresponding point group (Table 11.1). Additionally, graphic symbols (see Table 11.2) are used to indicate the symmetry elements, their directions, and locations in the unit cell. Below is a summary of space groups in 7 crystal systems:

Triclinic space groups: The combination of P lattice with point group 1 and $\bar{1}$ gives $P1$ and $P\bar{1}$.

Monoclinic space groups: We have two types of lattices in the monoclinic system, P and C and three point groups, 2, m, and $2/m$. There is only one characteristic symmetry direction, which is the b axis. By placing the point symmetry operations directly on every lattice points of those two lattices along b, we obtain the following symmorphic space groups, $P2$, $C2$, Pm, Cm, $P2/m$, and $C2/m$. Next, we consider replacing twofold rotation axes with 2_1 screw axes and mirrors with glide planes. In the latter case, the mirror is normal to b and the translation direction can be either along a or c. Since the only unique axis in the monoclinic system is the b axis, a and c axes can be selected arbitrarily. Thus, as a convention, we take the glide along c. In $P2_1$, the screw axes generate other parallel screw axes that bisect the unit cell. Similarly, combination of translations according to C centers with twofold rotation axes generate other screw axes, while C centers with 2_1 screw axes generate twofold rotation axes. Thus, $C2$ is equivalent to $C2_1$. Table 11.3 summaries the 13 unique space groups obtained this way.

Figure 11.5c shows the symmetry in the monoclinic space group $P2_1/c$ as viewed along the c-axis with b in horizontal direction. The 2_1 screw axis is along b as indicated in the figure, while the c glide plane is normal to b. In the figure, the open circle represents a structural unit, which can be an atom or a molecule or a group of atoms or molecules.

Figure 11.5a, c show the International Tables for Crystallography Volume A diagram for the space group $P2_1/c$ (number 14) when viewed along the b and c axes, respectively. The unit cell origin is now selected at the center of symmetry. This gives the c glide planes at the ¼ and ¾ positions as indicated by the label on the left and the dashed lines on the right.

Orthorhombic space groups: The orthorhombic system has three point groups (222, *mm2*, and *mmm*) and four possible lattice types (*P, C, I,* and *F*, see Fig. 11.4). So their combinations give twelve space groups, before we substitute for screw axes and glide planes. When all of this is done, we add 38 new space groups. Unlike the monoclinic system, there are three symmetry directions along a, b, and c in the orthorhombic system. We can have glide planes along (100), (010), and

Table 11.2 Space-group symmetry elements and their symbols

Symmetry element	Symbol
Center of symmetry	o $\bar{1}$
Rotation, screw and roto-inversion axes normal to plane	2 2_1 3 3_1 3_2 4 4_1 4_2 4_3 6 6_1 6_2 6_3 6_4 6_5 $\bar{6}$ $\bar{4}$ $\bar{1}$
Rotation axes in plane	⟶
Screw axes in plane	⟶
Mirror and glide planes in plane	*m* *b* *n* *e* *d*
Mirror and glide planes normal to plane	————— *m* – – – – – *a* ·············· *c* —·—·—·— *n* —··—··—··— *e* —·→·—·—·→·— *d*

Table 11.3 The monoclinic space groups tabulated according to point groups and P and C lattices

Class	*P* lattice	*C* lattice
2	*P2 P2$_1$*	*C2*
m	*Pm Pc*	*Cm Cc*
2/*m*	*P2/m P2$_1$/m P2/c P2$_1$/c*	*C2/m C2/c*

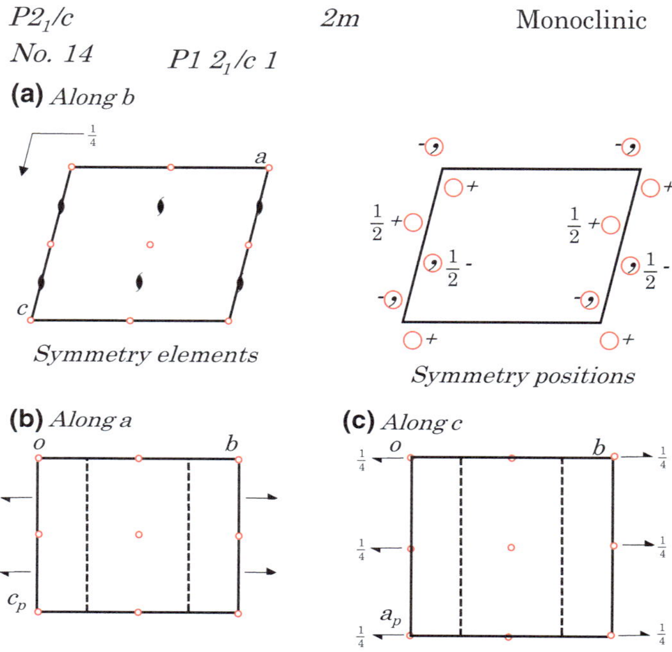

Fig. 11.5 Symmetry elements and positions of monoclinic space group $P2_1/c$ as viewed along b, a, and c axes, respectively. The graphical symbols used here are listed in Table 11.2

(001) because they are perpendicular to each other. The glide direction can be along a, b, c, n, or d. Since the three directions are mutually perpendicular to each other, there are 6 permutations for a, b, and c. Because of this, the space group of *Pnma* can have six different symbols: *Pnma*, *Pbnm*, *Pmcn*, *Pnam*, *Pmnb*, and *Pcmn*. This can be most confusing as different orientations were selected by past researchers, creating a large number of nonstandard space groups. Fortunately, their differences as well as their conversion are all specified in the International Tables (Hahn 2005).

A question that arises with multiple symmetry directions in whether rotation, roto-inversion, or screw axes intersect each other or not. This comes up first in the orthorhombic system, where there are three orthogonal symmetry directions. Take the space groups $P222$ (no. 16) and $P222_1$ (no. 17) as example (Fig. 11.6). In $P222$,

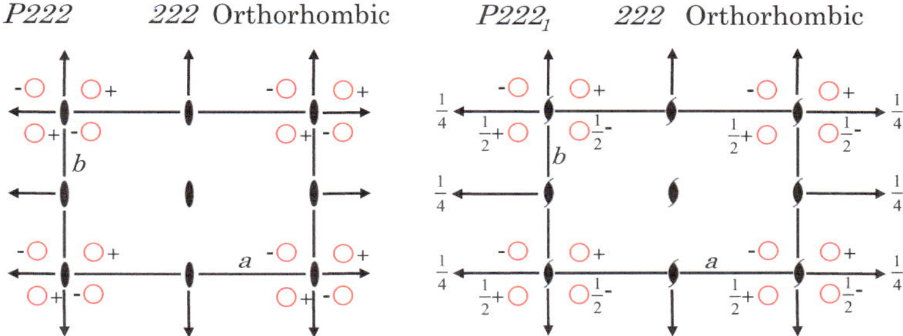

Fig. 11.6 Space groups of $P222$ (no. 16) and $P222_1$ (no. 17) with intersecting and nonintersecting axes. The diagram is along the c-axis. In $P222_1$, the twofold axes along a and b are offset along c by ¼

by having the twofold rotation axes along a and b intersecting each other, it generates another intersecting twofold axis in the normal direction, along c. Next, if we move the twofold axis along b to ¼ position along c as in $P222_1$, the combination of a and b twofold axes then generates a 2_1 screw axis along c in this case.

Tetragonal space groups: Unlike in the orthorhombic system, there is no ambiguity in the choice of coordinates here. The first symbol after the lattice type always refers to the c-axis with a fourfold symmetry. According to our discussion on point groups, there are seven point groups in this system. Together with two lattice types, P and I, they give 14 symmorphic groups. Another 52 space groups are created by substituting for screw axes and glide planes. Among these, the combination of the body-centered lattice (I) with fourfold screw axes of 4_1, 4_2, and 4_3 leaves only $I4_1$ as a distinct space group, while $I4_2$ has the same symmetry as $I4$ and $I4_1$ has both 4_1 and 4_3 screw axes.

Trigonal and hexagonal space groups: There are two lattice types in this case, P and R. The difference between trigonal and hexagonal systems is the first having a threefold axis, while the second with a sixfold axis, which is taken along c. The hexagonal system comprises 27 space groups (no. 168–194). All of these are primitive (P). The trigonal system is the tricky one, and its 25 space groups (no. 143–167) have either the hexagonal primitive lattice (P, 18 space groups) or the rhombohedral lattice (R, 7 space groups). There is no fixed rules for the rhombohedral lattice, and it can occur in any point group of the trigonal system.

Cubic space groups (also known as isometric space groups): What differentiates this set of space groups from the others is the 4 nonorthogonal threefold axes, which make it particularly difficult to visualize their symmetries using projected diagrams. The choice of axes is universal. The first symbol after the lattice is along the cubic axes, which is then followed by symmetry in the secondary and tertiary directions, in the same way as in the point group symbol. Thus, in the m-$3m$ system, we have Fd-$3m$ and Ia-$3d$, the d for diamond glide occurs in very different directions. Similar to the tetragonal system, we generate 3 more space groups by substitution of fourfold screw axes in $P432$, but there are only two distinct space groups in I and

F lattices for $I4_132$ and $F4_132$. Both $I23$ and $I2_13$ contain 2_1 screw axes, and the difference between these two is that all twofold axes intersect in $I23$, but not in $I2_13$. Also in $I2_13$, the 3_1 and 3_2 axes do not intersect. Cubic systems with nonintersecting threefold axes cannot have tetragonal site symmetry, since it requires 4 intersecting threefold axes.

A consequence of symmetry operations is the concept of asymmetric unit, which is the smallest volume in a unit cell that can be used to generate the entire unit cell volume by applications of symmetry operations. In crystal diffraction, the asymmetric unit can be approximated as comprised of atoms. Then, for a given atomic position at x, y, z, the symmetry operations place the same atom at a number of equivalent positions generated by symmetry. Figure 11.7 shows the example of $C2/c$ taken from International Tables, and there are 4 equivalent positions for a general position. There are also special positions, where x, y, and z assume special values. They are assigned with a number and symbol called Wyckoff positions. These positions have different degrees of point symmetry, and they are listed according to the symmetry with the highest symmetry at the bottom of the list. At the top of the listing is the coordinates, under which two are listed. These represent the lattice point position with (000) at the unit cell origin and (1/2, 1/2, 0) from the C centering. The + indicates that these coordinates add to the positions listed below. The bar on top of x, y, or z means −, e.g., $\bar{x} = -x$.

The space groups of some common simple crystal structures are as follows: the cubic close-packed metals, sodium chloride, calcium fluorite, $Fm3m$; hexagonal close-packed metals, $Im3m$; diamond, $Fd3m$; graphite, $P6_3/mmc$, cesium chloride, perovskite, $Pm3m$; sphalerite, $F\bar{4}3m$; wurzite, $P6_3mc$; rutile, $P4_2/mmm$; and spinel, $Fd3m$.

Positions

Multiplicity, Wyckoff letter, Site symmetry			Coordinates $(0,0,0)+ \quad (\tfrac{1}{2},\tfrac{1}{2},0)+$			
8	f	1	(1) x,y,z	(2) $\bar{x},y,\bar{z}+\tfrac{1}{2}$	(3) $\bar{x},\bar{y},\bar{z}$	(4) $x,\bar{y},z+\tfrac{1}{2}$
4	e	2	$0,y,\tfrac{1}{4}$	$0,\bar{y},\tfrac{3}{4}$		
4	d	$\bar{1}$	$\tfrac{1}{4},\tfrac{1}{4},\tfrac{1}{2}$	$\tfrac{3}{4},\tfrac{1}{4},0$		
4	c	$\bar{1}$	$\tfrac{1}{4},\tfrac{1}{4},0$	$\tfrac{3}{4},\tfrac{1}{4},\tfrac{1}{2}$		
4	b	$\bar{1}$	$0,\tfrac{1}{2},0$	$0,\tfrac{1}{2},\tfrac{1}{2}$		
4	a	$\bar{1}$	$0,0,0$	$0,0,\tfrac{1}{2}$		

Fig. 11.7 Special positions in the space group of $P2_1/c$

11.4 Symmetry Operation in Real and Reciprocal Spaces

A symmetry operation in general consists of two parts: a rotation R and a translation T. The rotation R can be represented by a 3×3 matrix, while T is a 3D vector in the real space. The operation is isometric, e.g., the distances between points are preserved. For a point in the real space, $\vec{r} = (x, y, z)$, the symmetry operation transforms it to $\vec{r}' = (x', y', z')$ according to

$$\vec{r}' = S\{\vec{r}\} = R\vec{r} + \vec{T}, \tag{11.2}$$

and in the matrix representation

$$\begin{pmatrix} x' \\ y' \\ z' \end{pmatrix} = \begin{pmatrix} R_{11} & R_{12} & R_{13} \\ R_{21} & R_{22} & R_{23} \\ R_{31} & R_{32} & R_{33} \end{pmatrix} \begin{pmatrix} x \\ y \\ z \end{pmatrix} + \begin{pmatrix} T_x \\ T_y \\ T_z \end{pmatrix}. \tag{11.3}$$

The R matrix is subjected to number constraints. First, it must be orthogonal

$$R^{-1} = R^{\mathrm{T}} = \begin{pmatrix} R_{11} & R_{21} & R_{31} \\ R_{12} & R_{22} & R_{32} \\ R_{13} & R_{23} & R_{33} \end{pmatrix}, RR^{-1} = I, \tag{11.4}$$

where I is the identity matrix. Secondly, R belongs to the symmetry group of the real space lattice. Thus, for a lattice translation vector T, RT also belongs to the lattice.

A convenient way to write the above matrix operations is to use the Seitz notation, with $S = \{R|T\}$, where R is substituted by a symbol for the rotation, and the fractional coordinates are used for T. The basic algebra of Seitz operators is

$$\{R_1|T_1\}\{R_2|T_2\} = \{R_1R_2|R_1T_2 + T_1\}.$$
$$\{R|T\}^{-1} = \{R^{-1}| - R^{-1}T\} \tag{11.5}$$

A listing of R and T for all symmetry operations can be found in Glazer et al. (2014). Below are some examples:

Twofold rotation along z, $\{2_{001}|0\}$:

$$\begin{pmatrix} x' \\ y' \\ z' \end{pmatrix} = \begin{pmatrix} -1 & 0 & 0 \\ 0 & -1 & 0 \\ 0 & 0 & 1 \end{pmatrix} \begin{pmatrix} x \\ y \\ z \end{pmatrix} + \begin{pmatrix} 0 \\ 0 \\ 0 \end{pmatrix}, \tag{11.6}$$

2_1 screw axis along z, $\{2_{001}|001/2\}$

$$\begin{pmatrix} x' \\ y' \\ z' \end{pmatrix} = \begin{pmatrix} -1 & 0 & 0 \\ 0 & -1 & 0 \\ 0 & 0 & 1 \end{pmatrix} \begin{pmatrix} x \\ y \\ z \end{pmatrix} + \begin{pmatrix} 0 \\ 0 \\ 1/2 \end{pmatrix}, \tag{11.7}$$

Mirror normal to z, $\{m_{001}|0\}$:

$$\begin{pmatrix} x' \\ y' \\ z' \end{pmatrix} = \begin{pmatrix} 1 & 0 & 0 \\ 0 & 1 & 0 \\ 0 & 0 & -1 \end{pmatrix} \begin{pmatrix} x \\ y \\ z \end{pmatrix} + \begin{pmatrix} 0 \\ 0 \\ 0 \end{pmatrix}. \tag{11.8}$$

The origin of the reciprocal lattice is fixed, and it is thus invariant under translation, e.g., $\{1|T\}\vec{g} = \vec{g}$. What we have is the point symmetry for the reciprocal lattice, which is the same as the point group symmetry of the real space lattice. To show this, we use the invariance of the dot product of $\vec{g} \cdot \vec{r}$ under rotation R

$$R\vec{g} \cdot R\vec{T} = \sum_i \sum_j R_{ji} g_j R_{ij} T_j = \sum_i g_j T_j = \vec{g} \cdot \vec{T} \tag{11.9}$$

Since $R\vec{T}$ is a vector of the real lattice, $R\vec{g}$ is equivalent to $\vec{g}$ by symmetry. One immediate result from having the same symmetry of the real lattice is that the reciprocal lattice must belong to one of the 14 Bravais lattices. However, the Bravais lattice type can be different.

11.5 Symmetry Determination Using Kinematic Diffraction Intensities

Under the kinematical approximation, the peak intensity of a Bragg reflection or the integrated intensity from a rotating crystal is proportional to the square of the crystal structure factor (Sect. 4.6), $|F_{hkl}|^2$. Since

$$F_{hkl} = \sum_{i=1}^{n} f_i e^{-2\pi i(hx_i + ky_i + lz_i)} = F^*_{\bar{h}\bar{k}\bar{l}}, \tag{11.10}$$

We have

$$|F_{hkl}|^2 = |F_{\bar{h}\bar{k}\bar{l}}|^2. \tag{11.11}$$

Thus, the intensity of the g reflection equals to that of the $-g$ reflection, if we ignore the anomalous dispersion. This is known as Friedel's law, which is applicable to Fourier transform of any real functions. To a large extent, X-ray diffraction can be described by kinematic approximation. Because of this, X-ray diffraction patterns would show a center of symmetry (unlike CBED patterns). For an acentric crystal, the diffraction pattern would be centric belonging to the point groups with the center of symmetry, which is a subset of the point groups (11 out of 32) that we discussed in Sect. 11.2. These groups are known as Laue groups (see Table 11.4 for the relationship between the two groups). In electron diffraction, dynamic diffraction or the presence of large anomalous absorption leads to the breakdown of Friedel's law and thus the center of symmetry can be determined directly using techniques such as CBED or anomalous X-ray diffraction.

In high-energy electron diffraction, the condition for kinematic diffraction is only met in very thin samples of light materials, such as graphene. Recent progress in precession electron diffraction, however, shows that the integrated diffraction intensity obtained with a large precession angle scales with the square of the electron structure factor for some reflections (Midgley and Eggeman 2015). To these, kinematic diffraction data analysis can be extended. We shall see in the next section, symmetry determination by CBED requires the recording of CBED patterns with a convergence angle, which can be difficult for crystals with a large unit cell or radiation (beam) sensitive or both (an example is protein crystals). In such cases, precession electron diffraction-based symmetry analysis could be helpful.

A popular method to observe pattern symmetry in X-ray diffraction is Laue diffraction, which is performed using a broad spectrum of X-ray wavelengths with the crystal oriented so the incident beam is along a symmetry axis. The range of wavelengths allows all reflections between the smallest and largest Ewald spheres to be recorded in the back scattering or transmission Laue pattern. Take the threefold rotation axis (or roto-inversion axis or screw axis) for example, a reflection recorded at point P will appear at P' and P'' at 120° apart because of the trigonal axis, and they are called equivalent reflections (other than the Friedel pair, $\pm(hkl)$). Similarly, equivalent reflections are recorded in a precession electron diffraction pattern along the symmetry axis for all reflections swept through by the Laue circle of the rotating electron beam.

Table 11.4 Laue groups and crystal classes

Laue group	Crystal classes			
$\bar{1}$	1	$\bar{1}$		
$2/m$	2	m	$2/m$	
mmm	222,	$mm2$	mmm	
$4/m$	4	$\bar{4}$	$4/m$	
$4/mmm$	422	$4mm$	$\bar{4}2m$	$4/mm$
$\bar{3}$	3	$\bar{3}$		
$\bar{3}\,m$	32	$3m$	$\bar{3}\,m$	
$6/m$	6	$\bar{6}$	$6/m$	
$6/mmm$	622	$6mm$	$\bar{6}m2$	$6/mm$
$m3$	23	$m3$		
$m3m$	432	$\bar{4}3m$	$m3m$	

The presence of rotation axes, mirrors, and screw axes can be determined by the intensity of equivalent reflections. Take the monoclinic system as example, an atom in a general position (x, y, z) generates the following pair of equivalent positions and the structure factor relationship:

(1) Twofold rotation: (x, y, z) and $(-x, y, -z)$

$$F_{hkl} = \sum_{i=1}^{N} f_i e^{-2\pi i (hx_i + ky_i + lz_i)}$$

$$= \sum_{i=1}^{N/2} 2f_i \cos 2\pi (hx_i + lz_i) e^{-2\pi i k y_i} = F^*_{h\bar{k}l}$$

(2) 2_1 c screw axis: (x, y, z) and $(-x, y, -z + 1/2)$

$$F_{hkl} = -\sum_{i=1}^{N/2} 2i f_i \sin 2\pi (hx_i + lz_i) e^{-2\pi i k y_i} = -F^*_{h\bar{k}l}$$

(3) Mirror plane: (x, y, z) and $(x, -y, z)$

$$F_{hkl} = \sum_{i=1}^{N/2} 2f_i \cos(2\pi k y_i) e^{-2\pi i (hx_i + lz_i)} = F_{h\bar{k}l}$$

In above three cases, we have

$$|F_{hkl}| = |F_{h\bar{k}l}| \tag{11.12}$$

and by the extension of Friedel's law

$$|F_{hkl}| = |F_{h\bar{k}l}| = |F_{\bar{h}k\bar{l}}| = |F_{\bar{h}\bar{k}\bar{l}}|. \tag{11.13}$$

The glide planes can be distinguished from mirrors, or rotation axes from screw axes, using the systematic absence observed in kinematic diffraction patterns. Take the c glide plane normal to b as example, atoms at (x, y, z) and $(x, -y, z + 1/2)$ are paired by the symmetry, and we have for even l number

$$F_{hkl} = \sum_{i=1}^{N/2} 2f_i \cos(2\pi k y_i) e^{-2\pi i (hx_i + lz_i)} e^{-\pi i l z_i},$$

and odd l number

$$F_{hkl} = \sum_{i=1}^{N/2} 2f_i \sin(2\pi k y_i) e^{-2\pi i(hx_i + lz_i)} e^{-\pi i l z_i},$$

which is 0 for $k = 0$. This is known as extinction conditions. International Tables for Crystallography A lists the extinction conditions for all space groups that can be used for identification of glide planes and screw axes.

Extinction also arises from lattice centering. In a fcc crystal, reflections of mixed Miller indices are extinct, while in a bcc crystal, the extinction condition is $h + k + l = 2n + 1$. For C centering, $F_{hkl} = 0$, for $h + k = 2n + 1$ for any l values.

The extinction due to lattice centering is also observed in electron diffraction even under dynamic diffraction since they arise from the presence of the primitive cell, which is smaller than the crystal unit cell. The primitive cell determines the dimensions of the reciprocal lattice. The extinction due to glide planes or screw axes is only observed under special conditions in dynamic diffraction. This will be discussed in the next section.

A major difficulty in using kinematic diffraction intensities to determine the crystal symmetry is to identify whether the crystal is centric or acentric. Because structure factors are real for centric crystals with the phase of 0 or π, while the structure factor phase of acentric crystals can be any value between 0 and 2π, the determination of the center of symmetry plays a critical role in phasing the measured structure factors for solving the crystal structure. For crystals with a large unit cell and number of atoms, the distribution of the structure factor amplitudes can be used to determine the center of symmetry. Details about this type of quantitative analysis can be found in the book by Woolfson (1997).

11.6 Symmetry Determination by CBED

A glance of CBED patterns recorded at major zone axes, like the one in Fig. 11.9 or Fig. 11.8, reveals certain symmetry about the center of the pattern. The pattern symmetry is determined by the symmetry of the crystal, the specimen, and electron diffraction. Thus, by observing the recorded diffraction symmetry under certain experimental conditions, it is possible to determine the crystal symmetry. Figure 11.9 illustrates the symmetry of an ideal CBED experiment. First, a parallel-sided sample is prepared with its surface normal along a principal symmetry direction, which is taken as the z-axis. In the horizontal direction, the exact sample shape is not important as long as its size is larger than the electron beam lateral coherence length (L_c) at the sample plane. Within the limit of coherent diffraction, the diffracting sample can be considered as a circular disk of thickness t and diameter L_c. A CBED pattern is recorded using a convergent beam formed by a circular condenser aperture, centered on the z-axis. Thus, the diffraction geometry

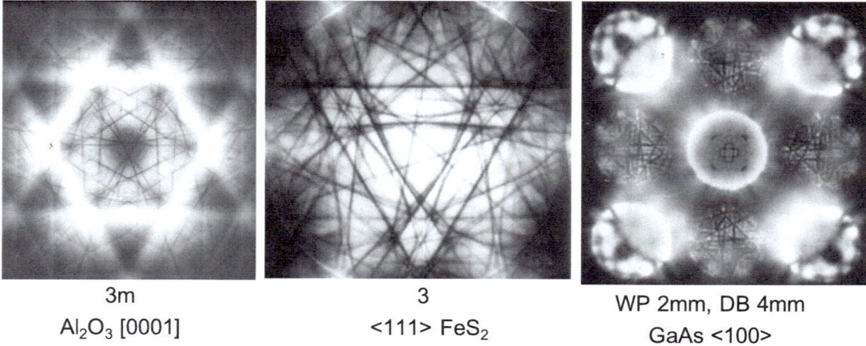

Fig. 11.8 Examples of observed symmetry in experimental CBED patterns. (Provided by John Steeds, Bristol University)

Fig. 11.9 The geometry of CBED used for symmetry determination. The CBED pattern shown was recorded from NiO at T = 106 K and 119.52 kV using a LEO 912 TEM with the in-column Ω-energy filter (Peng and Zuo 1999)

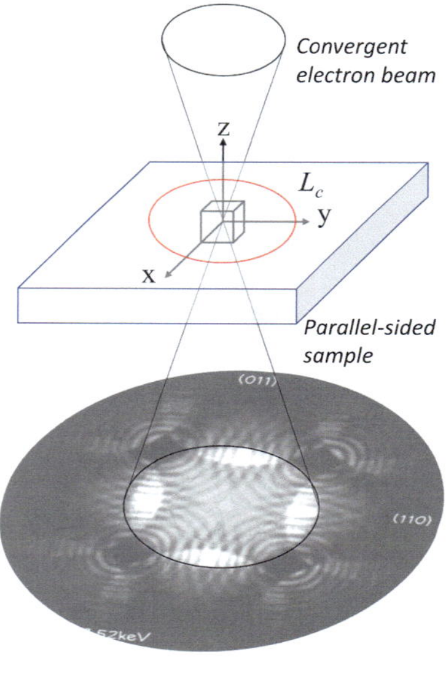

possesses the symmetry of a circular disk, including the cylindrical rotation axis and a horizontal mirror plane at the middle of the disk and twofold rotation axes in the horizontal mirror plane.

In practice, the sample is often tilted by a small angle. The tilt effect in dynamic theory is described by the renormalized eigenvalue equation (Eq. (5.8)) using the normal component of the wave vector K_n). For a small tilt angle θ, we have $K_n \approx K\left(1 - \theta^2/2\right) \approx K$. Thus, to the first-order approximation, the effect of a small sample tilt on symmetry determination can be neglected.

A CBED pattern records the diffracted intensity for a set of reflections as a function of the incident beam wave vector given by the direct beams (see Chap. 3 for details). At a point inside the disk of the reflection g, diffraction intensity is determined by the incident beam wave vector $\vec{K}$, the complex crystal interaction potential $U(\vec{r})$, other reflections $\vec{h}$, and the sample thickness or geometry $\Lambda(\vec{r})$. Under a symmetry operation S that belongs to both the crystal and the sample (e.g., both $U(\vec{r})$ and $\Lambda(\vec{r})$ stay the same under the operation), we have

$$
\begin{aligned}
S\left\{I_g\left(\vec{K},\vec{h},U(\vec{r}),\Lambda(\vec{r})\right)\right\} &= I_{S\{g\}}\left(S\{\vec{K}\},S\{\vec{h}\},S\{U(\vec{r})\},S\{\Lambda(\vec{r})\}\right) \\
&= I_{S\{g\}}\left(S\{\vec{K}\},S\{\vec{h}\},U(\vec{r}),\Lambda(\vec{r})\right)
\end{aligned}
, \quad (11.14)
$$

which we simplify as

$$
S\left\{I_g\left(\vec{K},\vec{h}\right)\right\} = I_{S\{g\}}\left(S\{\vec{K}\},S\{\vec{h}\}\right).
$$

For a 3D symmetry operation that includes a translation along the z-axis, the sample symmetry is broken in principle. However, for a thick sample, the amount of translation is much smaller than the sample thickness, and consequently the symmetry breaking is small and can be neglected.

11.6.1 Point Symmetry in Dynamic Diffraction

We have pointed out the need to have a thick sample in order to observe 3D symmetry. In general, CBED patterns are recorded from relatively thick samples in order to obtain the full symmetry information from dynamic diffraction. The Bloch wave theory described in Chap. 5 is a good starting point for examining the symmetry in dynamic diffraction. According to this theory, electron diffraction is determined by the following eigenvalue equation

$$
A(\vec{K})C(\vec{K}) = 2K_n\gamma(\vec{K})C(\vec{K}), \quad (11.15)
$$

where A is an n by n matrix with n beams included, dependent on the incident wave vector $\vec{K}$, namely

$$
A(\vec{K}) = \begin{pmatrix}
0 & U_{-g} & U_{-h} & \cdots \\
U_g & 2KS_g(\vec{K}) & U_{gh} & \cdots \\
U_h & U_{hg} & 2KS_h(\vec{K}) & \cdots \\
\vdots & \vdots & \vdots & \ddots
\end{pmatrix}, \quad (11.16)
$$

where

$$2KS_g\big(\vec{K}\big) = K^2 - \big(\vec{K} + \vec{g}\big)^2,$$

and

$$U_g = \frac{1}{V_c} \int\limits_{u.c.} U(\vec{r})e^{-2\pi i\vec{g}\cdot\vec{r}}d\vec{r}.$$

Under a symmetry operation, where we transform $\vec{r}$ to $R\vec{r} + \vec{T}$, $\vec{K}$ to $R\vec{K}$, and $\vec{g}$ to $R\vec{g}$, it can be shown

$$2KS_{Rg}\big(R\vec{K}\big) = 2KS_g\big(\vec{K}\big). \tag{11.17}$$

For the electron structure factor, by using Eq. (11.9) and $d\big(R\vec{r} + \vec{T}\big) = d\vec{r}$, we have

$$
\begin{aligned}
U_{Rg} &= \frac{1}{V_c} \int\limits_{u.c.} U\big(R\vec{r} + \vec{T}\big)e^{-2\pi iR\vec{g}\cdot\left(R\vec{r} + \vec{T}\right)}d\big(R\vec{r} + \vec{T}\big) \\
&= \frac{1}{V_c} \int\limits_{u.c.} U(\vec{r})e^{-2\pi i\vec{g}\cdot\left(\vec{r} + R^{-1}\vec{T}\right)}d\vec{r} \tag{11.18} \\
&= U_g e^{-2\pi i\vec{g}\cdot R^{-1}\vec{T}} = \tilde{U}_g.
\end{aligned}
$$

Thus, U_g and U_{Rg} are related by a phase shift, and $\tilde{U}_g$ is the electron structure factor of the crystal with its origin translated by $R^{-1}\vec{T}$. Combining this result with Eq. (11.17), we have

$$
\begin{aligned}
S\{A(\vec{K})\} &= \begin{pmatrix}
0 & U_{-Rg} & U_{-Rh} & \cdots \\
U_{Rg} & 2KS_{Rg}\big(R\vec{K}\big) & U_{Rgh} & \cdots \\
U_{Rh} & U_{Rhg} & 2KS_{Rh}\big(R\vec{K}\big) & \cdots \\
\vdots & \vdots & \vdots & \ddots
\end{pmatrix} \\
&= \begin{pmatrix}
0 & \tilde{U}_{-g} & \tilde{U}_{-h} & \cdots \\
\tilde{U}_g & 2KS_g\big(\vec{K}\big) & \tilde{U}_{gh} & \cdots \\
\tilde{U}_h & \tilde{U}_{hg} & 2KS_h\big(\vec{K}\big) & \cdots \\
\vdots & \vdots & \vdots & \ddots
\end{pmatrix},
\end{aligned}
$$

which has the same eigenvalues as $A(\vec{K})$ and the eigenvectors given by

$$\tilde{C}_g(\vec{K}) = C_g(\vec{K})\mathrm{e}^{-2\pi i \vec{g}\cdot R^{-1}\vec{T}}.$$

From this, we obtain the following relationship:

$$\phi_{Rg}(R\vec{K}) = \mathrm{e}^{-2\pi i \vec{g}\cdot R^{-1}\vec{T}} \sum_i c_i C_g^i \mathrm{e}^{2\pi i \gamma^i t} = \mathrm{e}^{-2\pi i \vec{g}\cdot R^{-1}\vec{T}} \phi_g(\vec{K}), \tag{11.19}$$

and

$$\left|\phi_g(\vec{K})\right| = \left|\phi_{Rg}(R\vec{K})\right|. \tag{11.20}$$

Electron diffraction patterns possess additional symmetry resulted from the symmetry in the Schrödinger wave equation (Eq. (2.17), Chap. 2), in which we have both

$$\phi^+(\vec{r}, t) = \exp\left[2\pi i \vec{k}\cdot \vec{r}\right] \exp(-\pi i \omega t)$$

and

$$\phi^-(\vec{r}, t) = \exp\left[-2\pi i \vec{k}\cdot \vec{r}\right] \exp(\pi i \omega t)$$

as the solutions of the Schrödinger equation. Thus, the wave equation is invariant with the coordinate transformation from $\left(\vec{r}, \vec{k}\right)$ to $\left(\vec{r}, -\vec{k}\right)$. An important result is the reciprocity theorem, which in terms of two points A and B and the scatter at P (see Fig. 11.10) can be stated as Pogany and Turner (1968)

Fig. 11.10 Reciprocity in electron diffraction

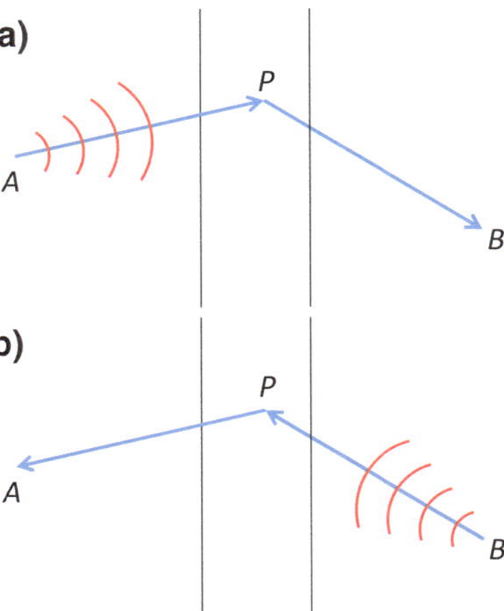

The amplitude at B of a wave originating from a source at A, and scattered by P, is equal to the scattered amplitude at A due to the same source placed at B.

A point source ϕ_o placed at $\vec{r}_A$ diffracted by the object gives the wave field at B as $\phi(\vec{r}_A, \vec{r}_B)$. According to reciprocity theorem, the wave field at A, $\phi(\vec{r}_B, \vec{r}_A)$, with the point source placed at B relates to $\phi(\vec{r}_A, \vec{r}_B)$ according to

$$|\phi(\vec{r}_B, \vec{r}_A)| = |\phi(\vec{r}_A, \vec{r}_B)|. \tag{11.21}$$

This holds when the entire wave field is included (with no aperture). It is also valid under the assumption where both the incident and scattered waves vanish outside the aperture. The latter case applies to electron diffraction since both the electron source and the detector are far from the sample and the incident and scattered wave fields are small compared to the microscope aperture. A formal proof of Eq. (11.21) can be obtained using the properties of Green's function (Pogany and Turner 1968 and Peng et al. 2004). Pogany and Turner also showed that the reciprocity principle holds for electron elastic scattering for complex optical potentials.

To see how reciprocity combines with the 3D symmetry of the sample, we consider first Fig. 11.11, where the sample has a mirror symmetry at the middle (m'). Take the incident beam marked as x, it gives rise to the diffracted beam marked as $\bullet$ in the dark-field pattern (DP) of a reflection g belonging to the ZOLZ. Next, we write the incident wave vector $\vec{K}$ in terms of its components parallel ($\vec{K}_z$) and normal ($\vec{K}_t$) to z as

$$\vec{K} = \vec{K}_z + \vec{K}_t = \vec{K}_z - \vec{g}/2 + \vec{\Delta}. \tag{11.22}$$

Here, we have taken $\vec{K}_t = -\vec{g}/2 + \vec{\Delta}$, where $\vec{\Delta}$ is a vector in the reciprocal space normal to z. Under the mirror symmetry operation in Eq. (11.8),

$$R\vec{K} = R\vec{K}_z - R\vec{g}/2 + R\vec{\Delta} = -\vec{K}_z - \vec{g}/2 + \vec{\Delta}.$$

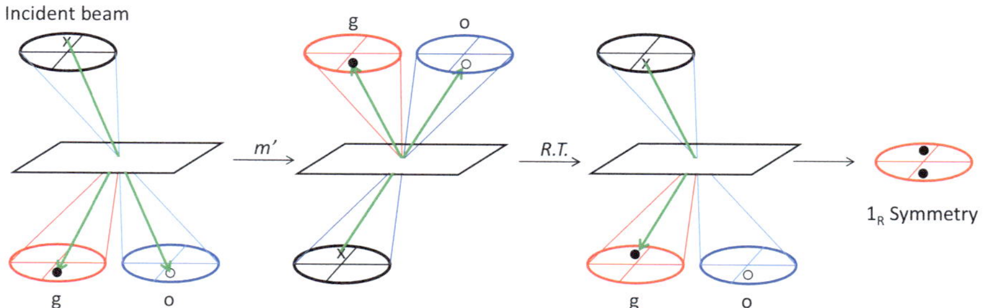

Fig. 11.11 Center of symmetry (1_R with R for reciprocity) in dark-field pattern (g) from the combination of a horizontal mirror (m', prime for 3D symmetry) and reciprocity theory (R.T.)

The diffracted beam in Fig. 11.11 has the wave vector

$$\vec{K}' = R\vec{K} + \vec{g} = -\vec{K}_z + \vec{g}/2 + \vec{\Delta},$$

and the wave function

$$\phi_{Rg}\left(R\vec{K}\right) = \phi_g\left(\vec{K}\right) = \phi_g\left(\vec{K}_z - \vec{g}/2 + \vec{\Delta}\right).$$

Using the reciprocity as shown in Fig. 11.11, we have

$$\left|\phi_{Rg}\left(R\vec{K}\right)\right| = \left|\phi_g\left(-\left(R\vec{K} + \vec{g}\right)\right)\right|$$

and thus

$$\left|\phi_g\left(\vec{K}_z - \vec{g}/2 + \vec{\Delta}\right)\right| = \left|\phi_g\left(\vec{K}_z - \vec{g}/2 - \vec{\Delta}\right)\right|, \qquad (11.23)$$

which gives rise to the twofold symmetry in the dark-field pattern of g as shown in
Fig. 11.11. Other symmetry obtained based on reciprocity are m_2, 2_R, and 4_R from a
horizontal twofold axis ($2'$), inversion center, and $\bar{4}$ parallel to z (Fig. 11.12). They
can be derived using the same procedure used for 1_R.

The point symmetry of CBED patterns belongs to the diffraction groups that are
constructed based on the ten symmetry elements of a parallel side perfect crystal,
which contain six 2D symmetry elements and 4 three-dimensional ones. The 2D
symmetries are one, two, three, four, and sixfold rotation axes and the mirror planes
m that are parallel to the z-axis. These symmetries are directly observed in the
diffraction pattern according to Eq. (11.20). They belong to the 10 2D symmetry
groups. (1, 2, 3, 4, 6, m, $2m(m)$, $3m$, $4m(m)$, and $6m(m)$, where the third symmetry in
the bracket is generated by the first two.) The four three-dimensional ones are the
normal mirror plane, $2'$ axes, inversion center, and $\bar{4}$ roto-inversion axis parallel to z,

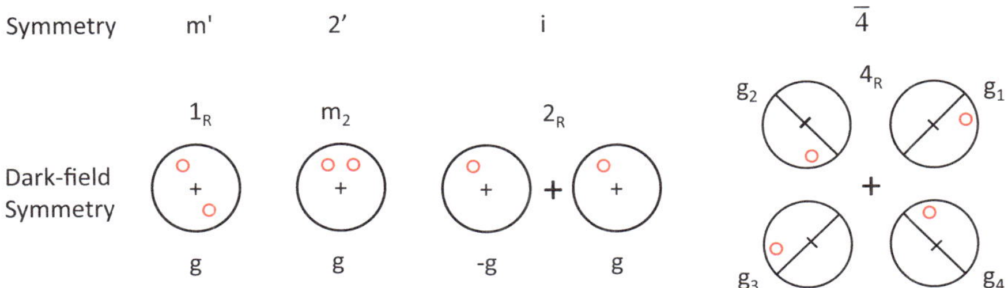

Fig. 11.12 Three-dimensional symmetries of dark-field patterns in CBED obtained from the
horizontal mirror plane m', the horizontal axis $2'$, the inversion center i, and the fourfold rotary
inversion $\bar{4}$ parallel to the z-axis

Table 11.5 Relationships between the diffraction groups and crystal point groups (Buxton et al. 1976)

Diffraction group	Tr 1	Tr $\bar{1}$	M 2	M m	M $2/m$	O 222	O $mm2$	O mmm	Te 4	Te $\bar{4}$	Te $4/m$	Te 422	Te $4mm$	Te $\bar{4}2m$	Te $4/mmm$	Tg 3	Tg $\bar{3}$	Tg 32	Tg $3m$	Tg $\bar{3}m$	H 6	H $\bar{6}$	H $6/m$	H 622	H $6mm$	H $\bar{6}m2$	H $6/mmm$	C 23	C $m3$	C 432	C $\bar{4}3m$	C $m3m$
$6mm1_R$																											×					
$3m1_R$																										×						
$6mm$																									×							
$6m_Rm_R$																								×								
61_R																							×									
31_R																						×										
6																					×											
6_Rmm_R																				×												×
$3m$																			×												×	
$3m_R$																		×												×		
6_R																	×												×			
3																×												×				
$4mm1_R$															×																	×
4_Rmm_R														×																	×	
$4mm$													×																			
$4m_Rm_R$												×																		×		
41_R											×																					
4_R										×																						
4									×																							
$2mm1_R$								×							×												×		×			×
2_Rmm_R							×						×	×											×						×	
$2mm$							×																			×						
$2m_Rm_R$						×						×		×										×				×		×		
$m1_R$					×						×												×									
m				×															×			×										
m_R				×															×													
21_R					×															×												
2_R			×						×	×																		×				
2			×															×														
1_R		×			×			×			×				×		×			×			×				×		×			×
1	×		×	×		×	×		×	×		×	×	×		×		×	×		×	×		×	×	×		×		×	×	

Crystal point groups

which as we have shown lead to the symmetry of $\bar{4}_R$, m_2, 2_R, and 4_R. Their combinations with 10 2D symmetry groups give 31 diffraction groups that are listed in Table 11.5 (Buxton et al. 1976).

11.6.2 Point Group Determination by CBED

Particular tests exist for individual symmetry elements. The following general rules apply and can be obtained by combining the reciprocity theorem with the symmetry elements of the crystal. Rotation axes are seen directly in CBED patterns when the beam is aligned with the rotation axis, as shown in Fig. 5.14. Note, for example, in such figures how the ZOLZ detail shows sixfold symmetry, which the diamond structure possesses only if it is projected along (111). The whole pattern, however, shows the true threefold symmetry of the three-dimensional lattice when account is taken of the detail in the HOLZ lines, both those crossing the central disk and those in the outer ring. This HOLZ detail results from diffraction events with a component along the beam path and is therefore sensitive to the three-dimensional crystal symmetry.

Mirror planes of symmetry in the crystal are also seen directly as mirror lines in the CBED pattern, if the beam lies in the mirror plane of symmetry. A vertical glide plane produces a mirror line in the CBED pattern. A horizontal twofold axis or twofold screw axis in the ZOLZ along g imparts a mirror line of symmetry onto disk g if it is at the Bragg condition, and this line runs normal to g.

Horizontal three, four, and sixfold axes produce no useful symmetries. (The four and sixfold axes, however, include a twofold axis.) A horizontal mirror plane or glide plane (strictly running through the midplane of the crystal slab) produces a centric distribution of intensity in every CBED disk at the Bragg condition. (This is given as the diffraction group symbol 1_R in Table 11.5) Such a horizontal mirror is always present in the projection approximation.

Unlike X-ray diffraction patterns, CBED patterns are very sensitive to the existence or absence of a center of symmetry (Goodman and Lehmpfuhl 1968 and Sect. 5.6.3). An early test for a center of symmetry was based on a comparison of the g and $-g$ disks recorded successively at the Bragg condition. [These distributions are related by translation (not rotation).] A simpler method, not requiring tilting, is based on determination of the diffraction group as described below, since no diffraction group can come from both a centered and a noncentered point group (Eades 1991). The central disk is a special case. Under general three-dimensional diffraction conditions, the presence of a center of symmetry in the crystal does not impose a centric distribution on the (000) disk. However, in the projection approximation, the (000) disk always has a center of symmetry, even in acentric crystals, due to the reciprocity theorem. The absence of a center in the (000) disk of a centric crystal may therefore be used as a test for three-dimensional scattering.

The analysis below will be based solely on patterns taken at zone axes, and a suitable tabulation is shown in Tables 11.5 and 11.6. It should be emphasized that

Table 11.6 Table showing the relation between the observed symmetries in CBED patterns and the 31 diffraction groups, which correspond to the 32 three-dimensional crystal point groups

Observed symmetry in zero-order zone	Projection diffraction group	Possible diffraction groups	Symmetries of high-order information	
			Whole pattern	Zero-order disk
1	1_R	1	1	1
		1_R	1	2
2	21_R	2	2	2
		2_R	1	1
		21_R	2	2
m	$m1_R$	m_R	1	m
		m	m	m
		$m1_R$	m	$2m$
$2mm$	$2mm1_R$	$2m_Rm_R$	2	$2mm$
		$2mm$	$2mm$	$2mm$
		2_Rmm_R	m	m
		$2mm1_R$	$2mm$	$2mm$
4	41_R	4	4	4
		4_R	2	4
		41_R	4	4
$4mm$	$4mm1_R$	$4m_Rm_R$	4	$4mm$
		4_Rmm_R	$4mm$	$4mm$
		$4mm1_R$	$2mm$	$4mm$
		$4mm1_R$	$4mm$	$4mm$
3	31_R	3	3	3
		31_R	3	6
$3m$	$3\,m1_R$	$3m_R$	3	$3m$
		$3\,m$	$3\,m$	$3m$
		$3m1_R$	$3m$	$6mm$
6	61_R	6	6	6
		6_R	3	3
		61_R	6	6
$6mm$	$6mm1_R$	$6m_Rm_R$	6	$6mm$
		$6mm$	$6mm$	$6mm$
		6_Rmm_R	$3m$	$3m$
		$6mm1_R$	$6mm$	$6mm$

Loretto (1984)

the method is sensitive to the symmetry of the physical crystal as a whole, including its boundaries and defects. In some cases (as shown in Sect. 11.6), inclined boundaries or defects may eliminate symmetry elements present in the infinite crystal, and the presence of defects must always be checked for using TEM imaging. (The use of a smaller probe minimizes the contribution from defects.) Since every disk at the Bragg condition is centric in the projection approximation (ZOLZ detail only), the absence of a center in such a disk may be used as a test for defects (Steeds 1979). (A twofold axis is equivalent to a center in two dimensions.) Under ultrahigh vacuum conditions, forbidden "termination" reflections will also produce misleading symmetries. Loosely speaking, these arise if the crystal contains incomplete unit cells at its surface. These expose the distinction between the symmetry of an infinite crystal and that of the finite slabs used in CBED (Ishizuka 1982).

A formal procedure based solely on zone-axis patterns would consist of three steps:

1. Determination of the symmetry of the projection diffraction group, using ZOLZ detail.
2. Determination of the diffraction group, using HOLZ detail.
3. Determination of the point group from the above information, using tables.

We discuss these steps in turn, with respect to the example of BeO given in Figs. Figure 11.13a–e.

(1)   A set of zone-axis patterns should be recorded at all the highest symmetry zones which can be found. From Table 11.5, we see that the higher the symmetry of the pattern, the fewer will be the possible point groups. Patterns will be needed at both large and small camera length, with exposures which reveal the HOLZ and ZOLZ detail separately. We thus assume that it is possible to separate the HOLZ features in a CBED pattern from the ZOLZ features. All two-dimensional patterns may be classified into one of ten classes, the ten two-dimensional point groups. These are listed in the first column of Table 11.6. Here, n denotes an n-fold rotation axis normal to the pattern, and m denotes a mirror line in the plane of the pattern. The first step is to determine into which of these classes the experimental pattern falls, when only the ZOLZ detail is considered. Thus, we ignore the fine HOLZ lines crossing the central disk and the outer HOLZ rings. Figure 11.13a shows one of many similar high-symmetry patterns which could be obtained at the same axis from our BeO crystal, using a 100 nm-diameter electron probe. A survey of many of these suggests that they contain two orthogonal mirror lines of symmetry and a sixfold axis. No one pattern, however, showed precisely this symmetry, due perhaps to the presence of defects or inclined surfaces. Our conclusion that the ZOLZ symmetry is therefore 6 mm rests on a subjective judgement based on experience. This emphasizes the important point that the CBED method, applied to real materials and using a "large" electron probe, is always likely to apparently underestimate the symmetry of the crystal, since it gives the true symmetry of the actual piece of material under the probe rather than that of the ideal crystal structure. Results closer to the ideal can be obtained using a smaller probe, which will minimize defect and boundary condition effects. The second column of Table 11.6 then indicates the name of the projection

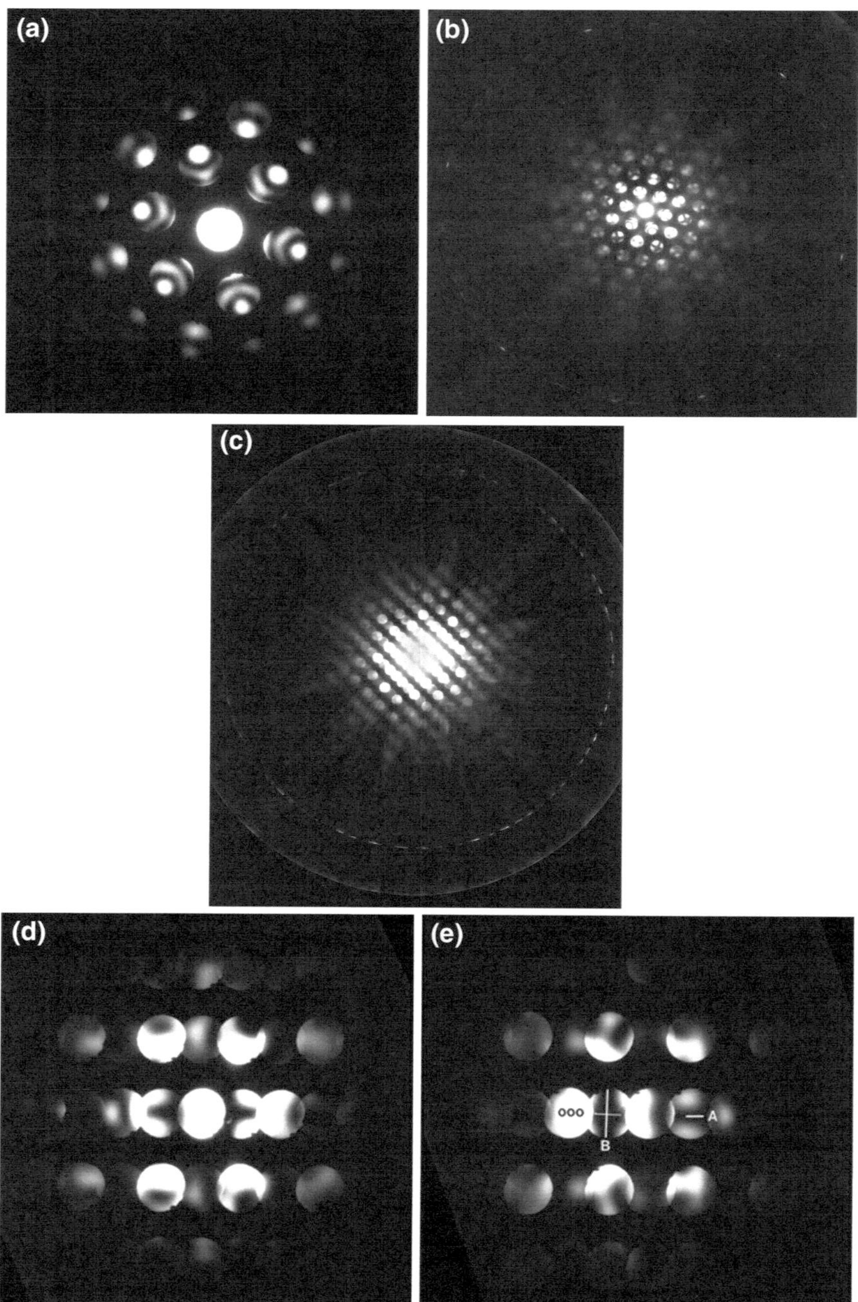

Fig. 11.13 **a** BeO CBED pattern recorded at 100 kV. The inner reflections only are shown. **b** similar to **a** but showing HOLZ ring observable only with cooling to −183 °C. **c** BeO CBED pattern recorded from a second high-symmetry zone axis (actually [1–100, normal to *c*-axis. Out HOLZ ring is seen. **d** Inner ZOLZ reflections from **c**. **e** BeO [10–10] zone axis, with first-order reflection (0001) at the Bragg condition, showing G-M black cross AB, and dark radial lines A in every second order along the *c*-axis

diffraction group. This label takes into account the additional effects of reciprocity (Pogany and Turner 1968) and horizontal mirror planes present in all ZOLZ patterns, and thus renames the classes in a systematic way (Buxton et al. 1976). There are more diffraction groups other than two-dimensional point groups because the diffraction group takes into account the internal symmetry of a CBED disk at the Bragg condition. The projection diffraction group for our example is thus 6 mm1_R.

(2)   The last four entries in column three of Table 11.6. Table 11.6 now indicates the possible diffraction groups for our pattern. These may be distinguished using the HOLZ information, taken separately in the zero-order disk (or outer HOLZ ring) and for the pattern as a whole. From Fig. 11.13b, we see that both the outer HOLZ ring (or the HOLZ lines in the central disk, not shown) and the whole pattern (including this HOLZ detail) have symmetry 6*mm*, so that the diffraction groups for this pattern may be either 6*mm* or 6*mm*1_R.

Our aim is to uniquely specify a point group on Table 11.5. The two diffraction groups we have identified now allow two possible point groups on Table 11.5, 6*mm* and 6/*mmm*. To distinguish these possibilities, we need a second CBED pattern taken from the same crystal in a different orientation. Figure 11.13c and d show a high-symmetry CBED pattern which was obtained with the beam normal to the direction used in Fig. 11.13a, b. A similar analysis from Table 11.6 shows that the diffraction group for this pattern is *m*1_R, since the outer HOLZ ring in Fig. 11.13d shows 2*mm* rather than m symmetry. In more complicated cases, it may be necessary to use a knowledge of the unit cell (giving the crystal system) to obtain more zone-axis patterns or to use the internal symmetries of CBED disks at the Bragg condition.

(3)   Turning to Table 11.5, we may now determine the point group for the crystal by tracing across from the diffraction group symbols on the left. Only the 6*mm* column intersects both the *m*1_R and 6*mm* rows. The point group is therefore 6*mm*.

In practice, when confronted with an unknown crystal, one looks initially for mirror lines of symmetry possibly in the Kikuchi pattern. Usually, something will be known about the crystal, and experienced workers use a combination of short-cuts, simple reasoning, and experience to isolate the possible symmetry groups after looking at "whole patterns" from as many high-symmetry axes as they can find. The *angle* between these axes, which can be read from the goniometer stage, provides additional important information. The internal symmetries of CBED disks at the Bragg condition may also be used.

When one has a crystal in which there is uncertainty about only a single symmetry element, then it is simply a matter of recording a CBED pattern along a suitable zone axis and observing the resulting symmetry of the whole pattern, including HOLZ detail. As an example, it was determined at an early stage by X-ray diffraction that the space group of $YBa_2Cu_3O_{7-\Delta}$ was either *Pmmm* (orthorhombic) or *P4mm* (tetragonal). Twinning and spatial variations in oxygen stoichiometry made further determination difficult. The matter was resolved using CBED patterns along the c-axis, with a probe size smaller than the twin spacing, for samples of differing stoichiometry. These clearly revealed the presence of a third mirror in the orthorhombic phase (x = 0), which is obtained by cooling from the high-temperature tetragonal phase (x = 1) (Eagelsham et al. 1987; Graham et al. 1987).

Other systematic procedures have been described which exploit symmetries in certain CBED disks at the Bragg condition ("dark-field disks") which may involve fewer steps and micrographs (Tanaka et al. 1983; Buxton et al. 1976), and the various methods have been compared (Howe et al. 1986). In particular, if the incident beam tilt controls are put under electronic control and scanned, it is possible to obtain a time-resolved hollow cone illumination mode. The use of this to provide "whole pattern" symmetry information is described by Tanaka and Terauchi (1985b). A wide-angle method has also been demonstrated which makes it possible to record simultaneously on a single micrograph the whole pattern, the central disk ("bright field"), and diffracted orders at the Bragg condition ("dark field") (Terauchi and Tanaka 1985).

Similar principles may be applied to the study of quasicrystals, provided a large probe is used. Methods for computing CBED patterns from quasicrystals are discussed in Cheng and Wang (1989), and experimental patterns can be found in Tanaka et al. (1988).

11.7 Bravais Lattice Determination

The next step in a general space-group determination should be to index the pattern, determine the Bravais lattice, and identify a unit cell. Worked examples of this procedure can be found in Raghavan et al. (1983), Ayer (1989), and Morniroli and Steeds (1992). The preceding analysis will have determined the crystal system (see Table 11.5). Identifying a cell and its centering will greatly reduce the number of possible space groups. The 14 Bravais lattices may be primitive (P), body-centered (I), face-centered (F), centered on two sides (C), or trigonal. The corresponding real space lattices are then defined by the vector relations given in Appendix C or by a Fourier transform. If there are no absent layers of reciprocal lattice points, it is usually possible to reconstruct the geometry of the three-dimensional reciprocal lattice by noting the relationship between the position of the HOLZ points and those in the ZOLZ "below." For this purpose, a CBED pattern recorded with a small condenser aperture along a high-symmetry axis should be used, using a small camera length. Lines may be drawn on a print or photocopy of the pattern, parallel to the ZOLZ rows of spots, across the CBED pattern, and connecting points in the HOLZ. The electron-optical distortions present in small camera length patterns can make this analysis difficult. These lines will pass either between the ZOLZ spots or through them. Two orthogonal sets need to be drawn, and their intersections will define the points in the FOLZ. In this way, it may be possible to tell if the reciprocal lattice is P, F, or I. The real space lattice is then P, I, or F, respectively. Indexed examples are given in Appendix D together with enough information to drawn the

HOLZ spots. Indices can then be assigned to the pattern starting with the shortest three-dimensional reciprocal lattice vector. The diameter of the FOLZ gives the height of the first layer of spots [Eq. (3.24) with $K_t = 0$] from which the three-dimensional reciprocal lattice vectors can be deduced. A worked example of this procedure can be found in Ayer (1989). Questions of accuracy clearly arise in distinguishing possibilites. For example, it has been pointed out by Eades (1991) that the error in a measurement of the c-axis (taken parallel to the beam) will differ from that of the other axes, because of the square-root dependence in Eq. (3.24). There may also be difficulties due to absent planes of reflections arising from translational symmetry. For example, in the reciprocal lattice for diamond, all structure factors in the sixth (nonzero) reciprocal lattice plane along [111] are zero due to the diamond glide symmetry element.

It is possible to write a computer program which will determine the Bravais lattice type, unit cell dimensions, and crystal system from a spot diffraction pattern rather than a CBED pattern (Page 1992). Even with the most careful calibration to allow for lens distortions, this approach may lead to errors. However, it is important to establish that it is possible in principle to find a systematic procedure for this purpose. Assuming that the Bravais lattice has been found, many choices of unit cell are then possible. A Buerger primitive cell has been defined (Buerger 1956) by the shortest three noncoplanar translations. This cell may be identified by a computer, but this choice is not unique. The Niggli cell, however, is unique and further allows the lattice type and crystal system to be immediately identified. A systematic procedure for determining the Niggli cell is given by Krivy and Gruber (1976). The accuracy for lattice identification can be improved by using HOLZ lines (Zuo 1993). The method described by Zuo allows the unique Niggli reduced cell to be determined from measurements of at least six HOLZ and ZOLZ lines in the central disk.

From the pattern shown in Fig. 11.13a, the lattice was found to be primitive. The possible space groups are now restricted to those listed in the International Tables for Crystallography (Volume A) which belong to the point group $6mm$ with a primitive lattice. (The point group is given as the third entry on the top line of the left-hand page for each space group.)

11.8 Space Groups

The aim of the final step in the analysis is to determine if either of the translational symmetry elements—screw axes or glide planes—are present, and if so, to find their orientation. When combined with the point group and lattice determination described in previous sections, this will usually provide enough information to determine the space group. Clear accounts of the practical procedure used are given in Tanaka et al. (1983) and Eades (1988). In X-ray diffraction, these translational symmetry elements introduce certain systematic absences as discussed in Sect. 11.5. The CBED method is based on the fact that, *for certain incident beam directions*, some of these reflections which are absent due to the presence of translational symmetry remain absent, despite multiple scattering, for all sample thicknesses and accelerating voltages. This causes a dark band or cross to be seen in CBED patterns known as a Gjønnes-Moodie or G-M line, as shown in Fig. 11.13e. A full explanation for this effect, first observed by Goodman and Lehmpfuhl (1968), was eventually given (following earlier papers by Fujimoto, Cowley, Miyake, Takagi, and others) by Gjonnes and Moodie (1965). An elegant derivation of the theory is also given using group theory in Portier and Gratias (1981).

Here, we provide a derivation and elaborate on the exceptions. Consider the beam direction approximately normal to the page (the z direction) in Fig. 11.14, and a screw axis running along the axis $(k, 0)$, or a glide plane that intersects the ZOLZ along $(k, 0)$. The glide translation is taken along k. Firstly, we consider only ZOLZ interactions in the projection approximation. In projection, the screw axis becomes a glide line and the projected two-dimensional symmetry becomes the plane group *pmg*. Then, the symmetry operation takes a position at (x, y) to (x', y') by

$$\begin{pmatrix} x' \\ y' \end{pmatrix} = \begin{pmatrix} 1 & 0 \\ 0 & -1 \end{pmatrix} \begin{pmatrix} x \\ y \end{pmatrix} + \begin{pmatrix} 1/2 \\ 0 \end{pmatrix}.$$

It can be shown that the above symmetry operation gives the structure factors that have the signs shown in Fig. 11.14 with (see Sect. 11.5 and Appendix C.2)

$$\begin{aligned} U(k, l) &= U(k, -l) && \text{for } k = 2n, \\ U(k, l) &= -U(k, -l) && \text{for } k = 2n + 1, \\ U(k, 0) &= 0 && \text{for } k = 2n + 1. \end{aligned} \tag{11.24}$$

Hence, alternate reflections along the screw axis (or glide) are kinematically forbidden. In addition, the sign of the structure factors alternates in the rows above and below the $(k, 0)$ axis as indicated in the figure. We shall see that these relationships among the structure factors lead to certain extinctions even in the presence of multiple scattering.

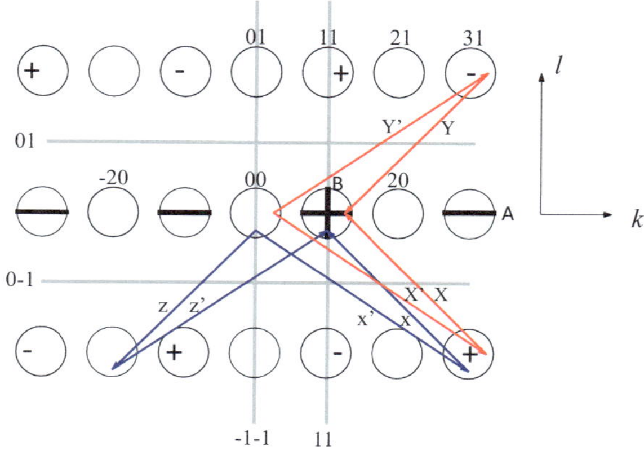

Fig. 11.14 The principle of dynamically forbidden reflections. Only ZOLZ (2D diffraction pattern) is shown. The black cross formed by lines A and B shows the locus of K_t along which all scattering path cancel in the first-order reflection (10) at the Bragg condition

In the coupled differential equations of the dynamical theory developed by Howie and Whelan, the wave function of a diffracted beam is given by (Sect. 5.2)

$$\frac{\mathrm{d}\phi_g}{\mathrm{d}z} = i\pi\lambda U_o\phi_g + i\pi\lambda \sum_{h\neq g} \phi_h U_{gh} \exp\left(2\pi i\left(\vec{S}_h - \vec{S}_g\right)\cdot\vec{r}\right) \tag{11.25}$$

where U_{gh} with $h \neq 0$ gives the multiple scattering from other diffracted beams, dependent on the excitation errors, $\vec{S}_g$ and $\vec{S}_h$. When the relationships of Eqs. (11.24) are imposed on Eq. (11.25), it is found that the resulting series of terms can be arranged in pairs with products of $U(k, l)$ of equal magnitude and opposite sign. This occurs only if $\vec{K}_t$ lies on the bold black cross in Fig. 11.14 in the projection approximation. Then, extinction along the radial line A occurs from the exact cancelation of all pairs of multiple scattering paths by the destructive interference, such as X and Y in Fig. 11.14. To prove this, we take $g = (2n + 1, 0)$ and two beams $h^+=(k, 1)$ and $h^- = (k, -1)$ above and below the center row. The wave functions of ϕ_{h+} and ϕ_{h-} are related according to Eq. (11.19) with

$$\phi_{h^-}(K_x, 0) = \mathrm{e}^{-\pi i k}\phi_{h^+}(K_x, 0), \tag{11.26}$$

while $S_{h^+} = S_{h^-} = S_h$ along the radial line A and

$$U_{gh^+} = U(2n + 1 - k, -l) = \mathrm{e}^{\pi i(1-k)}U_{gh^-}. \tag{11.27}$$

By combing Eqs. (11.25), (11.26), and (11.27), we obtain

$$\frac{\mathrm{d}\phi(2n+1,0)}{\mathrm{d}z} = i\pi\lambda U_o\phi(2n+1,0) + i\pi\lambda \sum_{h^+} \left(\phi_{h^+}U_{gh^+} + \phi_{h^-}U_{gh^-}\right)\exp\left(2\pi i\left(\vec{S}_h - \vec{S}_g\right)\cdot\vec{r}\right)$$

$$= i\pi\lambda U_o\phi(2n+1,0)$$

Thus, with the initial condition of $\phi(2n+1,0) = 0$ at the entrance surface, $\phi(2n+1,0)$ remains at 0 under dynamic diffraction.

Along the line B (where the Bragg condition is maintained for the (0, 1) reflection), the symmetry of paths such as x and Z gives rise to pairs of symmetrical and antisymmetrical Bloch waves that scatter destructively to the kinematically forbidden reflection. Valset and Tafto demonstrated this in a four beam case (Valset and Tafto 2011), and their conclusion is general and applicable to the many beam cases. Thus, within the projection approximation, experimental CBED patterns should show black lines in alternate orders along c and a black cross at the first-order reflection at the Bragg condition (or in any such diffracted beam that is tilted to the Bragg angle). Figure 11.13e shows our experimental pattern from BeO.

If HOLZ features are to be included in the analysis, the three-dimensional structure factors must be used. For a 2_1 screw axis along (010), the relationship linking them is

$$U(h, k, l) = (-1)^k U(-h, k, -l)$$

while for the same glide plane

$$U(h, k, l) = (-1)^k U(-h, k, l)$$

The screw axis might be, for example, a 6_3 screw axis. Repeated operation of this screw element and the 4_1, 4_3, 6_1, 6_3, and 6_5 screws show that they all include a 2_1 screw. (We note that 3_1, 3_2, 6_2, and 6_4 screws do not. Therefore, they do not produce G-M lines, which is one reason why 49 of the 230 space groups cannot be completely identified by G-M lines.)

If three-dimensional scattering is included, theory predicts that the screw axis produces only the B line of extinction across the $(0, 1)$ disk, while a glide plane produces only the radial line of extinction in the $(0, k)$ disks for k odd. The two cases may thus be distinguished in principle. However, at the high symmetry, low-index zone axes that are preferred for CBED study, it is usually the case that the three-dimensional effects are weak. Thus, in practice, the projection approximation applies and, for either a screw or a glide, there is extinction along A and B. In Tanaka's convention, we call the radial line of zero intensity A_2 if the scattering is two-dimensional, and A_3 under three-dimensional (HOLZ) conditions. A similar convention holds for the B lines.

For a glide plane normal to the beam, the period of the potential is halved by projection leading to the extinction of alternate reflections in the ZOLZ. (The FOLZ point pattern will then appear twice as dense as the ZOLZ allowing easy identification of this case.) In a CBED pattern, the theory thus predicts a point of extinction at the center of the black cross due to cancelation between paths Y and Z. This dark spot was observed by Tanaka's group (Tanaka et al. 1987).

In summary, if ZOLZ interactions only are considered, either the twofold screw axis or the glide plane described above will produce both the A and B G-M lines of extinction shown in Fig. 7.2. Under three-dimensional scattering conditions, the screw produces only the B line while the glide produces only the A lines.

These are the theoretical results. In practice, however, it may be difficult to ensure three-dimensional scattering in order to distinguish these two cases. An increase in thickness or a change in accelerating voltage may achieve this. Several other methods have therefore been suggested to distinguish screws from glides. If a glide plane is present, the A_2, B_2, and A_3 type G-M lines will remain if the crystal is rotated about h, but are destroyed if it is rotated about k. For a screw axis, A_2, B_2, and B_3 G-M lines are preserved if the crystal is rotated about k, but are destroyed by a rotation about h (Steeds and Vincent 1983). Second, it has been shown that the symmetry of the fine HOLZ lines which cross these disks can sometimes be used to distinguish the two cases (Tanaka 1989). If the HOLZ lines are symmetric about A_2 type G-M lines, a glide plane is present. If the lines are symmetric about the B_2 line, a screw axis exists. Finally, knowledge of the diffraction groups can be used to limit

Table 11.7 Origin of dynamical extinction in zero-layer reflections[a]

Single row of zero- layer extinctions	Perpendicular rows of zero-layer extinctons	Deduction	Tanaka symbol
m_R $2m_R m_R$	$2m_R m_R$ $4m_R m_R$	Screw axes parallel to each row of extinctions	A_2B_2 B_3
m $2m$	$2mm$ $4mm$	Glide planes parallel to the zone axis and each row of extinctions	A_2B_2 A_3
$2_R mm_R$	$4_R mm_R$	Glide, if parallel to whole pattern mirror, *or* Screw, if perpendicular to whole pattern mirror	A_2B_2 A_3 or A_2B_2 B_3
	$2_R mm_R$	Glide parallel *and* Screw perpendicular to whole pattern mirror	A_2B_2 A_3 and A_2B_2 B_3
$m1_R$ $2mm1_R$	$2mm1_R$ $4mm1_R$	Glide plane *and* Screw axis parallel to each line of extinctions[b]	A_2B_2 A_3B_3

[a]The table permits the proper interpretation of zero-layer extinctions when diffraction group is known

[b]This is the case for double diffraction routs in the zero layer. If the extinction is produced by double diffraction via HOLZ reflections, there are space groups for which the extinction can be produced by a glide or a screw alone (e.g., horizontal glide)

possibilities, since only certain combinations of the point and space-group elements are permitted (Eades 1988). These are indicated in Table 11.7. This table assumes that the diffraction group is known and then allows the correct deductions to be made concerning any screw or glide planes. An alternative and useful tabulation has also been provided by Tanaka et al. (1988), in which it is assumed that the point group is known. Then, his table shows the dynamical extinctions which are expected for each space group.

Perpendicular rows of black lines may also be seen if there is more than one translational symmetry element. No forbidden reflections occur if a screw axis is parallel to a twofold axis or if a glide is parallel to a mirror. The G-M lines are distinguishable from other zeros in intensity by the fact that they are present for all thicknesses and accelerating voltages. There are, in addition, some more subtle dynamical extinctions, such as those occurring in the outer HOLZ ring, or that due to a glide plane parallel to the surface, mentioned above. A vertical glide plane whose glide vector is not parallel to the crystal surface is strictly not a symmetry element of the layer groups for slab crystals—the implications of this are discussed in Ishizuka (1982). Here, it is shown that a vertical glide with a vertical translation does cause G-M lines if the HOLZ spacing is large. A symmetry classification for parallel-sided slab crystals has been given (the layer groups) (Goodman 1984) and may thus be used as the basis for space-group determination by CBED.

These procedures will identify 181 of the 230 space groups (For a discussion, see Eades 1988). Certain space groups (of which there are 23 sets) cannot be distinguished by this conventional CBED analysis. Two examples are the pairs ($I23$, $I2_13$) and ($I222$, $I2_12_12_1$). These differ by the absolute location of the symmetry elements in the unit cell and so could be distinguished by a real space image at atomic resolution. It has therefore been shown by Saitoh et al. (2001) that the coherent CBED method, using overlapping orders, can distinguish these space groups, since as we discuss in Sect. 14.8, the region in which coherent CBED disks overlap and interfere forms a kind of real space image, and so is sensitive to site symmetry, which depends on the phase of the structure factors. The coherent CBED patterns may form a probe smaller than the unit cell and will then show the local projected site symmetry of the crystal as reckoned about the center of the probe for a demonstration of this effect and its use for locating the beam in EELS spectroscopy see Spence and Lynch (1982). Left- and right-handed enantiomorphs (e.g., quartz) produce CBED patterns with the same symmetry, but different intensity distributions, so can only be distinguished by comparison with dynamical simulations for the two structures (Spence et al. 1994).

Table 11.8 lists those space-group pairs which are indistinguishable using the Gjonnes-Moodie extinction lines alone.

For the BeO pattern shown in Fig. 11.13e, we see that radial lines of absence A are obtained in every second reflection along c indicating a c glide. In addition, the first-order (0001) reflection is at the Bragg condition, and this shows a band B of extinction across the c-axis indicating a 2_1 screw axis along c. This is contained in the 6_3 screw element. From Fig. 11.13e alone, we may thus conclude only that that there exists either a screw axis along c, or a c glide, or both. Since the diffraction group for this pattern was determined to be $m1_R$, we can check from Table 11.7 that the observation here of a single row of zero-layer black lines and a cross indicates a "glide plane and screw axis parallel to each line of black crosses." The actual space group for the hexagonal wurtzite structure of BeO is $P6_3mc$.

The general procedure for an unknown crystal is as follows: First, find the point group, the crystal system, and index the pattern. Then proceed as follows. (1) List all possible space groups consistent with this information. (2) Use the table on page 162 of Tanaka and Terauchi (1985a) to determine the directions in which to look in

Table 11.8 Space groups indistinguishable by G-M extinction lines

1. $P3$, ($P3_1$, $P3_2$)	2. $P312$, ($P3_112$, $P3_212$)	3. $P321$, ($P3_121$, $P3_221$)
4. $P6$, ($P6_2$, $P6_4$)	5. $P622$, ($P6_222$, $P6_422$)	6. $P6_3$, ($P6_1$, $P6_5$)
7. $P6_322$, ($P6_122$, $P6_522$)	8. $P4$, $P4_2$	9. ($P4_1$, $P4_3$)
10. $P4/m$, $P4_2/m$	11. $P4/n$, $P42/n$	12. $P422$, $P4_222$
13. $P42_12$, $P4_22_12$	14. $I4$, $I4_1$	15. $I422$, $I4122$
16. $I23$, $I213$	17. $I222$, $I212121$	18. $P432$, $P4232$
19. ($P4132$, $P4332$)	20. $I432$, $I4132$	21. $F432$, $F4132$
22. ($P4122$, $P4322$)	23. ($P41212$, $P43212$)	

order to observe G-M lines for each possible space group. Check for screws which do not produce G-M lines (such as 3_1). Invaluable references include Eades (1988), Steeds and Vincent (1983), Tanaka (1989), and Tanaka et al. (1988).

The "chirality" of a crystal or its "handedness" can also be determined by CBED. (This corresponds to the difference between a right-handed and a left-handed screw. Space groups which differ in this way are known as enantiomorphs and are mirror images of one another.) The minimum strategy for distinguishing enantiomorphs by dynamical diffraction is described by Spence et al. (1994a). In the absence of anomalous dispersion, it is possible to determine the absolute hand of an enantiomorphic crystal by three-beam dynamical X-ray or electron diffraction in a general orientation only if a fourth noncoplanar reciprocal lattice point can be identified. Three-beam dynamical diffraction alone is unable to distinguish enantiomorphic forms. Identification is possible using four or more dynamical beams, in general, unless all relevant structure factors lie on a plane in reciprocal space passing through the origin. Applications to quartz are described in Goodman and Johnson (1977), and to MnSi in Tanaka and Terauchi (1985a). In both cases, the absolute orientation (i.e., the "handedness") can be determined. The effect of enantiomorphism is to reverse the sign of the three-phase invariant leading to observable effects.

11.9 Quantification of CBED Pattern Symmetry and Symmetry Mapping

Real crystals often have local symmetry dependent on sample position. An obvious case is the breakdown of symmetry to surface and interfacial stress and strain or the presence of defects. Another case is ferroelectric crystals, electric polarization removes the inversion symmetry and its direction coincides with the principle symmetry axis in tetragonal and rhombohedral crystals. Thus, measurement of local symmetry can be used to determine the polarization direction. There are many other examples, where local symmetry can help with phase identification and microstructure determination.

In using CBED for local symmetry determination, it is helpful to quantify the amount of symmetry recorded in CBED patterns (Mansfield 1985; Hu et al. 2000; Kim and Zuo 2013). The basic idea is to measure, from the diffraction intensities, the similarity between points inside the CBED disks that are related by symmetry. The similarity can be measured by the standard normalized cross-correlation coefficient (γ) (Lewis 1995). In order to measure the symmetry, regions in CBED patterns must be selected and are aligned. Figure 11.15 shows an experimental CBED pattern from the Si [110] zone axis. We use this pattern to demonstrate the image processing procedures employed for the dark-field symmetry quantification. The discussion below is specific to the mirror symmetry, but the principle also applies to rotational symmetry. First, two diffraction disks are selected on two sides of the mirror plane (marked by the yellow line) as shown in Fig. 11.15a. For the discussion, the selected

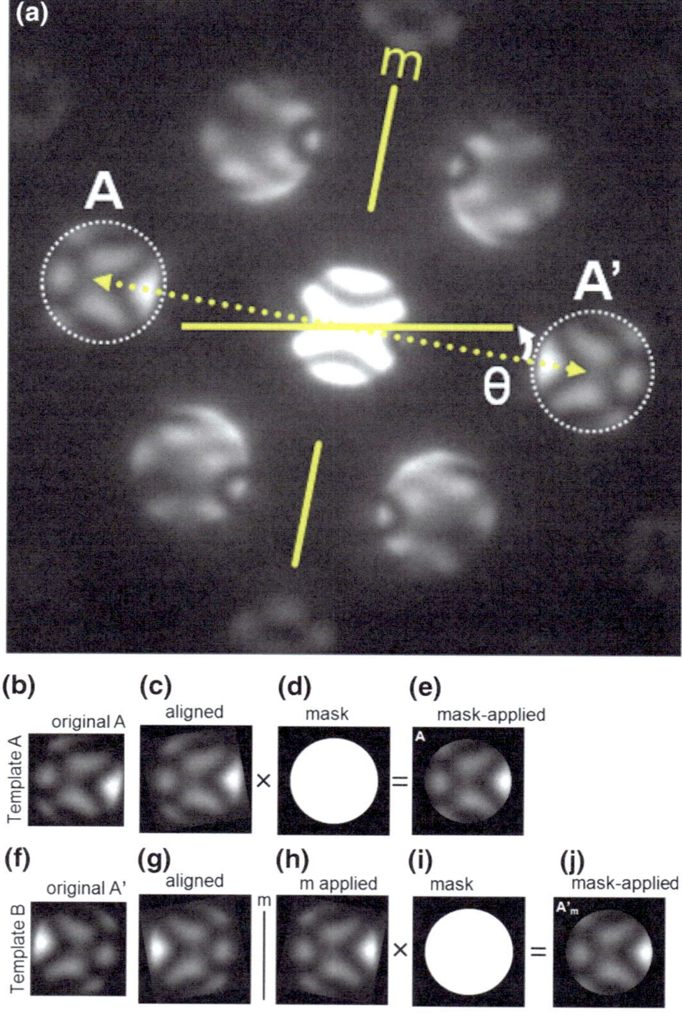

Fig. 11.15 Image processing procedures used for symmetry quantification. The example here is for the mirror symmetry. Two diffraction disks related by mirror are selected as indicated by the dotted circles A and A′ in the (**a**). The two disks are then processed to give two templates (A and B) as shown above

CBED disks are named as template A and template A′ (Fig. 11.15b, f), respectively. Each template is then rotated by an angle θ so that the mirror is aligned as shown in Fig. 11.15c, g. The template A is used as the reference motif so that the symmetry element is calculated by comparing with template A′. For the mirror operation, the template A′ is flipped to obtain a mirror image as shown in Fig. 11.15h. The mirror-applied image will be referred to as A'_m. For the rotational operation, the template A′ is rotated by 180°, 120°, 90°, and 60° for the two, three, four, and sixfold rotation, respectively. The rotated template A′ will be referred to as A′$_n$ (n = 2, 3, 4, 6). The circular mask shown in Fig. 11.15d, i is used to remove areas affected by CBED disk edge. Thus, the final templates are obtained by multiplying the mask image to the templates A and A′$_m$ as shown in Fig. 11.15e, j.

The normalized cross-correlation coefficient (γ) is used to quantify the similarity between A and B = A′$_{m \text{ or } n}$. The γ is calculated with the final templates of A and B based on pixel-by-pixel operation using

$$\gamma = \frac{\sum_{x,y}\left\{\left[I_A(x,y) - \overline{I_A}\right] \cdot \left[I_B(x,y) - \overline{I_B}\right]\right\}}{\sqrt{\left\{\sum_{x,y}\left[I_A(x,y) - \overline{I_A}\right]^2\right\} \cdot \left\{\sum_{x,y}\left[I_B(x,y) - \overline{I_B}\right]^2\right\}}}, \qquad (11.28)$$

where $\overline{I}_A$ and $\overline{I}_B$ are the mean values of two templates (Lewis 1995). From the Eq. (11.28), the numerator and denominator have the exact same values if the two templates are absolutely identical. By subtracting off the average intensity in the calculation of γ, it becomes less sensitive to the background intensity in the recorded patterns. For a pattern with perfect symmetry, $\gamma = 1$.

For the experimental pattern in Fig. 11.15, which was recorded using a JEOL 2100 LaB$_6$ TEM at 200 kV, the γ values range from 0.981 $\sim$ 0.991 for the mirror. A test of the robustness of the symmetry quantification procedure over 20 experimental Si [110] CBED patterns gave the γ values ranging from 0.981 to 0.991 for all quantification results.

The symmetry quantification method we have described can be combined with the scanning electron diffraction technique described in Sect. 10.4.7 for symmetry mapping. In scanning CBED, a series of CBED patterns are recorded and stored in a 4D dataset. The 4D dataset consists of m by n patterns; the m and n correspond to the number of sampling points along the two edges of the rectangular grid. Figure 11.16 shows an application of the symmetry mapping technique to a silicon crystal near a stacking fault. The medium magnification image presented in Fig. 11.16a shows the Bragg diffraction contrast from the stacking fault. The mirror selected for quantification is along the yellow line as indicated in Fig. 11.16b in the CBED pattern. The measurement used the A/A′, B/B′, and C/C′ disk pairs. From Fig. 11.16a, the symmetry distribution was mapped on 20 × 10 grid points. A probe of 7.8 nm in FWHM (full width half maximum) was used for scanning CBED with a step length of 8 nm. Thus, the physical dimension of the scanned area is 152 × 72 nm^2. Figure 11.16c shows a magnified image of the investigated area, and Fig. 11.16d is the calculated symmetry map using γ. A grid pixel in the symmetry map becomes bright as the intensity of two CBED disks matches by the selected symmetry (i.e., mirror). The dark contrast in the symmetry map indicates symmetry breaking from the selected mirror symmetry. For example, the profile of γ values is plotted along the line indicated in Fig. 11.16c and plotted in Fig. 11.16e. In the area of the stacking fault, the γ drops significantly from 0.98 to 0.18. Thus, symmetry breaking

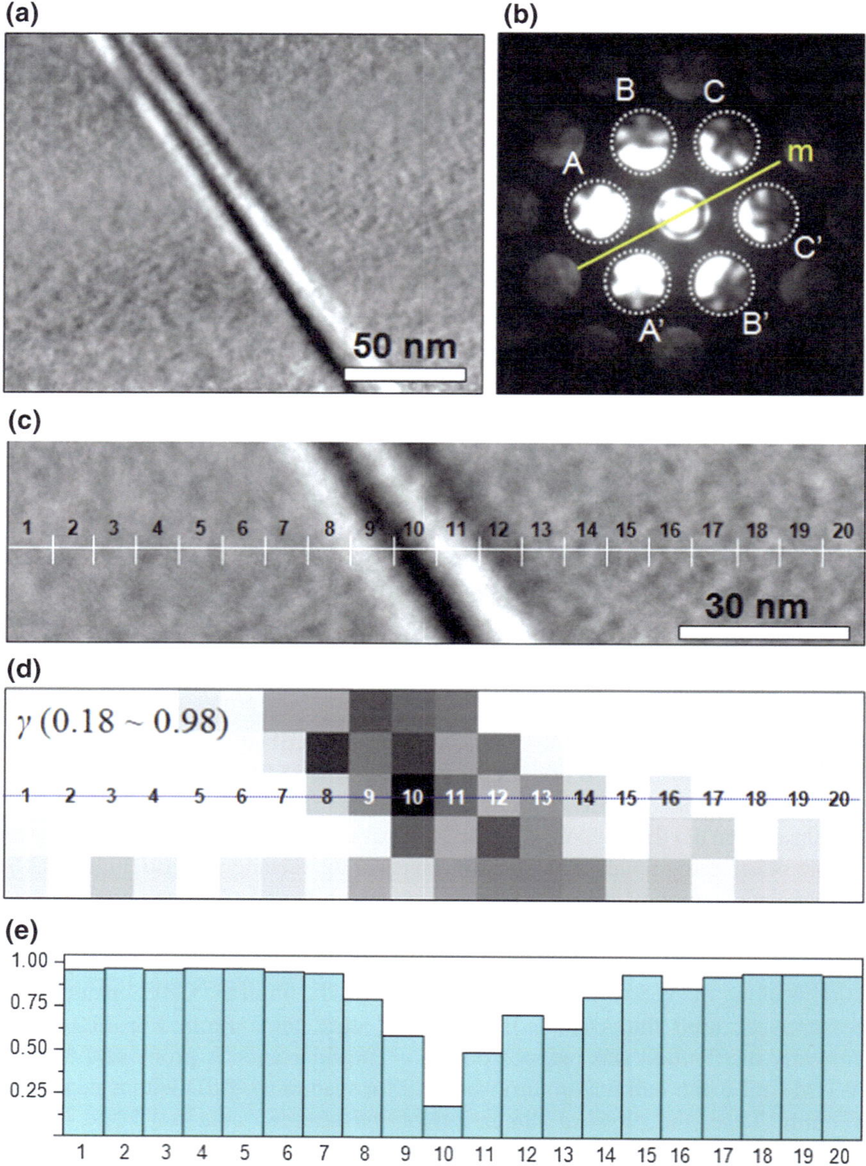

Fig. 11.16 **a** A medium magnification image of strained silicon showing a stacking fault, **b** a selected CBED pattern from the investigated area, and **c** a magnified image of the area investigated by scanning CBED. The symmetry maps using γ are shown in Fig. 7 (**d**). The **e** shows the γ profile across the stacking fault along the line indicated in the **c**

is detected across the stacking fault and near the stacking fault. The symmetry map also clearly shows that the symmetry breaking of the stacking fault is relatively localized because the rigid shift of the stacking fault introduces little strain unless it is terminated by a partial dislocation (Kim and Zuo 2013).

11.10   Symmetry and Polarization in Ferroelectric Crystals

Ferroelectric crystals exhibit spontaneous polarization. In perovskite ferroelectrics with the structure of calcium titanium oxide ($CaTiO_3$), polarization is characterized by a separation of positive and negative electric charges and its crystallographic direction, which can be poled by an applied external electric field. The spontaneous polarization is induced by phase transition from the high-temperature centric phase to a low-temperature acentric phase. The breakdown of inversion symmetry is accompanied by lattice distortion, which is dependent on the direction and magnitude of the spontaneous polarization. Thus, ferroelectric crystals are also piezoelectric as polarization gives rise to strain and polarization rotation changes strain.

Some ferroelectric crystals possess several low-temperature phases dependent on the range of temperatures, composition, or mechanical history. For example, barium titanate ($BaTiO_3$) is cubic and paraelectric above the Curie temperature ($T_c = 120\ °C$), but becomes tetragonal ($P4mm$) in the temperature range of 5–120 °C, orthorhombic ($Bmm2$) between −90 and 5 °C, and rhombohedral ($R3m$) below −90 °C. Above T_c, the dielectric constant follows a Curie-Weiss law of $\varepsilon = C/(T - T_o)$ with $T_o = 120\ °C$. The low-temperature phases differ in the direction of spontaneous polarization, which is along <100>, <110>, and <111> for tetragonal, orthorhombic, and rhombohedral phases, respectively. The stable polarization direction can take any one of crystallographic equivalent directions resulting in distinct polarization domains; each domain is characterized by the same polarization direction. Domain walls form at the interface where two domains of different polarization meet.

The interrelationship between the electromechanical, structural, and thermal properties of ferroelectric crystals can be obtained using the free energy model of Landau and Devonshire. The basic idea of this model is that the two phases in transition must be the same at their shared transition line in the phase diagram, and the transition is characterized in terms of order parameter, which is taken as polarization (P). The order parameter is uniform, and any local fluctuations must be small (Chandra and Littlewood 2006). During transition, the P changes from zero in the high-symmetry (disordered) phase to a finite value once the symmetry is lowered. The free energy, F, is assumed to be an analytic function of P, and it is then expanded as a power series in the vicinity of the transition.

A major unresolved question in the study of ferroelectric is the microscopic origin of polarization and the related transformation mechanism. Knowing the nature of polarization is a prerequisite for providing an accurate description of domains and interface and thus the microstructure. In $BaTiO_3$, phase transition is popularly explained by the displacive model (Kittel 1976; Cochran 1960), in which each Ti atom is positioned in the middle of the oxygen octahedron at equilibrium in the cubic phase, and the equilibrium position moves toward the $\langle 111 \rangle$, $\langle 011 \rangle$, or $\langle 001 \rangle$ directions for the R, O, and T ferroelectric phases, respectively. Experimentally, the X-ray absorption fine structure (XAFS) experiments (Ravel et al. 1998) show that the Ti atoms are displaced along various of the eight possible $\langle 111 \rangle$ directions in all phases. In X-ray diffraction, anomalous and strong diffuse scattering was reported in all but the rhombohedral phase (Comes et al. 1970). These and other results suggest a spontaneous symmetry breaking. It has been hypothesized that all Ti atoms are microscopically located in one of eight potential minima along the $\langle 111 \rangle$ directions for all crystal phases and phase transition occurs through an order-disorder transformation. Zhang et al. suggested that the stable rhombohedral phase has all distortions in phase, whereas higher temperature phases have antiferroelectric coupling in one, two, or three dimensions corresponding to orthorhombic, tetragonal, and cubic phases, respectively (Zhang et al. 2006). However, the observed entropy changes (≤ 0.52 J/mol) for each transition in $BaTiO_3$ is far smaller than the entropy change in an order-disorder transition (5.76 J/mol, see Comes et al. 1968 and Chaves et al. 1976). To explain this, a short correlation length between 5 to 10 nm has been postulated.

Direct evidence of rhombohedral nanostructures comes from CBED. Tsuda et al. (2012) observed nanoscale rhombohedral symmetry (a single diagonal mirror observed along the cubic axes, see Fig. 11.17) in the orthorhombic and tetragonal phases of $BaTiO_3$ using CBED. The symmetry of the orthorhombic phase is formed as the average of two rhombohedral variants with different polarizations and that of the tetragonal phase is formed as the average of four rhombohedral variants. Similar rhombohedral nanostructures were also found in the ferroelectric orthorhombic phase of $KNbO_3$ (Tsuda et al. 2013), while it was confirmed that the ferroelectric tetragonal phase of PbTiO3 does not have such rhombohedral nanostructures.

In technologically important piezoelectric materials of lead zirconium titanate ($PbZr_xTi_{1-x}O_3$ or $(1-x)PbTiO_3–xPbZrO_3$) or $(1-x)Pb(Mg_{1/3}Nb_{2/3})O_3–xPbTiO_3$

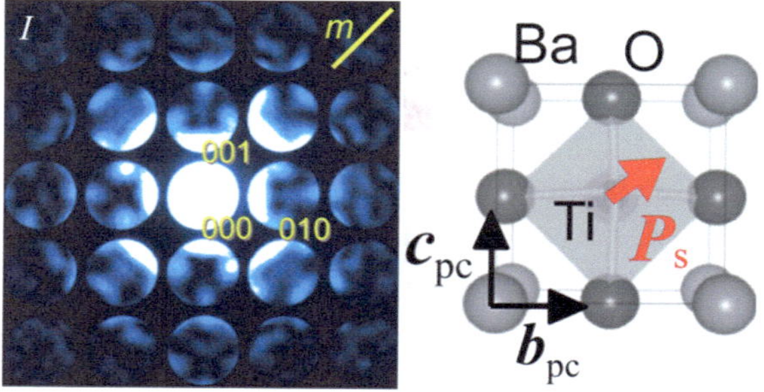

Fig. 11.17 CBED pattern recorded from low-temperature phase of $BaTiO_3$ along the high-temperature cubic axis ([001]), showing one of three mirrors of the 3 m point group (Provided by K. Tsuda, Tohuku Unversity, Japan)

(PMN–xPT), local chemical fluctuations are expected based on the substitution of multiple B-site cations. Chemical fluctuations induce local distortions that are governed by the nature of chemical bonds. Indeed, neutron total diffraction measurements revealed atomic pair distances that deviated significantly from that of the averaged crystal structure (Egami 1999, 2007). However, the extent of chemical fluctuations in real crystals is poorly understood and the development from short-range disordered structure to long-range ordered structure with well-defined symmetry is unknown. This knowledge gap has led to considerable debates about the nature of the symmetry in the morphotropic phase boundary (MPB) region, where the giant piezoelectric response is obtained.

TEM images reveal tweed-like contrast in single crystal PMN–xPT at the MPB region, which is common to relaxor-based ferroelectrics (Randall and Bhalla 1990; Viehland et al. 1995; Wang et al. 2006). The width of tweed-like contrast is on the order of a few tens of nm. Symmetry analysis of those nanodomains requires nanometer-sized probes, which can be achieved with the CBED technique. In the study reported by Kim et al., CBED patterns are recorded using different probe sizes to measure the local crystal symmetry. The amount of mirror symmetry in the CBED patterns is then quantified using Eq. (11.28) following the method described in Sect. 11.10. Figure 11.18a–e show the CBED patterns recorded from $[111]_C$ PMN–31 %PT after annealing at 500 °C in air with slow cooling using different probe sizes of 2.7, 5.2, 9, 15, and 24 nm, respectively. The CBED patterns are recorded from the same region within the crystal. For comparison, CBED patterns from a Si single crystal along the $[110]_C$ zone orientation are also recorded using similar electron probe sizes. These patterns are also shown in Fig. 11.18f–j. The $2mm$ whole pattern symmetry is preserved in Si for the different probe sizes ranging from 2.7 to 25 nm, whereas the patterns obtained from the ferroelectric relaxor are size dependent. Figure 11.18k illustrates the γ_m variations as a function of probe size. The γ_m values were quantified using the $(A/A')/(B/B')/(C/C')/(D/D')/(E/E')$ (Fig. 11.18) and $(A/A')/(B/B')/(C/C')$ (Fig. 11.18f) disk pairs for the PMN–31 %PT single crystal and the Si single crystal, respectively. The error bars plotted are the standard deviations measured from multiple CBED patterns recorded from different areas of similar thickness ($\sim$65 nm). The PMN–31 %PT $[111]_C$ CBED pattern has the lowest γ_m value of 27.3 % at 2.7 nm and rises thereafter to about 83.1 % as the probe size increases above 9 nm. By contrast with PMN–31 %PT, the γ_m values for Si $[110]_C$ were almost constant at $\sim$98 % for all probe sizes used. Thus, unlike silicon, where the symmetry of the pattern was almost perfect and constant down to $\sim$2.7 nm, the mirror symmetry of PMN–31 %PT was only obtained by averaging over regions greater than $\sim$15 nm. The highest symmetry obtained for PMN–31 %PT was $\sim$83 % in cross-correlation coefficient, so the mirror symmetry recorded in the CBED patterns was only approximate for the larger probe sizes.

The effect of volume averaging on symmetry is further examined in Fig. 11.19. The experimental CBED patterns are recorded on 3 × 3 grids in an area of 15 by 15 nm in size using a probe of 5 nm. Values for γ_m varied from 41.9 to 60.8 % for

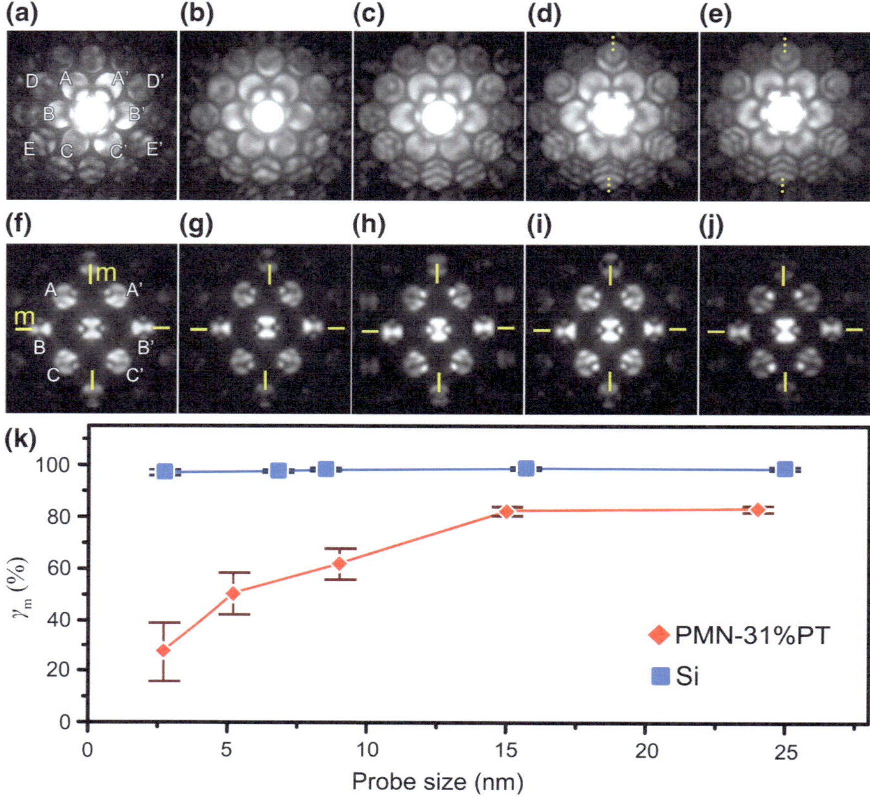

Fig. 11.18 Variations in CBED patterns obtained with different probe sizes ranging from 2.7 to 24 nm for (**a–e**) PMN–31 %PT and 2.5 to 25 nm for (**f–j**) Si single crystal. The cross-correlation coefficient for the mirror element (γ_m) is quantified along the indicated direction and plotted in (**k**). The error bars reflect symmetry variations obtained from measurements on different CBED patterns

these patterns. Figure 11.19b is a composition of the 9 CBED patterns. For comparison, Fig. 11.19c shows the experimental CBED pattern extracted from the 1st grid point in Fig. 11.19c, which is the same area given in Fig. 11.19b. By contrast with the individual CBED patterns obtained with the 5 nm probe, the averaged pattern increased in γ_m value up to 84 % close to the experimental γ_m value of 82 % recorded with the ~15 nm probe.

The mirror-like symmetry in CBED patterns when recorded using probes 15 nm or larger approximately belongs to space group *Cm* (Kim et al. 2012). According to structural data (Singh et al. 2006), the spontaneous polarization direction is along the [−u0w] ([−0.25, 0, 0.03]), which is within the mirror plane. The breakdown of symmetry detected at smaller probe sizes can be attributed to local ionic displacements.

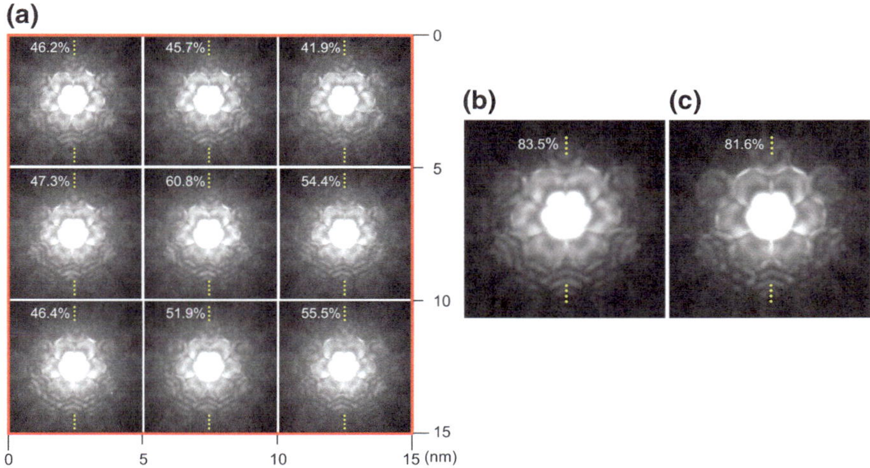

Fig. 11.19 **a** Experimental CBED patterns recorded on a 3×3 grid. The extracted CBED patterns are merged into **b** an averaged CBED pattern in order to compare with **c** an individual CBED pattern for the larger ~ 15 nm probe. The γ_m values are indicated along the dotted line. The dotted line in CBED patterns is originally orientated at $\sim 49°$ to the horizontal direction

In summary, symmetry determines the direction of polarization in ferroelectric crystals and CBED probes the local symmetry and thus the microscopic nature of polarization. Symmetry fluctuations were an important crystallographic occurrence on the local scale in both $BaTiO_3$ and complex ferroelectric crystals. Quantification of pattern symmetry provides a method by which symmetry fluctuations could be determined and measured by CBED.

References

Ayer R (1989) Determination of unit cell. J Electron Micr Tech 13:16–26

Buerger MJ (1956) Elementary crystallography. Wiley, New York

Buxton BF, Eades JA, Steeds JW, Rackham GM (1976) Symmetry of electron-diffraction zone axis patterns. Philos T Roy Soc A 281:171–194

Chandra P, Littlewood PB (2006) A Landau primer for ferroelectrics, arXiv:cond-mat/0609347v1

Chaves AS, Barreto FCS, Nogueira RA, Zěks B (1976) Thermodynamics of an eight-site order-disorder model for ferroelectrics. Phys Rev B 13(1):207–212

Cheng Y, Wang R (1989) Dynamical theory of electron diffraction for quasicrystals. Phys Status Solidi B 152:33–37

Cochran W (1960) Crystal stability and the theory of ferroelectricity. Adv Phys 9:387–423

Comes R, Lambert M, Guinier A (1968) Chain structure of $BaTiO_3$ and $KNbO_3$. Solid State Commun 6:715

Comes R, Lambert M, Guinier A (1970) Desordre lineaire dans les cristaux (cas du silicium, du quartz, et des perovskites ferroelectriques). Acta Crystallogr A 26:244–254

Eades JA (1988) Glide planes and screw axes in CBED: The standard procedure. Microbeam Analysis, San Francisco Press, San Francisco

Eades JA (1991) Personal communication

Eagelsham DJ, Humphreys CJ, Alford N, Clegg WJ, Harmer MA, Birchall JD (1987) EMAG 1987. In: Brown LM (ed) (Institute of Physics, London). I.O.P. 90:295

Egami T (1999) Microscopic model of relaxor phenomena in Pb containing mixed oxides. Ferroelectrics 222:421–428

Egami T (2007) Local structure of ferroelectric materials. Annu Rev Mater Res 37:297–315

Gjonnes J, Moodie AF (1965) Extinction conditions in the dynamic theory of electron diffraction. Acta Crystallogr 19:65–67

Glazer AM, Aroyo MI, Authier A (2014) Seitz symbols for crystallographic symmetry operations. Acta Crystallogr A 70:300–302

Goodman P (1984) A retabulation of the 80 layer groups for electron diffraction usage. Acta Cryst A 40:635–642

Goodman P, Johnson AWS (1977) Identification of enantiomorphically related space groups by electron diffraction—a second method. Acta Crystallogr A 33:997–1001

Goodman P, Lehmpfuhl G (1968) Observation of the breakdown of Friedel's law in electron diffraction and symmetry determination from zero-layer interactions. Acta Crystallogr A 24:339–347

Graham R, Ourmazd A, Spence JCH (1987) Unpublished work

Hahn T (2005) International tables for crystallography, volume A, space-group symmetry (5th), Springer

Howe JM, Sarikaya M, Gronsky R (1986) Space-group analyses of thin precipitates by different convergent-beam electron diffraction procedures. Acta Crystallogr A 42:368–380

Hu GB, Peng LM, Yu QF, Lu HQ (2000) Automated identification of symmetry in cbed patterns: a genetic approach. Ultramicroscopy 84:47–56

Ishizuka K (1982) Translation symmetries in convergent-beam electron diffraction. Ultramicroscopy 9:255–257

Jackson AG (1991) Handbook of crystallography, for electron microscopists and others. Springer, New York

Kim KH, Zuo JM (2013) Symmetry quantification and mapping using convergent beam electron diffraction. Ultramicroscopy 124:71–76

Kim K-H, Payne DA, Zuo J-M (2012) Symmetry of piezoelectric $(1-x)Pb(Mg_{1/3}Nb_{2/3})O_3$-$xPbTiO_3$ (x=0.31) single crystal at different length scales in the morphotropic phase boundary region. Phys Rev B 86:184113

Kittel C (1976) Introduction to solid state physics. Wiley, New York

Krivy I, Gruber B (1976) Unified algorithm for determining reduced (Niggli) cell. Acta Crystallogr A 32:297–298

Lewis JP (1995) Fast template matchig. Vision Interface 95:120–123

Loretto MH (1984) Electron beam analysis of materials. Chapman and Hall, London

Mansfield JF (1985) Error bars in CBED symmetry? Ultramicroscopy 18:91–96

McKie D, McKie C (1992) Essentials of crystallography. Blackwell Scientific Publications, Oxford

Midgley PA, Eggeman AS (2015) Precession electron diffraction—a topical review. Iucrj 2:126–136

Morniroli JP, Steeds JW (1992) Microdiffraction as a tool for crystal-structure identification and determination. Ultramicroscopy 45:219–239

Nye JF (1957) Physical properties of crystals. Clarendon Press, Oxford

Page YL (1992) Ab-initio primitive cell parameters from single convergent-beam electron diffraction patterns: A converse route to the identification of microcrystals with electrons. Microsc Res Tech 21:158–165

Peng LM, Zuo JM (1999) Anisotropic dispersion of the band structure and formation of ring patterns in CBED. Acta Crystallogr A 55:1026–1033

Peng LM, Dudarev SL, Whelan MJ (2004) High energy electron diffraction and microscopy. Oxford University Press

Pogany AP, Turner PS (1968) Reciprocity in electron diffraction and microscopy. Acta Crystallogr A 24:103

Portier R, Gratias D (1981) Extinction conditions in the dynamical electron diffraction: a symmetry approch. EMAG 1981. Inst Phys Conf Ser 61:275

Raghavan M, Koo JY, Petkovicluton R (1983) Some applications of convergent beam electron-diffraction in metallurgical research. J Met 35:44–50

Randall CA, Bhalla AS (1990) Nanostructural-property relations in complex lead perovskites. Japan J Appl Phys 29:327–333

Ravel B, Stern EA, Vedrinskii RI, Kraizman V (1998) Local structure and the phase transitions of $BaTiO_3$. Ferroelectrics 206:407–430

Saitoh K, Tsuda K, Terauchi M, Tanaka M (2001) Distinction between space groups having principal rotation and screw axes, which are combined with twofold rotation axes, using the coherent convergent-beam electron diffraction method. Acta Crystallogr A 57:219–230

Sands DE (1994) Introduction to crystallography (Revised edn), Dover

Singh AK, Pandey D, Zaharko O (2006) Powder neutron diffraction study of phase transitions in and a phase diagram of (1-x) $[Pb(Mg_{1/3}Nb_{2/3})O_3]$-xPbTiO$_3$. Phys Rev B 74:024101

Spence JCH, Lynch J (1982) STEM microanalysis by transmission electron-energy loss spectroscopy in crystals. Ultramicroscopy 9:267–276

Spence JCH, Qian W, Silverman MP (1994a) Electron source brightness and degeneracy from fresnel fringes in field emission point projection microscopy. J Vac Sci Technol A 12:542–547

Spence JCH, Zuo JM, Okeeffe M, Marthinsen K, Hoier R (1994b) On the minimum number of beams needed to distinguish enantiomorphs in x-ray and electron-diffraction. Acta Crystallogr A 50:647–650

Steeds JW (1979) Convergent beam electron diffraction. In: Hren JJ, Goldstein JI, Joy DC (eds) Introduction to analytic electron microscopy. Plenum Press, New York

Steeds JW, Vincent R (1983) Use of high-symmetry zone axes in electron diffraction in determining crystal point and space groups. J Appl Crystallogr 16:317–324

Stout GH, Jensen LH (1989) X-ray structure determination: a practical guide, 2nd edn. Wiley-Interscience, New York

Tanaka M (1989) Symmetry analysis. J Electron Micr Tech 13:27–39

Tanaka M, Terauchi M (1985a). Convergent beam electron diffraction. Tokyo, JEOL Company

Tanaka M, Terauchi M (1985b) Whole pattern in convergent-beam electron diffraction using the hollow-cone beam method. J Electron Microsc 34:52–55

Tanaka M, Sekii H, Nagasawa T (1983) Space-group determination by dynamic extinction in convergent-beam electron-diffraction. Acta Crystallogr A 39:825–837

Tanaka M, Terauchi M, Sekii H (1987) Observation of dynamic extinction due to a glide plane perpendicular to an incident beam by convergent-beam electron-diffraction. Ultramicroscopy 21:245–250

Tanaka M, Terauchi M, Kaneyama T (1988) Convergent beam electron diffraction II. JEOL Company, Tokyo, Japan

Terauchi M, Tanaka M (1985) Simultaneous observation of zone-axis pattern and ±g-dark-field pattern in convergent-beam electron-diffraction. J Electron Microsc 34:347–356

Tsuda K, Sano R, Tanaka M (2012) Nanoscale local structures of rhombohedral symmetry in the orthorhombic and tetragonal phases of $BaTiO_3$ studied by convergent-beam electron diffraction. Phys Rev B 86:214106

Tsuda K, Sano R, Tanaka M (2013) Observation of rhombohedral nanostructures in the orthorhombic phase of $KNbO_3$ using convergent-beam electron diffraction. Appl Phys Lett 102:051913

Valset K, Tafto J (2011) Bloch wave symmetries in electron diffraction: applications to friedels law, Gjonnes-Moodie lines and refraction at interfaces. Ultramicroscopy 111:854–859

Viehland D, Kim MC, Xu Z, Li JF (1995) Long-time present tweedlike precursors and paraelectric clusters in ferroelectrics containing strong quenched randomness. Appl Phys Lett 67: 2471–2473

Wang H, Zhu J, Lu N, Bokov AA, Ye ZG, Zhang XW (2006) Hierarchical micro-/nanoscale domain structure in m_c phase of $(1\text{-}x)Pb(Mg_{1/3}Nb_{2/3})O_3\text{-}xPbTiO_3$ single crystal. Appl Phys Lett 89:042908

Woolfson MM (1997) An introduction to x-ray crystallography (2nd edn). Cambridge University Press

Zhang Q, Cagin T, Goddard WA (2006) The ferroelectric and cubic phases in $BaTiO_3$ ferroelectrics are also antiferroelectric. Proc Natl Acad Sci USA 103:14695–14700

Zuo JM (1993) New method of Bravais lattice determination. Ultramicroscopy 52:459–464

Chapter 12
Crystal Structure and Bonding

Crystal structure is described by a lattice, symmetry, and atomic arrangements inside the asymmetric unit. Some crystals are further organized according to structural types, and empirical rules have been developed in solid-state chemistry regarding the classification of crystal structures according to atomic radii, electronegativity, packing, and tiling. In real crystals, atoms form bonds and lower the free energy providing the driving force for crystal growth and transformation. Bonding can be measured by the electron density from X-ray diffraction and quantitative CBED. Here, we will first examine some common crystal structural types and their representative features, and then introduce the concept of chemical bonding, followed by a study of electron density, its significance, experimental measurements, and their use for studying chemical bonding.

12.1 Description of Crystal Structure

We will use the example of one of the Cu-based high-T_c superconductors, yttrium barium copper oxide ($YBa_2CuO_{7-\delta}$), to introduce how crystal structures are described in the literature. Figure 12.1 shows the atomic structure of $YBa_2CuO_{6.91}$ as recorded in the ICSD database. The full dataset includes additional descriptions about the source of the data as well as the experimental methods used for structure determination. Here, we have selected only information related to the crystal structure. The first line in Fig. 12.1 is a record identification number in the database. This is followed by the chemical name (abbreviated as "Chem Name") and formula (given after "Structured"). The next is line 6, giving the unit cell constants a, b, c, α, β, and γ in units of Å and degrees, respectively. The space group that the crystal belongs to is specified next (line 8), which is *Pmmm*, orthorhombic with a primitive lattice in this case. Line 10 is the start of a description of the atoms in the asymmetric unit. This line specifies the information included in the description of each atom, starting with type of atoms ("Atom"), identification number ("#"), oxidation

© Springer Science+Business Media New York 2017

J.M. Zuo and J.C.H. Spence, *Advanced Transmission Electron Microscopy*,

DOI 10.1007/978-1-4939-6607-3_12

```
 1 *data for    ICSD #65582
 2 ...
 3 Chem Name    Yttrium Barium Copper Oxide (1/2/3/6.91)
 4 Structured   Y Ba2 Cu3 O6.91
 5 ...
 6 Unit Cell    3.8109(1) 3.8793(1) 11.6477(3) 90. 90. 90.
 7 ...
 8 Space Group  P m m m
 9 ...
10 Atom  #   OX    SITE       x           y           z          SOF      H
11  Ba   1  +2     2 t     0.5         0.5         0.1841(2)     1.       0
12  Y    1  +3     1 h     0.5         0.5         0.5           1.       0
13  Cu   1  +3     1 a     0           0           0             1.       0
14  Cu   2  +2     2 q     0           0           0.3549(1)     1.       0
15  O    1  -2     2 q     0           0           0.1589(1)     1.       0
16  O    2  -2     2 s     0.5         0           0.3779(1)     1.       0
17  O    3  -2     2 r     0           0.5         0.3776(2)     1.       0
18  O    4  -2     1 e     0           0.5         0          0.91(1)     0
19 Lbl   Type      B11         B22         B33         B12        B13         B23
20 Ba1   Ba2+  0.37(2)     0.37(2)     0.37(2)      0          0          0
21 Y1    Y3+   0.28(2)     0.28(2)     0.28(2)      0          0          0
22 Cu1   Cu3+  0.31(2)     0.31(2)     0.31(2)      0          0          0
23 Cu2   Cu2+  0.29(1)     0.29(1)     0.29(1)      0          0          0
24 O1    O2-   0.50(2)     0.50(2)     0.50(2)      0          0          0
25 O2    O2-   0.41(2)     0.41(2)     0.41(2)      0          0          0
26 O3    O2-   0.39(2)     0.39(2)     0.39(2)      0          0          0
27 O4    O2-   1.14(10)    0.12(7)     0.68(9)      0          0          0
28 ...
29 *end for     ICSD #65582
```

Fig. 12.1 Atomic structure of YBa_2CuO_7 with selected information from inorganic crystal structure database (ICSD)

state ("OX"), site symmetry in Wyckoff symbol ("site"), atomic position x, y, and z in fractional coordinates, site occupancy ("SOF"), and number of bonded hydrogen atoms ("H"). In what follows, the atoms in the asymmetric unit are listed with the above information. In $YBa_2Cu_3O_{7-\Delta}$, we have 1 Y, 1 Ba, 2 Cu, and 4 Oxygen atoms listed, and the rest are obtained by symmetry. For example, the first oxygen atom at the 2 q site, which has point symmetry mm^2 according to International Table for Crystallography A, and the two symmetry-related sites are at (00z) and (00-z). Only one of these two is specified. Starting from line 19, information on atomic thermal parameters is given. Here, the label ("Lbl") at line 19 specifies that they are specified by b_{ij} in the format of Eq. (12.20). This line is followed by a specification for each atom. It should be noted that not all structure determinations specify the thermal parameters, so they are included in the database record when available.

A qualitative, but vivid, description of a crystal structure results if we plot atoms in their equilibrium positions using computer-rendered drawings, true-to-scale, and with 3D perspective, in which atoms can be drawn simply as balls or as ellipsoids to represent atomic thermal vibration. The size of the atoms is chosen to represent their atomic or ionic radii or in the case of ellipsoids the probability of finding the atom. Often, atoms in close contact are drawn as connected using sticks to represent bonds or polyhedra to highlight atomic coordination. These different styles of

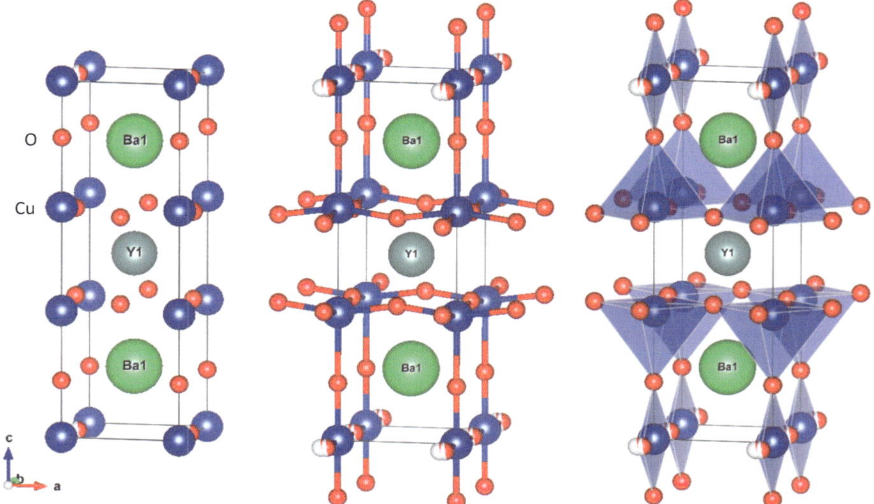

Fig. 12.2 Structure of $YBa_2Cu_3O_{7-\Delta}$ plotted using VESTA version 3.2.1 with balls representing atoms only, ball and stick, and ball and polyhedra

crystal structure rendering can be used to emphasize different aspects of a structure (e.g., see Fig. 12.2).

Further descriptions about the subtle features of crystal structure can be made using inter-atomic distances and angles. The inter-atomic distance is calculated from the equilibrium positions of two atoms, which are the average positions of the vibrating atoms. For example, in Fig. 12.1, the Cu_2–O_2 atoms are separated by a distance of 1.9237(2) Å, while the distance between Cu_2 and O_3 atoms is 1.9570(3) Å. The difference between these two is too small to show in a structural drawing. It is worth noting that a crystal structure determined by X-ray diffraction gives the center position of the electron density, which coincides with the positions of the nuclei to a good approximation, except for covalently bonded hydrogen atoms. To locate hydrogen atoms exactly, neutron diffraction is often used. Electron diffraction sees both electrons and nuclei and therefore is an alternative to neutron diffraction.

The immediate surrounding of an atom is characterized by the number of closest neighboring atoms (coordination). In deciding on the coordination number, both bond distances and types of atoms are considered. For example, the Cu_2 atom has four oxygen atoms at 1.9237 and 1.9570 Å distances away, another oxygen atom (apical oxygen) at 2.2767 Å distance, and Y and Ba atoms further away. The Cu_2 atom is said to be 4 coordinated with oxygen atoms in the CuO_2 plane because the apical oxygen is considered too far away (18 % longer distance) to be a coordinated atom. This may sound a bit arbitrary—it is! To address this issue, several schemes have been devised to provide a more quantitative description of atomic coordination, and for further details see the book by Müller (2006).

12.2 Common Structure Types

Crystals having the same atomic arrangements, but differing in atomic distances, belong to the same structure type. Examples include Cu and Au of the fcc structure or $BaTiO_3$ and $CaTiO_3$ of perovskite. In crystallography, a structure is often specified in the following manner: "X crystallizes in the Y type," where Y can be the name of a crystal to which the structure is linked (or initially identified). In such statements, small differences in atomic positions or difference in atomic distances are ignored. Thus, both Cu and Au belong to the fcc structure, and $BaTiO_3$ crystallizes in the structure of $CaTiO_3$, whereas in the latter case, details of the $BaTiO_3$ structure, such as different symmetries at different temperatures, are neglected.

Further distinctions are made if two ionic compounds have the same structure type, but the positions of cations in one compound are taken up by the anions of another and vice versa. Thus, we have an "exchange of cations and anions." Such cases are called "anti-types." Polytypism is a phenomenon by which a compound crystallizes in a variety of periodic and layered structures, called polytypes. The layer stacking changes in each polytype. A polytype is often characterized by the number or layer (n) stacking inside the unit cell. A well-known example is silicon carbide, which crystallizes in a large number of polytypes nH, n = 2, 4, 6 ..., where H stands for hexagonal. Crystals may adopt different structures for the same composition. This phenomenon is called polymorphism. The polymorphic forms of silica are known to form different crystal structures, including α-quartz, β-quartz, and cristobalite. We note also that some molecular crystals can crystallize in more than one crystal structure, retaining the same molecule within a different unit cell.

In the remainder of this section, we will introduce some common structure types and the principal features of these structure types. The lattice constants referred below are values obtained at room temperature.

Packing of spheres: In cases where atoms or molecules (C60 for example) can be approximated as rigid spheres of the same size, with a comparatively weak attractive force, their crystal structure can be understood based on packing of spheres, belonging to one of four basic structures: simple cubic (SC), body-centered cubic (bcc), hexagonal closest-packed (HCP), and cubic closest-packed (CCP) (Fig. 12.3). The SC structure is obtained by packing four spheres in contact with each other to form a square and extending this to generate a plane of spheres. Next, by stacking a second plane of spheres of the same pattern directly on top of the first and by repeating this process, we obtain a regular structure in which the lattice is primitive and cubic and the unit cell contains a single atom (see Fig. 12.3). Each atom bonds to its six nearest neighbors (coordination number of 6). Polonium is the only element known to support this structure. The bcc structure has the central sphere touching four spheres at the corners of a cube in the plane above and four more in the plane below, at the corners of a cube, which gives the coordination number of 8. The spheres in each plane also form a square pattern, but compared to the SC structure, there is a gap between the neighboring spheres.

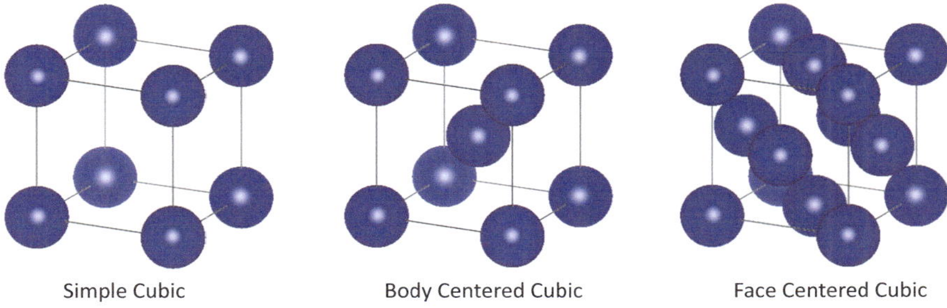

Simple Cubic   Body Centered Cubic   Face Centered Cubic

Fig. 12.3 Cubic structures formed by packing of spheres

Fig. 12.4 Close packing of spheres showing the planar close pack structure, AB stacking at *left* and ABC stacking at *right*

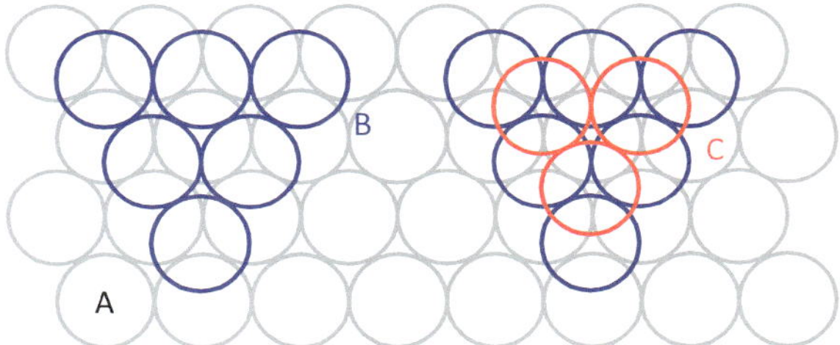

We take the atomic plane below as A, and the plane containing the center sphere as B, in which B is stacked on top of A in such a way that atoms in plane B sit in the middle of the squares in plane A. The crystal has the stacking sequence of ABABAB….

Another form of packing is to have three spheres in contact to form an equal-sided triangle (Fig. 12.4). This is known as close packing, since it gives the highest planar (and volume as shown below) packing density. The close packing creates two interstitial sites, by placing a second layer on top of these interstitial sites we create either AB or AC stacking. The fcc structure is created by ABCABC… stacking, while ABAB… stacking generates the hexagonal close-packed structure. Most of metallic elements adopt the fcc, bcc, and HCP (hexagonal close packing) structures. Examples of fcc metals include Ni, Cu, Rh, Pd, Ag, Ir, Pt, and Au. Mg, Ti, Co, and Zn are examples of the HCP structure. The bcc metals include Cr, Fe, Nb, Mo, Ta, and W. Table A6.1 lists the crystal structure of elements. Whether a metal adopts an fcc, bcc, or HCP structure is determined by electronic considerations. In the case of the bcc structure, the nearest-neighbor distance is about 3 % shorter than that of the fcc or HCP structure. The reduction in the atomic distances partially compensates for the poorer packing ratio in the bcc structure.

Diamond structure: The diamond structure (cubic with $a = 3.56679$ Å and space group $Fd3m$) is formed by having carbon atoms linked to four neighbors at the center of tetrahedral, which are corner shared. There are 2 atoms in the asymmetric

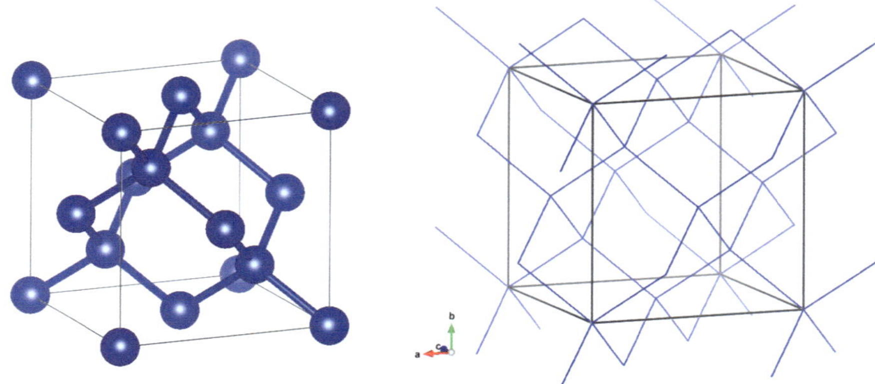

Fig. 12.5 Diamond structure showing one unit cell with tetrahedrally bonded atoms and the connection of the bonds

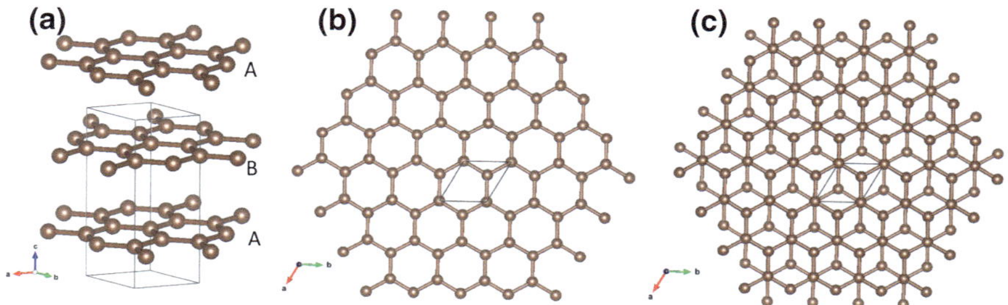

Fig. 12.6 Crystal structure of graphite structure (2H). **a** 3D rendering, **b** single atomic layer, and **c** double layers

unit at $\pm\left(\frac{1}{4},\frac{1}{4},\frac{1}{4}\right)$ positions in the cubic unit cell. The C–C bonds form a 3D 4-connected net Wells (1984) as shown in Fig. 12.5 that extends to the entire crystal. Other elements that form the same structure include Si ($a = 5.4307$ Å), Ge ($a = 5.65735$ Å) and tin (gray modification, $a = 6.4912$ Å).

Graphite: They are several structural forms (polytypes). What is shown in Fig. 12.6 is known as the 2H structure. All of the polytypes have the carbon atom linked to three neighboring atoms at the same bond distance of 1.421 Å, forming a network of regular hexagons in a flat layer honeycomb pattern (called graphene). The crystal is formed by stacking the layers of graphene at a distance 3.4 Å depending on layer stacking. In the 2H structure, half of the six vertices sit in the middle of the hexagons of the next layer in an ABAB… stacking, while the rhombohedral or 3R polytype has the layers stacked in an ABCABC… sequence.

NaCl structure: This is regarded as the prototypical ionic crystal with interpenetrating arrays of Na^+ and Cl^- ions (Fig. 12.7). The structure is constructed by placing a pair of Na^+ and Cl^- ions at the positions of (000) and (1/2, 1/2, 1/2),

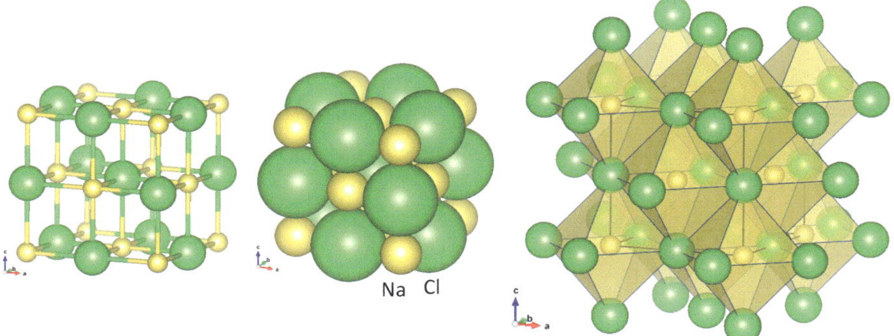

Fig. 12.7 NaCl structure with Na^+ ion at the corners and face centers and Cl^- ion at the middle of edges and body center. From *left* to *right*, rendered with reduced, full and reduced ionic radii. The *right figure* shows Na+ ions in the center of the octahedra formed by Cl^- ions

respectively, on all four lattice points of the fcc lattice ($a = 5.64$ Å). Table F.2 lists many MX-type (M = cation, X = anion) crystals that have the NaCl structure. The principle feature of the structure is the anion (Cl^- in NaCl) size, which are in close contact (see Fig. 12.7), and the cation (Na^+ in NaCl) occupies the interstitial space in the octahedron formed by Cl^- anions.

CsCl structure: This is SC with one molecular formula (MX) per unit cell. The structure is often specified by having M at (000) and X at the body center (1/2, 1/2, 1/2) positions. There are two kinds of substance which crystallizes in this structure. The first group is halides with large cations, Cs and Th for example, and the other are intermetallic compounds represented by beta-brass (CuZn). Table F.3 lists some MX compounds with the CsCl structure.

CaF$_2$ structure: Compounds of MX_2 with large-sized M atoms likely crystallize in the CaF_2 (fluorite) structure. The crystal structure is SC. The M atoms form a fcc lattice, and X atoms occupy the $\pm(1/4, 1/4, 1/4)$ positions (Fig. 12.8). Each M atom is surrounded by eight X atoms located at the corners of a cube. This coordination only occurs when the radius of M is at least 0.73 times that of X, e.g., $r(M)/r(X) \geq 0.73$. Compounds with the fluorite structure include halides (CaF_2), oxides, and sulfides of univalent alkalis (Li_2O), oxides of quadrivalent cations (HfO_2, ZrO_2), and intermetallic alloys ($AuAl_2$). The structure of Li_2O has the positions of the positive and negative ions reversed compared to the CaF_2 structure, and such an arrangement is called anti-fluorite.

ZnS structure: There are about 200 identified polytypes of ZnS, the two well-known being wurtzite and zincblende (sphalerite) structures, named after the minerals in which these structures were first identified. The zincblende structure has a cubic unit cell with $a = 5.4093$ Å and space group F-43m. Like NaCl, it has one pair of atoms [S at (000) and Zn at (¼, ¼, ¼)] at each fcc lattice points. Both Zn and S are tetrahedrally coordinated at the vertices of 4-connected net as in silicon. In fact, the structure is the same as diamond if the Zn and S are treated as indistinguishable. When viewed normal to one of $\langle 111 \rangle$ as shown in Fig. 12.9, the

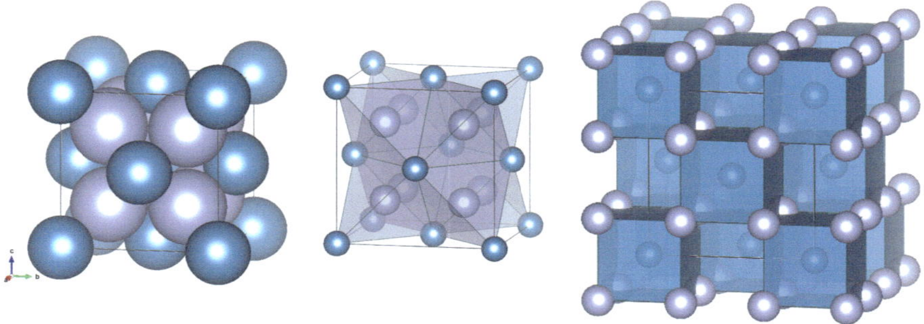

Fig. 12.8 CaF_2 structure with each F^- ion coordinated by four Ca^{2+} ions at the center of tetrahedron and each Ca^{2+} coordinated by eight F^- ions at the cubic corners

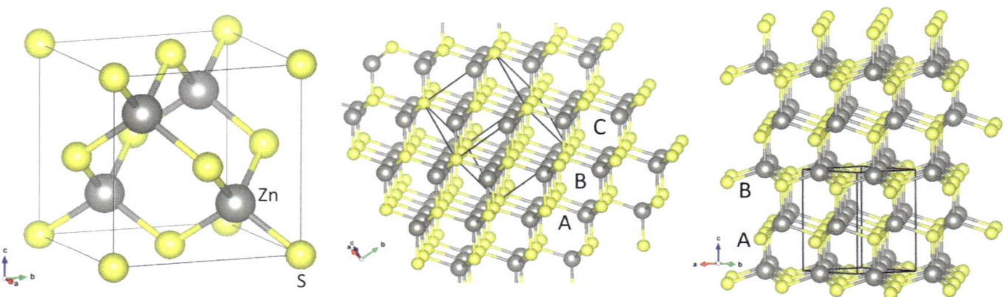

Fig. 12.9 From *left* to *right*, unit cell of cubic ZnS (Zincblende) structure, its ABCABC… stacking, and hexagonal ZnS (wurtzite) structure with ABAB… stacking

zincblende structure is formed by periodic stacking of A, B, and C layers. On the other hand, the wurtzite structure is hexagonal, with space group P6(3)mc (no. 186) with a = 3.8230 Å and c = 6.2565 Å. The crystal structure is obtained by having Zn at (1/3, 2/3, 0) and S at the (1/3, 2/3, 0.374) positions. Along the c-axis, the structure repeats every two layers in ABAB…. stacking. Several of the binary compounds that crystallize in the zincblende structure have industrial applications because of their physical properties. They include SiC, cubic BN, and the III–V semiconductor compounds.

Rutile: The lattice is tetragonal and primitive belonging to the space group (P4$_2$/mnm). There are two molecules in the unit cell, which has dimension of a = 4.5937 Å and c = 2.9581 Å for TiO_2. A number of dioxides of quadrivalent metals and fluorides of small divalent ions of formula MX_2 crystallize into this structure. The structure can be fully described by choosing M at 2a: (000) and (1/2, 1/2, 1/2) and X at 4f: $\pm(u, u, 0)$ and $\pm(u + 1/2, 1/2 - u, 1/2)$, where u is the only position parameter with a value close to 0.3 (u = 0.3053 for rutile). Each M atom is coordinated by six X atoms sitting at the corners of an octahedron. The octahedra are edge shared along the c-axis, and they are connected by corners in a snub square tiling pattern as shown along the c-axis (Figs. 12.10 and 12.11).

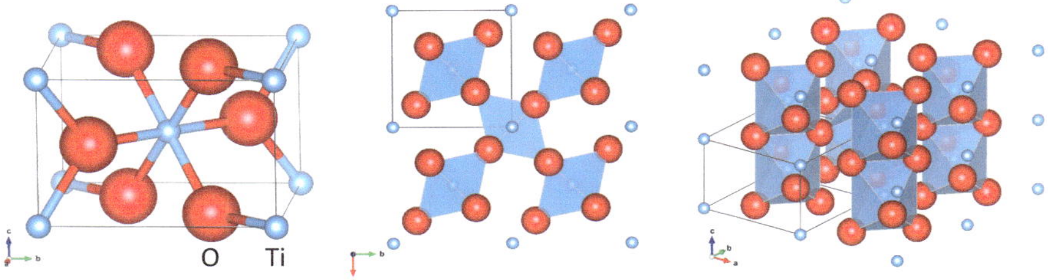

Fig. 12.10 Crystal structure of rutile (TiO$_2$). From *left* to *right*, single unit cell, viewed along the *c*-axis showing 4 unit cells, and a 3D view of the structure

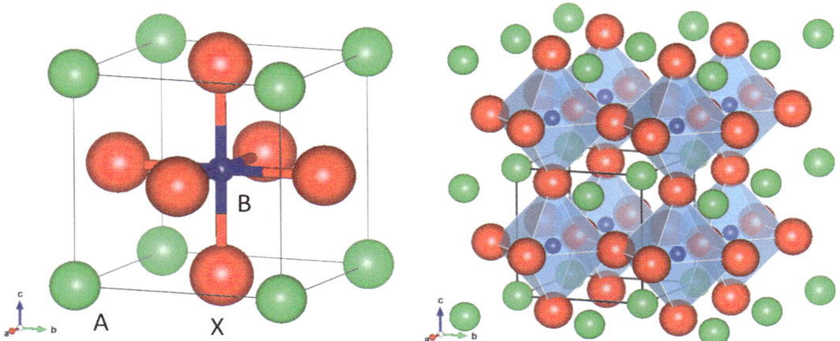

Fig. 12.11 The ideal structure of perovskite ABX$_3$

Perovskite (CaTiO$_3$): This is named after the Russian mineralogist, Count Lev Aleksevich von Perovski. This type of crystal structure is formed by a large group of ABX$_3$ (A and B for cations) compounds, especially a number of oxides (X=O) with diverse properties that collectively make the perovskites the most interesting structures for functional materials. The ideal perovskite structure is SC, with a unit cell containing a single molecule. The A atoms are shared at the cubic corners, X atoms are at the face centers of the cube, and a single B atom occupies the center. Each B atom is coordinated by six X atoms at the corners of an octahedron. These octahedra are corner-connected, instead of edge shared as in NaCl or edge and corner shared in rutile. Together, they form the cuboctahedral sites, where A atom has the coordination of 12. Alternatively, we could take the octahedron with B atom inside as a molecule of BX$_3$, and then, the structure becomes the same as that of CsCl in the form A(BX$_3$).

The actual structure of CaTiO$_3$ at room temperature is orthorhombic with $a = 5.3709$, $b = 5.4280$, and $c = 7.6268$ in Å, corresponding to a unit cell of $\sqrt{2}a_c \times \sqrt{2}a_c \times 2a_c$ where a_c is the cell length of the pseudocubic lattice. The octahedra formed by the oxygen atoms are tilted in both y- and z-axes (see Fig. 12.12), and the Ti atoms are at the center of the octahedra. Such tilts are found in other crystals too, such as GaFeO$_3$. In other structures such as BaTiO$_3$, the B

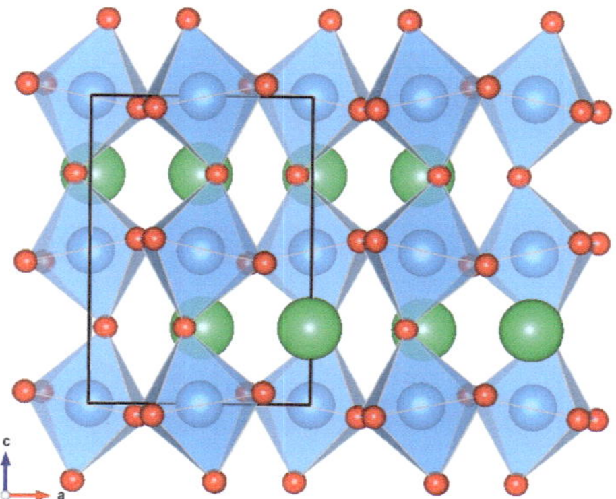

Fig. 12.12 Orthorhombic $CaTiO_3$ structure viewed along the b axis

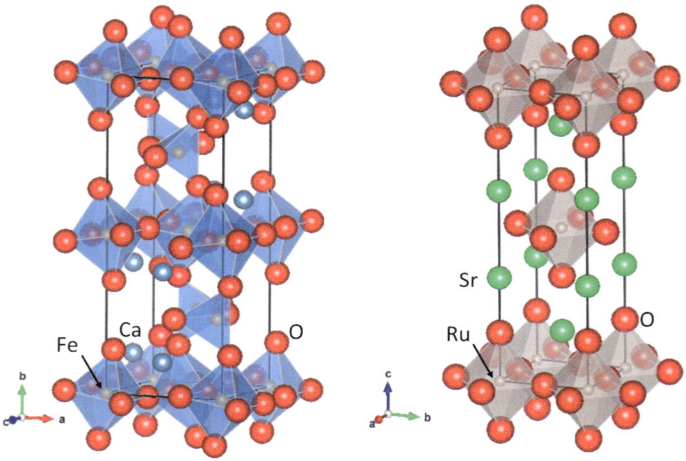

Fig. 12.13 Crystal structure of $Ca_2Fe_2O_5$ (*left*) and Sr_2RuO_4 (*right*)

cations move off center, lowering the symmetry and giving rise to ferroelectricity as discussed in Chap. 11.

There are a number of structures that are related to perovskite. Brownmillerite ($Ca_2Fe_2O_5$) is derived from the perovskite structure by introducing oxygen vacancies (1 per molecule). The vacancies are ordered in a pattern that produces alternating layers of octahedra and distorted tetrahedra (Fig. 12.13). Similarly, the high-T_c superconductor $YBa_2Cu_3O_{7-\Delta}$ structure derives from a tripled perovskite structure [$(YBa_2)Cu_3O_9$] with 2:1 layered ordering of Ba and Y and 2 oxygen vacancies in the Cu layers which lie between the Ba layers and in the Y layers. The first changes the coordination of the copper ions from octahedral to square planar. Further reduction in this layer with $x = 1$ leads to the formation of a linear Cu–O

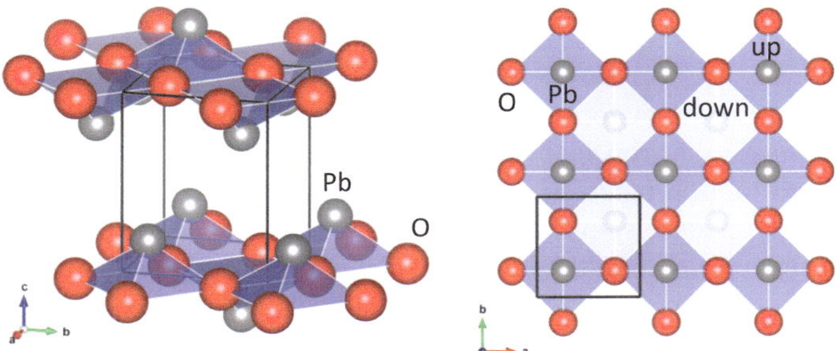

Fig. 12.14 The structure of tetragonal PbO

chain, which is common for Cu^+. The second type of oxygen vacancies gives rise to the Cu–O sheets adjacent to the Y layers, which are believed to be responsible for the high-temperature superconductivity of this compound and related structures. Other types of oxide structures are formed by layered perovskites comprised of infinite 2D slabs of the ABO_3-type structure, which are separated by another structural motif. The general formula of the perovskite slabs is as follows: $A_{n-1}B_nO_{3n+1}$, where n indicates the thickness of the 2D slabs in number of the BO_6 octahedron. The structure of Sr_2RuO_4 (Fig. 12.13) is an example, and here, the structure consists of alternating $n = 2$ and $n = 1$ layers. This is one of the crystal structures in the so-called Ruddlesden–Popper phases, where the perovskite slabs are separated by a layer of Sr_2, and offset by a (1/2, 1/2, 0) translation. Another example is the Aurivillius phases (($Bi_2O_2)A_{n-1}B_nO_{3n+1}$), which have the perovskite layers separated by $(Bi_2O_2)^{2+}$ layers.

PbO structure: PbO crystallizes in two polymorphs: one is tetragonal (litharge) and the other is orthorhombic (massicot). The tetragonal crystal structure, which is shown in Fig. 12.14, has two molecular units. Its principle feature is the square lattice formed by oxygen atoms and the checkerboard pattern formed by Pb atoms alternating in the up and down positions at the center of oxygen squares. This structure attracted considerable interest because it is found in pnictides with FeAs in the anti-PbO structure and Fe chalcogendies (FeSe and FeTe), which together form the third family of high-T_c superconductors (MgB_2 is the second after the discovery of copper-based high-T_c superconductors).

12.3   Chemical Bonding

Atoms are held together in a crystal by forces that are simply referred as the chemical bond. The nature of the chemical bond is quantum mechanical in origin, with the exception of the van der Waals forces. Here, we will introduce some basic concepts related to crystal bonding. Comprehensive treatments from the point of view of solid-state chemistry are given in a number of books by Hoffman (1989), Burdett (1995) and Pettifor (1995).

12.3.1   Bonding of a Diatomic Molecule

To start, we consider a diatomic molecule, comprised of two atoms A and B, where each has a single electron contributing to the bonding. As these two atoms are brought closer to each other, their atomic wave functions ϕ_A and ϕ_B overlap, and the electrons of opposite spin see the potential due to both atoms, while electrons of the same spins repel each other (due to the Pauli exclusion principle), giving rise to new electron wave functions. The same issue of delicately competing energies in balance arises even for a single atom, where Hund's rules give a consistent set of conditions for satisfying these requirements while adding electrons to an atom in a way which minimizes both the Coulomb and Pauli energies. For example, in carbon, we can ask whether the addition of a second electron to p orbitals should go in the same orbital as the first electron (maximizing Coulomb energy, but minimizing spin "repulsion") or to into a different p orbital (minimizing the proximity of the electrons and their Coulomb interaction, but, if they have the same spin, maximizing spin energy). In an approximation, known Hückel theory or local combination of atomic orbitals method, the new electron wave functions can be written as follows:

$$\phi_\pm = c_\pm(\phi_A \pm \varepsilon\phi_B), \tag{12.1}$$

which has the electron energy

$$E_\pm = 2c_\pm^2(E_{\mathrm{o}} \mp \varepsilon E_{AB}), \tag{12.2}$$

where $c_\pm = \sqrt{1 + \varepsilon^2 \pm 2\Delta}$, $\Delta = \mathrm{Re}\{\langle\phi_A|\phi_B\rangle\}$ is the normalization coefficient and

$$E_{\mathrm{o}} = \left(E_{AA} + \varepsilon^2 E_{BB}\right)/2. \tag{12.3}$$

The above energies are calculated using $E_{XY} = -\langle\phi_X|H|\phi_Y\rangle$ with

$$H = -\frac{\hbar}{2m_e}\nabla^2 - \frac{Z_A e^2}{4\pi\varepsilon_o\left|\vec{r} - \vec{R}_A\right|} - \frac{Z_B e^2}{4\pi\varepsilon_o\left|\vec{r} - \vec{R}_B\right|}. \tag{12.4}$$

Thus, E_{AB} comes from the orbital overlap that increases as the separation of atoms decreases.

For a homonuclear (B=A) molecule, $\varepsilon = 1$. In a heteronuclear molecule, ε is a measure of polarity. Its value ranges from 0 to infinite. A large value of ε means the B atom exerts a greater attractive force on the electron than the A atom. The two states obtained, ϕ_+ and ϕ_-, are called the bonding and antibonding state since the former is lower and the latter is higher in energy than the energy of separated atoms.

In a crystal, instead of a single bond in a diatomic molecule, there are an infinite number of bonds that connect the atoms, and the principle of lowering the energy by forming bonding states however is the same.

The ground-state many-electron wave function of the molecule is constructed as follows:

$$\phi = \phi_+(1)\phi_+(2) = c_+^2\,(\phi_A(1) + \varepsilon\phi_B(1))(\phi_A(2) + \varepsilon\phi_B(2))$$
$$= c_+^2\left[\phi_A(1)\phi_A(2) + \varepsilon^2\phi_B(1)\phi_B(2) + \varepsilon\phi_A(1)\phi_B(2) + \varepsilon\phi_B(1)\phi_A(2)\right] \quad (12.5)$$

The above wave function does not include the electron spin, which is antisymmetric for opposite spins.

When the atoms are further apart, the probability of having both electrons together in the same atom is small or, in other words, the electrons become more localized. Then, the first two terms in Eq. (12.5) no longer apply. It is more appropriate than to use the following many-electron wave function

$$\phi_{HL} = (\phi_A(1)\phi_B(2) + \phi_A(2)\phi_B(1))/\sqrt{2}, \quad (12.6)$$

which is known as Heitler–London (HL) wave function. It has a total energy of $E_{\mathrm{o}} = E_A + E_B$. Other states are formed when the electrons reside on the same atom, which gives

$$\phi'_{HL} = \phi_A(1)\phi_A(2) \text{ or } \phi'_{HL} = \phi_B(1)\phi_B(2) \quad (12.7)$$

These states have energy $E_{\mathrm{o}} = 2E_A + U_A$ and $E_{\mathrm{o}} = 2E_B + U_B$, respectively, where $U_X = \langle \phi_X(1)\phi_X(2)|r_{12}^{-1}|\phi_X(1)\phi_X(2)\rangle$ with $X = A$ or B is the electron–electron interaction energy which results from having both electrons on the same atom. The transition from localized to delocalized electron wave functions is one of most interesting areas of research in condensed matter physics, since tendency to electron localization leads to strong electron correlation.

The molecular binding energy can be obtained by plotting the electronic energy as a function of inter-nuclear separation. It is usually referred to as potential curve. It can be obtained by solving the Schrödinger equation. Since the exact solution is not possible, what is obtained is an approximation. The potential curve using wave function ϕ_- is repulsive, while ϕ_+ leads to a binding potential well with a minimum, corresponding to the equilibrium distance (r_{o}) of the molecule (Fig. 12.15). A repulsive force for ϕ_+ arises when the atomic distance is smaller than r_{o}. For closed-shell atoms or ions, the repulsion mainly comes from the mutual electrostatic repulsion of the electrons and the repulsion between electrons having the same spin (Pauli principle) with the latter contributing the principal part of the repulsion. Other contributions come from electrostatic forces between ions of the same sign, or in the case of the hydrogen molecules, the electrostatic repulsion between the atomic nuclei.

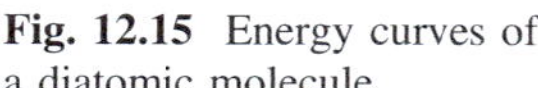

Fig. 12.15 Energy curves of a diatomic molecule

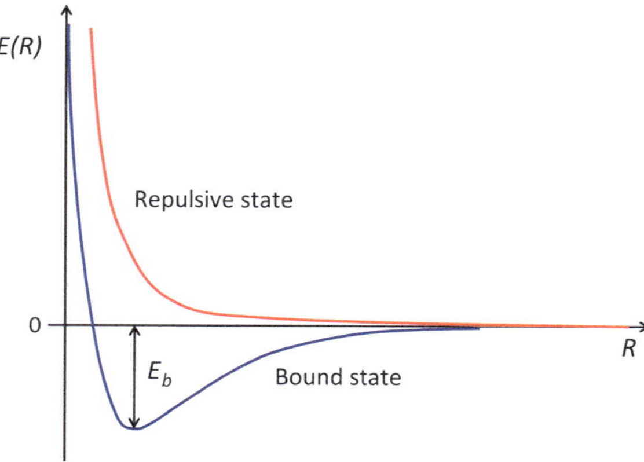

12.3.2   Atomic Sizes and Electronegativity

The equilibrium atomic distance allows a definition of the effective atomic size. The size of an atom, or ion, depends on the atomic number as well as the type of bonding since the attractive forces between the atoms differ significantly from one type of bonding to another. Thus, for every atom in the periodic table several atomic radii are assigned according to bonding type. Extensive experimental studies show that the atomic radius of an element is relatively constant for the same type of bonding in different materials. A further dependence is observed on the number of atoms which the atom bonds to (the coordination number), and this observation led to the concept of an effective ionic radius (Pauling 1960; Shannon 1976).

The polarity in a pair of atoms is measured by electronegativity X. This concept was first proposed by Pauling in his valence bond theory. In a diatomic molecule, the element with a low value of electronegativity is the electron donor and the other atom with a high value is the electron acceptor. The noble gas atoms are assigned with a value of 0 in electronegativity, because they neither accept nor donate electrons. To calculate the electronegativity, the difference between a pair of atoms of A and B is defined, according to Pauling, by

$$X_A - X_B = (\mathrm{eV})^{-1/2}\sqrt{E_d(AB) - [E_d(AA) + E_d(BB)]/2}, \qquad (12.8)$$

where the dissociation energy E_d of AB, AA, and BB bonds are given in volts. A value is assigned for each atom, in what is known as Pauling scale, by using hydrogen as the reference, which is taken as $X = 2.20$ (Fig. 12.16). The idea behind this is the hypothesis of resonance that a bond is formed as result of resonance between three hypothetical states, a covalent bond and two ionic bond states, which give the three energies in Eq. (12.8).

Periodic Table of Pauling's Electronegativity

Legend:
- Alkali metals
- Alkaline earth metals
- Transition metals
- Post-transition metals
- Metalloid
- Lanthanides
- Actinides

Period	1 IA/1A	2 IIA/2A	3 IIIB/3B	4 IVB/4B	5 VB/5B	6 VIB/6B	7 VIIB/7B	8	9 VIII	10	11 IB/1B	12 IIB/2B	13 IIIA/3A	14 IVA/4A	15 VA/5A	16 VIA/6A	17 VIIA/7A	18 VIIIA/8A
1	1 H 2.20																	2 He 0
2	3 Li 0.98	4 Be 1.57											5 B 2.04	6 C 2.55	7 N 3.04	8 O 3.44	9 F 3.98	10 Ne 0
3	11 Na 0.93	12 Mg 1.31											13 Al 1.61	14 Si 1.90	15 P 2.19	16 S 2.58	17 Cl 3.16	18 Ar 0
4	19 K 0.82	20 Ca 1.00	21 Sc 1.36	22 Ti 1.54	23 V 1.63	24 Cr 1.66	25 Mn 1.55	26 Fe 1.83	27 Co 1.88	28 Ni 1.91	29 Cu 1.90	30 Zn 1.65	31 Ga 1.81	32 Ge 2.01	33 As 2.18	34 Se 2.55	35 Br 2.96	36 Kr 3.0
5	37 Rb 0.82	38 Sr 0.95	39 Y 1.22	40 Zr 1.33	41 Nb 1.6	42 Mo 2.16	43 Tc 1.9	44 Ru 2.2	45 Rh 2.28	46 Pd 2.20	47 Ag 1.93	48 Cd 1.69	49 In 1.78	50 Sn 1.96	51 Sb 2.05	52 Te 2.1	53 I 2.66	54 Xe 2.60
6	55 Cs 0.79	56 Ba 0.89	*	72 Hf 1.3	73 Ta 1.5	74 W 2.36	75 Re 1.9	76 Os 2.2	77 Ir 2.20	78 Pt 2.28	79 Au 2.54	80 Hg 2.00	81 Tl 1.62	82 Pb 1.87	83 Bi 2.02	84 Po 2.0	85 At 2.2	86 Rn 2.2
7	87 Fr 0.7	88 Ra 0.9	**	104 Rf	105 Db	106 Sg	107 Bh	108 Hs	109 Mt	110 Ds	111 Rg	112 Cn	113 Uut	114 Fl	115 Uup	116 Lv	117 Uus	118 Uuo

Lanthanide Series*

57 La 1.1	58 Ce 1.12	59 Pr 1.13	60 Nd 1.14	61 Pm 1.13	62 Sm 1.17	63 Eu 1.2	64 Gd 1.2	65 Tb 1.1	66 Dy 1.22	67 Ho 1.23	68 Er 1.24	69 Tm 1.25	70 Yb 1.1	71 Lu 1.27

Actinide Series**

89 Ac 1.1	90 Th 1.3	91 Pa 1.5	92 U 1.38	93 Np 1.36	94 Pu 1.28	95 Am 1.13	96 Cm 1.28	97 Bk 1.3	98 Cf 1.3	99 Es 1.3	100 Fm 1.3	101 Md 1.3	102 No 1.3	103 Lr 1.3

Fig. 12.16 Periodic table of electronegativity

12.3.3 *Bonding in Infinite Crystals*

In an infinitely large crystal, the discrete energy levels of the molecules broaden into energy bands. Within each band, the lowest and highest energies correspond to the bonding and antibonding states, while intermediate energies are obtained with the electron wave functions corresponding to a mixture of these two states. Figure 12.17 illustrates this concept for a periodic 1D atomic chain made of the same atoms with each atom contributing a single valence electron. The electron wave function, in the same LCAO approximation that we used for the diatomic molecule, may be written as follows:

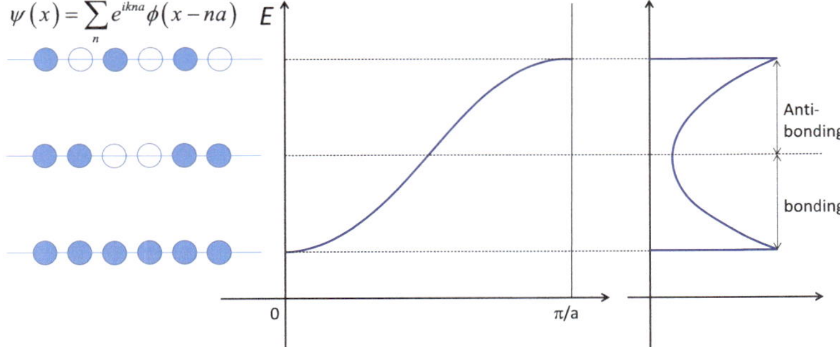

Fig. 12.17 Electron wave function, energy band, and density of states of a 1D atomic chain. The filled and open circles represent atomic wave functions of opposite signs

$$\psi(x) = \sum_n e^{ikna} \phi(x - na)$$

Here, $k = 2\pi/\lambda$. It belongs to the first Brillouin zone with $-\pi/a < k \leq \pi/a$, where a is the period of the atomic chain. The energy is obtained using the same method as the diatomic molecule assuming only the wave functions of two neighboring atoms overlap. This gives

$$E(k) = E_\mathrm{o} - 2E_{AB} \cos ka. \tag{12.9}$$

The plot of Eq. (12.9) gives the energy bands as shown in Fig. 12.17. In a 3D crystal, k is a three-dimensional vector, and it is then common to plot the energy along special high symmetry directions. The energies along these lines then give either the maximum or minimum energies for bands across the whole Brillouin zone. At a given range of energies, the number of states available per unit volume gives the density of states (DOS) $g(E)$, which is defined by

$$g(E)\mathrm{d}E = \text{number of electron states per unit volume between } E \text{ and } E + \mathrm{d}E. \tag{12.10}$$

Figure 12.17 shows the DOS for the 1D atomic chain. The DOS is filled with electrons up to the Fermi level. Above the Fermi level, the states are unoccupied. These states can be measured by electron energy loss spectroscopy (EELS) or X-ray absorption spectroscopy (XAS), while the occupied states can be probed by photoemission spectroscopy. In crystals with mixed bonding, EELS can be used to probe the local density of states and thus different types of bonds of the same atom.

12.3.4 Types of Bonds

The bonds seen in crystals in general are diverse. There are four ideal cases, corresponding to covalent, ionic, metallic, and Van der Waals bonds or molecular bonds. Another type of bond involves hydrogen, which is in its own category.

Covalent bond: Covalent bonds are formed when electrons are equally shared through the overlap of orbitals of adjacent atoms to achieve a stable outer-shell electron configuration. Since the atomic electron wave functions involved have different shapes and orientations, there is a strong directionality to the covalent bond.

Ionic bond: Ionic bonds are strong bonds formed between oppositely charged ions that attract each other by electrostatic forces. To form such bonds, atoms first loss or gain electrons and become ions. These ions are electronic stable with their completely filled outer electron shells and the gain in the electrostatic energy in a crystal. The prototypical ionic crystals are the alkali halides in which a group I atom with low electronegativity transfers an electron to a group VII atom with high electronegativity, resulting in two oppositely charged ions, each in the stable closed-shell configuration. The positive and negative charges are arranged next to each that creates an attractive force. The rock salt (NaCl) crystal structure offers the optimal, and alternating, packing of differently sized positive and negative ions.

Metallic bond: In metals, the valence electrons are highly delocalized; sharing among many atoms gives rise to metallic bonding. Thus, metallic bonds are largely nondirectional, and close-packed structures are thus mostly adopted.

Molecular bond: van der Waals bonding arises through the instantaneous induced dipole-induced dipole (dispersive) forces between atoms or molecules. The graphene layers in graphite are held together with van der Waals forces. Solids of rare gases at low temperatures are also formed with van der Waals attractive forces. The van der Waals forces are nondirectional and weak.

A common practice is to characterize a crystal by the presence of these types of bonds. Thus, graphite is covalently bonded in the graphene layer and van der Waals bonded between layers.

12.3.5   *Characteristics of Bonds*

Experimentally, chemical bonding can be studied by measuring the characteristics of the bonds, and these include the following:

1. Equilibrium distances as measured by bond lengths obtained from the crystal structure;
2. Cohesive energy as the energy that must be added to a crystal in order to separate its constituent atoms. Values in the unit of kcal per mole for elemental crystals are given in the book by Kittel. These values range from a fraction for rare gases to 7.37 kcal/mole for C;
3. Bulk modulus as defined by $B = \Omega\left(d^2E/d\Omega^2\right)$ with Ω for cell volume, which can be measured with a great accuracy and calculated by theory. The internal energy in the vicinity of the minimum is an analytical function of volume and bulk modulus according to thermodynamic considerations. A popular form is Murnaghan equation of state (Murnaghan 1944);
4. High-pressure-induced phase transformations give information about bonding and energy when atoms are forced to be close to each other;
5. Spectroscopy such as angle-resolved photoemission, EELS, or XAFS can be used to measure the occupied or unoccupied states through the excitation of electrons;
6. Direct measurement of electron density and its topology;
7. Mean crystal potential, or average value of the Coulomb potential.

Among the above, the measurement of charge density provides the most detailed and direct information on bonding. The basic idea can be illustrated using the above diatomic molecule model. If we were able to measure its charge density by diffraction, then we have in the independent electron and LCAO approximation

$$\rho = |c_+(\phi_A + \varepsilon\phi_B)|^2 = c_+^2\left(|\phi_A|^2 + \varepsilon^2|\phi_B|^2 + 2\varepsilon\mathrm{Re}\left(\phi_A\phi_B^*\right)\right)$$
$$= c_+^2\left(\rho_A + \varepsilon^2\rho_B + 2\varepsilon\mathrm{Re}\left(\phi_A\phi_B^*\right)\right) \tag{12.11}$$

Thus, the experimental charge density carries information about the atomic charge density (ρ_A and ρ_B), polarity ε, and the bonding charge ($2\varepsilon\mathrm{Re}\left(\phi_A\phi_B^*\right)$). What experimentally measured is the time-averaged real space crystal electron density $\rho(\vec{r}) = \left\langle|\psi(\vec{r}_1\ldots\vec{r}_n,t)|^2\right\rangle$, where ψ is the wave function of the many-electron system. Since both X-ray and electron diffraction use elastic scattering, the measured electron density is a ground-state property of the crystal.

In a monatomic crystal consisting of atoms on lattice sites, with unit cell volume V_c, its mean potential is given by

$$V_0 = \frac{|e|n_0}{3V_c\varepsilon_0}\left\langle r^2\right\rangle = \frac{|e|N}{3\varepsilon_0}\left\langle r^2\right\rangle \tag{12.12}$$

where n_0 is the number of atoms per cell, N is the number per unit volume, and

$$\left\langle r^2\right\rangle = \int \rho(\vec{r})r^2\mathrm{d}\vec{r} \tag{12.13}$$

is the mean-square radius of the atom. Thus, we see that V_0 has an important interpretation as a measure of the "size" of an atom (the other is a measure of diamagnetic susceptibility). It is thus the most sensitive of all the structure factors to the state of ionicity of atoms in a crystal and, because of the terms in r^2, depends strongly on the distribution of outer valence electrons. For example, a simple expression for V_0 in terms of the Gaussian expansion for atomic scattering factors is given by Eq. 4.51. For MgO, this atomic estimate (17.6 V) differs considerably from the experimental value of 13.6 V. The difference is due to the redistribution of charge in the solid state and to the sensitivity that Eq. (12.13) provides for this effect. Methods for evaluating V_0 in terms of either a multipole expansion of the crystal charge density or using an infinite sum of X-ray structure factors are described in Coppens (1997).

Values of V_0 have been measured by many workers, most commonly by using an electron biprism in an interference experiment. It may also be measured from RHEED experiments. Interferometry measurements provide values of the product $V_0 t$, so that thickness must be known accurately. This problem is avoided in the reflection geometry. In transmission work, the highest accuracy has been obtained by using cleaved crystalline wedges, for which the thickness is known accurately at each point (Gajdardziskajosifovska et al. 1993; Kruse et al. 2003). Table F.8 lists the experimental and theoretical mean potential for selected crystals. For further discussions on mean potential, see O'Keeffe and Spence (1994).

12.3.6   Charge Density as the Ground-State Property in Density Functional Theory

The ground-state electron density is also the fundamental property in the density functional theory (DFT) of Hohenberg and Kohn (1964). They proved the remarkable result that the total energy, E, of a nonspin polarized system of interacting electrons in the presence of an "external" potential $V(r)$ of nuclear charges (the atomic nuclei of a crystal, for example) is given exactly as a functional (meaning a function of a function) of the ground-state electron density $\rho(r)$:

$$E[\rho, V] = \int \mathrm{d}^3 r\, V(\vec{r}) \rho(\vec{r}) + F[\rho] \tag{12.14}$$

$E[\rho, V]$ is minimized by the ground-state electron density. Here, $F[\rho]$ is a universal, but unknown, functional of the electron density ρ only and does not depend on potential. Universal means that it takes the same functional form for all materials. It includes the electron correlation (the interaction due to correlated electron motion) and the exchange interactions for identical particles, needed to respect the anonymity of indistinguishable electrons. Using the variation principle, Kohn and Sham (1965) showed that the ground-state electron density can be obtained from a single particle-like wave function $\chi_i(\vec{r})$ and the occupation c

$$\rho(\mathbf{r}) = \sum_i c_i |\chi_i(\vec{r})|^2 \tag{12.15}$$

The wave function $\chi_i(\vec{r})$ is obtained by solving an independent electron-like Kohn–Sham equation:

$$\left(-\frac{h^2}{2m} \nabla^2 + V_N + V_e + V_{xc} \right) \chi_i(\vec{r}) = \varepsilon_i \chi_i(\vec{r}) \tag{12.16}$$

Since the potential depends on electron density, the Kohn–Sham equation for these quasiparticles must be solved self-consistently in an iterative process. Significant progress in solid-state physics has been made by replacing the unknown exchange and correlation energy or its corresponding potential using various approximations. Among these approximations, the most popular one is the local density approximation (LDA), which writes the exchange and correlation energy of the electron as an integral of the local electron density ρ times the exchange and correlation energy of a homogenous free-electron gas of density ρ. Here, we are replacing the energy contribution from some local value of charge density by that due to free-electron gas which has that constant density everywhere and which can be more simply estimated. Recent progress has been made by going beyond LDA. One approach is by introducing a gradient term to the exchange and correlation energy or potential. This has resulted in several approximations, which is generally known as the generalized gradient approximation (GGA). A comprehensive introduction to DFT and related methods is given by Martin (2008).

12.4   Experimental Measurement of Charge Density

In the following, we introduce the techniques for charge-density measurement in crystals, while the application of these techniques to the study of different types of bonds is discussed in the next section.

12.4.1   X-Ray Diffraction

X-rays are diffracted by electrons inside the crystal, not the nuclei. The structure factor, F_g, whose amplitude is measured by diffraction, is the Fourier transform of the time-averaged electron density, to a very good approximation:

$$F_g = \int_{\text{cell}} \langle \rho(\vec{r}) \rangle \exp[2\pi i \vec{g} \cdot \vec{r}] \mathrm{d}\vec{r} \tag{12.17}$$

The integration is over the unit cell volume, and $\vec{g} = h\vec{a}^* + k\vec{b}^* + l\vec{c}^*$ is the scattering vector corresponding to one of the reciprocal lattice vectors.

Conventional X-ray structure analysis approximates the crystal electron density by the superposition of spherical atoms or ions. Each atom is centered at its equilibrium position $\vec{R}$. In real crystals, atoms deviate from their equilibrium position by a small displacement $\vec{u}(t)$ because of thermal vibrations. The average crystal electron density is then given by

$$\langle \rho(\vec{r}) \rangle = \left\langle \sum_i \rho_i(\vec{r} - \vec{R}_i - \vec{u}(t)) \right\rangle \tag{12.18}$$

This is the independent atom model (IAM). In using a centered atomic electron density, it is assumed that the Born–Oppenheimer approximation holds, i.e., that electrons follow instantaneously around the nuclear positions because of the large mass difference between an electron and a nucleus. In crystals, while there is no unique way to divide the crystal electron density into atomic densities, the core electrons of each atom remain separate and the valence electrons, to a good approximation, can be expressed as a generalized nucleus-centered atomic electron density, by going beyond the spherical atom or ions approximation. Fourier transform of the averaged electron density gives

$$
\begin{aligned}
F_g &= \int_{\text{cell}} \left\langle \sum_i \rho_i(\vec{r} - \vec{R}_i - \vec{u}(t)) \right\rangle \exp[2\pi i \vec{g} \cdot \vec{r}] \mathrm{d}\vec{r} \\
&= \sum_i f_i(\vec{g}) \exp(2\pi i \vec{g} \cdot \vec{R}_i) \langle \exp(2\pi i \vec{g} \cdot \vec{u}_i(t)) \rangle \qquad (12.19) \\
&= \sum_i f_i(\vec{g}) T(\vec{g}) \exp(2\pi i \vec{g} \cdot \vec{R}_i)
\end{aligned}
$$

where $T(g)$ is the temperature factor. For harmonic oscillators, it can be shown that

$$
\begin{aligned}
T(\vec{g}) &= \langle \exp(2\pi i \vec{g} \cdot \vec{u}_i(t)) \rangle = \exp\left(-2\pi^2 \left\langle (\vec{g} \cdot \vec{u}_i)^2 \right\rangle \right) \\
&= \exp\left(-2\pi^2 \sum_{jk} \sigma_{jk} h_j h_k \right) \\
&= \exp\left(-\sum_{jk} \beta_{jk} h_j h_k \right) \qquad (12.20) \\
&= \exp\left(-2\pi^2 \sum_{jk} U_{jk} a_j^* a_k^* h_j h_k \right)
\end{aligned}
$$

where $\sigma_{jk} = \langle u_j u_k \rangle$, $\beta_{jk} = 2\pi^2 \langle u_j u_k \rangle$, and $U_{jk} = \beta_{jk}/2\pi^2 a_j^* a_k^*$ are the different expressions of the averaged, squared, thermal displacements relative to the reciprocal vector $\vec{g} = h_1 \vec{a}_1^* + h_2 \vec{a}_2^* + h_3 \vec{a}_3^*$. For isotropic vibrations, this reduces to the familiar Debye–Waller factor (Eq. 4.43). A comprehensive treatment for thermal vibrations, including anharmonic vibrations, can be found in Willis and Pryor (1975).

Thus, if both the amplitude and phase of the structure factor can be measured, then in principle crystal electron density can be synthesized by an inverse Fourier transform. In practice, electron density maps are obtained by modeling the crystal electron density using the multipole model (Sect. 12.4.4). Fourier synthesis is often used to form the deformation density or electron density difference map:

$$
\Delta\rho(\vec{r}) = \rho^{\text{exp}}(\vec{r}) - \rho^{\text{IAM}}(\vec{r}) = \sum_g \left(F_g^{\text{Exp}} - F_g^{\text{IAM}} \right) \exp(-2\pi i \vec{g} \cdot \vec{r}) \qquad (12.21)
$$

where $\rho^{\text{IAM}}(\vec{r})$ is a reference electron density calculated by superimposing independent atoms or spherical ions and F_g^{IAM} is the X-ray structure factor calculated from $\rho^{\text{IAM}}(\vec{r})$.

In the case of X-ray diffraction, the phase of the structure factor is not measured and the phase calculated from IAM is used. Because of this, Eq. (12.21) is only applicable to crystals with the center of symmetry, and the phases correspond to the signs of the structure factors. The phase can be measured very accurately using dynamical electron diffraction, which will be discussed in the following section.

The most popular technique for measuring X-ray structure factors is to use the intensity integrated over the angular profile of each Bragg reflection, assuming an ideally imperfect crystal made of mosaic blocks, each block of which diffracts coherently. Here, each slightly tilted block acts as a monochromator, picking out one component wavelength from the incident beam and sending it into a slightly different direction. The full theory of diffraction from a mosaic crystal is given in textbooks (Zachariasen 1967). To perform X-ray diffraction for electron density analysis, the integrated intensities are measured through a careful quantification of the diffraction peak profile and subtraction of the diffuse background. These intensities are then scaled and corrected for absorption, which varies with the beam path and can be monitored by measuring the direct beam intensity. For this reason, the preferred crystal shape is spherical. There are several reviews on quantification of X-ray diffraction intensities, and for details see Becker and Coppens (1974) and Zuo (2004). Other techniques that have been applied for charge-density analysis include the X-ray Pendellosung technique, which is similar to the thickness fringes in electron diffraction, and high-energy γ-ray diffraction. The former requires a large wedge-shaped perfect crystal, such as silicon. The later relies the high-energy γ-rays to reduce the extinction effects that limit the accuracy in conventional X-ray diffraction.

12.4.2   Electron Diffraction

Electrons interact with an atom through the Coulomb potential of the positive nuclei and that of all the electrons including the core electrons surrounding each nucleus. The relationship between this potential and electron density is given by Poisson's equation:

$$\nabla^2 V(\vec{r}) = -\frac{e[Z(\vec{r}) - \rho(\vec{r})]}{\varepsilon_0} \tag{12.22}$$

where $Z(r)$ is the nuclear charge density. For a 100 kV electron traveling at a speed of $\sim 1/3$ the speed of light, the time it spends inside a TEM specimen is $\sim 10^{-15}$ s, whereas the typical phonon frequency is $\sim 10^{12}$ Hz. Thus, the fast electrons see the instantaneous frozen configuration of a vibrating lattice, a snapshot. An experimentally recorded diffraction pattern at an exposure time of ~ 1 s is thus an average of the individual electron diffraction patterns from many static configurations of the crystal lattice. To a good approximation, the average experimental diffraction intensity can then be modeled by the electron interaction with an average potential

$$\langle V(\vec{r}) \rangle = \frac{|e|}{4\pi\varepsilon_0} \int d^3\vec{r}' \frac{\langle Z(\vec{r}') - \rho(\vec{r}') \rangle}{|\vec{r} - \vec{r}'|}. \tag{12.23}$$

The electron structure factor is obtained from

$$
\begin{aligned}
V_g &= \int d^3\vec{r} \exp(2\pi i\vec{g} \cdot \vec{r}) \langle V(\vec{r}) \rangle \\
&= \frac{|e|}{4\pi\varepsilon_0} \int d^3\vec{r}' \exp(2\pi i\vec{g} \cdot \vec{r}') \langle Z(\vec{r}') - \rho(\vec{r}') \rangle \int d^3\vec{r} \frac{\exp(2\pi i\vec{g} \cdot (\vec{r} - \vec{r}'))}{|\vec{r} - \vec{r}'|} \\
&= \frac{|e|}{4\pi\varepsilon_0 g^2} \left(Z_g - F_g \right)
\end{aligned}
$$

$$\tag{12.24}$$

where

$$Z_g = \sum_i Z_i T(\vec{g}) \exp(2\pi i\vec{g} \cdot \vec{r}) \tag{12.25}$$

and F_g is the average X-ray structure factor (Eq. 12.19). Equation (12.24) is used to convert the measured electron structure factor to the X-ray structure factor.

Electron structure factors are obtained from experimental CBED patterns by using the refinement method (Zuo and Spence 1991; Zuo 1999; Ogata et al. 2008; Nakashima and Muddle 2010). The refinement method works like the Rietveld method in powder X-ray or neutron diffraction, where the structure factors are treated as structural parameters, which together with other parameters are obtained by comparing experimental and theoretical intensities and optimizing for the best fit. Multiple scattering effects are taken into consideration by using dynamical theory to calculate diffraction intensities during the refinement. In this way, the failures of the kinematical approximation in electron diffraction are avoided. Further, electron interference due to coherent multiple scattering actually enhances the sensitivity of the diffracted intensities to the crystal potential and crystal thickness and thus improves the electron diffraction measurement accuracy.

The refinement is automated by defining a goodness-of-fit (GOF) parameter and using a numerical optimization routine to do the search in a computer. One of the most useful GOFs for a direct comparison between experimental and theoretical intensities is the value of χ^2 as defined by

$$\chi^2 = \frac{1}{n - p - 1} \sum_{i,j} \frac{1}{\sigma_{i,j}^2} \left(I_{i,j}^{\text{exp}} - c I_{i,j}^{\text{Model}}(a_1, a_2, \ldots, a_p) \right)^2 \tag{12.26}$$

Here, $I_{i,j}^{\mathrm{exp}}$ is the experimental intensity (in units of counts) measured from an energy-filtered CBED pattern, i and j are the pixel coordinate on the detector, n is the total number of data points, and a and p are the adjusted parameter and the number of parameters. $I_{i,j}^{\mathrm{Model}}$ is the model intensity calculated with parameters a_1 to $a_p,$ and c is the normalization coefficient. The other commonly used GOF is the R-factor:

$$R = \sum_{i,j} \left| I_{i,j}^{\mathrm{exp}} - c I_{i,j}^{\mathrm{Model}}(a_1, a_2, \ldots, a_p) \right| / \sum_{i,j} \left| I_{i,j}^{\mathrm{exp}} \right| \tag{12.27}$$

The optimum χ^2 has a value close to unity, which is obtained when the differences between theory and experiment are normally distributed and the variance σ is correctly estimated. The value of χ^2 smaller than 1 indicates an overestimation of the variance in the experimental data. The R-factor simply measures the residual difference in percentage and does not need the estimate of σ. Both criteria have been used in electron diffraction refinement techniques.

Experimental electron diffraction data are collected by recording diffraction patterns for each reflection at or near its Bragg condition. The diffraction pattern must be energy-filtered to remove the inelastic background [a thickness difference technique was developed by Nakashima that avoids this requirement (Nakashima and Muddle 2010)]. Experimental issues involved in the diffraction pattern recording include the geometric distortions of the diffraction patterns, the detector resolution and noise, the optimum sample thickness, and diffraction geometry. These are discussed in detail in the literature (Zuo 1998; Tsuda and Tanaka 1999; Friis et al. 2003; Nakashima 2005). The noise in the experimental data can be estimated using the measured detector quantum efficiency (DQE)

$$\mathrm{var}(I) = mgI/\mathrm{DQE}(I) \tag{12.28}$$

Here, I is the estimated experimental intensity, var denotes the variance, m is the area under the MTF, and g is the gain of the detector (Zuo 2000). Details of energy filters and electron detectors are given in Chaps. 9 and 10.

Three types of diffraction conditions have been used for electron structure-factor measurements. One is the systematic row diffraction condition (Fig. 12.18), in which a set of parallel lattice planes is close to the Bragg diffraction condition. The CBED pattern appears one-dimensional with relatively uniform intensity normal to the systematic direction, because diffraction is dominated by reflections belong to same set of lattice planes (a "systematic row"). The largest effect of a small change in the structure factor of a reflection on its diffraction intensity is near the Bragg condition, which can be accomplished for two reflections in a systematic row CBED pattern. Other choices of diffraction conditions include the symmetric zone-axis orientation and slightly off-zone-axis orientations (Saunders et al. 1995; Tsuda and Tanaka 1999). The advantage of these orientations is that a larger number of reflections can be refined and measured simultaneously. This is done at

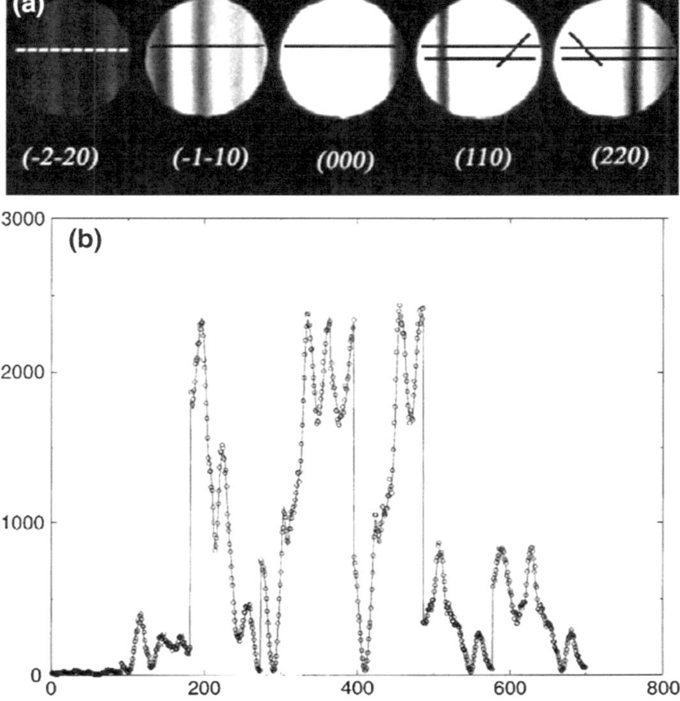

Fig. 12.18 **a** A systematic row CBED pattern of rutile (TiO$_2$), recorded at 120 kV with inelastic background removed by energy-filtering, after removing the detector PSF response by deconvolution. **b** The best fit, obtained from dynamical electron diffraction simulations and refinement, is also shown (Jiang et al 2003a, reproduced with permission)

the cost of increased complexity in comparing two-dimensional patterns and computing time because a large number of reflections contribute to diffraction in the zone-axis orientation.

To model diffraction intensities, both the detector response and the background intensity from thermal diffuse scattering must be included. A general expression for the model intensity including both factors is as follows:

$$I_{i,j}^{\text{Model}} = I^{\text{Theory}}(i,j) \otimes H'(i,j) + B(i,j)$$

Here, the theoretical intensity I^{Theory} is convoluted with the detector response function H$'$ plus the background B. The theoretical intensity is integrated over the area of a pixel. For a pixilated detector with a fixed size, the electron microscope camera length determines the resolution of the recorded diffraction patterns (see Chap. 9). At sufficiently large camera length, we can approximate H$'$ by a delta function for a deconvoluted diffraction pattern. The background intensity B, in general, is slowly varying, which can be subtracted or approximated by a constant for each reflection.

To calculate the theoretical intensities, an approximate model of the potential is needed. In the case of electron density measurements, the crystal structure (the atomic species and their coordinates, and the cell constants) is first determined very accurately. This is usually done by X-ray or neutron diffraction. The unknowns for electron refinement are the low-order structure factors (which are the most sensitive to bonding effects, as we have seen), the absorption coefficients, and the experimental parameters related to diffraction geometry and specimen thickness. The structure factors calculated from a spherical atom or ionic model can be used as a starting point. Absorption coefficients are estimated using the Einstein model with known Debye–Waller factors either from direct measurement of X-ray or neutron diffraction (Willis and Pryor 1975) or theory (see Sect. 4.7).

The actual refinement is divided into two steps. In the first step, the theoretical diffraction pattern is calculated based on the starting parameters. The pattern can be the whole or, part of, the experimental diffraction pattern. In some cases, such as a systematic row, a few line scans across the experimental diffraction pattern contain enough data points for the refinement purpose. In the second step, the calculated pattern is placed on top of the experimental pattern. The two patterns are matched by shifting, scaling, and rotating the theoretical pattern. Both steps are automated by optimization. The first step optimizes structural parameters, and the second step is for the experimental parameters. For the experimental parameters, we have the following:

(1) The zone-axis center (in practice, the tangential wave vector $\vec{K}_t$ for a specific pixel),
(2) The length and angle of the x-axis in the zone-axis coordinate used for simulations, and
(3) The specimen thickness, an intensity normalization coefficient, and the background intensity model.

The structural parameters include the following:

(1) Structure-factor amplitude and phase (in case of an acentric crystal) of the selected reflection (hkl at or near the Bragg condition).
(2) Fourier amplitude (and phase in some cases) of the absorption potential of the selected reflection for the same hkl as above.

Given a set of calculated theoretical intensities, their corresponding values in the experimental pattern can be found by adjusting some experimental parameters (e.g., orientation) without the need of dynamical calculations.

Figure 12.19 shows the χ^2 map as function of the structure factors of the (110) and (220) reflections, for a rutile (110) systematic refinement. It clearly shows that near the global minima, there is no other local minimum. This property ensures that the refinement program can find the true global minima. It is interesting to note that for the (110) reflection, the minimum point is almost independent of the (220) reflection.

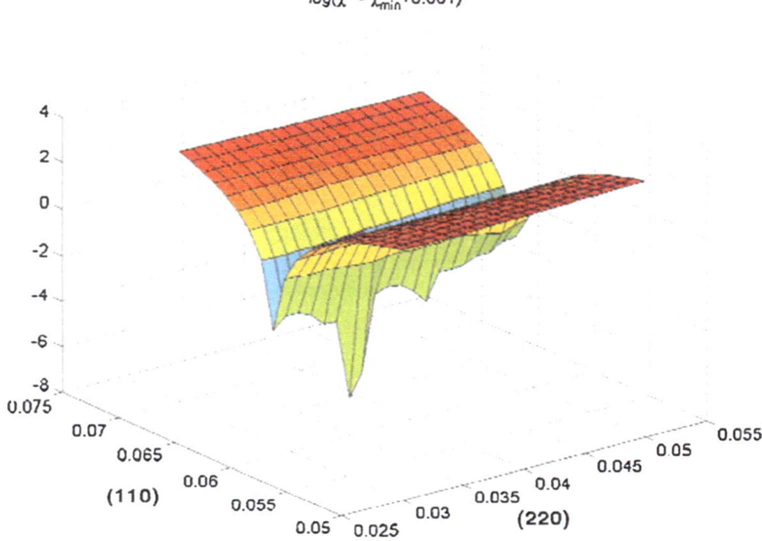

Fig. 12.19 χ^2 map plotted as function of structure factor amplitudes for the (220) and (110) reflections in Fig. 12.18

Estimates of errors in refined parameters can also be obtained by repeating the measurement. For CBED, this can be done by using different regions of the pattern or patterns recorded at different diffraction conditions and sample thicknesses. A test of electron diffraction accuracy has been reported by Saunders et al. (1995), Ren et al. (1997), and Ogata et al. (2008). They measured the low-order structure factors of silicon, which are known from X-ray Pendellösung measurements. The experimental data of Ren et al. (1997) has since then been re-refined using 279 beams, selected using a more stringent beam selection criteria. This lowered χ^2 to 1.58. The electron structure factors obtained are $U(\bar{1}11) = 0.04738(5)$ Å^{-2} and $U(\bar{2}22) = 0.00095(5)$ Å^{-2}. The X-ray Pendellösung measurements, converted to electron structure factor, give $U(\bar{1}11) = 0.04736(4)$ Å^{-2} and $U(\bar{2}22) = 0.000943(5)$ Å^{-2}. In this case, the different electron diffraction measurements and the best X-ray data agree within the experimental measurement error.

It is important to note that, additional accuracy is gained by converting the electron structure-factor to its X-ray value. The reason is the prefactor in Eq. (12.39). For small $s = \sin\theta/\lambda$, the prefactor is less than one, which demagnifies the error in the measured electron structure factor. This advantage eventually disappears as s increases.

12.4.3　Combined Electron and X-Ray Analysis

There are several reasons why it is advantageous to combine electron and X-ray diffraction measurements for experimental mapping of electron density. First, in such a combination, electron diffraction would be used to measure the structure factors of the low-order strong reflections (where it is most sensitive) and X-ray for weak and high-order reflections (where they are most sensitive). There is a gain in accuracy from converting the low-order electron structure factors to the X-ray ones, as we saw in the last section. Secondly, direct X-ray measurement of low-order strong reflections suffers from the extinction effects, which limits its accuracy (Zuo 2004). However, the effect of extinction generally is small for weak and high-order reflections. The structure factors of high-order reflections are needed to obtain accurate information on thermal vibrations and the consequent smearing of the electron density. High-order reflections also contribute to electron density in cases where localized electronic states, such as d electrons, are involved in bonding.

The practical issues involved in merging X-ray and electron data together are different sample temperatures, the temperature factors of nuclear vibrations, X-ray data scaling, and extinction correction. These mutually dependent issues are best treated during multipole model fitting, which will be addressed in the study of bonding in TiO_2 and Cu_2O.

12.4.4　Multipole Expansion of Electron Density

The difference between the true bonded crystal density and that generated by a superimposed spherical, independent atomic or ionic model electron density can be approximated by a set of atom-centered multipolar functions, which are consistent with the site symmetry of the atoms in the crystal. Using this difference, we can extract the charge density responsible for chemical bonding. In the formalism of Hansen and Coppens (1978) (also see Coppens 1997), the atomic electron density is divided into the three components, the core, the valence, and the deformed valence electrons, as:

$$\rho(\vec{r}) = \rho_c(\vec{r}) + P_v \kappa^3 \rho_v(\kappa\vec{r}) + \sum_{l=0}^{l_{max}} \kappa'^3 R_l(\kappa' r) \sum_{m=0}^{+l} P_{lm} d_{lm\pm}(\theta, \phi) \qquad (12.29)$$

Here, ρ_c and ρ_v are the spherical atomic core and valence electron densities, respectively, and P_v and P_{lm} are the electron populations for the spherical and deformed valence electron density. We will see that this decomposition will allow us to break up the bond charge into its ionic and covalent contributions. The valence electron density is allowed to expand or contract through the scaling coefficient κ. The third term is a polar expansion of the deformed valence electron density based on real spherical harmonics, $d_{lm\pm}$, which are normalized for density such that $\int |d_{lm\pm}| \sin\theta d\theta d\varphi = 2$ for $l > 0$ and $\int |d_{lm\pm}| \sin\theta d\theta d\varphi = 1$ for $l = 0$.

The function $d_{lm\pm}$ differs from the more familiar spherical harmonics $y_{lm\pm}$ only by the normalization coefficients. The expansion in Eq. (12.29) can be used to describe any arbitrary angular function, if it is taken to sufficiently high order, because the spherical harmonics are mutually orthogonal and form a complete basis set. The number of spherical harmonics can be reduced significantly for atoms with a high site symmetry. For example, atoms in the fcc metals have the cubic site symmetry and all terms up to $l = 4$ vanish, except for the monopole and a linear combination of hexadecapoles (known as Kubic harmonics, see von der Large and Bethe 1947). A general description of the index picking rules can be found in Kurkisuonio (1977) for different site symmetries.

The choice for the radial density of the deformed valence electron density, R_l, is somewhat arbitrary because there is no unique solution for decomposing a crystal electron density into atomic electron densities. In Eq. (12.29) from Hansen and Coppens (1978), the radial functions are taken to have the simple Slater functions (Clementi and Raimondi 1963):

$$R_l(r) = \frac{\alpha_l^{n_l + 3}}{(n_l + 2)!} r^{n_l} \exp(-\alpha_l r) \qquad (12.30)$$

Here, the coefficients, n_l, are selected from an examination of the products of hydrogenic orbitals, which give rise to a particular multipole. Values for the coefficient α_l may be taken from coefficients of isolated atoms that are available in the literature Clementi and Roetti (1974) or from other atomic calculations.

The aspherical atomic electron density of Eq. (12.29) leads to a generalized atomic scattering factor

$$f(\vec{s}) = f_c(s) + P_v f_v(s/\kappa) + 4\pi \sum_{l=0}^{l_{\max}} \sum_{m=0}^{+l} i^l \langle j_l \rangle P_{lm} d_{lm\pm}(\beta, \gamma) \qquad (12.31)$$

where f_c and f_v are the Fourier transforms of ρ_c and ρ_v, respectively, and j_l is the lth-order Fourier–Bessel transform of the radial function R_l

$$\langle j_l \rangle = \int j_l(2\pi s r) R_l(r) r^2 dr \qquad (12.32)$$

and β and γ are the polar coordinates in reciprocal space.

Equation (12.31) enables an extraction of experimental electron density from the measured structure factors by fitting the parameters in this multipole model. The multipole model also allows the recovery of phases up to a limit (Spackman and Byrom 1997), which are lost in X-ray diffraction intensity measurement. This is a considerable advantage of using the multipole model, compared to direct Fourier synthesis, since experimental data have missing phases and contain incomplete reflections and accuracies varying from one reflection to another. Because it imposes the atomicity constraint, the multipole model also allows the recovery of phases in crystals without inversion symmetry. The multipole approach also offers the flexibility of combining different datasets, such as measurements from electron and X-ray diffraction.

The fitting between experiment and model electron density is typically performed using the least-squares method, where the difference between experiment and model is measured and minimized:

$$\chi^2 = \frac{1}{n-p} \sum_g \frac{1}{\sigma_{F_g}^2} \left(\left| F_g^{\mathrm{Model}} \right| - \left| F_g^{\mathrm{Exp}} \right| \right)^2 \tag{12.33}$$

Here, F_g^{model} is calculated using Eq. (12.29) and $f(s)$ from Eq. (12.31). Computer programs for multipole model fitting are available in several distributed software packages for electron density analysis (Volkov et al. 2015).

12.5   Crystal Electron Density and Bonding

Electrons are the glue which holds atoms together in solids. But the redistribution of electron density which occurs when atoms come together to form a solid is an extremely small part of the total. For example, in GaAs, the chemical bonds between atoms which form to lower the total energy of the atoms in the condensed matter comprise only about 0.5 % (0.08 electrons per bond) of the number of electrons in the atom. Thus, the total energy of a crystal is extremely sensitive to the bond charge distribution. Yet this tiny effect is responsible for the form and properties of all the solid matter in our universe. Thus, contour maps of charge density for neutral atoms placed on lattice sites are almost indistinguishable from those of the fully bonded and relaxed crystal. Solid-state effects can only be seen in difference maps and may affect only low-order structure factors. In this section, we will review our understanding of bonding in crystals, treating each type in turn, and summarize recent research and methods.

12.5.1   Covalent Bonding in Diamond Structure

Among the group IV elements, C, Si, Ge, and Sn crystallize in the diamond structure with successively larger lattice constants. Each has four valence electrons, as separated atoms, and they occupy the outer-shell s and p orbitals. The atomic and electronic structure can be simply understood based on the energy diagram shown in Fig. 12.20. First, the s and p wave functions hybridize to form the tetrahedral sp^3 orbitals according to

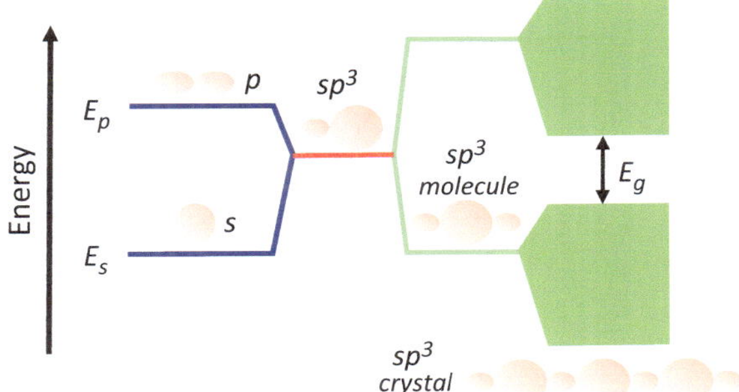

Fig. 12.20 Energy diagram of bond formation in diamond structures

$$
\begin{aligned}
\phi_1 &= \tfrac{1}{2}\left[\phi_s + \phi_{p_x} + \phi_{p_y} + \phi_{p_z}\right] \quad \text{along} \quad [111] \\
\phi_2 &= \tfrac{1}{2}\left[\phi_s + \phi_{p_x} - \phi_{p_y} - \phi_{p_z}\right] \quad\qquad\ [1\bar{1}\bar{1}] \\
\phi_3 &= \tfrac{1}{2}\left[\phi_s - \phi_{p_x} + \phi_{p_y} - \phi_{p_z}\right] \quad\qquad\ [\bar{1}1\bar{1}] \\
\phi_4 &= \tfrac{1}{2}\left[\phi_s - \phi_{p_x} - \phi_{p_y} + \phi_{p_z}\right] \quad\qquad\ [\bar{1}\bar{1}1]
\end{aligned}
\tag{12.34}
$$

The overlap of sp^3 orbitals on two adjacent atoms then splits the energy into two levels: one for a bonding and one for an antibonding state. For the bonding state, there are two electrons in each bond with opposite spins, and there are four bonds for each atom. The bonding and antibonding states broaden into bands in an infinite solid. The energy broadening is less than the energy difference between the bonding and antibonding states and leaves a gap, which makes them semiconductors.

The qualitative picture emerging from the above LCAO model is that of positive ion cores at the atomic positions of the diamond structure, they are connected by a net of covalent bonds formed a pair of electrons of opposite spins. The net charge on each atom is zero since the ion cores are surrounded by the shared covalent electrons for each atom.

The structure factors of Si have been measured using the highly accurate X-ray Pendellösung technique, which requires wedge-shaped perfect crystals that are available from silicon (Aldred and Hart 1973). Experimental structure factors up to the (880) reflection are available through 5 independent measurements by Aldred and Hart (1973), Teworte and Bonse (1984), and Saka and Kato (1986). Additional measurements for specific reflections have also been made with electron diffraction (Saunders et al. 1995; Ren et al. 1997). The consolidation of X-ray datasets by Cummings and Hart (1988) shows an averaged accuracy of ~ 3–5 me in the measured structure factors. A list of structure factors can be found in Zuo (2004).

Figure 12.21 shows the synthesized experimental deformation density using the measured experimental structure factors, and Eq. (12.21) in the (110) plane cutting through Si atoms and the high-resolution theoretical deformation maps calculated using the WIEN program (Blaha et al. 2001). The experimental deformation density uses the atomic density calculated by the Multiconfiguration Dirac-Fock (MCDF)

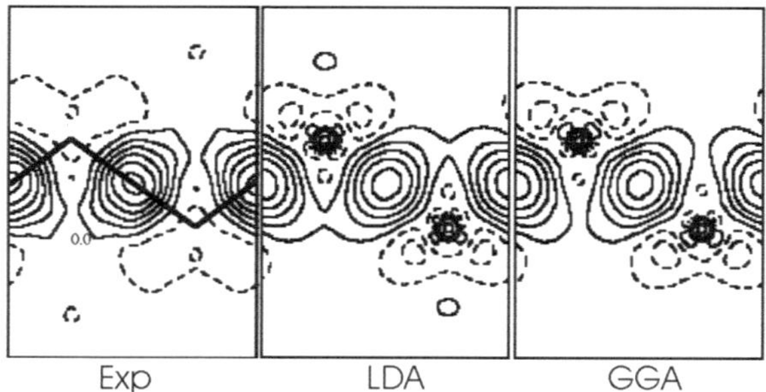

Fig. 12.21 Silicon electron density difference maps of the (110) plane which cuts through silicon atom, for experimental and theoretical electron densities. Two theoretical maps obtained by DFT are shown, one calculated with the local density approximation (LDA) and another using the generalized gradient approximation (GGA)

approximation as a reference (Rez et al. 1994, 1997), while the atomic densities calculated in the Dirac-Slater approximation using LDA or GGA for exchange and correlation potential and are used for the corresponding LAPW deformation maps, respectively. Both the LDA and the GGA maps reproduce the bonding electron density peak in the silicon–silicon bond with the peak value of 0.189 and 0.194 e/Å^3, respectively, compared to the experimental value of 0.213 $\pm$ 0.003 e/Å^3.

The theoretical difference density shows a modulation of core electron density that is absent in the experimental map. The question is, then, whether this feature is also present in the experimental data. To address this, multipole treatments of experimental structure factors are needed. Following the previous discussed multipole model, the structure factors of silicon are expanded in terms of Kubic harmonics that obey the cubic site symmetry:

$$F(h,k,l) \approx 8\cos\phi\{f_c + \delta_{c,4} - f_{a,3}\tan\phi\}\exp\left(-Bs^2\right) \tag{12.35}$$

The expansion includes the nonspherical terms in electron density up to the fourth order. Here, the phase $\phi = (h+k+l)\pi/4$. The f_c, $\delta_{c,4}$, and $f_{a,3}$ are the spherical, fourth-order and third-order Kubic harmonic part of the generalized "atomic" scattering factors of silicon, respectively. Only harmonic isotropic thermal vibration is considered. The spherical term f_c is taken as the scaled form of an atomic scattering factor:

$$f_c(\mathbf{s}) = \sum_{nl} f_{nl}(\mathbf{s}/k_{nl}) \tag{12.36}$$

where nl designates the shell and κ_{nl} is the scaling constant for each shell, which is usually taken as unit, except for the valence shell. The nonspherical third- and fourth-order terms are expressed by:

Table 12.1 Parameters obtained from the multiple model fitting of experimental and theoretical structure factors. $B = 0.4668$ Å^2

	k (L shell)	k (M shell)	O	H	α	R-factor (%)
Experiment	0.9998(5)	0.971(8)	0.37(2)	–0.14(2)	4.76 (10)	0.146
LDA	1.0013	0.970	0.340	–0.0794	4.70	0.093
GGA (PW91)	1.0013	0.967	0.355	–0.0841	4.67	0.094

$$f_{a,3} = O \frac{8\pi\alpha^7}{6!} \frac{hkl}{(h^2 + k^2 + l^2)^{3/2}} \int_0^\infty r^4 \exp(-\alpha r) j_3(4\pi s r)\mathrm{d}r$$

$$\delta_{c,4} = H \frac{8\pi\alpha^7}{6!} \frac{640}{27\sqrt{3}} \left[\frac{h^2 + k^2 + l^2}{(h^2 + k^2 + l^2)^2} - 3/5 \right] \int_0^\infty r^4 \exp(-\alpha r) j_4(4\pi s r)\mathrm{d}r$$

$$(12.37)$$

The scaling constant k_{nl}, the exponential component α, the occupation number O, and H are the fitting parameters in this multipole model.

Table 12.1 summarizes the multipole fitting results from Zuo et al. (1997). The overall agreement between the fitted structure factors and the original set is remarkably good, at the same level as the best agreement between first-principle theory and experiment. The nonspherical terms are also very similar for both experiment and theory. The M shell scaling differs significantly from 1.0. The scaling $k = 0.967$ of M shell in Table 12.1 was obtained with the 31 structure factors (Zuo et al. 1997). By comparison, the scaling obtained directly from the theoretical electron density using the GGA gives $k = 0.9992$. Thus, in the case of Si, the electron density deformation in the covalent bonds is robust and well reproduced by both the multipole model and the theory. Subtle effects such as contraction and expansion of Si core shells are still difficult to extract from experimental structure factors using the multipole model.

The charge densities of diamond and Ge follow a similar trend to Si (Lu et al. 1995). The excess covalent charge in the middle of the bond is significantly larger for diamond with $\Delta\rho_{\max} = 0.45$ e/Å^3 and a smaller value for Ge with $\Delta\rho_{\max} \sim 0.14$ e/Å^3, compared to that of Si at 0.213 ± 0.003 e/Å^3. DFT calculations give $\Delta\rho_{\max} = 0.099$ e/Å^3. The decrease in $\Delta\rho_{\max}$ comes with an increase in the metallicity (small band gap) of the bond (Harrison 1983) and increase in the bond length.

Tin crystallizes in one of two allotropic forms. At ambient pressure, the stable phase at low temperatures is α-Sn (gray tin), which has the diamond structure and is a zero-gap semiconductor (Fig. 12.22). Above $T_c = 13.2$ °C, the crystal transforms into the β phase (the metallic form or white tin), which is a body-centered tetragonal with a unit cell of $a = 5.8327$ Å and $c = 3.1825$ Å and four atoms per unit cell. Between the two structures, the bond length increases from 2.81 Å in the α phase to 3.02 Å in the β phase. The transition was known in medieval Europe as "tin pest" that turned white, shiny, church organ pipes into gray dust. What drives the phase transition is the entropy due to a difference in the vibrational properties of the two

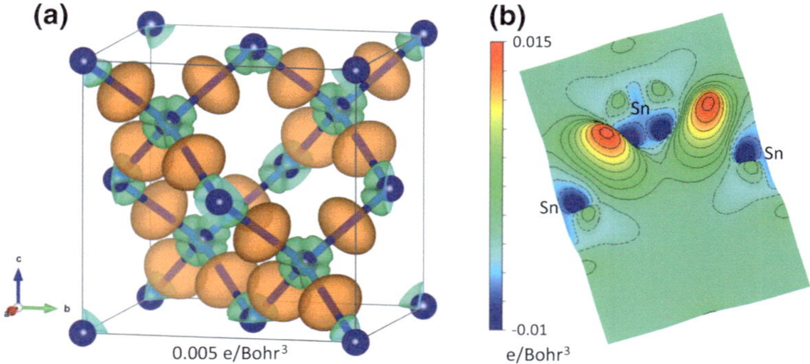

Fig. 12.22 Difference charge density of α-Sn obtained by DFT theory. **a** A 3D plot showing the isosurfaces at positive and negative $\Delta\rho_{max} = 0.005$ e/Bohr³. **b** The [110] section of the difference charge density map plotted as 2D surface with contours from -0.01 to 0.015 e/Bohr³ and increment of 0.002 e/Bohr³ (Provided by Jihwan Kwon, University of Illinois)

Fig. 12.23 Difference charge density of β-Sn obtained by DFT theory showing the isosurfaces at positive and negative $\Delta\rho_{max} = 0.005$ e/Bohr³ (Provided by Jihwan Kwon, University of Illinois)

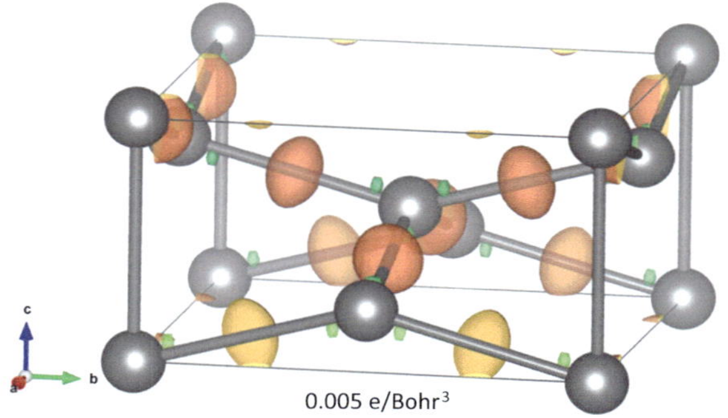

phases (Musgrave 1963; Pavone et al. 1998). Transformation to β tin structure also occurs in Si and Ge under high pressure. Compared to the Si difference charge density, the tetrahedral bonds of α-Sn is accompanied by noticeable polarization of Sn core electrons, with more electrons transferred along the bond (thus more positive) than between the bonds. This feature can be seen in the plot of isosurfaces and the surface plot of Fig. 12.22. In β-Sn, the covalent bond charge is diminished as evidenced in Fig. 12.23 with $\Delta\rho_{max} = 0.05$ e/Å³, which is half of that of α-Sn. Polarization around the Sn atoms is also much reduced. These features are consistent with the metallicity of β-Sn.

Thus, covalent bonding as seen in the diamond structures is characterized by excess bond charges in the middle of a bond accompanied by charge depletion near the atomic core and a modification of the core electron density. The amount of bond charge decreases with increasing bond length and correlates with the decrease in the band gap in these materials. The transition from the zero-gap α-Sn to metallic β-Sn is accompanied by a large reduction in the bond charge and polarization of Sn atoms. Together, these trends show the bond charge to be a major metric of covalent bonding that has been measured quantitatively by diffraction methods for C and Si and obtained by theory for Ge and Sn.

12.5.2 Ionic Bonding

12.5.2.1 The Case of MgO

Magnesium oxide is a prototype ionic crystal with the NaCl structure (unit cell $a = 4.2112$ Å). The electronic configurations of the neutral atoms are as follows: Mg: $1s^2 2s^2 2p^6 3s^2$ and O: $1s^2 2s^2 2p^4$. Ionic bonding is formed in MgO by transferring the two outer-shell ($3s$) electrons of Mg to O $2p$ to attain closed-shell configuration for both ions. The O^{2-} ion is stable only in the lattice potential of neighboring positive Mg^{2+} ions. This affects a few low-order structure factors because of the large radius of the O^{2-} ion, which have been accurately measured by CBED (Zuo et al. 1997).

The structure factors of MgO with its fcc lattice in an ionic model with Mg^{++} at (000) and O^{2-} at (1/2, 1/2, 1/2) is given by

$$F_{hkl} = 4\left[f^{Mg^{2+}}(s) + (-1)^{h+k+l}\left(f^O(s) + \Delta f(s) \right) \right],$$

where $\Delta f(s)$ is the scattering factor of the two transferred electrons (including its effects on the charge density of the neutral oxygen atoms). A particularly useful scheme for distinguishing different bonding models in ionic crystals is then to use the scattering factor of the transferred electrons Δf, which is given by

$$\Delta f(s) = (-1)^{h+k+l}\left[F_{hkl} - 4 f^{Mg^{2+}}(s) - (-1)^{h+k+l} 4 f^O(s) \right]/4. \qquad (12.38)$$

Figure 12.24b plots Δf for the experimental and theoretical structure factors obtained using the LAPW method and $Mg^{++}O^{2-}$, and obtained using the Dirac-Fock method and a Watson well (a spherical well of positive potential) of 1.2 Å radius (Zuo et al. 1997). The largest contribution to Δf comes from (111) and (200). It is clear that overall both the experimental and theoretical Δf resemble that of $Mg^{++}O^{2-}$.

Figure 12.24 shows the experimental map of the difference between the crystal electron density and that of superimposed neutral atoms on the (100) plane. The experimental map was constructed using parameters from a multipole fitting of the low-order electron structure factors together with high-order X-ray structure factors of Lawrence (1973). Both experiment and theory clearly indicate charge transfer from Mg to O. The amount of charge transfer depends on the partitioning model (Redinger and Schwarz 1981). The theoretical electron density was directly compared with models, and this favored a description with a charge transfer of two electrons per Mg atom (Redinger and Schwarz 1981; Mehl et al. 1988).

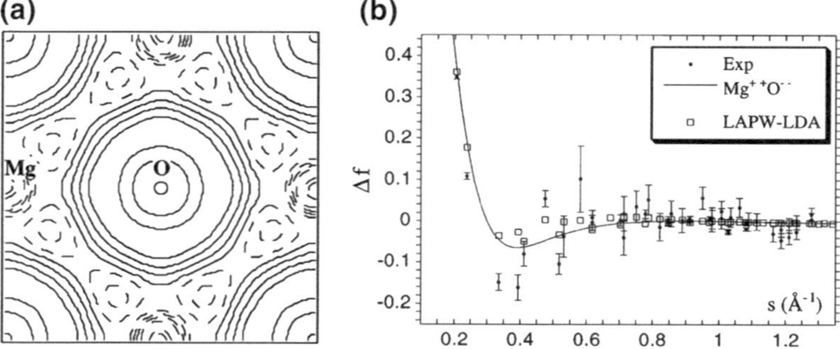

Fig. 12.24 a The (001) plane electron density deformation map between crystal and superposition of neutral atoms (Contour interval is 0.03 e/Å^3, full and *dash lines* for positive and negative difference charge, respectively). **b** Scattering factors (Δf) of the two Mg 3s electrons at the oxygen site obtained from experiment, the LAPW using LDA, and the Watson sphere model (Zuo et al. 1997)

The multipole model for the electron density of MgO is based on the charge transfer model:

$$F(h,k,l) = 4\left(f'_{Mg^{2+}} + (2-q)f_{3s} + \delta f^1\right)\exp\left(-B_{Mg}s^2\right)$$
$$+ 4(-1)^{h+k+l}\left(f'_O + qf_{2p} + \delta f^2\right)\exp\left(-B_O s^2\right) \tag{12.39}$$

Here, the scaled spherical Mg^{2+} and O electron densities are used as a reference, and f is the atomic scattering factor. The scaling is carried out using $f' = \sum_n f_n(s/\kappa_n)$, with scaling constant κ_n. For the deeply bounded 1s electrons, the influence of crystal fields is generally small, so that $\kappa_n = 1$. Ionic bonding is described by introducing the charge transfer q. Here, f$_{3s}$ is the scattering factor of the Mg 3s electrons, and f_{2p} is the difference scattering factor between O^{2-} and O. To a good approximation, f_{2p} is well described by the Fourier transform of the function $r^{n-2}e^{-\alpha r}$. Here, n and α are determined by fitting. For the nonspherical charge distortion, only the $l = 4$ Kubic harmonic term (the lowest *non*zero term) was included. The associated electron density in real space is as follows:

$$\rho(\mathbf{r}) = 13.68534 H N_h r^n \exp(-\beta r)\left[(x^4 + y^4 + z^4)/r^4 - 3/5\right] \tag{12.40}$$

with $N_h = \beta^{n+3}/(n+2)!$, and H and β are also determined by fitting. The experimental electron density is well described by the spherical-ion model, with scaling of the 2p electrons only in Mg and O. There is a small hexadecapole modulation. (The introduction of nonspherical terms in the model improves the R-factor by 0.03 % for the fit to experimental structure factors). The experimental electron density also agrees well with the full ionic model and a charge transfer of 2 electrons, and this is further supported by theory (Zuo et al. 1997).

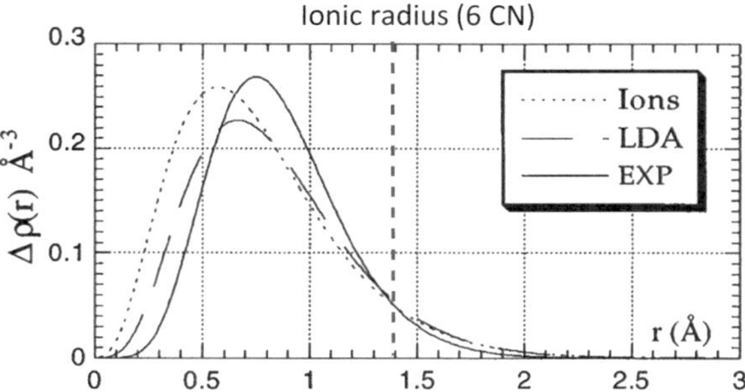

Fig. 12.25 Spherical averaged charge-density difference between O^{2-} and O for experiment, the LAPW (LDA), and the Watson sphere model (ions). The *dashed line* indicates the Shannon ionic radius for O^{2-} of 6 CN at 1.42 Å (Shannon 1976)

Figure 12.25 plots the spherically averaged charge-density difference between O^{2-} and O in MgO as found from the multipole model refinements for both experimental and LAPW-LDA structure factors. For comparison, the Watson model of O^{2-} is also plotted. As shown in Fig. 12.3, the experimental charge is significantly moved outward, indicating a more diffuse O^{2-} ion in the real crystal. The electron density of also extends beyond the classical definition of ionic radius, which is taken between 1.40 and 1.42 Å for O^{2-} of 8 coordination number.

12.5.2.2 Cohesive Energy of Ionic Crystals

The cohesive energy of a crystal is defined by the energy difference between bonded and unbounded atoms. In DFT, the total energy of Eq. (12.14) separates into four parts: the kinetic, Coulomb, exchange, and correlation energies. Among these, the Coulomb energy can be calculated directly from the experimental charge density according to

$$E_{\text{Coul}}[\rho] = \frac{e^2}{8\pi\varepsilon_o}\left[\sum_{i\neq j}\frac{Z_iZ_j}{|\vec{r}_i-\vec{r}_j|} - \sum_i Z_i\int \mathrm{d}^3r\,\frac{\rho(\vec{r})}{|\vec{r}-\vec{r}_i|} + \int \mathrm{d}^3r\mathrm{d}^3r'\,\frac{\rho(\vec{r})\rho(\vec{r}')}{|\vec{r}-\vec{r}'|}\right]$$

$$(12.41)$$

Here, the first and last terms describe the Coulomb repulsion between nuclear charges and the electrons, respectively, while the second term gives the attractive energy between positively charged nuclei and negatively charged electrons. General methods for evaluating Eq. (12.41) using the experimental structure factors are described in the book by Coppens (1997).

In the development of solid-state theory, early studies have shown that in simple ionic crystals, such as MgO, their cohesive energy can be accounted for mostly by a simple model which treats the approximately spherical ions as point charges with their electrostatic interactions. Taking the NaCl structure type as an example, the energy of a positive ion can be expressed in terms of its nearest-neighbor distance r according to

Table 12.2 Madelung constants A_m and nearest-neighbor coordination NN for selected ionic crystal structure types

Structure	$A_{m\,\mathrm{Mad}}$	NN
Cesium chloride	1.76	8
Sodium chloride	1.75	6
Wurtzite	1.64	4
Zincblende	1.64	4

$$\frac{1}{2}\sum_{n,m,l}\frac{q^2(-1)^{n+m+l}}{4\pi\varepsilon_0\left(n^2+m^2+l^2\right)^{1/2}r}, \tag{12.42}$$

where we have chosen the position of the positive ion as the origin and (n, m, l) as the index of the simple cubic lattice occupied alternatively by negative ions at sites ($n + m + l$ = odd numbers) and positive ions ($n + m + l$ = even numbers). The same energy is obtained for the negative ions by simply changing the sign of q. Together, they contribute to the attractive part of the cohesive energy

$$E_a = \frac{1}{2}\sum_{n,m,l}\frac{q^2(-1)^{n+m+l}}{4\pi\varepsilon_0\left(n^2+m^2+l^2\right)^{1/2}r} = -\frac{A_m q^2}{4\pi\varepsilon_0 r}, \tag{12.43}$$

where A_m comes from the lattice sum, is negative, and is completely determined by the crystal structure. It is known as the Madelung constant. The convergence of the sum is slow, thus making it difficult to calculate for a general crystal structure. Special methods have been developed, and for further details see Chap. 9 of Coppens (1997). Table 12.2 lists the Madelung constant for the common ionic crystal structure types.

The repulsive part of the cohesive energy coming from the forces that we have already discussed has the form as follows:

$$E_r = \frac{C}{r^n}. \tag{12.44}$$

Putting Eqs. (12.43) and (12.44) together, we have per molecule (M),

$$\frac{E}{M} = E_a + E_r = -\frac{A_m q^2}{4\pi\varepsilon_0 r} + \frac{C}{r^n}, \tag{12.45}$$

And its minimum gives the Madelung cohesive energy in the form

$$\frac{E_{\mathrm{Mad}}}{M} = \frac{A_m q^2}{4\pi\varepsilon_0 r}\left(1-\frac{1}{n}\right). \tag{12.46}$$

The cohesive energy depends on n: The larger the n, the steeper the repulsive force and the smaller its contribution to the cohesive energy. The value of n can be determined from the experimental modulus according to

Table 12.3 Experimental *NN* distance, Madelung electrostatic energy E^{Mad}, theoretical cohesive energy E^{coh} (theory), and experimental cohesive energy per molecule for a number of alkali halides of the NaCl structure type

Compound	c_o (exp) (Å)	$-E^{Mad}$ (eV)	E^{coh} (theory) (eV)	E^{coh} (exp) (eV)
LiF	2.01	11.8	10.83	11.45
LiCl	2.57	9.65	8.85	8.98
LiBr	2.75	9.28	8.51	8.39
LiI	3.01	8.64	7.92	7.66
NaF	2.32	10.49	9.62	9.96
NaCl	2.82	8.32	8.18	8.18
NaBr	2.99	8.52	7.81	7.72
NaI	3.24	7.39	7.32	7.13

From Scheffler et al. (2012)

$$B = \frac{(n-1)A_m q^2}{4\pi\varepsilon_o \times 18 r_o^2}.\tag{12.47}$$

Table 12.3 lists the Madelung energy obtained from the simple point charge model for a number of alkali halides and compares this with values obtained by theory and experiment. A large part of the cohesive energy is accounted for by the Madelung energy. The remaining 10–20 % of the cohesive energy can be attributed to the overlapping charge density and their nonspherical terms (Coppens 1998) plus a small contribution from the exchange and correlation energies. Thus, the point charge model captures the essential physics in these crystals. Both experimental studies of charge densities, as the example of MgO shows, and consideration of binding energies point to a small overlap of the ionic charge together with deviations from the charge density of the spherical ions. Thus, to a very good approximation, these systems can be simply considered as comprised of spherical ions.

The simple nature of chemical bonds in some of ionic crystals explains some structural trends exhibited by these systems. One of these trends is Pauling's so-called radius ratio rules that relate the relative size of the anion and cation in a crystal to the preferred structure types (ionic radii are obtained from atomic distances determined by experiment with, e.g., X-ray diffraction). Specifically, these rules give the intervals within which various structures are likely to occur:

$$
\begin{aligned}
1 &> \frac{R_+}{R_-} > 0.73 \qquad &\text{(CsCl structure)} \\[4pt]
0.73 &> \frac{R_+}{R_-} > 0.41 \qquad &\text{(NaCl structure)}. \\[4pt]
0.41 &> \frac{R_+}{R_-} > 0.23 \qquad &\text{(ZnS structure)}
\end{aligned}
\tag{12.48}
$$

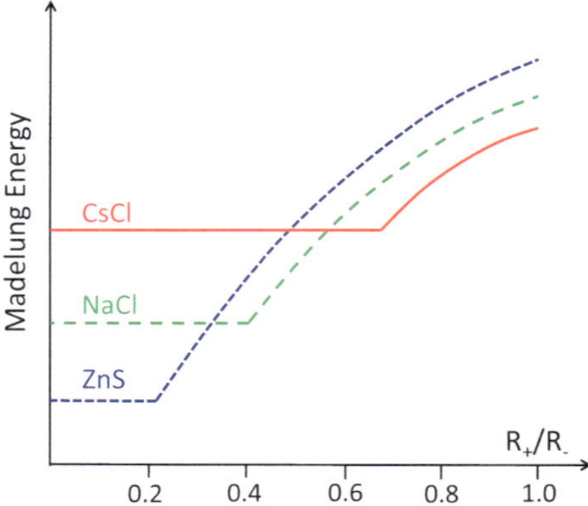

Fig. 12.26 The Madelung energy in ionic compounds as a function of the radius ratio for CsCl, NaCl, and cubic ZnS lattices (assuming the anion radius is held constant) (after Pettifor 1995)

The above trends are based on the most efficient ways of packing opposite charged spheres of different sizes in common binary crystal structures. A partial explanation of these trends is provided by the dependence of E_{Mad} as a function of the anion–cation ratio R_+/R_-, which is plotted in Fig. 12.26. For example, considering the CsCl to NaCl transition in Fig. 12.26, the nearest-neighbor distance in the CsCl structure is determined solely by the second nearest-neighbor anion–anion distance. As the size of cations reduce, $R_+/R_- = \sqrt{3}/2 = 0.73$, adjacent anions come into contact and no additional energy can be gained by shrinking the cation further. Thus, the Madelung energy remains constant when $R_+/R_- < 0.73$ and the cations simply "rattle" in the interstitial sites of anions in the CsCl structure. A gain is the Madelung energy can be made by adopting the NaCl structure as it allows a shorter cation–anion distance at the next highest coordination number (6). Further examples of these simple rules can be found in the book by Pettifor (1995).

12.5.2.3   Polarized Oxygen Anion in Corundum, α-Al$_2$O$_3$

Corundum, α-Al$_2$O$_3$, has a rhombohedral unit cell with the space group $R\bar{3}c$. The crystal structure can be described as approximately hexagonal close packing of oxygen atoms with Al occupying two-thirds of the octahedral sites as required by the cation and anion ratio (Fig. 12.27). Each Al is surrounded by six neighboring oxygen atoms, with two different bond lengths at 1.8551(2) and 1.9716(3) Å. The short and long bond lengths are formed by repulsion of two Al ions, which are brought close to each other in the two face-shared octahedra as marked by "]" in Fig. 12.27. This leads to the puckered layer of Al ions and also slight distortion of the layer of oxygen ions. Corundum is a high-temperature structural ceramic with excellent mechanical strength and corrosion resistance. The electron density of α-Al$_2$O$_3$ has been the focus of several electron density studies because both Al and O are light atoms and the corundum structure has a medium-sized unit cell featuring a number of low-order structure factors of both strong and weak reflections that are

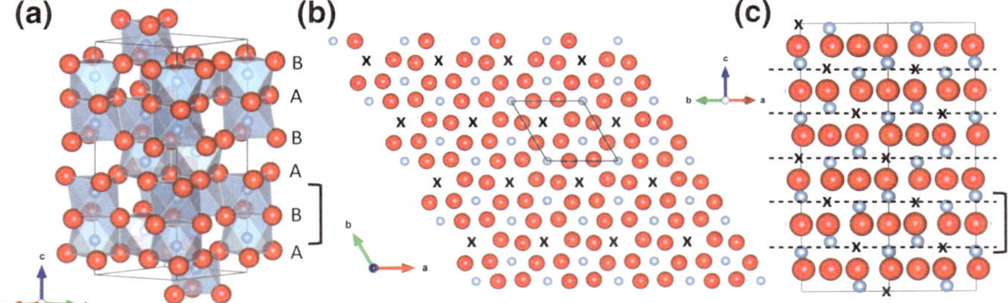

Fig. 12.27 a Structure of a-Al$_2$O$_3$ (corundum) showing face-shared oxygen octahedra and occupation by Al atoms. The oxygen atoms are hexagonally closed packed in ABAB... stacking. **b** A single layer of AB stacked oxygen atoms showing the Al atoms occupying 2/3 of octahedral sites. X marks the vacant sites. **c** Projection along [110] showing arrangements of Al atoms along *c*-axis. *Dashed lines* indicate the middle positions of the octahedral sites

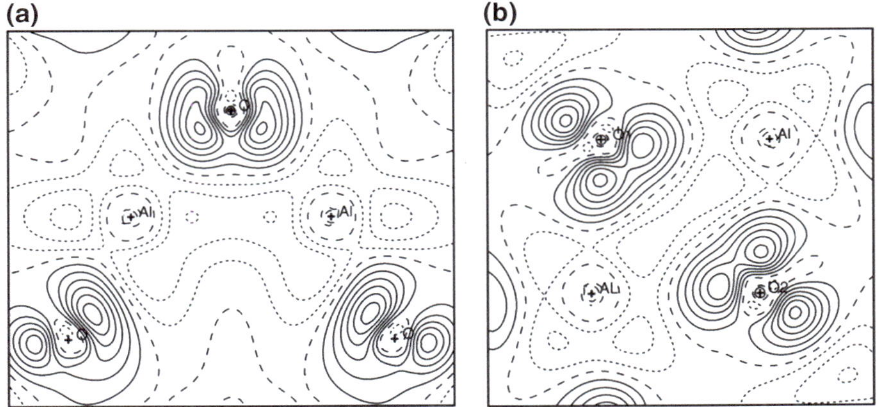

Fig. 12.28 Difference density maps obtained from the multipole refinement of a combined synchrotron X-ray and CBED dataset for α-Al$_2$O$_3$ in **a** the (010) plane and **b** the plane through the Al–O short and long bonds. The maps are plotted at contour interval of 0.05 e/Å^3 and the negative deformation density is shown as short *dashed lines*. Map borders are 6 × 5 and 4.5 × 4.5 Å, respectively (From Streltsov et al. 2003)

sensitive to bonding. For these reasons, α-Al$_2$O$_3$ is a good system for testing the limits of experimental techniques and multipole refinement (Pillet et al. 2001). From the chemical point of view, the difference in bond lengths and the low site symmetry of oxygen make α-Al$_2$O$_3$ an interesting system in which to investigate metal–oxygen bonding.

Figure 12.28 shows the difference charge-density map obtained by Streltsov, Nakashima et al. (2003) from a combined dataset of structure factors obtained by synchrotron X-ray diffraction and CBED. These maps show the transfer of electrons from Al atoms to O atoms with negative Δρ at the Al sites and positive Δρ at the oxygen sites. Compared to the difference density of MgO in Fig. 12.24, significant

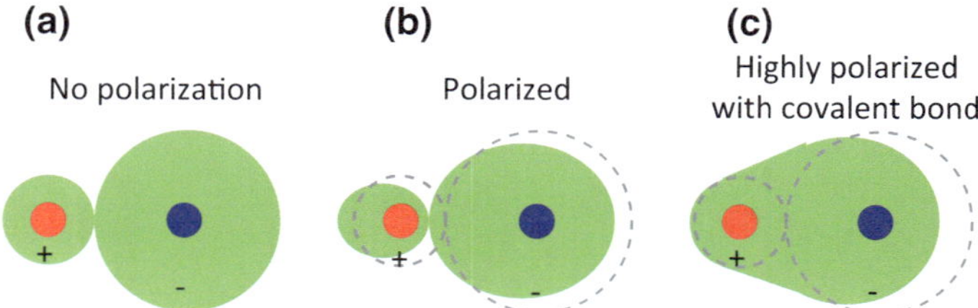

Fig. 12.29 Schematic comparison between ions with no polarization, polarization, and high polarization for the formation of covalent bond

polarization of the O atom electron density is seen toward the Al atoms along the Al–O bonds. Higher excess $\Delta\rho$ density peaks at 0.3 e/Å^3 along the shortest Al–O bond directions and a lower density at 0.25 e/Å^3 along the longer Al–O bond directions.

The distortion of the electron density of the oxygen anion and the higher electron density in the bond than in the corresponding ionic bond all show that the Al cations have high polarizing power than Mg cations in MgO. In these two cases, the polarizability of the oxygen anions is the same, and the differences between these two systems are as follows:

1. The ionic distance of Al$_2$O$_3$ (shortest $r = 1.85$ Å) is much shorter than that of MgO ($r = 2.11$ Å). This, together with the higher valence of Al (3 compared to 2 for Mg), contributes to a higher polarizing power for Al cations;
2. The amount of covalency in the bonds of Al$_2$O$_3$ is higher than for MgO based on the difference in the electronegativity between Al and O (1.83) and Mg and O (2.13);
3. The Al and Mg share the same core electron configuration with no d electrons.

In general, bonds between polarizing cations and polarizable anions are expected to have significant covalent character. Figure 12.29 illustrates the effect of polarizing cation on the ionic bond between a pair of atoms. To have a significant covalent character in a bond, the cation must be polarizing and the anion must be polarizable. The polarizability of the anion depends both on its size and on charge. Anions of lower charge density are more polarizable. For anions having the same charge, the larger anion is polarized to a greater extent. For anions of similar size, the more polarizable anion has a greater negative charge. According to these considerations, the strong polarizing power of the Al cations is the major factor here for the significant covalent character observed in the experimental charge density.

12.5.3   Metallic Bonding

A simple model of metallic bonding is that of a homogeneous electron gas embedded with positive ions. According to this model:

1. A homogeneous electron gas lacks any directionality. This is reflected in the mechanical properties of metals, which are elastic and ductile.
2. A large overlap of valence electron wave functions leads to a large energy dispersion and the disappearance of energy gaps in the DOS.
3. The Fermi level is determined by the filling of electrons into unoccupied states immediately above the highest occupied valence state. Because of this, metals are electrical conductors.
4. As the electrons are free to move in a metal, the highly delocalized valence electrons are shared by many atoms.
5. The Coulomb interaction between positive ions are largely canceled out by screening due to the free electrons. Contributions to the cohesive energy arise from delocalized electrons shared among many atoms that cannot be described by a sum of pair potentials as in the case of ionic crystals.

In metals that are a good approximation to a free-electron gas, the change in the total electron density due to bonding is subtle, making it difficult to determine. On the other hand, metals such as Al alone account for more than 40 % of world production in nonferrous metals. An important problem in the study of metals is the effect of inter-atomic bonding on the mechanical properties of metals. Eberhart (1996) has demonstrated a strong correlation between the topology of the total electron density and the anisotropy of elastic constants in metals. Specifically, bond directionality is associated with the intrinsic elastic behavior of metals and alloys. For example, single crystals of iridium, which is a high melting point fcc metal, fail by brittle cleavage at room temperature (Hecker et al. 1978). It has been suggested that cleavage in iridium is intrinsic, resulting from very strong and directed atomic binding forces. In case of Al, its charge density has been studied experimentally and theoretically by Nakashima et al. (2011). Their work shows that the redistribution of total electron density, as measured by the difference density method, ranges from -0.74 to 0.05 e/Å^3, with the negative peaks centered on the Al atoms and the positive peaks at the tetrahedral interstitial sites of the fcc lattice, with almost no bonding electron density in the octahedral interstice. The directional bonding is directly correlated to the anisotropic elastic constants of Al. The amount of bond charge is about ¼ of that in covalently bonded Si to the right of Al in the periodic table, and about the same as the metal β-Sn.

Transition metals in three rows of the periodic table are characterized by progressive filling of 3d, 4d, and 5d states. A subset is the noble metals that are resistant to corrosion and oxidation, comprising metals such as palladium, silver, platinum, and gold. The relative diffuse s and p electrons give rise to the free-electron-like sp band, while the d electrons are more localized leading to the narrow d band. As we move from the early to late transition metals, the arising and falling contributions from the d band to the DOS near the Fermi level have a large impact on the varying properties of the transition metal series. In the "rigid band model," where DOS stays the same, the number of d electrons increases from 1 (Sc for example) to 10 (Cu), and electron filling shifts the Fermi level further to the right. At the end of the transition metal series, with the d-states completely filled, the Fermi level lies in the s-like DOS above the d band.

Copper is an example of the end transition metal series. It has a full d shell electronic configuration, plus one 4s electron. The calculated cohesive energy using the 4s electron only is too low when compared to the experimental value (Kambe 1955). It is believed that the extra cohesion observed experimentally (evidenced by the high melting point) must be a result of hybridization of d electrons with those of 4s and 4p bands (Barrett and Massalski 1966).

Low-order structure factors for copper have been selectively measured by several groups (for a review and comparison see Friis et al. 2003). A systematic study was carried out by Jiang et al. (2004). These are summarized in Table 1 of Zuo (2004). Multipole refinement using the CBED data and γ-ray diffraction results (Schneider et al. 1981; Friis et al. 2003) for high-order reflections was carried out by Jiang et al. (2004).

The resulting electron density difference map (Fig. 12.30) shows a spherical electron deficiency region (0.9 Å in radius) around the copper atom and an electron surplus region between atoms. The electron surplus in the interstitial region is about 0.05 e/Å^3 or a 25 % increase in the valence electron density. The hexadecapole populations were found to be very small, and electron redistribution due to non-spherical deformation is less than 10^{-6} e/Å^3 between nearest-neighbor atoms. Thus, there is no evidence of a covalent contribution to bonding, contrary to an earlier speculation (Smart and Humphreys 1980). This finding agrees with the theoretical result of Ogata et al. (2002), who concluded that Cu has a homogeneous electron distribution with little bond directionality.

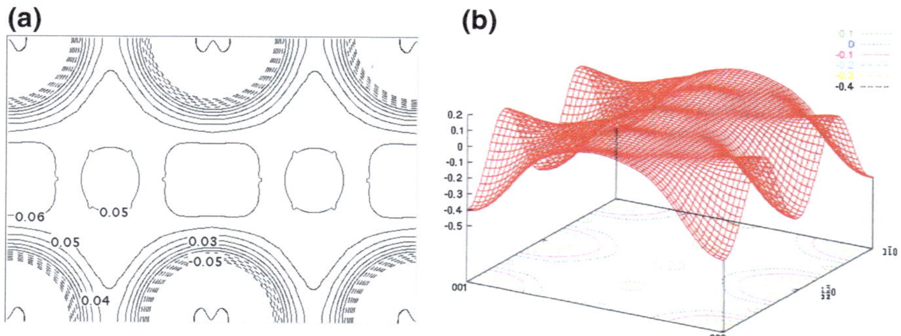

Fig. 12.30 Experimental electron deformation density map of the (110) with copper atoms at the corners and the middle of the top and bottom edges. **a** The map shows the difference between the experimental electron density and the IAM model as reference. (The IAM model is an artificial crystal made up of superimposed neutral atoms). The *dashed* and *solid lines* are contours with $\Delta\rho < 0$ and $\Delta\rho > 0$, respectively; the contour increment is 0.01 e/Å^3. **b** A surface plot of the deformation density map (From Jiang et al. 2004 and Friis et al. 2003)

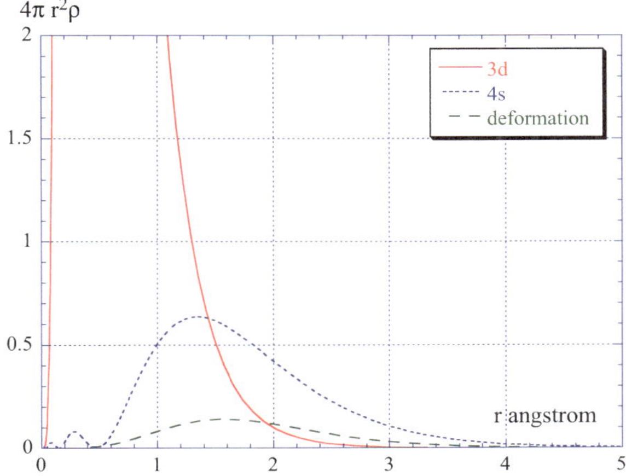

Fig. 12.31 Electron density of 3*d* and 4*s* and the deformation of 3*d* electrons. For details see text

Large negative difference density near copper atoms comes from the deformed d orbital because of 3d and 4s hybridization, $3d_{\text{deformed}} = 3d^{10-n}4s^{n}$, which was used by Jiang et al. (2004) to model the electron density in copper with

$$\rho^{\text{crystal}} = (10 - q)\rho_d + \rho_{4s} + qN_o r^4 \exp(-\alpha r) + P_4 N_4 r^4 \exp(-\alpha r) K_{l=4}(\theta, \varphi), \tag{12.49}$$

where a part of the d electron density takes the new form of $r^4 \exp(-\alpha r)$, which is introduced to account for a change in the d electron radial density from 3d-4s hybridization.

The main feature that emerges from this three parameters (q, α, and P_4) model is the amount of d electron density being pushed outward from 3d-4s hybridization (Fig. 12.31). The nonspherical deformation is very small. The deformation is well described by a single Slater orbital as the fitting demonstrates. In the model of Eq. (12.49), the number of electrons involved in deformation is significant ($q = 0.23$ e/Cu). Since the radial distribution of the deformation closely resembles that of 4 s electrons, their contribution to bonding must also be significant.

12.5.4   Transition Metal Oxides

This section discusses metal/oxygen and metal/metal bonding in transition metal oxides. This class of materials presents rich and complex interplay between charge, lattice, and magnetic spin, leading to a variety of physical properties including ferroelectricity, magnetism, magnetoresistance, superconductivity, and catalysis that are very much at the forefront of research in physics, chemistry and materials science. Theoretical treatment of this subject can be found in the books by Cox (2010) and Burdett (1995). Here, we examine bonding in rutile and Cu_2O as examples.

12.5.4.1   Ti–O Bonds in Rutile, TiO_2

The structure of Rutile is described in Sect. 12.2. At room temperature, its tetragonal unit cell has the dimensions of $a = b = 4.5937$ Å and $c = 2.9587$ Å. The Ti atoms occupy site (000), (1/2, 1/2, 1/2). Each Ti is coordinated by six oxygen at the sites of $(x, x, 0)$ and $(1/2 - x, 1/2 + x, 1/2)$ with $x = 0.30479$. There are two Ti–O bond lengths: two long apical Ti–O bonds and four short equatorial Ti–O bonds at the lengths of 1.9800 and 1.9485 Å, respectively. The long apical bonds are along $\pm[\bar{1}10]$ (see Fig. 12.10). The structure is considered to be ionic. The formal d^0 electron configuration of Ti^{4+} ions (in its highest oxidation state) has two $3d$ electrons and 2 $4s$ electrons removed from the Ti $[Ar]3d^24s^2$ electron configuration with an ionic radius of 0.605 Å.

Figure 12.32 shows the difference charge density determined by Jiang et al. (2003b) which were obtained using a combination of structure factors obtained by electron and X-ray diffraction (Restori et al. 1987). The plots taken in two different sections containing the short Ti–O bonds and long Ti–O bonds, respectively, reveal several features of Ti–O bonding in rutile:

(1)   The charge redistribution from deficiency at the Ti site to surplus at the O site indicates ionic bonding;
(2)   The anisotropic charge deficiency at the Ti site along the crystal c-axis direction has the characteristics of a d orbital hole;
(3)   The strong directional electron distribution around O atoms shows covalent bonding between O and Ti atoms;
(4)   There is a remarkable difference between the electron distributions of O atoms with shorter or longer bond distances to Ti. The nonspherical distortion is much stronger for the short bonds than the long ones (Fig. 12.32).

To understand bonding in rutile, we will consider next how Ti atoms interact with oxygen atoms. First, the Ti atoms are surrounded by six O atoms, forming a distorted octahedron. Because of the charge transfer, the O atoms are negatively charged and treated as point charges (ligands) first as an approximation. For an ideal octahedron with the m3m symmetry, the degenerate Ti atomic orbitals split into two levels under the influence of the ligand field, one for the e_g and one for the t_{2g} orbitals (Fig. 12.23). Even though the real symmetry at Ti is mmm in rutile, we will continue to use the notation (e.g., t_{2g}) by considering it as a good approximation. Next, we consider the O atoms, which are surrounded by three positive Ti ions in a perfect planar geometry. Such configuration is best described by the sp^2 hybridization of the O atom forming three bonds in the plane and one bond out of

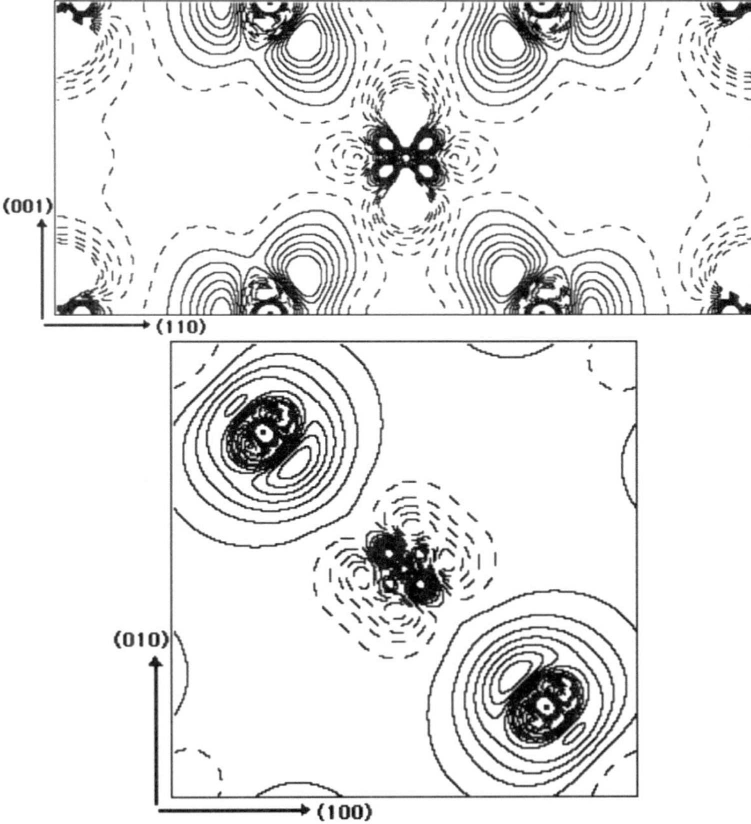

Fig. 12.32 Experimental electron density difference map of TiO_2. **a** The (1–10) plane with four short Ti–O bonds and **b** the (001) plane through Ti at the center of the unit cell with two long Ti–O bonds. The map was obtained from multipole model fitting of the combined electron and X-ray data. The neutral atom model was used as the reference. Negative electron density difference is plotted as *dashed lines*. The contour values range from –0.4 to 0.4 e/Å^3 with contour interval of 0.05 e/Å^3 (Taken from Fig. 3 of Jiang et al. 2003b)

the plane. According to the molecular orbital (MO) diagram of the TiO_6 octahedral complex (shown in Fig. 12.34), two Ti e_g orbitals combining with Ti sp^3 orbitals will form six σ-metal-ligand bonds pointing to the six O atoms and form molecular orbitals. The d_{yz} orbitals form a π bond with the apical oxygen p_z orbital, while the d_{xz} and d_{x2-y2} orbitals have σ nonbonding with O atoms or form σ or π bonds with the Ti atom in the x direction depending on the metal–atom distance. This Ti–Ti bonding is expected to be very weak owing to the large distance between Ti metal atoms. Figure 12.35 shows the decomposed DOS for Ti $3d$ and O $2p$ orbitals. The figure indicates the following features: (1) The band between −6 and 0 eV are mainly from O $2p$ orbitals, (2) the bands above Fermi energy are dominated by Ti

Fig. 12.33 d orbitals formed under octahedral ligand field

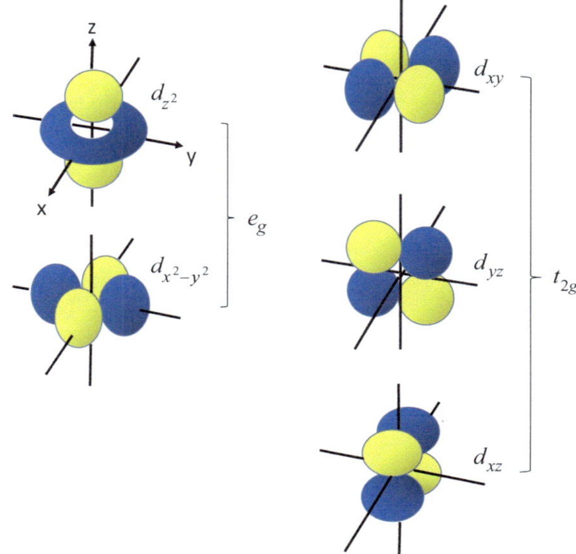

Fig. 12.34 Molecular orbital diagram of rutile (Ti_2O_4). The occupied energy levels are shaded in *gray*, and the orbitals are defined in the coordinate system of $x\|[001]$, $y\|[110]$ and $z\|[-110]$ (After Sorantin and Schwarz 1992)

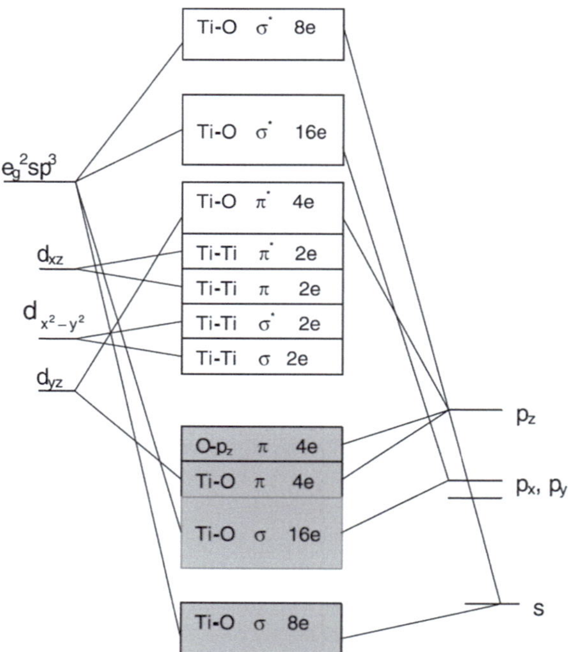

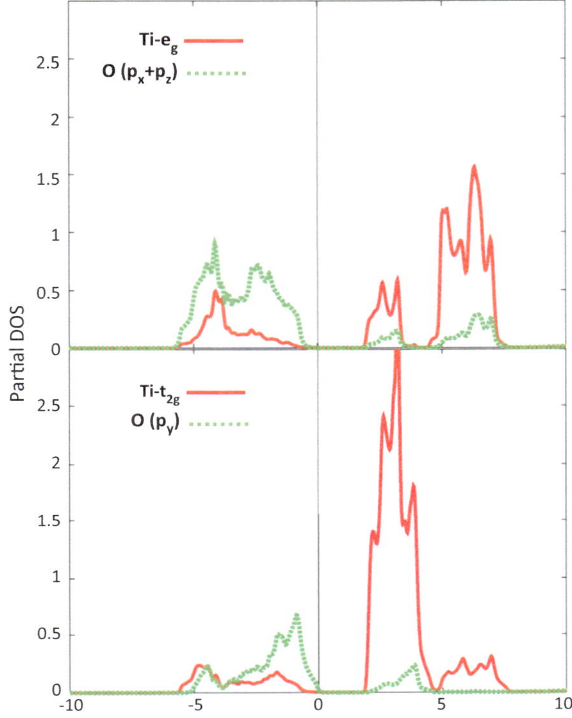

Fig. 12.35 Decomposition of theoretical TiO_2 DOS into the contributions from Ti t_{2g} and e_g and oxygen p_x, p_y, and p_z orbitals

$3d$ orbitals, and (3) orbital hybridization occurs mainly at the bottom of the valence band for Ti $3d$ and O $2p$ valence electrons.

The difference electron density peaks between the Ti–O short and long bonds at 0.41 and 0.3 e/Å³, respectively. Analysis based on the multipole models shows that among the e_g orbitals, d_{x2-y2} has more electron population than the d_z^2 orbital, thus forming the strongest bond with oxygen for the four short Ti–O σ bonds, and the d_{z^2} orbital forms the second strongest bond with the apical oxygen for the long Ti–O σ bond. The band theory calculation suggests that Ti e_g hybridize with the O $p_x + p_z$ to form covalent bonds, while t_{2g} with O p_y form weak π bonds.

Overall, both ionic and covalent bonding all play a role in Ti–O bonding, in agreement with the hypothesis of O'keeffe (1977). The interactions stem from the charge transfer from Ti to O atoms and the hybridization of Ti $3d$ and oxygen $2p$ orbitals as shown in Figs. 12.32 and 12.35.

12.5.4.2 d_{z^2} Holes and Cu–Cu Bonding in Cuprite: Cu_2O

The structure of cuprite is cubic with no free internal parameters (only Ag_2O is isostructural). The copper atoms form an fcc lattice, while oxygen atoms occupy two of the four tetrahedral sites (labeled as O in Fig. 12.37): (1/4, 1/4, 1/4) and

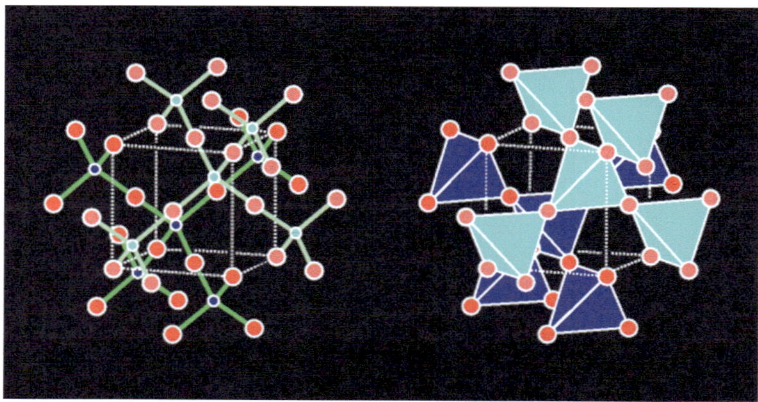

Fig. 12.36 *Left*, as a ball-and-stick model with O atoms *blue*, Cu atoms *red*, and bonds *green*. One network is colored darker than the other. Note that there are no bonds joining the two nets. *Right*, as corner-connected OCu_4 tetrahedra. *Dark* and *light* tetrahedra are on independent networks. In both sketches, *dotted white lines* outline a unit cell

(3/4, 3/4, 3/4) as rendered in Fig. 12.36 which has oxygen atoms in the corners and center of the cube. The size of the cell is $a = 4.2596$ Å at -170 °C. The Cu^+ in Cu_2O of the electronic configuration of d^{10} is linearly coordinated by two oxygen atoms. Together, the bonded Cu and O atoms form two sublattices, in the form of connected and interpenetrating networks as illustrated in Fig. 12.35. The simple theory of ionic crystals based on the closed-shell positive and negative ions predicts highest symmetry coordination for cations, such as the octahedral or tetrahedral coordination. Such theory is inadequate for cuprite, as evidenced by the Cu–O bond distance which at 1.84 Å is significantly less than that of 2.3 Å predicted based on the ionic radii of Cu^+ and O^{2-} ions (Shannon 1976). Furthermore, the two sublattices repel each other electrostatically, so that to account for their interpenetration some short-range Cu–Cu attractive interaction must be invoked (O'keeffe 1963). The closest approach of atoms of the two networks is a Cu–Cu distance of 3.02 Å— the shortest O–O distances are 3.70 Å.

The unusual linear coordination of cuprite have been explained (Orgel 1958) by invoking the participation in bonding of electronic orbitals of higher principal quantum number—that is, $(n + 1)s$ and $(n + 1)p$—accompanied by the creation of d orbital holes on the Cu^+ ion and the occurrence of metal–metal bonding despite their formal nd^{10} configuration. To test these hypotheses, Zuo et al. have used the quantitative CBED technique described in Sect. 12.4.2 combined with X-ray diffraction to map the charge-density distribution in cuprite, and the result is shown in Fig. 12.36 in the form of the difference charge density between the experiment (Zuo et al. 1999); the superimposed spherical O^{2-} and Cu^+ ions. The O^{2-} ion was calculated using a Watson sphere of 1.2 Å radius. The experimental charge density was obtained from the multiple model fitting of electron and X-ray structure factors (Fig. 12.37).

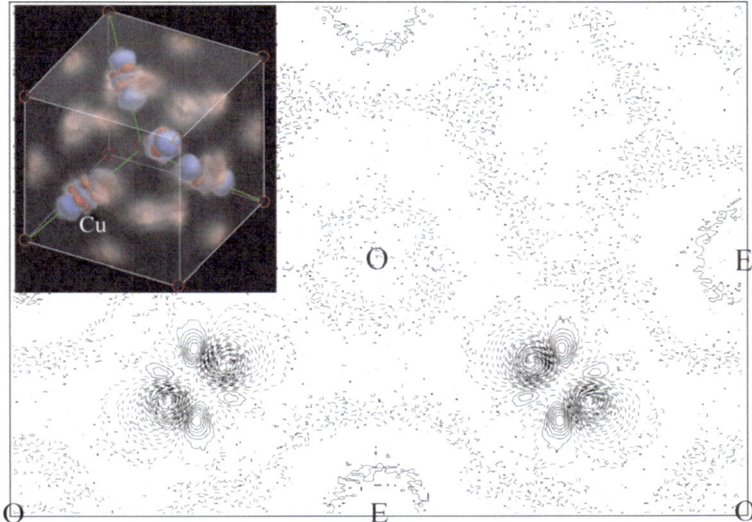

Fig. 12.37 Electron density difference map Cu_2O of (110) plane with oxygen at the center and corners of the rectangle (marked by O, E for empty site). The 3D rendering is shown in the inset with red and blue colors representing excess electrons and holes, respectively. The contour interval is 0.2 e/$Å^3$ with full line for positive differences (From Zuo 2004)

The difference electron density shown in Fig. 12.37 would be zero everywhere if cuprite were purely ionic (i.e., consisted of spherical ions). The difference confirms the earlier theoretical speculation (Orgel 1958) that a covalent contribution exists. Strong nonspherical charge distortion is seen around the copper atoms, with the characteristic shape of d orbitals, and excess charge in the interstitial region. The correspondence between the *experimental* map and the classical diagrams of d_z^2 orbitals sketched in textbooks is striking. There is little variation around oxygen in both the experimental and the theoretical results, which suggests that an O^{2-} anion description is valid. The experimental map also shows a large (~ 0.2 e/$Å^3$) positive peak in the unoccupied tetrahedral interstitial region of the 4 neighboring Cu atoms, which suggests a strong Cu^+–Cu^+ covalent bonding (marked as E in Fig. 12.37).

The nonspherical electron density around Cu^+ can be interpreted as due to the hybridization of d electrons with higher energy unoccupied s and p states (Fig. 12.38). Among these states, hybridization is only allowed for d_z^2 and $4s$ by symmetry, and when this happens, part of the d_z^2 state becomes unoccupied ("d hole"). These states are responsible for the spatial distribution of the deficiency in the map shown in Fig. 12.36. The complementary empty states are important for EELS, which probes empty states. The experimental studies reveal that the unoccupied states are predominately Cu-d character for the Cu $L_{2,3}$ edge; the theory shows that they originate from the hybridized d_z^2 orbital. This interpretation, based on the calculated partial DOS of the one-electron band structure, is supported by the

Fig. 12.38 Calculated partial density of states (DOS) for Cu. These spectra show the evidence of $3d$-$4s$ hybridization and the unoccupied states (holes) for $3d_z^2$ electrons

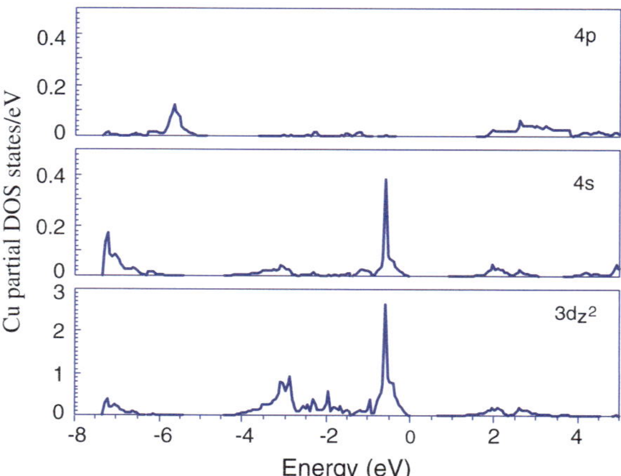

generally good agreement with experimental spectroscopy of both occupied and unoccupied states (Ghijsen et al. 1988). From the electron density, Zuo et al. estimated the hybridization coefficient between d_z^2 and $4s$, $|x| \sim 0.36$, so that about 0.22 electrons are removed from d_z^2 states.

References

Aldred PJE, Hart M (1973) Electron-distribution in silicon. I. Experiment. Proc Roy Soc London Ser A 332:223–238

Barrett CS, Massalski TB (1966) Structure of metals. McGraw-Hill, New York

Becker PJ, Coppens P (1974) Extinction within limit of validity of Darwin transfer equations 1. General formalisms for primary and secondary extinction and their application to spherical crystals. Acta Cryst AA 30:129–147

Blaha P, Schwarz K, Madsen G, Kvasnicka D, Luitz J (2001) In: K Schwarz K (ed) WIEN2k: an augmented plane wave plus local orbitals program for calculating crystal properties. Vienna University of Technology, Austria

Burdett JK (1995) Chemical bonding in solids. Oxford University Press, Oxford

Clementi E, Raimondi DL (1963) Atomic screening constants from SCF functions. J Chem Phy 38:2686–2689

Clementi E, Roetti C (1974) Roothaan-Hartree-Fock atomic wavefunctions. Basis functions and their coefficients for ground and certain excited states of neutral and ionized atoms. At Data Nucl Data Tables 14:177–478

Coppens P (1997) X-ray charge densities and chemical bonding. International Union of Crystallography

Coppens P (1998) Charge-density analysis at the turn of the century. Acta Cryst A54:779–788

Cox PA (2010) Transition metal oxides: an introduction to their electronic structure and properties. Oxford University Press, Oxford

Cummings S, Hart M (1988) Redetermination of absolute structure factors for silicon at room and liquid-nitrogen temperatures. Aust J Phys 41:423–431

Eberhart ME (1996) The metallic bond: elastic properties. Acta Mater 44:2495–2504

Friis J, Jiang B, Spence JCH, Holmestad R (2003) Quantitative convergent beam electron diffraction measurements of low-order structure factors in copper. Micros Microanal 9: 379–389

Gajdardziskajosifovska M, McCartney MR, Deruijter WJ, Smith DJ, Weiss JK, Zuo JM (1993) Accurate measurements of mean inner potential of crystal wedges using digital electron holograms. Ultramicroscopy 50:285–299

Ghijsen J, Tjeng LH, Vanelp J, Eskes H, Westerink J, Sawatzky GA, Czyzyk MT (1988) Electronic-structure of Cu_2O and CuO. Phys Rev B 38:11322–11330

Hansen NK, Coppens P (1978) Electron population analysis of accurate diffraction data.6. Testing aspherical atom refinements on small-molecule data sets. Acta Cryst A34:909–921

Harrison WA (1983) Theory of the two-center bond. Phys Rev B 27:3592–3604

Hecker SS, Rohr DL, Stein DF (1978) Brittle-fracture in iridium. Metall Trans A 9:481–488

Hoffman R (1989) Solids and surfaces: a chemist's view of bonding in extended structures. Wiley, New York

Hohenberg P, Kohn W (1964) Inhomogeneous electron gas. Phys Rev B 136:B864

Jiang B, Zuo JM, Jiang N, O'Keeffe M, Spence JCH (2003a) Charge density and chemical bonding in rutile, TiO_2. Acta Cryst A59:341–350

Jiang B, Zuo JM, Friis J, Spence JCH (2003b) On the consistency of QCBED structure factor measurements for TiO_2 (rutile). Micros Microanal 9:457–467

Jiang B, Friis J, Holmestad R, Zuo JM, O'Keeffe M, Spence JCH (2004) Electron density and implication for bonding in Cu. Phys Rev B 69:245110

Kambe K (1955) Cohesive energy of noble metals. Phys Rev 99:419–422

Kohn W, Sham LJ (1965) Self-consistent equations including exchange and correlation effects. Phys Rev 140:1133

Kruse P, Rosenauer A, Gerthsen D (2003) Determination of the mean inner potential in iii-v semiconductors by electron holography. Ultramicroscopy 96:11–16

Kurkisuonio K (1977) Symmetry and its implications. Isr J Chem 16:115–123

Lawrence JL (1973) Debye-waller factors for magnesium-oxide. Acta Cryst A29:94–95

Lu ZW, Zunger A, Deutsch M (1995) Electronic charge-distribution in crystalline germanium. Phys Rev B 52:11904–11911

Martin RM (2008) Electronic structure: Basic theory and practical methods. Cambridge University Press, Cambridge

Mehl MJ, Cohen RE, Krakauer H (1988) Linearized augmented plane-wave electronic-structure calculations for MgO and CaO. J Geophys Res-Solid 93:8009

Müller U (2006) Inorganic structural chemistry. Wiley, West Sussex

Murnaghan FD (1944) The compressibility of media under extreme pressures. Proc Natl Acad Sci USA 30:244

Musgrave MJ (1963) On relation between grey and white tin (alpha-Sn and beta-Sn). Proc Roy Soc London Ser A 272:503

Nakashima PNH (2005) Improved quantitative CBED structure-factor measurement by refinement of nonlinear geometric distortion corrections. J Appl Crystallogr 38:374–376

Nakashima PNH, Muddle BC (2010) Differential convergent beam electron diffraction: experiment and theory. Phys Rev B 81:115135

Nakashima PNH, Smith AE, Etheridge J, Muddle BC (2011) The bonding electron density in aluminum. Science 331:1583–1586

Ogata S, Li J, Yip S (2002) Ideal pure shear strength of aluminum and copper. Science 298: 807–811

Ogata Y, Tsuda K, Tanaka M (2008) Determination of the electrostatic potential and electron density of silicon using convergent-beam electron diffraction. Acta Cryst A64:587–597

O'Keeffe M (1963) Madelung constants for c3 and c9 structures. J Chem Phy 38(12):3035

O'Keeffe M (1977) Arrangements of ions in crystals. Acta Cryst A 33:924–927

O'Keeffe M, Spence JCH (1994) On the average coulomb potential and constraints on the electron density in crystals. Acta Cryst A50:33–45

Orgel LE (1958) Stereochemistry of metals of the b sub-groups .1. Ions with filled d-electron shells. J Chem Soc 4186–4190

Pauling L (1960) The nature of the chemical bond (3rd). Cornell University Press, New York

Pavone P, Baroni S, de Gironcoli S (1998) $\alpha\leftrightarrow\beta$ phase transition in tin: a theoretical study based on density-functional perturbation theory. Phys Rev B 57:10421–10423

Pettifor DG (1995) Bonding and structure of molecules and solids. Clarendon Press, Oxford

Pillet S, Souhassou M, Lecomte C, Schwarz K, Blaha P, Rerat M, Lichanot A, Roversi P (2001) Recovering experimental and theoretical electron densities in corundum using the multipolar model: IUCr multipole refinement project. Acta Cryst A57:290–303

Redinger J, Schwarz K (1981) Electronic charge-distribution of the polarizable O^{-2}-ion in MgO and CaO in contrast to the f^{-} ion in NaF. Z Physik B 40:269–276

Ren G, Zuo JM, Peng LM (1997) Accurate measurements of crystal structure factors using a FEG electron microscope. Micron 28:459–467

Restori R, Schwarzenbach D, Schneider JR (1987) Charge-density in rutile, TiO_2. Acta Crystallogr B 43:251–257

Rez D, Rez P, Grant I (1994) Dirac-Fock calculations of X-ray-scattering factors and contributions to the mean inner potential for electron-scattering. Acta Cryst A50:481–497

Rez D, Rez P, Grant I (1997) Dirack-fock calculations of X-ray scattering factors and contributions to the mean inner potential for electron scattering. Acta Cryst A53:522–522 (A50:481, 1994)

Saka T, Kato N (1986) Accurate measurement of the Si structure factor by the pendellosung method. Acta Cryst A42:469–478

Saunders M, Bird DM, Zaluzec NJ, Burgess WG, Preston AR, Humphreys CJ (1995) Measurement of low-order structure factors for silicon from zone-axis CBED patterns. Ultramicroscopy 60:311–323

Scheffler M, Tkatchenko A, Rinke P (2012) Theoretical material science. Fritz Haber Institute of the Max Planck Society, Berlin

Schneider JR, Hansen NK, Kretschmer H (1981) A charge-density study of copper by gamma-ray diffractometry on imperfect single-crystals. Acta Cryst A37:711–722

Shannon RD (1976) Revised effective ionic radii and systematic studies of interatomic distances in halides and chalcogenides. Acta Cryst A32:751–767

Smart D, Humphreys CJ (1980) EMAG. Int Phys IOP 52:211

Sorantin PI, Schwarz K (1992) Chemical bonding in rutile-type compounds. Inorg Chem 31: 567–576

Spackman MA, Byrom PG (1997) Retrieval of structure-factor phases in non-centrosymmetric space groups. Model studies using multipole refinements. Acta Crystallogr B 53:553–564

Streltsov VA, Nakashima PNH, Johnson AWS (2003) A combination method of charge density measurement in hard materials using accurate, quantitative electron and X-ray diffraction: the alpha-Al_2O_3 case. Micros Microanal 9:419–427

Teworte R, Bonse U (1984) High-precision determination of structure factors F_h of silicon. Phys Rev B 29:2102–2108

Tsuda K, Tanaka M (1999) Refinement of crystal structural parameters using two-dimensional energy-filtered CBED patterns. Acta Cryst. A55:939–954

Volkov A, Macchi P, Farrugia LJ, Gatti C, Mallinson P, Richter T, Koritsanszky T (2014) XD 2015: a computer program package for multipole refinement, topological analysis of charge densities and evaluation of intermolecular energies from experimental and theoretical structure factors. http://www.chem.gla.ac.uk/ ~ louis/xd-home/

von der Large FC, Bethe HA (1947) A method for obtaining electronic eigenfunctions and eigenvalues in solids with an application to sodium. Phys Rev 71:612

Wells AF (1984) Structural inorganic chemistry (5th). Oxford University Press, Oxford

Willis BTM, Pryor AW (1975) Thermal vibrations in crystallography. Cambridge University Press, Cambridge

Zachariasen WH (1967) A general theory of X-ray diffraction in crystals. Acta Crystallogr 23:558

Zuo JM (1998) Quantitative convergent beam electron diffraction. Mater Trans, JIM 39:938–946

Zuo JM (1999) Accurate structure refinement and measurement of crystal charge distribution using convergent beam electron diffraction. Microsc Res Tech 46:220–233

Zuo JM (2000) Electron detection characteristics of a slow-scan CCD camera, imaging plates and film, and electron image restoration. Microsc Res Tech 49:245–268

Zuo JM (2004) Measurements of electron densities in solids: a real-space view of electronic structure and bonding in inorganic crystals. Rep Prog Phys 67:2053–2103

Zuo JM, Spence JCH (1991) Automated structure factor refinement from convergent-beam patterns. Ultramicroscopy 35(3–4):185–196

Zuo JM, Blaha P, Schwarz K (1997a) The theoretical charge density of silicon: experimental testing of exchange and correlation potentials. J Phys C 9:7541–7561

Zuo JM, Okeeffe M, Rez P, Spence JCH (1997b) Charge density of MgO: implications of precise new measurements for theory. Phys Rev Lett 78:4777–4780

Zuo JM, Kim M, O'Keeffe M, Spence JCH (1999) Direct observation of d-orbital holes and Cu-Cu bonding in Cu_2O. Nature 401:49–52

Chapter 13
Diffuse Scattering

We have so far discussed Bragg diffraction and its use for the study of crystal symmetry and bonding. Other kinds of scattering recorded in electron diffraction patterns include inelastic scattering by phonon, plasmon or electron excitations, and elastic scattering by crystal imperfections. These contribute to the diffuse scattering that is observed between Bragg diffraction peaks. Thus, the study of diffuse scattering concerns its interpretation in relation to the crystal structure, beyond the periodically averaged crystal structure obtained from Bragg diffraction. It will be understood that, for example, the study of diffuse scattering in diffraction patterns as a function of temperature across a phase transition can provide a far simpler, and sometimes more informative experimental arrangement, covering a wider range of conditions, than temperature-dependent observations using atomic-resolution imaging.

The study of diffuse scattering has its origins in X-ray studies in the middle of the last century, where the relevant theory was developed. In the kinematic limit, the theory of electron diffuse scattering is the same as that for X-rays or neutrons. However, electron diffuse scattering does have its own unique aspects due to the short wavelength of high-energy electrons and strong electron interaction with matter. We therefore start the chapter with a discussion of electron diffuse scattering and then move on to the kinematical theory of thermal diffuse scattering, diffuse scattering from small defects, solid solutions, and modulated structures. We finish with the theory of multiple beam diffraction effects in electron diffuse scattering. Further reading on the subjects covered here can be found in the books by Krivoglaz (1996), Cowley (1995), Welberry (2010), Amoros (1968), and Peng et al. (2004).

To simplify the mathematical derivations in this chapter, we have adopted a different sign convention in Sects. 13.2–13.5, in which the structure factor is written as

$$U\left(\vec{G}\right) = \sum_n f_n e^{i\vec{G}\cdot\vec{r}_n},$$

© Springer Science+Business Media New York 2017

J.M. Zuo and J.C.H. Spence, *Advanced Transmission Electron Microscopy*,
DOI 10.1007/978-1-4939-6607-3_13

and

$$U(\vec{r}) = \sum_{\overrightarrow{G}} U\left(\overrightarrow{G}\right) e^{-i\overrightarrow{G}\cdot\vec{r}_n},$$

for the electron interaction potential. The reciprocal lattice vector $\overrightarrow{G}$ includes the 2π prefactor with $\overrightarrow{G} = 2\pi\vec{g}$ and an incident electron plane wave is given by $\phi = \exp(-i\vec{q}\cdot\vec{r})$, where $\vec{q} = 2\pi\vec{k}$. The minus sign here follows the crystallographic sign convention.

13.1 Electron Diffuse Scattering

The dominant diffuse scattering seen in an electron diffraction pattern comes from small-angle inelastic scattering, due to the excitation of plasmons (Howie 1963). The other forms of diffuse scattering are relatively weak in comparison. Thus, the study of elastic electron diffuse scattering from defects is greatly assisted by energy filtering for the purely elastic scattering. A procedure for recording this energy-filtered electron diffuse scattering was described by Zuo et al. (2000) using an in-column Omega filter and imaging plates for diffraction pattern recording. This combination offers a large acceptance angle, a large detector dynamical range, and high DQE at low electron signal levels, which are all beneficial for the recording of electron diffuse scattering. Unlike the energy filters commonly used in the STEM mode, this Omega filter filters every pixel of a TEM image or diffraction pattern simultaneously. The large dynamic range is especially useful since the intensity of Bragg peaks is very strong, while diffuse scattering seen between Bragg peaks is weak. The recent development of direct single-electron detection systems (McMullan et al. 2014), as used in cryo-electron microscopy, can be expected to be even better, if provision can be made to exclude the very strong Bragg reflections. Alternatively, a post-column energy filter and the CCD camera can be used. Prior to the development of these technologies, electron diffuse scattering was studied using films.

Figure 13.1 shows an example of electron diffuse scattering, recorded from a crystal of magnetite (Fe_3O_4, cubic spinel structure) at $T = 144$ K above its cubic-to-monoclinic phase transition temperature, known as the Verwey transition (Verwey et al. 1947). The diffraction pattern was recorded from a polycrystalline Fe_3O_4, chemically thinned to electron transparency, using a parallel electron beam of 1.5 μm in diameter formed using the Kohler illumination mode in a Zeiss LEO 912 TEM. A parallel beam is used here for its small convergence angle and thus high angular resolution in the diffraction pattern. Details of how to set up parallel beam illumination can be found in Chap. 10.

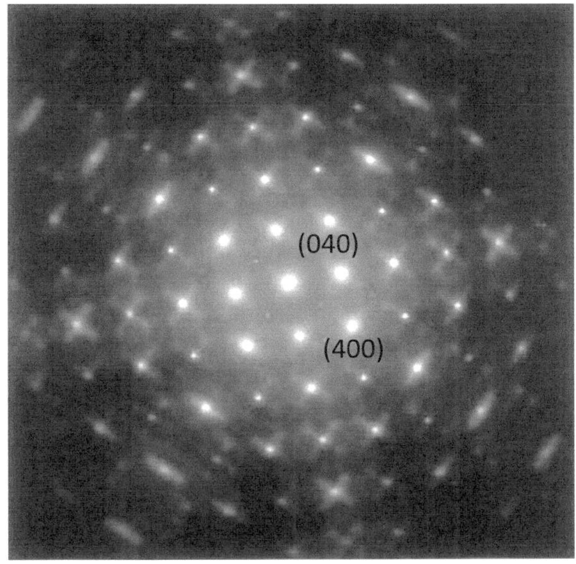

Fig. 13.1 Electron diffraction pattern showing diffuse scattering from magnetite (Fe_3O_4) along [001] at 144 K, which was recorded using a 15 eV energy-filtering window around the zero-loss peak, and an imaging plate detector

In the kinematic limit, what is recorded in an electron diffraction pattern is a cross section of diffuse scattering distributed in the three-dimensional reciprocal space. Because of the short electron wavelength, to a good approximation, we have a planar section that passes through the origin of the reciprocal lattice and the plane normal is the incident beam direction.

In electron diffraction, multiple scattering modifies the intensity of Bragg peaks as well as the diffuse scattering intensity by multiple beam diffraction effects that introduce different scattering paths to diffuse scattering. In thick crystals, Bragg diffraction of diffuse scattering gives rise to Kikuchi lines as discussed in Chap. 2. Multiple diffuse scattering broadens its distribution and hinders quantitative analysis. Fortunately, multiple scattering involving the excitation of plasmons or higher energy losses can be removed using an energy filter. Strong Bragg diffraction can be avoided to an extent by using an off-axis and off-Bragg diffraction condition, allowing the limitations of the kinematic approximation to be extended. Nevertheless, the quantitative analysis of electron diffuse scattering may still require a full treatment of dynamical diffraction.

The major advantages of using electrons to study diffuse scattering are as follows:

(1) Electron diffuse scattering can be recorded using a parallel electron beam, whose diameter ranges from a few tens of nm to a few microns. Such small probes enable the study of small crystals, polycrystalline materials, and thin films. The strong interaction means that electron diffuse scattering is much stronger than that produced by most X-ray or neutron sources.

(2) With the help of electron imaging, electron diffuse scattering can be recorded from regions free of other defects, and so allows us to separate the diffuse scattering from different types of defects;

(3) Electron diffuse scattering can be directly correlated with diffraction contrast imaging, for the study of nanodomains, or atomic-resolution images for the study of chemical disorder or structural modulations;

(4) Dynamical diffraction effects can be exploited to locate the source of electron diffuse scattering within the crystal unit cell Tafto and Spence (1982).

The disadvantages of using electron diffraction include the strong multiple scattering effects observed in thick crystals, especially near zone-axis orientations and the large inelastic background, which must be carefully filtered out using an energy filter.

The interpretation of electron diffuse scattering generally relies on physical models, such as phonon models for thermal diffuse scattering (which can generally not be excluded using an energy filter), or structural models for the study of crystal imperfections. The conventional crystallographic approach, based on the use of Patterson functions derived from diffraction patterns, requires a correction for dynamical effects on electron diffraction, which is difficult to achieve.

13.2   Thermal Diffuse Scattering

Much of our understanding of thermal diffuse scattering (TDS) can be traced back to the study of the effects of thermal vibration on X-ray scattering, with early contributions from Debye (1914), Waller (1923), Laval (1938, 1939), Zachariasen (1940), Born (1942) and James (1948). It was James who first gave the subject a comprehensive treatment in his book on X-ray diffraction that became the standard text for this subject. There are two major effects associated with the thermal vibration of atoms. The first is the introduction of a temperature factor $\exp(-2\,M)$, which reduces the intensity of Bragg reflections (Sect. 4.7) and so limits the information that can be extracted by Bragg diffraction. The second is TDS, whose relationship with lattice vibrations was first established by Laval (1938, 1939) and subsequently employed for the determination of phonon dispersion relations in simple metals and some compounds (see Xu and Chiang 2005 for a recent review). The cross section for TDS is small. Early experimental measurements of TDS performed using X-ray tubes were limited by the slow data acquisition rate, and the accuracy of the results was therefore limited. This situation has improved significantly with the development of synchrotrons and two-dimensional X-ray detectors such as image plates and CCD cameras that allow parallel measurements over a large solid angle. Compared with inelastic neutron diffraction, which is the standard technique for studying phonons, TDS is much better suited for small crystals; materials such as thin films or nanostructures can be studied by electron diffraction, while polycrystalline materials, and samples contained in high-pressure cells, can be best studied using focused X-ray beams in a synchrotron. By comparison, neutron scattering typically requires large single crystals whose size is about 1 cm.

The effect of thermal vibration on Bragg diffraction was first discussed in Sect. 4.7, Chap. 4. For a monoatomic crystal, the atom at a lattice site $\vec{R}_n$ oscillates around its equilibrium position and its position at any given time can be written as

$$\vec{r}_n(\tau) = \vec{R}_n + \vec{u}_n(t).\tag{13.1}$$

In a lattice, atomic vibration is determined by the equation of motion, which has special solutions or so-called normal modes or phonons in quantum mechanics in the form of a single frequency traveling wave

$$\vec{u}(t) = a_{\vec{k}}\vec{e}\cos\left[\omega\left(\vec{k}\right)t - \vec{k}\cdot\vec{R}_n + \phi\left(\vec{k}\right)\right],\tag{13.2}$$

where $\vec{k}$ is the phonon wavevector, $\omega(\vec{k})$ the phonon frequency, M the mass of the atom, $\phi(\vec{k})$ the initial phase, and $a_{\vec{k}}$ and $\vec{e}$ are the vibrational amplitude and direction, respectively.

At any given time t, the net atomic displacement is a sum of the displacements of all lattice waves, including different vibrational modes. There are 3 modes for a monoatomic crystal, two transverse acoustic (TA) waves, and one longitudinal acoustic (LA) wave. A sum over these modes and wavevectors gives

$$\vec{u}(t) = \sum_{\vec{k},j} a_{\vec{k},j}\vec{e}_j\cos\left[\omega_j\left(\vec{k}\right)t - \vec{k}\cdot\vec{R}_n + \phi_j\left(\vec{k}\right)\right].\tag{13.3}$$

Assuming displacements are small, we obtain from Eq. 4.42 (see Sect. 4.7) the following time-averaged diffraction intensity

$$\begin{aligned}
\langle I(t)\rangle &= f^2\sum_n\sum_m e^{i\vec{q}\cdot\left(\vec{R}_n-\vec{R}_m\right)} e^{-q^2 U_{qn}^2/2}e^{-q^2 U_{qm}^2/2}e^{q^2\langle U_{qn}U_{qm}\rangle}\\
&\approx f^2\sum_n\sum_m e^{i\vec{q}\cdot\left(\vec{R}_n-\vec{R}_m\right)} e^{-q^2 U_{qn}^2/2}e^{-q^2 U_{qm}^2/2}\left\{1 + q^2\langle U_{qn}U_{qm}\rangle + \cdots\right\}\\
&= I_0 + I_1 + \cdots.
\end{aligned}\tag{13.4}$$

The U_{qn} in the above equation represents the amplitude of atomic vibration along the scattering vector direction ($U_{qn} = \vec{U}_n\cdot\hat{q}$).

The expansion in Eq. (13.4) gives the time-averaged diffraction intensity as a sum of successively higher order terms, where the zero-order term I_0 corresponds to Bragg diffraction, which is followed by the first-order TDS, second-order TDS, etc. The I_0 contribution is given by

$$I_{\mathrm{o}} = f^2 \sum_n \sum_m e^{i\vec{q}\cdot\left(\vec{R}_n - \vec{R}_m\right)} e^{-2M}, \tag{13.5}$$

where

$$M = q^2 \left\langle U_{q_m}^2 \right\rangle / 2 = B(\sin\theta/\lambda)^2,$$

for an isotropic crystal. Here B stands for the Debye–Waller factor, which was introduced in Sect. 4.7.

The TDS is generally nonzero for any q and dominated by the first-order term, especially at low temperatures. To explore its relationship with phonon excitations in the crystal, we want to calculate $\left\langle U_{q_n} U_{q_m} \right\rangle$ in Eq. (13.4). Using the relationship

$$\cos x_1 \cos x_2 = \frac{1}{2}\left[\cos(x_1 - x_2) + \cos(x_1 + x_2)\right]$$

and $\langle \cos(\omega t + \phi)\rangle = 0$, we obtain

$$\left\langle U_{q_n} U_{q_m} \right\rangle = \frac{1}{2}\sum_{\vec{k},j} \left\langle a_{\vec{k},j}^2 \right\rangle (\vec{q}\cdot\vec{e}_j)^2 \cos\left[\vec{k}\cdot\left(\vec{R}_n - \vec{R}_m\right)\right]. \tag{13.6}$$

Substitution of Eq. (13.6) into (13.4) gives the first-order TDS

$$I_1 = \frac{f^2}{4} e^{-2M} \sum_n \sum_m \sum_{\vec{k},j} \left\langle a_{\vec{k},j}^2 \right\rangle (\vec{q}\cdot\vec{e}_j)^2 \left[e^{i(\vec{q}+\vec{k})\cdot\left(\vec{R}_n - \vec{R}_m\right)} + e^{i(\vec{q}-\vec{k})\cdot\left(\vec{R}_n - \vec{R}_m\right)} \right].$$

$$\tag{13.7}$$

Using the identity,

$$\sum_{n=1}^{N} e^{i\vec{k}\cdot\vec{R}_n} = N \sum_{h=1}^{N} \delta\left(\vec{k} - \vec{G}\right), \tag{13.8}$$

for large N, where $\vec{G} = 2\pi\vec{g}$ and the δ function is obtained from the summation over a very large lattice (see Sect. 4.4), we obtain

$$I_1 = \frac{f^2 N^2}{4} e^{-2M} \sum_{\vec{k},j} \left\langle a_{\vec{k},j}^2 \right\rangle (\vec{q}\cdot\vec{e}_j)^2 \left[\delta\left(\vec{q}+\vec{k}-\vec{G}\right) + \delta\left(\vec{q}-\vec{k}-\vec{G}\right)\right]. \tag{13.9}$$

This shows that TDS arises from correlated atomic motion. In each normal mode, atoms vibrate at a single frequency, and their motions are correlated with the lattice wave vector. The correlation produces TDS.

The result of Eq. (13.7) shows scattering by a lattice wave gives rise to two satellite peaks symmetrically located next to the Bragg peak at the $\overrightarrow{G} \pm \vec{k}$ positions. Since the excitation of lattice waves is random due to thermal fluctuations, we have the whole spectrum of lattice waves, which is continuous in a large crystal so that the consequent TDS distribution is also continuous. The TDS intensity is a sum of the contributions from all the different lattice normal modes, each contributing an amount which depends on (1) the angle between the scattering vector and the direction of lattice vibrations and (2) the amplitude of lattice vibrations. These sums produce streaks in the diffraction pattern, centered on the Bragg spots.

The mean-square amplitude $\left\langle a_{\vec{k},j}^2 \right\rangle$ is related to the phonon energy. To demonstrate this, we note that the mean total energy of phonons in the quantum mechanical theory is given by

$$\langle E \rangle = \sum_{\vec{k},j} \left(n_p + \frac{1}{2} \right) \hbar \omega_j\left(\vec{k}\right) = \sum_{\vec{k},j} \left(\frac{1}{e^{\hbar \omega_j/k_B T} - 1} + \frac{1}{2} \right) \hbar \omega_j\left(\vec{k}\right)$$

$$= \sum_{\vec{k},j} \frac{1}{2} \coth\left(\frac{\hbar \omega}{2 k_B T} \right) \hbar \omega\left(\vec{k}\right),$$

which in classical mechanics can be written as a sum of potential and kinetic energies

$$\langle E \rangle = \langle U \rangle + \langle T \rangle = 2\langle T \rangle = \sum_{m=1}^{N} 2 \left\langle \frac{1}{2} m_A \left(\frac{du_m}{dt} \right)^2 \right\rangle$$

$$= N m_A \sum_{\vec{k},j} \left\langle a_{\vec{k}}^2 \omega_j^2\left(\vec{k}\right) \sin^2\left[\omega_j\left(\vec{k}\right) t - \vec{k} \cdot \overrightarrow{R}_n + \phi_j\left(\vec{k}\right) \right] \right\rangle \qquad (13.10)$$

$$= N \frac{m_A}{2} \sum_{\vec{k},j} \left\langle a_{\vec{k}}^2 \right\rangle \omega_j^2\left(\vec{k}\right).$$

Using the above results, we obtain for monoatomic crystals the following expression for the first-order TDS intensity

$$I_1 = \frac{f^2 N}{2} e^{-2M} \sum_{\vec{k},j} \frac{\langle E \rangle}{m_A \omega^2\left(\vec{k}\right)} \left(\vec{q} \cdot \vec{e}_j\right)^2 \left[\delta\left(\vec{q} + \vec{k} - \overrightarrow{G}\right) + \delta\left(\vec{q} - \vec{k} - \overrightarrow{G}\right) \right]$$

$$= \frac{f^2 N}{4} e^{-2M} \sum_{\vec{k},j} \frac{\hbar}{m_A \omega\left(\vec{k}\right)} \coth\left(\frac{\hbar \omega}{2 k_B T} \right) \left(\vec{q} \cdot \vec{e}_j\right)^2 \left[\delta\left(\vec{q} + \vec{k} - \overrightarrow{G}\right) + \delta\left(\vec{q} - \vec{k} - \overrightarrow{G}\right) \right].$$

$$(13.11)$$

At high temperatures, the average phonon energy is expected to approach $k_B T$, and then I_1 is inversely proportional to the square of the phonon frequency $\omega^2\left(\vec{k}\right)$.

Because of this, the contributions to TDS come mostly from low-frequency acoustic phonons.

In a compound crystal, the unit cell has more than one atom. The atomic position is specified by $\vec{r}_{n,m} = \vec{R}_n + \vec{r}_m + \vec{u}_n(t)$, where $\vec{r}_m$ is the position vector of mth atom within a unit cell. Thus, the average diffraction intensity from the crystal is given by

$$\langle I \rangle = \left\langle \left| \sum_n \sum_m f_m e^{i\vec{q}\cdot\vec{r}_{n,m}} \right|^2 \right\rangle$$

$$= \sum_n \sum_m \sum_{n'} \sum_{m'} f_m f_{m'} e^{i\vec{q}\cdot\left(\vec{R}_n + \vec{r}_m - \vec{R}_{n'} - \vec{r}_{m'}\right)} \left\langle e^{i\vec{q}\cdot\left(\vec{u}_{nm} - \vec{u}_{n'm'}\right)} \right\rangle \qquad (13.12)$$

Using the same procedures for the monoatomic crystals, it can be shown (Xu and Chiang (2005)).

$$I_1 = \frac{N}{4} \sum_{\vec{k},j} \frac{\hbar}{\omega\left(\vec{k}\right)} \coth\left(\frac{\hbar\omega}{2k_B T}\right) \left| F_j\left(\vec{G}, \vec{q}\right) \right|^2 \left[\delta\left(\vec{q} + \vec{k} - \vec{G}\right) + \delta\left(\vec{q} - \vec{k} - \vec{G}\right) \right]$$

$$(13.13)$$

where

$$F_j\left(\vec{G}, \vec{q}\right) = \sum_n \frac{f_n}{\sqrt{m_n}} e^{-M_n} \left(\vec{q} \cdot \vec{e}_{\vec{q}-\vec{G}}\right) e^{-i\vec{G}\cdot\vec{r}_n}. \qquad (13.14)$$

The direct dependence of TDS intensity on the dispersion of phonon frequencies ($\omega(\vec{k})$ in Eqs. (13.11) or (13.13)) can be used indirectly to determine the phonon dispersion relations. This involves fitting the experimental TDS data using calculated intensities. Unlike neutron scattering or inelastic X-ray scattering methods which measures the phonon dispersion relations directly, fitting TDS data requires a theoretical model, such as the Born-von Karman force constant model (Born and Huang 1954), to calculate the phonon dispersion relations using a small number of parameters, which can be adjusted for best fit. The fitting requires two or more diffraction patterns recorded along different crystallographic directions in order to provide sufficient sampling of the TDS in reciprocal space. Figure 13.2 shows the two-dimensional TDS patterns recorded from Si on imaging plates using synchrotron radiation along [111] and [100] directions, and their theoretical fit, respectively. Each pixel in these diffraction patterns corresponds to a scattering vector q on the Ewald sphere cutting through reciprocal space. The bright spots in the TDS patterns correspond to points on the Ewald sphere close to reciprocal lattice points, where the acoustic phonon populations are high. For first-order scattering, the TDS intensity from each phonon mode is directly proportional to the thermal population of phonons (including the zero-point vibration effect). A low-frequency mode, such as an acoustic mode near a reciprocal lattice point, has a high thermal population and thus yields a high TDS intensity. The analysis leading to a determination of phonon dispersion curves was performed in the following manner. A Born-von Karman model with force constants up to the sixth neighbor was employed to calculate the phonon eigenvalues and eigenvectors, which were then used to calculate theoretical TDS patterns and fitted to the experimental patterns by a least-square procedure, with the force constants as fitting parameters. Figure 13.2c, d shows the results of the fit. A review of this work for X-ray scattering can be found in Xu and Chiang (2005). So far the application to electron diffraction has yet to be made, although the same principles apply.

13.3   Diffuse Scattering from Small Lattice Defects

Small lattice defects that give strong diffuse scattering in diffraction patterns fall into two categories: The first is characterized by deviations from the average composition on atomic sites. Examples include short-range ordering in otherwise disordered solid solution alloys, where diffuse scattering arises from fluctuations in atomic scattering. In the second category, diffuse scattering derives from disruption to the lattice. Examples here include atomic vacancies, nanodomains in ferroelectric or ferroelastic crystals, and charge fluctuations in ionic crystals. In all these cases, an averaged lattice can be defined. The two cases can be distinguished immediately from the form of the diffuse scattering—in the first case (substitutional disorder), the diffuse is peaked at the origin of reciprocal space, while in the second it falls to zero around the origin, as for thermal diffuse scattering. The small defects belong to the type I defects defined by Krivoglaz (type II defects are defined by the lack of distinct Bragg diffraction spots) (Krivoglaz 1996). Scattering from small lattice defects thus can be treated based on deviation from the average periodic interaction potential

$$\Delta U(\vec{r}) = U(\vec{r}) - \langle U(\vec{r}) \rangle, \tag{13.15}$$

where $U(\vec{r})$ is the interaction potential of the real crystal and $\langle U(\vec{r}) \rangle$ is the periodic, averaged potential. Dependent on the type of defects, $\Delta U(\vec{r})$ can be approximated as (A) localized at the defect site or (B) extended involving a large number of atoms. The periodically averaged potential is obtained by adding every unit cell of the actual crystal into one cell, and then periodically extending this.

In what follows, we will discuss case B, involving the modification of the interaction potential of multiple atoms, such as the case of atomic vacancies where the missing atom(s) causes neighboring atoms to displace over an extended spatial region. A treatment of case A is given in the next section.

In the kinematical approximation, the electron diffraction intensity from a nonperiodic potential is given by

$$I(\vec{q}) = |F(\vec{q})|^2 = \left| \int e^{i\vec{q}\cdot\vec{r}} [\langle U(\vec{r}) \rangle + \Delta U(\vec{r})] d\vec{r} \right|^2, \tag{13.16}$$

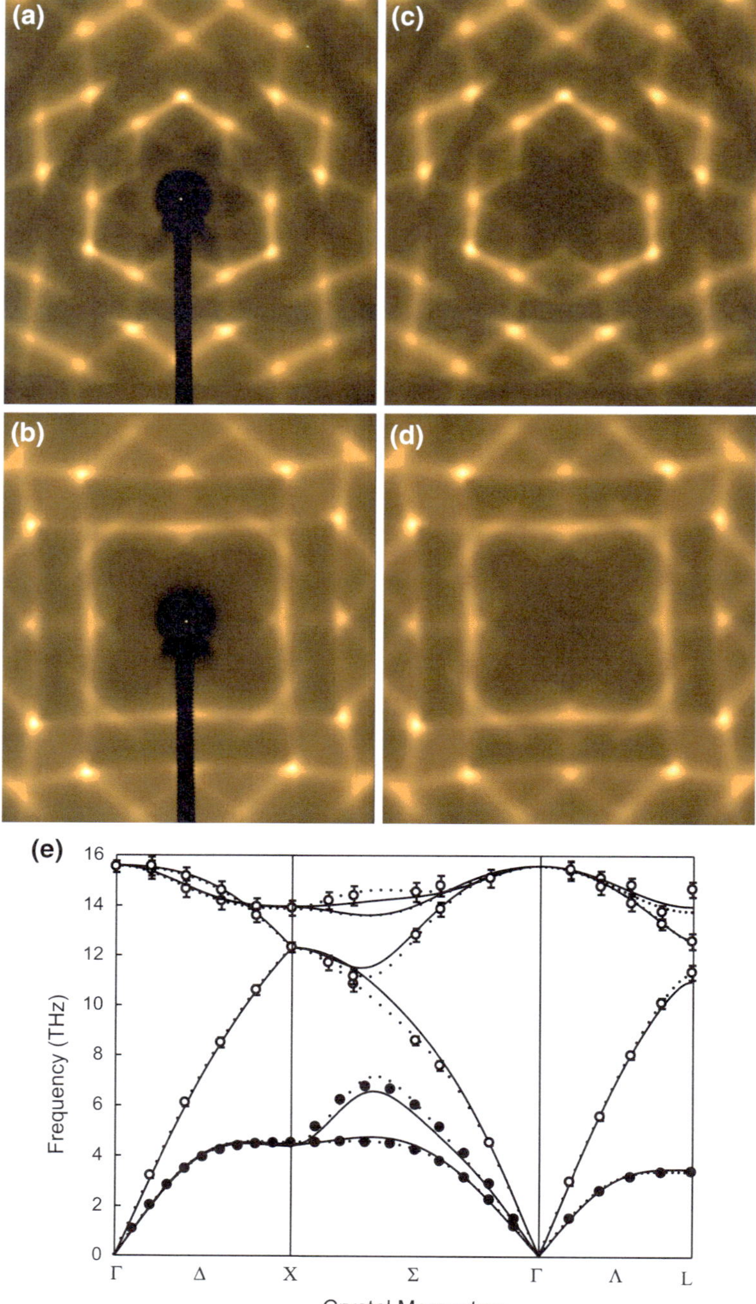
(a)
(b)
(c)
(d)
(e)
16
14
12
10
8
6
4
2
0
Frequency (THz)
Γ   Δ   X   Σ   Γ   Λ   L
Crystal Momentum

◀ **Fig. 13.2** Experimental X-ray TDS scattering for (**a**) Si [111] and (**b**) Si [100] as well as theoretical first-order TDS images for (**c**) Si(111) and (**d**) Si(100). The theoretical images are calculated from a sixth-nearest-neighbor Born-von Karman force constant model. (**e**) Phonon dispersion curves of Si. Circles are neutron scattering data. Solid curves are results derived from best fits to the TDS patterns. (Provided by T. C. Chiang, University of Illinois, Urbana-Champaign)

where $\vec{q} = 2\pi\left(\vec{K} - \vec{K}_\mathrm{o}\right)$ is the scattering vector. For small defects that are much smaller than the size of the electron probe, diffraction intensity is obtained over a large number of defects. In such cases, what is recorded in a diffraction pattern is the spatially averaged diffraction intensity, which is given by

$$
\begin{aligned}
\langle I(\vec{q})\rangle = \left\langle |F(\vec{q})|^2 \right\rangle &= \int\int e^{i\vec{q}\cdot(\vec{r}_1-\vec{r}_2)}\mathrm{d}\vec{r}_1\mathrm{d}\vec{r}_2\langle U(\vec{r}_1)U(\vec{r}_2)\rangle \\
&= \int\int e^{i\vec{q}\cdot(\vec{r}_1-\vec{r}_2)}\mathrm{d}\vec{r}_1\mathrm{d}\vec{r}_2[\langle U(\vec{r}_1)\rangle\langle U(\vec{r}_2)\rangle + \langle \Delta U(\vec{r}_1)\Delta U(\vec{r}_2)\rangle] \\
&= I_\mathrm{o}(\vec{q}) + VS(\vec{q})
\end{aligned}
\tag{13.17}
$$

where

$$
I_\mathrm{o}(\vec{q}) = \left| \int e^{i\vec{q}\cdot\vec{r}}\langle U(\vec{r})\rangle\mathrm{d}\vec{r} \right|^2
\tag{13.18}
$$

and

$$
\begin{aligned}
S(\vec{q}) &= \frac{1}{V}\int\int e^{i\vec{q}\cdot(\vec{r}_1-\vec{r}_2)}\mathrm{d}\vec{r}_1\mathrm{d}\vec{r}_2\langle \Delta U(\vec{r}_1)\Delta U(\vec{r}_2)\rangle \\
&= \frac{1}{V}\int\int e^{i\vec{q}\cdot(\vec{r}_1-\vec{r}_2)}\mathrm{d}\vec{r}_1\mathrm{d}\vec{r}_2 S_2(\vec{r}_1,\vec{r}_2),
\end{aligned}
\tag{13.19}
$$

with V for the volume of the sample under the electron probe. The function

$$
S_2(\vec{r}_1,\vec{r}_2) = \langle \Delta U(\vec{r}_1)\Delta U(\vec{r}_2)\rangle = \langle U(\vec{r}_1)U(\vec{r}_2)\rangle - \langle U(\vec{r}_1)\rangle\langle U(\vec{r}_2)\rangle
\tag{13.20}
$$

is known in statistics as Ursell function or connected correlation function.

To simplify the mathematics, we will assume a uniform atomic displacement within a crystal unit cell. This approximation is based on the elastic response of a crystal lattice to defects, which works very well except at the center of the defect where there is a net force. It is also sufficient as long as we focus on the diffuse scattering near the Bragg peak, whose intensity is largely determined by atomic displacements away from the defect. Under the above approximation, we have

$$
U\left(\vec{r}\right) = U_\mathrm{uc}(\vec{r}) \otimes \sum_m \delta\left(\vec{r} - \vec{R}_m - \vec{u}_m\right).
\tag{13.21}
$$

Here the first term $U_{\mathrm{uc}}(\vec{r})$ describes the potential in the crystal unit cell resulting from the contribution of each atom in the cell. Using this, we have

$$
\int\int e^{i\vec{q}\cdot(\vec{r}_1-\vec{r}_2)}d\vec{r}_1 d\vec{r}_2 \langle U(\vec{r}_1)U(\vec{r}_2)\rangle
$$

$$
= |F(\vec{q})|^2 \int\int e^{i\vec{q}\cdot(\vec{r}_1-\vec{r}_2)}d\vec{r}_1 d\vec{r}_2 \left\langle \sum_m \sum_{m'} \delta(\vec{r}-\vec{R}_m-\vec{u}_m)\delta(\vec{r}-\vec{R}_{m'}-\vec{u}_{m'})\right\rangle
$$

$$
= |F(\vec{q})|^2 \sum_m \sum_{m'} e^{i\vec{q}\cdot(\vec{R}_m-\vec{R}_{m'})}\left\langle e^{i\vec{q}\cdot(\vec{u}_m-\vec{u}_{m'})}\right\rangle.
$$

and

$$
\int\int e^{i\vec{q}\cdot(\vec{r}_1-\vec{r}_2)}d\vec{r}_1 d\vec{r}_2 \langle U(\vec{r}_1)\rangle\langle U(\vec{r}_2)\rangle
$$

$$
= |F(\vec{q})|^2 \sum_m \sum_{m'} e^{i\vec{q}\cdot(\vec{R}_m-\vec{R}_{m'})}\left\langle e^{i\vec{q}\cdot\vec{u}_m}\right\rangle\left\langle e^{i\vec{q}\cdot\vec{u}_{m'}}\right\rangle.
$$

Substitution of the above results into Eq. (13.19) gives the following equation for the diffuse scattering

$$
S(\vec{q}) = |F(\vec{q})|^2 \sum_m \sum_{m'} e^{i\vec{q}\cdot(\vec{R}_m-\vec{R}_{m'})}\left\{\left\langle e^{i\vec{q}\cdot(\vec{u}_m-\vec{u}_{m'})}\right\rangle - \left\langle e^{i\vec{q}\cdot\vec{u}_m}\right\rangle\left\langle e^{i\vec{q}\cdot\vec{u}_{m'}}\right\rangle\right\}. \tag{13.22}
$$

This equation shows that the diffuse scattering intensity is obtained from the difference between the first and second terms, which gives the total scattering and the Bragg intensities, respectively.

A defect at the lattice site n causes the atom at the lattice site m to be displaced by a vector $\vec{t}_{m,n}$. The net displacement $\vec{u}_m$ of the atom is a sum of contributions from all defects in the form

$$
\vec{u}_m = \sum_n c_n \vec{t}_{m,n}, \tag{13.23}
$$

where $c_n = 1$, if the lattice site n is occupied by a defect or $c_n = 0$ otherwise. We will assume that defects are randomly distributed; then, c_n is a random number with the following properties: $\langle c_n\rangle = c$, $\langle c_n^2\rangle = \langle c_n\rangle = c$, etc., where c is the average concentration of defects per lattice site. The atomic displacement $\vec{t}_{m,n}$ can be further separated into a local component $\vec{t}_{m-n} \propto 1/\left|\vec{R}_m-\vec{R}_n\right|^2$ due to the force exerted by the defect and a slow varying component $\vec{\tau}_{m,n}$ that can be attributed to the image forces resulting from a finite crystal (Dederich 1971). These two together contribute to an "expanded" lattice, where the average lattice vector is given by

$$\left\langle \vec{R}_m \right\rangle = \vec{R}_m + \langle \vec{u}_m \rangle = \vec{R}_m + c \sum_n \left(\vec{t}_{m-n} + \vec{\tau}_{m,n} \right). \tag{13.24}$$

To evaluate Eq. (13.22) using the above the results, we make use of the following approximation (Dederich 1971)

$$\left\langle e^{i\vec{q}\cdot\vec{u}_m} \right\rangle = \prod_n \left\langle e^{i\vec{q}\cdot c_n \vec{t}_{m,n}} \right\rangle = \prod_n \left[1 + c\left(e^{i\vec{q}\cdot\vec{t}_{m,n}} - 1 \right) \right] = \exp\left\{ \sum_n \ln\left(1 + c\left(e^{i\vec{q}\cdot\vec{t}_{m,n}} - 1 \right) \right) \right\}$$

$$\approx \exp\left\{ c \sum_n \left(e^{i\vec{q}\cdot\vec{t}_{m,n}} - 1 \right) \right\},$$

$$\tag{13.25}$$

where the approximation is for small c ($c \ll 1$), e.g., at a low concentration of defects.

By combining Eqs. (13.24) and (13.25) with Eq. (13.22), we obtain the following equation

$$S(\vec{q}) = |F(\vec{q})|^2 \sum_m \sum_{m'} e^{i\vec{q}\cdot\left(\langle\vec{R}_m\rangle - \langle\vec{R}_{m'}\rangle\right)}$$

$$\left\{ \begin{array}{c} \exp\left\{ c \sum_n \left[e^{i\vec{q}\cdot(\vec{t}_{m-n} - \vec{t}_{m'-n})} - 1 - i\vec{q}\cdot(\vec{t}_{m-n} - \vec{t}_{m'-n}) \right] \right\} \\[2mm] - \exp\left\{ c \sum_n \left[e^{i\vec{q}\cdot\vec{t}_{m-n}} + e^{-i\vec{q}\cdot\vec{t}_{m'-n}} - 2 - i\vec{q}\cdot(\vec{t}_{m-n} - \vec{t}_{m'-n}) \right] \right\} \end{array} \right\}. \tag{13.26}$$

At a large separation distances with $\left|\vec{R}_m - \vec{R}_{m'}\right| >> R_o$, where R_o defines the radius of the impact by the defect, there is no longer any correlation between the displacements at the two lattice sites and the scattering is determined by the average lattice. Because of this, the bracket in Eq. (13.26) goes to zero, and thus, there are no sharp peaks in $S(\vec{q})$ as for Bragg diffraction. The scattering vector $\vec{q}$ can be written as

$$\vec{q} = \vec{G} + \vec{Q}, \tag{13.27}$$

where $\vec{G}$ is the nearest Bragg reflection and $\vec{Q}$ is the diffuse scattering vector within the Brillouin zone. Significant diffuse scattering is expected near $\left|\vec{Q}\right| \sim 2\pi/R_0$. R_0 is much larger the unit cell dimensions for many inorganic crystals. Considering the above factors, the double summation over m and m' can be replaced by an integration over $\vec{R}$ for all lattice points, which gives the following expression for a sample volume containing N lattice points

$$S\left(\vec{G}+\vec{Q}\right) = \frac{N}{V}\left|F\left(\vec{G}+\vec{Q}\right)\right|^{2}\int e^{i\left(\vec{G}+\vec{Q}\right)\cdot\vec{R}}\left\{\exp\left[\phi\left(\vec{R}\right)-\phi(\infty)\right]\right\}d\vec{R},$$

(13.28)

where

$$\phi\left(\vec{R}\right) = \frac{c}{V_c}\int\left[e^{i\left(\vec{G}+\vec{Q}\right)\cdot\left[\vec{t}\left(\vec{r}+\vec{R}\right)-\vec{t}(\vec{r})\right]} - 1 - i\left(\vec{G}+\vec{Q}\right)\cdot\left[\vec{t}\left(\vec{r}+\vec{R}\right)-\vec{t}(\vec{r})\right]\right]d\vec{r},$$

(13.29)

and

$$\begin{aligned}
\phi(\infty) &= \frac{c}{V_c}\int\left[e^{i\left(\vec{G}+\vec{Q}\right)\cdot\vec{t}\left(\vec{r}+\vec{R}\right)} + e^{-i\left(\vec{G}+\vec{Q}\right)\cdot\vec{t}(\vec{r})} - 2 - i\left(\vec{G}+\vec{Q}\right)\cdot\left[\vec{t}\left(\vec{r}+\vec{R}\right)-\vec{t}(\vec{r})\right]\right]d\vec{r} \\
&= -\frac{2c}{V_c}\int\left[1-\cos\left(\vec{G}+\vec{Q}\right)\cdot\vec{t}(\vec{r})\right]d\vec{r} \\
&= -2M_{\vec{G}+\vec{Q}}.
\end{aligned}$$

(13.30)

The term $2M_{\vec{G}+\vec{Q}}$ in Eq. (13.30) contributes to the Debye–Waller factor; thus, a further reduction of Bragg diffraction intensity from atomic displacements is induced by defects in addition to contribution from thermal vibrations. These two can be separated because the defect contribution is approximately constant at different temperatures.

Using the result in Eq. (13.30), it can be shown that Eq. (13.29) can be rewritten as

$$\phi\left(\vec{R}\right) = -2M_{\vec{G}+\vec{Q}} + \frac{c}{V_c}\int\left\{e^{i\left(\vec{G}+\vec{Q}\right)\cdot\vec{t}\left(\vec{r}+\vec{R}\right)} - 1\right\}\left\{e^{-i\left(\vec{G}+\vec{Q}\right)\cdot\vec{t}(\vec{r})} - 1\right\}d\vec{r}.$$

(13.31)

In the limit of small defect concentrations with $c < \!\!< 1$, we have, to first order,

$$e^{\phi(\infty)} \approx 1 - 2M_{\vec{G}+\vec{Q}}$$

(13.32)

and

$$e^{\phi\left(\vec{R}\right)} \approx 1 - 2M_{\vec{G}+\vec{Q}} + \frac{c}{V_c}\int\left\{e^{i\left(\vec{G}+\vec{Q}\right)\cdot\vec{u}\left(\vec{r}+\vec{R}\right)} - 1\right\}\left\{e^{-i\left(\vec{G}+\vec{Q}\right)\cdot\vec{u}(\vec{r})} - 1\right\}d\vec{r}.$$

(13.33)

The integral in Eq. (13.33) follows the general form of

$$P\left(\vec{R}\right) = \int f^*(\vec{r})f\left(\vec{r}+\vec{R}\right)d\vec{r},$$

(13.34)

which is known as Patterson function or autocorrelation function $\left(P(\vec{R}) = f^*(\vec{r})*\right.$
$\left. f(-\vec{r})\right)$ in signal processing. It has a special property in crystallography as its
Fourier transform gives the kinematical diffraction intensity. Thus, by substituting
the above result into (13.28), we obtain

$$S\left(\vec{G}+\vec{Q}\right) \approx \frac{Nc}{V}\left|F\left(\vec{G}+\vec{Q}\right)\right|^2\left|D\left(\vec{G}+\vec{Q}\right)\right|^2,$$

(13.35)

where

$$D\left(\vec{G}+\vec{Q}\right) = \frac{1}{V_c}\int e^{i\left(\vec{G}+\vec{Q}\right)\cdot\vec{r}}\left\{e^{i\left(\vec{G}+\vec{Q}\right)\cdot\vec{u}(\vec{r})} - 1\right\}d\vec{r}$$

(13.36)

for a low concentration of defects.

Diffuse scattering from crystals containing dislocation loops has been calculated
by Ehrhart et al. (1982). Characteristic differences of diffuse scattering are shown
between perfect and faulted loops and between vacancy- and interstitial-type loops.
Such results can be directly compared with experimental results. Figure 13.3 shows
an example. Using a nanometer-sized coherent electron parallel beam, Kirk et al.
(2005) has demonstrated that electron diffuse scattering can be recorded from a
single nanometer-sized defect. The pattern in Fig. 13.3 is from a Frank loop of size
near 5 nm in an electropolished gold foil, irradiation by 1 meV Kr^{2+} to a fluence of
3×10^{10} ions/cm^2. The microscope used for electron diffraction was the JEOL
2010F (200 kV) in the Center for Microanalysis of Materials in the Frederick Seitz
Materials Research Laboratory of the University of Illinois, Champaign–Urbana.
This microscope produces a small, nearly parallel, coherent electron beam, which
was used together with a Gatan Imaging Filter and CCD camera for zero-loss
energy filtering and digital data acquisition. By comparing the experimental diffuse
scattering pattern with that calculated by Ehrhart et al. (1982), Kirk et al. (2005)
concluded that the nature of the loop is most likely to be vacancy.

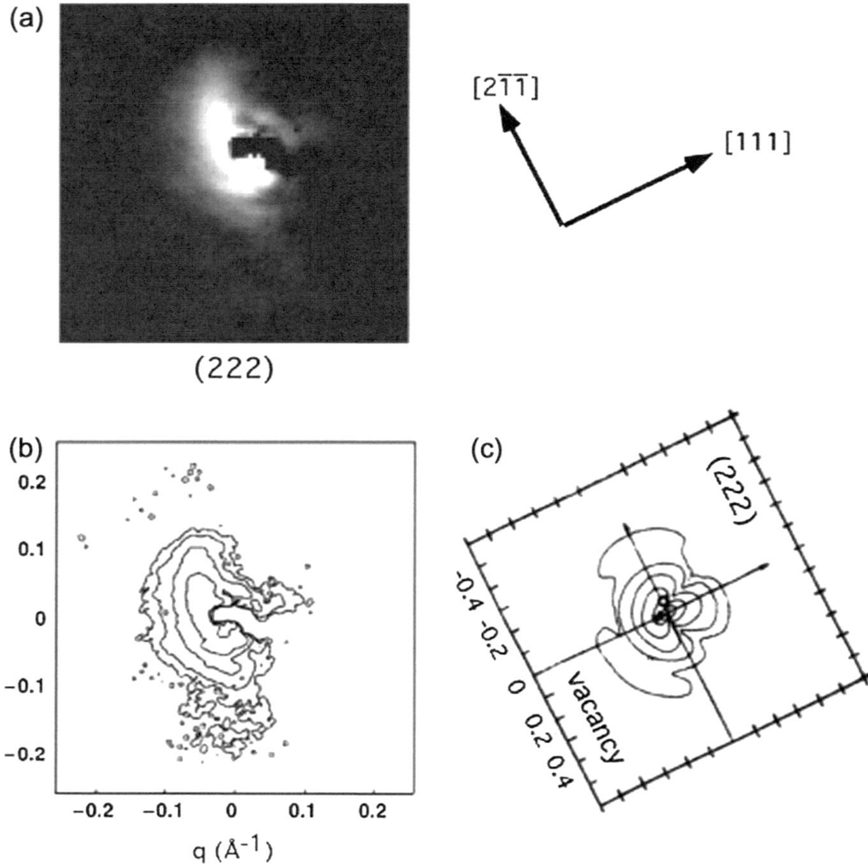

Fig. 13.3 Experimental diffuse scattering (panels (**a**) and (**b**)) from a Frank loop. Panel (**c**) is the calculated X-ray diffuse scattering for another specific Frank loop (on plane) from Ehrhart et al. (1982) (from Kirk et al. 2005, reproduced with permission)

13.4   Scattering by Solid Solutions

The simplest examples of a solid solution are binary alloys (A_xB_{1-x}) such as β-brass (CuZn) or CuAu. In the CuZn alloy, above a critical temperature (T_c), the two sites of a body-centered lattice are occupied by the Cu and Zn atoms in a random (disordered) way such that the probability of finding one of them at a particular site (probability of occupancy) is proportional to its atomic fraction. Below T_c, partial ordering develops over many unit cells where the probability of occupancy depends on the composition of neighboring atomic sites. In alloys, ordering is an important parameter determining their physical properties and thus, the development of ordering in the microstructure has great significance in commercial applications.

In complex oxides or sulfides, ordering is related to the nonstoichiometry and phase transitions. The development of diffuse scattering associated with charge ordering in magnetite (Fe_3O_4, see Fig. 13.1) is an example.

Scattering by a solid solution may be treated following the example of crystals with small defects. The scattering potential is separated into an average, or perfect, crystal potential comprised of averaged periodically arranged "atoms" and a deviation potential representing inhomogeneity in the crystal. There are two contributions to the crystal inhomogeneity: One comes from the difference in the atomic scattering factors of different atoms and the other from atomic displacements. Since different atoms have different sizes, the local interatomic distances are determined by local composition; e.g., whether the A atom is near other A atoms, or whether a B atom is near other B atoms. In a partially ordered alloy, a distribution of local displacements occurs because it is then energetically favorable for the atoms to relax from their average positions in a periodic crystal. This gain of energy, called "relaxation energy," is important for the thermodynamics of many alloys (Ducastelle 1991, pp. 439–442). The combination of atomic scattering factor difference and atomic displacements gives rise to following scattering effects:

(1) A shift in the Bragg peak position,
(2) Weakening of Bragg diffraction intensity,
(3) The appearance of diffuse scattering.

The shift in Bragg peak position corresponds to the "expanded" lattice as defined in Eq. (13.24), while weakening of the Bragg diffraction intensity can be attributed to the contribution of atomic displacements to the static Debye–Waller factor (Eq. (13.30) and further discussed below). In the case that the difference in atomic sizes is large in an alloy, or the atomic sizes are coupled with charge in an ionic crystal, atomic displacements extend to a small region around each solute atom and the "atomic size effect" is similar that of small lattice defects, which causes a broad diffuse intensity to appear asymmetrically around the Bragg peaks. In what follows, we will consider the case where atomic displacements are rather small, and diffuse scattering is dominated by modulations in composition.

To develop a scattering model for solid solutions, we will consider a binary solid solution of A and B atoms in a crystal with N unit cells, and there are v number of atomic sites within the unit cell. The nth atomic site inside the mth unit cell is occupied by an atom with the atomic potential

$$c_{mn}U_A + (1 - c_{mn})U_B, \qquad (13.37)$$

where $c_{mn} = 1$ or 0, when occupied by A or B atom. Combining this result with the δ-function to represent the atomic position, as we have done previously, gives the following mathematical form for the crystal potential of a solid solution

$$U(\vec{r}) = \sum_{m=1}^{N}\sum_{n=1}^{v} [c_{mn}U_A(\vec{r}) + (1 - c_{mn})U_B(\vec{r})] * \delta\left(\vec{r} - \vec{R}_m - \vec{r}_n - \vec{u}_{mn}\right). \quad (13.38)$$

In the above equation, $\vec{R}_m$ and $\vec{r}_n$ denote the unit cell and the atomic position inside the unit cell, respectively, and $\vec{u}_{mn}$ is introduced to represent the atomic displacement due to atomic substitution-induced distortion.

For the Bragg reflection, according to Eq. (13.18), we have

$$
\begin{aligned}
I_0 &= \left| \sum_{m=1}^{N} \sum_{n=1}^{v} \int e^{i\vec{q}\cdot\vec{r}} \left\langle [c_{mn}U_A(\vec{r}) + (1 - c_{mn})U_B(\vec{r})] * \delta\left(\vec{r} - \vec{R}_m - \vec{r}_n - \vec{u}_{mn}\right) \right\rangle d\vec{r} \right|^2 \\
&= \left| \sum_{m=1}^{N} \sum_{n=1}^{v} e^{i\vec{q}\cdot\left(\vec{R}_m + \vec{r}_n\right)} \left\{ \left\langle c_{mn}f_A e^{i\vec{q}\cdot\vec{u}_{mn}} \right\rangle + \left\langle (1 - c_{mn})f_B e^{i\vec{q}\cdot\vec{u}_{mn}} \right\rangle \right\} \right|^2 \\
&= \left| \sum_{m=1}^{N} \sum_{n=1}^{v} e^{i\vec{q}\cdot\left(\vec{R}_m + \vec{r}_n\right)} \left[cf_A e^{-M_A} + (1 - c)f_B e^{-M_B} \right] \right|^2 \\
&= \left| N \sum_{\vec{G}} F(\vec{q})\delta(\vec{q} - \vec{G}) \right|^2,
\end{aligned}
$$

$$\tag{13.39}$$

where

$$
\begin{aligned}
F(\vec{q}) &= \sum_{n=1}^{v} e^{i\vec{q}\cdot\vec{r}_n} \left[c_n f_A e^{-M_A} + (1 - c_n)f_B e^{-M_B} \right] \\
&= \sum_{n=1}^{v} \bar{f}_n e^{i\vec{q}\cdot\vec{r}_n}.
\end{aligned}
$$

$$\tag{13.40}$$

Thus, the Bragg diffraction in a solid solution is described by the structure factor obtained with the average, site-specific, atomic scattering factor $\bar{f}_n = c_n f_A e^{-M_A} + (1 - c_n)f_B e^{-M_B}$ with c_n for the volume-averaged composition, $e^{-M_A} = \left\langle e^{i\vec{q}\cdot\vec{u}_{mn}} \right\rangle_{A,n}$ and $e^{-M_B} = \left\langle e^{i\vec{q}\cdot\vec{u}_{mn}} \right\rangle_{B,n}$ where the averaging is carried over all atomic sites of n occupied by A and B atoms, respectively. The M_A and M_B thus are the static Debye–Waller factors introduced by atomic distortions. They are different in general in both cases of different atoms occupying the same atomic sites or same atoms occupying different atomic sites.

The above results show that for Bragg diffraction a solid solution takes the form of an ideal crystal. The diffraction intensity is determined by the averaged atomic scattering factor at each atomic position inside the unit cell and an addition to the atomic Debye–Waller factor from atomic distortions. The shift in the Bragg peak position comes from an overall "expansion" of the lattice, when $\langle \vec{u}_{mn} \rangle = \vec{x}_m$ leads to a shift in the unit cell position.

Next, we evaluate the integral in Eq. (13.19) for the diffuse scattering. We start with

$$I(\vec{q}) = \int \int e^{i\vec{q}\cdot(\vec{r}_1 - \vec{r}_2)} d\vec{r}_1 d\vec{r}_2 \langle U(\vec{r}_1) U(\vec{r}_2) \rangle$$

$$= \left\langle \left| \sum_{n=1}^{v} e^{i\vec{q}\cdot\vec{r}_n} \sum_{m=1}^{N} [c_{mn} f_A + (1 - c_{mn}) f_B] e^{i\vec{q}\cdot\vec{R}_m} e^{i\vec{q}\cdot\vec{u}_{mn}} \right|^2 \right\rangle.$$

$$(13.41)$$

In an alloy exhibiting significant deviations from random occupation, a preference of one type of atom over another develops, depending on the neighboring atoms. Under such circumstances, the composition fluctuation can be represented by a Fourier series

$$c_{mn} - c_n = \sum_{\vec{k}} C_n\!\left(\vec{k}\right) e^{-i\vec{k}\cdot\vec{R}_{mn}}$$

$$(13.42)$$

where

$$C_n\!\left(\vec{k}\right) = \sum_{m=1}^{N} (c_{mn} - c_n) e^{i\vec{k}\cdot\vec{R}_{mn}}.$$

$$(13.43)$$

Similarly for the atomic displacement, we write

$$\vec{u}_{mn} = \sum_{\vec{k}} \vec{U}_n\!\left(\vec{k}\right) e^{-i\vec{k}\cdot\vec{R}_{mn}}$$

$$(13.44)$$

and

$$\vec{U}_n\!\left(\vec{k}\right) = \sum_{m=1}^{N} \vec{u}_{mn} e^{i\vec{k}\cdot\vec{R}_{mn}}.$$

$$(13.45)$$

The distortion introduced by atomic substitution in alloys is small, compared to other defects such as interstitials or atomic vacancies. For relatively small scattering vectors, to a good approximation we can use the linear approximation

$$e^{i\vec{q}\cdot\vec{u}_{mn}} \approx 1 + i\vec{q}\cdot\vec{u}_{mn}.$$

Substituting the above results into Eq. (13.41), we obtain

$$I(\vec{q}) \approx \left\langle \left| \sum_{n=1}^{v} e^{i\vec{q}\cdot\vec{r}_n} \sum_{m=1}^{N} \left[\bar{f}_{no} + \sum_{\vec{k}} \left[(f_A - f_B) C_n\!\left(\vec{k}\right) + i\vec{q}\cdot\vec{U}_n\!\left(\vec{k}\right) \bar{f}_o \right] e^{-i\vec{k}\cdot\vec{R}_{mn}} \right] e^{i\vec{Q}\cdot\vec{R}_m} \right|^2 \right\rangle,$$

$$(13.46)$$

Here $\bar{f}_{no} = c_n f_A + (1 - c_n)f_B \approx \bar{f}_n$ and $\vec{Q} = \vec{q} - \vec{G}$ (see Eq. (13.27)). The approximation is made by only keeping the first-order terms involving $C_n\left(\vec{k}\right)$ and $\vec{U}_n\left(\vec{k}\right)$.

In the limit of a large crystal,

$$\sum_{m=1}^{N} e^{i\left(\vec{Q}-\vec{k}\right)\cdot\vec{R}_m} \approx N\delta\left(\vec{Q}-\vec{k}\right).$$

Thus, we have two terms from Eq. (13.46): One corresponds to Bragg diffraction (I_0) with $\vec{Q} = 0$ and other with $\vec{Q} = \vec{k}$ gives the diffuse scattering

$$I_1\left(\vec{G}+\vec{Q}\right) = N^2 \left\langle \left| \sum_{n=1}^{v} e^{i\vec{G}\cdot\vec{r}_n} \left[(f_A - f_B)C_n\left(\vec{Q}\right) + i\left(\vec{G}+\vec{Q}\right)\cdot\vec{U}_n\left(\vec{Q}\right)\bar{f}_0 \right] \right|^2 \right\rangle.$$

$$(13.47)$$

The second term inside [] in Eq. (13.47) describes the contribution from atomic distortions. It is asymmetric around the Bragg reflection with a change of the sign for $\vec{Q}$. The first term is proportional to the difference in atomic scattering factors of A and B atom and the extent of ordering as described by the Fourier coefficient $C_n\left(\vec{k}\right)$. These two terms interfere in a solid solution that exhibit short-range ordering and the related atomic distortion.

Equation (13.47) is greatly simplified when there is only one atom per unit cell $(v = 1)$. For example, both fcc and bcc crystals can be described by a primitive unit cell. The atomic distortion is proportional to the composition in this case; thus, we have

$$I_1\left(\vec{G}+\vec{Q}\right) = N^2 \left\langle \left| C\left(\vec{Q}\right) \right|^2 (f_A - f_B) + i\left(\vec{G}+\vec{Q}\right)\cdot\vec{A}\left(\vec{Q}\right)\bar{f}_0 \right|^2 \right\rangle \qquad (13.48)$$

where $\vec{A}\left(\vec{Q}\right)$ is the proportional vector that relates the compositional term $C_n\left(\vec{Q}\right)$ to the atomic displacement term, $\vec{U}_n\left(\vec{Q}\right)$. In the case of very small distortions, $\vec{A}\left(\vec{k}\right) \approx 0$; then,

$$I_1\left(\vec{G}+\vec{Q}\right) = N^2(f_A - f_B)^2 \left\langle \left| C\left(\vec{Q}\right) \right|^2 \right\rangle.$$

To calculate $\left\langle \left| C\left(\vec{Q}\right) \right|^2 \right\rangle$, we recall Eq. (13.43) and obtain

$$\left\langle \left| C_n\left(\vec{Q}\right) \right|^2 \right\rangle = \frac{1}{N^2}\left[\sum_m \left\langle (c_{mn} - c_n)(c_{mn} - c_n) \right\rangle + \sum_{m \neq m'} \left\langle (c_{mn} - c_n)(c_{m'n} - c_n) \right\rangle e^{i\vec{Q}\cdot(\vec{R}_m - \vec{R}_{m'})} \right]$$

Since c_{mn} has the value of 1 or 0 only, $\left\langle c_{mn}^2 \right\rangle = c_{mn}$ and thus,

$$\left\langle (c_{mn} - c_n)(c_{mn} - c_n) \right\rangle = c_n(1 - c_n),$$

and

$$\begin{aligned}
\left\langle \left| C_n\left(\vec{Q}\right) \right|^2 \right\rangle &= \frac{1}{N^2}\left[\sum_m c_n(1 - c_n) + \sum_{m \neq m'} \left[\left\langle c_{mn} c_{m'n} \right\rangle - c_n^2 \right] e^{i\vec{Q}\cdot(\vec{R}_m - \vec{R}_{m'})} \right] \\
&= \frac{1}{N^2}\left[\sum_m c_n(1 - c_n) + \sum_{m \neq m'} \left[P_{mn,m'n}^{AA} - c_n^2 \right] e^{i\vec{Q}\cdot(\vec{R}_m - \vec{R}_{m'})} \right],
\end{aligned}$$

$$(13.49)$$

where $P_{mn,m'n}^{AA}$ is the probability of finding A atom at the m'n site with A atoms at the mn site, or vice versa. Using the symmetry of $P_{mn,m'n}^{AA}$, Eq. (13.49) can be simplified as

$$\begin{aligned}
\left\langle \left| C_n\left(\vec{Q}\right) \right|^2 \right\rangle &= \frac{1}{N^2}\left[\sum_m c_n(1 - c_n) + \sum_{m \neq m'} \left[P_{mn,m'n}^{AA} - c_n^2 \right] e^{i\vec{Q}\cdot(\vec{R}_m - \vec{R}_{m'})} \right] \\
&= \frac{1}{N}\left\{ c_n(1 - c_n) + \sum_{\vec{R} \neq 0} \varepsilon(\vec{R}) \cos\left[\vec{Q} \cdot (\vec{R}_m - \vec{R}_{m'}) \right] \right\}
\end{aligned}$$

where $\varepsilon\left(\vec{R}\right)$ is the correlation parameter. Using this result, we obtain

$$I_1\left(\vec{G} + \vec{Q}\right) = N(f_A - f_B)^2 \left\{ c_n(1 - c_n) + \sum_{\vec{R} \neq 0} \varepsilon(\vec{R}) \cos\left[\vec{Q} \cdot (\vec{R}_m - \vec{R}_{m'}) \right] \right\}.$$

And (see Krivoglaz 1996, p. 164)

$$\varepsilon\left(\vec{R}\right) = \frac{v}{8\pi^3} \int_{1/v} \frac{I_1}{N}(f_A - f_B)^{-2} \cos \vec{Q} \cdot \vec{R} \, d\vec{Q},$$

where the integration is carried out over area υ within the unit cell of the reciprocal lattice.

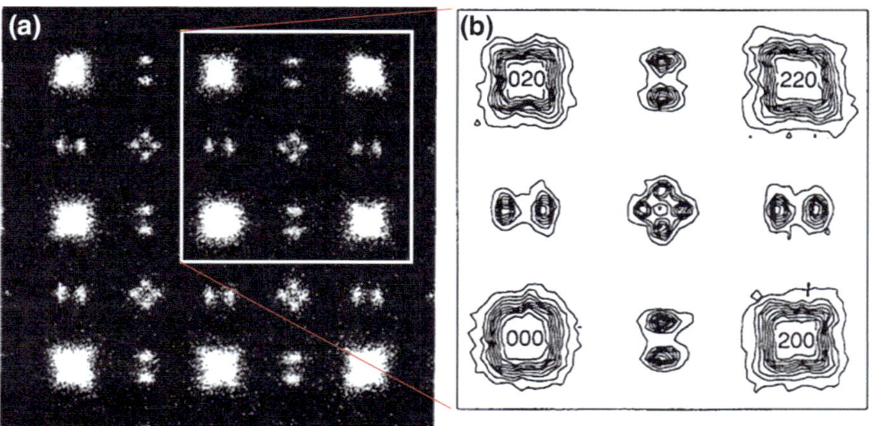

Fig. 13.4 Energy-filtered electron diffraction pattern recorded from a Cu-27.5 at %Pd alloy crystal of 281 nm thick. (From Ikematsu et al. (2000))

Figure 13.4 shows the recorded elastic energy-filtered diffuse scattering from a Cu–Pd alloy at 27.5 at.% of Pd by Ikematsu et al. (2000). The characteristic twofold and fourfold splitting diffuse maxima around {100} and {110} in the diffraction pattern are observed in the Pd concentration range from about 13.0 to 60.0 at.% (Ohshima and Watanabe 1973a, b). This diffuse scattering comes from several types of superlattice structure as illustrated in Fig. 13.5. They are built from the unit cell of the ordered Cu_3Au type ($L1_2$ structure). Along the ordering direction (c-axis) in the superlattice, the composition alternates between Cu–Pd and Cu–Cu and the repeat is interrupted by the anti-phase domain boundary. The superstructures form below the composition-dependent critical temperature T_c. Above T_c, the Cu–Pd alloy system forms a continuous series of solid solutions with the fcc structure (Massalski and Okamoto 1990, p 1454). The alloys with Pd concentrations from 18 to 28 at.% have periodic one-dimensional or two-dimensional anti-phase domain structures in the ordered state, depending on the composition and temperature.

The diffraction pattern can be qualitatively understood from Fig. 13.5. Compared to the disordered phase, which has the fcc structure, additional weak diffraction spots appear at the {100} and {110} positions for the ordered Cu_3Pd phase (Fig. 13.5 a, c). In the superlattice structure model of Fig. 13.5c, the unit cell has the approximate dimensions of a, a, and $2Ma$, where M is half of the number of Cu3Pd unit cells along the superlattice direction. The domain boundary is characterized by a shift of (1/2, 1/2, 0), which gives a body-centered lattice with the extinction condition of $h + k+l = 2n + 1$. With the electron beam along [010], we have diffraction peaks only at $l = 2n$ with h even. When h is odd, we have diffraction peaks when l is odd. They are separated by an odd multiple of $1/2\,M$, which gives the distance of the two nearest peaks at $1/M$. In a real alloy, the superlattice can be along any of the three cubic axes. Further, there are several

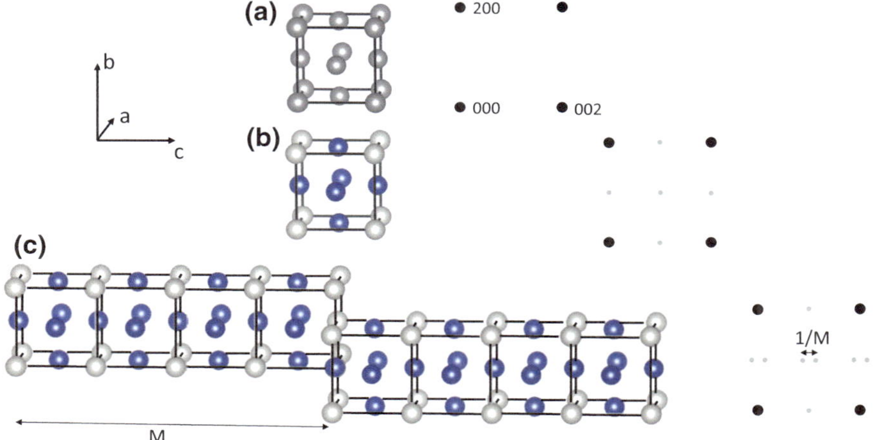

Fig. 13.5 Model structure of Cu_xPd **a** solid solution, **b** Cu_3Pd ($L1_2$ structure), and **c** superlattice with two blocks of 4 unit cells of Cu_3Pd with a domain boundary, and their corresponding diffraction patterns

variants of superlattice structures associated with the domain boundaries occurring at different intervals, and the diffraction pattern is a sum of equivalent distributions of these variants and directions.

13.5   Modulated Structures

Sharp diffraction spots are observed in modulated structures with long-range ordering and a single or few harmonic modulation frequencies. Examples of such cases include crystals exhibiting periodic modulation of electron charge density (charge-density wave or CDW) accompanied by distortion in the lattice, incommensurate structure in the Bi-based high-T_c superconductors ($Bi_2Sr_2Ca_{n-1}Cu_nO_{2n+4}$), and superstructures observed in $La_{1-x}Ca_xMnO_3$. CDWs can be found in 1D chain compounds such as $NbSe_3$ or layered 2D materials such as RTe_3 (R for rare earth elements). The incommensurate structure in $Bi_2Sr_2Ca_{n-1}Cu_nO_{2n+4}$ is attributed to the extra oxygen atoms in the Bi–O layer (Petricek et al. 1990), while superstructures in $La_{1-x}Ca_xMnO_3$ are attributed to ordering of Mn^{3+} and the related lattice distortion (see Fig. 13.6 for examples of electron diffraction patterns). In each of these cases, the materials' electrical conductivity is dramatically influenced by the change in structure. To understand the mechanism of the CDW transition or the properties in the modulated state, it is important to know the structure of the incommensurate-modulated phases.

In a modulated crystal structure, atoms are subjected to substitutional and/or positional fluctuations. If the period of modulation matches an integer ratio of the period of the averaged crystal structure, then a superstructure results, such as the

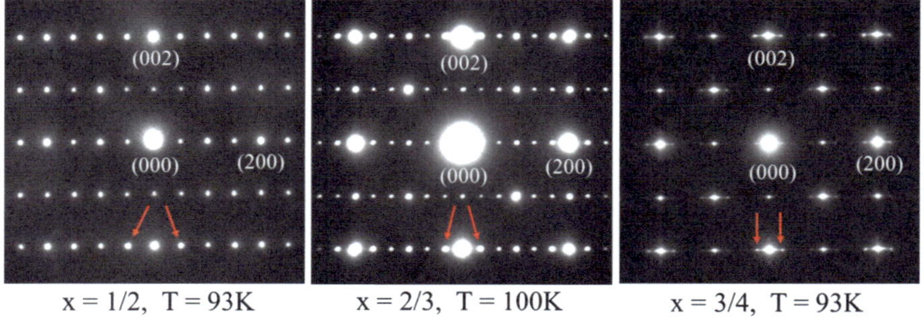

Fig. 13.6 Electron diffraction patterns recorded at low temperatures from $La_{1-x}Ca_xMnO_3$ of perovskite structure with $x = 1/2$, $2/3$ and $\frac{3}{4}$. The arrows indicate sharp diffraction spots from low-temperature-modulated structures

superlattices observed in the binary alloy of $Cu_{1-x}Pd_x$. Otherwise, an incommensurate-modulated structure is obtained. A common feature of incommensurate-modulated structures is that they do not have three-dimensional periodicity. In crystallography, incommensurate-modulated structures are mathematically described as the three-dimensional projection of a 4- or higher-dimensional periodic structure (Dewolff 1974; Janner and Janssen 1977; Yamamoto 1996). The symmetry in the higher-dimensional space allows a determination of incommensurate-modulated structures using direct phasing.

The modulated structure is one of the three kinds of quasiperiodic structures to which higher-dimensional crystallography is applicable. The other two are the so-called composite crystals discovered in the 1970s (Makovicky and Hyde 1981) and quasicrystals found in 1984 (Shechtman et al. 1984). A quasicrystal shows no translational periodicity in three-dimensional space and has noncrystallographic rotational symmetry, including fivefold axes. But its diffraction patterns give sharp Bragg diffraction spots, suggesting translational symmetry. This combination of ten sharp spots around the origin therefore appears inexplicable ("ten" indicating 5-fold symmetry, and "sharp" indicating periodicity), until we understand that the Fourier transform of the nonperiodic Fibonacci series describing the crystal plane spacing along one direction converges to delta functions as the number of planes increases. The basis for understanding the structure of quasicrystals is the same as incommensurately modulated structures, namely higher-dimensional crystallography, since the structures are periodic in higher dimensions ($n > 3$), but not when projected onto three dimensions (Goldman and Kelton 1993).

We will first consider a modulated structure where the atoms are displaced by a sinusoidal wave with wavevector $\vec{Q}$ and phase ϕ_n. For an atom at site n in unit cell l, its displacement is given by

$$\vec{u}_{ln} = \vec{U}_n\left(\vec{Q}\right) \sin\left(\vec{Q} \cdot \vec{R}_{ln} - \phi_n\right),$$

where $\vec{U}_n\left(\vec{Q}\right)$ is the amplitude of the modulation wave as a vector, representing displacements along the three axes of the crystal. The structure factor of the modulated crystal is given by

$$F(\vec{q}) = \sum_l \sum_n f_{ln}\,\mathrm{e}^{i\vec{q}\cdot\left(\vec{R}_{ln}+\vec{u}_{ln}\right)}$$

$$= \sum_l \sum_n f_n\,\mathrm{e}^{i\vec{q}\cdot\vec{R}_{ln}}\,\mathrm{e}^{i\vec{q}\cdot\vec{U}_n\left(\vec{Q}\right)\sin\left(\vec{Q}\cdot\vec{R}_{ln}-\phi_n\right)}$$

Using the Jacobi–Auger relation

$$\mathrm{e}^{ix\sin\theta} = \sum_m J_{-m}(x)\mathrm{e}^{-im\theta}$$

We obtain

$$F(\vec{q}) = \sum_n f_n\,\mathrm{e}^{i\vec{q}\cdot\vec{R}_n}\sum_m J_{-m}\left(\vec{q}\cdot\vec{U}_n\left(\vec{Q}\right)\right)\mathrm{e}^{im\phi_n}\sum_l \mathrm{e}^{i\vec{q}\cdot\vec{R}_l}\mathrm{e}^{-im\vec{Q}\cdot\vec{R}_l}$$

$$= \sum_n f_n\,\mathrm{e}^{i\vec{q}\cdot\vec{R}_n}\sum_m (-1)^m J_m\left(\vec{q}\cdot\vec{U}_n\left(\vec{Q}\right)\right)\mathrm{e}^{im\phi_n}\sum_{\vec{G}}\delta\left(\vec{q}-\vec{G}-m\vec{Q}\right).$$

Thus, diffraction peaks occur at $\vec{q} = \vec{G}+m\vec{Q}$. The main diffraction peaks from the average crystal structure are obtained by letting $m = 0$, while satellite peaks are expected around each main diffraction peak $(\vec{G})$ with $m \neq 0$.

The three-dimensional modulation wave vector $\vec{Q}$ can be written as components of the reciprocal lattice vector

$$\vec{Q} = x_1\vec{a}^* + x_2\vec{b}^* + x_3\vec{c}^*.$$

When the modulation is incommensurate, at least one of the three components is not a rational fraction; then, the satellite peaks do not overlap and we obtain

$$F\left(\vec{G},m\right) = (-1)^m\sum_n f_n\,\mathrm{e}^{i\left(\vec{G}+m\vec{Q}\right)\cdot\vec{R}_n} J_m\left(\vec{q}\cdot\vec{U}_n\left(\vec{Q}\right)\right)\mathrm{e}^{im\phi_n}.$$

Many modulated structures found in materials have several modulation waves that are superimposed, where

$$\vec{u}_{ln} = \sum_{\vec{Q}}\vec{U}_n\left(\vec{Q}\right)\sin\left(\vec{Q}\cdot\vec{R}_{ln}-\phi_n\right)$$

Additionally, the atomic displacement is accompanied by a modulation to the atomic scattering factor

$$\delta f_{mn} = \sum_{\vec{Q}} \bar{f}_n C_n\left(\vec{Q}\right) e^{-i\vec{Q}\cdot\vec{R}_{mn}}, \tag{13.50}$$

Here the average atomic scattering factor of site n is $\bar{f}_n$, and $C_n(\vec{k})$ is the modulation amplitude. Since the atomic scattering factor is a real quantity (in the absence of absorption), the complex amplitude has the property $C_n\left(\vec{k}\right) = C_n\left(-\vec{k}\right)^*$. Substitution of the above expressions into the structure factor formula leads to a complex formula involving products of Bessel functions—the details are given in Withers (2008).

Figure 13.7 shows an example of an electron diffraction study of modulated structures. The diffraction patterns were recorded from the low-temperature-modulated structures of $La_{0.33}Ca_{0.67}MnO_3$ using a parallel electron beam of $\sim$ 100 nm in diameter. $La_{0.33}Ca_{0.67}MnO_3$ at room temperature (RT) has an orthorhombic-distorted perovskite structure (space group Pnma, $a \approx c \approx b/\sqrt{2} \approx \sqrt{2}a_c$ with a_c for high-T cubic cell). A check on the valence sum shows that the ratio of Mn^{3+} and Mn^{4+} in this crystal is 1:2. The structure undergoes a continuous incommensurate–commensurate structural transition below $T_{co} \sim$ 260 K. It has been suggested that the structural transition is accompanied by ordering of Mn^{3+} and Mn^{4+}. The low-temperature superstructure is approximately two-dimensional with the b-axis as the invariant direction. It induces a lattice distortion through the electron–lattice coupling manifested by the Jahn–Teller effect. Figure 13.11b shows a model of the lattice modulation. The modulation period is along the a-axis, while the modulation amplitude is mostly along the c-axis. Strong modulation peaks are observed around the reflections (002) and ($\pm$202) in Fig. 13.7.

For atomic displacement-type diffuse scattering, the important selection rule is $\left(\vec{G} + \vec{Q}\right)\cdot\vec{\Delta}$, where Δ is the atomic displacement associated with the modulation wave. Diffuse scattering is strongest for reflections parallel to the atomic displacement. In Fig. 13.7, modulation peaks are very weak in the (h, 0, 0) row at 261 K, and strong superlattice peaks are associated with strong fundamental reflections in the $l = 2n$ rows. Both suggest that the modulation peaks originate predominately from small atomic displacements parallel to the c-axis.

Multiple beam diffraction effects have a strong effect on the intensity of modulations peaks (see Sect. 13.6). Information that can be obtained directly from the electron diffraction patterns includes the modulation period, coherence length, and the relative peak intensity. In Fig. 13.8, the first-order modulation peak is at (h $\pm$ q, 0, 1). At 98 K, q $\approx$ 0.33, thus indicating a nearly commensurate CO structure in this system at low temperature. The modulation reflection is characterized by three variables: the wave vector q, the width, and the intensity. During the phase transition, all three change continuously with temperature. The temperature dependence

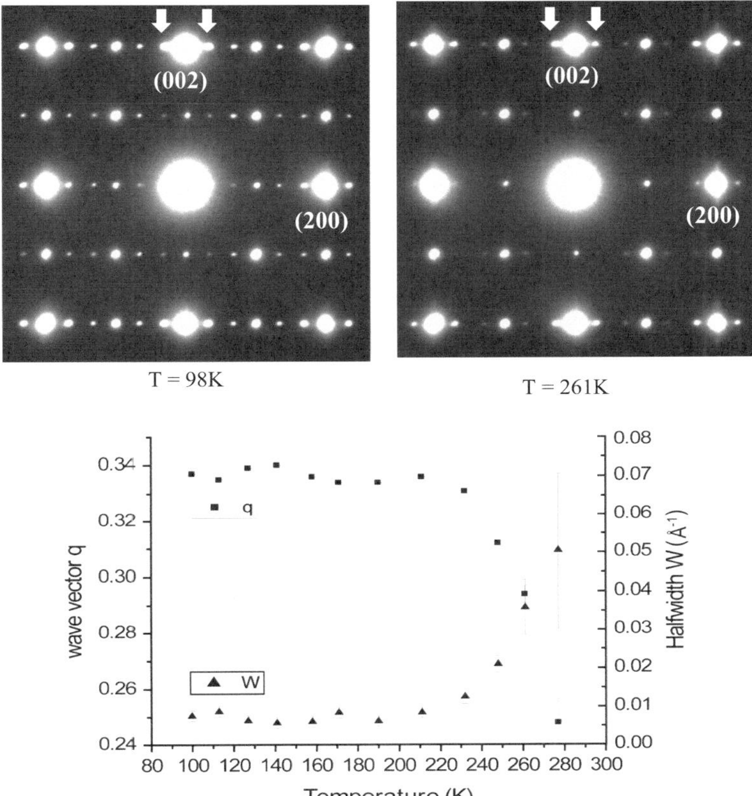

Fig. 13.7 (*Top*) Two examples of electron diffraction patterns are shown recorded from $La_{0.33}Ca_{0.67}MnO_3$ at T = 98 K and 261 K. (*Below*) The temperature dependence of the first-order charge-ordering reflection q, its half-width, and amplitude. The data were obtained from electron diffraction patterns recorded from a single domain. Reflections (pointed out by the arrows) from structural modulation become clearly visible at 261 K. (From Jing Tao, Ph. D. Thesis, University of Illinois, Urbana-Champaign)

of q and the peak width during cooling are shown in Fig. 13.7, which can be summarized as follows: The positions of the superlattice peaks move toward the fundamental reflections (h, 0, l) and appear at (h $\pm$ [1/3 - ε], 0, l), where ε is the incommensurability. The incommensurate to commensurate phase transition is observed from the wave vector curve. The modulation peaks disappear completely at room temperature. The half-width of the modulation peak, which gives the inverse of the average size of coherent charge-ordering regions, decreases with decreasing temperature. The average size of charge-ordered regions increases gradually from 16 nm at T = 260 K to $\sim$80 nm at T = 100 K.

Locally modulated structure in an otherwise disordered structure gives rise to characteristic diffuse scattering. Figure 13.8 shows three examples. The diffuse

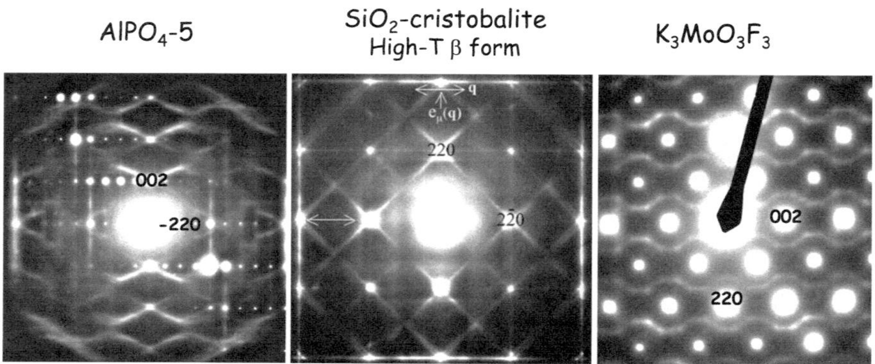

Fig. 13.8 Diffuse scattering observed in electron diffraction patterns recorded from (*left* to *right*) the microporous aluminophosphate (AlPO$_4$-5), high-temperature SiO$_2$ cristobalite (β form), and K$_3$MoO$_3$F$_3$ (Provided by Ray Withers, Australian National University, Canberra)

scattering seen here is spread out rather uniformly on well-defined reciprocal space surfaces (their intersections with the Ewald sphere give rise to the diffuse streaks seen in these diffraction patterns). While the broadly distributed diffuse scattering in the example of K$_3$MoO$_3$F$_3$ suggests local short-range order, the sharp diffuse streaks that are observed in cristobalite above 275 °C underlie much longer range order (of the order of tens of nanometers, or more in some cases, see Withers (2015)). Both cristobalite and AlPO$_4$-5 have the tetrahedral corner-connected framework structures. They are dynamically disordered at high temperatures due to simultaneous excitation of several zero-frequency rigid unit modes (RUM) of distortions involving rotations of individual tetrahedral and changes in their relative orientations. The sharp diffuse scattering maps out these RUM modes of distortion in these crystals. In the high-temperature cristobalite diffraction pattern, strong diffuse streaks are observed with $\vec{Q}\|\langle 110\rangle$ running through {440} reflections, which requires $\vec{U}\left(\vec{Q}\right)$ to be perpendicular to $\vec{Q}$. In this case, the diffuse distribution takes the form of {110} sheets of diffuse intensity perpendicular to the six $\langle 110\rangle$ directions, the Fd-3 m average structure of b cristobalite. The atomic displacements responsible for these diffuse sheets are these of Si atoms along $\langle 110\rangle$ directions. The Si atoms are also the dominant scatterers in SiO$_2$.

13.6 Multiple Scattering Effects in Diffuse Scattering

So far we have discussed the relation between the scattering potential and diffraction intensities in the kinematic approximation. In electron diffraction, this relationship is complicated by multiple beam diffraction effects. In electron microscopy, it has been known for a long time (Kamiya and Uyeda 1961; Cundy et al. 1969) that the diffuse scattering, including inelastic scattering, carries some of the phase relationships among Bragg reflections and so can be used to form a lattice image. The theory of the interaction between diffuse and Bragg scattering has been discussed for the contrast of Kikuchi lines (Kainuma 1955) and for inelastic scattering by Fujimoto and Kainuma (1963) and Howie (1963). An n-beam dynamical treatment based on scattering matrix was outlined by Gjonnes (1966). Application of this approach was made by Gjonnes and Watanabe (1966) for inelastic scattering in MgO involving relatively few beams. A formulation based on the multislice method was described by Cowley and Pogany (1968), and this approach was subsequently used by Doyle (1969) for detailed calculations on thermal diffuse and plasmon scattering and by Cowley and Murray (1968) for short-range order scattering. A multislice method for scattering from defects which allows multiple Bragg scattering coupled to single kinematic diffuse scattering has been described by Spence (1978). A more recent treatment of thermal diffuse scattering using the same multislice approach was given by Wang (2003). The same multislice method using the independent atomic vibration model (Einstein model) has also been used successfully to simulate the atomic-resolution Z-contrast images. Separately, the subject of dynamic effects in diffuse scattering was treated in the book by Peng et al. (2004), where a description of the distorted wave Born approximation and kinematic equation can be found for electron diffraction.

Here, we will provide a general treatment based on a combination of the scattering matrix method of Gjonnes and the distorted wave Born approximation (DWBA). As will become apparent below, the use of DWBA extends the thickness of the coherent volume where diffuse scattering occurs, compared to the original perturbation treatment of Gjonnes based on the kinematic approximation. The scattering matrix approach used here has a number of advantages over the multislice method. First, the scattering matrix method can be applied to any crystal orientations. Secondly, the theory can be applied to nonperiodic objects and closed form solutions obtained for the case of diffraction involving few beams. Note that we have reverted to the mathematical symbols and sign convention used in other chapters of the book here.

In the cases we have discussed so far, Bragg diffraction is strong while diffuse scattering is relatively weak. A good approximation then is to treat ΔU as a perturbation in the single diffuse scattering model as illustrated in Fig. 13.9. This model considers four scattering paths (A to D) in addition to the transmitted beam, which are single diffuse scattering (A), Bragg diffraction including multiple Bragg diffraction (B), single diffuse scattering followed by Bragg diffraction (C), and Bragg diffraction followed by diffuse scattering (D). The single diffuse scattering approximation is justified by the weak deviation potential ΔU. Using this, the sample then separates into three slices: I, II, and III, where diffuse scattering takes

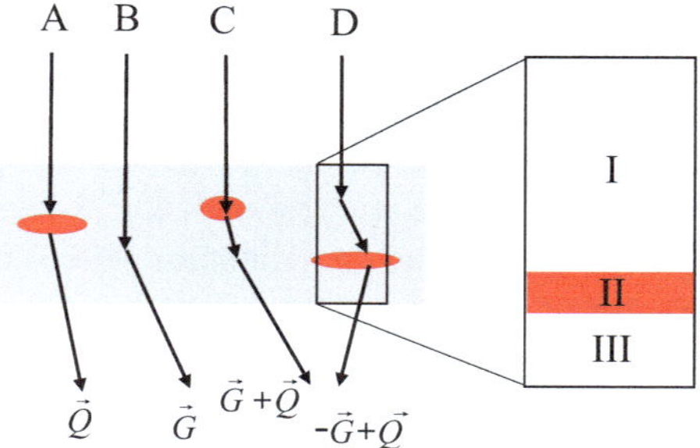

Fig. 13.9 Electron multiple scattering involving single diffuse scattering

place in region II. Scattering in slices I and III is described by the scattering matrix (S) of the average crystal structure (Eq. 5.16)

$$
\begin{pmatrix} \phi_{\mathrm{o}}\!\left(\vec{k},t\right) \\ \phi_{g}\!\left(\vec{k},t\right) \\ \vdots \end{pmatrix}
=
\begin{pmatrix}
S_{\mathrm{oo}}\!\left(\vec{k},t\right) & S_{\mathrm{og}}\!\left(\vec{k},t\right) & \cdots \\
S_{\mathrm{go}}\!\left(\vec{k},t\right) & S_{gg}\!\left(\vec{k},t\right) & \vdots \\
\cdots & \cdots & \ddots
\end{pmatrix}
\begin{pmatrix} \phi_{\mathrm{o}}\!\left(\vec{k},0\right) \\ \phi_{g}\!\left(\vec{k},0\right) \\ \vdots \end{pmatrix},
$$

Or

$$
\Phi_{\vec{k}}(t) = S_{\vec{k}}\Phi_{\vec{k}}(0) \tag{13.51}
$$

where $\vec{k}$ is the wavevector of the direct beam (with the 2π prefactor), t is the thickness of the slice, and $\Phi_{\vec{k}} = \left(\phi_{\mathrm{o}}\!\left(\vec{k},t\right), \phi_{g}\!\left(\vec{k},t\right), \cdots\right)^{T}$ is an column vector of N dimension. With N beams of $\left(0, \vec{G}, \cdots\right)$ included, S is an $N \times N$ matrix that relates the incident waves to the exit waves of the slice. In region II, in addition to Bragg diffraction, scattering by the deviation potential $\Delta U(\vec{r}) = U(\vec{r}) - \langle U(\vec{r})\rangle$ gives a second set of beams of $\left(0 + \vec{Q}, \vec{g} + \vec{Q}, \cdots\right)$. The relationship between incident waves (which are beams exiting slice I) and exit waves can be similarly described using the scattering matrix

$$
\begin{bmatrix} \Phi_{\vec{k}}(t_{II}) \\ \Phi_{\vec{k}+\vec{Q}}(t_{II}) \end{bmatrix}
=
\begin{bmatrix} S_{\vec{k}}(t_{II})\Phi_{\vec{k}}(t_{I}) \\ S_{\vec{Q}}(t_{II})\Phi_{\vec{k}}(t_{I}) \end{bmatrix},
$$

where $S_{\vec{Q}}(t_{II})$ describes the diffuse scattering. Combining these results, we have for the exit waves of the sample

$$
\begin{bmatrix} \Phi_{\vec{k}}(t_{III}) \\ \Phi_{\vec{k}+\vec{Q}}(t_{III}) \end{bmatrix}
=
\begin{bmatrix} S_{\vec{k}}(t_{III}) & 0 \\ 0 & S_{\vec{k}+\vec{Q}}(t_{III}) \end{bmatrix}
\begin{bmatrix} S_{\vec{k}}(t_{II})S_{\vec{k}}(t_{I})\Phi_{\vec{k}}(0) \\ S_{\vec{Q}}(t_{II})S_{\vec{k}}(t_{I})\Phi_{\vec{k}}(0) \end{bmatrix}
$$

and

$$\Phi_{\vec{k}+\overrightarrow{Q}}(t_{\mathrm{III}}) = S_{\vec{k}+\overrightarrow{Q}}(t_{\mathrm{III}})S_{\overrightarrow{Q}}(t_{\mathrm{II}})S_{\vec{k}}(t_{\mathrm{I}})\Phi_{\vec{k}}(0). \tag{13.52}$$

Here $\mathbf{S}_{\vec{k}+\overrightarrow{Q}}(t_{\mathrm{III}})$ describes Bragg diffraction among the beams of $\left(0+\overrightarrow{Q}, \vec{g}+\overrightarrow{Q}, \cdots\right)$, which can be simply obtained using the Bloch wave method with the diffuse scattering wave vector added to the incident beam.

The diffraction intensity from the above scattering processes for the beam of $\overrightarrow{G}+\overrightarrow{Q}$ is obtained by

$$I_{\vec{k}+\overrightarrow{Q}}(t_{\mathrm{I}},t_{\mathrm{II}},t_{\mathrm{III}}) = \left| \phi_{\vec{k}+\overrightarrow{Q}}(t_{\mathrm{I}},t_{\mathrm{II}},t_{\mathrm{III}}) \right|^2 .$$

Assuming that the diffuse scattering is random and can occur at any thickness and any region of the sample under the illumination, then

$$I_{\vec{k}+\vec{Q}}(t) = \left\langle \int_0^t \left| \phi_{\vec{k}+\vec{Q}}(t_{\mathrm{I}},t_{\mathrm{II}},t_{\mathrm{III}}) \right|^2 dt_{\mathrm{II}} \right\rangle. \tag{13.53}$$

Next, we look into how to calculate the diffuse scattering matrix $\mathbf{S}_{\vec{Q}}(t_{\mathrm{II}})$. For this purpose, we use the distorted wave Born approximation (DWBA) by starting with the time-independent Schrödinger equation (Eq. 2.13) in the form

$$\frac{1}{4\pi^2}\nabla^2\phi + k^2\phi + U_1(\vec{r})\phi = -\Delta U(\vec{r})\phi \tag{13.54}$$

where $U_1(\vec{r}) = \langle U(\vec{r})\rangle$ and $\Delta U(\vec{r})$ is the deviation potential defined in Eq. (13.15). Following the example of Sect. 4.1, Eq. (13.54) can be transformed into the Lippmann–Schwinger integral equation:

$$\phi(\vec{r}) = \phi_1(\vec{r}) - \int d^3\vec{r}'G_1(\vec{r},\vec{r}')\Delta U(\vec{r}')\phi(\vec{r}') \tag{13.55}$$

Here $\phi_1(\vec{r})$ is a solution of the modified homogeneous wave equation

$$\frac{1}{4\pi^2}\nabla^2\phi_{1,\vec{k}_o} + k_o^2\phi_{1,\vec{k}_o} + U_1(\vec{r})\phi_{1,\vec{k}_o} = 0 \tag{13.56}$$

and $G_1(\vec{r},\vec{r}')$ is the solution of

$$\frac{1}{4\pi^2}\nabla^2 G_1(\vec{r},\vec{r}') + k_o^2 G_1(\vec{r},\vec{r}') + U_1(\vec{r})G_1(\vec{r},\vec{r}') = \delta(\vec{r}-\vec{r}'). \tag{13.57}$$

Compared to the potential-free case discussed in the kinematical theory of electron diffraction (Sect. 4.1), the differences here are the reference potential $U_1(\vec{r})$ that appears in the Schrödinger equation, the $\Delta U(\vec{r})$ that substitutes $U(\vec{r})$ as the inhomogeneous scattering potential in the Lippmann–Schwinger equation, and ϕ_1 as the homogeneous solution instead of the incident plane wave. The Born approximation to Eq. (13.55) is given by

$$\phi(\vec{r}) \approx \phi_{1,\vec{k}_o}(\vec{r}) - \int d^3\vec{r}' G_1(\vec{r},\vec{r}')\Delta U(\vec{r}')\phi_{1,\vec{k}_o}(\vec{r}'). \tag{13.58}$$

To satisfy the boundary conditions of a diffraction experiment, the homogeneous solution $\phi_{1,\vec{k}_o}$ must also approximate as a plane wave plus an outgoing wave in the far field limit

$$\phi_{1,\vec{k}_o}(\vec{r}) = e^{2\pi i \vec{k}_o \cdot \vec{r}} + \frac{1}{r}e^{2\pi i k r}f_1\left(\vec{k},\vec{k}_o\right) \tag{13.59}$$

and

$$f_1\left(\vec{k},\vec{k}_o\right) = \pi \int d^3\vec{r}\, e^{-2\pi i \vec{k}\cdot\vec{r}} U_1(\vec{r})\phi_{1,\vec{k}_o}(\vec{r}), \tag{13.60}$$

which is simply the scattering amplitude for the potential $U_1(\vec{r}')$. The total scattering amplitude in the presence of the deviation potential is then given by

$$f\left(\vec{k},\vec{k}_o\right) = f_1\left(\vec{k},\vec{k}_o\right) + \Delta f\left(\vec{k},\vec{k}_o\right), \tag{13.61}$$

and

$$\Delta f\left(\vec{k},\vec{k}_o\right) = \pi \int d^3\vec{r}\,\phi_{1,-\vec{k}}(\vec{r})\Delta U(\vec{r})\phi_{1,\vec{k}_o}(\vec{r}). \tag{13.62}$$

Here, $\phi_{1,\vec{k}_o}(\vec{r}')$ and $\phi_{1,-\vec{k}}(\vec{r}')$ are the solutions of Eq. (13.56) for the incident waves of $\exp\left(2\pi i \vec{k}_o \cdot \vec{r}\right)$ and $\exp\left(-2\pi i \vec{k} \cdot \vec{r}\right)$, respectively. Equations (13.61) and (13.62) give the diffracted electron waves in the DWBA.

Equation (13.62) can be evaluated by expanding the functions in the integral in Fourier series

$$\Delta U(\vec{r}) = \sum_{\vec{Q}} \Delta U_{\vec{Q}}\, e^{2\pi i \vec{Q}\cdot\vec{r}},$$

$$\phi_{1,\vec{k}_o}(\vec{r}) = \sum_i c_i\left(\vec{k}_o\right)e^{2\pi i[\vec{k}_o + \gamma_i(\vec{k}_o)\vec{n}]\cdot\vec{r}} \sum_{\vec{g}} C_{\vec{g}}^i\left(\vec{k}_o\right)e^{2\pi i \vec{g}\cdot\vec{r}}. \tag{13.63}$$

$$\phi_{1,-\vec{k}}(\vec{r}) = \sum_i c_i\left(-\vec{k}\right)e^{2\pi i[-\vec{k} + \gamma_i(-\vec{k})\vec{n}]\cdot\vec{r}} \sum_{\vec{g}} C_{\vec{g}}\left(-\vec{k}\right)e^{2\pi i \vec{g}\cdot\vec{r}}$$

Here c_i, γ_i, $C_{\vec{g}}^i$ are the excitation, eigenvalue, and eigenvector of the ith Bloch wave and $\vec{n}$ is the sample surface normal vector. Using the above results, we obtain

$$\Delta f\left(\vec{k},\vec{k}_o\right) = \pi \int d^3\vec{r} \sum_i \sum_j c_i\left(\vec{k}_o\right) c_j\left(-\vec{k}\right) e^{2\pi i\left[\vec{k}_o + \gamma_i(\vec{k}_o)\vec{n}\right]\cdot\vec{r}} e^{2\pi i\left[-\vec{k} + \gamma_j(-\vec{k})\vec{n}\right]\cdot\vec{r}}$$

$$\cdot \sum_{\vec{h}} \sum_{\vec{g}} C_{\vec{g}}^i\left(\vec{k}_o\right) C_{\vec{h}}^j\left(-\vec{k}\right) e^{2\pi i(\vec{g}+\vec{h})\cdot\vec{r}} \sum_{\overrightarrow{Q}} \Delta U_{\overrightarrow{Q}} e^{2\pi i \overrightarrow{Q}\cdot\vec{r}}$$

Next, we will take $\vec{n}$ to be parallel to $\vec{z}$, $\vec{k} - \vec{k}_o = \vec{s}_\| + \vec{s}_z$ and $\vec{s}_\|$, $\vec{g}$, and $\vec{h}$ are normal to $\vec{z}$. Then we obtain

$$\Delta f\left(\vec{k},\vec{k}_o\right) = \pi \sum_{\vec{q}} \sum_i \sum_j c_i\left(\vec{k}_o\right) c_j\left(-\vec{k}\right) \Gamma(s_z, Q_z, t)$$

$$\cdot \sum_{\vec{h}} \sum_{\vec{g}} C_{\vec{g}}^i\left(\vec{k}_o\right) C_{\vec{h}}^j\left(-\vec{k}\right) \Delta U_{\overrightarrow{Q}} \delta\left(\vec{s}_\| - \vec{g} - \vec{h} - \overrightarrow{Q}_\|\right) \tag{13.64}$$

where

$$\Gamma(s_z, Q_z, t) = \int_0^t e^{2\pi i\left[Q_z - s_z + \gamma_i(\vec{k}_o) + \gamma_j(-\vec{k})\right]z} dz. \tag{13.65}$$

To gain further insight into dynamical effects in electron diffuse scattering, we consider the two-beam case as illustrated in Fig. 13.10. One Bragg reflection, g, is excited by the incident beam. The scattering matrix $S_{\overrightarrow{Q}}(t_{\mathrm{II}})$ is a 2×2 matrix that is given by

$$S_{\overrightarrow{Q}}(t_{\mathrm{II}}) = \begin{bmatrix} f\left(\vec{k}_o + \overrightarrow{Q}, \vec{k}_o\right) & f\left(\vec{k}_o + \overrightarrow{Q}, \vec{k}_o + \vec{g}\right) \\ f\left(\vec{k}_o + \overrightarrow{Q} + \vec{g}, \vec{k}_o\right) & f\left(\vec{k}_o + \overrightarrow{Q} + \vec{g}, \vec{k}_o + \vec{g}\right) \end{bmatrix}.$$

By substituting this into Eq. (13.52), we obtain

$$\begin{aligned}
\phi_Q &= S_{\mathrm{co}}\left(\vec{k}_o + \overrightarrow{Q}, t_{\mathrm{III}}\right) S_{\mathrm{oo}}\left(\vec{k}_o, t_{\mathrm{I}}\right) f\left(\vec{k}_o + \overrightarrow{Q}, \vec{k}_o\right) \\
&+ S_{\mathrm{oo}}\left(\vec{k}_o + \overrightarrow{Q}, t_{\mathrm{III}}\right) S_{\mathrm{go}}\left(\vec{k}_o, t_{\mathrm{I}}\right) f\left(\vec{k}_o + \overrightarrow{Q}, \vec{k}_o + \vec{g}\right) \\
&+ S_{\mathrm{og}}\left(\vec{k}_o + \overrightarrow{Q}, t_{\mathrm{III}}\right) S_{\mathrm{oo}}\left(\vec{k}_o, t_{\mathrm{I}}\right) f\left(\vec{k}_o + \overrightarrow{Q} + \vec{g}, \vec{k}_o\right) \\
&+ S_{\mathrm{og}}\left(\vec{k}_o + \overrightarrow{Q}, t_{\mathrm{III}}\right) S_{\mathrm{go}}\left(\vec{k}_o, t_{\mathrm{I}}\right) f\left(\vec{k}_o + \overrightarrow{Q} + \vec{g}, \vec{k}_o + \vec{g}\right) \\
&= Af\left(\vec{k}_o + \overrightarrow{Q}, \vec{k}_o\right) + Bf\left(\vec{k}_o + \overrightarrow{Q} + \vec{g}, \vec{k}_o + \vec{g}\right) \\
&+ Cf\left(\vec{k}_o + \overrightarrow{Q}, \vec{k}_o + \vec{g}\right) + Df\left(\vec{k}_o + \overrightarrow{Q} + \vec{g}, \vec{k}_o\right).
\end{aligned} \tag{13.66}$$

The four terms in Eq. (13.66) are illustrated in Fig. 13.10. Among these, the first term corresponds to diffuse scattering from the direct beam and its coefficient A describes the amplitude of the direct beam before diffuse scattering and the subsequent transmission through the rest of the sample. The calculation of the amplitudes of A to D can be made based on the two-beam results in Chap. 5.

The electron diffuse scattering theory outlined in the previous section was applied to determine the structure of nanometer-sized charge-ordered domains in $La_{2/3}Ca_{1/3}MnO_3$ (Zuo (2002)). The neutron measurements show that the superlattice reflections from these domains are elastic with timescale $t > 1$ ps. The electron diffraction experiment was carried out using the energy filtering LEO-912 Omega electron microscope (120 kV). Single-crystal ED patterns were recorded from a thin area of $La_{2/3}Ca_{1/3}MnO_3$ above 300 nm in diameter using a parallel beam and the Fuji 25 μm image-plate detector. Figure 2 shows an example along the [010] zone-axis orientation. The most remarkable feature in the diffraction pattern are the nearly commensurate satellite peaks with half indices of (h + 1/2, 0, 1); these are superstructure reflections with a period of twice the original a-axis. They disappear at low temperature. The superstructure reflection intensity changes with temperature, and they are most visible near $T \sim 270$ K. The superstructure reflection also sits on a broad anisotropic diffuse background, especially near strong fundamental reflections. Inspection of the [001] and [100] zone axes revealed anisotropic diffuse scattering, but no superstructure reflections.

The need to apply incoherent scattering theory became apparent from the two coherence lengths seen in the diffraction pattern. The broadening of superstructure reflections corresponds to structural domains of about 36 ± 2 Å in diameter, whereas the beam coherence is about >7 times larger.

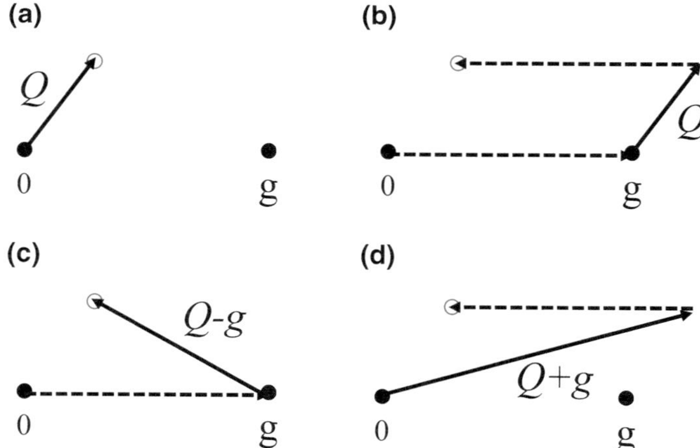

Fig. 13.10 Contributions to diffuse scattering from multiple beam paths in a two-beam case. The *dashed* and *full lines* are for Bragg diffraction and diffuse scattering, respectively

A structural model was constructed based on atomic displacements estimated from the X-ray bond-length measurements. Atomic pair distributions from X-ray results indicated three bond lengths (~ 1.85–1.9 Å, ~ 1.97 Å and ~ 2.1–2.15Å) at 300 K and a single bond length at 20 K for this crystal. In Fig. 3, the shift of oxygen atoms in the opposite direction around Mn(A) gives two short and two long Mn–O bonds in the a-c plane, differing by $\sim \Delta/\sqrt{2}$, with Δ standing for the transverse atomic displacement. We expect the shortest bond of ~ 1.9 Å for the Mn^{4+}-O^{2-} bond and the longer two bond lengths for Mn^{3+}-O^{2-}. To fit these bond lengths, we need an additional contraction (δ) for the Mn^{4+}-O^{2} bond in the a-c plane and a consequent dilation for the Mn^{3+}-O^{2-} bond in the same plane. By fitting the X-ray data, we obtain $\Delta \sim 0.14$ Å and $\delta \sim 0.06$ Å.

To see whether the model proposed above fits the experimental data, full dynamical calculations were carried out using the scattering matrix method described above and the Bloch wave method. The calculation was done by integrating Eq. 2 over the sample thickness, which we estimated to be 840 Å from the intensities of the fundamental reflections. The superstructure reflection structure factors were calculated using $\Delta = 0.14$ Å (they are not affected by the contraction δ). The thickness of nanodomains, t, was treated as a parameter. The calculation was compared with 21 independent superlattice reflections in the [010] zone axis. The best fit between theory and experiment was obtained with $t \sim 15$ Å, giving $\chi^2 \sim 1.$ 4. The comparison between our model and experiment is shown in Fig. 13.11 c, d. The agreement is good, considering that the model only contains two structural parameters. From the intensity of the superstructure reflections, we estimate the volume percentage of nanodomains to be a few percent at 272 K with $\Delta = 0.14$ Å. This estimate is inversely proportional to Δ^2.

The above example demonstrates the importance of multiple beam diffraction effects in the quantitative study of electron diffuse scattering. The main differences between the dynamic and kinematic diffraction intensities are the weak (..2n + 1) reflections and the strong (..2) reflections. In the kinematic model, the intensities of (..2n + 1) reflections are very small, while the dynamic intensities of these same reflections are weak but of measurable intensity. Their origin can be attributed to the redistribution of diffuse intensity by multiple diffractions. The difference between the strong dynamical effects of (..2) reflections and the much smaller effect on the (..4) reflections is a result of the strong dynamical scattering nearing a zone axis and the much weaker multiple scattering away from the zone-axis orientation.

The results presented here also demonstrate that quantitative structural information can be obtained from electron incoherent scattering. This was made possible by (1) an accurate measurement of electron diffraction intensities using the electron energy filter and imaging plates and (2) by taking into account electron multiple scattering in comparing theoretical and experimental intensity, using the scattering matrix method presented here.

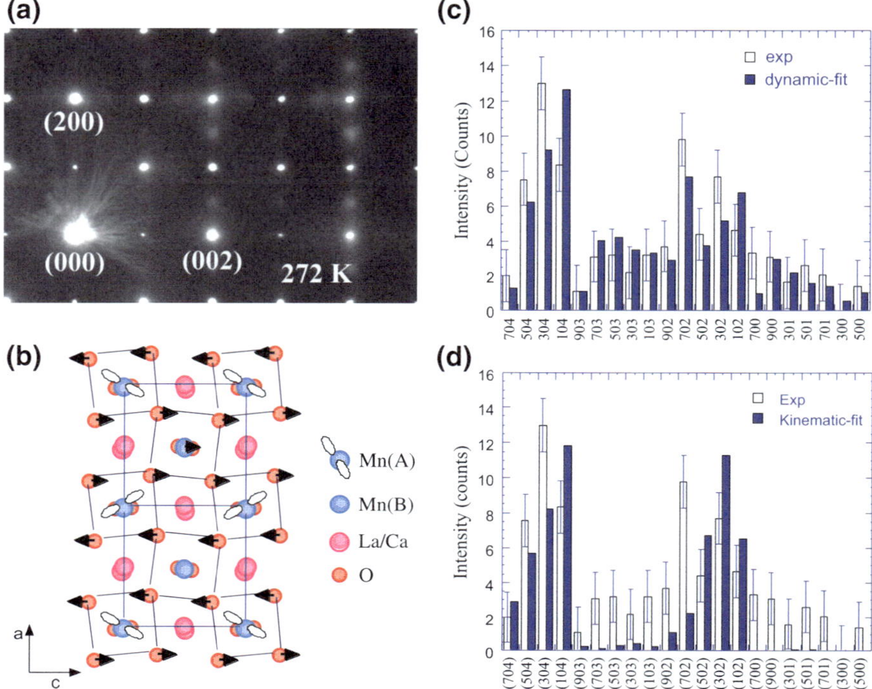

Fig. 13.11 **a** Evidence of the temperature-dependent superstructure in $La_{2/3}Ca_{1/3}MnO_3$. The diffraction pattern shown was recorded in the [010] zone-axis orientation at 272 K, displayed on an absolute intensity scale in units of counts (1 counts corresponding to about 1.2 beam electrons). **b** Atomic structure model of lattice distortions proposed for the superstructure. In this model, atoms shift by $\pm\Delta$ or 0 depending on their position (as indicated by the arrow). The displacement pattern resembles a transverse wave with $k = (1/2, 0, 0)$. **c** and **d** A comparison between the experimental and theoretical diffraction intensities of superstructure reflections. The reflection index (x-axis) is based on the supercell. Theoretical intensities were calculated using the structure model proposed in b), and the domain thickness of 15 Å based on the dynamic and kinematic theory (from Zuo 2002)

References

Amoros JL (1968) Molecular crystals: their transforms and diffuse scattering. Wiley, New York

Born M (1942) Theoretical investigations on the relation between crystal dynamics and X-ray scattering. Rep Prog Phys 9:294–333

Born M, Huang K (1954) Dynamical theory of crystal lattices. Clarendon Press, Oxford

Cowley JM (1995) Diffaction physics, 3rd edn. Elsevier Science, NL

Cowley JM, Pogany AP (1968) Diffuse scattering in electron diffraction patterns. I. General theory and computational methods. Acta Crystallogr A24:109

Cundy SL, Howie A, Valdre U (1969) Preservation of electron microscope image contrast after inelastic scattering. Philos Mag 20:147

Debye P (1914) Interference of Röntgen rays and heat motion. Ann Phys 43:49–95

Dederich PH (1971) Diffuse scattering from defect clusters near Bragg reflections. Phys Rev B 4:1041

Dewolff PM (1974) Pseudo-symmetry of modulated crystal-structures. Acta Crystallogr A 30: 777–785

Doyle PA (1969) Dynamical calculation of thermal diffuse electron scattering. Acta Crystallogr A 25:569

Ducastelle F (1991) Order and phase stability in alloys. Amsterdam North-Holland

Ehrhart P, Trinkaus H, Larson BC (1982) Diffuse-scattering from dislocation loops. Phys Rev B 25:834–848

Fujimoto F, Kainuma Y (1963) Inelastic scattering of fast electrons by thin crystals. J Phys Soc Jpn 18(12):1792

Gjonnes J (1966) Influence of Bragg scattering on inelastic and other forms of diffuse scattering of electrons. Acta Crystallogr 20:240

Gjonnes J, Watanabe D (1966) Dynamical diffuse scattering from magnesium oxide single crystals. Acta Crystallogr 21:297–302

Goldman AI, Kelton RF (1993) Quasicrystals and crystalline approximants. Rev Mod Phys 65:213–230

Howie A (1963) Inelastic scattering of electrons by crystals. I. Theory of small-angle inelastic scattering. Proc Roy Soc London Ser A 271:268

Ikematsu Y, Shindo D, Oikawa T (2000) Quantitative analysis of short-range order diffuse scattering in Cu-27.5 at %Pd alloy with energy-filtered electron diffraction. Mater Trans JIM 41:238–241

Janner A, Janssen T (1977) Symmetry of periodically distorted crystals. Phys Rev B 15:643–658

Kainuma Y (1955) The theory of Kikuchi patterns. Acta Crystallogr 8:247–257

Kamiya Y, Uyeda R (1961) Effect of incoherent waves on electron microscopic images of crystals. J Phys Soc Jpn 16:1361

Kirk MA, Davidson RS, Jenkins ML, Twesten RD (2005) Measurement of diffuse electron scattering by single nanometre-sized defects in gold. Philos Mag 85:497–507

Krivoglaz MA (1996) Diffuse scattering of x-rays and neutrons by fluctuations. Springer, New York

Laval J (1938) Crystallography. On the diffusion of X rays by a crystal. C R Hebd Seances Acad Sci 207:169–170

Laval J (1939) Diffusing x-rays with crystals outside the bearing of selective reflection. C R Hebd Seances Acad Sci 208:1512–1514

Makovicky E, Hyde B (1981) Non-commensurate (misfit) layer structures. Inorg Chem Struct Bond 46:101–170

Massalski TB, Okamoto H (1990) Binary alloy phase diagrams. ASM International, Ohio

McMullan G, Faruqi AR, Clare D, Henderson R (2014) Comparison of optimal performance at 300 keV of three direct electron detectors for use in low dose electron microscopy. Ultramicroscopy 147:156–163

Ohshima K, Watanabe D (1973a) Electron-diffraction study of short-range-order diffuse scattering from disordered Cu-Pd and Cu-Pt alloys. Acta Cryst A 29:520

Ohshima K, Watanabe D (1973b) Electron-diffraction study of short-range-order diffuse scattering from disordered Cu-Pd and Cu-Pt alloys. Acta Cryst A 29:520

Peng LM, Dudarev SL, Whelan MJ (2004) High energy electron diffraction and microscopy. Oxford University Press, Oxford

Petricek V, Gao Y, Lee P, Coppens P (1990) X-ray-analysis of the incommensurate modulation in the 2-2-1-2 Bi-Sr-Ca-Cu-O superconductor including the oxygen-atoms. Phys Rev B 42: 387–392

Shechtman D, Blech I, Gratias D, Cahn JW (1984) Metallic phase with long-range orientational order and no translational symmetry. Phys Rev Lett 53:1951–1953

Spence JCH (1978) Approximations for dynamical calculations of microdiffraction patterns and images of defects. Acta Cryst A34:112–116

Tafto J, Spence JCH (1982) Atomic site determination using the channeling effect in electron-induced X-ray-emission. Ultramicroscopy 9:243–247

Verwey EJ, Haayman PW, Romeijn FC (1947) Physical properties and cation arrangement of oxides with spinel structures. 2. Electronic conductivity. J Chem Phys 15:181–187

Waller I (1923) On the question of the influence of thermal motion on the interference of X-rays. Z Physik 17:398–408

Wang ZL (2003) Thermal diffuse scattering in sub-angstrom quantitative electron microscopy—phenomenon, effects and approaches. Micron 34:141–155

Welberry TR (2010) Diffuse scattering and models of disorder. Oxford University Press, IUCr

Withers RL (2008) "Disorder": structured diffuse scattering and local crystal chemistry. In: Hawkes PW (ed) Advances in imaging and electron physics, vol 152. Elsevier Academic Press Inc, San Diego, pp 303–337

Withers R (2015) A modulation wave approach to the order hidden in disorder. IUCRj 2:74–84

Xu RQ, Chiang TC (2005) Determination of phonon dispersion relations by X-ray thermal diffuse scattering. Z Kristallogr 220:1009–1016

Yamamoto A (1996) Crystallography of quasiperiodic crystals. Acta Cryst A52:509–560

Zachariasen WH (1940) A theoretical study of the diffuse scattering of X-rays by crystals. Phys Rev 57:597–602

Zuo JM (2002) Quantitative electron incoherent scattering and application to nanometre-sized charge ordering in $La_{2/3}Ca_{1/3}MnO_3$. J Electron Microsc 51:S67–S72

Zuo JM, Pacaud J, Hoier R, Spence JCH (2000) Experimental measurement of electron diffuse scattering in magnetite using energy-filter and imaging plates. Micron 31:527–532

Chapter 14
Atomic Resolution Electron Imaging

This chapter introduces the principles of high-resolution electron microscopy (HREM) and scanning transmission electron microscopy (STEM). These are the two major techniques for atomic resolution imaging of nanostructures, certain types of crystal defects and interfaces, as well as metrology in electronic and magnetic devices when atomic resolution is needed. In STEM, atomic resolution imaging using an annular dark-field detector can be combined with analytical techniques, such as energy-loss spectroscopy (EELS, which gives similar information to soft X-ray absorption spectroscopy), energy dispersive X-ray spectroscopy (EDS), and nanodiffraction, for composition and crystallographic analysis.

We start with a brief history and introduction of HREM and STEM. After this, we introduce Abbe's theory of coherent imaging based on wave optics, which will be used as the basis for further discussion of HREM and probe formation in STEM. The major topics covered under HREM include linear imaging theory, the contrast transfer function, and partial coherence. For STEM, we discuss probe formation, coherence and bright-field STEM contrast, ronchigrams, dark-field STEM, and aberration-corrected STEM. The chapter finishes with a section on simulation methods. Further reading on the subjects covered here can be found in the books by Spence (2013), Pennycook and Nellist (2011), Cowley (1995), Peng et al. (2004) and Kirkland (2010).

14.1 Introduction and a Brief History

HREM can be said to begin with publication of the first lattice image of a thin crystal by Menter in 1956. This was a two-beam interference image of copper phthalocyanine, with a spacing of 1.2 nm. Menter achieved a resolution of 12 Å. This started the field that came to be known as high-resolution electron microscopy. Theoretical work had begun much earlier, notably with the important paper of Scherzer (1949) defining the wave-front aberration function and the resolution of

© Springer Science+Business Media New York 2017

J.M. Zuo and J.C.H. Spence, *Advanced Transmission Electron Microscopy*,

DOI 10.1007/978-1-4939-6607-3_14

the electron microscope, and with early work on electron holography (Haine and Mulvey 1952). Throughout the sixties, improvements in electronic and mechanical stability lead to instruments with a "line" resolution of less than 1 Å before the end of the decade, but it was widely held that these few beam images provided more information about the electron microscope than they did about the sample. They were analyzed using the two-beam dynamical theory (Cowley 1959). Three breakthroughs occurred in the early seventies which gave renewed confidence to the field—the observation of individual atoms by STEM (Crewe et al. 1970), the first observations of two-dimensional lattice images containing useful information on crystal chemistry and defects (Iijima 1971; Allpress and Sanders 1973), and the widespread adoption of Thon's "diffractograms" to measure lens aberrations and focus settings (see Spence 2013 for details).

The first STEM was designed and constructed by Manfred von Ardenne in Berlin in 1937–1938 soon after the electron microscope itself was invented in 1931 by Ruska and Knoll. The HREM image formed in a TEM is obtained from very thin samples illuminated by a broad electron beam. The objective aperture is opened up so that more than one diffracted beam can reach the image plane, where they interfere to form lattice fringes. In contrast, the STEM operates, as for an SEM, by scanning a fine probe over sample; however, the sample is sufficiently thin that electrons transmitted through the sample and scattered, and may be detected and shown on a raster display, which is synchronized with the probe scan position. The instrument built by Crewe for high-resolution STEM used a cold field emission gun, which was first developed by A. Crewe and coworkers at the University of Chicago (Crewe 1966), and an annular dark-field (ADF) detector. Crewe and his coworkers used it to obtain the first electron microscope images of individual atoms in 1970 (Crewe et al. 1970). These images of isolated heavy atoms on a thin carbon film were obtained at 30 kV. While the resolution of STEM is entirely competitive (or better) than that obtainable by HREM, early STEM images often suffered from distortions due to movements of the sample relative to the probe during scanning (or distortion in the scan raster), and a stationary beam may quickly drill a hole in the sample if it is uncontaminated.

In STEM, various forms of image contrast can be obtained according to the signals selected by different detectors, which are designed for BF (bright field), ABF (annular bright field), ADF-STEM, or high-angle ADF (HAADF) imaging. The BF detector in STEM measures the intensity of the direct beam and some of the scattered electrons at low angles. A BF-STEM image has contrast similar to the BF HREM image. The ADF detector collects scattered electrons over a range of scattering angles between its inner and outer radius (Sect. 9.1). The ADF imaging mode provides the most important contrast mechanism for image formation in a STEM.

The major advantage of ADF-STEM relates to the distribution of inelastic scattered electrons, which dominates low-angle scattering and falls off much faster than electrons scattered by the atomic nucleus (e.g., elastically scattered electrons). Since inelastic scattering by the outer-shell electrons is less localized than elastic scattering, a significant gain in image contrast is expected by detecting high-angle elastically scattered electrons. The use of the ADF detector also allows simultaneous imaging together with EELS, since the portion of the beam which passes through the central hole in the detector may be passed to the detector below. When ADF-STEM is performed using a high inner cutoff angle, it is capable of sub-angstrom spatial resolution, and this allows atomic resolution images to be obtained in registry with EELS spectra from individual columns of atoms.

A major conceptual advance in ADF-STEM is the use of a high inner angle or the high-angle ADF (HAADF) detector for the so-called Z-contrast imaging. During work at the Cavendish in the late 1970s on imaging metal catalyst particles on a crystalline substrate (Treacy et al. 1978), Howie proposed that a sufficiently high-angle detector would minimize the unwanted lower-angle Bragg scattering, giving a STEM HAADF image with strong atomic number contrast (proportional to Z^2) according to the Rutherford formula for scattering from an unscreened nucleus (Treacy and Gibson 1993). An inner hole diameter of about 40 mrad was suggested (Howie 1979), corresponding to the angle at which the corresponding d-spacing becomes comparable with the thermal vibration amplitude. An important development was the first publication of atomic-scale images using high-angle scattering (Pennycook and Boatner 1988). A full treatment of this problem requires the inclusion of both Bragg and thermal diffuse scattering (TDS), and, with the development of sufficiently fast computers, these treatments began to appear in the 1990s, notably from the Cornell group (Hillyard and Silcox 1995) and elsewhere.

The use of field emission electron sources was critical for STEM for forming a small and intense beam. As in HREM, the main resolution limit was the spherical aberration of the objectives. Thus, advances in the technology of aberration correctors that started in late 1990s (Chap. 7) have brought significant improvement in the image resolution for both HREM and STEM from about 0.34 nm (a common spacing in graphite) in the late 1970s to 0.05 nm today, sufficient to resolve individual columns of atoms in most thin crystals. The use of charge-coupled device (CCD) detectors played a critical role here (Spence and Zuo 1988). Together with dramatic increases in computing power, these detectors have made image simulation (including all multiple scattering effects) routine, enabled the tomography mode in biology (where digital images of projections must be combined into 3D), and made the automated alignment and control of electron microscopes possible. These advances also allow much more quantitative analysis of high-resolution images and the more accurate measurement of the experimental parameters on which image simulations depend.

14.2   Abbe's Theory of Coherent Imaging

The starting point for understanding HREM is the theory of coherent image formation, first formulated by Ernst Abbe in 1873 in Jena, Germany, for an optical system. Geometrical optics adequately predicts image positions, geometry, and magnifications, but does not provide any explanation of image quality, especially image resolution and contrast. Abbe realized that the diffraction pattern formed in the back focal plane of the objective lens is the critical and deterministic factor in image formation. He proposed that for a microscope objective lens equipped with a fixed numerical aperture and illumination (wavelength), coherent interference between the zero-order and higher order diffracted beams from the specimen produces image contrast and determines the maximum spatial resolution.

To examine the above idea further, we consider a diffraction grating made of alternating opaque and transparent strips as the object. Illumination of a parallel beam produces an array of diffraction spots normal to the strips, which can be indexed according to their diffraction angles with two first-order diffracted beams appear immediately next to the incident beam (zero order), followed by the second, third order and so on. In order to have interference, at least two beams must enter the objective lens, independent of the lens magnification used.

It is then instructive to examine Fig. 14.1. When illuminated with a plane wave, the diffracted beams are also plane waves. Since sharp diffraction spots are formed in the back focal plane, the lens must perform a wave-front transformation that turns an incident plane wave into a spherical wave that converges to a point on the back focal plane. For the image plane, each diffraction spot acts as a separate source of secondary waves. Interference of these waves gives rise to an image. In the case of only zero-order diffraction passing through the objective lens, no image of the optical grating is formed since there is no interference.

An analogy to the optical lens can be made for electrons. An electron lens has the same properties as the optical lens and electron wave propagate in a space free of electromagnetic fields except in the sample and the lens. All focusing actions of the lens can be assumed to occur within the boundary of the lens. Such a model is valid because the lens magnetic field is macroscopic, weak, and relatively uniform

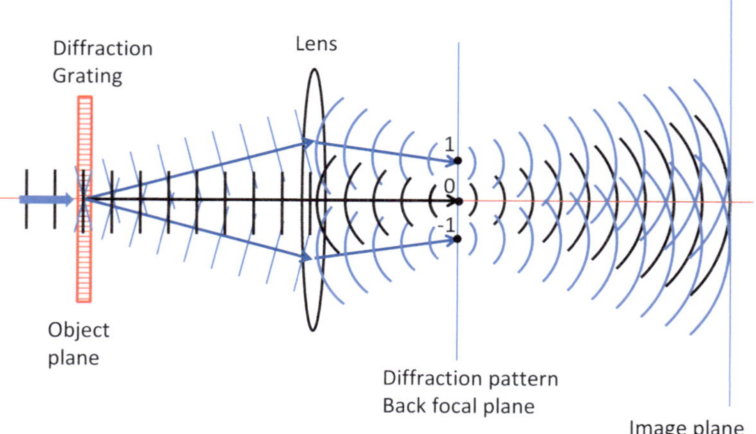

Fig. 14.1 Illustration of Abbe's theory of image formation

as the field varies on a length scale far greater than the electron wave length. In what follows, we will formulate the imaging process based on Fresnel propagation at small angles.

14.3   Coherent Imaging in an Ideal Lens

In an ideal lens, a parallel beam corresponding to a plane wave on the object side along the optical axis is focused by the lens to the back focal point (f in Fig. 14.2) and thus transformed into a spherical wave. This spherical wave is represented by the circle centered at f and in contact with the principle plane (H_2) in Fig. 14.2. Both the phase and amplitude are constant on this circle. Compared to this, the wave exiting the principle plane (H_2) has a phase shift determined by the path difference δ as shown in Fig. 14.2:

$$\alpha = -\frac{2\pi\delta}{\lambda} = \frac{2\pi}{\lambda}f\left(1 - \frac{1}{\cos\theta}\right) \approx -\frac{\pi}{\lambda}f\theta^2 \approx -\frac{\pi}{\lambda f}\left(x^2 + y^2\right), \tag{14.1}$$

where x and y are the coordinates in the plane normal to z as illustrated in Fig. 14.2. The approximation taken in (14.1) is for small θ, the same approximation that was used in the paraxial approximation in electron optics. Thus, the wave function on H_2 can be written as

$$\phi_{H_2} = e^{-\frac{\pi i}{\lambda f}\left(x^2 + y^2\right)} \tag{14.2}$$

By combining Eq. (14.2) and (2.58), we have:

$$\phi_f(x, y) = -i\frac{e^{2\pi i k f}}{f\lambda} e^{\frac{\pi i}{\lambda f}\left(x^2 + y^2\right)} \iint e^{-\frac{2\pi i}{\lambda f}(xX + yY)} \mathrm{d}X\mathrm{d}Y$$

$$= -ie^{2\pi i k f}\delta(k_x)\delta\left(k_y\right) \tag{14.3}$$

where

$$k_x = x/\lambda f \quad \text{and} \quad k_y = y/\lambda f. \tag{14.4}$$

The delta function $\delta(x)$, which has the following property

$$\delta(x - a) = \begin{cases} \infty & x = a \\ 0 & x \neq a \end{cases}$$

In deriving Eq. (14.3), we have used a property of the Fourier transform (Appendix E). The above results show that lens action can be described by a parabolic phase shift increasing with distance from the optical axis. For an incoming plane wave, this phase shift leads to a focused point at f by Fresnel propagation.

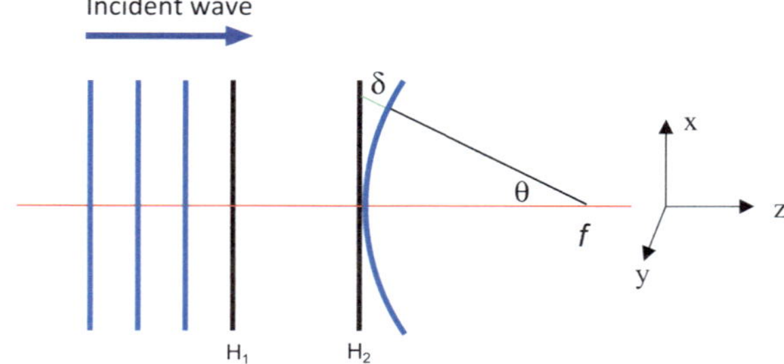

Fig. 14.2 Lens action on an incident plane wave (*left*) and transform into a converging spherical wave (*right*)

Next, we consider coherent imaging of an object at distance U from the principle plan H_1, and the image at V distance, for a focal distance f. We start with the exit-wave function of the object

$$\phi_o = \phi_e(x, y)$$

The wave function at the principle plane H_1 is obtained by propagating the exit wave according to Eq. 2.58, which can be written in the mathematical form of convolution (see Appendix E for details):

$$\phi_{H_1}(x, y) = \phi_e(x, y) * P_U(x, y) \tag{14.5}$$

where

$$P_z(x, y) = -i\frac{e^{2\pi ikz}}{z\lambda}\, e^{\frac{\pi i}{\lambda z}\left(x^2 + y^2\right)} \tag{14.6}$$

is the so-called Fresnel propagator. At H_2, the lens action introduces a phase shift according to (14.2). From this, we obtain

$$\phi_{H_2}(x, y) = \phi_{H_1}(x, y)e^{-\frac{\pi i}{\lambda f}\left(x^2 + y^2\right)} \tag{14.7}$$

Similarly, the wave function at the image plane is given by propagating electron wave function from the back focal plane to the image plane:

$$\phi_i(x, y) = \phi_f(x, y) * P_{V-f}(x, y) \tag{14.8}$$

Using the same procedure as in Eq. (14.3), we calculate the wave function at the back focal plane of the lens

$$\phi_f(k_x, k_y) = -i\frac{e^{2\pi ikf}}{f\lambda}\, e^{\pi i\lambda f\left(k_x^2 + k_y^2\right)}\mathrm{FT}\left[\phi_{H_1}(x, y)\right] \tag{14.9}$$

where k_x and k_y are defined according to Eq. (14.4); together they specify the coordinate in the back focal plane. To further evaluate Eq. (14.9), we first examine the Fourier transform of the Fresnel propagator

$$\mathrm{FT}\left(e^{\frac{\pi i}{\lambda z}x^2}\right) = \int\limits_{-\infty}^{\infty} e^{\frac{\pi i}{\lambda z}x^2} e^{2\pi i k x}\,\mathrm{d}x = \sqrt{\frac{\lambda z}{2}}(1+i)e^{-\pi i k^2 \lambda z},$$

which gives

$$\mathrm{FT}[P_z(x,y)] = e^{2\pi i k z} e^{-\pi i \left(k_x^2 + k_y^2\right)\lambda z} \tag{14.10}$$

Using this result and the property that the Fourier transform of two convoluted functions is the product of the Fourier transform of each function, we obtain

$$\phi_f\left(k_x, k_y\right) = -i\frac{e^{2\pi i k(f+U)}}{f\lambda} e^{-\pi i \frac{\lambda f}{M}\left(k_x^2 + k_y^2\right)} \phi_e\left(k_x, k_y\right) = C\phi_e\left(k_x, k_y\right) \tag{14.11}$$

where

$$\phi_e\left(k_x, k_y\right) = \iint \phi_e(x,y) e^{-2\pi i\left(k_x x + k_y y\right)}\,\mathrm{d}x\mathrm{d}y,$$

and C is a complex coefficient. In obtaining (14.11), we have used the relationship between the magnification and distances from the geometric optics:

$$\frac{f}{U-f} = \frac{V-f}{f} = M.$$

From Eq. (14.11), we see that the wave function at the focal plane is the Fourier transform of the object wave function with an additional, angle dependent, phase shift.

Using the same procedures for wave propagation between the back focal plane and the image plane, we obtain for an ideal lens

$$\phi_i(x,y) = -\frac{1}{M} e^{2\pi i k(U+V)} e^{\frac{\pi i}{\lambda M f}\left[x^2 + y^2\right]} \phi_e\left(-\frac{x}{M}, -\frac{y}{M}\right) \tag{14.12}$$

Thus, the wave function at the image plane is a magnified and inverted copy of the object exit-wave function, modified by additional phase term.

14.4 Coherent Imaging in a Real Lens

According to the results in Sect. 7.2, the effect of lens aberrations, including defocus, is described by introducing an additional phase factor to the wave function at the back focal plane, in the form of

$$\phi_f = C\phi_e\left(k_x, k_y\right)e^{-i\chi\left(k_x,k_y\right)}. \tag{14.13}$$

The relationship between the diffraction coordinate of $\left(k_x, k_y\right)$ used here and the angle coordinate $\left(\theta_x, \theta_y\right)$ in Chap. 6 is simply $k_x \approx k\theta_x$ and $k_y \approx k\theta_y$. For a round magnetic lens, we have

$$\chi(K_t) = \pi\lambda K_t^2\left(\frac{C_s\lambda^2}{2}K_t^2 + \Delta f\right) = \frac{\pi}{\lambda}\theta^2\left(\frac{C_s}{2}\theta^2 + \Delta f\right), \tag{14.14}$$

where $K_t = \sqrt{k_x^2 + k_y^2} \approx \theta/\lambda$ is the amplitude. In a TEM equipped with an aberration corrector, C_s is reduced to the order of micrometers (compared to millimeters for an uncorrected objective lens). Its contribution to $\chi\left(k_x, k_y\right)$ is thus no longer the largest, compared to other forms of aberrations; these must now be considered. The general form of $\chi\left(\theta_x, \theta_y\right)$ was given in Chap. 7 for describing image formation in a C_s corrector up to 5th order.

An objective aperture is used to limit the contributions of higher order aberrated beams in image formation. The objective lens aperture is placed in the vicinity of the back focal plane and its effect can be effectively described by imposing a top-hat function on the back focal plane

$$A\left(k_x, k_y\right) = \begin{cases} 1 & \sqrt{k_x^2 + k_y^2} < k_{\text{cutoff}} \\ 0 & \text{otherwise} \end{cases}. \tag{14.15}$$

This represents a circular aperture. The aperture radius defines the maximum spatial frequency $\left(k_{\text{cutoff}}\right)$ that can be recorded in an image.

By combining Eqs. (14.13), (14.12), and (14.15), we obtain the image wave function in a real lens

$$\phi_i(x, y) = C\iint A\left(k_x, k_y\right)\phi_e\left(k_x, k_y\right)e^{-i\chi\left(k_x,k_y\right)}e^{-2\pi i\left[k_x x + k_y y\right]/M}dk_x dk_y. \tag{14.16}$$

It is often convenient to assume unit magnification and without inversion ($M = -1$) in image calculations. Then the image wave function, aside from a quadratic phase factor, is given by

$$\phi_i(x,y) = \iint \phi_e(k_x, k_y) H(k_x, k_y) e^{2\pi i(k_x x + k_y y)} \mathrm{d}k_x \mathrm{d}k_y = \phi_e(x,y) * h_C(x,y),$$

$$(14.17)$$

where

$$H(k_x, k_y) = A(k_x, k_y) e^{-i\chi(k_x, k_y)} \tag{14.18}$$

and

$$h_C(x,y) = \mathrm{FT}^{-1}\left[H(k_x, k_y)\right] = \mathrm{FT}^{-1}\left[A(k_x, k_y) e^{-i\chi(k_x, k_y)}\right] \tag{14.19}$$

is a complex function, called the lens transfer function. The coherent image intensity is then given by

$$I_i(x,y) = |\phi_i(x,y)|^2 = |\phi_e(x,y) * h_C(x,y)|^2. \tag{14.20}$$

Thus, under the coherent imaging condition, the image intensity is obtained by convoluting the exit-wave function with the lens transfer function.

14.5   Linear Imaging Theory and Contrast Transfer Function (CTF)

To examine the effects of lens aberrations, aperture, and defocus on coherent imaging, we will consider a very thin object using the weak phase object (WPO) approximation (Sect. 4.2), where the electron exit wave relates directly to the projected object potential

$$\phi_e(x,y) = 1 + i\pi\lambda U(x,y), \tag{14.21}$$

where $\phi_o(\vec{r}) = 1$ for the incident wave and

$$U(x,y) = \int_0^t U(x,y,z)\mathrm{d}z \tag{14.22}$$

is the projected potential. Furthermore, the imaginary part of the electron-optical potential can be neglected for a very thin object, and thus $U(x,y)$ can be taken as a real function. By Fourier transform, we obtain

$$\phi_e(k_x, k_y) = \delta(k_x, k_y) + i\pi\lambda U(k_x, k_y). \tag{14.23}$$

According to Eqs. (14.13) and (14.17), the wave functions at the back focal plane and the image plane are given by

$$\phi_f\left(k_x, k_y\right) \approx \left\{\delta\left(k_x, k_y\right) + i\pi\lambda U\left(k_x, k_y\right)\right\} \exp\left(-i\chi\left(k_x, k_y\right)\right)$$
$$= \delta\left(k_x, k_y\right) + \pi\lambda U\left(k_x, k_y\right) \sin\chi\left(k_x, k_y\right) + i\pi\lambda U\left(k_x, k_y\right) \cos\chi\left(k_x, k_y\right),$$

$$(14.24)$$

And the image wave function is obtained by Fourier transform, which gives

$$\phi_i(x, y) = 1 + \pi\lambda U(x, y) * h(x, y) + i\pi\lambda U(x, y) * h'(x, y). \tag{14.25}$$

Here $h(x, y)$ and $h'(x, y)$ are the Fourier transform of $\sin\chi\left(k_x, k_y\right)$ and $\cos\chi\left(k_x, k_y\right)$, respectively. The image intensity is given by the amplitude square of $\phi_i(x, y)$

$$I(x, y) \approx 1 + 2\pi\lambda U(x, y) * h(x, y) + \left[\pi\lambda U(x, y) * h'(x, y)\right]^2, \tag{14.26}$$

and by ignoring the second-order terms, we obtain

$$I(x, y) \approx 1 + 2\pi\lambda U(x, y) * h(x, y). \tag{14.27}$$

Thus, under the WPO approximation, the image intensity to first order is given by the projected potential convoluted with $h(x, y)$. If the values of the defocus and aberration coefficients are optimized such that $h(x, y)$ contains a single sharp peak, then the image gives the projected potential at the resolution defined by the width of the sharp peak. The image resolution thus depends on the $\sin\chi\left(k_x, k_y\right)$ function.

The linear relationship obtained between the image intensity and the object function in Eq. (14.27) is known as the linear image model, which has the general form

$$I(\vec{r}) = f(\vec{r}) * h(\vec{r})$$

where $h(\vec{r})$ with $\vec{r} = (x, y)$ is the point spread function (PSF) of the instrument. In frequency space, the linear image model gives the following product

$$I(\vec{K}_t) = F(\vec{K}_t) H(\vec{K}_t),$$

where $\vec{K}_t = \left(k_x, k_y\right)$, and $H(\vec{K}_t)$ is the Fourier transform of $h(\vec{r})$. In this model, if the object is a sharp point, then the image of this sharp point gives the PSF of the instrument. Since $h(\vec{r})$ is independent of the object in the linear image model, it provides a convenient mechanism for understanding the imaging process and image contrast.

In electron imaging, it is a common practice to describe image formation in frequency space based on $H(\vec{K}_t)$. To relate $H(\vec{K}_t)$ to the image contrast, we consider a sinusoidal WPO, giving rise to the following exit-wave function

$$\phi_e(x) \approx 1 + i\varepsilon \cos(2\pi K_t x). \qquad (14.28)$$

According to Eq. (14.27), the image intensity is given by

$$I(x) = 1 + 2\varepsilon \cos(2\pi K_t x) \sin \chi(K_t) \qquad (14.29)$$

The difference between the maximum and minimum intensity gives the image contrast. The maximum contrast in an ideal image is obtained by setting $\sin \chi(K_t) = 1$ or $\chi(K_t) = \pi/2$, in which case, $I_{Max} - I_{Min} = 4\varepsilon$. In a nonideal image, the contrast transfer function is the ratio of the difference with $\sin \chi(K) \neq 1$ and 4ε, defined by:

$$CTF = \frac{I_{Max} - I_{Min}}{(I_{Max} - I_{Min})_{MAX}} = \frac{|4\varepsilon \sin \chi(K_t)|}{4\varepsilon} = |\sin \chi(K_t)|. \qquad (14.30)$$

Thus, for an object with many frequencies, the CTF defines the contribution of each frequency to the final image, which is determined by $|\sin \chi(K_t)|$ for a weak phase object.

Ideally, the amplitudes and phase of the image Fourier spectrum should closely resemble that of the object by keeping the $\sin \chi(K_t)$ function relatively flat and constant in sign over the range of spatial frequencies that are enclosed by the objective aperture. To meet this requirement, the oscillations of the sine function, which cause serious problems because it changes sign, should be excluded. In a TEM, the controls that are easily accessed are the defocus and the objective aperture, once the wavelength is determined by the microscope acceleration voltage. The value of C_s is fixed in a conventional TEM and corrected using a C_s corrector.

To compute for the optimum defocus and objective aperture for a conventional TEM, we first introduce the reduced scattering angle

$$\theta^* = \theta(C_s/\lambda)^{1/4} \qquad (14.31)$$

and defocus:

$$\Delta f^* = \Delta f / \sqrt{C_s \lambda} \qquad (14.32)$$

Substituting these into the Eq. (14.14), we have

$$\chi(\theta^*)/2\pi = \frac{\theta^{*4}}{4} + \frac{\theta^{*2}}{2} \Delta f^*, \qquad (14.33)$$

with only defocus and the third-order spherical aberration included.

The maximum contrast is obtained when $|\sin \chi(\theta^*)| = 1$. In order to represent the object in the image as closely as possible, $|\sin \chi(\theta^*)|$ should be close to 1 for as many frequencies as possible. These two conditions can be met by having a minimum in the aberration function and having the minimum $\chi_{\min}(\theta^*) = \pi(2n+1)/2$ by using a negative defocus (Fig. 14.3). This is obtained by setting the derivative of $\chi(\theta^*)$ to zero, which yields

$$\frac{\partial}{\partial \theta^*}[\chi(\theta^*)/2\pi] = \theta^{*^3} + \theta^* \Delta f^* = 0$$

$$\theta^* = \sqrt{|\Delta f^*|}$$

and $\Delta f^* < 0$.

At the minimum,

$$[\chi(\theta^*)/2\pi]_{\min} = -\frac{\Delta f^{*^2}}{4}.$$

The phase shift of $\pi(2n + 1)/2$ is obtained with

$$\Delta f^* = -\sqrt{2n+1},$$

with $n = 0, 1, 2, \ldots$. At the optimum defocus,

$$[\chi(\theta*)]_{\min} = -n\pi - \frac{1}{2}\pi, \tag{14.34}$$

and

$$\left|\sin[\chi(\theta^*)]_{\min}\right| = 1.$$

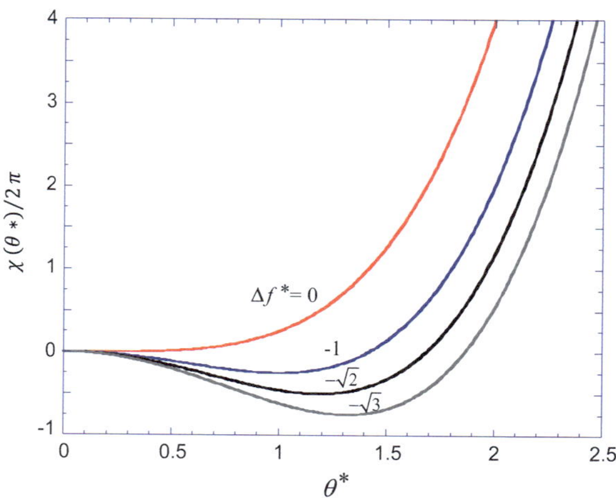

Fig. 14.3 Aberration as function of the reduced scattering angle and defocus

At $n = 0$, we obtain

$$\Delta f_{\text{opt}} = -(C_s \lambda)^{1/2} \tag{14.35}$$

Figure 14.4 plots $\sin \chi(\theta^*)$ for $\Delta f^* = -1$ and -1.22. At $\Delta f^* = -1$, a flat minimum with $\sin \chi(\theta^*) = -1$ is obtained at $\theta^* = 1$. As θ^* increases, $\sin \chi(\theta^*)$ turns upward and crosses zero at

$$\chi(\theta^*) = 0 \quad \text{and} \quad \theta^* = \sqrt{2}.$$

Further increase in θ^* leads to progressively rapid oscillations. The optimum imaging condition is to cut off these oscillations by placing an objective aperture with the following cutoff angle:

$$\theta_{\text{opt}} = \frac{\sqrt{2}}{(C_s/\lambda)^{1/4}} \approx 1.41 \left(\frac{\lambda}{C_s}\right)^{1/4}. \tag{14.36}$$

The lower bound of the image resolution that can be achieved under such condition is obtained from

$$d > 1/k_{\text{max}} = \lambda/\theta_{\text{opt}} = 0.71 \left(\lambda C_s^3\right)^{1/4}. \tag{14.37}$$

The resolution can be further improved by using the so-called Scherzer focus with

$$\Delta f^* = -\sqrt{1.5} \approx -1.22 \quad \text{and} \quad \Delta f = -\sqrt{3/2}(C_s \lambda)^{1/2} \tag{14.38}$$

which adds a $-\pi/4$ phase to $[\chi(\theta^*)]_{\text{min}}$ and pushes the zero crossover further out (see Fig. 14.4) leading to

$$\theta_{\text{opt}} = \frac{\sqrt{2\sqrt{1.5}}}{(C_s/\lambda)^{1/4}} = \left(\frac{6\lambda}{C_s}\right)^{1/4}, \tag{14.39}$$

and

$$d > 1/k_{\text{max}} = \lambda/\theta_{\text{opt}} = 0.64 \left(\lambda C_s^3\right)^{1/4}. \tag{14.40}$$

Together, Eqs. (14.38) and (14.39) are referred to in HREM as the Scherzer conditions.

In a C_s corrector without the parasitic aberrations, C_s can be adjusted by the corrector while C_5 is the performance limiting aberration. In 1970, Scherzer described the optimum for the defocus, the third-order spherical aberration C_s (C_3 as defined in Sect. 7.2), the objective aperture θ_{opt}, and the lower bound of

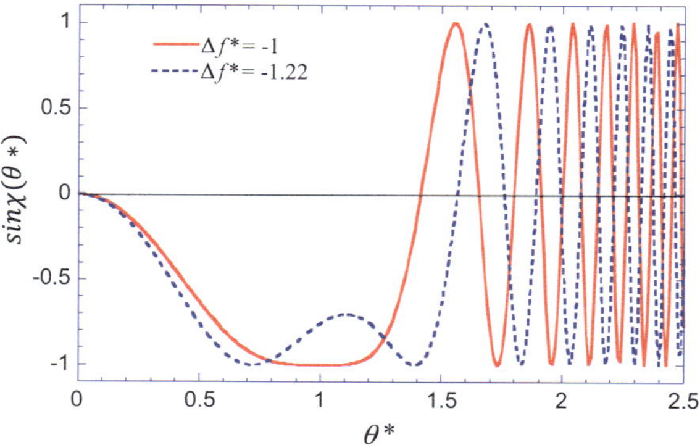

Fig. 14.4 $\sin \chi(\theta^*)$ plotted for $\Delta f^* = -1$ and -1.22. The latter is known as the Scherzer focus

resolution by optimizing a phase shift of $\pi(1 \pm 1/2)/2$. These parameters are given by Haider et al. (2000)

$$\Delta f = 2.0 \sqrt[3]{\lambda^2 C_5}$$

$$C_s = -3.2 \sqrt[3]{\lambda^2 C_5^2} \tag{14.41}$$

$$\theta_{\text{opt}} = \frac{7}{4} \sqrt[6]{\frac{\lambda}{C_5}}.$$

We note that the optimal value of C_s is thus not zero, rather a small negative value in an aberration-corrected TEM. Its value together with defocus is determined by C_5, which is assumed uncorrected and positive. For example, with $C_5 = 5$ mm, Eq. (14.41) gives $\Delta f = 6.31$ nm, $C_s = -12.71$ µm, and $\theta_{\text{opt}} = 49.3$ mrad at 200 kV. These values here serve only as a guide to operating aberration-corrected TEMs, as their performance is also limited by the lower order, nonspherical, aberrations, which were not taken into consideration in Eq. (14.41). However, there are experimental benefits having a negative C_s, which was demonstrated by Lentzen et al. (2002) and Jia et al. (2003). Once the corrector is adjusted and aligned, the best imaging conditions are obtained by optimizing Δf and twofold astigmatism (A1, see Sect. 7.2), while θ_{opt} is fixed at around 25 mrad at 200 kV (see Fig. 14.5 for example).

The WPO approximation is valid as long as the sample is thin enough and the electron acceleration voltage is high enough. To account for "not so weak" objects, the second-order term in Eq. (14.26) must be kept. This contribution is proportional to the square of the potential that is smeared out by $h'(\vec{r})$. The smearing effect of $h'(\vec{r})$ is approximately the same as multiplying the diffraction pattern by $A(K_t) \cos \chi(K_t)$. Because of this, $\cos \chi(K_t)$ is also known as the amplitude contrast term. At the optimum imaging condition, $h'(\vec{r})$ is rather broad and featureless, and it contributes to the background intensity in HREM images. At medium magnification, low-angle

Fig. 14.5 Calculated $\sin \chi(\theta_x, \theta_y)$ for a third-order C_s corrector at 200 kV using experimentally adjusted aberration coefficients and the optimization of Δf and twofold astigmatism (Wen et al. 2010). The *marked circle* has a radius of 25 mrad

scattering becomes important for the image contrast; the amplitude contrast dominates as $\cos \chi(K_t)$ approaches 1 while $\sin \chi(K_t)$ becomes zero.

For strongly scattering objects, full multiple scattering calculations must be made, and it is no longer possible to use the simple concept of a transfer function or even to define resolution simply, because it then becomes a property of the sample and the microscope, rather than of the microscope alone. Nevertheless, although an intuitive interpretation of the image in terms of sample structure is lost, it is still possible to compare simulated images for model structures with experimental images and to seek the best-fitting model structure. This topic is treated more fully in Kirkland (2010).

14.6   The Effects of Electron Energy Spread and Partial Coherence

So far, we have assumed perfect illumination by a monochromatic, parallel, and coherent beam and thus a single focus for the magnetic lens. These assumptions break down in practice for the following reasons:

(1)   The electron beam carries a small electron energy spread (ΔE) from the electron source that ranges from ~ 0.3 eV for a cold field emitter to 2–3 eV for a thermionic source. In addition, fluctuations in the accelerating voltage $(\Delta \Phi)$ contribute to the further broadening of the electron energy distribution.

The energy spread can be reduced using an electron monochromator (see Chap. 10) and by improving the stability of the high voltage supply.

(2) Instead of a single incident wave, there are many waves coming from the effective electron source. The lateral coherence length of these waves is defined by the source convergence angle α_s according to Eq. (10.7).

(3) The magnetic lens focal length changes with the lens current. Thus, the current instability (ΔI) gives rise to a distribution of focal lengths.

The effects of these factors all can be represented as time-dependent variations in the focal length variation

$$\Delta_f = C_c f_r \sqrt{\left(\frac{\Delta E}{E_o}\right)^2 + \left(2\frac{\Delta I}{I}\right)^2 + \left(\frac{\Delta\Phi}{\Phi}\right)^2}. \tag{14.42}$$

where

$$f_r = \frac{1 + e\Phi/m_e c^2}{1 + 2e\Phi/m_e c^2}.$$

To take account of the effects of an extended source, we assume that successive fast electrons are independent and any interaction between them (such as the Boersch effect) is neglected. We assign a separate wavevector and direction to each incident electron. Two electrons with the same wavevector would arrive at the specimen at different times. The intensity at a point in the final image $I(\vec{r})$ is obtained by summing the intensities due to each fast electron. Thus,

$$I(\vec{r}) = \iint \left|\phi_i(\vec{r}, \vec{K}_t, \Delta f)\right|^2 p(\vec{K}) p(\Delta f) \mathrm{d}^2\vec{K}_t \mathrm{d}\Delta f \tag{14.43}$$

where $\phi_i(\vec{r}) = \phi_e(\vec{r}) * h(\vec{r})$ according to Eq. (14.17); $p(\Delta f)$ and $p(\vec{K})$ are the probability of finding an electron with the momentum between $\vec{K}_t$ and $\vec{K}_t + \mathrm{d}\vec{K}_t$ and at the focus setting between Δf and $\Delta f + \mathrm{d}\Delta f$, respectively. Both $p(\Delta f)$ and $p(\vec{K}_t)$ are normalized such that the overall probability is one. The probability $p(\Delta f)$ is approximately Gaussian distributed because of the random nature of the electron emission, high voltage, and the objective lens current fluctuations. The exact profile of $p(\vec{K}_t)$ can be measured from an intensity profile taken across the central diffraction spot formed by a parallel beam as long as a short exposure time is used to avoid the detector saturation. In an approximation, $p(\vec{K}_t)$ can also be described as a Gaussian function.

The integration in Eq. (14.43) may be computed exactly, using the result of a numerical solution for $\phi_i(\vec{r}, \vec{K}_t, \Delta f)$. This method makes no approximation but requires a separate dynamical calculation for each incident plane wave $(\vec{K}_t)$. Early results of such calculations, exploring the effect of variations in coherence, can be found in O'Keefe and Sanders (1975). Alternatively, to avoid the need for many dynamical calculations, an approximation valid for small beam divergence may be adopted. This requires a single dynamical calculation for $\phi_i(\vec{r}, \vec{K}_t, \Delta f)$. For very thin crystals, we may further assume the weak phase object approximation, which gives the analytical expression of partial coherence effects in the form of a multiplicative transfer function. This will be shown below.

We start with the sinusoidal WPO discussed in Eq. (14.28), which gives the following image intensity for a tilted incident beam $\vec{K}_{to}$

$$I(x) = 1 + \varepsilon \cos(2\pi K_t x)\left[\sin \chi(\vec{K}_t + \vec{K}_{to}) + \sin \chi(\vec{K}_t - \vec{K}_{to})\right]. \tag{14.44}$$

For small values of $\vec{K}_{to}$, the phase shift $\chi(\vec{K}_t \pm \vec{K}_{to})$ can be expanded in a Taylor series,

$$\chi(\vec{K}_t \pm \vec{K}_{to}) = \chi(\vec{K}_t) \pm \nabla\chi(\vec{K}_t) \cdot \vec{K}_{to} + \cdots,$$

where ∇ is the differential operator. By keeping the first-order term in Eq. (14.44), we have

$$I(x) = 1 + 2\varepsilon \cos(2\pi K_t x) \sin \chi(\vec{K}_t) \cos\left[\nabla\chi(\vec{K}_t) \cdot \vec{K}_{to}\right]. \tag{14.45}$$

Substituting this result into Eq. (14.43) gives the following expression (Reimer and Kohl 2008)

$$I(x) = 1 + 2\varepsilon \cos(2\pi K_t x) \sin \chi(\vec{K}_t) E_c(\Delta_f, K_t) E_s(\alpha_s, K_t), \tag{14.46}$$

where $E_c(\Delta_f, K_t)$ and $E_K(\alpha_s, K_t)$ are two envelope functions obtained from the integration of the Gaussian function with $\cos\left[\nabla\chi(\vec{K}_t) \cdot \vec{K}_{to}\right]$. They are given by

$$E_c(\Delta_f, K_t) = \exp\left[-\left(\frac{\pi\lambda K_t^2 \Delta_f}{4\sqrt{\ln 2}}\right)^2\right] \tag{14.47}$$

and

$$E_s(\alpha_s, K_t) = \exp\left[-\frac{\left(\pi C_s \lambda^2 K_t^3 - \pi\Delta f K_t\right)^2 \alpha_s^2}{\ln 2}\right], \tag{14.48}$$

where Δ_f is defined by Eq. (14.42) and α_s is the full-width at Half Maxima (FWHM) of the effective source angle.

A comparison of the two equations in (14.46) and (14.29) shows that in the WPO approximation, the partial coherence for the C_s limited objective lens in the absence of the twofold astigmatism is accounted for by modifying the Fourier transform of the PSF

$$H(K_t) = \sin \chi(K_t) E_c\left(\Delta_f, K_t\right) E_s(\alpha_s, K_t) = \sin \chi(K_t) E(K_t). \tag{14.49}$$

The envelope function $E(K_t)$ damps out the oscillations in $\sin \chi(K_t)$ at high spatial frequencies (see Figs. 14.6 and 14.7 for example), while it has little effect at low frequencies because the exponential term increases to the fourth power of the spatial frequency in $E_c(\Delta_f, K)$. In HREM, the envelop function defines the extent of information transferred by the lens. The ultimate limit to information transfer is placed by the focal spread (Eq. 14.47), which falls to $\exp(-2)$ or 13.5 % at

$$\left(\frac{\pi \lambda K_t^2 \Delta_f}{4\sqrt{\ln 2}}\right)^2 = 2 \tag{14.50}$$

or

$$\Lambda_{\min} = \frac{1}{K_{t\max}} = \left(\frac{\pi \lambda \Delta_f}{4\sqrt{2 \ln 2}}\right)^{1/2}. \tag{14.51}$$

Other information limiting factors in the imaging process are from the sample, electron inelastic scattering, and the electron detector. Specifically,

(1) any sample movement or instability during imaging taking translates into a degradation of the image contrast, and its effect is a sum of snapshots recorded almost instantaneously by incident electrons.
(2) the recorded image intensity is convoluted with the PSF of the detector; Boothroyd et al. showed that small difference in the experimentally measured PSFs using different methods can lead to sensitive changes in image contrast (Boothroyd and Dunin-Borkowski (2004), also see Chap. 9 for a discussion on detector characterization).

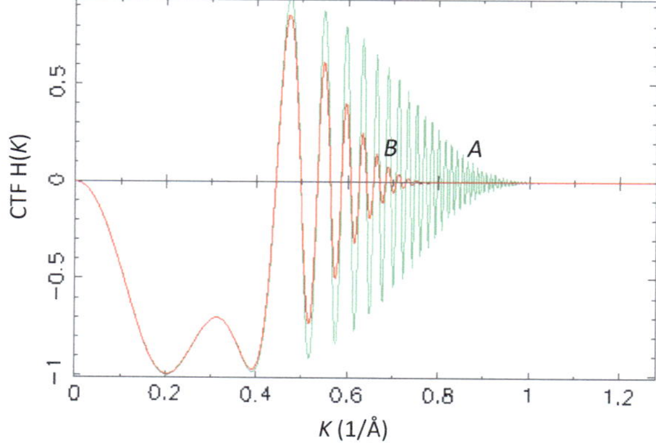

Fig. 14.6 Weak phase object contrast transfer function taking into account partial coherence with (*B*) and without the source convergence angle of 0.25 mrad (*A*). Parameters used for calculations are $C_s = 1$ mm, $C_c = 1.2$ mm, $\Delta f = -61.3$ nm, $\Delta E = 0.7$ eV, $\Delta V = \Delta I = 0$

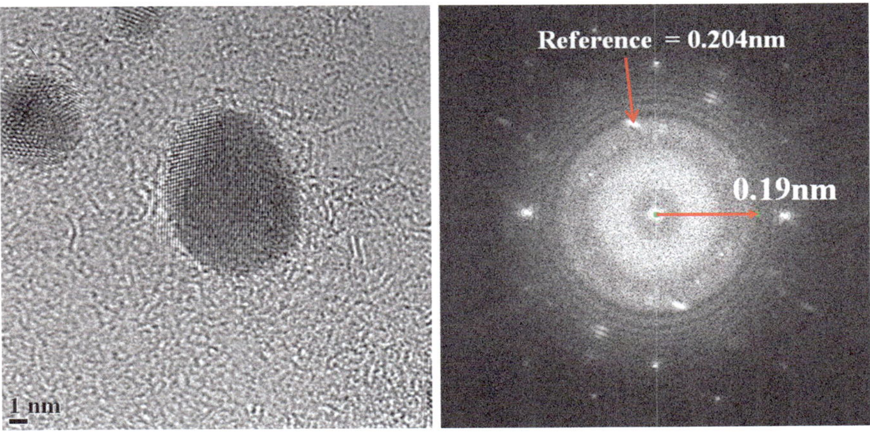

Fig. 14.7 *Left* HREM image of Au nanoparticles supported on amorphous Ge thin film, *Right* power spectrum of the *left image* showing the first zero of the contrast transfer function and the subsequent oscillations

(3)   inelastic scattered electrons contribute to high-resolution image contrast due to subsequent elastic scattering, even though inelastic scattering process is incoherent. Because the contribution from inelastic scattering comes from all thickness within the sample and at different focuses due to the chromatic aberration, the contrast from inelastic scattered electrons is distinctly different from that of elastic scattered electrons. The inelastic contribution is significant even from the thinnest sample (Boothroyd 1998).

(4)   electron TDS, which is distributed over a large range of scattering angles, contributes to a constant background in the image to the first-order approximation.

14.7   Electron Probes for High-Resolution STEM and Analysis

The premise of STEM imaging is the detection of scattering by a measurement of diffraction intensities, which occurs to a focused probe as it traverses through the sample. To achieve a sufficient level of image contrast, the focused probe must be small to meet the resolution requirement and bright enough to provide the signal. The smallest probe is obtained by making source contribution to the probe size (d_s) small in Eq. (10.5) by combining a small physical source with a large demagnification. Then, the probe formation becomes diffraction limited or aberration limited, and the illumination [from Eq. (10.9)] becomes necessarily coherent ($2L > 2R_a$, Sect. 10.3.5). Under perfectly coherent conditions, the probe on the sample is formed by a converging wavefront emerging from the lens. The wavefront is defined by the angle and lens aberration-dependent phase, $\chi(\vec{K}_t)$. Considering these factors together, the electron probe wave function is given by (also see Eq. 10.16)

$$\phi_P(\vec{r}, E) = \int_{-\infty}^{\infty} \phi_S(-M\vec{K}_t) A(\vec{K}_t) \exp\left[-i\chi(\vec{K}_t)\right] \exp(2\pi i \vec{K}_t \cdot \vec{r}) \mathrm{d}^2\vec{K}_t$$

$$= \phi_S(-\vec{r}/M) \otimes h(\vec{r}, E)$$

$$\approx h(\vec{r}, E). \tag{14.52}$$

where $h(\vec{r}, E)$ is the lens resolution function as defined in (14.19) for a specific electron energy. The approximation of $\phi_P(\vec{r}, E) \approx h(\vec{r}, E)$ is made for $M \gg 1$. The intensity distribution of the probe at the sample is given by

$$I(\vec{r}) = \int p(E) \phi_P(\vec{r}, E) \phi_P^*(\vec{r}, E) \mathrm{d}E, \tag{14.53}$$

where $p(E)$ is the probability of finding an electron with an energy between E and $E + \mathrm{d}E$. The conditions to obtain the "most compact" probe in a conventional TEM are given in Eqs. 10.18 and 10.19, and the probe size containing 70 % intensity is given by Eq. 10.20 (Fig. 14.8).

Different operating modes of STEM instruments require different focusing conditions. For analytical microscopy of a thick sample, the scattering of the probe distribution $I(\vec{r})$ is treated by solving a transport equation using, for example,

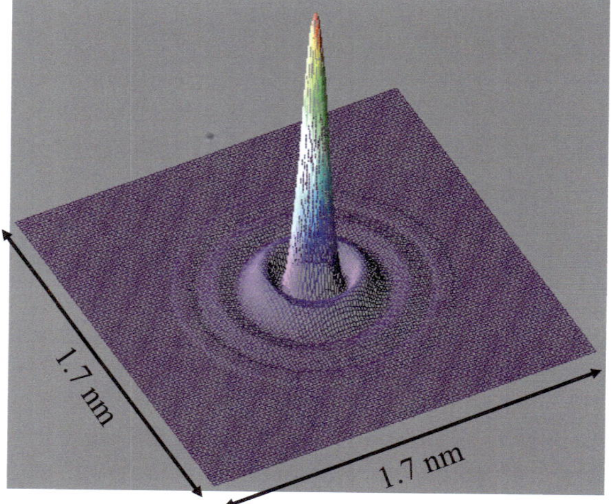

Fig. 14.8 Electron probe formed at following conditions: $C_s = 1$ mm, $C_c = 1.2$ mm, $\Delta E = 0.7$ eV, $\Delta f = -61.3$ nm and $\theta_c = 10.03$ mrad. The diameter containing 50 % of the probe intensity is 2.7 Å

Monte Carlo methods. Thus, although the probe formation process is treated coherently, the multiple scattering within the sample is treated incoherently, using the Boltzmann transport equation rather than the Schrödinger equation.

14.8   Probe Size and Resolution in Bright-Field STEM

For STEM imaging, dependent on the detector, a different focusing condition is required to produce the most faithful image of crystal. To examine this, we first consider the diffraction geometry used in BF-STEM as illustrated in Fig. 14.9. If CBED orders overlap, and if the angular range over which the illumination is coherent exceeds the Bragg angle, it becomes possible to form a phase-contrast STEM lattice image. The relevant theory is given in Spence and Cowley (1978) and a review in Spence (2013). The image is formed by detecting part of the CBED pattern (the overlap region).

To start, we assume a point source, and therefore complete coherence, and a conventional STEM. The focused probe is a diffraction-limited image of this point source, as formed by an imperfect lens. The probe wavefunction is given by Eq. (14.52), and we may think of the phase factor in Eq. (14.52) as filling the exit pupil of the probe-forming (objective) lens. We see that the orders just touch if $\theta_c = \theta_B$. If the orders do overlap slightly, as shown, and if the illumination aperture

Fig. 14.9 Lattice imaging in STEM. The illumination angle is larger than θ_B, which allows a small overlap of transmitted and diffracted disks, so that a detector at D records a two-beam lattice image as the probe is scanned. Interchanging S and D shows that this is equivalent to inclined illumination TEM imaging

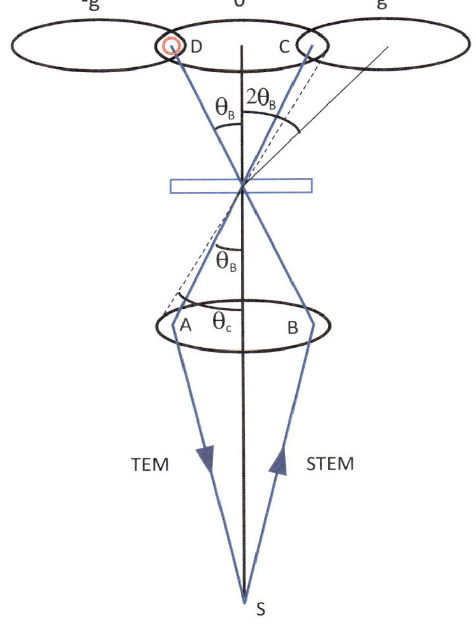

is coherently filled across A–B, it is possible for radiation from two different points A and B to reach the same detector point D following the paths of BD for the direct beam (ϕ_o) and ACD for the diffraction beam (ϕ_{-g}). The interference at D is given by

$$
\begin{aligned}
I_D &= \left| \phi_\mathrm{o} e^{-i\chi(-\theta_B)} + \phi_{-g} e^{-i\chi(\theta_B)} e^{-2\pi i g x} \right|^2 \\
&= |\phi_\mathrm{o}|^2 + |\phi_{-g}|^2 + 2|\phi_\mathrm{o}||\phi_{-g}| \cos\left[2\pi g x + \chi(\theta_B) - \chi(-\theta_B) + \alpha_{-g}\right].
\end{aligned}
\tag{14.54}
$$

where x is the probe position along the direction of g and α_{-g} is the phase angle between ϕ_o and ϕ_{-g}. The recorded intensity at D will therefore depend on the probe position and vary sinusoidally in this case with the period $c = 1/g = d = \lambda/2\theta_B$ of the lattice. The position depends on the cancelation of $\chi(\theta_B) - \chi(-\theta_B)$. It also depends on the phases of the crystal structure factors, so that experiments with overlapping CBED orders may be used to measure these quantities (Nellist et al. 1995).

However, we can see immediately that a lattice image will be formed, by using the theorem of reciprocity (see Fig. 11.12). On interchanging the point source S and a point detector at D, we have exactly the arrangement used to form two-beam lattice images in a TEM instrument, using inclined illumination from D. (It may be useful to imagine the crystal being scanned under a stationary probe to see this more clearly.) We note in passing that if $C_s = 0$ and $\Delta f = 0$, then the full width at half maximum height (FWHM) of the Bessel function probe would be

$$
d_\mathrm{FW} = 0.61\lambda/\theta_c
$$

Using the Bragg law and setting $d_\mathrm{FW} = d_{hkl}$, we find that the FWHM for the probe just equals the lattice spacing d_{hkl} if

$$
\theta_c = 1.2\theta_B
\tag{14.55}
$$

In this sense, it becomes possible to resolve the lattice in a STEM instrument when the probe size becomes comparable to the d-spacing of interest, and this occurs with a 20 % overlap of the orders. In practice, the effects of spherical aberration, tip vibration, and electrical instabilities cannot be neglected. It is also more useful to think of lattice resolution in STEM as limited by the angular range over which radiation incident on the sample is coherent.

To form a BF axial image, it is necessary to use the geometry shown in Fig. 14.10, with an axial detector, since, by the reciprocity theorem, such an image is identical to that which would be formed in a high-resolution TEM instrument if reciprocal aperture, source, and detector sizes are used. It is for this reason that the illumination aperture in STEM is known as the objective aperture, since it may be thought of (by reciprocity) as limiting the number of "beams" which contribute to a STEM lattice image. The Scherzer focus condition therefore gives the most faithful

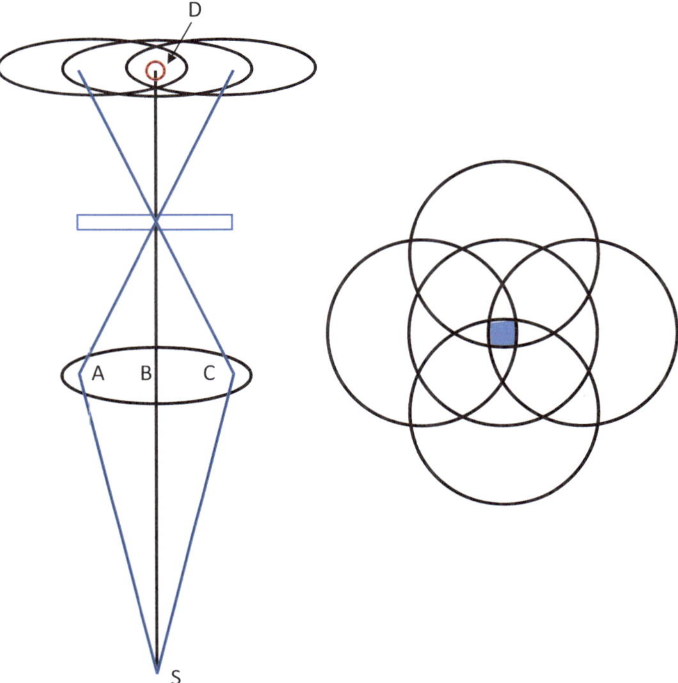

Fig. 14.10 Axial three-beam lattice imaging in STEM. By opening up the illumination angle to twice the Bragg angle or larger, three disks overlap at the axial detector D. The appearance of a two-dimensional coherent CBED pattern used for axial five-beam lattice imaging is shown at the *right*

representation of a weakly scattering sample in STEM. The lattice images for STEM and TEM may therefore be computed in the same way, with the integration over the source required in TEM becoming the integration over the detector in STEM.

Figure 14.11 shows an experimental coherent CBED patterns obtained as a Å-sized probe is positioned on single crystal silicon. Close examination of the figure shows the presence of both 1D and 2D interference patterns formed for overlapping disks of different orders. The overlapping portions of the patterns change as the probe is moved. In a STEM, the probe may be stopped at any point in the image and used to form a CBED pattern. Instrumentation requirements for this are described in Sect. 10.3 We will see that the central portion of the pattern, with a stationary probe, is a faithful lattice shadow-image of the crystal.

A dark-field or minimum phase-contrast focus condition also can be defined (Cowley 1995), for which the numerical factor in Eq. (10.17) becomes -0.44 instead of -0.75. The optimum focus for annular dark-field STEM imaging is found to be midway between this condition and the Scherzer condition (Mory et al. 1987).

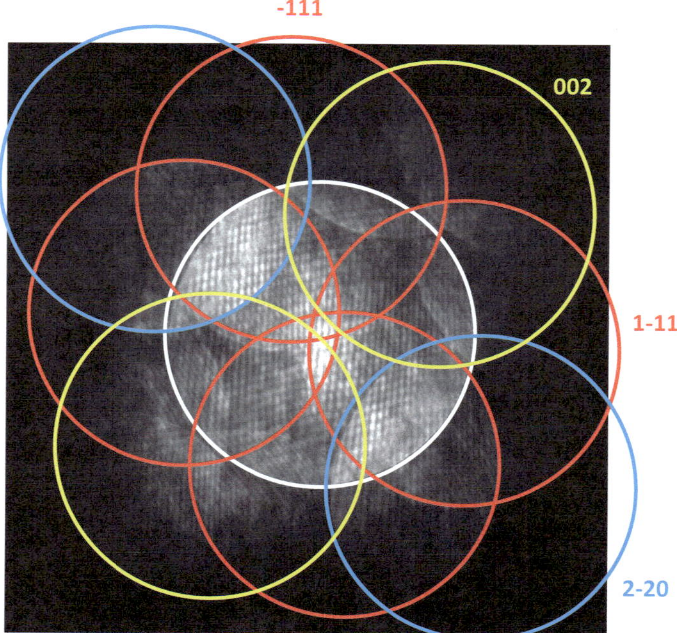

Fig. 14.11 CBED pattern of Si [110] recorded in the STEM mode of a JEOL2200FS (S)TEM equipped with the CEOS C_s corrector at 200 kV. Coherence can be observed in the areas where disks overlap

14.9   Ronchigrams

A ronchigram is a coherent CBED pattern recorded with a very large objective aperture, or with the aperture removed entirely. If the sample is crystalline, gross overlap of the CBED orders results. These patterns are called ronchigrams because the geometry used to obtain them is identical to that used to test optical lenses and mirrors (Malacara 1978; Ronchi 1964). They have a number of interesting properties and uses for STEM (Lupini 2011), as follows:

(1)   They may be used to measure the aberration coefficients and defocus setting of the probe-forming lens. These quantities are required for the alignment of the probe aberration corrector, optimization of resolution in any resulting STEM images, and interpretation of BF image contrast.
(2)   They are used to align the probe-forming lens and to correct astigmatism. This is often critical for STEM using Å-sized probes.
(3)   If the sample is crystalline, the central portion of this pattern will be shown to consist of a high magnification lattice image of the sample (Cowley 1979). Thus, a lattice image can be formed without scanning.

(4) For very thin sample, the ronchigram is an in-line Gabor electron hologram. All the image-processing techniques previously developed for holography in optics may therefore be applied to the interpretation of these patterns (Lin and Cowley 1986a, b).

(5) An understanding of these patterns is basic to the interpretation of bright- and dark-field STEM lattice images, and to the design of special detectors for this purpose.

We commence with a simple geometric explanation of the formation of projection lattice images and Fourier imaging, then summarize the relevant theory, and give experimental examples. Figure 14.12 shows a ray diagram for coherent CBED with an out of focus probe focused in front of a thin crystal. Bragg diffraction of the incident cone generates additional cones deflected by multiples of twice the Bragg angle. If $\theta_c > 2\theta_B$, these will overlap and interfere at the center of the detector. By tracing these deflected cones back toward the source, a set of virtual sources may be defined, which are necessarily coherent with the physical source. These virtual sources lie on the reciprocal lattice, and the situation is thus identical to that found in the back focal plane of a TEM when used for lattice imaging. If an ideal point source were available, the arrangement would therefore produce an unaberrated lattice image of the crystal, without using either lenses or scanning. The magnification of the lattice image is defined by the projection

$$M = (r - r_s)/r_s \approx r/r_s, \tag{14.56}$$

where r and r_s are the distances from the source to the detector and sample, respectively. Thus, a very large magnification is obtained by placing the probe close to the sample. The magnification changes from positive to negative as the probe changes from overfocus to under-focus conditions with the magnification at infinite at the zero focus.

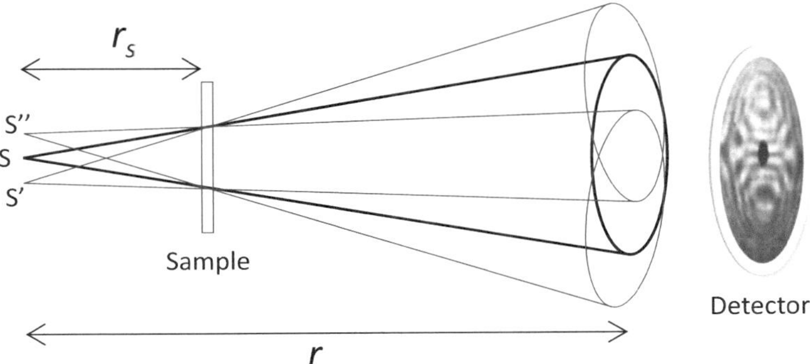

Fig. 14.12 Coherent overlapping orders produced by a point source focused in front of a thin crystal and virtual sources which result from Bragg diffraction if $\theta_c > 2\theta_B$, in a underfocused case

This point-projection method for electron lattice imaging was first proposed by Cowley and Moodie (1957a, b), who provided the theory and experimental results using coherent light. They named the resulting images "Fourier images." It is important to note that only strictly periodic detail is faithfully imaged by this technique, since it relies on Bragg diffraction. Defects are not seen in Fourier images [see Spence (1988) for a review]; however, if the object is weakly scattering (or compact), the patterns may be interpreted as holograms of defects. Figure 14.13 shows ronchigrams obtained from a thin crystal of beryl, using an out-of-focus 0.3-nm-diameter probe. According to the preceding discussion, the central region shows an (aberrated) lattice image of the 0.8-nm crystal planes. The magnification of the image is seen to depend on the focus. No scanning was used to obtain these images.

For experimental ronchigrams obtained in a STEM, the previous description must now be modified to take into account the aberrations of the lens and the finite source size. First, we treat the general case of a nonperiodic object, which provides a useful practical method of electron-optical alignment. The transmission function of the sample is taken to be (Eq. 4.10)

$$q(\vec{r}) \approx \exp[i\pi\lambda U(\vec{r})t] \tag{14.57}$$

where $\bar{U}(\vec{r})$ is the projected sample potential, and the sample is assumed sufficiently thin so that there is no variation in the intensity of the rocking curves within each CBED order. Equation (14.57) includes multiple scattering effects within the approximation of a "flat" Ewald sphere. Then, if $Q(\vec{K})$ is the Fourier transform FT $\{q(\vec{r})\}$, the intensity distribution at the detector is

$$I(\vec{K}_t) = \left| Q(\vec{K}_t) * A(\vec{K}_t)e^{-i\chi(\vec{K}_t)} \right|^2 \tag{14.58}$$

Here $\vec{K}$ is a two-dimensional coordinate in the illumination aperture plane. The asterisk denotes convolution. Numerical computations based on this expression are in excellent agreement with experimental ronchigrams. The reader is referred to Lupini (2011) for more details. These characteristic patterns are obtained by placing the probe on the edge of a noncrystalline sample and removing the objective aperture. They may be used for the final stages of lens alignment and astigmatism correction for STEM.

We now consider in more detail the special case of a thin extended crystal. From the interior of a thin crystal, patterns such as that shown in Fig. 14.13 are obtained. These may also be simulated with the aid of Eq. (14.58) using a periodic potential. We now show that the resulting pattern is a true image, and discuss its aberrations (Cowley 1995). It is convenient to transform to new spatial coordinates x and y in the detector plane. Let $q(\vec{r})$ have Fourier coefficients $F_{h,k}$. Then, $Q(\vec{K}_t)$ consists of a set of delta functions on reciprocal lattice sites. If the objective aperture has been removed, we may take $A(\vec{K}_t) = 1$. The convolution in Eq. (14.58) then becomes

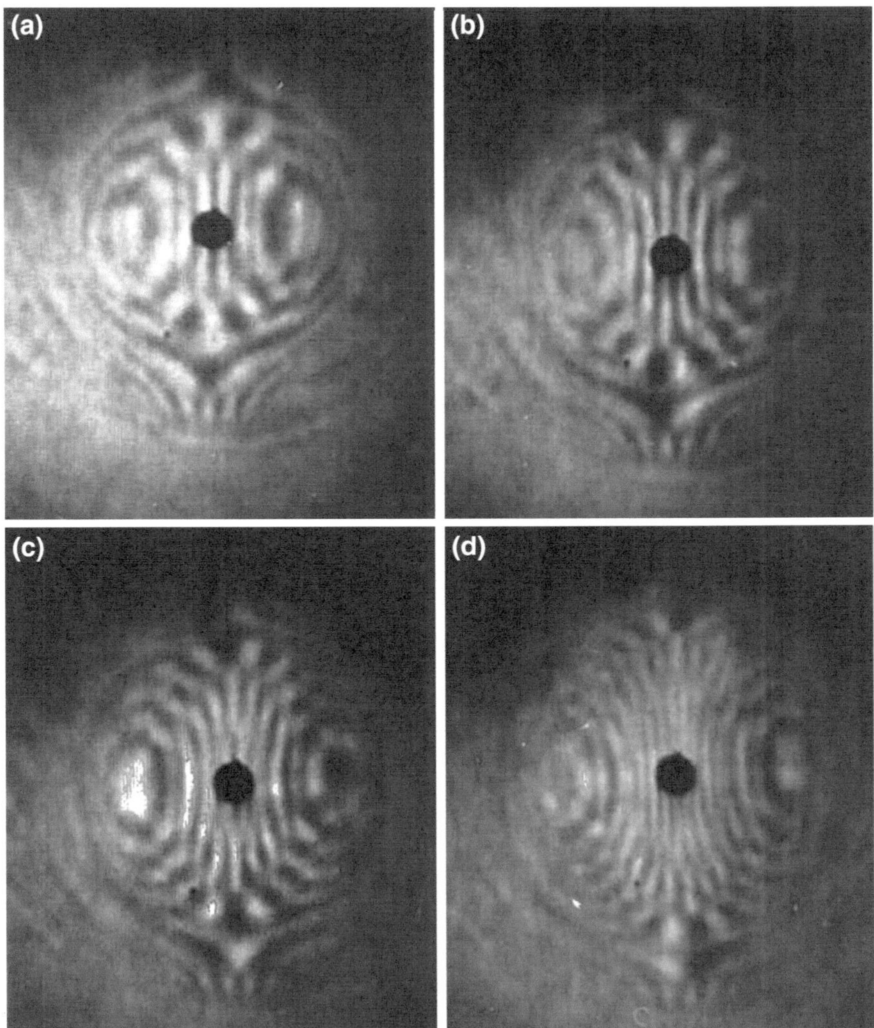

Fig. 14.13 Ronchigrams obtained with a 0.3 nm probe for four focus settings (**a–d**) from a thin crystal of the mineral beryl. The fringes in the center are an aberrated point projection of lattice image of the crystal. The lattice spacing is 0.8 nm. Outer ellipses allow measurement of the spherical aberration constant and defocus for the probe-forming lens (Lin and Cowley 1986a)

$$\phi(x, y) = \sum_{h,k} F_{h,k} \exp\left[-i\chi\left(\frac{x}{R\lambda} - \frac{h}{a}, \frac{y}{R\lambda} - \frac{k}{b}\right)\right] \tag{14.59}$$

where R is the distance from sample to detector, a and b are the cell constants for the projected unit cell, and h and k are integers.

To show that Eq. (14.59) produces a faithful, magnified image of the sample, we consider first points near the optical axis such that the effects of spherical aberration are negligible. Then, with $C_s = 0$, and using Eqs. (14.14), (14.59) becomes, in one dimension,

$$\phi(x) = e^{i\alpha} \sum_h F_h \exp\left(i\frac{2\pi h x}{Ma}\right) \exp\left(i\frac{\pi\lambda\Delta f h^2}{a^2}\right) \tag{14.60}$$

where α is a constant phase factor and $M = R/\Delta f$. For the special infinite set of focus settings

$$\Delta f_n = 2na^2/\lambda \tag{14.61}$$

the last term in Eq. (14.60) becomes unity since n and h are integers. The intensity distribution at the detector therefore becomes

$$I(x) = \left|\sum_h F_h \exp\left(i\frac{2\pi h x}{Ma}\right)\right|^2 \tag{14.62}$$

which is a magnified copy of the modulus squared of the transmission function of the object $q(\vec{r})$. The magnification $M = R/\Delta f$ is that of a geometric point-projection image, projected from the defocused probe position above (or below) the sample. These Fourier images also occur in two dimensions if certain restrictions are placed on the ratio of the projected cell dimensions and angle. In general, the images are periodic in probe defocus Δf, and it may be shown that additional half-period and reversed contrast images also occur at intermediate focus settings. The reader is referred to the original papers for more details (Cowley 1979; Cowley and Moodie 1957a, b). In fact, for the particular transmission function used in this example (a phase object), the "in-focus" Fourier image at Δf_n will show no contrast, since the modulus of Eq. (14.57) is unity. It has been shown, however, that contrast is obtained near the in-focus setting (Cowley and Moodie 1960).

For experimental ronchigrams, the effects of spherical aberration cannot be neglected. The preceding discussion explains the appearance of the inner lattice fringes in Fig. 14.13. Moving away from the center horizontally, we see two "eyes," within which the intensity is relatively constant. This may be understood with reference to Fig. 14.14, which provides a geometrical interpretation of Eq. (14.59) for the axial case of the three beams. Here the lens aberration function χ has been sketched across the illumination aperture. The on-axis image intensity at O arises from interference along paths AO and A'O (which are Bragg scattered in passing through the sample), together with the axial ray. An off-axis point such as P involves paths CP and C'P, and therefore samples the aberration function at different points. We have seen that for points near the axis, lattice fringes are produced. For points further from the axis, the effects of spherical aberration become dominant. The "eyes" in the experimental pattern (Fig. 14.13) result from

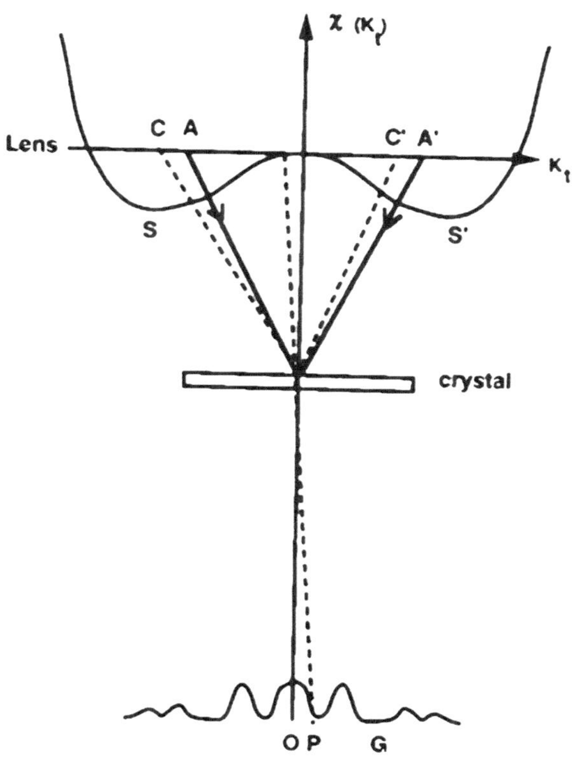

Fig. 14.14 Ray diagram showing formation of ronchigram and point-projection lattice images. The on-axis image point results from interference between paths AO and $A'O$. An off-axis image point P selects rays originating at C and C' if only Bragg scattering is allowed. Stationary-phase points are shown in S and S'. The wave-front aberration function for the probe-forming lens is $\chi(K_t)$

rays leaving near the stationary-phase turning points S and S'. Then there is relatively little phase change with change in detector position. For a given lattice spacing d_{hkl}, this defocus occurs for

$$-\Delta f = C_s \lambda^2 / d_{hkl}^2 \qquad (14.63)$$

The experimental pattern is also seen to consist of a set of outer ellipses. These ellipses are more visible in the simulated ronchigrams as shown in Fig. 14.16 from a simulated thin sample consisting of randomly distributed Ge atoms. The simulation parameters are $C_s = 1$ mm, $\Delta f = -1500$ nm, and $E_o = 300$ kV. Two distinctive rings are seen: One is made of radially stretched patterns and another consists of azimuthal rotated patterns.

A basic understanding of aforementioned features from a thin amorphous sample can be obtained based on the projection geometry as illustrated in Fig. 14.15. Diffraction by the thin sample is weak and its effects are neglected; thus, only the direct beam (source S) is retained in Fig. 14.15. Under this approximation, the ronchigram is treated as the geometrical shadow of a mostly transparent object (Cowley 1979). We will assume that the ronchigram is recorded with a large magnification with the detector placed at a distance far greater than the probe to

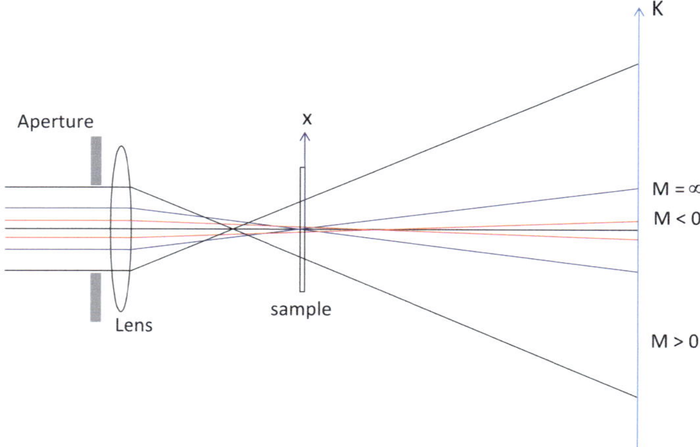

Fig. 14.15 Geometric illustration of ronchigram formation in an aberrated lens showing three regions of negative, infinite, and finite magnifications

sample distance. To a good approximation, the detector position gives the ray angle with

$$\vec{\theta} = (\theta_x, \theta_y) = \left(\frac{X}{R}, \frac{Y}{R}\right) \quad \text{and} \quad \vec{K}_t = (\theta_x/\lambda, \theta_y/\lambda) = (u, v)$$

where X and Y are the detector position. A particular ray at $\vec{K}$ goes through the sample at position $\vec{r} = (x, y)$. In the absence of aberrations,

$$\vec{r}_{\mathrm{o}} = (R_s\theta_x, R_s\theta_y) = (R_s\lambda u, R_s\lambda v).$$

The aberration deflects the ray by the angle $\nabla\chi(u, v)/\lambda$ and it intersects the sample at

$$\vec{r} = R_s\nabla\chi(u, v)/\lambda + \vec{r}_{\mathrm{o}}$$

A second ray recorded at the detector position $\vec{K} + \mathrm{d}\vec{K}$ intersects the sample at

$$\vec{r} + \mathrm{d}\vec{r} = R_s\nabla\chi(u + \mathrm{d}u, v + \mathrm{d}v)/\lambda + \vec{r}_{\mathrm{o}} + \mathrm{d}\vec{r}_{\mathrm{o}}.$$

Using the expansion

$$\nabla\chi(u + \mathrm{d}u, v + \mathrm{d}v) \approx \nabla\chi(u, v) + \frac{\partial^2}{\partial u^2}\chi(u, v)\mathrm{d}u + \frac{\partial^2}{\partial u\partial v}\chi(u, v)\mathrm{d}v$$
$$+ \frac{\partial^2}{\partial v\partial u}\chi(u, v)\mathrm{d}u + \frac{\partial^2}{\partial v^2}\chi(u, v)\mathrm{d}v,$$

we have

$$d\vec{r} = H_\chi d\vec{K}_t + d\vec{r}_o, \tag{14.64}$$

where H_χ is a 2×2 matrix in the form

$$H_\chi = \frac{R_s}{\lambda} \begin{pmatrix} \frac{\partial^2}{\partial u^2} \chi(u,v) & \frac{\partial^2}{\partial u \partial v} \chi(u,v) \\ \frac{\partial^2}{\partial v \partial u} \chi(u,v) & \frac{\partial^2}{\partial v^2} \chi(u,v) \end{pmatrix}. \tag{14.65}$$

Alternatively, the sample vector $d\vec{r}$ is projected onto the detector according to

$$d\vec{K}_t = H_\chi^{-1}(d\vec{r} - d\vec{r}_o) \tag{14.66}$$

Two important cases are obtained directly from this formulation (Lupini 2011)

(1)   The matrix has no inverse where its determinant is equal to 0. The magnification at those locations is undefined, or infinite.
(2)   The magnification matrix as measured by dK_t/dz with $t = x$ or y at a point in the ronchigram is related to the second derivatives of the aberration function at that corresponding angle after the calibration of the source and detector distances. Thus, a measurement of the local magnification gives the local second derivatives. When this measurement is made at several points in the ronchigram, it is possible to fit the whole aberration function from the obtained second derivatives and to measure the microscope aberrations.

Let us consider case 1 first for a round objective lens as simulated in Fig. 14.16. The aberration function is determined by C_s (C_3) and defocus (C_1) only and we have

$$H_\chi = \frac{R_s}{\lambda} \begin{pmatrix} C_1 + C_3(3u^2 + v^2) & C_3 2uv \\ C_3 2uv & C_1 + C_3(u^2 + 3v^2) \end{pmatrix}. \tag{14.67}$$

By taking the determinant to zero, we obtain the following equation:

$$\begin{aligned} \left(C_1 + C_3\left(3u^2 + v^2\right)\right)\left(C_1 + C_3\left(u^2 + 3v^2\right)\right) - \left(C_3 2uv\right)^2 \\ = \left(C_1 + 3C_3\left(u^2 + v^2\right)\right)\left(C_1 + C_3\left(u^2 + v^2\right)\right) = 0. \end{aligned} \tag{14.68}$$

This equation gives two solutions for the radius

$$r_r = \sqrt{\frac{C_1}{3C_3}} \quad \text{and} \quad r_a = \sqrt{\frac{C_1}{C_3}},$$

where $r = \lambda\sqrt{u^2 + v^2}$. At r_r, the magnification tends to infinity along the radial direction, and r_a gives the radius of the larger, azimuthal, circle of infinite magnification. At r_r, the caustic intersects the sample plane, giving rise to the infinite

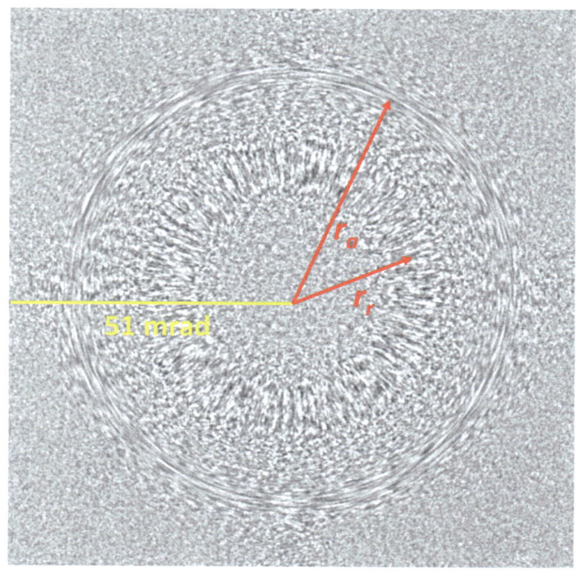

Fig. 14.16 *Top* A simulated ronchigram from a thin sample containing randomly distributed Ge atoms with $C_s = 1$ mm, $\Delta f = -1500$ nm and electron energy of 300 keV

magnification. At r_a, the rays at the same radius go through the same point on the sample, giving rise to infinite azimuthal magnification.

There are several practical uses of the second case where magnification is nonuniform but finite. First, for experimentalists, a very useful result here is that the change of magnification recorded in the ronchigram depends on the aberration function, or more precisely its second-order derivatives. Thus, significant insight can be gained from ronchigrams about the dominant aberrations, regarding their symmetry and approximate magnitude. Secondly, using samples with distinct features such as small nanoparticles, it is possible to determine the magnification by shifting the probe position by a known amount and measure the resulting apparent shifts in the ronchigram at different locations. The measurement of shift can be done by numerically using the cross-correlation of small patches of the ronchigrams. The determined magnifications are then used to determine the aberration function. This forms the basis of the so-called Nion method to measure aberrations (Dellby et al. 2001; Krivanek et al. 1999).

14.10   Coherence in STEM

In STEM, Fourier transform of the detector shape function $D(\vec{K})$ plays the same role as the coherence function $\gamma(\vec{r})$ (Sect. 14.6) in HREM. To show this, we consider the case of an ideal point field emitter, where the STEM probe wave function is just the lens resolution function $h(\vec{r}, E)$ of the probe-forming lens (Eq. 14.52 and Fig. 14.8), and the probe formation process (as opposed to overall image formation) is thus ideally coherent. The detector shape, by controlling the degree to which scattering from different atoms can interfere, determines the degree of coherence for scattering within the sample. For a probe positioned at $\vec{r}_p$ and object transmission function $q(\vec{r})$, the exit face wave function is given as:

$$\phi_e(\vec{r}, \vec{r}_p) = h(\vec{r} - \vec{r}_p, E)q(\vec{r})$$

and the wavefunction at a distant detector is the Fourier transform of this:

$$
\begin{aligned}
\phi_e(\vec{K}_t, \vec{r}_p) &= \int h(\vec{r} - \vec{r}_p, E)q(\vec{r})e^{-2\pi i\vec{K}_t\cdot\vec{r}}\mathrm{d}^2\vec{r} \\
&= H(\vec{K}_t, E)e^{-2\pi i\vec{K}_t\cdot\vec{r}_p} * Q(\vec{K}_t) \\
&= \int H(\vec{K}'_t, E)e^{-2\pi i\vec{K}'_t\cdot\vec{r}_p}Q(\vec{K}_t - \vec{K}'_t)\mathrm{d}^2\vec{K}'_t \\
&= -h(-\vec{r}_p, E) * q(\vec{r}_p)e^{-2\pi i\vec{K}_t\cdot\vec{r}_p}
\end{aligned}
\tag{14.69}
$$

where the asterisk denotes convolution. The intensity collected by a detector is described by an integration over the detector shape (or sensitivity) function $D(\vec{K})$ and the electron beam energy

$$I(\vec{r}) = \int p(E)\mathrm{d}E \int_{\text{detector}} D(\vec{K}_t)\left|\int H(\vec{K}'_t, E)e^{-2\pi i\vec{K}'_t\cdot\vec{r}}Q(\vec{K}_t - \vec{K}'_t)\mathrm{d}^2\vec{K}'_t\right|^2 \mathrm{d}^2\vec{K}_t.
\tag{14.70}$$

We can compare this with the corresponding expression for HREM (Eq. 14.43) using an instrument with the same lens resolution function if we allow for an incoherent source, so that the intensities of images may be added for each illumination direction. Equation (14.43) describes the effects of plane wave illumination inclined from direction $\vec{K}$ on a sample. Then, since for a tilted incident beam $\vec{K}$

$$\phi_e(\vec{r}) = q(\vec{r})e^{2\pi i\vec{K}\cdot\vec{r}},
\tag{14.71}$$

the HREM image is given by a convolution of the impulse response with the object transmission function for inclined illumination,

$$\phi_i(\vec{r}) = q(\vec{r})e^{2\pi i\vec{K}_t\cdot\vec{r}} * h(\vec{r})$$

which is identical to Eq. (14.69) for STEM for a round symmetrical lens. The integration of intensity over an incoherently filled illumination source $p(\vec{K}_t)$ for HREM is therefore also described by Eq. (14.70), so that the partial coherence is given by either the illumination source for HREM or the detector shape in STEM. The key approximation made (Eq. 14.71) holds only for a thin sample. Under this condition, we therefore expect that the use of a bright-field detector in the shape of a small disk in STEM will produce images with similar coherence properties as those obtained under TEM. For example, in the case of a very small STEM BF detector, $D(\vec{K}_t) = \delta(\vec{K}_t)$, and we obtain, from Eq. (14.70),

$$I_{\mathrm{STEM}}^{\mathrm{BF}}(\vec{r}) = \int p(E)\mathrm{d}E \int_{\mathrm{detector}} D(\vec{K}_t)\left|\int H(\vec{K}_t', E)\mathrm{e}^{-2\pi i\vec{K}_t'\cdot\vec{r}}Q(\vec{K}_t - \vec{K}_t')\mathrm{d}^2\vec{K}_t'\right|^2 \mathrm{d}^2\vec{K}_t$$

$$= \int |h(-\vec{r}, E) * q(\vec{r})|^2 p(E)\mathrm{d}E,$$

$$(14.72)$$

which is identical to the expression for bright-field HREM imaging using a coherent point source.

The use of an annular detector in STEM is therefore equivalent to conical or hollow-cone illumination achieved in a TEM. In both cases, it was found that the degree of coherence fell more slowly for atoms separated in the z-direction than in the transverse direction (Fertig and Rose 1977). This effect, for example, has been used effectively to suppress interference effects in BF high-resolution imaging of amorphous thin films by using a hollow-cone illumination (Gibson and Howie 1979). Incoherent imaging is achieved by having a large annular detector that covers a sufficiently wide angular range to detect all of the scattering. To show this, we can take the integration over the detector in Eq. (14.70) to cover all scattering angles for a uniform detector $(D(\vec{K}) = 1)$

$$I_{\mathrm{STEM}}^{\mathrm{ALL}}(\vec{r}, E) = \int\left|\int q(\vec{r}')h(\vec{r}' - \vec{r}, E)\mathrm{e}^{-2\pi i\vec{K}_t\cdot\vec{r}'}\mathrm{d}^2\vec{r}'\right|^2 \mathrm{d}^2\vec{K}_t$$

$$= \int\int\int q(\vec{r}')q*(\vec{r}'')h(\vec{r}' - \vec{r}, E)h^*(\vec{r}'' - \vec{r}, E)\mathrm{e}^{-2\pi i\vec{K}\cdot(\vec{r}'-\vec{r}'')}\mathrm{d}^2\vec{r}'\mathrm{d}^2\vec{r}''\mathrm{d}^2\vec{K}_t$$

$$= \int\int q(\vec{r}')q*(\vec{r}'')h(\vec{r}' - \vec{r}, E)h^*(\vec{r}'' - \vec{r}, E)\delta(\vec{r}' - \vec{r}'')\mathrm{d}^2\vec{r}'\mathrm{d}^2\vec{r}''$$

$$= \int q(\vec{r}')q*(\vec{r}')h(\vec{r}' - \vec{r}, E)h^*(\vec{r}' - \vec{r}, E)\mathrm{d}^2\vec{r}',$$

which gives

$$I_{\mathrm{STEM}}^{\mathrm{ALL}}(\vec{r}, E) = |q(\vec{r})|^2 * |h(\vec{r}, E)|^2.$$

$$(14.73)$$

The second term on the right of Eq. (14.73) is the intensity distribution in the probe, and the object function is convoluted with this. This relationship describes the ideal incoherent imaging mode, such as that used, for example, by a camera used to image "self-luminous" objects. Compared to the phase contrast, where contrast can be reversed by focus, the change of focus in the incoherent imaging mode only contributes to the blurring of the image from the broadening of the lens resolution function.

For a phase grating including the effect of absorption potential, we have

$$q(\vec{r}) = \exp[-i\sigma V(\vec{r})]\exp[-\sigma V'(\vec{r})]$$

And thus for the elastic scattering

$$I_{\text{STEM}}^{\text{Elastic}}(\vec{r}, E) = \exp[-2\sigma V'(\vec{r})] * |h(\vec{r}, E)|^2, \qquad (14.74)$$

where $V(\vec{r})$ and $V'(\vec{r})$ are the projected real and imaginary (absorption) potential. The absorption potential describes the loss of electrons to inelastic scattering. Thus, detection of all scattered electrons gives no contrast in phase objects. The absorption contrast is observed with energy filtering. Since the absorption potential is only a fraction of the elastic scattering potential, only weak contrast is produced by energy filtering.

The ideal of incoherent imaging breaks down with the introduction of a hole at the center of the ADF detector, which could produce strong contrast for both phase grating and multiple scattering objects. This can be seen at once by noting that the total scattering is a constant, i.e., $I_{\text{elastic}} + I_{\text{inelastic}} = 1$, independent of probe position, so that the scattering contribution to a detector with a very small hole must be the complement of the bright-field STEM image. Thus

$$I_{\text{STEM}}^{\text{DF}}(\vec{r}) = 1 - I_{\text{STEM}}^{\text{BF}}(\vec{r}). \qquad (14.75)$$

At high resolution, scattering of the electron probe by neighboring atoms produces interference effects, which modify the distribution of scattered electrons around the central hole of the annular detector and give the BF-STEM contrast as discussed before. Because of this, the assumption that the collected ADF-STEM image intensity under the incoherent imaging condition is proportional to the total scattered electrons with a large annular detector is no longer valid. In order to detect all scattered electrons to form incoherent images at atomic resolution, the detector inner angle must be reduced and the same time a smaller objective aperture must be used in order to avoid the detection of transmitted electrons. The small objective aperture necessarily reduces the imaging resolution (Ade 1977). This is known as the hole-in-the-detector problem (Cowley 1976; Ade 1977; Jesson and Pennycook 1993).

We have so far assumed phase objects for electron scattering. In thick samples, coherent electron multiple scattering and incoherent inelastic scattering also contribute to image contrast. The inelastic scattering part can be separated out according to its distribution in diffraction patterns. Early work by Crew and his coworkers demonstrated remarkable high resolution in micrographs recorded using an ADF detector that detects 60–80 % of the elastically scattered electrons and lets about 90 % of the inelastically scattered electrons passing through a large hole in the ADF detector (see Fig. 14.17 for example). The image contrast is dominated by Rutherford scattering with a $Z^{3/2}$ dependence (Crewe et al. 1975).

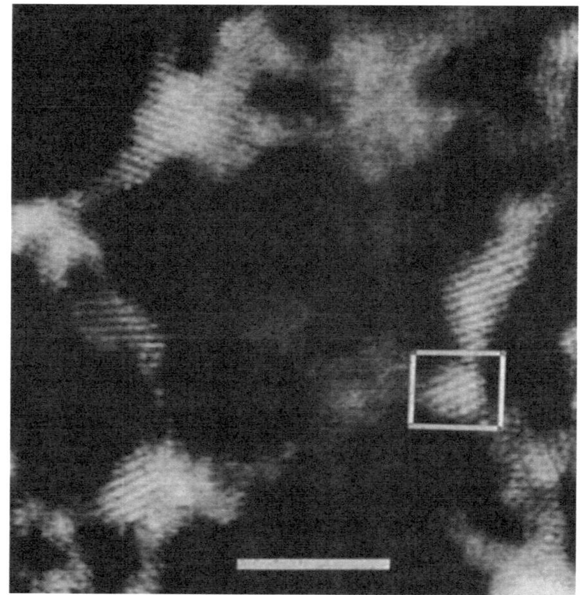

Fig. 14.17 A micrograph of a thorium specimen recorded at University of Chicago using a 33.5 keV STEM and a silicon annular dark-field detector with an inner and outer cutoff angles of 20 and 200 mrad, respectively. Several thorium crystallites are resolved at resolution close to 3 Å. The scale bar indicates 50 Å (from Wall et al. Proc. Nat. Acad. Sci., 71, pp. 1–5, 1974, produced with the author's permission)

14.11　HAADF-STEM (Z-Contrast) Imaging

When imaging crystals or partially ordered structures in STEM using an ADF detector, both coherent elastic (Bragg diffraction) and TDS contribute to the detected signal. TDS dominates at larger angles. Thus, the ADF detector will collect the coherent diffraction signal as well as the thermal diffuse signal. Howie (1979) suggested that a sufficiently high-angle detector (40 mrad or greater) would minimize the unwanted Bragg scattering, giving a ADF-STEM image with strong atomic number contrast (proportional to Z^2) according to the Rutherford formula for scattering from an unscreened nucleus. The reasoning is that the corresponding d-spacing at the 40 mrad scattering angle is comparable with the thermal vibration amplitude. Above this angle, TDS contribution becomes increasingly more important, and eventually dominates the HAADF signal (Pennycook and Jesson 1991). A systematic ADF-STEM imaging study of ion-implanted heavy impurity atoms in single crystal silicon as function of the ADF inner cutoff angle was carried out by Pennycook et al. (1986). Their results showed that contrast due to thickness variations, diffraction, and channeling effects can be avoided using a large inner cutoff angle (~ 110 mrad), and the resulting image contrast quantitatively reflects variations in impurity concentration. This form of Z-contrast imaging led to the idea of attempting to resolve crystal structure directly (Pennycook et al. 1986).

A full treatment of HAADF-STEM requires the inclusion of both Bragg and TDS. Contributions from the TDS to the HAADF intensity can be approximately calculated using the Einstein model which treats atomic vibrations in a solid as independent quantum harmonic oscillators. Using this approximation, the scattering into the HAADF detector can be described by an atomic potential of $V_a^{\text{HAADF}}(\vec{r})$, whose atomic scattering factor is approximately given by Pennycook and Jesson (1991), Ishizuka (2002) (also see Sect. 4.9)

$$
f_a^{\text{HAADF}}(s, M_a) = \frac{4\pi\hbar}{m_e v} \int\limits_{\text{detector}} f_a(s')f_a(|\vec{s} - \vec{s}'|)\left[1 - \exp\left\{-2M_a\left(s'^2 - \vec{s}\cdot\vec{s}'\right)\right\}\right]\mathrm{d}^2\vec{s}'
\tag{14.76}
$$

where M_a is the Debye–Waller factor of the atom, m_e and v stand for electron mass and velocity, respectively, and $\vec{s} = \vec{S}/2 = \vec{k} - \vec{k}_\mathrm{o}$ is half of the electron scattering vector. For a HAADF detector with a large inner cutoff angle, s' is large and for small s including $s = 0$, $f_a^{\text{HAADF}}(s, M_a)$ approaches a constant of

$$
f_a^{\text{HAADF}}(M_a) = \frac{4\pi\hbar}{m_e v} \int\limits_{\text{detector}} f_a^2(s')\left[1 - \exp\left(-2M_a s'^2\right)\right]\mathrm{d}^2\vec{s}'
\tag{14.77}
$$

Since $f_a(s) \propto Z/s^2$ for large s, Eq. (14.77) gives the Z^2 dependence that is qualitatively observed in experimental HAADF-STEM images in thin specimens. This simple relationship allows a direct interpretation, and Z-contrast, of images recorded in HAADF-STEM (see Fig. 14.18 for example).

As s increases and approaching the inner cutoff angle of the HAADF detector, $f_a^{\text{HAADF}}(s, M_a)$ falls off but the falloff rate is slower than that of electron atomic scattering factor. This indicates that HAADF atomic scattering potential is more localized than the electron elastic scattering potential. Thus, in principle, resolution in HAADF-STEM can surpass the resolution of electron images formed based on electron elastic scattering.

The intensity recorded by the HAADF detector in the above approximations is simply an integration of the product of the probe intensity and the HAADF potential with the probe placed at according to

$$
I^{\text{HAADF}}(\vec{r}) = \frac{2}{\hbar v}\sum_i \int \left|\phi_p(\vec{r}' - \vec{r}, z)\right|^2 V_i^{\text{HAADF}}(\vec{r}' - \vec{r}_i, z - z_i)\mathrm{d}^2\vec{r}'\mathrm{d}z
\tag{14.78}
$$

where the sum is over atoms in the solid. In the limit of very thin samples, the electron probe wave function can be approximately assumed as that of the incident electrons; then, the resolution of the recorded HAADF-STEM image is simply defined by that width of the electron probe intensity distribution. In the same approximation, the intensity is proportional to thickness as the number of atoms under the electron probe scales with the specimen thickness. These relationships break down as the thickness increases as electron multiple scattering modifies the probe intensity distribution as well as the distribution of scattered electrons.

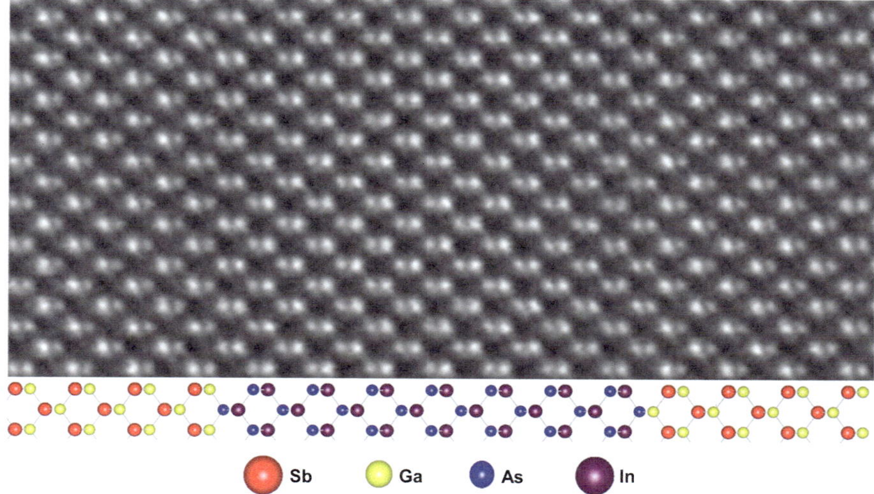

Fig. 14.18 Atomic resolution Z-contrast image of an InAs/GaSb superlattice recorded along [110] using the FEI aberration-corrected STEM (Titan Ultimate) at 300 kV. The difference in the atomic number between In and As and Ga and Sb gives rise to differences in the image intensity of the atomic columns (From Honggyu Kim, University of Illinois, Urbana-Champaign, USA and Jean-Luc Rouviere, CEA, INAC, France)

Specifically, as discussed below, electrons in the focused probe initially channel along the atomic columns when they first enter the sample and are scattered by the atomic columns, which leads to electron dechanneling. The electron channeling effect allows imaging of relatively thicker crystalline specimens in HAADF-STEM than in HREM. Also, change in focus affects the probe intensity distribution and thus the image resolution in HAADF-STEM; unlike in HREM change in focus could lead to a reversal of image contrast.

14.12 Aberration-Corrected STEM

The objective lens spherical aberration limits the performance of high-resolution STEM in two ways. The first is the smallest probe that can be formed and thus the optimum imaging resolution that can be obtained in STEM. Secondly, the size of the objective aperture must be small in order to obtain the smallest probe. Figure 14.19 shows the optimized probe size as function of the aperture size for a lens with $C_s = 1$ mm and $C_c = 1.2$ mm and $\Delta E = 0.7$ eV. The smallest probe is obtained with a 10 mrad objective aperture. Further increase in the objective aperture size leads to a rapid rise in the probe size. Because of the small objective aperture, the probe intensity for high-resolution STEM imaging is low and the signal-to-noise ratio in the recorded STEM images is also low. A major benefit of aberration correction is thus a significant increase in the probe intensity from the use of a larger objective aperture. The corrector for STEM is placed between the condenser and the objective pre-field (probe corrector) (Krivanek et al. 1999), and thus, it corrects aberrations in the illumination. Figure 14.20 shows a typical ronchigram from an amorphous Ge film after tuning the probe using the probe corrector in a JEOL2200FS STEM. This microscope has a Schottky field emission gun with a maximum accelerating voltage of 200 kV and a third-order CEOS aberration corrector based on two hexapole lenses on the probe-forming side (Haider et al. 2008). The objective lens pole piece has a 2-mm gap with a specified point-to-point resolution of 0.18 nm in TEM imaging mode and 0.1 nm in STEM mode with aberration correction. As shown in the figure, the ronchigram at the Gaussian focus ($\Delta f = 0$) shows a sixfold symmetry due to the off-axis distortion of the hexapoles. The half angle of the nearly spherical aberration-free "flat" area reaches 49 mrad. The inner ring indicates a 26 mrad half angle corresponding to the size of a 30-µm condenser aperture.

The Schottky field emission gun of the JEOL 2200FS TEM produces ~ 160 µA of emission current at the recommended gun settings. The probe current is ~ 30 pA using a 30-µm condenser aperture (26.5 mrad semi-convergence angle) and ~ 13 pA using a 20-µm aperture (17 mrad semi-convergence angle). The measurements showed that the full width half maximum (FWHM) of the zero loss peak is 0.77 eV at an emission current of 170 µA and 0.5 eV at a low emission current of 30 µA. At these conditions, spatial resolution better than 0.1 nm can be obtained at a large camera length of 60 cm and the inner cutoff angle of the ADF detector about 100 mrad. The probe current can be further increased for analytical work by selecting a spot size setting with a larger probe size and higher probe current using the 30-µm aperture, ~ 160 pA probe current. The large probe current delivers a reasonably good spatial resolution of 0.11 nm. It is significant to note that the probe current is increased by fivefold, while the probe size is increased by only 0.03 nm with aberration correction.

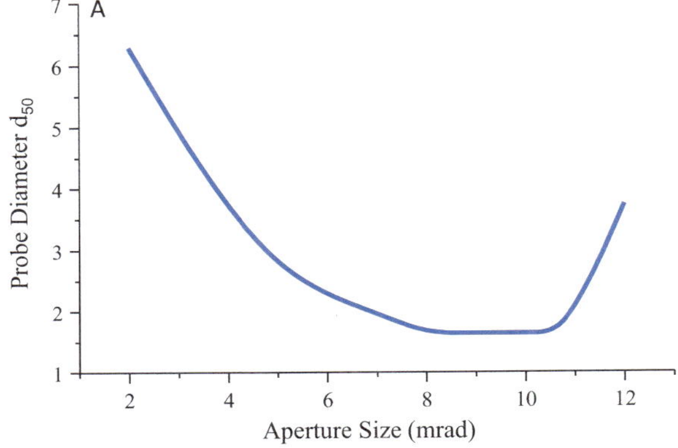

Fig. 14.19 Probe diameter (d_{50} containing 50 % probe intensity) plotted as function of aperture size for an objective lens with $C_s = 1$ mm

The spatial resolution of the microscope in STEM mode can be assessed in two ways. One is to examine the information transfer in the power spectrum of a recorded STEM image and the other is to resolve closely spaced atoms in a selected sample. In the first case, a common practice is to investigate the microscope performance by imaging a Si single crystal along the <110> zone axis. Figure 14.21 shows a representative HAADF image of Si <110> using the probe of 30 pA and 26.5 mrad convergence angle. Si atomic columns with 0.136 nm separation ("dumbbells") are clearly resolved. The average contrast of peak to valley intensity as defined by $(I_{max} - I_{min})/(I_{max} + I_{min})$ in the intensity profile across the dumbbell Si atomic column pairs is 43 %. This contrast ratio is consistently larger than 33 % over entire image, and in some locations, the contrast is as high as 60 % in some areas. The FWHM of each Si column is also smaller than 0.1 nm, suggesting a sub-0.1 nm effective probe size. The 2D fast Fourier transform (FFT) of the STEM image shows several spots corresponding to Si diffraction (335, 440, and 444) with d-spacing of less than 0.1 nm. This demonstrates information transfer better than 0.1 nm. The spatial resolution using the smaller condenser aperture of 17 mrad is slightly lower than when using the 26.5 mrad aperture, due to the diffraction-limited broadening of the smaller condenser aperture. This condition, however, is quite useful for high-resolution imaging of beam-sensitive specimens with less than half the probe current of the 26.5 mrad aperture. A sub-0.1 nm probe size is further verified by resolving two Ga columns separated by 0.104 nm in a GaN thin film grown on a sapphire substrate. The valley-to-peak ratio is ~ 80 %, which satisfies the Raleigh 81 % criteria.

The probe-corrected JEOL2200FS described above belongs to the first generation of aberration-corrected TEM/STEM using the technology of the CEOS hexapole corrector (see Chap. 7). Subsequent improvements were made to the corrector technology in the USA, in the Department of Energy Transmission Electron Aberration-Corrected Microscope (TEAM) project, which was in collaboration with FEI (Portland, USA) and CEOS (Heidelberg, Germany), and the Core

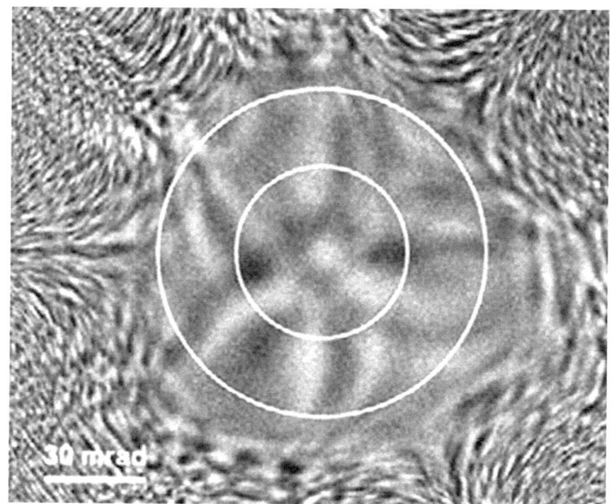

Fig. 14.20 A typical ronchigram from an amorphous Ge thin film recorded using a third-order CEOS probe aberration corrector. The half angle of the flat area is 49 mrad. The *inner white circle* indicates the angle when using the 30 µm (26.5 mrad semi-convergence angle) condenser aperture, which is typically used for high-resolution STEM imaging

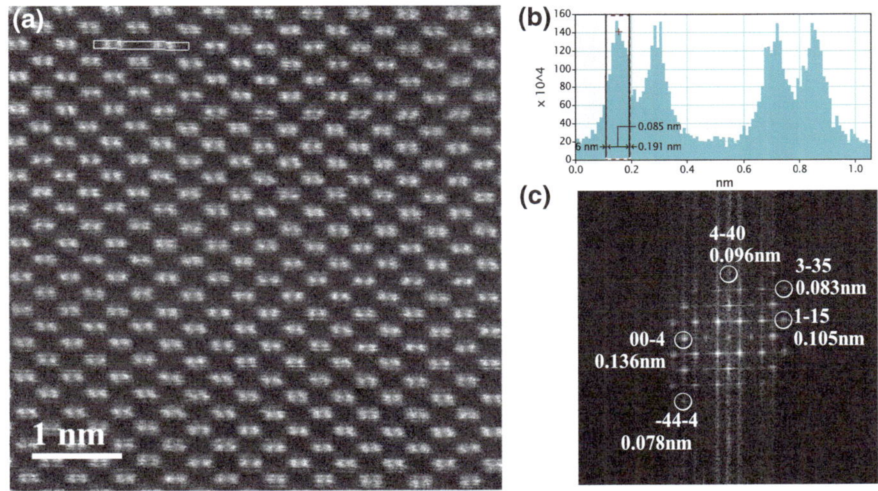

Fig. 14.21 Performance of a 200 kV aberration-corrected STEM for imaging silicon dumbbells along Si[110]

Research for Evolutional Science and Technology (CREST) project in Japan in collaboration with JEOL (Tokyo, Japan) (for details, see Müller et al. 2006 and Sawada et al. 2010). These improvements enabled the correction of fifth-order spherical aberration and sixfold astigmatism and achieved sub-angstrom resolution of 0.63 Å for STEM at 300 kV accelerating voltage.

14.13   Three-Dimensional Imaging in STEM

The improvements brought by aberration correction also open up the opportunity for three-dimensional imaging, using depth (or "optical") sectioning. Along the optical axis, the probe intensity falls off from the Gaussian focus, which defines the depth of focus. To examine how the probe intensity varies with the focus, we consider a diffraction-limited electron probe where the probe scattering effect is small. If the probe is focused at position $(\vec{R}, z)$, we obtain from Eq. (14.52)

$$\phi_P(\vec{r}, \vec{R}, z, \Delta f) = \int_{-\infty}^{\infty} A(\vec{K}_t) \exp\left[-i\pi\lambda K_t^2(\Delta f + z)\right] \exp\left(2\pi i \vec{K}_t \cdot (\vec{r} - \vec{R})\right) d^2\vec{K}_t,$$

where the integration is carried out over the transverse momentum up to the limit defined by the aperture radius, K_{max}. The probe intensity distribution is axially symmetric since both the aperture function and the phase introduced by defocus have rotation symmetry. Thus, it suffices to describe the full 3D electron intensity based on its dependence on the perpendicular radius r and axial distance z. At $z = 0$, $\Delta f = 0$ and zero probe displacement $(\vec{R} = 0)$, we have

$$\phi_P(r,0,0,0) = \int\limits_0^{K_{max}} \int\limits_0^{2\pi} e^{2\pi i K_t r \cos\theta} K\mathrm{d}K\mathrm{d}\theta = \int\limits_0^{K_{max}} 2\pi J_0(2\pi K_t r) K_t \mathrm{d}K_t$$

$$= \frac{K_{tmax}}{r} J_1(2\pi K_{tmax} r),$$

(14.79)

and thus

$$I_P(r,0,0,0) = \left| \frac{K_{tmax}}{r} J_1(2\pi K_{tmax} r) \right|^2$$

(14.80)

Along the optical axis,

$$I_P(0,0,z,0) = \left| 2\pi \int\limits_0^{K_{max}} \exp\left[-i\pi\lambda K_t^2 z\right] K_t \mathrm{d}K_t \right|^2 = \left| \pi K_{tmax}^2 \sin c\left(\frac{\pi\lambda K_{tmax}^2 z}{2}\right) \right|^2.$$

(14.81)

These results give the following radial and axial FWHM (D'Alfonso et al. 2007)

$$r_{\mathrm{FWHM}} = \frac{0.5145}{K_{tmax}} = \frac{0.5145\lambda}{\alpha}$$

$$z_{\mathrm{FWHM}} = \frac{1.772}{\lambda K_{tmax}^2} = \frac{1.772\lambda}{\alpha^2},$$

(14.82)

where α is the probe convergence angle, which is same as the objective lens numerical aperture. Thus, the improvement in the probe width with aberration is proportional to the inverse of the numerical aperture allowed by the aberration corrector, while the improvement in the depth of focus is much larger since it is inversely proportional to the square of the numerical aperture. In an aberration-corrected STEM, the depth of focus may be reduced to just a few nanometers. Among the most commonly studied crystals, the unit cell sizes are typically less than a nanometer, and having a depth of focus of 3 nm does not enable depth sectioning at the resolution of a unit cell. However, as pointed out by D'Alfonso et al. (2007), in extracting quantitative depth information, the precision is not entirely dependent upon probe depth resolution. There is a clear distinction between the precision to which one can determine, for example, the depth of sparse dopant atoms and the depth resolution of a probe. With the help of good experimental signal-to-noise ratio and accurate theoretical simulations of experimental images, it is possible to perform structure determination from probes having relatively poor depth resolution.

There are two approaches to obtaining 3D structural information in STEM. One simply uses the HAADF detector to record a focal series of Z-contrast images, thereby forming a 3D stack. The use of incoherently scattered particles for imaging is similar to the confocal scanning optical microscope (CSOM), where a 3D image of the sample is obtained by raster scanning a focused light beam across all dimensions and scattered light is collected using the objective lens. To enable optical slicing or sectioning of the sample in a CSOM, a small pinhole aperture is placed just prior to the detector. This aperture excludes the out-of-focus scattered light. The depth of focus observed in the image is determined by the size of the pinhole and the axial resolving power of the objective lens. In HAADF-STEM imaging, the depth-elongated probe is scattered by out-of-focus objects, which nonetheless contribute to the image intensity. To remove the out-of-focus intensity, several electron microscope adaptations of the CSOM detection technique have been proposed (Zaluzec 2003; Einspahr and Voyles 2006). Confocal STEM is implemented using both illumination lenses and post-specimen imaging lenses. The illumination optics is used to focus and position the beam within the sample. Post-specimen lenses are used to image the probe to an aperture placed in front of the detector. This aperture works in the same way as the pinhole in a CSOM by removing scattering from points in the sample away from the so-called confocal point. For depth resolutions of a few nanometers, aberration-corrected optics is required in both the pre- and post-specimen lenses. It thus requires a double-corrected (S)TEM with probe and image lens correctors. Additionally, to establish a confocal trajectory in such a double aberration-corrected microscope requires the simultaneous operation of the pre- and post-specimen optics in accurate mutual alignment, including the pre- and post-specimen fields of the objective lens. The challenge, as well as a method to establish a confocal geometry in a double aberration-corrected TEM, is discussed by Nellist et al. (2008). A successful application example of confocal STEM for imaging 3D structures was reported by Hashimoto, Shimojo et al. (2009) for carbon helices.

The best resolution in electron depth imaging so far was demonstrated by focal series HAADF-STEM. Figure 14.22 shows an example study of the Au/TiO$_2$ interfacial structure. The Z-contrast images were recorded in a focal series for depth sectioning. The principle of depth sectioning using STEM Z-contrast imaging is illustrated in Fig. 14.22a. The electron probe was formed by converging the electron beam onto the specimen; the probe intensity peaks where the probe is smallest. For the 26 mrad aperture and 200 kV electrons that were used here, *the depth of focus is* 6.5 nm. Figure 14.22b shows a series of the Z-contrast images from the interfacial region of an Au nanocrystal, recorded at 9 different probe focuses. The images were aligned using the cross-correlation method. After the alignment, at each focus setting, three intensity profiles were taken along (a) the interfacial layer, (b) the Au layer immediately adjacent to the interfacial layer, and (c) the TiO layer next to the interface, respectively (the positions of these layers are marked by red lines in Fig. 14.22b). These intensity profiles were then used to form the depth-sectioning images shown in Fig. 14.22c for the Au, TiO, and interfacial layers. In both Au and TiO layers, atomic columns give rise to vertical intensity

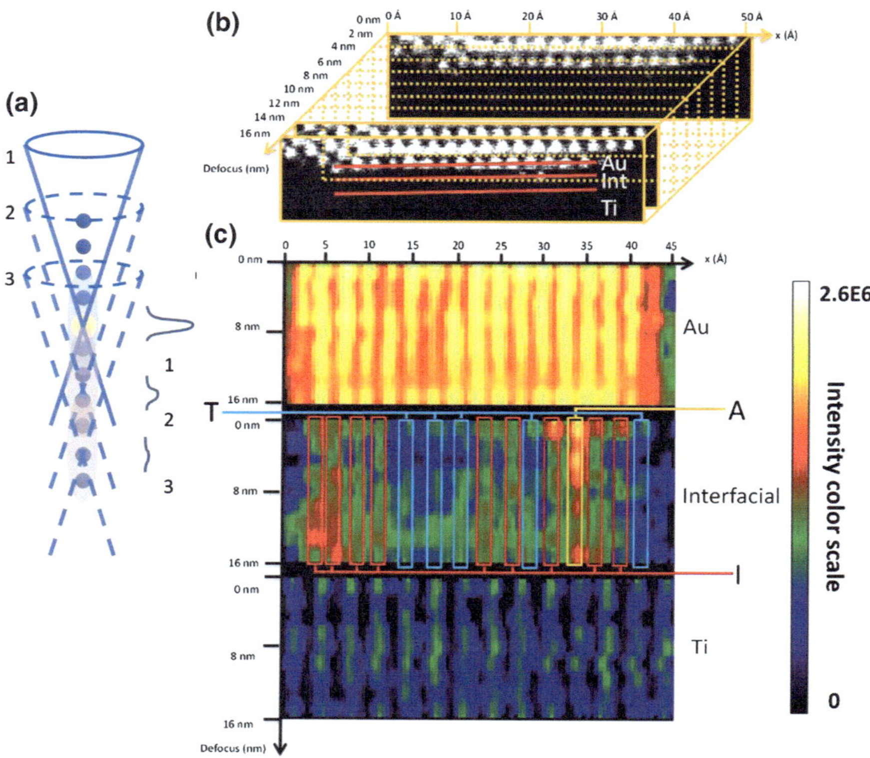

Fig. 14.22 An illustration of the principle of depth sectioning using STEM Z-contrast imaging. **a** The electron probe is formed by converging the electron beam onto the specimen; the probe intensity peaks where the probe is smallest. **b** The focal series of Z-contrast images recorded from the Au nanocrystal near the interfacial region shown in (**c**). The images were aligned using the cross-correlation method. These intensity profiles were then used to form the depth-sectioning images shown in (**c**) for the Au, TiO, and interfacial layers, respectively. The intensity bands were marked using the *color boxes* to identify the column as an Au-column, Ti-column, or an interfacial-column. From Gao et al. (2015)

bands along the focus axis in the depth-sectioning images. Each band corresponds to an atomic column, which is resolved horizontally along the scanning direction. Within the atomic column, the atoms are not resolved in the depth direction because of the limited depth resolution. Notably, at the interface, the depth-sectioning image shows intermediate intensities between these of Ti- and Au-columns. These were attributed to interfacial Au atoms with the help of image simulations. Upon interacting with a real solid, the electron probe is modified by electron scattering. Simulations are therefore required in the interpretation of depth-sectioning data. Using the principle described above, an atom can be located in 3D depending on how well the intensity distribution can be resolved and interpreted (Gao et al. 2015).

14.14 Channeling, Bound States, and Atomic Strings

A question directly related to the interpretation of HAADF images including depth sectioning is whether an object function, independent of thickness, can be usefully defined in the presence of strong multiple elastic scattering? The answer is strictly no as there is no simple expression for the probe wave function taking into account the electron multiple scattering effects. But some understanding of this problem can be obtained based on the quantum mechanical channeling theory.

Channeling has been defined as the tendency of charged particles to run along paths of low potential energy in crystals. These paths are the nuclear columns for electrons and the interatomic spaces for positrons (see Spence (1992) for a review). Electrons can be encouraged to travel along the nuclear strings either by focusing the wave field entirely onto one such string (as in STEM) or by exciting a set of laterally infinite Bloch waves in a crystal, in which case those which concentrate flux onto the atomic strings will be preferentially absorbed. Coherent probe propagation inside a perfect crystal in the zone axis "channeling" condition has been studied by Fertig and Rose (1981) and by Loane et al. (1988). Remarkable electron focusing effects were reported in the channeling condition, and the penetration of the probe was found to be different for probes focused onto atomic columns containing different species.

The experimental geometry and theory of electron channeling in STEM are essentially the same as those of zone axis CBED patterns, except that in STEM we are concerned with much smaller sample thicknesses and overlapping CBED orders. Since the main theoretical approach used to analyze the elastic portion of electron channeling has also proven useful in analyzing CBED patterns (Buxton and Tremewan 1980; Vincent et al. 1984), we give it here briefly as a basis for understanding channeling in STEM and in order to relate it to the previous many-beam treatment (Chap. 5).

The elastic contribution to electron channeling patterns is normally obtained by expressing the solution to Eq. (5.1) in the form (Howie 1966)

$$\phi(\vec{r}) = \sum_j \alpha^j \exp\left(2\pi i k_z^j z\right) B^j(x, y) \tag{14.83}$$

This separates the energetic forward free-particle motion of the electron (now described by a plane wave) from its transverse motion, described by the lateral eigenfunction $B^j(x, y)$. This separation is valid if all the important reciprocal lattice vectors lie in a plane normal to the beam, that is, if the variation of crystal potential in the beam direction can be neglected and the "projection" or ZOLZ approximation made. HOLZ effects are ignored. The Bloch wave excitation amplitude α^j is determined by matching the wave in the crystal to the incident wave at the boundary. From Eq. (5.1), we then obtain

$$\frac{1}{4\pi^2}\left[\frac{\partial^2}{\partial x^2} + \frac{\partial^2}{\partial y^2}\right]B^j + U(x,y)B^j = -\left(K_o^2 - k_z^{j2}\right)B^j \tag{14.84}$$

where $U(x,y)$ is the crystal interaction potential (in Å^{-2}) averaged in the z-direction. This averaging is known as the projection or ZOLZ approximation in electron diffraction. The solution of Eq. (14.84) gives the eigenvalues ε^j corresponding to the transverse eigenstates $B^j(x,y)$, which in unit of energy are thus

$$\varepsilon^j\left(\vec{K}_t\right) = \frac{h^2}{2m}\left(K_o^2 - k_z^{j2}\right) \tag{14.85}$$

where $\vec{K}_t$ in the plane normal to the optical axis specifies the incident-beam direction. Since ε^j is an eigenvalue, with the units of energy, we may imagine that the transverse motion of the electron is described by bound (or free) states $B^j(x,y)$ within the transverse crystal potential energy well given by $U(x,y)$. For bound states ε^j is also negative, while for free states it lies above the maximum value of the crystal potential energy. We note that these energies can be related to the dispersion surfaces of dynamical electron diffraction theory using $k_z^j(K) = K_z + \gamma^j$ and thus

$$\varepsilon^j\left(\vec{K}_t\right) \approx -\frac{h^2 k}{2m}\left(\gamma^j\left(\vec{K}_t\right) - \frac{K_t^2}{2k}\right) - eV_o. \tag{14.86}$$

where $k = 1/\lambda$ and V_o is the mean crystal potential. In the axial orientation with the crystal zone axis aligned with the optical axis $\vec{K}_t = 0$. In addition, the dispersion $\gamma^j\left(\vec{K}_t\right)$ is taken to be along the zone axis. Several branches of the dispersion surfaces are involved in channeling, with the uppermost dispersion surface corresponding to the deepest bound state.

In the zone axis STEM imaging condition, the incident electron wave is a wave packet containing a range of incident-beam directions with the central beam along the zone axis direction. Thus, more than one set of Bloch states are excited. The wave packet channeling depends on thickness even in the absence of absorption and the channeling may run down one or more atomic columns. The wave packet may be channeled and it also may be dechanneled. To examine this, we consider a perfect crystal using the Bloch wave method. For one component of the coherent probe or wave packet with the incident-beam direction K_t, the wavefunction inside the crystal is given by

$$\phi\left(\vec{r},\vec{K}_t\right) = \sum_j \sum_g c_j\left(\vec{K}_t\right) C_g^j\left(\vec{K}_t\right) \exp\left\{2\pi i\left[\vec{K} + \gamma^j\left(\vec{K}_t\right)\hat{z} + \vec{g}\right]\cdot\vec{r} + i\alpha\left(\vec{K}_t\right)\right\}$$

$$\tag{14.87}$$

Here the usual symbols and the same sign convention as Chap. 5 are used, and $\alpha(\vec{K}_t)$ is the initial phase of

$$\alpha(\vec{K}_t) = \chi(\vec{K}_t - \vec{K}_{to}) + 2\pi(\vec{K}_t - \vec{K}_{to}) \cdot \vec{r}_p. \tag{14.88}$$

Here also $\vec{K}_{to}$ is the value of $\vec{K}_t$, at the center of the CBED disk. The electron probe inside the crystal is the integration of each wave component

$$\phi(\vec{r}) = \int \phi(\vec{r}, \vec{K}_t) d^2\vec{K}_t$$

$$= \sum_j \int \sum_g c_j(\vec{K}_t) C_g^j(\vec{K}_t) \exp\{2\pi i[\vec{K} + \gamma^j(\vec{K}_t)\hat{z} + \vec{g}] \cdot \vec{r} + i\alpha(\vec{K}_t)\} d^2\vec{K}_t$$

$$\tag{14.89}$$

The excitation coefficients and the eigenvalues and eigenvectors may be obtained through diagonalization for each wave component.

Figure 14.23 shows a 1D scattering example where we have evaluated Eq. (14.89) for each dispersion surface branch separately. The crystal and orientation used is Si (220) systematics with the central beam perpendicular to the systematics direction. To differentiate elastic effects from absorption effects, absorption was not included. The Si (220) systematic dispersion surfaces are plotted in Fig. 14.22a; these are labeled according to the convention used by Buxton et al. (1978). Figure 14.22b shows the integrated wave amplitude for the top three dispersion surface branches for different crystal thicknesses. The wave amplitudes for branch 1 are plotted as full lines, and the other two as dashed lines. The contribution from branch 3 is very small, barely visible in Fig. 14.22. As it travels deeper into the crystal, the wave associated with branch 1 first is focused or channeled into the atomic columns with most of its energy in a single atomic column. At a thickness of 543 Å, electrons start to dechannel into neighboring atomic columns. The wave associated with the second branch is channeled between the atomic columns with much less concentration in the center of the probe, and the energy spreads away from the probe center as it travels deeper into the crystal. Thus, the wave associated with branch 1 is much more localized than branch 2.

Besides the dechanneling due to elastic scattering in the zone axis orientation, absorption will also further reduce channeling. The dispersion surface has an imaginary part when absorption is included. For branch 1, the imaginary dispersion is positive and we expect the wave to dissipate as it travels deeper into the crystal, while the wave associated with branch 2 increases because of the negative imaginary dispersion.

In ADF-STEM using a detector with a large hole, we enhance the contribution from the more localized Bloch wave states, which scatter into larger angles. In addition, these states are the least dispersive, and hence are not washed out by the large beam divergence used in STEM. In such a Bloch wave model, it may then be argued that the image is formed predominantly from the most localized 1 s

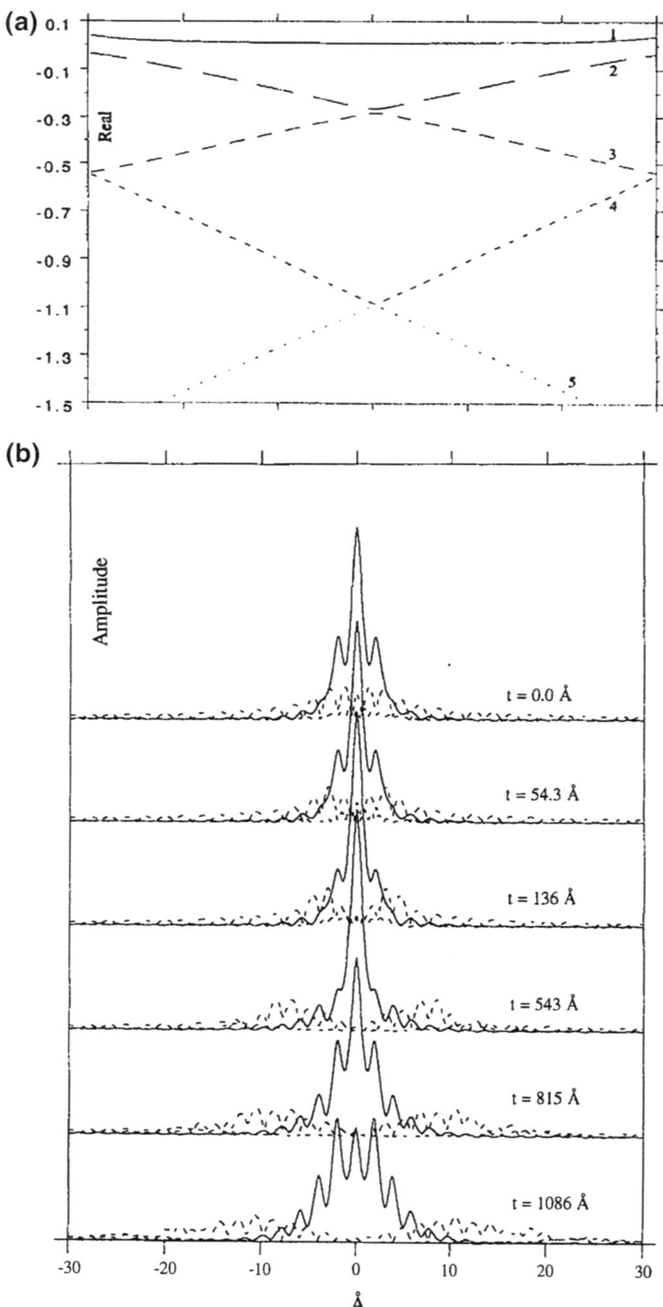

Fig. 14.23 **a** Dispersion surfaces (real part) for silicon (220) systematics calculated by the Bloch wave method with nine systematic beams included. The *vertical axis* is γ in units of 1/Å. **b** Integrated wave amplitude at different thicknesses for dispersion surface branch 1 (*full line*), 2 and 3 (*dashed lines*) according to Eq. (14.89). For details, see text (from Zuo and Spence 1993)

transverse eigenstates (Pennycook and Jesson 1991). These have the highest transverse kinetic energy, occupying regions of the lowest potential energy. An expression for the ADF image may be derived along similar lines to Eq. (14.78), and an approximation developed which includes the probe coordinate and the predominant 1 s state. The images do depend on thickness and may best be understood by following the propagation of the probe down atomic columns (Fertig and Rose 1981; Hillyard and Silcox 1995). The propagation is dependent on the composition of atomic columns. For example, in InP, Hillyard and Silcox (1993) find that when the probe is located over an Indium atomic column, a channeling peak quickly appears, disappearing after a thickness of about 10 nm. Locating the probe over a P column, however, generates a peak up to a thickness of 59 nm.

14.15 Image Simulation Using the Multislice Method

The starting point is to simulate the interaction of an electron beam with matter. For this purpose, two approaches are in use: those intended for the thinnest samples and based on the weak phase object approximation (Chap. 4), and those based on the dynamical theory intended for thicker samples in which multiple scattering cannot be neglected. The Bloch wave theory developed in Chap. 5 is still the method of choice for crystal diffraction calculations. But the multislice formulation developed by Cowley and Moodie and others (Cowley 1995; Ishizuka 1982b) has the advantage that it can treat both crystals and nonperiodic structures, including amorphous structures. The text by Spence and Zuo (1992) contains the complete FORTRAN source code for the Bloch wave and multislice methods, and the book by Kirkland (2010) contains an excellent analysis of HREM and STEM image simulation and related C code. This book is strongly recommended.

The multislice method solves the time-independent Schrödinger equation by forward-scattering approximation. The scattering angle in the forward direction is small, considering the electron short wavelength. Thus, we have in the near field

$$|\vec{r} - \vec{r}'| = \sqrt{(z-z')^2 + (x-x')^2 + (y-y')^2} \approx |z-z'| + \frac{(x-x')^2 + (y-y')^2}{2|z-z'|}$$

Applying this to the kinematic solution of Schrödinger equation for a thin sample (Eq. 4.5), we obtain

$$\phi(\vec{r}) \approx \phi_{\mathrm{o}}(\vec{r}) + \pi i\lambda \int d^3\vec{r}' P(\vec{r} - \vec{r}')U(\vec{r}')\phi_{\mathrm{o}}(\vec{r}') \qquad (14.90)$$

where $\phi_{\mathrm{o}}(\vec{r}) = e^{2\pi i\vec{k}_{\mathrm{o}}\cdot\vec{r}}$ is the incident wave and

$$P(\vec{r} - \vec{r'}) = \frac{e^{2\pi i k|z-z'|} e^{\pi i k\left[(x-x')^2 + (y-y')^2\right]/|z-z'|}}{i\lambda|z - z'|}$$

is the Fresnel propagator, coming from the Fresnel integral that we introduced in Sect. 2.12.3 (also Eq. 14.6).

Next, we consider electron scattering by a thin atomic layer as illustrated in Fig. 14.23. The thin layer is bounded by the top and bottom surfaces at z_o and z_e. To compute Eq. (14.90), we take a thin slice of thickness dz at the depth z. The thickness is small enough so the potential inside the thin slice is constant. This slice contributes to a part of the scattered wave at z_e by

$$d\phi(\vec{R}, z_e) \approx \left\{\left[\pi i\lambda U(\vec{R}, z)dz\right]\phi_o(\vec{R}, z_o) * P(\vec{R}, z - z_o)\right\} * P(\vec{R}, z_e - z)$$
$$\approx \left[\pi i\lambda U(\vec{R}, z)\phi_o(\vec{R}, z_o)dz\right] * P(\vec{R}, z_e - z_o) \tag{14.91}$$

where we have used the method of stationary phase and one of the properties of the Fresnel propagator (Ishizuka and Uyeda 1977)

$$P(\vec{R}, z_e - z_o) = P(\vec{R}, z - z_o) * P(\vec{R}, z_e - z).$$

Integration over z in Eq. (14.91) gives

$$\phi(\vec{R}, z_e) = \left[1 + \pi i\lambda\bar{U}(\vec{R})\right]\phi_o(\vec{R}, z_o) * P(\vec{R}, z_e - z_o)$$

where

$$\bar{U}(\vec{R}) = \int\limits_{z_o}^{z_e} U(\vec{R}, z)dz$$

is the projected potential. The above equation is obtained under the WPO approximation for a thin sample. In case of heavy atoms, a better approximation is the phase grating approximation, which is attained by following substitution

$$1 + \pi i\lambda\bar{U}(\vec{R}) = e^{i\pi\lambda\bar{U}(\vec{R})} = e^{i\sigma\bar{V}(\vec{R})} \tag{14.92}$$

where $\sigma = 2\pi me\lambda/h^2$ is called interaction constant. Using this, we obtain the basic iterative equation for the multislice method that relates the exit-wave function to the incident wave function at the entrance surface according to

$$\phi(\vec{R}, z_e) \approx \left\{e^{i\sigma\bar{V}(\vec{R})}\phi_o(\vec{R}, z_o)\right\} * P(\vec{R}, z_e - z_o). \tag{14.93}$$

In the above equation, the term inside the bracket { } describes a modification to the phase of the electron wave by the thin sample's projected potential.

Transmission of electrons through the thin sample is represented as a two-dimensional phase object, which takes place at the entrance surface. Then, Fresnel propagation takes the modified wave function to the exit surface as in vacuum. This form of description is completely consistent with the more conventional quantum mechanical descriptions in the limit of a WPO. The relationship between the multislice algorithm, Feynman's path-integral method, the Bloch wave method derived from the Schroedinger equation, and other multiple scattering methods, such as the Howie–Whelan equations, is given in Spence (2013).

For a sample of finite thickness, we model the forward propagation of the electron waves through a successive thin slice of potentials as illustrated in Fig. 14.24. Equation (14.93) then provides the relationship between the incident wave $\phi_n(x,y)$ and the exit wave $\phi_{n+1}(x,y)$ of the nth slice (Fig. 14.25)

$$\phi_{n+1}(x,y) = \{\phi_n(x,y)\exp[i\sigma\bar{V}_n(x,y)]\} * P(x,y,\Delta z_n) \tag{14.94}$$

The projected potential here is an integration of the potential over the slice thickness:

$$\bar{V}_n(x,y) = \int_{\Delta z_n} V(x,y,z)\,dz. \tag{14.95}$$

The propagation between two waves over a short distance is described by the Fresnel propagator:

$$P(x,y,\Delta z_n) = \frac{1}{\Delta z_n}\frac{1}{\lambda i}\exp\left(\pi i\frac{x^2+y^2}{\lambda\Delta z_n}\right)\exp\left[2\pi i(k_x x + k_y y)\right] \tag{14.96}$$

where the phase factor $\exp\left[2\pi i(k_x x + k_y y)\right]$ is used to take account of a tilt in illumination.

Thus, the multislice method divides the specimen into a sequence of thin rectangular slices. The slice is in a plane perpendicular to the optical axis of the electron microscope (along z) as in Fig. 14.24. The slice thickness (Δz) must be thin enough

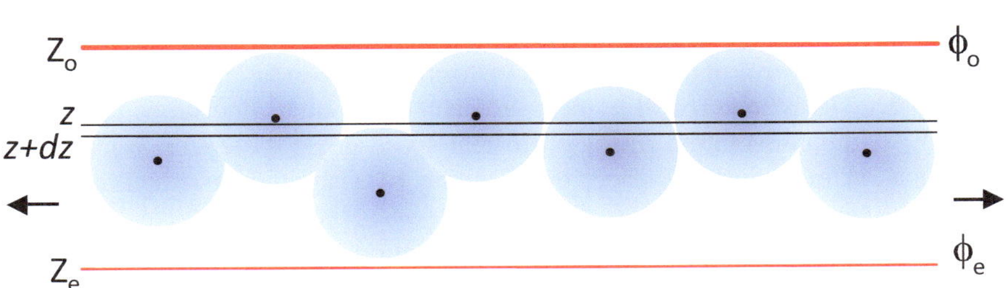

Fig. 14.24 Electron scattering by a thin sample with the incident and exit waves at the *top* and *bottom* of the sample

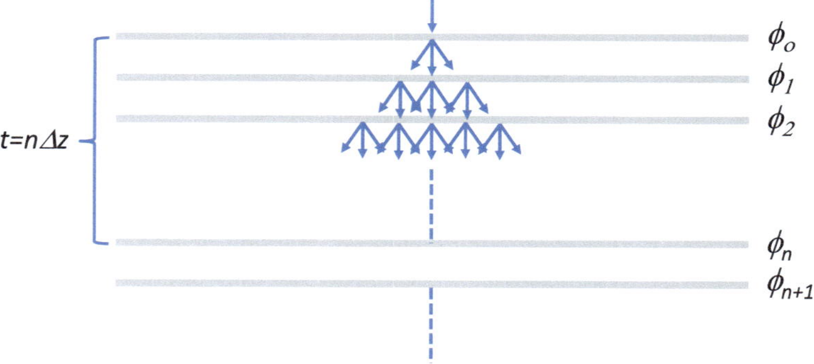

Fig. 14.25 A schematic illustration of the multislice method. The potential is divided into slices of thickness Δz and averaged along z for each slice. The choice of slice thickness affects the numerical convergence of the calculation and accuracy of higher order Laue zone reflections. Along the x- and y-directions, the potential is sampled in discrete points in pixels

to be a phase object. In a crystalline sample, because of its periodic structure, it is then possible to identify the slices with the atomic layers in the specimen when the specimen is aligned such that these layers are perpendicular to the electron beam direction. In fact, several structure types can be described as a repeat of a small number of identical layers. For example, the (111) projection of a fcc structure has a stacking sequence of *abcabc...* with three repeating layers. Ideally, each atomic layer should be sampled along Z in smaller slices to allow a good representation of the potential for HOLZ reflections. However, in most multislice implementations, it is convenient to have all of the atoms within z to $z + \Delta z$ compressed into a flat plane or slice at z according to Eq. (14.94). Such approximation limits the simulation accuracy of HOLZ reflections.

For STEM or electron nanodiffraction, the incident electron wave is set to the electron probe function as described in Sect. 14.7, e.g.,

$$\phi_{\mathrm{o}}(x, y) = \phi_{\mathrm{P}}(x, y).$$

The electron exit wave, $\phi_{\mathrm{exit}}(x, y)$, is obtained by applying Eq. (14.94) sequentially from the first to the last slices.

Equation (14.94) is computed numerically using the fast Fourier transform (FFT). The FFT performs the Fourier transform of a periodic and discretely sampled object. It is one of the most efficient computer algorithms available. Using FFT, the convolution operation in real space in Eq. (14.94) is replaced by multiplication in reciprocal space, thus leading to a significant reduction in the computer time.

The idea of sampling is based on the information theory, which was first formulated by Nyquist in 1928 and further developed by Shannon (1949). The Nyquist–Shannon theorem states that (Shannon 1949): "if a function $f(x)$ vanishes outside the points $x = \pm a/2$, then its Fourier transform $F(k)$ is completely specified by the values which it assumes at the points $k = 0, \pm 1/a, \pm 2/a, \ldots$." The minimum sampling frequency of $1/a$ is called the Nyquist frequency.

Discrete sampling is used in digital representation of the object or diffraction patterns, where a continuous 2D object of dimensions a and b is approximated by discrete points denoted by $x_n = (n - 1)a/N$ and $y_m = (m - 1)b/M$ with $n = 0, \ldots, N - 1$ and $m = 0, \ldots, M - 1$. Fourier transform is carried out via summation for a set of discrete frequencies denoted by $k_j = (j - 1)/a$ and $k_k = (k - 1)/b$ with $j = 0, \ldots, N - 1$ and $k = 0, \ldots, M - 1$:

$$F(j,k) = \text{FFT}(f(x,y)) = \sum_{n=0}^{N-1}\sum_{m}^{M-1} f(n,m)\exp(-2\pi ijn/N)\exp(-2\pi ikm/M)$$

$$(14.97)$$

$$f(n,m) = \text{FFT}^{-1}\left(F(k_x,k_y)\right)$$
$$= \frac{1}{NM}\sum_{j=0}^{N-1}\sum_{k=0}^{M-1} F(j,k)\exp(2\pi ijn/N)\exp(2\pi ikm/M) \qquad (14.98)$$

where Eqs. (14.97) and (14.98) denote the forward and inverse Fourier transform, respectively. The smallest frequency in Fourier transform is $1/a$. Since $\exp(-2\pi i(N - j)n/N) = \exp(2\pi ijn/N)$, k_j and k_{N-j} thus belong to the same frequency with the opposite signs. Because of this, the maximum frequency represented in the discrete Fourier transform is $N/2a$ and $M/2b$ along the reciprocal a- and b-axes.

The use of discrete Fourier transform requires that the slice must be periodic in the plane of the slice. If the slices do not obey periodic boundary conditions in the x- and y-directions, serious artifacts may be generated in the image due to the so-called wrap-around error. There is no requirement on the periodicity of slices along the optical axis (the z-direction), although periodicity along the z-direction can save considerable computation time.

To implement the multislice method using FFT, a computational unit cell (CUC) containing an isolated object or unit cells of periodic crystal is constructed with a- and b-axes that define the slicing plane, and the c-axis usually assumed to be perpendicular to the ab plane. This unit cell is divided into n_s slices along the c-axis. For a periodic structure, the CUC can be repeated along with the c-axis to represent crystals of different thicknesses. For a nonperiodic object, the length of c represents the thickness of the object. The CUC contains N_a atoms with each atom specified by its atomic number (Z), valence (for ionic crystals), Debye–Waller factor (B), occupancy (o), and atomic position (x, y, z), where x, y, and z are the fractional coordinates, e.g.,

$$\vec{r} = x\vec{a} + y\vec{b} + z\vec{c}.$$

Table 14.1 describes the basic steps for calculating the exit-wave function for HREM or STEM images. Step 1 takes care of the information that is required for the simulations, while step 2 is the start of computation. The atoms in the CUC are separated into slices in this step. For each slice, the projected potential is calculated using one of two approaches available: One is to calculate the Fourier coefficients of the potential and transform back to potential in real space using FFT, and the other approach is to calculate the projected atomic potential in real space for each of the atoms and place the atomic potentials onto the 2D plane. The Fourier coefficient of potential is obtained by summing the contributions from individual atoms and this is repeated for each sampling point in the reciprocal space. This approach becomes computationally intensity as the CUC size and the number of atoms increases. In such cases, the second approach becomes much more efficient.

The Fourier coefficient of the Fresnel propagator is calculated using the following formulas

$$P(j,k) = \exp(2\pi i S(H,K)\Delta z)$$

$$S(H,K) = -\lambda/2\left[H'Ha^{*2} + K'Kb^{*2} + (H'K + HK')a^*b^* \cos\gamma\right]$$

$$H = j - N/2, \quad K = k - M/2$$

$$H = 2H_o + H, \quad K' = 2K_o + K$$

<table>
<tr><td>Table 14.1 Major steps in a multislice algorithm for HREM and STEM</td><td>

1. Separate the computational unit cell into thin slices and compute the complex projected potential ($\bar{V}(n, \mathrm{m})$) for every sampling points in each of the slices with the real and imaginary parts for the Coulomb and absorptive potentials, respectively

</td></tr>
<tr><td></td><td>

2. Use the potential to calculate the phase grating for each of the slices and save it in memory

</td></tr>
<tr><td></td><td>

3. Compute the Fresnel propagator in reciprocal space ($P(j, k)$) and save it in memory

</td></tr>
<tr><td></td><td>

4. For STEM, move the electron probe and calculate the electron probe wave function ϕ_o. For TEM, take $\phi_o = 1$. Assign the current slice (i) to 1 and $j = 1$

</td></tr>
<tr><td></td><td>

5. Assign the current slice projected potential to $\bar{V}(n, \mathrm{m})$ and compute the product of $\phi_1(n, m) = \phi_o(n, m)\exp[i\sigma\bar{V}(n, \mathrm{m})]$ in real space

</td></tr>
<tr><td></td><td>

6. Compute FFT of $\phi_1(n, m)$, multiplied with $P(j, k)$ and perform inverse FFT

</td></tr>
<tr><td></td><td>

7. For ADF-STEM, compute electron scattering into the ADF detector from the current slice

</td></tr>
<tr><td></td><td>

8. Take $\phi_o(n, m) = \phi_1(n, m)$ and $i = i + 1$, go back to step 5 if $i < n_s$ (total number of slices)

</td></tr>
<tr><td></td><td>

9. For STEM, take $j = j + 1$, go back to step 4 if $j < n_p$ (total number of probe positions)

</td></tr>
<tr><td></td><td>

10. Save the exit-wave function for HREM

</td></tr>
</table>

where H_o, K_o are used to describe the beam tilt and $S(H, K)$ is essentially the excitation error. Step 6 calculates the phase grating in real space while the Fresnel propagation carried out in step 7 is performed in the reciprocal space. These two steps are repeated in a loop (step 9) until the propagation of waves is carried out through the entire CUC.

For HREM image simulation, the calculated exit-wave function from the multislice simulation (or Bloch wave method, for a perfect crystal) is Fourier transformed and multiplied by the lens aberration function $\exp(-i\chi(H, K))$ and then transformed back to obtain the electron wave function at the detector plane, according to Eq. (14.16), from which the image intensity is calculated.

Different considerations arise when using a CUC for either image or diffraction pattern simulation. Consider the case of a glassy amorphous film for which the periodicity of the CUC in real space creates unphysical discontinuities at the boundaries. Because of the forward-scattering nature of high-energy electron diffraction, a high-resolution image simulation of the center of this CUC will be correct, since the image there depends only on the local potential in a thin column erected about that point along the beam direction (the "column approximation"). For the diffraction pattern, however, the discontinuities will generate a small unphysical contribution to scattering across the entire pattern.

For ADF-STEM simulation, an additional loop is added to calculate the probe wave function and iterate over different probe positions as in steps 5 and 10 of Table 14.1. The inelastic scattering into the HAADF detector can be calculated using the so-called frozen phonon model, where atoms are randomly displaced to represent the instantaneous atomic positions in the sample, or using the approximation to represent the inelastic phonon scattering using an inelastic atomic potential as described in Sect. 14.11.

The main limitation of the multislice method is the number of atoms that can be included realistically in a simulation. The limitation comes from the atomic potential sampling considerations. The 3D sample potential in a multislice calculation is represented in a 2D numerical array for each slice along the beam direction. As this slice is made thinner, three-dimensional multiple scattering, including HOLZ scattering along the beam direction, is included with increasing accuracy. The representation of the atomic potentials requires a minimum number of sampling points. For example, a minimum of 5 points are required to represent the center, the size, and the gap of the atomic potential. For a 1 Å sized atom, the spacing between these points is 0.2 Å defines a minimum pixel size in the real space. A 1 k $\times$ 1 k in this case represents a sample area of 20×20 nm^2. Figure 14.26 shows an example, where a spacing of 0.161 Å was used to sample a large Pt nanoparticle of 30 nm in diameter. With 2048×2048 sampling points, the field of view in the simulated image is 33 nm.

Fig. 14.26 An example of a simulated HREM image of a Pt icosahedral nanoparticle of 30 nm in diameter with 738,221 atoms. The model was obtained by molecular dynamics simulations, which was kindly provided by Liang Qi, University of Michigan. The multislice calculation was carried out using the Zmult program developed at University of Illinois with 2048 × 2048 sampling points and 150 slices over a field of view of 33 nm

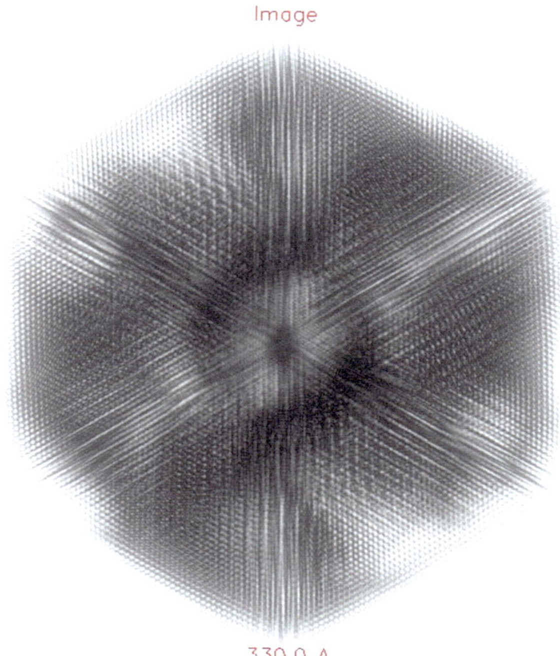

References

Ade G (1977) Incoherent imaging in scanning-transmission electron-microscope (STEM). Optik 49:113–116

Allpress JG, Sanders JV (1973) The direct observation of the structure of real crystals by lattice imaging. J Appl Cryst 6:165–190

Boothroyd CB (1998) Why don't high-resolution simulations and images match? J Microsc 190:99–108

Boothroyd CB, Dunin-Borkowski RE (2004) The contribution of phonon scattering to high-resolution images measured by off-axis electron holography. Ultramicroscopy 98(2–4): 115–133

Buxton BF, Tremewan PT (1980) Atomic-string approximation in cross-grating high-energy electron-diffraction.I. Dispersion surface and Bloch waves. Acta Cryst A 36:304–315

Buxton BF, Loveluck JE, Steeds JW (1978) Bloch waves and their corresponding atomic and molecular-orbitals in high-energy electron-diffraction. Philos Mag A 38:259–278

Cowley JM (1959) The electron-optical imaging of crystal lattices. Acta Crystallogr 12:367

Cowley JM (1976) Scanning-transmission electron-microscopy of thin specimens. Ultramicroscopy 2:3–16

Cowley JM (1979) Adjustment of a STEM instrument by use of shadow images. Ultramicroscopy 4:413–418

Cowley JM (1995) Diffaction physics, 3rd edn. Elsevier Science, NL

Cowley JM, Moodie AF (1957a) Fourier images: II—the out-of-focus patterns. Proc Phys Soc B 70:497

Cowley JM, Moodie AF (1957b) The scattering of electrons by atoms and crystals. I. A new theoretical approach. Acta Crystallogr 10:609–619

Cowley JM, Moodie AF (1960) Fourier images IV: the phase grating. Proc Phys Soc 76:378

Crewe AV (1966) Scanning electron microscopes—is high resolution possible. Science 154: 729–738

Crewe AV, Wall J, Langmore J (1970) Visibility of single atoms. Science 168:1338

Crewe AV, Langmore JP, Issacson MS (1975) Resolution and contrast in the scanning transmission electron microscope. In: Siegel BM, Beaman DR (eds) Physical aspects of electron microscopy and microbeam analysis. Wiley, New York

D'Alfonso AJ, Findlay SD, Oxley MP, Pennycook SJ, Van Benthem K, Allen LJ (2007) Depth sectioning in scanning transmission electron microscopy based on core-loss spectroscopy. Ultramicroscopy 108:17–28

Dellby N, Krivanek OL, Nellist PD, Batson PE, Lupini AR (2001) Progress in aberration-corrected scanning transmission electron microscopy. J Electr Microsc 50:177–185

Einspahr JJ, Voyles PM (2006) Prospects for 3d, nanometer-resolution imaging by confocal STEM. Ultramicroscopy 106:1041–1052

Fertig J, Rose H (1977) Reflection on partial coherence in electron-microscopy. Ultramicroscopy 2:269–279

Fertig J, Rose H (1981) Resolution and contrast of crystalline objects in high-resolution scanning-transmission electron-microscopy. Optik 59:407–429

Gao WP, Sivaramakrishnan S, Wen JG, Zuo JM (2015) Direct observation of interfacial Au atoms on TiO_2 in three dimensions. Nano Lett 15:2548–2554

Gibson JM, Howie A (1979) Investigation of local-structure and composition in amorphous solids by high-resolution electron-microscopy. Chem Scr 14:109–116

Haider M, Uhlemann S, Zach J (2000) Upper limits for the residual aberrations of a high-resolution aberration-corrected STEM. Ultramicroscopy 81:163–175

Haider M, Muller H, Uhlemann S (2008) Present and future hexapole aberration correctors for high-resolution electron microscopy. Adv Imaging Electron Phys 153:43–119

Haine ME, Mulvey T (1952) The formation of the diffraction image with electrons in the Gabor diffraction microscope. J Opt Soc Am 42:763–773

Hashimoto A, Shimojo M, Mitsuishi K, Takeguchi M (2009) Three-dimensional imaging of carbon nanostructures by scanning confocal electron microscopy. J Appl Phys 106:086101

Hillyard S, Silcox J (1993) Thickness effects in ADF STEM zone-axis images. Ultramicroscopy 52:325–334

Hillyard S, Silcox J (1995) Detector geometry, thermal diffuse-scattering and strain effects in ADF STEM imaging. Ultramicroscopy 58:6–17

Howie A (1966) Diffraction channelling of fast electrons and positrons in crystals. Philos Mag 14:223–237

Howie A (1979) Image-contrast and localized signal selection techniques. J Microsc 117:11–23

Iijima S (1971) High-resolution electron microscopy of crystal lattice of titanium-niobium oxide. J Appl Phys 42:5891

Ishizuka K (1982) Multislice formula for inclined illumination. Acta Cryst A 38:773–779

Ishizuka K (2002) A practical approach for STEM image simulation based on the FFT multislice method. Ultramicroscopy 90:71–83

Ishizuka K, Uyeda N (1977) New theoretical and practical approach to multislice method. Acta Cryst A 33:740–749

Jesson DE, Pennycook SJ (1993) Incoherent imaging of thin specimens using coherently scattered electrons. P Roy Soc Lond A 441:261–281

Jia CL, Lentzen M, Urban K (2003) Atomic-resolution imaging of oxygen in perovskite ceramics. Science 299:870–873

Kirkland EJ (2010) Advanced computing in electron microscopy, 2nd edn. Springer, New York

Krivanek OL, Dellby N, Lupini AR (1999) Towards sub-angstrom electron beams. Ultramicroscopy 78:1–11

Lentzen M, Jahnen B, Jia CL, Thust A, Tillmann K, Urban K (2002) High-resolution imaging with an aberration-corrected transmission electron microscope. Ultramicroscopy 92:233–242

Lin JA, Cowley JM (1986a) Calibration of the operating parameters for an HB5 STEM instrument. Ultramicroscopy 19:31–42

Lin JA, Cowley JM (1986b) Reconstruction from in-line electron holograms by digital processing. Ultramicroscopy 19(2):179–190

Loane RF, Kirkland EJ, Silcox J (1988) Visibility of single heavy-atoms on thin crystalline silicon in simulated annular dark-field STEM images. Acta Cryst A 44:912–927

Lupini AR (2011) The electron ronchigram. In: S Pennycook, P Nellist (eds) Scanning transmission electron microscopy. Springer, New York

Malacara D (1978) Optical shop testing. Wiley, New York

Menter JW (1956) The direct study by electron microscopy of crystal lattices and their imperfections. Proc Roy Soc Lond Ser A 236:119

Mory C, Colliex C, Cowley JM (1987) Optimum defocus for STEM imaging and microanalysis. Ultramicroscopy 21:171–177

Müller H, Uhlemann S, Hartel P, Haider M (2006) Advancing the hexapole Cs-corrector for the scanning transmission electron microscope. Micros Microanal 12:442–455

Nellist PD, McCallum BC, Rodenburg JM (1995) Resolution beyond the information limit in transmission electron-microscopy. Nature 374:630–632

Nellist PD, Cosgriff EC, Behan G, Kirkland AI (2008) Imaging modes for scanning confocal electron microscopy in a double aberration-corrected transmission electron microscope. Micros Microanal 14:82–88

Nyquist H (1928) Certain topics in telegraph transmission theory. Trans AIEE 47:617

O'Keefe MA, Sanders JV (1975) N-beam. Lattice images, VI. Degradation of image resolution by a combination of incident-beam divergence and spherical aberration. Acta Crystallogr A 31:307–310

Peng LM, Dudarev SL, Whelan MJ (2004) High energy electron diffraction and microscopy. Oxford University Press, Oxford

Pennycook SJ, Boatner LA (1988) Chemically sensitive structure-imaging with a scanning-transmission electron-microscope. Nature 336:565–567

Pennycook SJ, Jesson DE (1991) High-resolution Z-contrast imaging of crystals. Ultramicroscopy 37:14–38

Pennycook S, Nellist P (eds) (2011) Scanning transmission electron microscopy, imaging and analysis. Springer, New York

Pennycook SJ, Berger SD, Culbertson RJ (1986) Elemental mapping with elastically scattered electrons. J Microsc 144:229–249

Reimer L, Kohl H (2008) Transmission electron microscopy, 4th edn. Springer, Berlin

Ronchi V (1964) Forty years of history of a grating interferometer. Appl Opt 3:437–451

Sawada H, Sasaki T, Naruse M, Honda T, Hambridge P, Hartel P, Haider M, Hetherington C, Doole R, Kirkland A, Hutchison J, Titchmarsh J, Cockayne D (2010) Higher-order aberration corrector for an image-forming system in a transmission electron microscope. Ultramicroscopy 110:958–961

Scherzer O (1949) The theoretical resolution limit of the electron microscope. J Appl Phys 20:20–29

Shannon CE (1949) Communication in the presence of noise. Inst Radio Eng 37:10

Spence JCH (1988) Experimental high-resolution electron microscopy. Oxford University Press, New York

Spence JCH (1992) Electron channelling. In: Cowley JM (ed) Techniques of electron diffraction, vol 1. Oxford University Press, Oxford

Spence JCH (2013) High resolution electron microscopy, 4th edn. Oxford University Press, Oxford

Spence JCH, Cowley JM (1978) Lattice imaging in STEM. Optik 50:129–142

Spence JCH, Zuo JM (1988) Large dynamic-range, parallel detection system for electron-diffraction and imaging. Rev Sci Instrum 59:2102–2105

Spence JCH, Zuo JM (1992) Electron microdiffraction. Plenum, New York

Treacy MMJ, Gibson JM (1993) Coherence and multiple-scattering in "Z-contrast" images. Ultramicroscopy 52:31–53

Treacy MMJ, Howie A, Wilson CJ (1978) Z-contrast of platinum and palladium catalysts. Philos Mag A 38:569–585

Vincent R, Bird DM, Steeds JW (1984) Structure of AuGeAs determined by convergent-beam electron-diffraction. I. Derivation of basic structure. Philos Mag A 50:745–763

Wen JG, Mabon J, Lei C, Burdin S, Sammann E, Petrov I, Shah AB, Chobpattana VG, Zhang J, Ran K, Zuo JM, Mishina S, Aoki T (2010) The formation and utility of sub-angstrom to nanometer-sized electron probes in the aberration-corrected transmission electron microscope at the University of Illinois. Micros Microanal 16:183–193

Zaluzec NJ (2003) The scanning confocal electron microscope. Microscopy-Today 6:8–12

Zuo JM, Spence JCH (1993) Coherent electron nanodiffraction from perfect and imperfect crystals. Philos Mag A 68:1055–1078

Chapter 15
Imaging and Characterization of Crystal Defects

This chapter describes the theory of imaging and characterization of defects in crystals by electron microscopy. We start with an overview. This is followed by an introduction to atomic displacements and strain in crystals. The following sections then discuss the kinematic theory of diffraction contrast imaging, the weak-beam imaging technique, and the dynamical theory of electron diffraction from crystal defects. This is followed by a review of diffraction-based defect characterization methods, using CBIM or LACBED. The last section describes the determination of atomic structure of defects using HREM and STEM.

The reading of this chapter can be helped significantly by having some basic knowledge of crystal defects, which can be found in the excellent books by Kelly and Knowles (2012), Hirth and Lothe (1983), Sutton and Balluffi (1997) and Howe (1997). Further discussions on the TEM characterization of defects can be found in the books by De Graef (2003), Edington (1975, 1976), Hirsch et al. (1977), Head et al. (1973) and Amelinckx et al. (1978).

15.1 Overview

Defects in crystals, such as dislocations, stacking faults, and aggregates of point defects, disrupt coherent Bragg scattering. Under suitable diffraction conditions, the disruption is large enough to give rise to "diffraction contrast" in TEM images formed from a single Bragg beam, which maps out the strain field around a defect. At high resolution, using many beams, the atomic arrangement of defects can be observed directly. Together, diffraction contrast imaging and HREM had contributed much of our knowledge of defects in real materials.

© Springer Science+Business Media New York 2017
J.M. Zuo and J.C.H. Spence, *Advanced Transmission Electron Microscopy*,
DOI 10.1007/978-1-4939-6607-3_15

Crystal defects are usually classified according to their dimensions (see Fig. 15.1). Solute or impurity atoms or vacancies alter the crystal lattice at a single site and so are considered to be point, or 0-dimensional, defects. Dislocations are line defects or 1-dimensional defects, consisting of a crystal plane which terminates inside the crystal along a line of atoms, as shown in Fig. 15.1b. The repetition of the crystal lattice breaks down across the dislocation line. Surfaces or interfaces, include stacking faults or grain boundaries, are considered to be 2-dimensional defects, where distinct crystal lattices terminate or are joined together. 3-dimensional defects change the crystal pattern over a finite volume. These include precipitates, large voids, or inclusions of second-phase particles. In all cases, defects are accompanied by strain fields, which are usually 3-dimensional.

The basis for the study of extended defects, such as dislocations and stacking faults, at medium resolution, is the recognition of characteristic contrast patterns produced by the transmitted and diffracted beams under certain diffraction conditions. The theory for such work was largely developed by Hirsch, Howie, Whelan, and others in the 1960s for the interpretation of these characteristic contrast patterns (Hirsch et al. 1977). They developed an important approximation, the so-called column approximation, discussed in more detail in Sect. 15.3.1. This assumes that away from the core of defects the strain field changes slowly. In addition, electron scattering angles in TEM are very small. As a result, the diffracted intensity at each point in a single-beam dark-field image can be expressed in terms of the local diffraction conditions at that point. Given a model of the strain field, and an approximation for the dynamic scattering, the column approximation provides a useful device for calculating medium resolution images that allow the character of a defect to be determined without knowing the details of its atomic structure—only the form of the strain field is needed. In addition, if the atomic displacements due to strain all lie in the planes used to form the image, the strain field will be invisible. This applies, for example, to the planes normal to the dislocation line AA' in Fig. 15.1b, so that this dislocation would not be visible when imaged using Bragg reflection from these planes. We will see that this invisibility condition can be used to determine the Burgers' vector of a dislocation.

The absence of an extended strain field from point defects, such as impurities or interstitial atoms, or vacancies, means that their detection must rely on either amplitude or phase contrast, or significantly improved sensitivity in strain field detection. Even a single impurity atom of large Z in a column of light atoms can provide sufficient amplitude contrast for its detection in atomic-resolution Z-contrast imaging, as long as the depth of focus is small enough, or the crystal is thin enough. It is entirely different for the detection of atomic vacancies, where the scattering power of a vacancy is similar to that of a single atom, not taking into account the strain associated with such a defect. Even if there is sufficient sensitivity in the image contrast, an unavoidable difficulty is to distinguish a vacancy from a

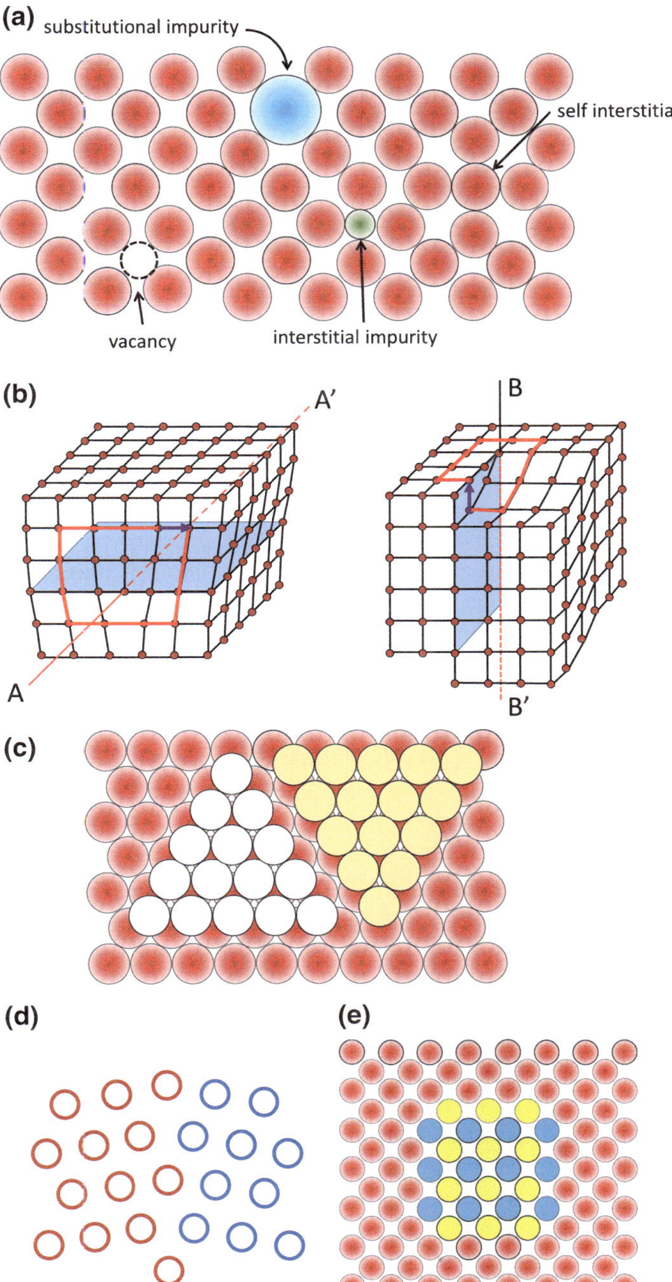

Fig. 15.1 Types of defects, from 0- to 3-dimensional as illustrated in **a** point defects, **b** edge and screw dislocations with *AA'* and *BB'* marking the dislocation *line* and *arrows* for Burger's vector, **c** stacking fault, **d** grain boundary and **e** volume defect as represented by a coherent precipitate

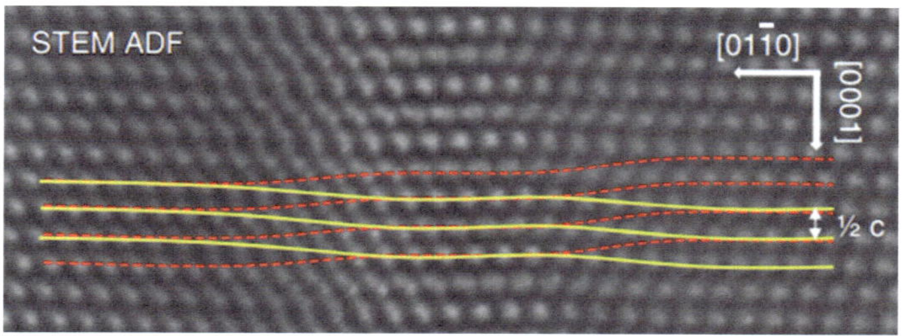

Fig. 15.2 Experimental and simulated ADF image of a dissociated dislocation whose line direction runs up the page, perpendicular to the electron beam (normal to the page). The screw displacements associated with each of the partial dislocations are indicated by the *solid* and *dashed lines* on top of atomic peaks. $C = 5.18$ Å for GaN (from Yang et al. 2015)

missing surface atom—these are indistinguishable in the projection approximation. In general, it is very difficult to remove all other sources of contrast due to surface roughness, oxidation, and contamination even in a perfect thin crystal. And so far, no independent method exists for determining the presence or absence of these artifacts. The recognition of a distinctive pattern of contrast is thus again necessary to detect some of the smaller defects.

Progress in aberration-corrected electron microscopy has now advanced to a level where atomic-resolution electron microscopy may determine atomic arrangements in the cores of dislocations, in fault planes, and at grain boundaries. Since they control the strength of ductile materials, the cores of dislocations, especially, have been of long-standing interest in materials science. In semiconductors, they possess important electronic properties, due, for example, to reconstruction of the chain of atoms running along the edge AA' of the terminating plane of atoms in Fig. 15.1b, which could then act as a one-dimensional conducting wire, to short out a device. These properties were previously amenable only to theoretical simulations and indirect experimental methods. Using the principles of optical sectioning based on ADF-STEM (Sect. 14.11), Yang et al. imaged atomic displacements for screw dislocations whose line lay in a plane transverse to the electron beam, for a mixed $[a + c]$ dislocation in GaN. This allowed direct imaging of a dissociated screw dislocation with a distance of 1.65 nm between the partial dislocations (Fig. 15.2). When the faults are planar or linear and parallel to the incident beam, the atom positions in the fault can be seen directly. This is the case, for example, in Fig. 15.3, where a 60° dislocation is observed at the strained SiGe/Si interface using aberration-corrected STEM and Z-contrast imaging.

Fig. 15.3 Observation of the 60° dislocation at the Si/SiGe interface, which is dissociated into two partial dislocations separated by a stacking fault. The atomic structure of the dislocation is directly determined and the step at the SiGe/Si interface introduced by the dislocation is also clearly visible. The *arrows* indicate the start and end of the dislocation (provided by Jean-Luc Rouvière, also see Rouvière et al. 2013)

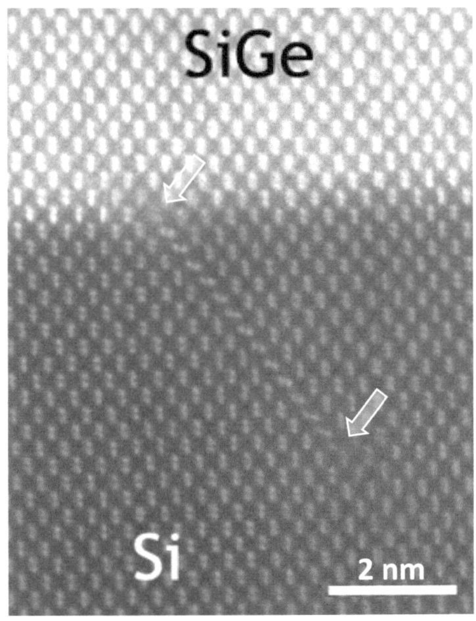

15.2   Atomic Displacements, Strain, and Stress

For a simple elastically isotropic material, Hooke's law $\sigma = E\varepsilon$ relates a small applied stress σ (with the units of pressure) to strain ε (a dimensionless fractional displacement) via the Young's modulus E. The velocity of sound in the medium is then $V_s = \sqrt{K/\rho}$, where K is the bulk modulus and ρ is the density of the medium. A long-wavelength acoustic phonon of wave vector q then has dispersion relation $\omega = V_s\, q$.

Locally, in a statically deformed crystal, atoms are also displaced from their ideal lattice positions around defects. The atomic displacement is determined by the following difference vector

$$\vec{u}(x, y, z) = \vec{r}'(x, y, z) - \vec{r}(x, y, z),$$

Here, $\vec{r}(x, y, z)$ and $\vec{r}'(x, y, z)$ are the positions of the atom before and after deformation, and (x, y, z) is its Cartesian coordinate. Away from the core of a defect, the deformation is specified on a unit cell basis by assuming the same atomic displacement for all atoms within the unit cell. The unit cell position is specified by its lattice vector with $(x, y, z) = (n, m, l)$ and $\vec{r}(x, y, z) = n\vec{a} + m\vec{b} + l\vec{c}$. Figure 15.4 illustrates the case of a screw dislocation. The crystal shear as defined by the Burger's vector $\vec{b}$ is accompanied by crystal deformation and displacements of some atoms from their ideal positions (dashed circles). The atomic displacements

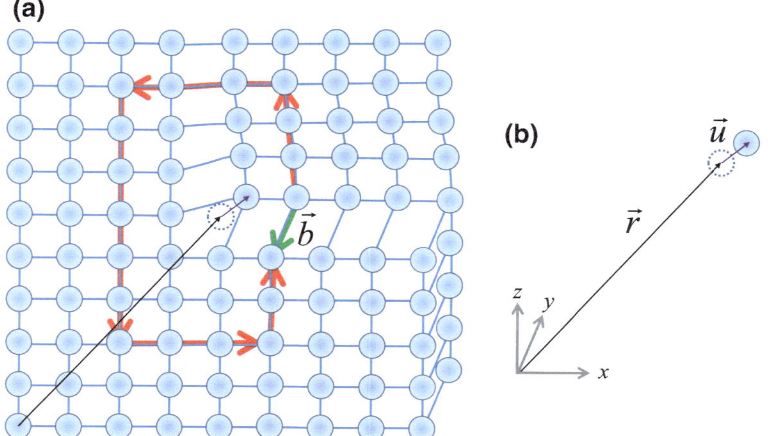

Fig. 15.4 Atomic displacements by crystal defects. **a** A schematic illustration of atomic displacement in a screw dislocation and **b** atomic position of the displaced atom in the *xyz* coordinate

on each lattice site constitute a vector field, or the displacement field, which is inhomogeneous and characteristic of the defect.

The inhomogeneous displacement field modifies atomic bonds from their normal bond lengths and directions, including broken bonds due to dislocations or vacancies. Atomic bonds can be deformed for several reasons, including (i) pressure or forces on the surface of the material, (ii) the presence of defects, which can be point defects (vacancies, interstitials or dopant/impurity atoms), linear (mainly dislocations), or planar (stacking faults or grain boundaries or interfaces, (iii) change of temperature. This last parameter is particularly important in thin-film devices because complex structures are always composed of several materials that have different thermal expansion coefficients. When the temperature in the material changes, for instance after growth, the lengths of their bonds change in different ways, and forces can build up at the materials interface and propagate into the material. While certain defects are specific to crystals, other types of bond deformation can occur in both crystals and in amorphous materials. In fact, because of the lack of translational symmetry, atomic scale strain is intrinsic to amorphous materials.

Since rigid translations and rotations do not affect atomic bonds, strain is introduced to describe the relative deformation. In linear elasticity, the nine components of strain are defined by the first-order derivatives of the displacement components according to Kelly and Knowles (2012)

$$\varepsilon_{xx} = \frac{\partial u_x}{\partial x}, \quad \varepsilon_{yy} = \frac{\partial u_y}{\partial y}, \quad \varepsilon_{zz} = \frac{\partial u_z}{\partial z} \tag{15.1}$$

and

$$\varepsilon_{xy} = \varepsilon_{yx} = \frac{1}{2}\left(\frac{\partial u_x}{\partial y} + \frac{\partial u_y}{\partial x}\right), \quad \varepsilon_{xz} = \varepsilon_{zx} = \frac{1}{2}\left(\frac{\partial u_x}{\partial z} + \frac{\partial u_z}{\partial x}\right),$$
$$\varepsilon_{yz} = \varepsilon_{zy} = \frac{1}{2}\left(\frac{\partial u_y}{\partial z} + \frac{\partial u_z}{\partial y}\right)$$

$$(15.2)$$

The first three are normal strain s, representing the fractional change in length along the x, y, and z direction, while the other six are shear strains with the first-order derivatives inside the () bracket corresponding to the shear angles. The magnitude of strain for each of the nine components is assumed to be small ($\ll 1$). A small volume V is changed by strain to $V + \Delta V = V(1 + \varepsilon_{xx})(1 + \varepsilon_{yy})(1 + \varepsilon_{zz}) \approx V(1 + \varepsilon_{xx} + \varepsilon_{yy} + \varepsilon_{zz})$.

Forces are present in the deformed atomic bonds, which sum to zero at the atomic nucleus. Stress is the physical quantity that allows a measurement of these internal forces. To quantify the state of stress at a point P, consider the point as a small cube. The stress acting on each of the six sides of the cube is resolved into three components normal and parallel to the two sides of the face. This is illustrated in Fig. 15.5. The stress σ_{ij}, where $i, j = x, y, z$, is defined by the force along j acting on the surface direction along i divided by the surface area and thus has the unit of force per area. Since there are three different surfaces, a complete description of the stresses therefore requires the following 3×3 matrix (second-order tensor)

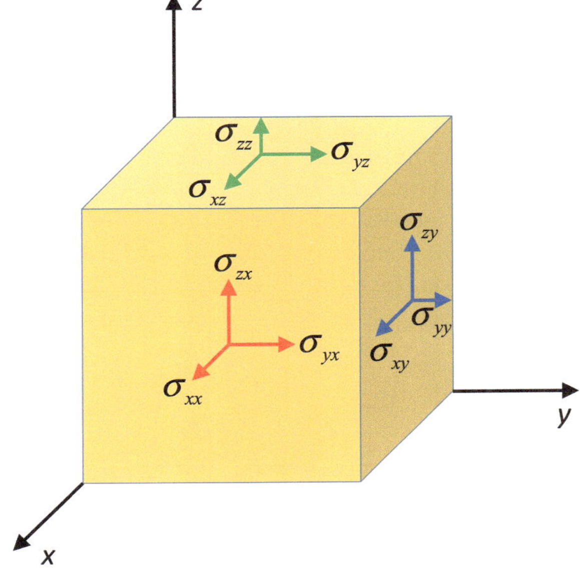

Fig. 15.5 Stresses acting on the surfaces of a cube. Only the stress vectors on 3 sides are shown, the other three sides have exactly the opposites of the illustrated stress vectors

$$\sigma = \begin{pmatrix} \sigma_{xx} & \sigma_{yx} & \sigma_{zx} \\ \sigma_{xy} & \sigma_{yy} & \sigma_{zy} \\ \sigma_{xz} & \sigma_{yz} & \sigma_{zz} \end{pmatrix}. \tag{15.3}$$

The matrix is symmetrical with $\sigma_{ij} = \sigma_{ji}$, which is required in order to have a zero net moment. Together, there are six components that completely specify the state of stress at any points of the continuum. The components with $i \neq j$ are the shear stresses. They are often abbreviated as τ. In dislocation theory, τ without an index is used to represent the shear stress acting on the slip plane in the slip direction of a crystal. The three remaining components σ_{xx}, σ_{yy}, σ_{zz} are the normal stresses. And positive and negative normal stresses give rise to tension and compression, respectively. An effective pressure acting on a volume element is given by the negative average of three normal stresses.

Stress and strain are directly related and one does not exist without the presence of the other. Their relationship is linear according to Hooke's law in the case of small deformations. In its most general form, for anisotropic materials, Hooke's law reads

$$\sigma_{ij} = \sum_k \sum_l c_{ijkl}\varepsilon_{kl} \quad \text{or simply} \quad \sigma_{ij} = c_{ijkl}\varepsilon_{kl} \tag{15.4}$$

where i, j, k, $l = x$, y, z. Since both i and j have three components, there are 9 equations in total. The fourth-rank tensor c_{ijkl} is called the stiffness tensor with 81 coefficients. However, due to the symmetry of the stress and strain tensor, there are only 36 independent coefficients with

$$c_{ijkl} = c_{jikl} = c_{ijlk} = c_{jilk}.$$

Further, the existence of a unique strain energy potential requires that

$$c_{ijkl} = c_{klij},$$

which reduces the number of independent coefficients in the stiffness tensor to 21. It is helpful to further simplify the $3 \times 3 \times 3 \times 3$ stiffness tensor into a 6×6 matrix using the Voigt notation. The stress and strain tensors are vectorized, i.e., converted into vectors of ε_I and σ_I with $I = 1$–6 and $I = 1, 2, 3$ for xx, yy, zz and 4, 5, 6 for yz, zx, and xy. This gives the following stress and strain tensor

$$\begin{pmatrix} \sigma_1 & \sigma_6 & \sigma_5 \\ \sigma_6 & \sigma_2 & \sigma_4 \\ \sigma_5 & \sigma_4 & \sigma_3 \end{pmatrix} \quad \text{and} \quad \begin{pmatrix} \varepsilon_1 & \varepsilon_6/2 & \varepsilon_5/2 \\ \varepsilon_6/2 & \varepsilon_2 & \varepsilon_4/2 \\ \varepsilon_5/2 & \varepsilon_4/2 & \varepsilon_3 \end{pmatrix}.$$

Note the shear strain is divided by 2 here. In this notation, for example, $\varepsilon_6 = 2\varepsilon_{xy} = \gamma_{xy}$, where γ_{xy} is the engineering shear strain. In the simplified notation, we have

$$\sigma_I = C_{IJ}\varepsilon_J \text{ or } \varepsilon_I = S_{IJ}\sigma_J$$

where C_{IJ} and S_{IJ} are the 6×6 elastic stiffness and compliance matrix, respectively. Both are symmetrical with $C_{IJ} = C_{JI}$ or $S_{IJ} = S_{JI}$, leaving $30/2 + 6 = 21$ coefficients for the general anisotropic linear elastic solid. For a crystal, these coefficients must conform to the crystal symmetry and thus can be simplified significantly as required by Neumann's principle (Chap. 13), details can be found in the book by Nye (1957). For a cubic crystal, the elastic stiffness matrix simplifies to

$$\begin{bmatrix} C_{11} & C_{12} & C_{12} & 0 & 0 & 0 \\ C_{12} & C_{11} & C_{12} & 0 & 0 & 0 \\ C_{12} & C_{12} & C_{11} & 0 & 0 & 0 \\ 0 & 0 & 0 & C_{44} & 0 & 0 \\ 0 & 0 & 0 & 0 & C_{44} & 0 \\ 0 & 0 & 0 & 0 & 0 & C_{44} \end{bmatrix}, \qquad (15.5)$$

with only three independent elastic constants. For isotropic materials, there are two independent elastic constants with

$$C_{44} = (C_{11} - C_{12})/2.$$

The anisotropy of a cubic crystal is defined by the Zener anisotropy ratio

$$A = 2C_{44}/(C_{11} - C_{12}).$$

Aluminum and tungsten have values of A at 1.23 and 1.0, while Cu and Au have high anisotropy ratio close to 3. In dislocation theory, crystals are often treated as isotropic. The elastic constants of an isotropic material are given by the Young's modulus E, the bulk modulus K, and Poisson's ratio v, Lamè constants (λ and μ) with

Young's modulus: $E = 1/S_{11}$

Shear modulus: $G = 1/S_{44} = 1/2(S_{11} - S_{12})$

Poisson's ratio: $v = -S_{12}/S_{11}$ $\qquad\qquad\qquad\qquad\qquad\quad$ (15.6)

Lamè constants: $\mu = C_{44} = (C_{11} - C_{12})/2 = 1/S_{44} = G, \quad \lambda = C_{12}$

Bulk modulus: $K = E/3(1 - 2v)$

Using Eqs. (15.5) and (15.6), we obtain for isotropic solids

$$\sigma_1 = C_{11}\varepsilon_1 + C_{12}\varepsilon_2 + C_{12}\varepsilon_3 = (2\mu + \lambda)\varepsilon_1 + \lambda\varepsilon_2 + \lambda\varepsilon_3,$$
$$\sigma_2 = C_{12}\varepsilon_1 + C_{11}\varepsilon_2 + C_{12}\varepsilon_3 = \lambda\varepsilon_1 + (2\mu + \lambda)\varepsilon_2 + \lambda\varepsilon_3,$$
$$\sigma_3 = C_{12}\varepsilon_1 + C_{12}\varepsilon_2 + C_{11}\varepsilon_3 = \lambda\varepsilon_1 + \lambda\varepsilon_2 + (2\mu + \lambda)\varepsilon_3,$$
$$\sigma_4 = \frac{1}{2}(C_{11} - C_{12})\varepsilon_4 = \mu\varepsilon_4 \tag{15.7}$$
$$\sigma_5 = \frac{1}{2}(C_{11} - C_{12})\varepsilon_5 = \mu\varepsilon_5$$
$$\sigma_6 = \frac{1}{2}(C_{11} - C_{12})\varepsilon_6 = \mu\varepsilon_6.$$

Typical values of E and v for metals and ceramic solids are in the range of 40–600 GNm^{-2} and 0.2–0.45, respectively.

15.3   Diffraction Contrast Imaging

The term "diffraction contrast" has been used with different meanings in different communities. Here, we take it to mean the change in intensity of a Bragg beam due local changes in diffraction conditions within a crystal, usually due to strain, composition, or sample thickness. (By contrast, "phase contrast" is taken to refer to near-field interference effects described by the weak-phase approximation, focusing, and Fresnel fringe effects.) In the simplest case, a diffracted beam is selected to form a dark-field image using a small aperture in the back focal plane of the objective lens (Fig. 15.6). Similarly, a bright-field image is formed using the transmitted or undeflected beam. Diffracted waves inside the aperture act as effective sources, and their Fourier transform generates the image according to Abbe's image formation

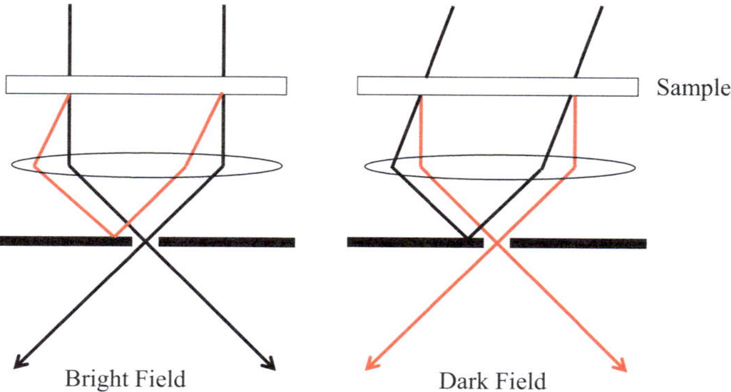

Fig. 15.6 Schematics illustrating the bright- and dark-field imaging in a TEM

theory, so that resolution remains limited by the aperture size, even for these "single-beam" images. Each local region of crystal is taken to support a constant strain and so define a particular diffraction condition (i.e., excitation error). We can then parameterize a single-beam image in terms of the local value of the excitation error, which defines the local rotation of the crystal out of the Bragg condition at one point in the dark-field image. "Bend contours" are the intensity oscillations seen in these single-beam images as a bent crystal is seen to rotate across the exact Bragg condition in the image, and intensity oscillations are also seen due to thickness variations. The diffracted wave is only partially coherent since diffraction contrast imaging is often carried out using relative thick samples and over a field view larger than the lateral coherence length of the electrons. Because of this, diffraction contrast imaging is often treated by taking account of diffraction intensity alone at the sample exit surface.

To set up the dark-field diffraction contrast imaging properly, the incident beam is tilted first using the illumination deflection coils (Sect. 10.3) so the diffracted beam is along the optical axis (Fig. 15.6). Then, the crystal is rotated to a specific diffraction condition. For imaging, a small objective aperture is centered on the diffracted beam. The size of the aperture defines the diffraction-limited optical resolution. The beam tilt is performed using the dark tilt in the TEM. Using bright tilt brings the incident beam back to the optical axis, together the dark and bright tilts provide for complementary bright- and dark-field imaging. Alternatively, dark-field diffraction contrast images can be obtained by quickly placing the objective aperture around an off-axis diffracted beam. Such images suffer from astigmatism because of the off-axis optical aberrations.

A uniformly illuminated perfect crystal of uniform thickness gives no contrast since diffraction intensity is uniform. Thus, diffraction contrast must come from local deviations from the Bragg condition or thickness variations within the sample, including small-angle scattering that occurs within the sample, for the following reasons:

1. Crystal bending. The excitation error of the diffracted beam varies continuously in a bent crystal. Extinction contours or bent contours are observed in the bright-field image as it records the same rocking curve information as the transmitted beam in LACBED. The dark-field image of a bent crystal is similar to the dark-field LACBED.
2. Shape of crystals. A direct image of the projected crystal shape is made in the diffraction contrast image. For large crystals, along the beam direction, variations in the crystal thickness give rise to thickness fringes, which can be related directly to the crystal thickness.
3. Change of composition. The diffraction intensity is dependent on the crystal structure factor, which in turn depends on composition. Thus, structure-factor contrast can be observed by using a composition sensitive reflection for imaging.

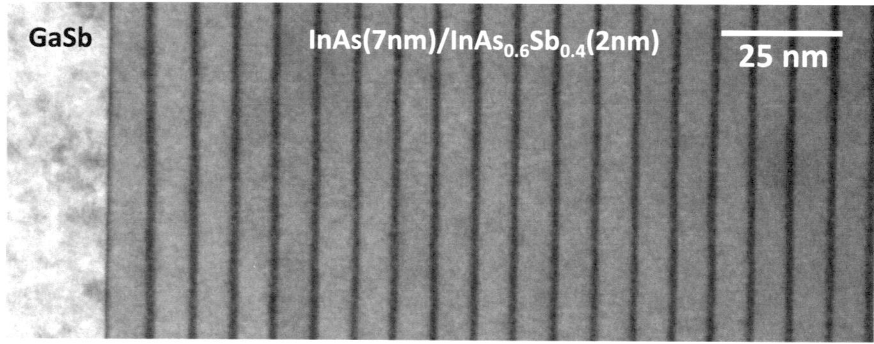

Fig. 15.7 Cross-sectional TEM of MOCVD Grown an InAs/InAsSb superlattice on GaSb Substrate. The image was formed using the (002) reflection in the two-beam diffraction condition. (Image was taken by Honggyu Kim, University of Illinois from the sample provided by Russell D Dupuis, Georgia Institute of Technology)

4. Lattice strain. The presence of defects or interfaces introduces local lattice strain and bending of lattice planes and thus changes in diffraction intensity giving rise to the strain contrast.

Figure 15.7 shows an example. The image was recorded from a cross section of a MOCVD grown InAs (7 nm)/InAs$_{0.6}$Sb$_{0.4}$ (2 nm) superlattice on top of the GaSb substrate. The image was formed using the (002) reflection in the two-beam diffraction condition. Both GaSb and InAs are III–V semiconductors having the zincblende structure (Sect. 12.1). The (002) structure factor is given by the difference between the scattering factors of the group III and V elements, which make it composition sensitive. In the image, the InAs$_{0.6}$Sb$_{0.4}$ layers appear in dark contrast because of the substitution of As with Sb brings their combined scattering factor closer to that of In.

Diffraction contrast effects also appear in STEM images recorded using an ADF detector of a small inner cutoff angle. The elastically scattered electrons detected by the ADF detector give rise to the diffraction contrast. Interpretation of such images is complicated by the contributions from multiple diffracted beams.

15.3.1   Column Approximation

The interpretation of diffraction contrast images is usually made under the column approximation (Hirsch et al. 1960). The crystal is divided into columns, whose z-axis is taken to lie along the direction of the diffracted beam. Within the column, atomic displacements vary only with z. The premise is thus that these columns can be chosen narrow enough that displacements due to the presence of defects is uniform normal to z, and yet they are wide enough so that electrons enter at the top of the sample are not diffracted out of these columns (Humphreys 1979b). The column approximation works well in thin foils at an image resolution of about 1 nm. In this case, the diffracted wave comes mostly from the incident beam, whose amplitude at thickness t is determined by Fresnel propagation, with 50 % contribution from the first Fresnel zone (Chap. 2). Thus, an approximation can be made that the diffracted wave at the exit surface of the foil comes from a column of diameter $D \approx \sqrt{\lambda t}$. For 200 kV electrons and $t = 100$ nm, D is about 0.5 nm. Under strong two-beam diffraction condition, the direction of the electron flux is parallel to the lattice planes, which is taken to be the column direction. The column direction is not affected by defects since their scattering occurs mostly at small angles. The column under strong two-beam condition has a diameter of $D \approx 2\theta_B t = \lambda t / d_{hkl}$, which is about 1.3 nm under these conditions with $d_{hkl} = 2$ Å.

15.3.2   Thickness Fringes and Bend Contours

The diffraction contrast of thin crystals can be predicted using the dynamical theory of Chap. 5 under the column approximation. For example, the two-beam theory gives (Eq. 5.20)

$$I_g(t) = \left| \phi_g(t) \right|^2 = \frac{1}{1 + \omega^2} \sin^2 \left\{ \frac{\pi t}{\xi_g} \sqrt{1 + \omega^2} \right\} \tag{15.8}$$

where $\omega = \xi_g S_g$. Thus, a dark-field image of a crystal wedge recorded under the two-beam condition is dark at the edge of the wedge corresponding to $t = 0$, followed by a broad bright contour parallel to the sample edge (Fig. 15.8). The bright-field image formed by the direct beam shows complementary contrast with $I_0 = 1 - I_g$ in the two-beam approximation and in the absence of absorption. The dark contours in the dark-field image are obtained under the extinction condition $t = n\xi_g / \sqrt{1 + \omega^2}$. For this reason, they are known as thickness fringes. The variation of the intensity with thickness is known as "Pendellosung," or thickness fringe oscillations. At the Bragg condition, the oscillation period is the largest and equals ξ_g. Equation (15.8) does not take account of the absorption effect. With absorption included, the contrast of thickness fringes decreases as the crystal thickness increases and eventually disappears in the thick part of the crystal (Sect. 5.5). Additionally, the overall intensity of both bright-field and dark-field images decreases in the thicker part of the crystal because of the mean absorption (Chap. 2).

In a bent crystal of uniform thickness, bend contours that resemble a LACBED zone-axis pattern are observed near the crystal area near the zone-axis orientation. Bend contours are observed in samples that are under stress. In high-resolution imaging, they offer a means of locating the area that is near a zone axis. Figure 15.9

Fig. 15.8 Thickness fringes observed at the grain boundaries of polycrystalline AlN (courtesy of Changqiang Chen)

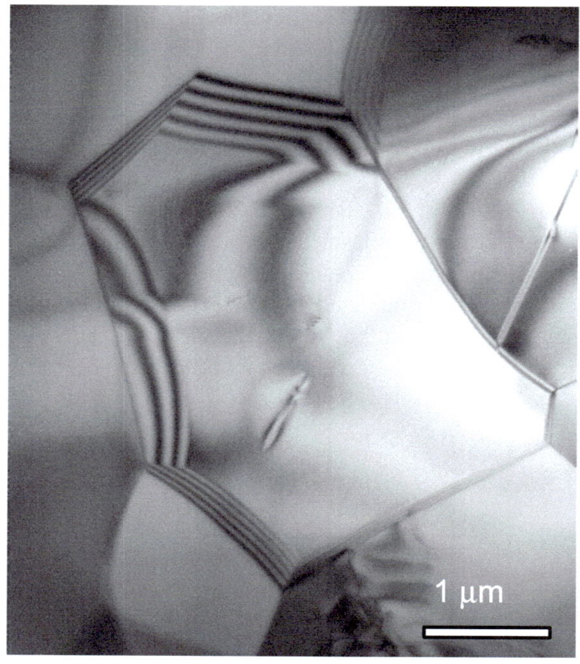

Fig. 15.9 A schematic diagram of bent contour formation in bright-field TEM. The image shown below is recorded from a silicon crystal (after Jean-Paul Morniroli, France)

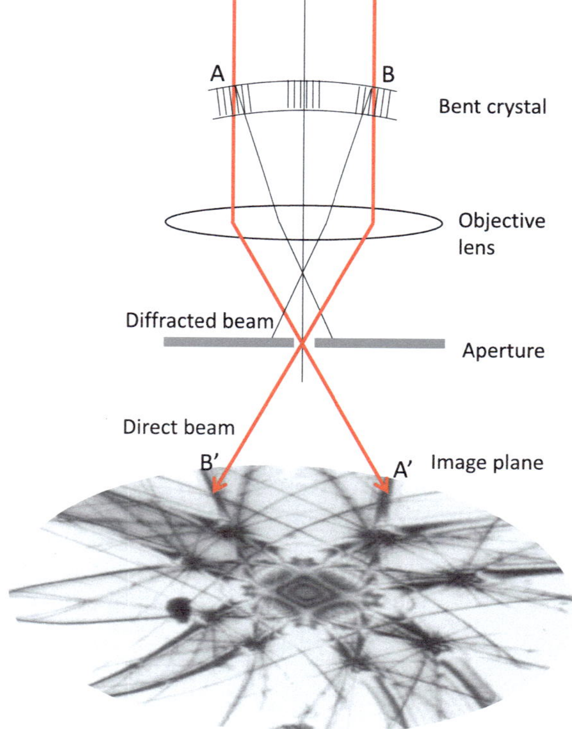

shows an example. Because of the change of excitation error due to the change in the crystal orientation, the ω in Eq. (15.8) varies, and extinction occurs at

$$\omega = \pm\sqrt{\left(n\xi_g/t\right)^2 - 1}.$$

This applies to both $+g$ and $-g$ diffraction conditions. Thus, the bend contours occur in pairs just like Kikuchi patterns. Near the zone axis, where many beams are excited, the intensity is predicted by the many-beam theory as in CBED. The diffraction conditions of a bent crystal are illustrated in Fig. 15.9. At the center of bend contours, $S_g < 0$, which becomes more positive along the direction of g and more negative in the opposite direction. As S_g increases, Bragg diffraction, e.g., $S_g = 0$, occurs at the sample position A. The exact distance depends inversely on the sample curvature. For the $-g$ reflection, the direction of change is opposite and $S_{-g} = 0$ occurs on the other side of the center at the position B. The BF image as shown in Fig. 15.9 records the intensity of the transmitted beam, whereas only the intensity of the selected reflection is recorded in a DF image. The BF image is symmetrical, while the DF image is asymmetrical about the center. By comparing the recorded BF and DF images, it is possible to determine whether the bend is upward or downward.

15.3.3 Diffraction Contrast from Lattice Defects

Defects such as stacking faults and dislocations interrupt the crystal lattice and introduce local atomic displacements and strain fields. To deal with diffraction contrast imaging of strain fields, here we first introduce the kinematic theory of diffraction for crystals with defects (Hirsch et al. 1960) and then describe the scattering matrix method for multiple beam diffraction.

Starting with the Howie–Whelan equation (Eq. 5.16), we have for the diffracted beam

$$\begin{aligned}
\frac{d\phi_g(z)}{dz} &= i\pi\lambda U_o\phi_g(z) + i\pi\lambda \sum_{h\neq g} \phi_h(z)U_{gh}(z)\exp\left(2\pi i\left(\vec{S}_h - \vec{S}_g\right)\cdot\vec{r}\right) \\
&\approx i\pi\lambda\phi_o U_g(z)\exp\left(-2\pi iS_g z\right),
\end{aligned}$$

$$(15.9)$$

here the approximation is made for kinematic diffraction with $|\phi_o| \gg |\phi_h|$ or $|\phi_g|$. The z-axis is taken to be along the column direction. In a deformed crystal, at the depth of z, the crystal is displaced by $\vec{R}(z)$. The displacement introduces an additional phase to the electron structure factor in the form of

$$U_g(z) = U_g \exp\left(-2\pi i \vec{g} \cdot \vec{R}(z)\right).$$

By combining the above results together and using $\xi_g = 1/\lambda U_g$ and $\phi_o = 1$, we obtain the following equation

$$\frac{d\phi_g(z)}{dz} = i\frac{\pi}{\xi_g}\exp\left\{-2\pi i\left[\vec{g} \cdot \vec{R}(z) + S_g z\right]\right\}. \tag{15.10}$$

Under the column approximation, the intensity under the column at x in a real crystal is equal to the intensity of the Bragg beam diffracted by an equivalent crystal in which the strain is given by $\vec{R}(z)$ for all x. Only horizontal shearing is permitted in the equivalent crystal, and there is one such equivalent crystal for each image point.

Equation (15.10) provides an important rule that in order to image lattice defects the scalar product $\vec{g} \cdot \vec{R}(z)$ must be nonzero. When $\vec{g} \cdot \vec{R}(z) = 0$, the crystal diffracts like the perfect crystal as if the atomic displacements are nonexistent. This criterion breaks down in the presence of multiple beam scattering involving a nonsystematic reflection or reflections. In practice, to observe a defect, the magnitude of $\vec{g} \cdot \vec{R}(z)$ must be sufficiently large and its accumulated effect on the diffraction intensity must be above the background noise level. For example, for a stacking fault, a rule of thumb is that if $\vec{g} \cdot \vec{R} \leq 1/3$, there is no visible contrast associated with $\vec{R}$.

Equation (15.10) can be used to calculate the image contrast from defects such as dislocation loops, small precipitates, stacking faults, domain boundaries, grain, and interphase boundaries. The result is valid under the weak-beam imaging condition. Under strong diffraction conditions, dynamic theory must be applied in order to account for multiple scattering effects. Nevertheless, the kinematic theory, Eq. (15.10), explains the essential features observed from these defects.

Let us consider the cases of a screw dislocation and a stacking fault in the diffraction geometry of Fig. 15.10. In both cases, the solution of Eq. (15.10) can be obtained analytically. First, along column 1, $\vec{g} \cdot \vec{R}(z) = 0$, thus

$$\phi_g(z) = i\frac{\pi}{\xi_g}\int_{-t/2}^{t/2} e^{-2\pi i S_g z}dz = i\frac{\pi}{\xi_g}\frac{\sin\left(\pi S_g t\right)}{\pi S_g}, \tag{15.11}$$

where we have to take the origin of the z coordinate ($z = 0$) at the middle of the thin foil. Taking the square of Eq. (15.11) gives the diffracted beam intensity

$$I_g(z) = \frac{\sin^2\left(\pi S_g t\right)}{\left(S_g \xi_g\right)^2}. \tag{15.12}$$

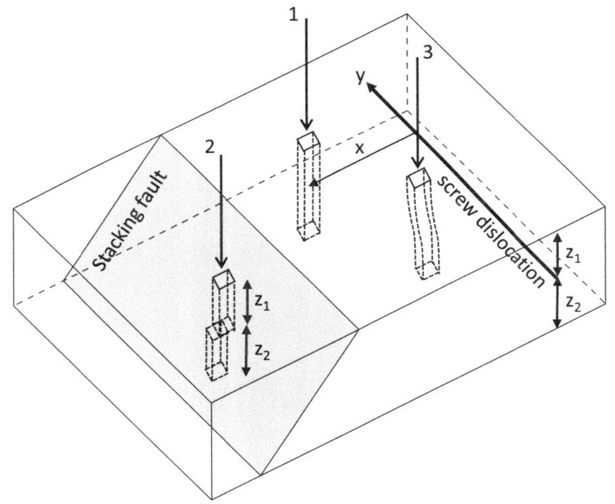

Fig. 15.10 The electron diffraction geometry of a thin crystal of thickness t containing a stacking fault and a screw dislocation. Three columns are marked as *1*, *2*, and *3*. The column *1* is in an area of sample unaffected by the defects. In columns *2*, the crystal lattice in the lower and upper parts of the column is displaced by the vector $\vec{R}$, while the column *3* shows the continuous displacement in the presence of a screw dislocation

This result is equal to the two-beam theory (Eq. 15.8) in the limit of $|\omega| = |S_g|\xi_g \gg 1$ or $|S_g| \gg 1/\xi_g$ or at very small thickness with $|S_g| \sim 0$ and

$$I_g(z) \approx \pi^2 t^2 / \xi_g^2$$

where $t \ll 1/\pi\xi_g$ (which together with $|S_g| \gg 1/\xi_g$ define the limits of the kinematic approximation). Since diffraction contrast imaging is typically carried out in thick crystals, the large excitation error condition is more appropriate. This condition is approximately obtained in weak-beam imaging.

Along column 2 in Fig. 15.10, which intersects with a stacking fault, we have

$$2\pi\vec{g} \cdot \vec{R}(z) = \begin{cases} 0 & -t/2 \leq z < z_1 \\ \alpha & z_1 \leq z \leq t/2 \end{cases}, \tag{15.13}$$

and

$$\phi_g(z) = i\frac{\pi}{\xi_g} \int_{-t/2}^{z_1} e^{-2\pi i S_g z}\mathrm{d}z + i\frac{\pi}{\xi_g}e^{-i\alpha} \int_{z_1}^{t/2} e^{-2\pi i S_g z}\mathrm{d}z. \tag{15.14}$$

The possible values of α depend on the crystal structure. In a fcc crystal, the {111} planes are close-pack planes with the stacking sequence ABCABC... (Sect. 12.2). A stacking fault is formed if one part of the crystal is shifted from one of the two interstitial sites to another, corresponding to the displacement vector $\vec{R} = a\{112\}/6$. For example, a displacement of $\vec{R} = a\{1\bar{1}2\}/6$ moves a B layer into a C layer and the new sequence is ABCA|CABC..., where the line marks the stacking-fault position. The stacking fault can also be formed by removing a layer and by adding an additional layer; they are known as intrinsic and extrinsic stacking faults, respectively. For an fcc crystal, the Miller indices of the diffracted beam (hkl) must be all even or all odd. This gives a phase shift of $\alpha = 0$ or $2\pi n/3$. Only the values $\alpha = 0$, $\pm 2\pi/3$ lying between $-\pi$ and π need be considered. With $\alpha = 0$, the stacking fault is invisible because Eq. (15.14) becomes the same as Eq. (15.11) without the stacking fault. In a hexagonally packed crystal and an alloy with anti-phase boundaries (Chap. 11), $\alpha = \pi$ has to be considered as well.

The integration of Eq. (15.14) can be carried out directly, which gives

$$
\begin{aligned}
I_g(z) = \left|\phi_g\right|^2 = \frac{1}{\left(\xi_g S_g\right)^2}\Bigg[&\sin\left(\pi t S_g + \frac{\alpha}{2}\right) + \sin^2\left(\frac{\alpha}{2}\right) \\
&-2\sin\left(\frac{\alpha}{2}\right)\sin\left(\pi t S_g + \frac{\alpha}{2}\right)\cos\left(2\pi S_g z\right)\Bigg]
\end{aligned}
\tag{15.15}
$$

where $z = 0$ at the middle of the thin sample. Among the three terms inside the square bracket, only the last term depends on z in the form of a cosine function. Thus, for a stacking fault that runs obliquely across the sample as shown in Fig. 15.10, where the fault position (z) changes from $-t/2$ to $t/2$, Eq. (15.15) gives a set of parallel fringes whose direction is determined by the intersection of the stacking fault with the sample surface, as observed in Fig. 15.11. The contrast of these fringes is symmetrical according to the kinematic theory. They are separated by a period of $1/s$. Near the Bragg condition as in the case of Fig. 15.11, S_g is replaced by $S_g^{eff} = \sqrt{S_g^2 + 1/\xi_g^2}$ in the two-beam theory. At the Bragg condition, $S_g^{eff} = 1/\xi_g$ and the fringes are observed only when the sample thickness is greater than a few extinction distances. The number of fringes increases with increasing foil thickness or $\left|S_g\right|$. At large $\left|S_g\right|$, the contrast difference between the dark and bright fringes is less as determined by the factor of $1/S_g^2$ in Eq. (15.15). The contrast extinction is observed at specific thicknesses or tilts for which the term $\sin\left(\pi t S_g + \alpha/2\right)$ becomes zero. Extinction occurs at all thicknesses and all tilts when $\alpha = 0$. Additionally, stacking faults parallel to the sample surface give no fringe contrast since z is constant. In Fig. 15.11, the surface normal is [111]; thus, the observed fringes belong to the stacking faults along $[\bar{1}11]$, $[1\bar{1}1]$ or $[11\bar{1}]$. Among these, the stacking faults with the vector $\vec{R}$ equal to $\pm[112]/6$ and $\pm[11\bar{2}]/6$ are extinct when observed with $\vec{g} = (2\text{−}20)$. The direction of the displacement vector $\vec{R}$ can be determined by finding two different g, with help of sample tilting, at which the stacking-fault contrast disappears.

The integration of Eq. (15.14) can be also carried out graphically using the amplitude and phase diagram illustrated in Fig. 15.12, which provides a direct visualization of the solution (Hirsch et al. 1960). The contribution to the diffracted wave by a thin slice of the column lying between depths z and $z + dz$ has an amplitude proportional to dz and a phase angle $2\pi S z$. This complex quantity is represented by a small vector in the amplitude-phase diagram. The length of the vector is constant, while its angle increases linearly, as the slice position changes

Fig. 15.11 *Bright* and *dark-field* images of stacking faults in a [111] silicon single crystal deformed under hydrostatic pressure recorded using a JEOL2010F TEM operated at 200 kV and on Imaging Plates. The *inset* shows the diffraction condition used to record the images. The sample was provided by Jerome Pacaud, University of Poitier, France

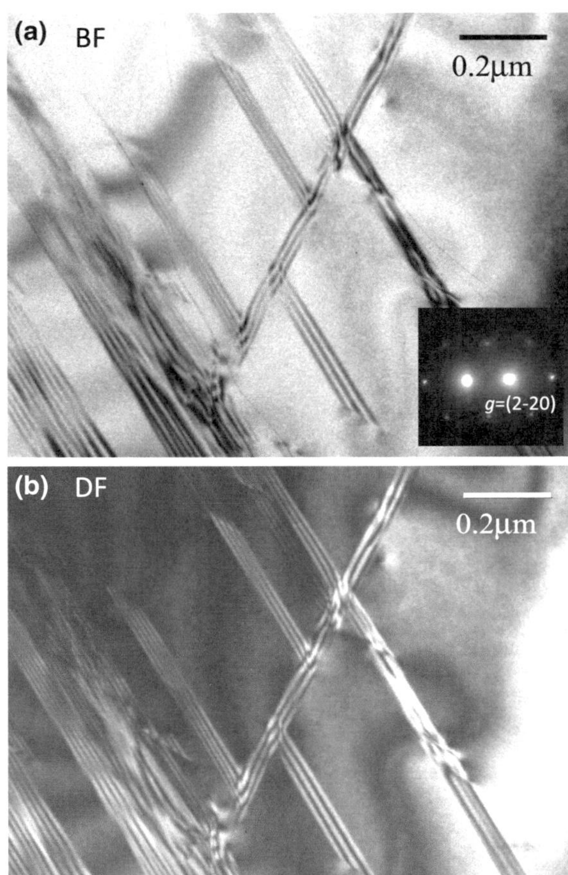

from the top to the bottom of the column. The sum of all vectors follows a circle of radius $(2\pi S)^{-1}$. If we choose the zero phase to be at the middle of the column for the coordinates that we have adopted, the complex wave diffracted by the top and bottom halves of the crystal are represented by the vectors of PO and OP' in Fig. 15.12. Their sum gives PP', whose length gives the amplitude of the diffracted wave. As the thickness t varies, PP' will oscillate between 0 and $(\pi S)^{-1}$ for a perfect crystal. In the presence of a stacking fault, the change of phase by α takes the integration on the initial circle for the top part of the column to the final circle for the bottom part of the column. The change occurs at the point Q on the initial circle corresponding to the depth at which the column intersects the fault plane and an abrupt change of phase of α ($-120°$ as illustrated in Fig. 15.12) occurs. Thus, we have the top part and bottom part of the column above and below the fault contributing PQ and QP' to the diffracted wave, respectively. Their sum gives the diffracted wave amplitude corresponding to the length of PP'', which is very different

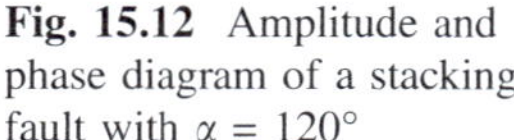

Fig. 15.12 Amplitude and phase diagram of a stacking fault with $\alpha = 120°$

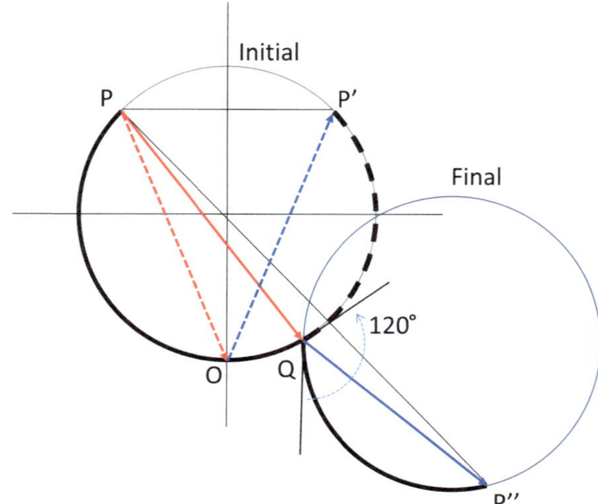

from PP', the amplitude from the perfect crystal. For a stacking fault that intersects the sample surface at an oblique angle, we have P fixed and both Q and P'' and thus PP'' varying as the column intersects the fault at different positions.

A dislocation, when viewed from its side, has the atoms above and below displaced by an amount equal to half the Burgers vector, e.g., $\vec{b}/2$. Thus, in a thin crystal, if the column passes through the center of the dislocation an abrupt change of phase occurs, and the resulting contrast for this column is similar to that for a stacking fault. Away from the dislocation center, the column is continuously deformed and consequently the phase angle $\alpha = 2\pi\vec{g} \cdot \vec{R}(z)$ is a continuous function of position. The atomic displacements are in opposite directions on the two sides of the dislocation. Because of this, the displacement phase $\alpha = 2\pi\vec{g} \cdot \vec{R}(z)$ adds to the thickness phase $(2\pi S_g z)$ on one side of the dislocation and subtracts from it on the other side. The addition of phase takes the crystal column further away from Bragg diffraction condition, while phase subtraction brings the column nearer to the reflecting condition. Since the amount of displacement falls off away from the dislocation, the dislocation effectively creates a locally strong diffraction condition on one side of the dislocation. Which side depends on the sign of α and S_g. Therefore, contrast is expected at the center and one side of the dislocation. This observation is general and applicable to dislocations of all types.

Figure 15.13 provides an example of diffraction contrast imaging of dislocations in silicon (Fig. 15.14 is an example of heavily deformed W). The reflection used for imaging is $(02\bar{2})$ in an off-Bragg diffraction condition tilted away from the [111] foil normal direction (see inset diffraction pattern in Fig. 15.13). Dislocations that run approximately parallel to the sample surface appear as dark lines in the BF image and bright lines in the DF image, while inclined dislocations show the dotted-line or zigzag contrast. These contrasts are observed in sample areas of thicknesses greater than the extinction distance $(t \gg \xi_g)$. Figure 15.13 also

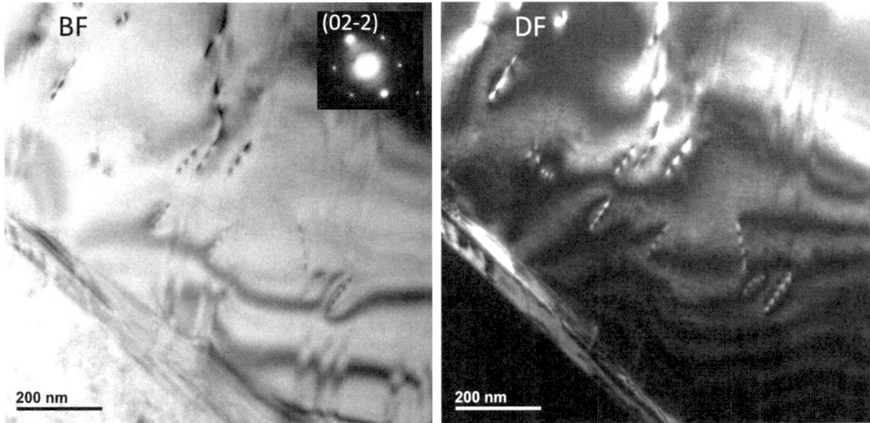

Fig. 15.13 *Bright* and *dark* field of dislocations in a thin deformed silicon sample imaged in a 200 kV TEM. Dislocations parallel to the surface show dark and bright contrast in the BF and DF images, respectively, inclined dislocations show complementary alternating contrast

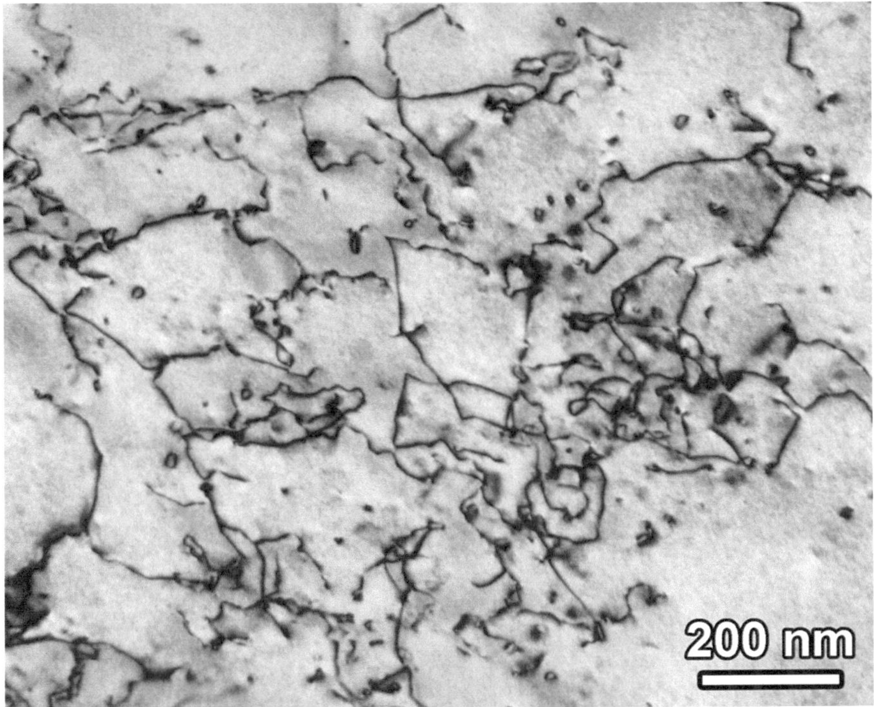

Fig. 15.14 *Dark*-field image of dislocations in a heavily deformed tantalum foil (courtesy of Changqiang Chen)

provides a striking example of the complementary contrast between the BF and DF images. In general, the contrast observed strongly depends on the diffraction condition; noncomplementary contrast (e.g., both BF and DF images show dark contrast) is observed when another reflection is operating.

The contrast observed using a reflection with large excitation errors can be largely accounted for using the kinematical theory, whereas the dynamical theory is necessary near the Bragg diffraction condition. To calculate diffraction contrast from a dislocation, the local displacement vector must be known. Such information can be obtained approximately as the sum of displacements from small straight segments (Saldin and Whelan 1979), whose solution is known analytically (Hirth and Lothe 1983). Here, we will consider the simplest case of a screw dislocation parallel to the surface of a thin foil and at distances of z_1 and z_2 from the top and bottom surfaces. The screw dislocation is formed by crystal shear, the crystals above and below the shear plane is displaced by the Burger's vector ($\vec{b}$), while the edge of the shear plane defines the core of the screw dislocation. The Burger's vector is parallel to the unit vector ($\vec{u}$) along the dislocation line. In an isotropic media, a circle centered on the dislocation line, on a lattice plane perpendicular to the dislocation line and starting from the shear, is continuously displaced and finishes after a 2π rotation with a shift of $\vec{b}$. The displacement vector of a screw dislocation is then given by

$$\vec{R} = \vec{b}\,\frac{\alpha}{2\pi} = \frac{\vec{b}}{2\pi}\arctan\left(\frac{z}{x}\right), \tag{15.16}$$

and

$$\alpha = 2\pi\vec{g}\cdot\vec{R} = \vec{g}\cdot\vec{b}\,\arctan\left(\frac{z}{x}\right) = n\,\arctan\left(\frac{z}{x}\right). \tag{15.17}$$

In general, the Burger's vector of dislocations is associated with lattice vectors; the most stable ones are those with shortest lattice vectors, and they are called perfect dislocations. On the other hand, the Schockley partial dislocations observed in the fcc crystals have the Burger's vector of $\{112\}a/6$, which does not belong to the fcc lattice. Interaction of perfect dislocations with stacking fault in a fcc crystal leads to the decomposition of a perfect dislocation into two Schockley partials. For a perfect dislocation, $\vec{g}\cdot\vec{b}$ is an integer since $\vec{g}$ is a reciprocal lattice vector. The integer may take positive and negative values and zero. The latter case ($n = 0$) is obtained when $\vec{g}$ and $\vec{b}$ are perpendicular to each other. When this happens, the dislocation is invisible for a screw dislocation. In diffraction contrast imaging, low-order reflections are typically used to produce good contrast. Because of this, the typical values of n are small. For example, for $\vec{g} = (220)$ and $\vec{b} = [101]/2$, $n = 1$ and $n = 2$ for $\vec{g} = (202)$.

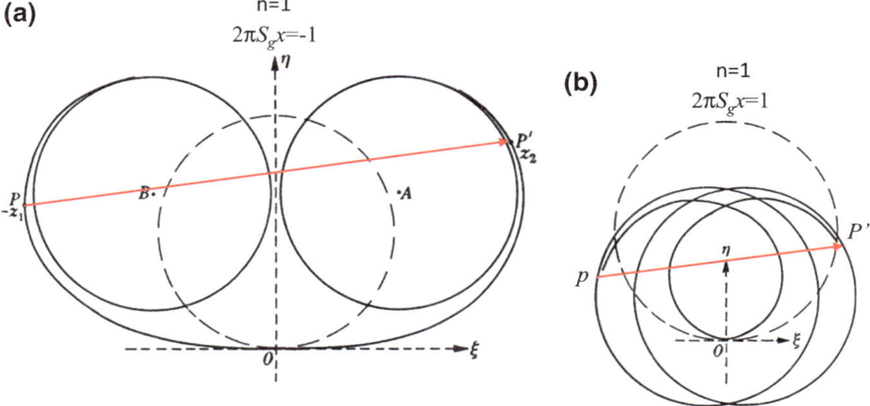

Fig. 15.15 Amplitude-phase diagrams for the column 3 close to a screw dislocation (Fig. 15.10). Two cases are illustrated here with $n = 1$: **a** $2\pi S_g x = -1$ and **b** $2\pi S_g x = 1$, respectively. The separation between the points P and P' measured along the curves is equal to the crystal thickness (from Hirsch et al. 1960)

Substituting Eq. (15.17) into Eq. (15.10) gives the following integral

$$\phi_g(z) = i\frac{\pi}{\xi_g} \int_{-z_1}^{z_2} e^{-i\left(2\pi i S_g z + n\arctan(z/x)\right)}\,dz = i\frac{\pi}{\xi_g} \int_{-z_1}^{z_2} e^{-i\varphi}\,dz. \tag{15.18}$$

The integration of Eq. (15.18) is illustrated in Fig. 15.15 using the amplitude-phase diagram for $n = 1$. Two cases are illustrated corresponding to (a) $2\pi S_g x = -1$ and (b) $2\pi S_g x = 1$, respectively. The dashed circles in Fig. 15.15a, b represent the perfect crystal case with the radius of $(2\pi S)^{-1}$. The point O corresponds to $z = 0$, at which the dislocation is located. In case of $S_g x < 0$, the defect phase subtracts from the thickness phase and consequently the curvature of the amplitude-phase diagram is reduced and less than $1/r$ (r radius or the circle). For a thick crystal, at the top of the column, the crystal is nearly perfect and the amplitude of the diffracted wave from this part of the crystal is asymptotic to a circle of radius $(2\pi S)^{-1}$ centered at B in Fig. 15.15a. Similarly, diffraction from the bottom of the crystal approaches the asymptotic circle centered at A. The complete amplitude-phase diagram shown in Fig. 15.15a is a double-spiral curve connecting the two asymptotic circles. In case of $S_g x > 0$, the situation is reversed with the defect phase adding to the thickness phase. The diffracted wave amplitude ϕ_g is proportional to the length L_{PP}, while curve length from z_1 to z_2 equals to the crystal thickness t. The intensity of the diffracted wave is $I_g = (\pi/\xi_g)^2 L_{PP'}^2$. There are two contributions to L_{PP}—a steady part coming from the distance of BA and a oscillating part depending on the exact depth position of the dislocation and the crystal thickness. Both contributions to the diffracted wave amplitude also change with

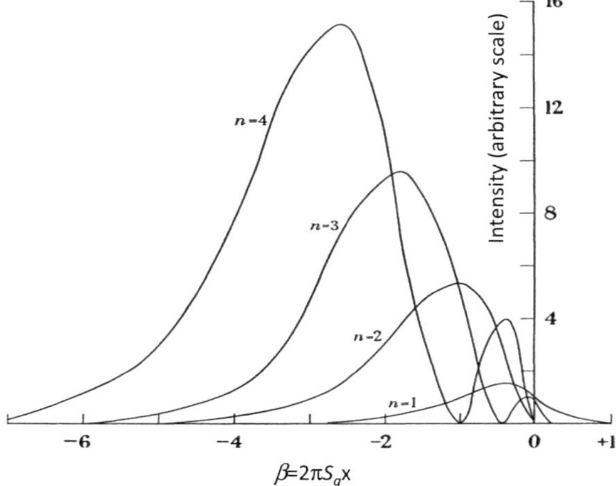

Fig. 15.16 Intensity profiles of images of a screw dislocation plotted for various values of *n* as a function of the normalized dislocation coordinate β. The dislocation is centered at $x = 0$ (from Hirsch et al. 1960)

$$\beta = 2\pi S_g x$$

horizontal position x and the excitation error S_g. These oscillations tend to average out in an experimental image. Thus for practical applications, the steady part of L_{PP}. is used to represent the average kinematic image intensity. This is calculated and plotted in Fig. 15.16 for various values of n across a screw dislocation.

Next we consider the case of an edge dislocation. In a coordinate of $\vec{x} \| \vec{b}$ and y parallel to the dislocation line, the displacements due to an edge dislocation in an infinite medium are

$$R_x = \frac{b}{2\pi} \left[\arctan\left(\frac{z}{x}\right) + \frac{xz}{2(1-v)(x^2+z^2)} \right]$$

$$R_z = -\frac{b}{2\pi} \left[\frac{1-2v}{4(1-v)} \ln(x^2+z^2) + \frac{x^2-z^2}{4(1-v)(x^2+z^2)} \right]. \tag{15.19}$$

$$R_y = 0$$

Here, v is Poisson's ratio. The slip plane is the xy plane. When it is parallel to the surface of the foil, R_x is the displacement which gives rise to the contrast. For $v = 1/3$, to a reasonable approximation, we have (Hirsch et al. 1960)

$$\alpha = 2\pi \vec{g} \cdot \vec{R} = n \left[\arctan\left(\frac{z}{x}\right) + \frac{3xz}{4(x^2+z^2)} \right] \approx n \arctan\left(\frac{2z}{x}\right). \tag{15.20}$$

Thus, in this case, the contrast is same as for screw dislocations for the same S_g except that the contrast of the edge dislocation is twice as wide than a screw dislocation. When the Burger's vector is normal to the foil, the phase factor for this displacement does not change sign with x. Secondly, most of the contrast effect should come from the $\ln(r)$ term in the phase factor since the second-term changes sign with z. Because of these factors, we expect a symmetrical line contrast relative to the dislocation. The line contrast is also very narrow since the $\ln(r)$ varies extremely slowly at large distances.

In summary, the simple kinematic theory predicts the following results about the diffraction contrast of dislocations:

1. If $\vec{g} \cdot \vec{b} = 0$, no image is produced, and thus, the dislocation is invisible or extinct.
2. The peak contrast of the dislocation is observed at one side of the dislocation core by a distance similar to the peak half-width (Fig. 15.16), which changes with the sign of S_g.
3. For $\vec{g} \cdot \vec{b} > 2$, the kinematic theory predicts the double-peak contrast with the weaker peak placed closer to the dislocation core.
4. The image of an edge dislocation with the slip plane parallel to the foil surface is similar to a screw dislocation, but wider by a factor of about two due to the different strain fields.

The above discussions concern a pure screw or edge dislocation, whose Burger's vector can be determined with the help of extinction condition and the intensity profile of the dislocation. In case of a mixed dislocation, displacements are seen in all directions; then, there is no diffracting vector that can render the dislocation invisible. The Burgers vector can only be found by detailed comparison of calculated profiles with experimental images. Additionally, we have assumed isotropic elasticity so far. The results, including the important extinction rule, do not apply to materials in which elastic anisotropy is important. Again, for such materials a detailed comparison of calculated and experimental images must be made.

15.3.4   Weak-Beam Imaging

The weak-beam imaging technique refers to the use of a low-order diffracted beam with a large excitation error (S_g) for diffraction contrast imaging. This technique allows individual dislocations to be imaged as relatively intense, narrow peaks (1–1.5 nm wide), positioned very close to the dislocation core. Consequently, the resolution of dislocation detail which can be obtained is greatly increased. A demonstration of the improved resolution is the imaging of dissociated dislocations separated by a distance of a few nanometers (also see Fig. 15.17). Under the two-beam diffraction condition, dislocation image widths are of the order of $\xi_g/3$ to $\xi_g/5$. With $\xi_g \sim 50$ nm, the best resolution is ~ 10 nm. Compared to the broad width obtained with the two-beam condition, weak-beam imaging improves to the resolution to such extent to permit imaging of fine dislocation structures in the object, such as narrowly separated partial dislocations.

A qualitative explanation of the weak-beam technique is as follows. Away from the defect, as the crystal is tilted far away from the Bragg position, the intensity in the diffracted beam is very weak, becoming a "weak beam." When the diffraction is

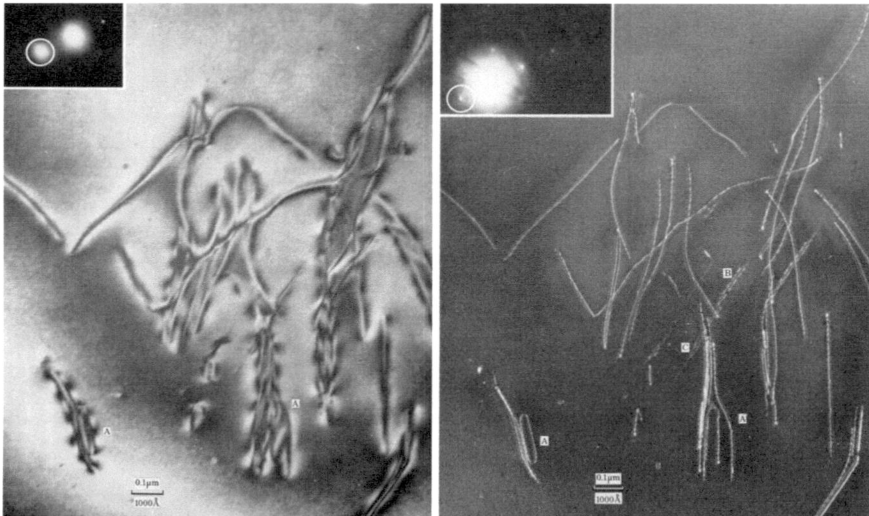

Fig. 15.17 Dislocations in an area of heavily deformed silicon imaged in a strong 220 diffracted beam. The *inset* shows the diffracting conditions used to form the image. A weak-beam 220 dark-field image of the same area as in **a** showing a considerable increase in the resolution of dislocation detail. The *inset* shows the diffracting conditions used to form the image (from Ray and Cockayne (1971), reproduced with permission)

carefully controlled, so there are no other diffracted beam at Bragg diffraction, the scattering of this weak beam can be treated using the kinematic approximation of Eq. (15.10). If the crystal contains a dislocation positioned at z_o, the lattice planes around the dislocation are locally tilted and the displacement can be approximated by

$$\vec{R}(z) \approx \vec{R}(z_\mathrm{o}) + \left[\frac{\partial}{\partial z}\vec{R}(z)\right]_{z=z_\mathrm{o}} (z - z_\mathrm{o}). \tag{15.21}$$

Substituting this result into Eq. (15.10), we obtain

$$\frac{\mathrm{d}\phi_g(z)}{\mathrm{d}z} = i\frac{\pi}{\xi_g}\exp\left\{-2\pi i S_g^{\mathrm{eff}}(z - z_\mathrm{o})\right\}\exp\left\{-2\pi i\vec{g}\cdot\vec{R}(z_\mathrm{o})\right\}\exp\left\{-2\pi i S_g z_\mathrm{o}\right\}.$$

Near the dislocation core, the local lattice rotation becomes large enough and on one side of the dislocation $S_g^{\mathrm{eff}} = 0$, and hence, the lattice planes are locally tilted back into the Bragg position. The intensity scattered locally from the region of crystal that satisfies the Bragg condition is strong. Below the dislocation, the lattice is again far away from the Bragg condition, so the loss of the diffracted intensity Bragg diffraction is small and the dislocation is imaged as a narrow, intense peak against a relatively uniform background. The image peak occurs at

$$S_g^{\text{eff}} = S_g + \vec{g} \cdot \left[\frac{\partial}{\partial z}\vec{R}(z)\right]_{z=z_0} = 0. \tag{15.22}$$

For the screw dislocation, Eq. (15.22) together with Eq. (15.16) gives the predict the image peak at the distance from the dislocation line

$$x = -\frac{\vec{g} \cdot \vec{b}}{2\pi S_g}. \tag{15.23}$$

In comparison, the full kinematical intensity calculation for the $\vec{g} \cdot \vec{b} = 2$ case gives the distance $x = -1/2\pi S_g$, which half of the result in Eq. (15.23). The distance of the image peak to the dislocation decreases as S_g increases. At a large S_g value, the dislocation image comes from crystal regions close to the dislocation core. The discussion so far is based on the argument of diffraction condition only. Experimentally, it has been found that the image peak distance also depends on the sample thickness and the dislocation depth. The change in the image peak distance from these effects is ~ 1 nm. This and the difference from kinematical theory define the limit that Eq. (15.22) can be used to predict the peak position.

The experimental procedures for weak-beam imaging were established by Cockayne et al. (1969), Cockayne (1981). These procedures allow the position of the dislocation core to be determined with an accuracy better than 1 nm under the following conditions:

1. $\vec{g} \cdot \vec{b} \leq 2$. For $\vec{g} \cdot \vec{b} > 2$, kinematical theory predicts 2 or more peaks (see Fig. 15.16 for an example). Thus, it is no longer possible to associate each image peak to a separate dislocation.
2. $S_g > 2 \times 10^{-2}$ 1/Å. This threshold was established in the early experimental studies from a compromise between the image intensity, which falls with $1/S_g^2$, the short distance to the dislocation, which decreases with $1/S_g$ and the dislocation peak width, which also falls off with $1/S_g$. The use of CCD camera and the high brightness gun such as the FEG have significantly improved the lowest image intensity that can be recorded, and thus, a higher value of S_g can be used for weak-beam imaging using new instruments.
3. No other reflection has $S_g \approx 0$, this avoids spurious peaks due to diffraction from an accidental strong reflection at the Bragg condition to the beam used for imaging.

A preferred setup for weak-beam imaging is to use the systematic diffraction condition and with help of Kikuchi lines as illustrated in Fig. 15.18. A first-order reflection g is used for imaging. First, the diffracted beam g is tilted onto the optical axis using the dark tilt, and then, the crystal is rotated, so the beam ng is close to (but not at) the Bragg condition. The excitation error of the g beam is approximately given by

Fig. 15.18 Diffraction condition for weak-beam imaging

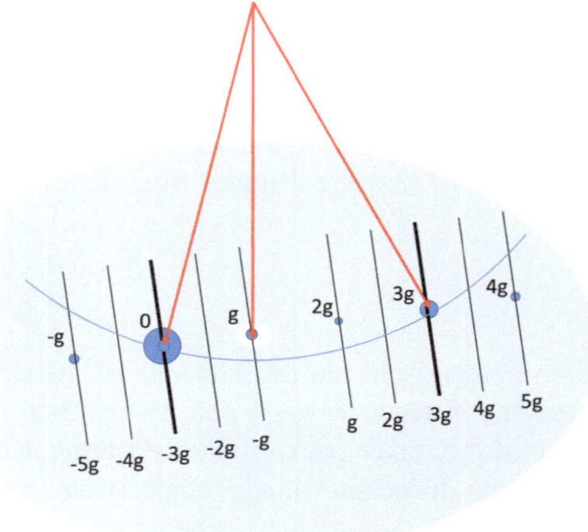

$$S_g = (n - 1)g^2\lambda/2. \tag{15.24}$$

The above diffraction is called $g(ng)$ condition. The example shown in Fig. 15.18 is the so-called $g(3g)$ condition with 3g at the Bragg condition. Take $g = 2.0$ Å^{-1} and $\lambda = 0.025$ Å at 200 kV for example, $S_g = 0.1$ Å^{-1}, which 5 times the minimum value set in the above condition 2. To satisfy the condition (3), $g(3.1g)$ condition is preferred.

Major applications of the weak-beam technique are the measurement of stacking-fault energies, and imaging of small defects, such as dislocation loops and point defect clusters. It has been demonstrated experimentally that the visibility of very small clusters of size <5 nm is usually better under weak-beam diffraction conditions, because of the improved contrast and resolution compared with other imaging conditions. Figure 15.19 is an example resolving the stair-rod dislocation

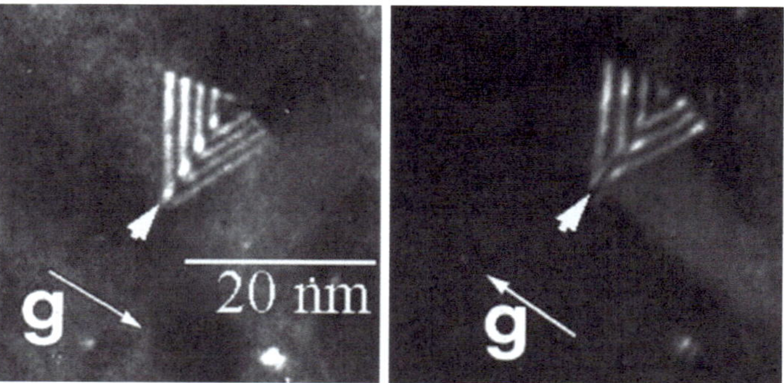

Fig. 15.19 Weak-beam images of stacking fault tetrahedron in quenched gold imaged with $g = \pm(220)$ and $S_g \approx 2 \times 10^{-2}$ Å^{-1}. The *short arrows* identify the contrast arising from the stair-rod dislocation (from Jenkins 1994, reproduced with permission)

in the stacking fault tetrahedral in gold. The improved image resolution also makes it possible to carry out an accurate and direct measurement of the stacking-fault energy γ by measuring the separation of Shockley partial dislocations as a function of dislocation line orientation (Cockayne et al. 1969; Ray and Cockayne 1971). The separation of two image peaks is interpreted using anisotropic elasticity theory; from this, values for the stacking-fault energies of these metals are obtained. Using this technique, the stacking-fault energy has been measured in a large number of materials, for the understanding of the plastic deformation in metals and defect formation in semiconductors.

15.4 Howie-Basinski Equations and the Dynamical Theory of Electron Diffraction from Crystal Defects

So far, we have provided a qualitative explanation of diffraction contrast of defects based on the kinematical theory. This theory is adequate for a number of applications, such as the identification of dislocations and stacking faults, which depends on observation of the diffraction conditions under which they become invisible and these conditions, as well as basic image features, are obtained from the kinematical theory. Applications replying on quantitative measurements of image intensities, in order to extract information about the detailed structure of defects, however, require the dynamical theory. For large defects imaged under the two-beam or similar strong diffraction conditions, the image contrast can be successfully simulated by solving the Howie–Whelan equations under the column approximation. However, the column approximation may fail near the core of defects for weak-beam images, which as discussed before are obtained at much higher spatial resolution. Here, we first introduce the Howie–Basinski equations (Howie and Basinski 1968) and then show how various additional approximations can be made. We finish with the scattering matrix solution to the Howie–Whelan equations for defect scattering under the column approximation. Examples, where dynamic theory is required, include the determination of the atomic arrangements in the dislocation core or in a small dislocation loop, or near clusters of point defects.

The Howie–Basinski equations give approximate solutions of the time-independent Schrödinger equation in the forward-scattering approximation. Similar to the derivation of Howie–Whelan equations in Sect. 5.2, the electron wave function is written as a sum over the diffracted beams $\phi_g(\vec{r})$ in the Bloch wave form

$$\phi(\vec{r}) = \sum \phi_g(\vec{r}) \exp\left(2\pi i \left(\vec{K} + \vec{g} + \vec{S}_g\right) \cdot \vec{r}\right), \tag{15.25}$$

where $\vec{K}$ is the incident electron wave vector. The crystal potential with defects is constructed using the deformable ion approximation

$$U(\vec{r}) = \sum_g U_g \exp\left(-2\pi i \vec{g} \cdot \vec{R}(\vec{r})\right) \exp(2\pi i \vec{g} \cdot \vec{r}), \tag{15.26}$$

with U_g as the Fourier coefficient of the perfect crystal potential and $\vec{R}(\vec{r})$ the atomic displacement. The constructed potential takes account of the modifications in the structure factors by strain, but not composition. Thus, it does not give the structure-factor contrast. The function $\vec{R}(\vec{r})$ can be obtained using elasticity or lattice models, such as the ones used in molecular dynamics simulations. In both cases, $\vec{R}(\vec{r})$ is represented as a continuous field of atomic displacements around the defect. The deformable ion approximation fails near the dislocation core when the field of displacements is calculated using elasticity due to singularities in its solutions. More realistic results obtained using lattice models show that deformations remain smooth and continuous even in the core of small defects (e.g., see Hudson et al. 2004). A practical solution is to eliminate singularities in elasticity by introducing a small regularizing correction (Zhou et al. 2006).

By substituting Eqs. (15.25) and (15.26) into the Schrödinger equation and by neglecting the second-order derivative $\nabla^2 \phi$, we obtain the Howie–Basinski equations of dynamical electron diffraction for crystals with a defect

$$\left(\vec{K}+\vec{g}\right)_x \frac{\partial \phi_g(\vec{r})}{\partial x} + \left(\vec{K}+\vec{g}\right)_y \frac{\partial \phi_g(\vec{r})}{\partial y} + \left(\vec{K}+\vec{g}+\vec{S}_g\right)_z \frac{\partial \phi_g(\vec{r})}{\partial z}$$
$$= i\pi U_o \phi_g(z) + i\pi \sum_{h \neq g} \phi_h(z) U_{gh}(z) e^{2\pi i\left(\vec{h}-\vec{g}\right)\cdot \vec{R}(\vec{r})} e^{2\pi i\left(\vec{S}_h-\vec{S}_g\right)\cdot \vec{r}}. \tag{15.27}$$

Following the example of Zhou et al. (2006), we apply the following gauge transformation to Eq. (15.27)

$$\phi_g(\vec{r}) = \Phi_g(\vec{r}) e^{-2\pi i \vec{g} \cdot \vec{R}(\vec{r})} e^{-2\pi i \vec{S}_g \cdot \vec{r}} e^{-\pi i\left[U_o/\left(\vec{K}+\vec{g}+\vec{S}_g\right)_z\right]z}. \tag{15.28}$$

Since Φ_g and ϕ_g here differ only by a phase factor, the gauge transformation does not affect the amplitude of the wave functions. After transformation, the Howie–Basinski equations become

$$\left(\vec{K}+\vec{g}\right)_x \frac{\partial \Phi_g(\vec{r})}{\partial x} + \left(\vec{K}+\vec{g}\right)_y \frac{\partial \Phi_g(\vec{r})}{\partial y} + \left(\vec{K}+\vec{g}+\vec{S}_g\right)_z \frac{\partial \Phi_g(\vec{r})}{\partial z}$$
$$= 2\pi i\left(\vec{K}+\vec{g}+\vec{S}_g\right)_z S_g^R \Phi_g(z) + i\pi \sum_{h \neq g} \Phi_h(z) U_{gh}, \tag{15.29}$$

where

$$S_g^R = S_g + \frac{\left(\vec{K}+\vec{g}\right)_x}{\left(\vec{K}+\vec{g}+\vec{S}_g\right)_z}\,\vec{g}\cdot\frac{\partial\vec{R}(\vec{r})}{\partial x} + \frac{\left(\vec{K}+\vec{g}\right)_y}{\left(\vec{K}+\vec{g}+\vec{S}_g\right)_z}\,\vec{g}\cdot\frac{\partial\vec{R}(\vec{r})}{\partial y} + \vec{g}\cdot\frac{\partial\vec{R}(\vec{r})}{\partial z}.$$

$$(15.30)$$

Thus, we have the derivatives of $\vec{R}(\vec{r})$ (the distortion field) in the transformed Howie–Basinski equations instead of the displacement field, which appeared in the original equations (Eq. 15.27). In general, the displacement field depends on the choice of reference lattice; thus, it may not be a well-defined quantity. The main benefit for having the distortion field is it falls off from the defect much more quickly than the displacement field.

Under the column approximation, the wave function and the strain are uniform normal to the column direction; thus, the x and y components of $\nabla\Phi_g$ are neglected as well as the derivatives of $\vec{R}(\vec{r})$ in these directions. We then obtain the Howie–Whelan equations for a crystal with a defect

$$\frac{\partial\Phi_g(\vec{r})}{\partial z} = 2\pi i S_g^R \Phi_g(\vec{r}) + i\frac{\pi}{\left(\vec{K}+\vec{g}+\vec{S}_g\right)_z}\sum_{h\neq g}\Phi_h(\vec{r})U_{gh}, \qquad (15.31)$$

with

$$S_g^R \approx S_g + \vec{g}\cdot\frac{\partial\vec{R}(\vec{r})}{\partial z}.$$

The accuracy of the Howie–Whelan equations can be improved for small defects by retaining the x and y components of the derivatives of $\vec{R}(\vec{r})$. Equations (15.31) together with Eq. (15.30) are called the modified Howie–Whelan equations, whose solution has been programmed for several types of defects (TEMACI).

Equation (15.31) can be solved by using the scattering matrix method (Sturkey 1957; Hirsch et al. 1977). The crystal is divided into columns along the diffracted beam direction; each column is then separated into parallel slices. Within each slice, $\Phi_g(\vec{r})$ and S_g^R are constant. The number of slices, n, is selected to give a good sampling of the distortion field along the beam direction. The electron wave function in reciprocal space at thickness t, $\Psi = \left(\Phi_o(t), \Phi_g(t), \ldots\right)^T$ (T for transpose), is related to the incident wave $\Psi_o = (1, 0, \ldots)^T$ through

$$\Psi = S_n S_{n-1}\ldots S_1\Psi_o$$

where S_n stands for the scattering matrix of the nth slice of the imperfect crystal and S can be calculated using the above Bloch wave method (Spence 1992) using

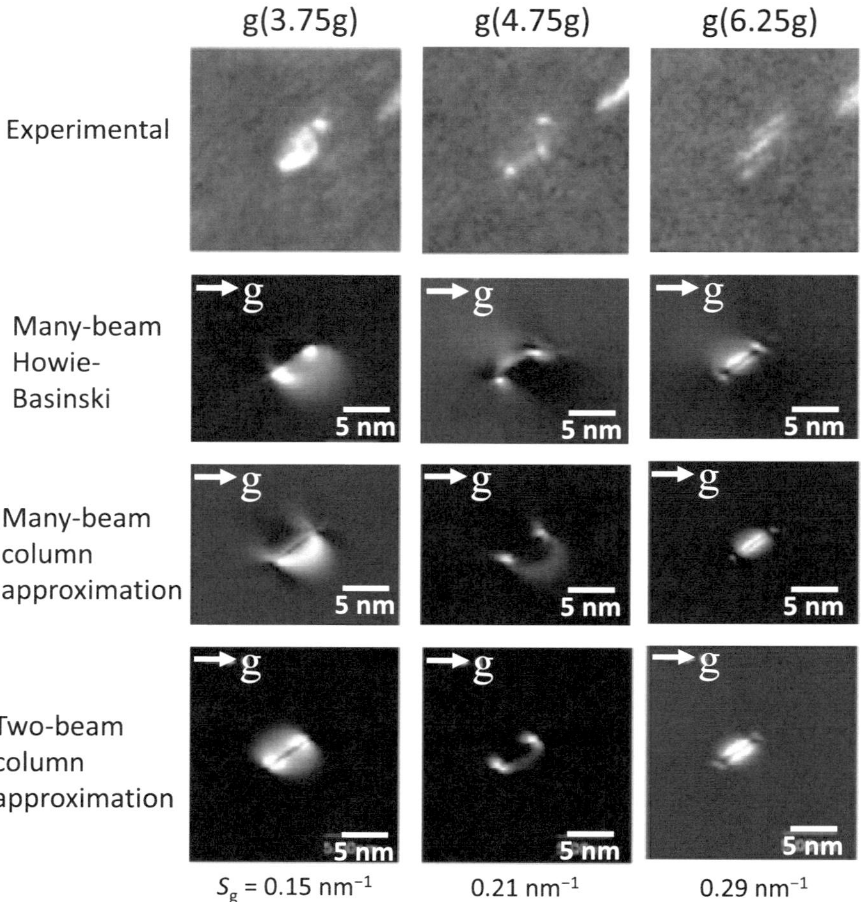

$$S_g = 0.15\ \mathrm{nm}^{-1} \qquad 0.21\ \mathrm{nm}^{-1} \qquad 0.29\ \mathrm{nm}^{-1}$$

Fig. 15.20 Simulated and experimental weak-beam images of dislocation loops in copper viewed edge-on at the electron energy of 100 keV and obtained with $g = (002)$ at three diffraction conditions as labeled at the *top* and *bottom*. Simulations were carried out for the loop diameter 5 nm, foil thickness 60 nm, loop depth 30 nm using the many-beam approximation (including eight beams, from 0 to 7 g), with and without the column approximation, and the two-beam approximation (from Zhou et al. 2006)

$$S = C \Upsilon C^{-1}$$

Here, C is the Bloch wave eigenvector matrix obtained from equation and Υ is the diagonal matrix

$$\Upsilon = \left\{ \exp\left(2\pi i \gamma^i \Delta z\right) \right\}$$

with Δz as slice thickness.

The various approximations described above can be used to simulate diffraction contrast images of any defect as long as its distortion field can be found. Figure 15.20 shows an example of a 5-nm dislocation loop in ion-irradiated copper viewed edge-on under weak-beam diffraction conditions (Zhou et al. 2006). The experimental images shown at the top were taken under well-controlled diffraction conditions (Jenkins et al. 1999). The observed defects had geometries based on Frank loops. For the foil orientation of (110) and diffraction vector $g = (200)$, all Frank loops have $\left| \vec{g} \cdot \vec{b} \right| = 2/3$ and in contrast. The edge-on Frank loops have $\vec{b} = [1\bar{1}1]/3$ and $[\bar{1}11]/3$. The images were simulated using (1) the full Howie–Basinski approach, including eight systematic reflections; (2) as (1) but using the column approximation to solve the modified Howie–Whelan Eq. (15.30) rather than the Howie–Basinski equations; and (3) as (2) but in the two-beam approximation (Zhou et al. 2006). For the simulations, linear, isotropic elasticity theory was used to obtain the distortion fields of dislocation loops using expressions derived from the displacement fields of angular dislocations given by Yoffe (1960).

15.5   Defect Analysis Using LACBED, Defocused CBED, and CBIM

The LACBED technique described in Sect. 10.6.4 using a defocused probe records a projected image of the sample inside the diffraction pattern. Similarly, images are formed in every CBED disks by using a defocused probe on the sample. There are several advantages in using LACBED or defocused CBED patterns for defect analysis (Cherns and Preston 1989; Tanaka et al. 1994). First, the spatial information is obtained by defocusing the image of an extended source, to form a shadow image in every beam. Since information on detailed atomic positions is not sought, the penalty in loss of spatial resolution is unimportant. Secondly, by separating the spatial and orientation information, these techniques also reveal the dislocation line direction, and it is also then possible to determine $\vec{b}$ and its sign. Finally, since effects on HOLZ lines can be seen, one obtains three-dimensional information on the strain field (Carpenter and Spence 1982).

The interpretation of LACBED patterns from defects is based on the analogy with images of dislocations crossing a bend contour (Hirsch et al. 1977). From Fig. 15.21 (also Fig. 10.21) we see that each local region of sample is illuminated from a different direction as a result of the probe focus defect. The simplest case to consider is a systematic row of reflections with a screw dislocation running parallel to this row. Then, $\vec{b}$ and $\vec{u}$ are parallel to $\vec{g}$. We take an image coordinate $\vec{x}$ running normal to $\vec{g}$ and the beam direction. Each point in the sample then defines an incident beam direction and a local value of the (depth-dependent) strain. For a high-order reflection, we may use the two-beam or kinematic theory, together with the column approximation. A single rocking curve calculated for one value of x therefore

Fig. 15.21 LACBED for
defect analysis

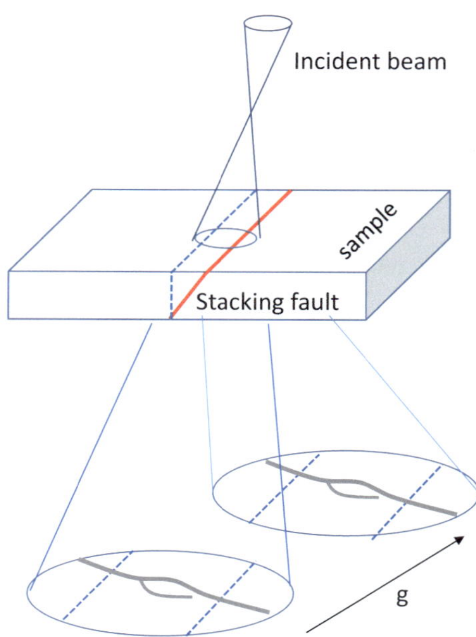

corresponds to a diffraction pattern computed for a probe size about equal to the
column width of the column approximation (typically a few angstroms). The effects
of larger probes can be incorporated using a local average over x.

Using the theory of linear elasticity for the strain field $\vec{R}(z)$ around the defect,
diffraction intensity can be calculated using the kinematic theory for high-order
reflections or dynamic theory in other cases. Figure 15.22 shows the results of
calculations based on two-beam theory. Since many orders are observed simultaneously in CBED patterns, a high-order reflection may be used for which the
two-beam theory is a good approximation. In Fig. 15.22, the dislocation core runs
up the page in the center of each pattern. Thus, for a suitably oriented defect, the
LACBED method can provide a one-dimensional real space shadow image along x,
together with an orthogonal one-dimensional rocking curve at each point x. Note
that the "splitting" of the intensity distribution seen here is not the same effect as
that which occurs on HOLZ lines in perfect crystals due to the degeneracy of the
dispersion surfaces (Sect. 5.4), which occurs on a much coarser scale. In Fig. 15.
22, the effects of using a larger, in-focus probe can be understood by integrating
over x. More extensive calculations and experimental images can be found for
partial dislocations and dislocation loops in Tanaka et al. (1988). The determination
of the direction and sense of **b** from the geometric distortions in CBED patterns are
further discussed by Wen et al. (1989).

We see from Fig. 15.22 that the value of $\vec{g} \cdot \vec{b} = n$ can be determined directly from
the patterns. It is equal to one more than the number of subsidiary maxima between
the dark lines entering from each side. If the sign of $\vec{g} \cdot \vec{b} = n$ is reversed, the effect is
to reflect the patterns about a horizontal central mirror line at $w = 0$ on the diagrams.

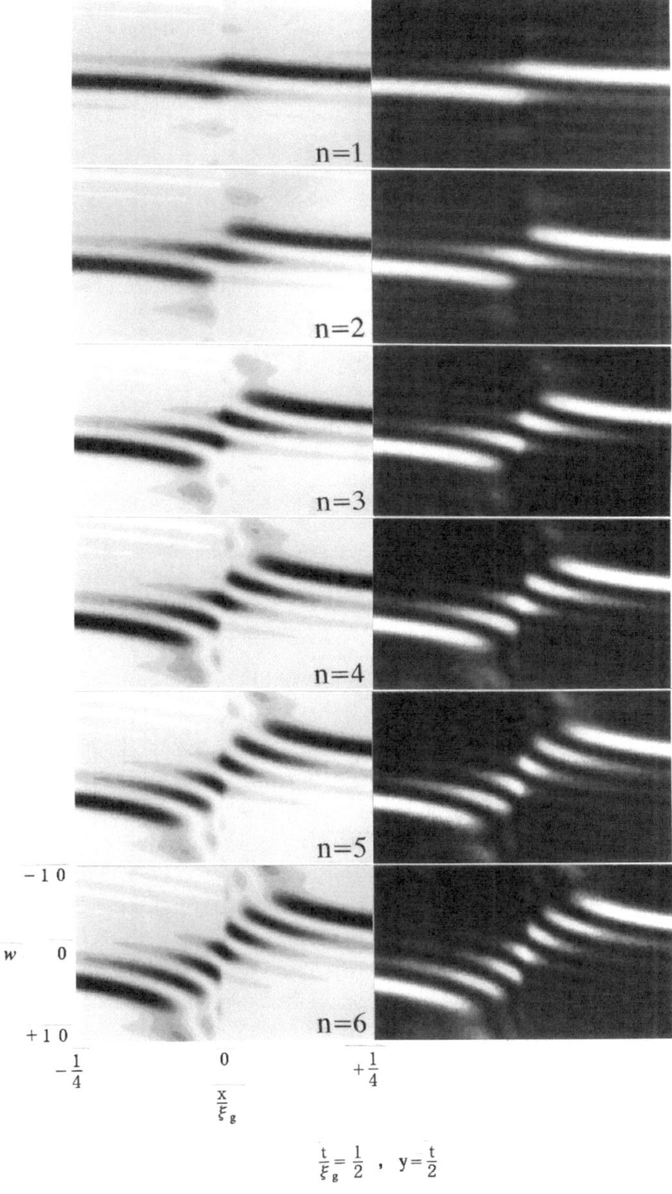

Fig. 15.22 Bright and dark field two calculations for LACBED disks from a crystal containing undissociated screw dislocation. The vertical axis corresponds to the deviation parameter $w = S_g \xi_g$, while the horizontal axis corresponds to distance x from the dislocation, which runs up the page at $x/\xi_g = 0$ (Tanaka et al. 1988)

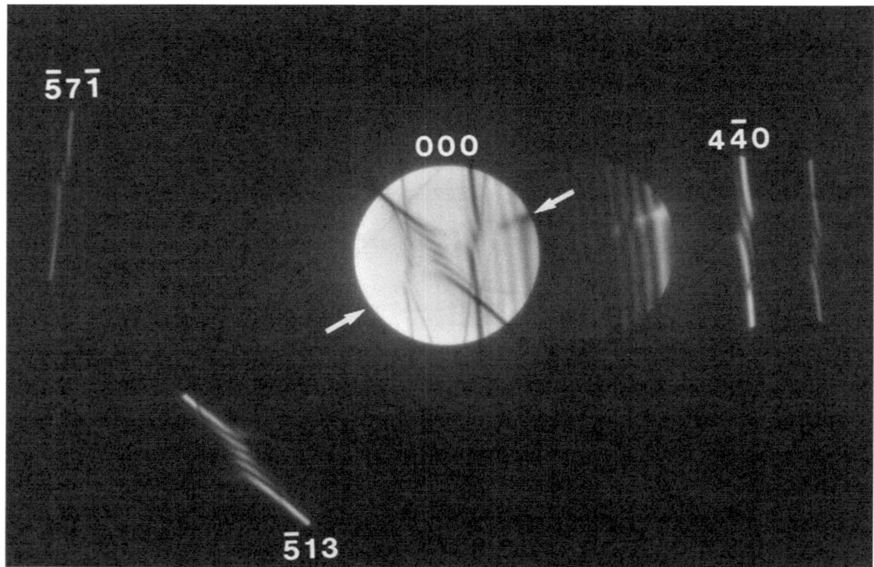

Fig. 15.23 Experimental CBED pattern from silicon containing a dislocation recorded with 200 kV electrons. Probe size about 200 nm (from Tanaka et al. 1988)

Again, contrast is best for a dislocation in the middle of the sample, and the two-beam results given above fit best for thicknesses of less than half an extinction distance— thus, first-order reflections at the Bragg condition are avoided. The $\vec{g} \cdot \vec{b} = n$ rule is found to be insensitive to the depth of the dislocation. If, in Fig. 8.14, we consider the intensity along a vertical line at a constant value of x (corresponding to a fixed position of a very small probe), we see that the position of the Bragg maximum shifts in angle (w) as the probe moves closer to the dislocation core. This is similar to the effect discussed for HOLZ line shifts in Sect. 5.4, although here the effect is solely due to strain and no composition change is involved.

Figure 15.23 shows an experimental pattern from silicon containing a dislocation running in the direction of the arrows. Three noncolinear weak reflections have been excited, and the (000) disk shows the clearest image of the dislocation. (A weaker "reflex" image showing the same dislocation line direction is formed in every order.) Three-beam conditions (where Bragg lines cross) are avoided. Thus, three values of $\vec{g} \cdot \vec{b}$ can be found, which is sufficient to solve for $\vec{b}$. In the case of Fig. 15.23, we have $\vec{g} \cdot \vec{b} = -2, 4$ and 2 for (4−40), (−513) and (−57−1) reflections, respectively, which gives $\vec{b} = (-101)/2$.

Similar principles apply to the study of planar faults and interfaces. Expressions for the diffracted intensity $I_g(S_g, x)$ as discussed before for diffraction contrast imaging may be used directly to interpret LACBED images. A complete two-beam analysis can be found in Wang and Wen (1989), together with experimental LACBED patterns from stacking faults in austenitic stainless steel. For example, conventional single-beam diffraction contrast images show fringes running parallel to the fault trace for an inclined fault, as the intensity varies with fault depth. A horizontal stacking fault therefore produces no intensity variation in an image. In CBED, however, we observe the variation of $I_g(S_g, x)$ with S_g and therefore see fringes from a horizontal fault. For an inclined fault observed by LACBED, we again have an image appearing in every diffracted order. Consider the two-beam case where the trace of the fault (its intersection with the surface) is parallel to $\vec{g}$. Then, one may examine the intensity variation $I_g(S_g, x)$ along $\vec{g}$, giving the rocking curve for constant fault depth. The intensity variation normal to $\vec{g}$ gives the intensity for constant excitation error as a function of fault depth. A study of these rocking curves shows several methods for deducing the fault type from the patterns (Cherns and Preston 1989; Tanaka 1986). The use of HOLZ lines appears to be the best method. A HOLZ line in the (000) disk usually shows a single sharp minimum in perfect crystal. The introduction of a stacking fault causes two minima to appear, one deeper than the other. The deepest minima occur to either side of that which occurs in the perfect crystal, depending on the sign of α. For $\alpha = 2\pi/3$ ($\alpha = -2\pi/3$), the deepest minima lie on the side for which $\omega < 0$ ($\omega > 0$). This rule is independent of the depth of the fault; however, faults at the mid-plane of the sample again give the most pronounced splitting. Since HOLZ reflections have large extinction distances, this rule is also usefully independent of sample thickness, since the required condition $t/\xi_g < 3/5$ will always be satisfied in practice. By noting which lines are split and which are not in the central disk, it is therefore possible to determine $\vec{R}$, as for the dislocation case. Worked examples can be found in Tanaka et al. (1988). By comparison with imaging, the CBED method has the advantage of allowing faults lying parallel to the surface to be solved. A clearer recording of the HOLZ lines is obtained using the hollow cone method, since this removes all ZOLZ detail from the central disk.

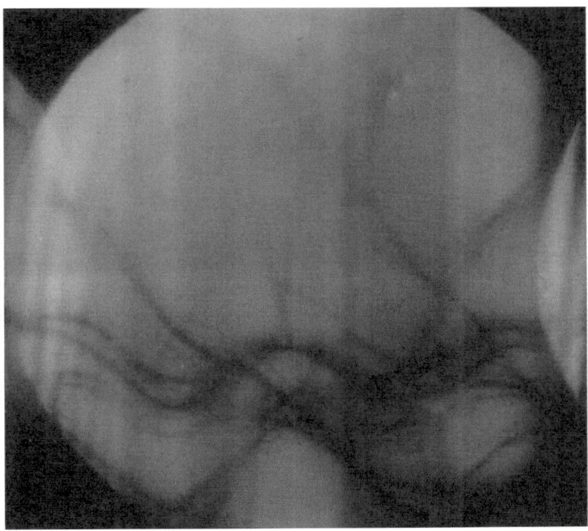

Fig. 15.24 CBIM image of alternating Ge_xSi_{1-x} and Si layers, showing HOLZ line oscillating from strain relaxation

A second CBED shadow image method known as convergent beam imaging, or CBIM, has been developed by Humphreys et al. (1988). We may distinguish this method from the LACBED method in an oversimplified way as follows. In the CBIM method, the microscope is focused onto the specimen plane—that is, the detector is conjugate to the sample. In the LACBED method, the detector screen is conjugate to the diffraction pattern in the back focal plane of the post-specimen lens. In CBIM, since the sample is in focus, the spatial resolution is given by the normal performance of the electron microscope, while the angular width of HOLZ lines depends on the probe size. The angular resolution is estimated to be typically 0.1 mrad. Conversely, in LACBED, the angular width of HOLZ lines is limited only by any diffraction limits imposed by the post-specimen lens (normally negligible) since the diffraction pattern is in focus, while the spatial resolution is limited by the probe size. Thus, there are contributions to a particular HOLZ line from an area at least as large as the defocused probe and extended throughout the thickness of the sample. Experimentally, however, LACBED patterns are usually obtained at some intermediate focus condition, giving compromise performance.

Figure 15.24 shows an experimental CBIM pattern from a series of interfaces between Ge_xSi_{1-x} and Si (Eaglesham 1989). The HOLZ lines are seen to curve as the lattice rotates at the interface, perhaps as a result of strain relaxation. Applications to the Si–SiO_2 interface are also described in this paper. Similar thin-film relaxation of a tetragonally distorted SiGe alloy has been observed by Humphreys et al. (1988) to cause a slowly varying rotation of the [013] zone axis, estimated to be about 0.001 rad. The authors discuss the advantages of cooling samples to improve HOLZ line visibility.

All these shadow-imaging methods have the disadvantage that strains due to composition variations cannot be distinguished from those due to sample bending as a result of elastic relaxation. From a CBIM image alone, it is not possible to distinguish HOLZ line displacements due to lattice parameter variations from those due to bending. (These bending effects are, however, much less than those present in the much thinner samples used for lattice imaging.) It has been suggested that best results are obtained by using a shadow image to locate the probe, followed by in-focus CBED patterns for the strain analysis.

15.6   Atomic Structure Determination of Defects from High-Resolution Electron Images

Crystal defects can be regarded as low-dimensional structures that retain some forms of periodicity and symmetry of the crystal lattice. Point defects are like molecules, involving the modification of chemical bonds of few to tens of atoms. A dislocation is then periodic along specific crystallographic directions. Real dislocations are curved. The curvature is formed by segments of straight dislocations in directions with low Peierls potential connected by kinks or jogs (Fig. 15.25). Kinks lie in the dislocations slip plane, while jogs are out of the slip plane. As an

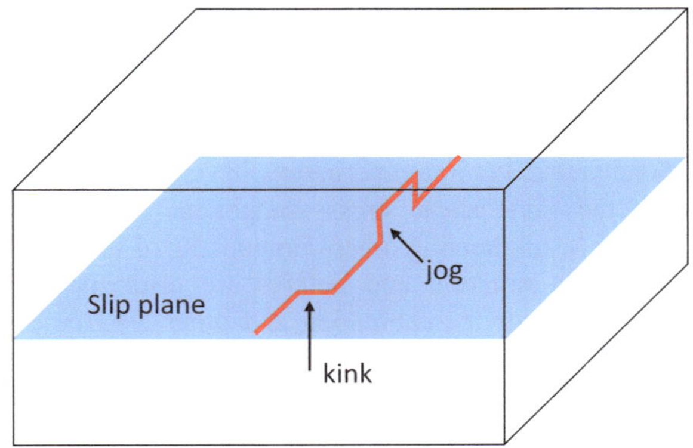

Fig. 15.25 Kink and jog on a dislocation line. A kink causes a shift of the dislocation line. The jogs of opposite sign and the region between the jogs has moved up from the original slip plane

example of 2D defects, a grain boundary is formed by the meeting of two surfaces; each has its own 2D periodicity. In general, a 2D surface can be characterized by its terraces, steps, or ledges and kinks in the so-called TSK or TLK model. The terraces are often low-index surfaces with low surface energy. The overlap of two 2D lattices is the basis for describing high-angle grain boundary structures.

The crystal chemical bonds are disrupted along the dislocation line or at grain boundaries. Because of this, reforming of chemical bonds, which is known as reconstruction in surface science, accompanies the formation of defects as well as their movements. This effect is especially important in covalent semiconductors and ionic crystals involving opposite charges.

The local stress created by defects also drives the migration of impurity atoms near these defects and their segregation to the defect. The characteristics of the segregation depend on the chemical nature of solute atoms. For example, the compressive stress above the dislocation line attracts the smaller substitutional atoms to this region, while interstitial solute atoms tend to migrate to the tensile stressed region below the dislocation line. This interaction between a dislocation and impurity atoms is known as a Cottrell atmosphere, e.g., the formation of almost rod-like clouds of impurity atoms by the segregation along the dislocation line. Similarly, grain boundary segregation can lead to a factor of 2–3 and sometimes orders of magnitude concentration of the solutes on the grain boundary than the grain interior, dependent on the bulk solubility of the solute. The smaller bulk solubility leads to the higher enrichment at the grain boundaries. Dislocation lines and grain boundaries can also act as fast pathways for enhanced mass transport. Fast diffusion along dislocation lines is known as pipe diffusion.

15.6.1   Atomic Structure of Dislocation Cores

The crystallography and dislocation core structure, effects not considered in the continuum elasticity theory, are responsible for the friction that the crystal structure offers to dislocation motion (the Peierls stress), and are indicated directly by several phenomena. These signatures of atomic granularity include the failure of Schmid's law (in which critical resolved shear stress does not depend on slip system), and of a strong temperature and orientation dependence of yield stress and dislocation velocity. Indentation experiments have shown that loops on different glide planes with the same shear stress component on their glide plane travel at different velocities, implicating other components of the stress tensor, contrary to Schmid's law. Core structure and defects are also important in generating deep states responsible for recombination and other electronic properties in semiconductors, with life-time limiting effects on devices. So far we have based our observation of dislocations on their extended strain field in diffraction contrast imaging, for which the continuum elasticity theory suffices. While long-range elastic interactions between dislocations are important in controlling the interaction between dislocations (and store most of the energy), direct evidence of dislocation core structure is needed to address above identified issues. Most of the evidence comes from HREM and STEM, while evidences of Cottrell atmosphere have been provided by atom probe tomography (for example, see Thompson et al. 2007). Questions directly related to the dislocation core structure are

1. What is the atomic structure of kinks and other defects on dislocations?
2. What are the atomic structures of jogs?
3. What are the atomic structures of the common complex extended line defects in compounds, such as intermetallic alloys and oxides? Are there any rules governing these structures?
4. Do atoms reconstruct at the defect, if yes, in what structural forms?
5. Do kinks collide and annihilate?

There are obvious needs for dislocation core structure determination. First foremost, atomic-resolution imaging would allow an observation of dislocation kinks and jogs and thus answering several major questions regarding what limits dislocation velocity at the atomic scale, and do kinks collide. Images of kinks in motion would yield information on pinning energies, and on any correlations between kink motions on different partials. A review of what is known about dislocation core structure, particularly from electron microscopy, is given in Spence (2007).

Dislocations are observed in high-resolution electron microscopy along the end-on or the normal directions along line AA' or BB' as illustrated in Fig. 15.1. Figure 15.26 shows a HREM image of the core structure edge dislocations formed along a 6° [001] symmetrical tilt grain boundary in the fcc aluminum. The image was taken from a section of the grain boundary that appears straight and roughly parallel to the median of the (110) planes of both crystals. A Burgers circuit drawn around the dislocation indicates a Burgers vector $b = [-110]/2$, which is the shortest lattice vector and therefore the most likely Burgers vector for perfect dislocations in the fcc structure. The (−110) planes perpendicular to b have a twofold stacking sequence ABAB... as shown in Fig. 15.26 and the "extra

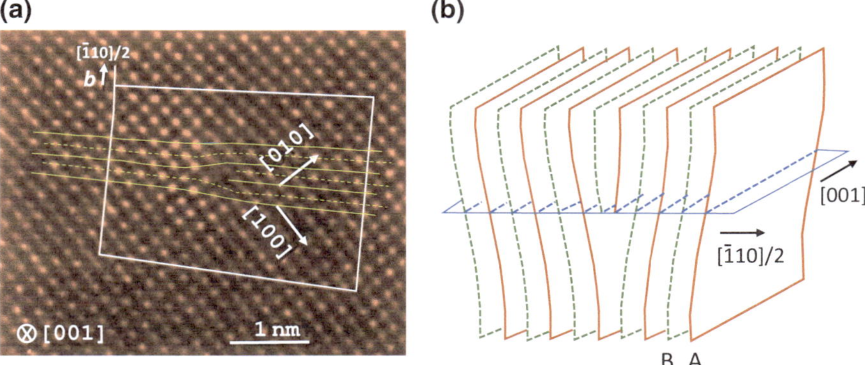

Fig. 15.26 Edge dislocation observed along a 6° [001] symmetrical tilt grain boundary in Al. The sample is an aluminum bicrystal prepared according to details described in (M. Shamsuzzoha et al., Scripta Metall, 24, p. 1611 (1990)). High-resolution electron microscopy was performed with a JEM 4000EX electron microscope operated at 400 kV. All images were obtained and recorded near the optimum defocus and atomic columns are dark (HREM image reproduced with permission from David Smith)

half-plane" of the edge dislocation is made of two (−110) half planes (A and B). This unit dislocation likes to glide on the {111} planes in fcc metals and the dislocation line is typically along <112> directions, rather than <001> . The dislocation glide retains the fcc crystal structure. The fact that this dislocation is observed along [001] implies that it only exist at this portion of the grain boundary.

Compared to edge dislocations, the end-on observation of a screw dislocation gives very weak contrast because in an infinite isotropic solid the displacements are along the atomic column direction parallel to the dislocation line. Such displacements can be only detected using HOLZ reflections and for a very thin sample where HOLZ reflections are weak, the screw dislocation would be invisible when viewed end-on. However, the screw dislocation becomes visible in thin samples where there are displacements normal to the dislocation caused either by surface relaxation or by core dissociation. The displacements caused by surface relaxation for a screw dislocation along z normal to an elastically isotropic foil of thickness t are known as the Eshelby twist. This deformation field is characterized by counter-rotations of the upper and lower parts of the foil, which for an isotropic material is given in polar coordinates (ρ, θ) by Eshelby and Stroh (1951)

$$R_\theta = -\frac{b_z}{2\pi} \sum_{n=0}^{\infty} (-1)^n \left\{ \frac{\rho}{(nt+z) + \left[(nt+z)^2 + \rho^2\right]^{1/2}} - \frac{\rho}{\left(\overline{n+1}t - z\right) + \left[\left(\overline{n+1}t - z\right)^2 + \rho^2\right]^{1/2}} \right\}$$

where b_z is the length of the Burgers vector of the screw dislocation and R_θ is the displacement vector, which is normal to ρ, in the plane normal to z. The displacements are much larger than the core displacements in the bulk. HREM and STEM image simulation studies (Groger et al. 2011; Cosgriff et al. 2010) suggested that the twist causes apparent displacements in the peak positions in the simulated images. However, the displacements obtained from the image peak positions do not reflect accurately the structure at the surface. Thus, in order to obtain information of the core displacements of the screw dislocation in the bulk, quantitative interpretation of experimental images with simulations using structural models is essential. Furthermore, the magnitude of the edge component of the core displacement in the bulk (~ 0.01 nm) is smaller than the currently demonstrated precision to which peak positions have been measured in a typical HREM experiment. For HADDF-STEM, the Z-contrast image of the dislocated crystal consists of an inner relatively bright region one or two atoms wide around the screw, surrounded by a somewhat wider darker region, as seen in the experimental image for a screw dislocation viewed end-on in GaN (Arslan et al. 2006; Lozano et al. 2014).

The most productive recent investigations of dislocation core structures by atomic-resolution microscopy come from the study of compound semiconductors and oxides including complex oxides (Zhang et al. 2002, 2003; Jia et al. 2003; Paulauskas et al. 2014). The improvement in resolution provided aberration correction resolves finer details of dislocation core structures. The benefits of having 1 Å compared to 2 Å resolution in conventional TEM/STEM for the characterization of semiconductors extend much further than a "mere" factor of two increase in resolution might suggest. The most common observation direction for diamond structure is along (110), where the projected atomic columns form dumbbells at distances of 1.36 and 1.41 Å for Si and Ge, respectively. Thus, the improved resolution by aberration correction allows a separation of atomic columns. The smaller probe formed with a larger condenser aperture improves signal-to-noise ratio and consequently improved image contrast.

Two approaches have been developed to image dislocations normal to the dislocation line, one uses forbidden reflections of bulk crystals and the other uses depth-sectioning based on HAADF-STEM (see Fig. 15.2 for a latter example).

The principle of using forbidden reflections for lattice image method is illustrated in Fig. 15.27a. Along the [111] projection of a fcc crystal, we have a hexagonal unit cell containing three atoms at A, B, and C sites at (000), (1/3,1/3,1/3), and (2/3,2/3,2/3) corresponding the ABC stacking of close-packed layers. The structure factor of such unit cell is given by

$$F(h,k,l) = f\left\{ 1 + e^{2\pi i (h+k+l)/3} + e^{4\pi i (h+k+l)/3} \right\}$$

In the zero-order Laue zone ($l = 0$), we have $F(h,k,l) = 0$, if $h + k = 3n \pm 1$ as forbidden reflections. The first-order reflections $\{100\}$ correspond to $\{2\!-\!42\}/3$ of the fcc lattice as shown in Fig. 15.27b, which do not exist. For such forbidden reflections, the structure factor of a perfect crystal column parallel to [111],

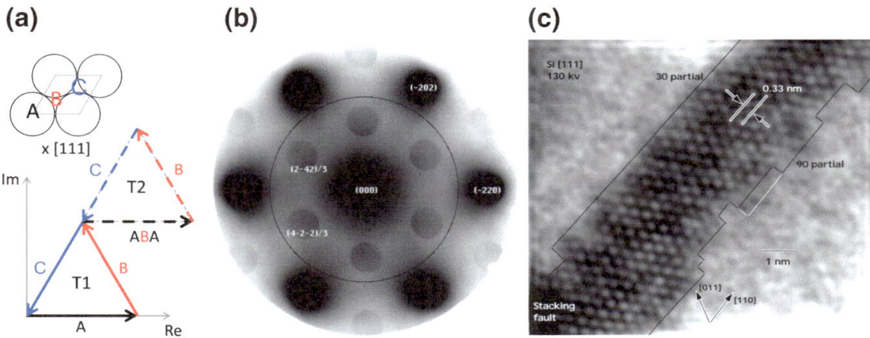

Fig. 15.27   **a** Argand diagram for a {−422}/3 reflection in fcc crystals with Re and Im for real and imaginary axes. **b** Experimental CBED pattern from an intrinsic stacking fault on (111) in silicon lying normal to the beam. **c** TEM image of dissociated 60° dislocation in silicon. The bright diagonal band of regular dots is six-membered rings in the ribbon of SF separating 30° and 90° partial dislocation lines. *Black lines* run along cores of the two partial dislocations. Fine *white line* shows typical alternative boundary used to estimate error in counting kinks. The partial dislocations can be identified by Burger's vector analysis

containing *N* close-packed layers, but not necessarily an integral number of unit cells, may be written as

$$F(h,k,l) = f \sum_{n=0}^{N-1} e^{2\pi i n/3},$$

where the sum follows the triangle *T*1 in the Argand diagram with $n = 3m$, $3m + 1$, and $3m + 2$ (*m* integer) contribute the sides of *A*, *B*, and *C*. Thus, for a perfect crystal column $F(h,k,l) = 0$, only if $N = 3\,m$. $F(h,k,l)$ is also nonzero if the column contains a stacking fault, as shown in Fig. 15.27a, in which case the missing *C* layer in the stacking of *ABA* takes the sum from *T*1 to *T*2, and the resulted amplitude of $F(h,k,l)$ is greater than or equal to *f*. TEM dark-field images were first formed with these reflections by Cherns in (1974), who used them to image monatomic surface steps on (111) gold films Cherns (1974). With the development of ultra-high vacuum transmission electron microscopy, these same termination reflections could be identified in transmission patterns from thin (111) silicon crystals with (7 × 7) reconstructed surfaces Tanishiro et al. (1986) and analyzed (Spence (1983)). They were first observed as additional spots in micro-diffraction patterns from stacking faults in 1986, using a field emission STEM probe narrower than the ribbon of SF separating two partial dislocations (Alexander et al (1986)). Figure 15.27b shows such a pattern, obtained by H. Kolar using convergent beam electron diffraction (CBED). Since the edges of the SF ribbon define partial dislocation cores, an image formed with the inner six of these "termination" or forbidden reflections in the (111) zone provides a lattice image of the SF alone, and its boundary at the dislocation core. The *d*-spacing for the

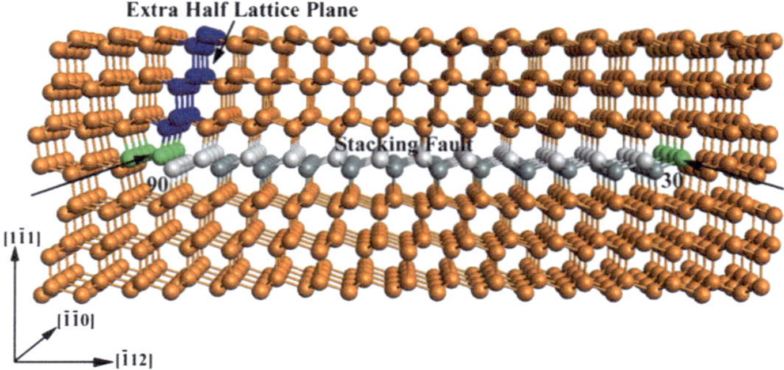

Fig. 15.28 Atomic structure model of a dissociated 60° dislocation in silicon after relaxation. The ribbon of stacking fault is bounded by the 30° and 90° partial dislocation lines and an extra half lattice plane as marked by *arrows* (from Christopher Koch)

"forbidden" planes is $d_{(422)/3} = h = 0.33$ nm, or one Peierls valley wide. These valleys run along the tunnels in the diamond structure, orthogonal to $(42{-}2)/3$. Real crystals have atomically rough surfaces, and these effects must also be considered, in addition to other sources of background in the images, which limit contrast. Large atomically flat surface islands produce sharp forbidden reflections unless $N = 3m$.

Figure 15.27c shows an example image, in which the 30°/90° dislocation partials have moved apart toward their equilibrium separation during an in situ heating experiment Kolar et al. (1996). Image calculations Alexander et al. (1986) show that the bright diagonal band of regularly spaced dots is a lattice image of the double layer of atoms which form the stacking-fault plane. Pairs of atoms appear as a single dark spot, bright spots are centered on the sixfold rings of a single double layer. The borders of this band of regular dots form the partial dislocation cores, as shown in the model of Fig. 15.28. The white scale lines indicate one Peierls valley, 0.33 nm wide, and the average stacking-fault width corresponds to a stress on the partials of 275 MPa. Although the accurate determination of kink density is complicated by the effects of surface roughness, the higher density on one partial (seen also in larger fields of view and in many different cases) suggests that surface effects are not dominant. In addition, monatomic surface islands are not seen outside the stacking fault—these would produce similar (but lower) contrast to the stacking fault. Video recordings at 600 °C showed kink motion (and pinning) at the edges of the lattice image of the stacking fault. The effects of beam induced motion were minimized by turning the beam off during dislocation motion (except during the pinning studies), and using low-dose techniques in conjunction with image plates and CCD recording. From this work, the kink unpinning energy could be found, together with the single kink formation energy (0.73 eV) and the migration energy (1.24 eV). The formation energy was obtained by applying the Hirth $\pm$ Lothe nucleation and growth rate equations to the observed concentration of saddle-point kink pairs under high stress conditions. The conclusion from this work is that, due

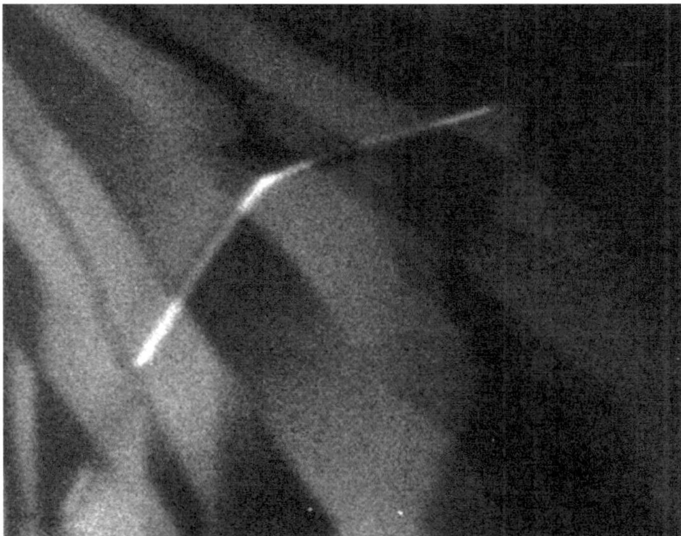

Fig. 15.29 Image from silicon observed along [111]. The sample was prepared in an ultra-high vacuum TEM with atomically smooth surfaces. The narrow bright band is a dissociated dislocation, while the *darker gray* shaded areas are where the sample thickness changes by a single double-layers of atoms (from Spence 2007)

to the deep Peierls valleys in covalently-bonded semiconductors, the atomic mechanism which controls the ductile strength of (warm) silicon is the kink migration energy, rather than kink formation energy, unlike the situation expected in metals.

The effects of surface roughness on the atomic scale and of surface islands spanning a partial dislocation can be eliminated by using an ultra-high vacuum electron microscope, which allows silicon surfaces to be formed which are atomically flat over large areas, by the method of controlled oxygen etching. This can be done at a low enough temperature to prevent glide of all dislocations out of the sample, and it was found that many dislocations remain pinned during this heating and etching to remove surface steps. Figure 15.29 shows an image recorded using the UHV TEM in the laboratory of Dr F. Ross at IBM using a $(-422)/3$ forbidden reflection in silicon. The broad bands running across the image run between two-atom high surface steps on either surface (sometimes crossing on different surfaces), while the thin bright line running down the page is a ribbon of stacking fault between straight 60/90 partial dislocation in the middle of the slab. We see how, according to the preceding equations, the occurence of an unfavorable termination, when combined with the effect of the stacking fault, can extinguish the fault contrast. Lattice images like these, formed using all six of the forbidden reflections (similar to Fig. 15.27c) would clearly reveal kinks, allowing the study of kink dynamics as the sample is heated to induce dislocation motion.

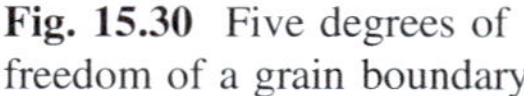

Fig. 15.30 Five degrees of freedom of a grain boundary

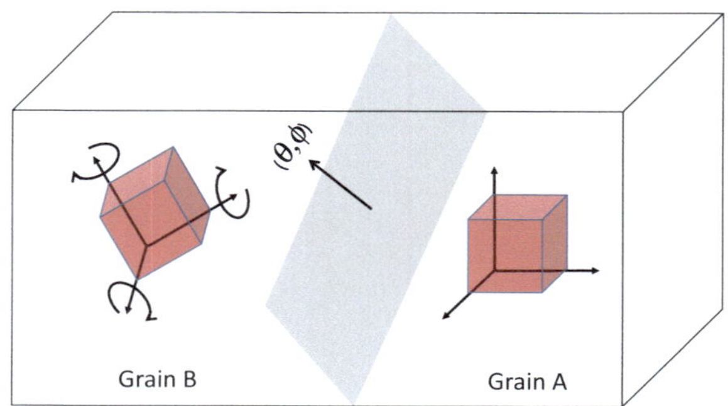

15.6.2   Grain Boundaries

Most crystalline materials except single crystals are polycrystals or polycrystalline, composed of many crystallites or grains of varying size, shape, and orientation. The interface between two grains forms the grain boundary (GB). The atomic structure of grain boundaries has been the subject of intense interest in electron microscopy, as well as in materials science, for the simple reason that grain size has the largest impact on materials properties. Studies have shown that the atomic structure of GBs influences GB segregation, diffusion, GB mobility, and sliding. A comprehensive review of interface structure and properties can be found in the text by Sutton and Balluffi (1997). In metals, GBs can either weaken through intercrystalline fracture, stress corrosion cracking or strengthen polycrystalline metallic materials, which is known as Hall–Petch effect.

A grain boundary has five degrees of freedom (DOF) as illustrated in Fig. 15.30 for a bicrystal as the simplest polycrystal containing two grains, A and B. The orientation of grain A is used as the reference. Three DOF are used to describe the rotation of grain B relative to A. The orientation relationship can be described by Euler angles (φ, θ, ψ), which transform the (x_B, y_B, z_B) coordinate into (x_A, y_A, z_A), where φ is a rotation around the z_B-axis, θ a rotation around the new x_B-axis and ψ around the z_A-axis. The grain boundary adjoining grain A and B is represented by a 2D surface, whose surface normal, (n_x, n_y, n_z), is determined by another two DOF. The transformation from (x_B, y_B, z_B) to (x_A, y_A, z_A) is described by a 3×3 rotation matrix, which can be determined by electron diffraction using Kikuchi lines or CBED (see Randle (1993), as well as Sects. 3.10 and 10.7).

Alternatively, the orientation relationship can be specified by using a common axis $\vec{O}$ for the two coordinates and a rotation θ around $\vec{O}$ to bring them together. It takes two DOF to describe the common axis $\vec{O}$; thus, there is no change in the DOF. Using this approach, grain boundaries may be categorized into three different types, twist, symmetrical, and asymmetrical tilt boundaries. The common axis is perpendicular to the grain boundary plane for the twist boundary (Fig. 15.31a). Because of this, the grain boundary plane is exactly defined and independent of the rotation angle. Compared to twist boundaries, the tilt boundaries have the common

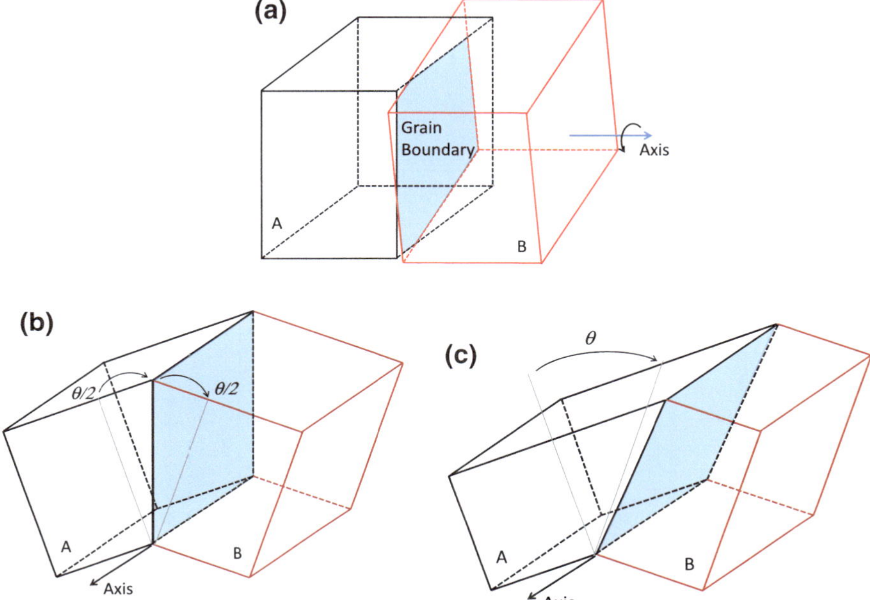

Fig. 15.31 Three types of grain boundaries: **a** twist, **b** symmetrical tilt, and **c** asymmetrical tilt boundaries

axis aligned parallel to the grain boundary plane (Fig. 15.31). For the symmetrical tilt boundaries, the boundary plane represents the plane of the mirror symmetry of the crystal lattices of two grains. There is no such symmetry for the asymmetrical tilt boundary. Consequently, there are an infinite number of possible grain boundary planes for a given rotation angle. Grain boundaries containing both twist and tilt components are called mixed. Even in such cases, it is useful to identify segments of boundaries of pure twist or tilt.

A grain boundary can also be described by the surface normal of the two boundary planes ($\vec{n}_1$ and $\vec{n}_2$, each with two DOFs), and their angle θ. If we examine the possible combinations of two planes, each has its 2D lattice. Now, let us consider how two lattices can meet relative to each other. By rotating and moving one lattice against the other, we will find some special orientations where a set of lattice points belonging to lattice 1 coincide exactly with another set of lattice points belonging to lattice 2. Thus, a kind of superstructure, called coincidence site lattice (CSL), develops. The density of coincidence sites, in its reciprocal form (Σ), is used to define the CSL with

$$\Sigma = \frac{\text{number of coincidence sites in the unit cell}}{\text{total number of lattice sites in the unit cell}},$$

Fig. 15.32 Faceted <1-10> tilt boundary ($q = 50°$, $\Sigma = 11$) showing (113)(113) vertical facet at bottom-left followed by (225)(441) and (557)(771) facets (experimental image from Merkle 1994)

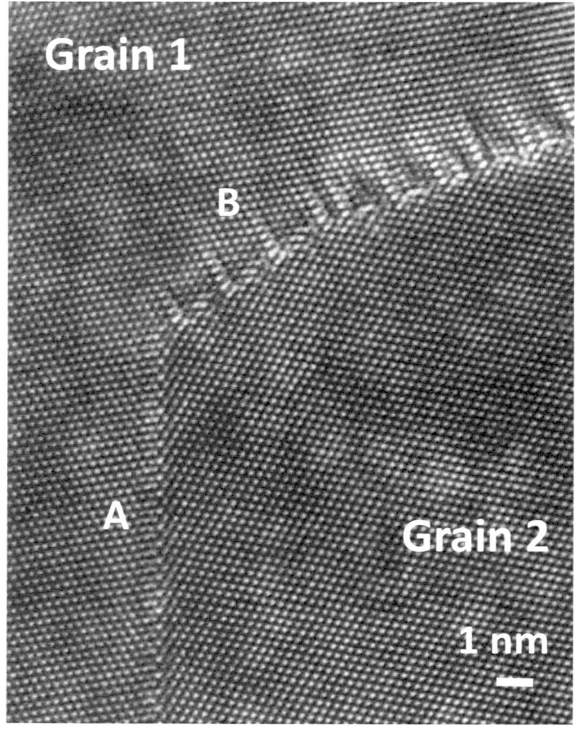

where the unit cell belongs to the CSL. Grain boundaries with low Σ tend to have special properties, e.g., low energy. For example, in fcc crystals, a measurement of tilt boundaries rotating along <110> shows pronounced minima in energy at $\Sigma 3$ and $\Sigma 11$. However, this tendency is not clear-cut, e.g., there is no direct correlation between energy and Σ values. Some grain boundaries of low Σ value might have especially low-energy values, whereas others have energies not so different from a random orientation. The abundance of special boundaries in some materials has also been observed, for example, X-ray diffraction microscopy measurements revealed peaks in the grain neighbor misorientation angle distribution at $60°$ ($\Sigma 3$) and $39°$ ($\Sigma 9$) with a resolution-limited width of $0.14°$ FWHM in well-annealed nickel (Hefferan et al. 2009).

Atomic-resolution electron microscopy studies of grain boundaries are typically performed on tilt GBs of low-index tilt axes, such as <100> and <110> in a fcc crystal. Since these GBs are also the ones that have low-energy, atomic-resolution electron microscopy can provide considerable insight regarding the GB atomic scale structure. Figure 15.32 is an example. The HREM image was recorded from an Au bicrystal prepared from thin (110) gold films grown epitaxially on NaCl in a UHV system and then pairwise sintered together at the required misorientation angle, using a modified method of Schober and Balluffi (1970) and Tan et al. (1976).

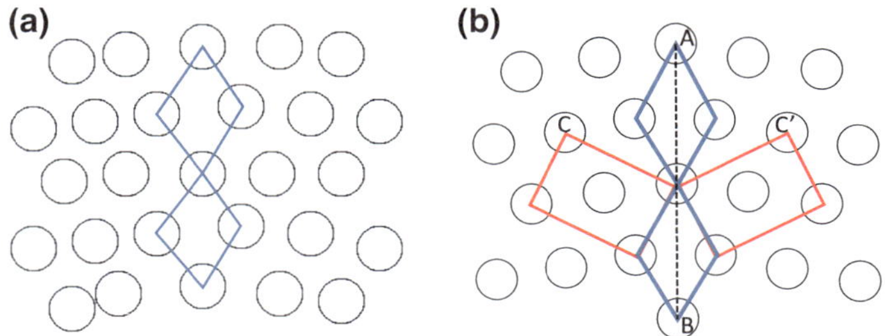

Fig. 15.33 Atomic structure of (113) symmetrical <1-10> tilt boundary, Σ = 11. **a** Experimental structure determined from HREM image. **b** equilibrium structure obtained from molecular dynamics simulation (Rittner and Seidman 1996)

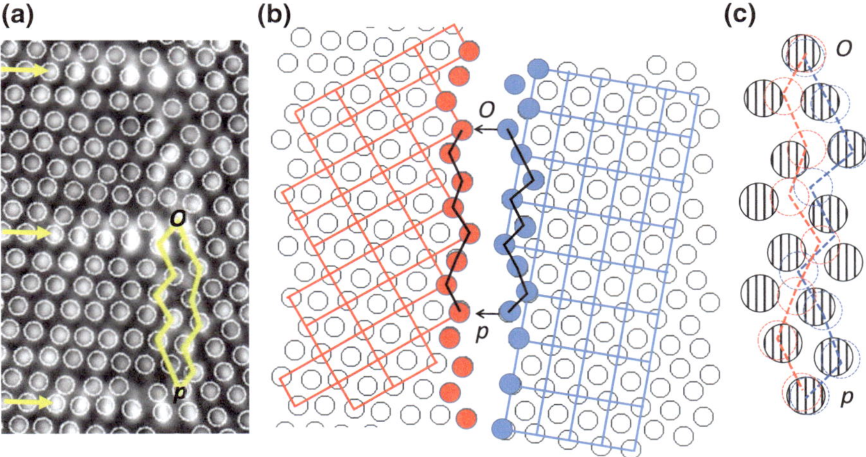

Fig. 15.34 Atomic structure of (225)(441) asymmetrical <1-10> tilt boundary, Σ = 11. **a** Experimental structure determined from the HREM image. **b** Structure model constructed based on the facets of (225) (*left side*) and (441) (*right side*), where filled *circles mark* atoms on the surface facets, and **c** superposition of the experimental structure (*shaded disks*) and the model (*dashed circles*)

The images were taken at a magnification of 700k× for several defocus values near-optimum defocus, utilizing a H9000 high-resolution electron microscope, operated at 300 kV. Part of a small grain at $\theta = 50°$ (Σ = 11) is shown in Fig. 15.33. The GB planes change from (113)(113) to (225)(441) and (557)(771). The boundary marked as A is the symmetrical (113)(113) GB. Regular arrangement of atomic columns is seen, which produces an optimum match between the two lattices, with no evidence of lattice strain. This is contrasted by the asymmetrical GB marked as B in Fig. 15.34, which the HREM image clearly show extended regions of strain over ∼5 atomic columns long, commensurate to the structural period, and on one side of the interface due to strongly asymmetrical relaxations in this boundary. It is interesting to note that the corner of the grain boundaries, where two the facets of A and B join, also appear to be well matched atomically, with no evidence for additional strain associated with forming a corner.

The projected grain boundary atomic structure can be determined directly by atomic-resolution HREM. Figures 15.33 and 15.34 are examples. In both cases, the atomic positions were determined using peak detection based on template matching (Zuo et al. (2014)), with the atomic peak as template. For the (113) symmetrical $[1-10]$ tilt boundary, the grain boundary has a diamond-like structural unit instead of the centered rectangular lattices, which has mirror symmetry at the grain boundary. The experimental structure agrees with the equilibrium structure obtained from molecular dynamics simulations, except some distortions, which can be attributed to strain associated with the small grain observed in the experiment. The top and bottom corners of the marked diamonds are the matching coincidence sites. The distance between matching sites is shortest in (113)(113). In the (225)(441) asymmetrical $\langle 1-10 \rangle$ tilt boundary, the shortest matching distance is OP as marked in Fig. 15.34. The atomic structure of this GB can be understood based on the structure of (225) and (441) facets as illustrated in Fig. 15.34b. The lattice sites of these two lattices overlap at O and P, which represents the shortest period. By bringing the facet atoms together at O and P, the structural unit of this GB is formed. A superposition of the model unit and the structural unit as determined by HREM shows the relaxation of atoms that avoids the atomic overlaps and also results in a larger grain boundary volume.

In general, the HREM experiments show that GBs can assume a great variety of different atomic structures, depending on both the CSL misorientation and the inclination of the GB plane. In a few instances, multiple GB structures for the same macroscopic GB parameters have also been found for both symmetrical (Merkle 1994) and asymmetrical GBs.

References

Alexander H, Spence JCH, Shindo D, Gottschalk H, Long N (1986) Forbidden-reflection lattice imaging for the determination of kink densities on partial dislocations. Philos Mag A 53: 627–643

Amelinckx S, Gevers R, Van Landuyt J (1978) Diffraction and imaging techniques in materials science. North-Holland, Amsterdam

Arslan I, Bleloch A, Stach EA, Ogut S, Browning ND (2006) Using EELS to observe composition and electronic structure variations at dislocation cores in GaN. Philos Mag 86:4727–4746

Carpenter RW, Spence JCH (1982) Three-dimensional strain-field information in convergent-beam electron diffraction patterns. Acta Cryst A38:55–61

Cherns D (1974) Direct resolution of surface atomic steps by transmission electron-microscopy. Phil Mag 30:549–556

Cherns D, Preston AR (1989) Convergent beam diffraction studies of interfaces, defects, and multilayers. J Electron Micr Tech 13:111–122

Cockayne DJH (1981) Weak-beam electron microscopy. Annu Rev Mater Sci 11:75–95

Cockayne DJH, Ray ILF, Whelan MJ (1969) Investigations of dislocation strain fields using weak beams. Philos Mag 20:1265–1270

Cosgriff EC, Nellist PD, Hirsch PB, Zhou Z, Cockayne DJH (2010) ADF STEM imaging of screw dislocations viewed end-on. Philos Mag 90:4361–4375

De Graef M (2003) Introduction to conventional transmission electron microscopy. Cambridge University Press, Cambridge

Eaglesham DJ (1989) Applications of convergent beam electron diffraction in materials science. J Electron Micr Tech 13:66–75

Edington JW (1975) Practical electron microscopy in materials science, 3. Interpretation of transmission electron micrographs. Philips Technical Library, Eindhoven

Edington JW (1976) Practical electron microscopy in materials science, 4. Typical electron microscope investigations. Philips Technical Library, Eindhoven

Eshelby JD, Stroh AN (1951) Dislocations in thin plates. Lond Edinb Dubl Phil Mag 42:1401–1405

Groger R, Dudeck KJ, Nellist PD, Vitek V, Hirsch PB, Cockayne DJH (2011) Effect of Eshelby twist on core structure of screw dislocations in molybdenum: atomic structure and electron microscope image simulations. Philos Mag 91:2364–2381

Head AK, Humble P, Clarebrough LM, Morton AJ, Forwood CT (1973) Computed electron micrographs and defect identification. North-Holland Publishing Company, Amsterdam

Hefferan CM, Li SF, Lind J, Lienert U, Rollett AD, Wynblatt P, Suter RM (2009) Statistics of high purity nickel microstructure from high energy X-ray diffraction microscopy. CMC-Comput Mater Continua 14:209–219

Hirsch PB, Howie A, Whelan MJ (1960) A kinematical theory of diffraction contrast of electron transmission microscope images of dislocations and other defects. Philos T R Soc Lond A 252:499

Hirsch P, Howie A, Nicolson RB, Pashley DW, Whelan MJ (1977) Electron microscopy of thin crystals. Robert E. Krieger Publishing Company, Malabar

Hirth JP, Lothe J (1983) Theory of dislocations. Krieger Publishing Company, Malabar

Howe JM (1997) Interfaces in materials: Atomic structure, thermodynamics and kinetics of solid-vapor, solid-liquid and solid-solid interfaces. Wiley, New York

Howie A, Basinski ZS (1968) Approximations of dynamical theory of diffraction contrast. Philos Mag 17:1039

Hudson TS, Dudarev SL, Sutton AP (2004) Confinement of interstitial cluster diffusion by oversized solute atoms. Proc R Soc A 460:2457–2475

Humphreys CJ (1979b) STEM imaging of crystals and defects. In: Introduction to analytical electron microscopy. J. J. Hren, J. I. Goldstein and D. C. Joy. Plenum, New York

Humphreys CJ, Maher DM, Fraser HL, Eaglesham DJ (1988) Convergent-beam imaging—a transmission electron-microscopy technique for investigating small localized distortions in crystals. Philos Mag A 58:787–798

Jenkins ML (1994) Characterization of radiation-damage microstructures by TEM. J Nucl Mater 216:124–156

Jenkins ML, Kirk MA, Fukushima H (1999) On the application of the weak-beam technique to the determination of the sizes of small point-defect clusters in ion-irradiated copper. J Electron Microsc 48:323–332

Jia CL, Lentzen M, Urban K (2003) Atomic-resolution imaging of oxygen in perovskite ceramics. Science 299:870–873

Kelly AA, Knowles KM (2012) Crystallography and crystal defects, 2nd edn. Wiley, West Sussex

Kolar HR, Spence JCH, Alexander H (1996) Observation of moving dislocation kinks and unpinning. Phys Rev Lett 77:4031–4034

Lozano JG, Guerrero-Lebrero MP, Yasuhara A, Okinishi E, Zhang S, Humphreys CJ, Galindo PL, Hirsch PB, Nellist PD (2014) Observation of depth-dependent atomic displacements related to dislocations in GaN by optical sectioning in the STEM. J Phys Conf Ser 522:012048

Merkle KL (1994) Atomic-structure of grain-boundaries. J Phys Chem Solids 55:991–1005

Nye JF (1957) Physical properties of crystals. Clarendon Press, Oxford

Paulauskas T, Buurma C, Colegrove E, Stafford B, Guo Z, Chan MKY, Sun C, Kim MJ, Sivananthan S, Klie RF (2014) Atomic scale study of polar Lomer-Cottrell and Hirth lock dislocation cores in CdTe. Acta Cryst A70:524–531

Randle V (1993) The measurement of grain boundary geometry. CRC Press, Boca Raton

Ray ILF, Cockayne DJ (1971) Dissociation of dislocations in silicon. Proc R Soc Lond Ser A 325:543

Rittner JD, Seidman DN (1996) <110> symmetric tilt grain-boundary structures in fcc metals with low stacking-fault energies. Phys Rev B 54:6999–7015

Rouvière JL, Prestat E, Bayle-Guillemaud P, Hertog MD, Bougerol C, Cooper D, Zuo J (2013) Advanced semiconductor characterization with aberration corrected electron microscopes. J Phys Conf Ser 471:012001

Saldin DK, Whelan MJ (1979) Construction of displacement-fields of dislocation loops and stacking-fault tetrahedra from angular dislocation segments. Philos T R Soc A 292:513–521

Schober T, Balluffi RW (1970) Quantitative observation of misfit dislocation arrays in low and high angle twist grain boundaries. Philos Mag 21:109

Spence JCH (1983) High-energy transmission electron-diffraction and imaging studies of the silicon(111) 7x7 surface-structure. Ultramicroscopy 11:117–124

Spence JCH (1992) Electron channelling. In: Cowley JM (ed) Techniques of electron diffraction, vol 1. Oxford University Press, Oxford

Spence JCH (2007) Experimental studies of dislocation core defects. In: Nabarro FRN, Hirth JP (eds) Dislocations in solids. Elsevier, Amsterdam

Sturkey L (1957) The use of electron-diffraction intensities in structure determination. Acta Crystallogr 10:858

Sutton AP, Balluffi RW (1997) Interfaces in crystalline materials. Clarendon Press, Oxford

Tan TY, Hwang JCM, Goodhew PJ, Balluffi RW (1976) Preparation and applications of thin-film specimens containing grain-boundaries of controlled geometry. Thin Solid Films 33:1–11

Tanaka M (1986) Conventional transmission-electron-microscopy techniques in convergent-beam electron diffraction. J Electron Microsc 35:314–323

Tanaka M, Terauchi M, Kaneyama T (1988) Convergent beam electron diffraction II. JEOL Company, Tokyo

Tanaka M, Terauchi M, Tsuda K (1994) Convergent beam electron diffraction III. JEOL Company, Tokyo

Tanishiro Y, Takayanagi K, Yagi K (1986) Observation of lattice fringes of the Si(111)-7x7 structure by reflection electron-microscopy. J Microsc 142:211–221

TEMACI, http://www.Materials.Ox.Ac.Uk/research/rippublications/temaci.Html

Thompson K, Flaitz PL, Ronsheim P, Larson DJ, Kelly TF (2007) Imaging of arsenic Cottrell atmospheres around silicon defects by three-dimensional atom probe tomography. Science 317:1370–1374

Wang R, Wen J (1989) Effects of a stacking fault on higher-order diffraction fringes. Acta Cryst A45:428–431

Yang H, Lozano JG, Pennycook TJ, Jones L, Hirsch PB, Nellist PD (2015) Imaging screw dislocations at atomic resolution by aberration-corrected electron optical sectioning. Nat Commun 6:7266

Yoffe EH (1960) The angular dislocation. Philos Mag 5:161–175

Zhang ZL, Sigle W, Ruhle M (2002) Atomic and electronic characterization of the a 100 dislocation core in SrTiO$_3$. Phys Rev B 66:094108

Zhang ZL, Sigle W, Phillipp F, Ruhle M (2003) Direct atom-resolved imaging of oxides and their grain boundaries. Science 302:846–849

Zhou Z, Jenkins ML, Dudarev SL, Sutton AP, Kirk MA (2006) Simulations of weak-beam diffraction contrast images of dislocation loops by the many-beam Howie-Basinski equations. Philos Mag 86:4851–4881

Zuo J-M, Shah AB, Kim H, Meng Y, Gao W, Rouviére J-L (2014) Lattice and strain analysis of atomic resolution Z-contrast images based on template matching. Ultramicroscopy 136:50–60

Chapter 16
Strain Measurements and Mapping

The concepts of strain and stress were introduced in Chap. 15. Direct measurements of strain and stress always rely on the measurement of deformation in a structure. An example is the case of stress measurement based on wafer curvature. From the Stoney formula (Stoney 1909), the measurement is made on the deformation of the substrate, i.e., its curvature, and the use of an analytical formula (the Stoney formula or its modified form) that links the stress to the curvature allows one to determine the stress present in a thin film. Alternatively, an indirect measurement of strain and stress can be made, for instance, Raman spectroscopy determines how the vibrations of atoms bonded together change when stress is present. The strain of uniform two-dimensional films can be measured using well-established micro-mechanical methods (Guckel et al. 1985; Nix 1989). For strain mapping, the techniques developed include X-ray diffraction, micro-Raman spectroscopy, and electron beam-based techniques. X-ray diffraction measures the lattice parameter by Bragg diffraction with an accuracy of 10^{-4}–10^{-5} Å. Micro-Raman spectroscopy has a spatial resolution on the order of 0.5 μm. High-resolution X-ray diffraction is now possible with the development of small synchrotron probes of tens to hundreds nanometers in diameter (Keckes et al. 2012). Presently, the best techniques capable of measuring strain with a spatial resolution between 1 Å and a few nanometers, and with good sensitivity (10^{-4}), are TEM-based techniques, which is the topic of this Chapter.

16.1 Local Lattice Parameters and Strain

What is measured by the diffraction or high-resolution imaging are the local lattice parameters. A 3D-crystal lattice is characterized by its lattice vectors $\vec{a}$, $\vec{b}$, and $\vec{c}$ (for a 2D measurement, only two vectors are considered), and measuring strain then comes down to the measurement of the local lattice vectors. Let $\vec{a}_{o}$, $\vec{b}_{o}$, and $\vec{c}_{o}$ be the

© Springer Science+Business Media New York 2017

J.M. Zuo and J.C.H. Spence, *Advanced Transmission Electron Microscopy*,
DOI 10.1007/978-1-4939-6607-3_16

lattice vectors of the undeformed crystal in a reference area and $\vec{a}$, $\vec{b}$, and $\vec{c}$ be the lattice vectors in the deformed area. By expressing these vectors in a Cartesian coordinate (x, y, z), we define the lattice matrix by

$$A = \begin{pmatrix} a_x & b_x & c_x \\ a_y & b_y & c_y \\ a_z & b_z & c_z \end{pmatrix}. \tag{16.1}$$

From this, a gradient deformation matrix F can be defined as

$$F = A\,A_o^{-1}, \tag{16.2}$$

where A_o is the lattice matrix defined for the reference structure. In the case of a small rotation and small deformation (i.e., rotation angle less than $10°$ and strain smaller than 4 %), strain is obtained by taking the symmetric part of F, and rotation is the antisymmetric part of F:

$$\begin{aligned} \varepsilon &= \frac{1}{2}\left(F + F^{\mathrm{T}}\right) - I \\ R &= \frac{1}{2}\left(F - F^{\mathrm{T}}\right) + I, \end{aligned} \tag{16.3}$$

where F^{T} is the transpose of the F matrix. In the case of larger rotations and strains, more accurate formulas can be obtained (Martin et al. 2016), but in most TEM applications Eq. (16.3) is sufficient.

The strain defined in Eq. (16.3) using a fixed reference is called the Lagrange strain. Depending on where the reference is taken, different definitions of strain can be made. As an example, we consider the case represented in Fig. 16.1 where a Ge layer is epitaxed on a Si substrate. Let us take a_{Ge} and a_{Si} as the equilibrium lattice

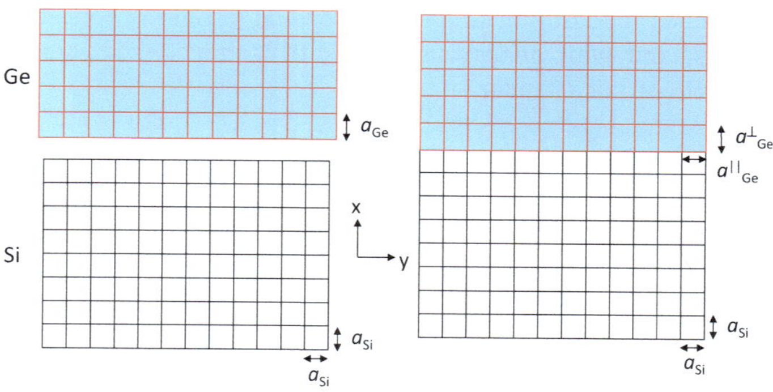

Fig. 16.1 Illustration of epitaxial strain of Ge on the Si substrate

parameters of Ge and Si, respectively. If a is the local lattice parameter along the x-axis in the Ge layer, the strain is defined by

$$\varepsilon_{xx}^{\text{mat}} = (a - a_{Ge})/a_{\text{Ge}} = a a_{\text{Ge}}^{-1} - 1 \tag{16.4}$$

We call $\varepsilon_{xx}^{\text{mat}}$ the material strain in order to differentiate it from the Lagrange or Euler strain, which are defined, respectively, as

Lagrange strain:

$$\varepsilon_{xx}^{\text{La}} = (a - a_{Si})/a_{Si} = a a_{\text{Si}}^{-1} - 1 \tag{16.5}$$

Euler strain:

$$\varepsilon_{xx}^{\text{Eu}} = (a - a_{Si})/a = 1 - a_{Si} a^{-1} \tag{16.6}$$

Lagrange strain takes the reference in the substrate (in our example, the silicon), while Euler strain uses the local lattice parameter as the reference. In the material strain, the reference is the lattice parameter of the material in its undeformed state, and this reference changes when the material changes from Si to Ge or when an alloy is formed between Si and Ge. With the definition of Lagrange strain, the reference is the same for different materials in the measured sample.

The material strain in the Ge layer differs significantly from the Lagrange and Euler strains, the latter two to first order are equal $((\varepsilon_{xx}^{\text{La}} + 1) = a a_{\text{Si}}^{-1} = 1/(1 - \varepsilon_{xx}^{\text{Eu}}) \approx \varepsilon_{xx}^{\text{Eu}} + 1)$. Introducing the misfit, $m = (a_{\text{Ge}} - a_{\text{Si}})/a_{\text{Si}} = a_{\text{Ge}} a_{\text{Si}}^{-1} - 1$ between Si and Ge, one obtains

$$(\varepsilon_{xx}^{mat} + 1) = a a_{Ge}^{-1} = (\varepsilon_{xx}^{La} + 1)/(m + 1) \tag{16.7}$$

Thus, to first order in strain, we have $\varepsilon_{xx}^{\text{mat}} \approx \varepsilon_{xx}^{\text{La}} - m$.

The above discussions use Ge on Si as an example. However, these concepts are general and applicable to other systems (e.g., A on B). To calculate stress, we need to know the material strain and thus the undeformed state.

16.2 Electron Beam-Based Strain Measurement Techniques

The dark-field diffraction contrast imaging introduced in the last chapter is a classical technique used to image the variation of strain. Diffraction contrast obtained from a specimen of constant thickness and composition can be sensitive to strain on the order of 10^{-3} at a resolution of few nanometers. A constant strain field, giving rise to uniform contrast, cannot be interpreted using this technique. Furthermore, because of dynamical diffraction, the recorded image intensity from a thin crystalline sample is sensitive to the crystal rotation, its thickness, and composition. Thus, analysis also requires accurate diffraction models and theory. A tool for performing such analysis was reported by Janssens et al. (1995).

CBED enables the recording of high-order Laue zone (HOLZ) lines using elastically diffracted electrons. The early use of CBED patterns for lattice parameter measurement was reported in the 1970s and 1980s (Jones et al. 1977; Kaufman et al. 1986; Twigg et al. 1987). Computer algorithms were developed for automated lattice parameter refinement in 1990s (Zuo 1992; Rozeveld and Howe 1993; Kramer et al. 2000). The measurement is often based on kinematical electron diffraction, which requires a tilt of the sample to a high-order zone axis (Armigliato et al. 2005; Diercks et al. 2010). This constraint can be removed by the use of the pattern matching method for lattice parameter measurement using dynamically calculated CBED patterns (Zuo et al. 1998; Kim et al. 2004).

The line width of a HOLZ line depends sensitively on strain variations and crystal rotations inside the crystal volume probed by the electron beam. A highly strained sample when chemically or otherwise thinned for TEM exhibits surface strain relaxation (Gibson and Treacy 1984). The relaxation leads to strain variations across the sample thickness, and HOLZ line splitting from such samples has been reported by a number of groups (Clement et al. 2004; Houdellier et al. 2006). Methods have been proposed to extract the thickness-dependent displacement field distribution from split HOLZ lines (see Sect. 16.4.6, Carpenter and Spence 1982; Spessot et al. 2007).

Scanning electron nanodiffraction (SEND) is a general technique for the characterization of nanostructures (Zuo and Tao 2011). It can be used to map strain by scanning a small electron probe across the specimen and recording the diffraction patterns directly on a two-dimensional detector for each probe position (Armigliato et al. 2008; Uesugi et al. 2011; Beche et al. 2013; Baumann 2014). This technique is associated with several names, including nanobeam electron diffraction (NBED) (Beche et al. 2009) or nanobeam diffraction (NBD) (Armigliato et al. 2008) using a focused, nanometer-sized, electron probe, and nanoarea electron diffraction (NAED) (Zuo et al. 2004) for a defocused and parallel electron beam.

The recorded electron diffraction patterns contain contributions from both elastic and inelastically scattered electrons. In CBED, the visibility of HOLZ lines can be significantly improved by removing inelastic scattering using energy filtering, e.g., by using a small slit placed around the zero loss peak in the electron energy loss spectrum (Chap. 10 and Rose 1995). Energy filtering also helps in improving the sharpness of diffraction spots recorded in NBD and thus their measurement precision.

EBSD is another diffraction-based technique. It determines crystallographic information from small volumes of material in a SEM using back scattered electrons collected using an area detector (Dingley 2004). The backscattered electrons are diffracted by crystal lattice planes giving rise to diffraction patterns similar to Kikuchi patterns recorded in TEM from a thick crystal. By placing the EBSD detector close to the crystal, diffraction patterns can be recorded with a large angular range. In EBSD, the sample is usually tilted toward the detector by 60°–80° in order to increase the quality of the diffraction patterns. EBSD maps are obtained by extracting crystallographic information from the recorded EBSD patterns using an automated Kikuchi line indexing procedure. Recent developments in high-resolution EBSD have made it capable of measuring strain with a good sensitivity (Wilkinson et al. 2006b). The strain is measured relative to a reference strain-free area. For example, the SiGe/Si interface strain has been studied by several groups, and good sensitivity has been reported (Wilkinson et al. 2006a; Vaudin et al. 2008; Villert et al. 2009). However, the spatial resolution of EBSD is degraded by the highly tilted incident beam.

The development of high-resolution electron microscopy has provided the impetus for the development of several quantitative methods for lattice, atomic displacement, and strain measurement. A popular method is the geometrical phase analysis (GPA) (Hytch et al. 1998, 2003), which was developed by Hytch and his collaborators in Toulouse, France. This method utilizes a set of lattice planes recorded in a HREM image to measure the so-called geometrical phase. Dark-field electron holography (DFEH) is another phase-based technique (Hytch et al. 2011). The principle of DFEH is based on Moire fringes observed between two lattices, which is well known in electron microscopy to provide a sensitive measurement of strain (Hirsch et al. 1977). By using a biprism in off-axis holography, the lattice from the substrate can be made to interfere with the lattice in the device region, for strain mapping. With the development of aberration correctors, recent publications have demonstrated that atomic positions can be measured with picometer precision (Kimoto et al. 2010; Zuo et al. 2014). This represents the ultimate spatial resolution achievable in a 2D strain measurement.

In both cases of HREM and DFEH, the information recorded comes from the electron exit wave function due to electron scattering inside the specimen. A direct correlation between the phase and crystal potential can only be made in a very thin specimen, and this approximation breaks down rapidly in the real specimens to which these techniques have been applied. In the absence of a full theory, several comparative studies based on measurements on well-designed test samples have been carried out using various experimental techniques and theoretical modeling using the finite element method. The results can be found in several journal publications (Beche et al. 2013; Denneulin et al. 2014; Glowacki et al. 2014).

In Table 16.1, different TEM-based techniques for strain measurements are compared based on measurement sensitivity, precision, accuracy, resolution, and field of view. Accuracy is determined with the help of a standard or calibrated sample. The precision can be estimated by repeated measurements or from the same measurement for different points inside a uniform specimen area. Sensitivity describes the smallest change that can be detected.

For diffraction-based techniques, the electron probe size and the beam direction relative to the strain axes are often presented as resolution in Table 16.1. The actual spatial resolution depends on several factors: the electron probe size, electron wave propagation inside the specimen due to electron scattering and its effects on the measurements, and the direction relative to the strain axes in a 2D projection. Theoretical considerations based on the column approximation suggest that the resolution in the zone-axis orientation can be as small as a few nanometers (Sect. 15.3.1, also see Zuo and Spence 1993). In the case of a thick sample, the resolution is limited by probe broadening for a focused beam (Chuvilin and Kaiser 2005). For atomic

Table 16.1 Comparison of different TEM-based techniques for strain analysis (with data from Zuo et al. 2016)

	CBED	NBED without or with precession (PED)	HRTEM-STEM	Dark-field holography
Sensitivity	2×10^{-4}	1×10^{-3}	1×10^{-3}	2.5×10^{-4}
Accuracy	2×10^{-4}	1×10^{-3} to 5×10^{-2}	1×10^{-3}	
Precision		6×10^{-4} to 1×10^{-3} (NBED) 9×10^{-5} to 2×10^{-4} (PED)	$1–3 \times 10^{-3}$	2×10^{-3} to 2×10^{-4}
Spatial resolution (nm)	1×7 for [340] 1×10 for [230][110]	2.5–10	2–4	4–6
Field of view (nm × nm)	map capable, limited by time	map capable, limited by time	image 150×150	image 500×1500
Optimum specimen thickness (nm)	~ 200	~ 50–100	~ 50 or less for HREM ~ 50-100 for STEM	~ 100
TEM hardware	Analytical TEM (ATEM)	ATEM, plus a small condenser aperture, Precession setup	HRTEM, probe or image aberration corrector for atomic resolution strain mapping	Schottky or FEG emitters electrostatic biprism and Lorentz lens

resolution imaging, the spatial resolution is determined by the signal localization, whether it is the exit wave function in HREM or the diffuse scattering recorded in HAADF-STEM. Strain can be measured on an individual atomic column basis using atomic resolution electron images.

The precision (or accuracy with calibration) of strain measurements can be improved by averaging over a number of measurements and thus becomes dependent on the measurement spatial resolution for imaging-based techniques. The same argument also applies to DFEH. In these cases, the higher the spatial resolution, the less accurate the measurement. In diffraction, sharp diffraction spots or HOLZ lines are obtained only when the number of lattice planes under the electron probe is sufficiently large (>10). For HOLZ reflections, the normal to these planes will not lie in the plane of the sample. Increasing the electron probe size beyond the limit of the electron beam lateral coherence does not improve the sharpness of the diffraction spots.

Field of view is a parameter for imaging-based strain measurement techniques. It refers to the dimensions of the specimen area in a single measurement with the targeted measurement resolution and sensitivity. In diffraction-based measurement techniques, the field of view is determined by the number of measurements and the extent of beam scanning, which is not constrained by the requirement of resolution and sensitivity, but rather it is limited by the experimental time.

16.3 Limitations of Electron Beam Techniques

The major limitation of electron beam techniques for strain measurement is related to the destructive process involved in the specimen preparation. The spatial resolution is obtained in the cross-sectional geometry. There are two effects that need to be addressed related to the cross-sectional sample preparation process. The first effect is the stress and strain introduced by the new surfaces created during the sample preparation. For example, Ar beam irradiation produces an amorphous layer on the surfaces. Bana et al. showed that the amorphous layer thickness decreases with the ion beam energy (Barna et al. 1998); for Silicon, the measured thickness is about 5 nm using 3 kV Ar ions (this number appears on the large side of values reported). The damage is also dose and incident angle dependent. The stress induced by the surface amorphous layer can be detected by CBED as demonstrated by Vincent et al. (1999). Jacob and Lefebvre (2003) demonstrated that chemical etching is effective for making samples with little or no surface damage. Mechanical polishing using a precision polisher to prepare thin wedges is suitable for TEM observation since it requires only a small amount of ion beam polishing. For the FIB method, various strategies have been developed for reducing the surface damage (Mayer et al. 2007).

The second effect of TEM sample preparation is strain relaxation when a 3D strained structure is prepared as a cross-sectional specimen. This is especially true for films containing a large strain. An example is the Eshelby twist that we discussed in Sect. 15.3.1. Gibson and Treacy (1984) described in details of the relaxation of shear stresses of modulated structures in thin TEM samples and its impact on the interpretation of HREM images. It can be said as a general statement that surface relaxation introduces 3D strain and complicates the experimental measurement. A solution to this is to combine the experimental measurement with FEM simulations and use the simulation results to infer the original strain/stress (Clement et al. 2004; Houdellier et al. 2006). The 3D strain in the prepared cross-sectional samples can be also measured using information provided by HOLZ splitting (Vincent et al. 1999; Spessot et al. 2007). It should be noted that surface relaxation is not a serious issue in the case of the small strain expected in the memory devices as the work by Kim et al. showed (Kim et al. 2004). In case of device characterization, the technology trend favors electron diffraction; for the FinFET transistors, they are so small, the whole device is examined by TEM.

In favorable cases (Zhang et al. 2006) for strain mapping within thinned semi-conductor devices, the direction of sample thinning may be perpendicular to the direction along which the wanted strain component dominates.

16.4 Electron Diffraction-Based Strain Measurement Techniques and Applications

In electron diffraction, the lattice d-spacing in the sample volume illuminated by the electron beam can be measured directly by Bragg diffraction. While the principle of electron diffraction is similar to X-ray diffraction, the spatial resolution of electron diffraction is much higher because of the small Bragg angle and the small electron probes formed inside a modern TEM. This section describes the principles and applications of NBD and CBED for strain measurements.

16.4.1 Nanobeam Diffraction

In NBD, a nanometer-sized electron beam is formed using the microscope's condenser lenses and a small condenser aperture; the beam can be scanned using the electron deflection coils built in the TEM illumination system (Tao et al. 2009), for strain mapping using the SEND technique (Beche et al. 2009). Local lattice parameters, in principle, can be measured directly by applying Bragg's law of $2d\sin\theta = \lambda$. A major advantage of NBD is that it can be used for strain analysis along the zone-axis direction of silicon parallel to the device interfaces. Thus, it gives the highest spatial resolution for mapping strain in a silicon device. The [110] or [100] zone axis is difficult for CBED since HOLZ lines are not visible at room temperature.

Figure 10.9 shows an example of a small focused probe illuminated on a Si (110) crystal with a FWHM of 2.7 nm. The diffraction pattern recorded using this probe consists of small diffraction disks with a convergence semi-angle of 0.37 mrad.

16.4.2 Diffraction Geometry

Within the recorded diffraction patterns, diffraction spots (in the case of a parallel illumination, for a convergent beam the discussion below refers to the center of the disk) are expected, where the transmitted and diffracted electron beams intersect with the detector. Thus, the basis of strain analysis is the crystal reciprocal lattice and the Laue diffraction condition (Fig. 16.2a):

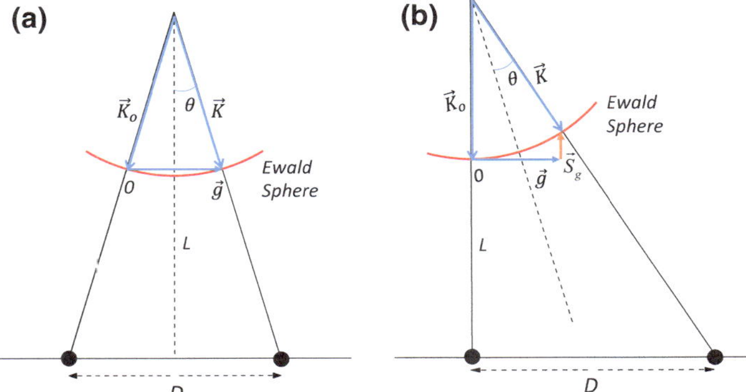

Fig. 16.2 **a** Bragg diffraction geometry with incident beam at Bragg angle to the optical axis (*vertical*), **b** zone-axis geometry with incident beam along the optical axis. *L* marks camera length and *D* is the distanced measured on the detector between the incident and diffracted spots

$$\vec{S} = \vec{k} - \vec{k}_o = \vec{g} \tag{16.8}$$

where $\vec{g}$ is a reciprocal lattice vector, and $\vec{k}_o$ and $\vec{k}$ are the wave vectors of the incident and diffracted beams inside the crystal. The d-spacing (d) can be obtained by measuring the length of $\vec{g}$ in the experimental diffraction pattern (D, see Fig. 16.2a) at the Bragg diffraction geometry using:

$$\frac{D}{2L} = \tan \theta \tag{16.9}$$

and

$$d \approx \frac{L\lambda}{D} \left(1 + \frac{1}{8} \left(\frac{D}{L} \right)^2 \right) \tag{16.10}$$

where L is the experimental camera length λ, which can be calibrated using a sample with known d-spacings. For 200 kV electrons, θ is 5 mrad for $d = 2.5$ Å. For such small angles, $g \approx D/L\lambda$ in Eq. (16.10) to an accuracy of 3×10^{-6}.

In the zone-axis orientation, the diffracted beam is not at the Bragg condition. To account for this, the scattering vector is taken as

$$\vec{S} = \vec{k} - \vec{k}_o = \vec{g} + \vec{S}_g. \tag{16.11}$$

Because of this, the relationship between the measured distance and the d-spacing has to be modified for the zone-axis orientation according to

$$d \approx \frac{L\lambda}{D} \left(1 + \frac{1}{2}\delta^2 + \frac{3}{8} \left(\frac{D}{L} \right)^2 \right) \tag{16.12}$$

where δ is the deviation angle from the Bragg diffraction condition and $\delta = 0$ at the exact zone-axis orientation.

16.4.3   Strain Mapping

Using the TEM deflection (or STEM scan) coils, SEND (scanning electron nanodiffraction) patterns can be recorded from an area of the specimen to provide spatially resolved strain information for strain mapping. The double deflection coils are placed before the objective lens. Scanning electron diffraction can be carried out by first selecting a region of interest (ROI) and dividing this area into a number of pixels and then placing the electron probe at each of these pixels to record diffraction patterns using a camera. The diffraction peak positions are determined from the recorded diffraction patterns and used to map the local strain in real space. Additionally, from the recorded diffraction patterns, bright- and dark-field STEM images can be obtained from SEND by integrating the diffraction intensities of the direct beam or the diffracted beams, respectively, which can be used to register the strain map to these images. SEND can be performed in either the STEM or TEM mode. When NBD is performed in STEM, the STEM ADF detector can be used to visualize the ROI by collecting the scattered electrons to large angles during scanning. In this way, diffraction can be correlated with the probe location during data collection. There is also a major difference in the configuration of condenser lenses between TEM and STEM, for details, see Chap. 10.

Using the above techniques, NBD can be performed in the scanning mode, and a series of diffraction patterns can be acquired along a line or a ROI. For strain mapping, a series diffraction patterns are acquired and compared to a reference diffraction pattern from an unstrained region. The strain is then calculated as the relative lattice mismatch by determination of the exact position of the diffraction spots positions. Using the FEI-Titan microscope operated in the u-Probe STEM mode with the 2.7 nm probe size and 0.5 mrad convergence angle, a precision of 6×10^{-4} was reportedly achieved (Beche et al. 2009). The accuracy depends on the sample preparation and related strain relaxation and is estimated to be about 10^{-3}.

The projected 2D strain ε can be obtained using the 2D deformation matrix (D) obtained from two measure reciprocal vectors, $\vec{g}_1$ and $\vec{g}_2$. They can be taken as the basis vectors for the zone-axis diffraction pattern or any two nonparallel vectors recorded in the diffraction pattern. Each vector is defined by its components along the x and y directions perpendicular to the zone axis. They give the following G-matrix:

$$G = \begin{pmatrix} g_{1x} & g_{2x} \\ g_{1y} & g_{2y} \end{pmatrix} \tag{16.13}$$

and D is simply given by

$$D = \left(G^{\mathrm{T}}\right)^{-1} G_{\mathrm{o}}^{\mathrm{T}} - I \tag{16.14}$$

where T represents transverse and I is a unit diagonal matrix and $G_{\mathrm{o}}^{\mathrm{T}}$ is the transverse of the G matrix of the reference crystal. The strain and crystal rotation is obtained from D using

$$\varepsilon = \frac{1}{2}(D + D^{t}) \quad \text{and} \quad \omega = \frac{1}{2}(D - D^{t}). \tag{16.15}$$

In NBD, dynamical effects can lead to rapid changes in the diffraction spot intensities with thickness and changes in the crystal orientation due to rotation. By recording electron diffraction patterns with the incident electron beam in precession, PED is able to provide the integrated electron diffraction intensity across the Bragg condition for many reflections. The same principle of reducing dynamic effects by precession can also be used to improve measurements of strain in nanostructures (Rouviere et al. 2013). PED has been shown to greatly improve the quality and robustness of electron diffraction strain analysis. It can work with a larger convergence angle and thus a small probe size, which offers increased flexibility in the experimental conditions.

Rouviere et al. reported an implementation of precession NBD on a FEI-Titan TEM (Rouviere et al. 2013). The scan coils of the STEM unit were used to precess the incident beam and to perform descan of the diffracted beams at a precession speed of 0.1 s. Diffraction patterns were recorded on a 2 × 2 k GATAN ultrascan CCD camera with acquisition times of ∼1 s. The incident beam used for strain measurements had a convergence angle of 1.8 mrad and a size of 2.4 nm. Thus, the diffraction patterns consisted of small disks. By using PED, the intensities within a given diffraction disk are made more uniform than the CBED like patterns recorded without precession. Diffracted beam positions were measured by detecting the edges of the diffraction disks instead of their peak intensity. Because of the improvements in the intensity distribution within the diffraction disk, better measurement accuracy could be obtained with the help of precession. Precession also helps by spreading the intensity across to high-index diffraction spots and making them more amenable for detection. Additionally, precession improves the robustness of measurements by reducing the crystal misorientation effects by averaging over the precessed incident beam directions at a cost of slightly larger beam diameter and increased crystal volume from the tilted incident beam and its precession.

Figure 16.3b shows the measured strain profiles from a Si/SiGe multilayer grown by RP-CVD (reduced pressure chemical vapor deposition) using NBED with and without precession. The sample contains four SiGe layers, each of 11 nm thick and of different Ge compositions: 20, 31, 38, and 45 %. The composition has been determined using Secondary Ion Mass Spectroscopy (SIMS). The SiGe layers are biaxially strained by the Si substrate and only in the direction x perpendicular to the layers the lattice parameter is larger than the reference substrate. Figure 16.3 also plots the strain profile obtained by finite element simulations of the structure to take into account the strain relaxation in the thin TEM lamella. The profile was plotted by averaging the strain along the $\langle 011 \rangle$ beam direction and convoluting the obtained profiles with the measured electron beam size, i.e., 2.5 nm. This reduced slightly the strain in the layers; this effect is greater in the layer having the higher Ge concentration where the strain is reduced from 2.76 % down to 2.6 %. As seen in Fig. 16.3, in the SiGe layers, the difference between the measurement of NBD with and without precession is small. At the center of the SiGe layers, the difference between the calculated and the experimental strain obtained with precession is about 0.1 % for the three layers that have the lowest Ge concentration. Large differences are observed inside the silicon, the profile obtained with precession is slightly negative inside silicon, which fits very well with the simulation result, while the NBED measurement without precession give far larger negative strain than the simulation result indicates, especially near the Si-SiGe layer interface. Away from the SiGe layers, the strain profile obtained with precession is very smooth with a root mean square of fluctuations of 1.5×10^{-4}.

Application of NBD with precession is demonstrated in Fig. 16.4 for the analysis of a transistor having recessed Si $_{0.65}$Ge $_{0.35}$ source and drain. Figure 16.4a is a bright-field image of the device. Strain and rotation maps have been obtained by using Eq. (16.15). The DPs used in the analysis were obtained using a beam semi-convergence angle of 1.8 mrad, a precession angle of 0.5°, and a beam diameter of 2.5 nm. The root mean square of strain in the Si substrate, far from the layers, can be used to measure the strain measurement precision, i.e., the

Fig. 16.3 a HAADF-STEM image of the observed sample composed of four SiGe layers deposited on a (001) silicon substrate. b ε xx strain profiles obtained from N-PED, NBED, and simulations (Provided by Jean-Luc Rouviere, CEA, Grenoble, France)

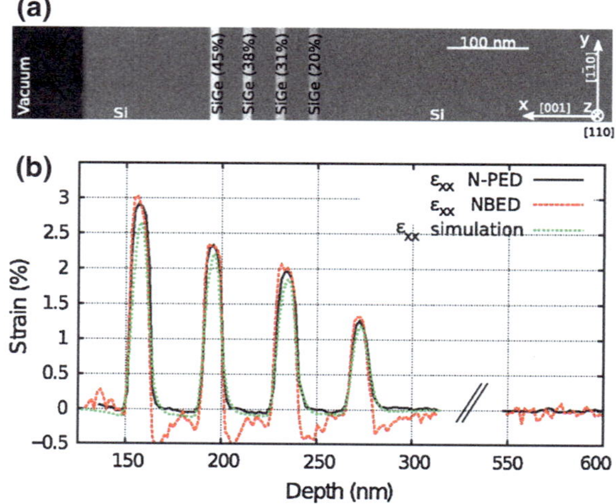

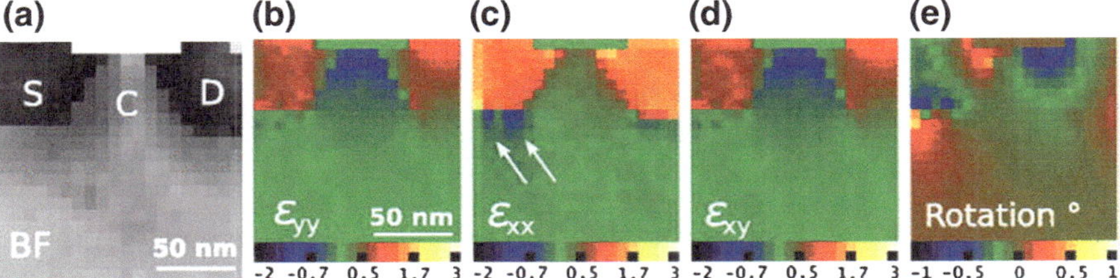

Fig. 16.4 a Bright-field image of a transistor having a 22 nm channel (*C*) and recessed SiGe source (*S*) and drain (*D*). Some dislocations are indicated by arrows. **b–e** The different strain and rotation maps obtained from the analysis of N-PED patterns (from J.-C. Rouviere, CEA, Grenoble, France)

reproducibility of the technique. For NBD, the strain precision can be as low as 6×10^{-4}. The best precision experimentally obtained on NBD with precession was 9×10^{-5} with a probe size of 2.5 nm. In the above examples, only small maps, 20×20 pixels, were obtained. With improved precession speed, larger memory and shorter camera acquisition time, larger maps can be realized. Overall, initial results demonstrate that NBD with precession is a very efficient technique to measure strain in nanostructures.

16.4.4   Convergent Beam Electron Diffraction (CBED)

The sensitivity of the CBED technique to lattice parameters is illustrated in Fig. 16.5. A small probe of electrons is used to measure the local lattice parameters from the positions of high-order Laue zone (HOLZ) lines recorded inside the diffraction disks from relative thick crystals (~ 100–300 nm using 200 kV electrons). The lines are formed by Bragg diffraction of high-order reflections in upper Laue zones, which are associated with small d-spacings and very sensitive to local changes in lattice parameters and thus strain.

The sensitivity comes from the large scattering angle. This can be seen in the case of a cubic crystal for which $\theta \approx g\lambda/2 = \sqrt{h^2 + k^2 + l^2}\,\lambda/2a$. A small change in a gives $\delta\theta \approx 0.5g\lambda\delta a/a$. The change in the Bragg angle is proportional to the length of g. The lines move relative to each other when the lattice parameters change. This effect can be used for accurate measurement of lattice parameters without the need of a reference (except for calibration of electron acceleration high voltage or wavelength).

In the analysis of CBED patterns, it is useful to express HOLZ lines using line equations in an orthogonal zone-axis coordinate system (x, y, z), with z parallel to the zone-axis direction (Sect. 3.6). The x direction can be taken along the horizontal direction of the experimental pattern and y is normal to x. The Bragg diffraction condition expressed in this coordinate system is given by

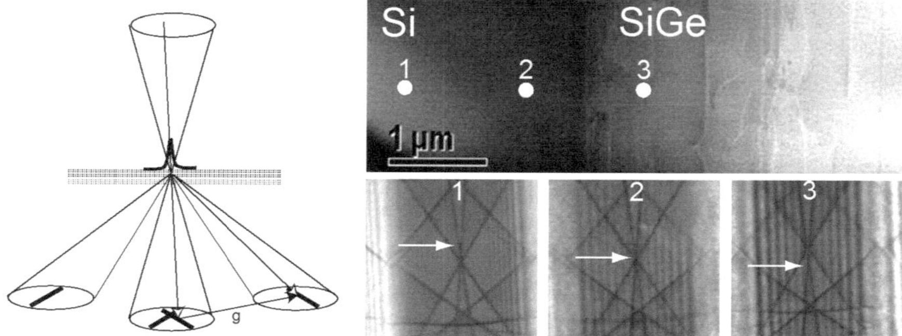

Fig. 16.5 *Left* A schematic diagram of convergent beam electron diffraction for strain mapping. A nanometer-sized electron beam is scanned across the device and used to recorded diffraction patterns. Bragg scattering in the form of lines changes with local lattice parameters and is used for strain mapping. *Right* An example showing three patterns recorded from positions of 1–3 from Si and SiGe buffer layer grown on top of Si. The change in the HOLZ positions is indicated by the *arrows*

$$k_y = -\frac{g_x}{g_y}k_y + \frac{g_z}{g_y}k_z - \frac{g^2}{2g_y} \tag{16.16}$$

and

$$|k_z| = \sqrt{k_o^2 - k_x^2 - k_y^2} \approx k_o = 1/\lambda \tag{16.17}$$

Equation (16.16) defines the so-called kinematical HOLZ line based on Bragg diffraction condition, which is only an approximation of real electron scattering in a relative thick crystal where sharp HOLZ lines are observed. The Bragg condition breaks down in electron diffraction in general because the phase of the incident electron wave changes nonlinearly as it propagates through the crystal due to multiple scattering (dynamic diffraction) (Sect. 5.6.4). The dynamically corrected HOLZ line equation is given by

$$k_y = -\frac{g_x}{g_y}k_y + \frac{g_z}{g_y}\left(k_z - \frac{k\gamma}{g_z}\right) - \frac{g^2}{2g_y} \tag{16.18}$$

It differs from the kinematic expression [Eq. (16.16)] only by the term in γ, called dispersion in dynamical theory. The dependence of γ over the incident wave vector gives the so-called dispersion surface (Chap. 5). Over a small region of the dispersion surface, we may assume that the dispersion γ is approximately constant in the zone axis of the incident beam direction. Then the effects of dynamical dispersion may be thought of as a correction to the accelerating voltage and accommodated by a change in the term k_z in Eq. (16.16). However, this correction to the high voltage differs from zone to zone because of the weighting g_z.

The value of γ decreases as one moves away from the center of a high-symmetry zone axis. For the [230] and [340] zone axes selected for silicon device characterization, Armigliato et al. show that good accuracy and strain sensitivity can be obtained in applications to shallow trench isolation structures by using this dynamical shift correction (Armigliato et al. 2005).

The method employed to measure strain from CBED patterns is to fit the position of the HOLZ lines to theoretical predictions. Two approaches have been developed: the first is based on the kinematic approximation, in which HOLZ lines are approximated by straight lines (Zuo 1992). The second is based on pattern matching using dynamic diffraction simulations. The fitting in the second case is achieved by using pattern matching between the experimental and simulated diffraction patterns (Zuo et al. 1998). Kinematic fitting involves several steps, starting with (1) measuring the experimental lines from the recorded CBED pattern, followed by (2) indexing the experimental lines, and finally (3) carrying out the fitting. The first step is achieved through the following automated functions (Zuo 1992; Senez et al. 2003): (1) preprocessing of diffraction patterns for line detection and (2) line detection using the Hough transformation. Using these techniques, CBED can measure lattice parameters to an accuracy of better than a change of 10^{-4} Å by detecting the shifts of HOLZ lines.

Pattern matching using dynamically calculated CBED patterns takes account of electron multiple scattering effects in the pattern simulations. Using a silicon crystal as a test, Kim et al. measured the Si lattice parameter at different orientations and sample positions. Their standard deviation is 0.0012 Å, which corresponds to 0.02 % (2×10^{-4}) of the Si lattice parameter. The major advantage of dynamical fitting is that this method can be applied to any crystal orientations where HOLZ lines are visible. A drawback of this technique is that it is computationally intensive compared to kinematic fitting. The error bar in kinematic fitting includes a systematic error due to the approximation of ignoring dynamical scattering. This error can be reduced by using the "effective" kinematic orientations (Buxton 1976; Lin et al. 1989), such as the [340] and [230] zone axes in silicon (Armigliato et al. 2003).

16.4.5 3D Strain and Deformation Gradient Matrix

CBED measures lattice constants in three dimensions (3D) by including HOLZ lines in the fitting. It is the only electron beam technique capable of providing 3D strain information at high resolution (EBSD is another technique for 3D strain mapping but at lower resolution). A convenient way to relate the measured lattice parameters to strain is to use the so-called deformation matrix defined in Sect. 16.1 by using the lattice constants of undeformed crystal as a reference. The volume strain of the lattice is given by

$$V_\varepsilon = \det(F) \tag{16.19}$$

Thus, determination of local deformation comes down to the determination of 9 parameters in the F matrix.

Using a high-index [651, 441, and 31] zone axis close to the kinematic diffraction condition, Martin found that the two parameters f_{zx} and f_{zy} had almost no effect on the positions of the HOLZ lines (Martin et al. 2016). Additionally, there is an almost linear dependence of the retrieved f_{zz} over $f_{xx} + f_{yy}$ for deficient HOLZ lines (within the zero disk). This ambiguity is removed when the excess HOLZ lines are included in the fitting. Thus, fitting without the excess HOLZ lines does not allow one to retrieve unique values of the F tensor diagonal terms, i.e., the volume of the lattice cannot be determined. Determination of f_{zx} and f_{zy} required an additional CBED pattern recorded in another zone axis. Thus, all 9 parameters of F matrix can be determined using CBED patterns recorded along two different directions and by including both excess and deficient HOLZ lines in the fitting.

16.4.6   HOLZ Line Splitting from 3D Strain

Those methods for strain measurement using CBED require sharp, well-defined, HOLZ lines. A sharp line is obtained where the spacing of the associated lattice plane under the illuminating electron probe is uniform and constant. In the presence of a 3D strain field, or strain relaxation when the sample is prepared as a cross-sectional TEM lamella, the assumption of a constant lattice under the electron probe is no longer valid. The effect of a deformed lattice on the HOLZ line intensity, to a good approximation, can be predicted using the kinematical diffraction theory of deformed crystals described in the previous chapter, which gives the diffracted wave $\vec{g}$ in the form of an integration over the crystal thickness along the incident beam direction (z).

$$\phi\left(S_g, t\right) \approx \frac{i\pi}{\xi_g} \int_0^t \exp\left(-2\pi i S_g z\right) \exp(-2\pi i \vec{g} \cdot \vec{u}(z))\mathrm{d}z \tag{16.20}$$
$$= FT[\exp(-2\pi i \vec{g} \cdot \vec{u}(z)) \times \Lambda(z)]$$

where the crystal deformation is described by a z-dependent displacement vector $\vec{u}(z)$ and $\Lambda(z) = 1$ for $0 \leq z \leq t$ and $\Lambda(z) = 0$, otherwise. The diffraction intensity is simply given by

$$I\left(S_g, t\right) = \left|\phi\left(S_g, t\right)\right|^2 = \phi\left(S_g, t\right)\phi^*\left(S_g, t\right) \tag{16.21}$$

Inside the CBED disk, S_g goes from positive to negative across the Bragg condition of $\vec{g}$ along the $\vec{g}$ direction. The slope of change is $|\vec{g}|$. Thus, the range of

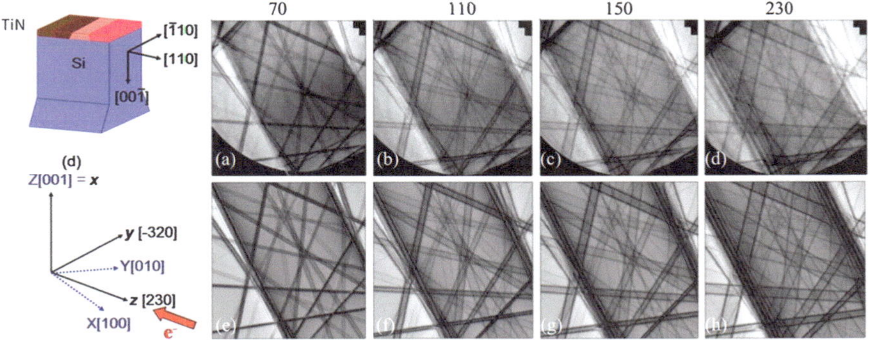

Fig. 16.6 Experimental (*top*) and simulated (*below*) CBED patterns from strained Si showing HOLZ splitting dependent on the electron probe position (labeled on *top*). From Spesspot, Ph.D Thesis

S_g recorded in CBED is far greater for a HOLZ reflection than for a low-order reflection. Secondly, the nonzero deformation phase, $2\pi\vec{g} \cdot \vec{u}(z)$, increases with $|\vec{g}|$. These two factors together make the intensity of HOLZ lines very sensitive to $\vec{u}(z)$ (Fig. 16.6). The ZOLZ reflections recorded in a zone-axis pattern in NBED are not sensitive to $u_z(z)$. By detecting the peak position not the intensity, measurements made in NBD thus is not sensitive to $\vec{u}(z)$.

The analysis of split HOLZ lines requires us to take account of the average strain and the strain variation under the electron beam. There are two approaches that have been developed to achieve this; one uses modeling and the other is through inversion. The finite elemental method (FEM) can be used to model strain in nanodevices. For a given strain model, electron diffraction patterns can be simulated using the scattering matrix method (Houdellier et al. 2006; Jacob et al. 2008) and compared with experimental diffraction patterns. This approach was demonstrated by Houdellier et al. (2006) for a strained SiGe epitaxial layer. Vincent et al. (1999) proposed that the strain profile can be inverted from the diffraction intensity profile. In this method, the z-dependent displacement parallel to g, R_g, is obtained directly by inverting an intensity line profile taken across the HOLZ line using an iterative phasing procedure. Since the intensity of the HOLZ line comes from the Fourier transform of the displacement if the amount of displacement is small and the kinematical approximation applies. The displacement is zero outside the sample, which provides the so-called support for iterative phasing (Fienup 1982; Spence et al. 2002). In this case, the diffraction intensities measured from the CBED pattern contains the necessary phase information and can be reconstructed using the iterative algorithms, such as the Fienup's Hybrid Input and Output (HIO) method (Fienup 1982), which was developed based on the Gerchberg-Saxton algorithm (Gerchberg and Saxton 1972). The feasibility of using phase retrieval for measuring vertical displacements was demonstrated by Vincent et al. (1999) for a Si thin film capped with surface amorphous layers created by Ar ion milling. The method has a significant advantage since it does not require modeling. While applications of this method to nanodevices have not been demonstrated, it appears general and deserves further attention. Alternatively, the model-based method developed by Spessot et al. can be used to extract the thickness-dependent displacement field distribution from split HOLZ lines (Spessot et al. 2007).

16.5   Electron Imaging-Based Strain Measurement Techniques and Applications

The crystal structure, projected along the beam direction, can be also determined from atomic resolution images. Both HREM and STEM have been employed for measuring strain arising from defects or interfaces. Information is recorded about the local atomic structure at a large magnification (M $\sim$ 1 million or higher at atomic resolution) in the form of lattice fringes in HREM, or resolved atomic columns in STEM and a Z-contrast image obtained using a high angle annular dark-field detector (HAADF). Thus, the (S)TEM-based techniques can provide the highest spatial resolution for strain mapping.

The atomic resolution images can be analyzed by locating the positions of the atomic columns recorded in a zone-axis orientation or by using the method of geometrical phase analysis (GPA). The first approach is the most direct; once the atomic positions are located, the strain can be calculated directly using the methods such as LADIA (Lattice Distortion Analysis) (Du et al. 2002), PPA (Peak Pair Analysis) (Galindo et al. 2007), and TeMA (Zuo et al. 2014). In HREM, atomic contrast is obtained only in very thin specimens, which is unrealistic for the characterization of the device structures. GPA is a method that has been developed for strain analysis without the requirement for atomic resolution.

16.5.1   Strain Mapping Using GPA

The geometrical phase analysis (GPA) is based on the simple relationship between the phase $P_g(\vec{r})$ and displacement $\vec{u}(\vec{r})$:

$$P_g(\vec{r}) = -2\pi\vec{g} \cdot [\vec{r} + \vec{u}(\vec{r})].\tag{16.22}$$

where the vector $\vec{r} = (x, y)$ marks a specimen position normal to the incident beam. The phase $-2\pi\vec{g} \cdot \vec{u}(\vec{r})$ is the same phase introduced by atomic displacements in the kinematical theory of defect scattering. A single phase image gives only the component of the displacement field in the direction of g. Two phase images, P_{g1} and P_{g2} where $g1$ and $g2$ are noncollinear, are required to determine the two-dimensional displacement field, $\vec{u}(\vec{r})$, from Eq. (16.22). To examine how the phase can be used to measure strain, it is convenient to express the phase as measured by sample coordinate (x, y) in the form

$$P_g(x, y) = -2\pi g_x x - 2\pi g_y y - 2\pi g_x u_x(x, y) - 2\pi g_y u_y(x, y). \tag{16.23}$$

By taking derivatives of Eq. (16.23), we obtain a relationship between the gradients of the measured phase and the strain

$$
\begin{aligned}
\frac{\partial P_g}{\partial x} &= -2\pi g_x - 2\pi g_x \frac{\partial u_x}{\partial x} - 2\pi g_y \frac{\partial u_y}{\partial x} \\
\frac{\partial P_g}{\partial y} &= -2\pi g_y - 2\pi g_x \frac{\partial u_x}{\partial y} - 2\pi g_y \frac{\partial u_y}{\partial y}
\end{aligned}
\tag{16.24}
$$

Using this for two independent g vectors, the four derivatives of the local displacement can be determined directly from their phase gradients:

$$
\begin{aligned}
e &= \begin{pmatrix} e_{xx} & e_{xy} \\ e_{yx} & e_{yy} \end{pmatrix} = \begin{pmatrix} \frac{\partial u_x}{\partial x} & \frac{\partial u_x}{\partial y} \\ \frac{\partial u_y}{\partial x} & \frac{\partial u_y}{\partial y} \end{pmatrix} = \frac{-1}{2\pi} \begin{pmatrix} g_{x1} & g_{y1} \\ g_{x2} & g_{y2} \end{pmatrix}^{-1} \begin{pmatrix} \frac{\partial P_{g1}}{\partial x} + 2\pi g_{1x} & \frac{\partial P_{g1}}{\partial y} + 2\pi g_{1y} \\ \frac{\partial P_{g2}}{\partial x} + 2\pi g_{2x} & \frac{\partial P_{g2}}{\partial y} + 2\pi g_{2y} \end{pmatrix} \\
&= \frac{-1}{2\pi} A P'
\end{aligned}
\tag{16.25}
$$

The matrix of e can be decomposed into symmetric and nonsymmetric parts, corresponding to strain (ε) and rotation (ω) according to

$$\varepsilon = \frac{1}{2}\left(e + e^{\mathrm{T}}\right); \quad \omega = \frac{1}{2}\left(e - e^{\mathrm{T}}\right). \tag{16.26}$$

The procedures for calculating the strain from the HREM image follow these major steps: (1) Fourier transform of the HREM image, (2) masking of g in the Fourier spectrum and applying inverse transform to obtain $P_g(\vec{r})$ for two independent reflections, (3) use of a reference region in the calculate phase map to obtain the matrix $A,$ and (4) evaluation of Eq. (16.25) and (16.26) to obtain the strain and rotation maps. Details of the implementation of the GPA method can be found in the references of Hytch et al. (1998), Rouviere and Sarigiannidou (2005).

Figure 16.7 shows an example of GPA analysis of a pure edge Lomer dislocation in silicon by Hytch et al. (2006). The dislocation has Burger's vector $\vec{b} = \tfrac{1}{2}[110]$ seen end-on in the $[1\bar{1}0]$ orientation. The HREM image (Fig. 16.7a) shows uniform contrast and little variation in the amplitude of the lattice fringes, which is critical for phase analysis Hÿtch et al. (1998). For the analysis, the phase images were calculated first using the (111) and the $(11\bar{1})$ lattice fringes. The phases were then converted using Eq. (16.23), into displacement parallel (x-axis) and perpendicular (y-axis) to the Burgers vector (Fig. 16.7b). The strain fields are

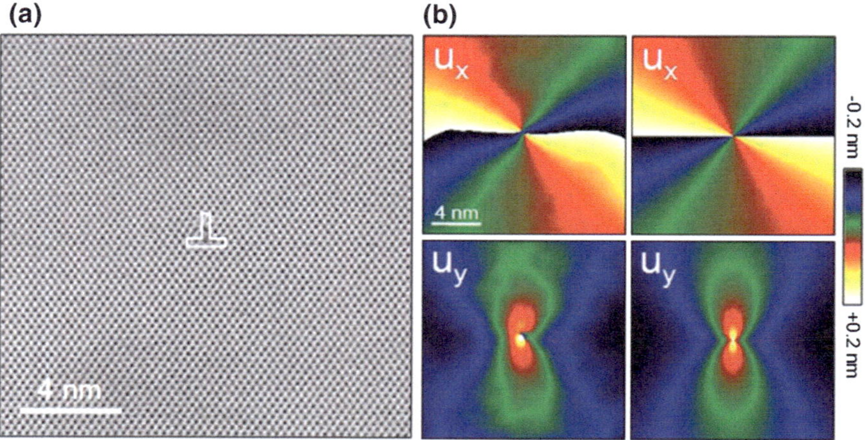

Fig. 16.7 Geometric phase analysis of an edge dislocation in silicon: **a** HREM image recorded in a conventional TEM at 200 kV; **b** in-plane displacement field measured from the experimental image by GPA and theoretical displacement field calculated from linear anisotropic elastic theory. The x-axis is parallel to Burgers vector and spatial resolution 2–3 nm. Contours in the strain map are every 0.5 % from −2.5 to +2.5 % strain (from Hytch et al. 2006)

determined directly from the phase images, according to Eq. (16.25). Results are shown in Fig. 16.7b. In order to compare the results with theory, the displacement field was calculated using anisotropic elastic theory for a dislocation in an infinite medium and using the bulk elastic constants of silicon. Theoretical phase images for the same lattice fringes were calculated and processed using the identical procedures as the experimental case to determine the strains (Fig. 16.7b). Larger values of strain occur in the immediate core region.

16.5.2   STEM and Its Application for Strain Measurements

In using STEM for strain mapping, images of crystals recorded using a HAADF detector provide the relatively uniform image contrast peaked at atomic columns and dependent on atomic number (Z-contrast) (see Chap. 14). The best contrast is obtained using the smallest electron probe. Most important for strain measurements, the peak intensity at the atomic column position in Z-contrast images shows an almost monotonic dependence over a large range of sample thicknesses (LeBeau et al. 2008, 2010b). Atomically centered contrast is also available in high-resolution electron microscopy (HREM), or bright-field STEM, but only when the sample is thin and at the right defocus (Spence 2013). Using these properties, the position of atomic columns can be determined from images recorded in samples of varying thickness using a peak finding technique, followed by a real space analysis of atomic displacements and the related strain.

The standard numerical techniques used for peak finding that have found applications in electron image processing include peak fitting using a model peak distribution, locating peak maximum using the parabolic curve fitting or fitting with other curve functions (Van Aert et al. 2009), the center of gravity method, and by fitting cross correlation function (Zuo et al. 2014).

To measure lattice displacements, a reference lattice can be defined using the measured atomic peak positions in a region of minimal distortion and away from strained regions. The lattice is defined using 6 parameters: base vectors $\vec{a}$ and $\vec{b}$ with its x and y components, and the lattice origin (x_o and y_o). These parameters can be obtained by minimizing the distance between the measured and the reference lattice. The displacement at each lattice point is obtained simply by the difference between the measured peak position and the reference lattice:

$$u_x(h,k) = x_{h,k} - ha_x - kb_x - x_o$$
$$u_y(h,k) = y_{h,k} - ha_y - kb_y - y_o$$

$$(16.27)$$

Both the displacements (u) and the peak position (x, y) are in unit of pixels, whose physical length is determined by the magnification in STEM. The local lattice strain can be determined from the derivative of the measured displacements. For this purpose, it is convenient to introduce the $2 \times 2\ e$ matrix of Eq. (16.25) to calculate the strain and crystal rotation using Eq. (16.26).

A major issue in using STEM image for strain measurements is the scan errors and scan noise introduced into the recorded image. As image intensities are acquired sequentially or pixel-by-pixel in STEM, the recorded image is susceptible to the movements of the probe and sample during image acquisition, as well as systematic and random noise in the probe scan. Random scan noise is introduced during the scan fly back time that has little effect on the atomic peak position detection. The systematic noise leads to atomic displacements that can arise from sample drift, environmental interference, and deviations from the scan generator saw-tooth voltages that were applied to the deflection coils, or any systematic scan noise introduced during the scan fly back time. All these sources contribute to scan errors in the recorded Z-contrast images. A number of numerical methods have been introduced to correct for scan noise and scan errors in as-recorded STEM images (Nakanishi et al. 2002; Rouvière et al. 2011; Braidy et al. 2012; Sang and LeBeau 2014; Ophus et al. 2016). In certain cases, scan errors can be calibrated simply by measuring the displacements of an undistorted lattice. The accuracy of scan error calibration can be improved by averaging the displacements obtained from the undistorted lattice at the same y-scan position (Rouvière et al. 2011). Once the calibration is made, scan error can be corrected in other parts of the image.

Two factors should be taken into consideration in using HAADF-STEM for strain mapping, one is the field of view and the second is the measurement precision. The field of view is determined by the image size times the size of the image pixels, which is determined by magnification. The precision of locating atomic peak shifts depends on the signal/noise ratio in the recorded images, for example, standard deviation of 0.3–0.4 pixels for the peak position determination have been demonstrated without scan noise correction, and a standard deviation of 0.21–0.24 pixels with correction, using a GaSb crystal viewed along [110] as an example (Zuo et al. 2014). The measurement accuracy can be further improved by reducing the noise in the experimental image. A number of groups have reported the use of averaging over multiple STEM images recorded with a shorter exposure times. Using this technique, Kimoto et al. (2010) has demonstrated few picometer precision in peak position by deconvoluting averaged ADF images.

16.6   Off-Axis Electron Holography

In off-axis electron holography, two electron waves of the same frequency are brought together to form an interference pattern. One of the two waves is used as the reference. The interference effect allows a measurement of the phase difference between the two waves. With the phase of the reference wave known, the phase of the second wave or the object wave can thus be determined. This is the basic principle of off-axis electron holography. The basic requirement for performing off-axis electron holography is coherence of the electron waves. In bringing together the electron waves from two separate points on the sample, interference is only obtained when these points fall within the lateral coherence width of the illuminating electrons. Secondly, the interference is only formed between elastically scattered electrons (Lichte 1995; Verbeeck et al. 2011).

Off-axis electron holography is achieved experimentally using an electron biprism that is placed typically between the intermediate and projection lenses. The design of the electrostatic biprism, invented by Möllenstadt and Düker, consists of a thin charged wire placed in between two grounded plates. The wire is positively biased. The potential of this wire for electrons is given by (Matteucci et al. 1998)

$$V(r) = -\frac{\sigma}{2\pi\varepsilon_o}\ln\left(\frac{\pi r^2}{2D}\right) \tag{16.28}$$

where σ is line charge density and D the distance between the two plates. Electrons traveling through this potential field are deflected by an angle of α, the deflection can be described by a phase shift proportional to the distance, x:

$$\beta_o(x) = \mp\frac{\pi}{2\lambda\Phi}\frac{\sigma}{\varepsilon_o}x = \mp\pi k_x x = \mp\pi k\alpha x \tag{16.29}$$

where the plus and minus are for electrons on the left and right side of the charged wire, respectively. Their overlap gives rise to the interference pattern.

The interference of an incident wave of $\varphi_i = \exp(2\pi i k_z z)$ with an object wave of $\varphi_i = A(x)\exp(2\pi i k_z z + \beta(x))$ gives the following interference pattern:

$$I(x) = 1 + A^2(x) + 2A(x)\cos[2\pi k_x x + \beta(x)] \qquad (16.30)$$

To extract the object wave function amplitude $A(x)$ and the relative phase $\beta(x)$, Fourier transform is applied to the electron hologram. The cosine function in Eq. (16.30) gives two major carrier frequencies at $\pm k_x$, where the spectrum of $FT\{A(x)\exp[i\beta(x)]\}$ is centered at $\pm k_x$. Thus, the hologram can be simply reconstructed by applying a mask around one of carrier frequency (For further details, see the review by Midgley 2001).

The phase measured in off-axis electron holography is the phase of the exit wave function from the specimen. In principle, this phase can be used to measure strain following the same methods employed in the GPA of HREM images. In practice, such application is limited because to achieve the same resolution in HREM requires a higher magnification to satisfy the sampling requirement for the electron holograms, which severely limits the field of view in electron holography. To overcome this limit, Hytch and coworkers at CEMES in Toulouse (Hytch et al. 2008, 2011) developed the dark-field electron holography (DFEH) technique. Instead of using an electron wave that has passed through the vacuum as the reference wave as in off-axis electron holography, a diffracted wave from an unstrained reference region is brought into interference with the diffracted electron wave that has passed through a strained ROI in DFEH. The diffracted wave is selected in the back focal plane of the objective lens by placing an objective aperture around the diffraction spot of an identified set of lattice planes. What is recorded in the electron hologram is the phase difference of

$$\Delta P_g(\vec{r}) = P_g(\vec{r}) - P_g^{ref}(\vec{r}) = -2\pi \vec{g} \cdot \vec{u}(\vec{r}), \qquad (16.31)$$

where we assume that the thickness of the reference region and the region of interest is the same, and we have neglected the contribution of a constant phase difference between the two regions. Thus, the reconstructed phase image from DFEH corresponds to the displacement field for the selected lattice planes. In order to determine a 2D displacement field, at least two sets of lattice planes of different directions are required for DFEH. Once the phase is obtained, the strain can then be readily calculated using the GPA method described in Sect. 16.5.1.

DFEH is carried out in the dark-field illumination mode (Beche et al. 2011). A selected diffracted beam is oriented along the optical axis. This achieved by tilting the incident beam and by rotating the specimen so it is oriented to the Bragg condition for the selected diffracted beam. The phase of the diffracted wave is strongly affected by the diffraction condition and specimen thickness. So specimens prepared for DFEH must be flat and uniform in thickness. The electrostatic biprism is set up to produce holographic fringe spacings of a few nm and a hologram width of several hundreds of nm dependent on the size of the 2D detector (Beche et al. 2011; Kasama et al. 2011). The field of view in the reconstructed amplitude and

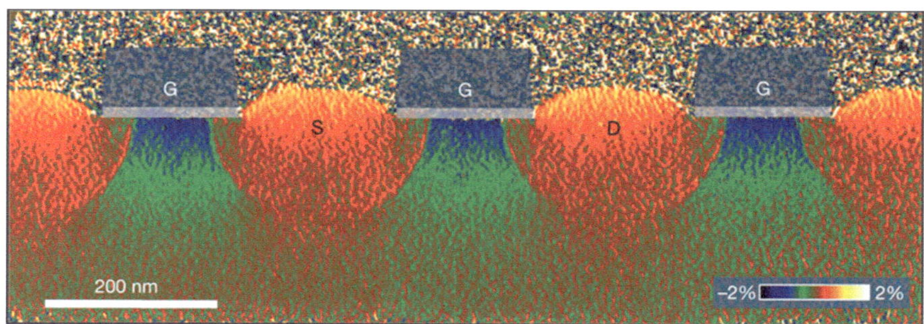

Fig. 16.8 An example of strain in transistor device measured by DFEH. From ref. Hytch et al. (2008)

phase map is determined by the hologram width. Figure 16.8 shows an example of the strain map measured by DFEH and reported by Hytch et al. (2008). The experiment was performed using holographic fringes spaced at 2 nm apart and an overlap width of 250 nm. The holograms were recorded at a nominal magnification of 20,000, on a 2 k slow-scan CCD camera. Digital sampling densities were 0.566 nm per pixel.

Phase and phase distribution can be measured with high accuracy using electron holography. The phase resolution in off-axis electron holography is determined by the following expression (Harscher and Lichte 1996; Lichte 2008):

$$\delta\beta_{\min} = \frac{1}{C}\sqrt{\frac{2}{nN_e\eta(q_c)}} \tag{16.32}$$

where C stands for the fringe visibility or contrast, n for the binning factor used in hologram reconstruction, N_e for the number of electrons per pixel in a recorded electron hologram and $\eta(q_c)$ for the frequency-dependent signal transfer efficiency as defined by the ratio of the output signal/noise ratio (SNR) and input SNR at the characteristic spatial frequency of the holographic fringes. In a typical experimental setup, N_e is about 100–500 for a hologram acquisition time of 2–8 s (Kasama et al. 2011). A longer acquisition time can be used to increase N_e as long as the biprism wire as well as the specimen is stable enough to maintain the fringe visibility. At large fringe spacings, the hologram fringe contrast (C) is determined by the lateral coherence of the electron illumination (which is fixed by the choice of microscope and condenser lens settings) and the quality of the specimen. As the fringe spacing decreases with increasing biprim biasing voltage, the fringe contrast decreases because of the point spread function of the electron detector.

The spatial resolution (r_s) achievable in the reconstructed phase image in DFEH is determined by the largest of following: three times the spacing of the hologram fringes and the size of crystal column which contributes most of the diffracted electron wave. In the device applications of DFEH, the specimen is thick enough to be treated as a strong amplitude and phase object. Lichte suggested masking out the sideband completely from the central band in such a case, then at least 3 fringes have to sample one reconstructed period. It also does not make sense to go much beyond this sampling requirement for two reasons. First, using a higher sampling rate requires enhanced microscope stability that is not always possible, and secondly it takes up more pixels on the electron-camera and thus limits the field of view. As a practical example, we may take 6 pixels used to record each fringe on a 1024 by 1024 pixel CCD for a field of view of 250 nm (see Fig. 16.8 for an example), which gives a fringe spacing of 1.5 nm. The spatial resolution of the phase images, and hence the strain maps, taken to be three times the fringe spacing is then 4.5 nm. The column size in the column approximation is approximately $\sqrt{\lambda t}$, which is 0.7 nm for 200 kV electrons and $t = 200$ nm. Thus, the resolution is determined by the fringe spacing in this case. The same spatial resolution will lead to a field of view of 0.5 μm when a 2048 by 2048 pixel CCD is used for electron hologram recording.

References

Armigliato A, Balboni R, Carnevale GP, Pavia G, Piccolo D, Frabboni S, Benedetti A, Cullis AG (2003) Application of convergent beam electron diffraction to two-dimensional strain mapping in silicon devices. Appl Phys Lett 82:2172–2174

Armigliato A, Balboni R, Frabboni S (2005) Improving spatial resolution of convergent beam electron diffraction strain mapping in silicon microstructures. Appl Phys Lett 86:063508

Armigliato A, Frabboni S, Gazzadi GC (2008) Electron diffraction with ten nanometer beam size for strain analysis of nanodevices. Appl Phys Lett 93:161906

Barna A, Pecz B, Menyhard M (1998) Amorphisation and surface morphology development at low-energy ion milling. Ultramicroscopy 70:161–171

Baumann FH (2014) High precision two-dimensional strain mapping in semiconductor devices using nanobeam electron diffraction in the transmission electron microscope. Appl Phys Lett 104:262102

Beche A, Rouviere JL, Clement L, Hartmann JM (2009) Improved precision in strain measurement using nanobeam electron diffraction. Appl Phys Lett 95:123114

Beche A, Rouviere JL, Barnes JP, Cooper D (2011) Dark field electron holography for strain measurement. Ultramicroscopy 111:227–238

Beche A, Rouviere JL, Barnes JP, Cooper D (2013) Strain measurement at the nanoscale: Comparison between convergent beam electron diffraction, nano-beam electron diffraction, high resolution imaging and dark field electron holography. Ultramicroscopy 131:10–23

Braidy N, Le Bouar Y, Lazar S, Ricolleau C (2012) Correcting scanning instabilities from images of periodic structures. Ultramicroscopy 118:67–76

Buxton BF (1976) Bloch waves and higher-order Laue zone effects in high-energy electron-diffraction. P Roy Soc Lond A 350:335–361

Carpenter RW, Spence JCH (1982) Three-dimensional strain-field information in convergent-beam electron diffraction patterns. Acta Cryst A38:55–61

Chuvilin A, Kaiser U (2005) On the peculiarities of CBED pattern formation revealed by multislice simulation. Ultramicroscopy 104:73–82

Clement L, Pantel R, Kwakman LFT, Rouviere JL (2004) Strain measurements by convergent-beam electron diffraction: The importance of stress relaxation in lamella preparations. Appl Phys Lett 85:651–653

Denneulin T, Cooper D, Rouviere JL (2014) Practical aspects of strain measurement in thin SiGe layers by (004) dark-field electron holography in Lorentz mode. Micron 62:52–65

Diercks DR, Kaufman MJ, Irwin RB, Jain A, Robertson L, Weijtmans JW, Wise R (2010) Using a <670> zone axis for convergent beam electron diffraction measurements of lattice strain in strained silicon. J Microsc 239:154–158

Dingley D (2004) Progressive steps in the development of electron backscatter diffraction and orientation imaging microscopy. J Microsc 213:214–224

Du K, Rau Y, Jin-Phillipp NY, Phillipp F (2002) Lattice distortion analysis directly from high resolution transmission electron microscopy images—the LADIA program package. J Mater Sci Technol 18:135–138

Fienup JR (1982) Phase retrieval algorithms—a comparison. Appl Opt 21:2758–2769

Galindo PL, Kret S, Sanchez AM, Laval JY, Yanez A, Pizarro J, Guerrero E, Ben T, Molina SI (2007) The peak pairs algorithm for strain mapping from hrtem images. Ultramicroscopy 107:1186–1193

Gerchberg RW, Saxton WO (1972) A practical algorithm for the determination of the phase from image and diffraction plane pictures. Optik 35, 237

Gibson JM, Treacy Michael Matthew John (1984) The effect of elastic relaxation on the local structure of lattice-modulated thin films. Ultramicroscopy 14:345–349

Glowacki E, Le Royer C, Morand Y, Pedini JM, Denneulin T, Cooper D, Barnes JP, Nguyen P, Rouchon D, Hartmann JM, Gourhant O, Baylac E, Campidelli Y, Barge D, Bonnin O, Schwarzenbach W (2014) Ultrathin (5 nm) SiGe-on-insulator with high compressive strain (−2 GPa): from fabrication (Ge enrichment process) to in-depth characterizations. Solid-State Electron 97:82–87

Guckel H, Randazzo T, Burns DW (1985) A simple technique for the determination of mechanical strain in thin-films with applications to polysilicon. J Appl Phys 57:1671–1675

Harscher A, Lichte H (1996) Experimental study of amplitude and phase detection limits in electron holography. Ultramicroscopy 64:57–66

Hirsch P, Howie A, Nicolson RB, Pashley DW, Whelan MJ (1977) Electron microscopy of thin crystals. Robert E Krieger Publishing Company, Malabar

Houdellier F, Roucau C, Clement L, Rouviere JL, Casanove MJ (2006) Quantitative analysis of HOLZ line splitting in CBED patterns of epitaxially strained layers. Ultramicroscopy 106: 951–959

Hytch MJ, Snoeck E, Kilaas R (1998) Quantitative measurement of displacement and strain fields from HREM micrographs. Ultramicroscopy 74:131–146

Hytch MJ, Putaux JL, Penisson JM (2003) Measurement of the displacement field of dislocations to 0.03 angstrom by electron microscopy. Nature 423:270–273

Hytch MJ, Putaux JL, Thibault J (2006) Stress and strain around grain-boundary dislocations measured by high-resolution electron microscopy. Philos Mag 86:4641–4656

Hytch M, Houdellier F, Hue F, Snoeck E (2008) Nanoscale holographic interferometry for strain measurements in electronic devices. Nature 453:1086

Hytch MJ, Houdellier F, Huee F, Snoeck E (2011) Dark-field electron holography for the measurement of geometric phase. Ultramicroscopy 111:1328–1337

Jacob D, Lefebvre A (2003) Interpretation of unexpected rocking curve asymmetry in lacbed patterns of semiconductors. Ultramicroscopy 96:1–9

Jacob D, Zuo JM, Lefebvre A, Cordier Y (2008) Composition analysis of semiconductor quantum wells by energy filtered convergent-beam electron diffraction. Ultramicroscopy 108:358–366

Janssens KGF, Vanderbiest O, Vanhellemont J, Maes HE, Hull R, Bean JC (1995) Localized strain characterization in semiconductor structures using electron-diffraction contrast imaging. Mater Sci Tech 11:66–71

Jones PM, Rackham GM, Steeds JW (1977) Higher-order Laue zone effects in electron-diffraction and their use in lattice-parameter determination. P Roy Soc Lond A 354:197

Kasama T, Dunin-Borkowski RE, Beleggia M (2011) In: Ramírez FAM (ed) Electron holography of magnetic materials. Holography—different fields of application. InTech

Kaufman MJ, Pearson DD, Fraser HL (1986) The use of convergent-beam electron-diffraction to determine local lattice-distortions in nickel-base superalloys. Philos Mag A 54:79–92

Keckes J, Bartosik M, Daniel R, Mitterer C, Maier G, Ecker W, Vila-Comamala J, David C, Schoeder S, Burghammer M (2012) X-ray nanodiffraction reveals strain and microstructure evolution in nanocrystalline thin films. Scripta Mater 67:748–751

Kim M, Zuo JM, Park GS (2004) High-resolution strain measurement in shallow trench isolation structures using dynamic electron diffraction. Appl Phys Lett 84:2181–2183

Kimoto K, Asaka T, Yu XZ, Nagai T, Matsui Y, Ishizuka K (2010) Local crystal structure analysis with several picometer precision using scanning transmission electron microscopy. Ultramicroscopy 110:778–782

Kramer S, Mayer J, Witt C, Weickenmeier A, Ruhle M (2000) Analysis of local strain in aluminium interconnects by energy filtered CBED. Ultramicroscopy 81:245–262

LeBeau JM, Findlay SD, Allen LJ, Stemmer S (2008) Quantitative atomic resolution scanning transmission electron microscopy. Phys Rev Lett 100:206101

LeBeau JM, Findlay SD, Allen LJ, Stemmer S (2010) Standardless atom counting in scanning transmission electron microscopy. Nano Lett 10:4405–4408

Lichte H (1995) In: Tonomura A, Allard LF, Pozzi G, Joy DC, Ono YA (eds) Electron holography

Lichte H (2008) Performance limits of electron holography. Ultramicroscopy 108:256–262

Lin YP, Bird DM, Vincent R (1989) Errors and correction term for HOLZ line simulations. Ultramicroscopy 27:233–240

Martin Y, Zuo JM, Favre-Nicolin V, Rouviere JL (2016) Measuring lattice parameters and local rotation using convergent beam electron diffraction: one step further. Ultramicroscopy 160:64–73

Matteucci G, Missiroli F, Pozzi G (1998) Electron holography of long-range electrostatic fields. Adv Imag Elect Phys 99:178–240

Mayer J, Giannuzzi LA, Kamino T, Michael J (2007) TEM sample preparation and FIB-induced damage. MRS Bull 32:400–407

Midgley PA (2001) An introduction to off-axis electron holography. Micron 32:167–184

Nakanishi N, Yamazaki T, Recnik A, Ceh M, Kawasaki M, Watanabe K, Shiojiri M (2002) Retrieval process of high-resolution HAADF-STEM images. J Electron Microsc 51:383–390

Nix WD (1989) Mechanical-properties of thin-films. Metall Trans A 20:2217–2245

Ophus C, Ciston J, Nelson CT (2016) Correcting nonlinear drift distortion of scanning probe and scanning transmission electron microscopies from image pairs with orthogonal scan directions. Ultramicroscopy 162:1–9

Rose H (1995) In: Reimer L (ed) Energy-filtering transmission electron microscopy. Springer, Berlin

Rouviere JL, Sarigiannidou E (2005) Theoretical discussions on the geometrical phase analysis. Ultramicroscopy 106:1–17

Rouviere JL, Beche A, Martin Y, Denneulin T, Cooper D (2013) Improved strain precision with high spatial resolution using nanobeam precession electron diffraction. Appl Phys Lett 103:241913

Rouvière JL, Mouti A, Stadelmann P (2011) Measuring strain on HR-STEM images: Application to threading dislocations in $Al_{0.8}In_{0.2}N$. J Phys: Conf Ser 326:012022

Rozeveld SJ, Howe JM (1993) Determination of multiple lattice-parameters from convergent-beam electron-diffraction patterns. Ultramicroscopy 50:41–56

Sang XH, LeBeau JM (2014) Revolving scanning transmission electron microscopy: correcting sample drift distortion without prior knowledge. Ultramicroscopy 138:28–35

Senez V, Armigliato A, De Wolf I, Carnevale G, Balboni R, Frabboni S, Benedetti A (2003) Strain determination in silicon microstructures by combined convergent beam electron diffraction, process simulation, and micro-raman spectroscopy. J Appl Phys 94:5574–5583

Spence JCH (2013) High resolution electron microscopy, 4th edn. Oxford University Press, Oxford

Spence JCH, Weierstall U, Howells M (2002) Phase recovery and lensless imaging by iterative methods in optical, x-ray and electron diffraction. Philos T Roy Soc A 360:875–895

Spessot A, Frabboni S, Balboni R, Armigliato A (2007) Method for determination of the displacement field in patterned nanostructures by TEM/CBED analysis of split high-order Laue zone line profiles. J Microsc 226:140–155

Stoney GG (1909) The tension of metallic films deposited by electrolysis. Proc Royal Soc Lond A 82:172–175

Tao J, Niebieskikwiat D, Varela M, Luo W, Schofield MA, Zhu Y, Salamon MB, Zuo JM, Pantelides ST, Pennycook SJ (2009) Direct imaging of nanoscale phase separation in $La_{0.55}Ca_{0.45}MnO_3$: relationship to colossal magnetoresistance. Phys Rev Lett 103:097202

Twigg ME, Chu SNG, Joy DC, Maher DM, Macrander AT, Chin AK (1987) Relative lattice-parameter measurement of submicron quaternary (InGaAsP) device structures grown on inp substrates. J Appl Phys 62:3156–3160

Uesugi F, Hokazono A, Takeno S (2011) Evaluation of two-dimensional strain distribution by STEM/NBD. Ultramicroscopy 111:995–998

Van Aert S, Verbeeck J, Erni R, Bals S, Luysberg M, Van Dyck D, Van Tendeloo G (2009) Quantitative atomic resolution mapping using high-angle annular dark field scanning transmission electron microscopy. Ultramicroscopy 109:1236–1244

Vaudin MD, Gerbig YB, Stranick SJ, Cook RF (2008) Comparison of nanoscale measurements of strain and stress using electron back scattered diffraction and confocal raman microscopy. Appl Phys Lett 93:193116

Verbeeck J, Bertoni G, Lichte H (2011) A holographic biprism as a perfect energy filter? Ultramicroscopy 111:887–893

Villert S, Maurice C, Wyon C, Fortunier R (2009) Accuracy assessment of elastic strain measurement by EBSD. J Microsc 233:290–301

Vincent R, Walsh TD, Pozzi M (1999) Iterative phase retrieval from kinematic rocking curves in CBED patterns. Ultramicroscopy 76:125–137

Wilkinson AJ, Meaden G, Dingley DJ (2006a) High resolution mapping of strains and rotations using electron backscatter diffraction. Mater Sci Technol 22:1271–1278

Wilkinson AJ, Meaden G, Dingley DJ (2006b) High-resolution elastic strain measurement from electron backscatter diffraction patterns: new levels of sensitivity. Ultramicroscopy 106: 307–313

Zhang P, Istratov AA, Weber ER, Kisielowski C, He H, Nelson C, Spence JC (2006) Direct strain measurement in a 65 nm node strained silicon transistor by convergent-beam electron diffraction. Appl Phys Lett 89:161907

Zuo JM (1992) Automated lattice-parameter measurement from HOLZ lines and their use for the measurement of oxygen-content in $YBa_2Cu_3O_{7-\Delta}$ from nanometer-sized region. Ultramicroscopy 41:211–223

Zuo JM, Spence JCH (1993) Coherent electron nanodiffraction from perfect and imperfect crystals. Philos Mag A 68:1055–1078

Zuo JM, Tao J (2011) Scanning electron nanodiffraction and diffraction imaging. In: Pennycook S, Nellist P (eds) Scanning transmission electron microscopy. Springer, New York

Zuo JM, Kim M, Holmestad R (1998) A new approach to lattice parameter measurements using dynamic electron diffraction and pattern matching. J Electron Microsc 47:121–127

Zuo JM, Gao M, Tao J, Li BQ, Twesten R, Petrov I (2004) Coherent nano-area electron diffraction. Microsc Res Tech 64:347–355

Zuo J-M, Shah AB, Kim H, Meng Y, Gao W, Rouviére J-L (2014) Lattice and strain analysis of atomic resolution Z-contrast images based on template matching. Ultramicroscopy 136:50–60

Zuo JM, Zhang J, Rouviere J-L (2016) Transistor strain measurement using electron beam techniques. In: ZMADG Seiler (ed) Characterization and metrology for nanoelectronics and nanostructures. Pan Stanford Publishing Pte Ltd (in print)

Chapter 17
Structure of Nanocrystals, Nanoparticles, and Nanotubes

This chapter introduces the study of crystallinity in nanostructures, i.e., nanocrystallography. Since the field of nanostructure study is vast and our knowledge about nanostructures is still emerging, we will focus instead on the fundamentals of the manifestation of chemical bonds in nanostructures and their study based on atomic-resolution imaging and electron diffraction. For this purpose, we discuss specifically the studies of fcc nanocrystals, nanoparticles, and carbon nanostructures, which represent the outstanding, as well as the most studied, examples of nanostructures.

17.1 Nanostructures and Nanoscale Phenomena

Nanostructures are broadly defined as materials having a critical length (L^*) between ~ 1 and 100 nm in at least one dimension. As defined, nanostructures are intermediate structures between individual atoms (or small molecules) and bulk materials. Nanostructures are categorized according to their dimensions:

(1) Zero-dimensional nanoclusters or nanoparticles for which $L < L^*$ in all three dimensions,
(2) One-dimensional nanowires or nanorods where $L < L^*$ in two dimensions,
(3) Two-dimensional atomic layers, where the thickness in the order of two times the atomic radius,
(4) Thin films where $L < L^*$ in the thickness direction.

Although $L^* = 100$ nm is often cited in the literature, its definition depends on the characteristic length of the physical property of the nanostructure. Significant properties arising in nanostructures are associated with the following phenomena.

Surface/interface effects: A significant fraction (or the majority) of atoms are located on or near the surfaces or interfaces in a nanostructure. Since the chemical bonds of surface or interface atoms can differ significantly from interior (or bulk)

© Springer Science+Business Media New York 2017

J.M. Zuo and J.C.H. Spence, *Advanced Transmission Electron Microscopy*,

DOI 10.1007/978-1-4939-6607-3_17

atoms, interfacial atoms give rise to distinct chemical, mechanical, thermodynamic, electronic, magnetic, and optical properties. The surface/interface effects are expected to increase as the ratio of surface to bulk atoms increases.

Quantum confinement: Electrons in a nanostructure are confined in a small space along the critical length direction(s), creating a quantum box or a quantum well. This leads to the quantization of the energy levels. The critical length is then defined by the exciton Bohr radius according to Pauli's exclusion principle. In a semiconductor nanostructure, the difference in energy between the highest valence band level and the lowest conduction band level increases as the nanostructure size decreases. Thus, the energy needed to excite a valence electron, and concurrently, the energy released by an excited electron when it returns to its ground state can be tuned experimentally by controlling the nanostructure size. An example is the synthesis of quantum dots, which can emit any color of light from the same material simply by changing the dot size. In metallic nanoparticles, the same quantum confinement effects are observed, however, only when the nanoparticles are made of about 100 atoms (Halperin 1986).

Topology of atomic bonds: New atomic structures result from a high surface/bulk atomic ratio. An example is the C60 molecule, made of 20 hexagons and 12 pentagons. The sp^2-bonded carbon atom has three bonds along the three edges of the polygon with the atom at the polygon vertex, with a short bond length (1.391) and long bond (1.455) for hexagons and pentagons, respectively. The high percentage of pentagons is due to the large curvature of C60 molecules, which is not allowed in 2D graphene. Another example is metal nanoparticles with special fivefold symmetry.

Nanoscale alloying: Nanosized alloys can form between elements that are immiscible in bulk, such as the archetypical case of Pt and Au (Wanjala et al. 2010; Petkov et al. 2012). The introduction of a surface alters the bulk free energy. In nanostructured crystalline materials, solute atoms with little solubility in the crystallite lattice often segregate to the crystalline interfaces and result in the reduction in the free energy (Gleiter 2000).

Nanostructures are synthesized *from the bottom-up* or *from the top-down*. The bottom-up approaches include the self-assembly of the materials' basic units (down to atoms), leading to nanostructure formation. During self-assembly, the operating physical or chemical forces are used to combine the basic units into a larger stable structure. A typical example is the formation of nanoparticles from a colloidal dispersion. The *top-down* approaches starts with a large initial structure and obtains the structure by processing. Examples include etching through the use of optical masks, ball milling, and the applications of severe plastic deformation. Thus, on the one hand, the nanostructures of interest can be an assembly of macro-molecules or nanoparticles. On the other hand, we have solid-state devices, such as the state-of-the-art transistors, with a characteristic length of a few to tens of nanometers.

Characterization plays a critical role in the study of nanostructures. X-ray diffraction, scanning probe microscopy (SPM) (including scanning tunneling microscopy (STM), atomic force microscopy (AFM), and their variants), and scanning and transmission electron microscopy are the most commonly used techniques. However, the determination of surface or interfacial atomic structure in a nanostructure is a challenge. SPM is surface sensitive, but does not have the penetrating power required for the atomic structure determination, and the field-ion microscope can be useful for 3D imaging of some structures in the form of needles (with species identification) at atomic resolution. The combination of electron diffraction and high-resolution imaging in principle makes the TEM- or STEM-based characterization techniques very powerful for nanostructure characterization, sometimes limited by radiation damage, but still there remains no general solution to the three-dimensional atomic-resolution structure determination problem. However, there are several promising techniques under the development toward this goal (Miao et al. 1999; Zuo et al. 2003; Pfeifer et al. 2006; Lu et al. 2015; Xu et al. 2015).

In what follows, we will describe the atomic structure of nanocrystals, nanoparticles, and carbon nanostructures. Through these examples, we will illustrate the applications of state-of-the-art electron diffraction and imaging techniques.

17.2 Structure of Nanocrystals

The structure of a nanocrystal is defined by (1) the arrangement of its interior atoms, (2) its surfaces and interfaces, which also define the shape of the nanocrystal, (3) any surface or interfacial atomic reconstruction, and (4) defects.

Catalysis is one specific field wherein the ability to determine the nanocrystalline structure of noble metals is critical because their catalytic response can differ markedly depending on the nature of the exposed surface (Bratlie et al. 2007; Chen et al. 2009). It has been frequently proposed, for example, that vertex and edge atom sites bind more strongly to molecules and reactive intermediates than other surface atoms, doing so in ways that broadly impact catalysis (Huang et al. 2008; Sanchez et al. 2009a, b; Bratlie et al. 2007).

17.2.1 Nanocrystal Equilibrium and Kinetic Shapes

The presence of surface gives rise to the surface energy (γ). For a crystal, the surface energy is defined as the excess free energy per unit area for a particular crystallographic face, which can be expressed as a function of the face angle (θ, φ). The total surface free energy of a shaped crystal can be written as follows:

$$G = \int \gamma(\theta, \varphi) \mathrm{d}A,$$

where the integral is over the whole surface. The equilibrium shape of a nanocrystal is determined by minimizing its total surface energies. For amorphous solids or liquids, the surface energy is isotropic and total surface energy can be lowered simply by decreasing the amount of surface area corresponding to a given volume. The resulting particle shape is a perfect sphere. Crystals on the other hand possess different surface energies for different crystal faces. This anisotropy results in stable morphologies where free energy is minimized.

Wulff (1901) has shown that the equilibrium shape of a crystal, which gives the minimum surface energy, can be constructed by the following scheme:

$$\frac{h_i}{h_j} = \frac{\gamma_i}{\gamma_j} = \text{const.} \tag{17.1}$$

where h_i and γ_i represent the distance from the particle center to facet i and the surface energy of that facet. The resulting shape minimizes the quantity G, e.g.,

$$d\sum_i \gamma_i A_i = \sum_i \gamma_i dA_i = 0 \tag{17.2}$$

where "i" represents each facet and A_i is the facet area. The equilibrium shape can then be obtained from the following construction:

(1) Starting from a point, draw vectors normal to all possible crystallographic faces.
(2) The vector length is taken as proportional to γ. Together, these vectors form the Wulff plot.
(3) At every point on the Wulff plot, draw a tangent surface (line in 2D) perpendicular to the normal vector.
(4) The inner envelope of tangent surfaces is the equilibrium shape.

If there are cusps in the Wulff plot, this construction gives faceted equilibrium shapes as illustrated in Fig. 17.1. Thus, crystals are bound by the crystal planes of low surface energies, which tend to be the low-index planes that exhibit closest atomic packing. Metals with the fcc structure tend to have the following low-energy surfaces: {111}, {100}, and {110}, with {111} surfaces having the lowest surface energy (Vitos et al. 1998). By considering only these planes, the essential features of the equilibrium crystal shape can be found. The exceptions are when nanocrystals have strong interactions with their environment. In such case, high-index surfaces that contain steps and kinks can offer better binding to the adsorbates which lower their surface energy.

There is a distinction between the equilibrium shape, which is only attained after careful thermal annealing (Sivaramakrishnan et al. 2010), and the shapes of nanocrystals synthesized under kinetic growth conditions, which have yet to achieve equilibrium (Marks and Peng 2016). In crystal growth, the fastest growing planes terminate during the early stages of growth, and consequently, the slowly growing planes dominate the crystal shape. This fact is well known in colloidal

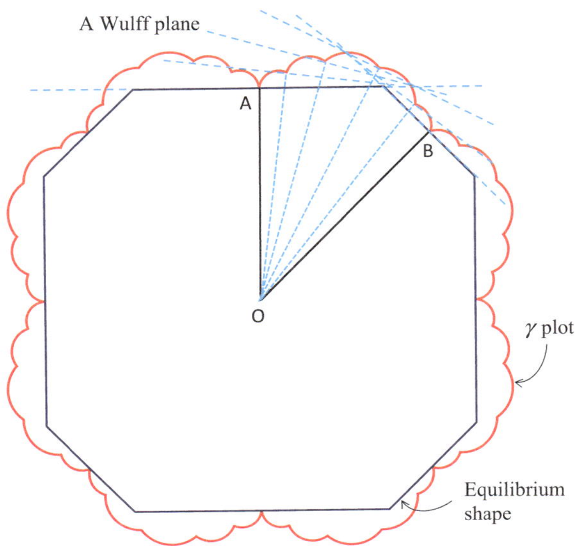

Fig. 17.1 Wulff plot. The length of OA represents the surface energy γ of a plane whose normal is along OA. A plot of γ for all planes whose normal is in the page gives a section of the γ surface (γ plot). Wulff planes are those normal to, and at the end of, the line from O to the γ surface. Wulff planes at the cusp, such as A and B, give the inner envelope and the equilibrium shape (after H. Lüth, Surfaces and Interfaces of Solid, Springer, 1993)

syntheses. For example, by employing molecular capping agents that selectively adsorb to specific crystal planes, shape control at the crystallographic level can be achieved, including anisotropic shapes, such as plates and rods (Murphy et al. 2005). The general strategy to generate different shapes is to stabilize a particular facet through the facet-dependent molecular interactions, e.g., strong binding is used to limit growth, while growth is promoted on crystal planes where binding is weak.

17.2.2 Nanocrystal Facet Determination

The determination of nanocrystal shapes requires the identification of crystal facets and crystallographic planes. For a convex nanocrystal, the outmost facets along the beam direction are captured in the electron image. These facets then can be identified by measuring the facet angles or indexed by measuring the d-spacing of the facet plane, using lattice fringes recorded from an HREM image. The indexing process can also be helped by Fourier transform. For example, the lattice spacing of the dominant facets in Fig. 17.2 is measured at 1.9 Å, which belongs to the $\{200\}$ of Pt. The Fourier spectrum shows streaks running along the $\{200\}$ directions associated with the facets.

The above imaging method works for convex nanocrystals in a few low-index zone-axis orientations. In concave nanocrystals, the image projection hides inward-facing facets. Also high-index facets are difficult to identify using the imaging-based method because of their smaller d-spacing. An alternative approach

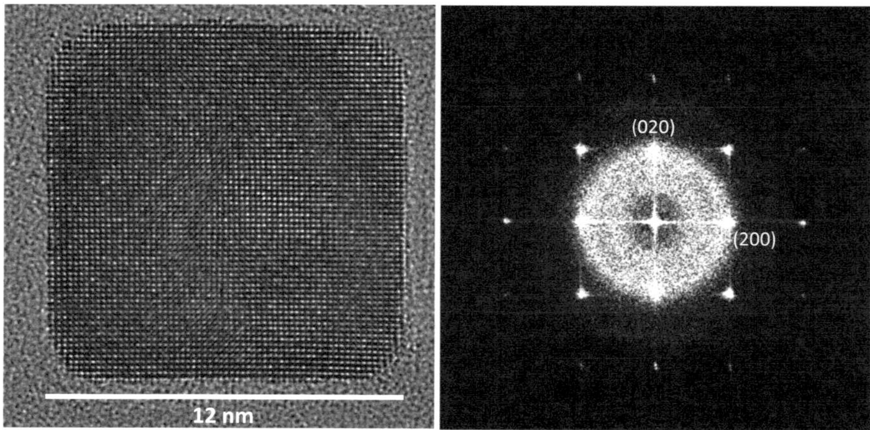

Fig. 17.2 HREM image of a cubic Pt nanocrystal showing the {100} facets and rounding at corners. The nanocrystal was synthesized by J.B. Wu and the HREM image taken by J.G. Wen and W.P. Gao

to overcome these difficulties is to use coherent nanoarea electron diffraction (NAED) described in Sect. 10.6.2. Figure 17.3 illustrates the principle. A nanocrystal is illuminated with a small parallel beam of coherent length L. This length covers at least one crystal facet, or multiple facets in the case of a small nanocrystal. In the limit of the kinematical approximation and a large coherent length, the diffraction intensity is proportional to the square of the nanocrystal structure factor, which is the Fourier transform of the nanocrystal potential

$$F\left(\vec{S}\right) = \mathrm{FT}\{V(\vec{r})\} = \mathrm{FT}\{s(\vec{r})\} \otimes \mathrm{FT}\left[V^{\mathrm{Crystal}}(\vec{r})\right], \qquad (17.3)$$

where $s(\vec{r})$ is the shape of the nanocrystal and $V^{\mathrm{Crystal}}(\vec{r})$ is the periodic crystal potential and

$$\mathrm{FT}\{V(\vec{r})\} = \int_{-\infty}^{\infty} V(\vec{r})e^{2\pi i \vec{S}\cdot\vec{r}}\,d\vec{r}. \qquad (17.4)$$

The FT of $V^{\mathrm{Crystal}}(\vec{r})$ gives rise to sharp diffraction peaks, whose position and kinematic intensity are described in Chap. 4. The nanocrystal shape introduces the shape function $\mathrm{FT}\{s(\vec{r})\}$. The convolution in Eq. (17.3) places the shape function onto each reciprocal lattice point. For a faceted crystal, the reciprocal lattice points of an infinite crystal transform into shape functions consisting of reciprocal rods for each facet. The Ewald sphere intersects the reciprocal lattice rods, and the diffraction pattern is a projection of that 2D slice of reciprocal space. Figure 17.3 illustrates a rectangular crystal and its Fourier transform. The diffraction pattern

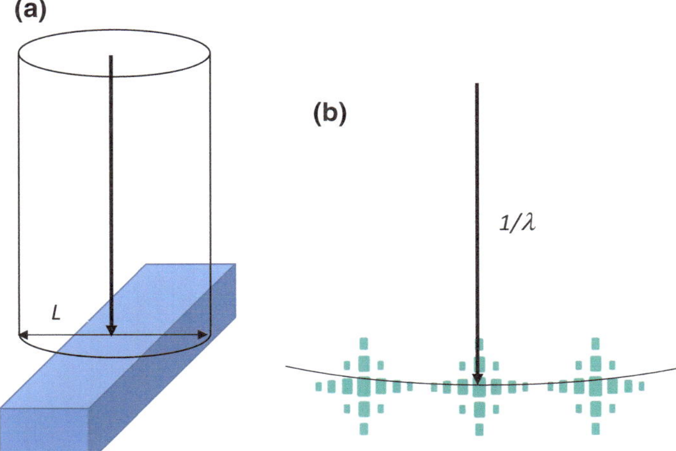

Fig. 17.3 Principle of facet determination using coherent nanoarea electron diffraction. **a** A parallel beam of coherent length L illuminates a 3D crystal. The resulting diffraction pattern is a slice of the Ewald sphere through the reciprocal lattice, producing a 2D diffraction pattern as illustrated in **b**

recorded is the intersection of the Ewald sphere with the shape function at each reciprocal lattice point. For high-energy electrons, the Ewald sphere is approximately flat because of the small electron wavelength. Thus, the reciprocal lattice rod of a nanocrystal facet intersects the Ewald sphere almost tangentially when the facet is normal to the incident electron beam. Measurement of the recorded reciprocal lattice rods, together with the electron diffraction pattern, allows a determination of the facets. For large nanocrystals of size greater than the coherence length, it is possible to determine individual facets by placing the coherent beam on these facets. It should be noted that electron multiple scattering causes the redistribution of diffraction intensities, but does not affect the direction of reciprocal rods.

Figure 17.4 displays a diffraction pattern from a 16 nm × 41 nm Au nanorod oriented to the [1 1 0] zone axis. The diffraction pattern shows a well-ordered spot pattern indicating the nanorod is single crystalline. The weak diffuse rings in addition to sharp diffraction spots are due to diffraction of the amorphous carbon, which is used as support. Each reflection has a streak pointed to the (0 1 −1) direction in reciprocal space. These (0 1 −1) streaks come from intersection of the Ewald sphere with reciprocal lattice rods of (0 2 −2) planes. Streaks for (0 0 2) planes in the [0 0 1] direction are weakly observed; they are much shorter and broader than those in the [0 1 −1] direction.

Figure 17.5 displays a diffraction pattern from a 44 nm trisoctahedron (TOH) oriented to the [1 1 0] zone axis, with an inset of (0 0 −4) to show the streaks more clearly. Around each of the Bragg reflections, we observe 8 strong steaks. In fcc crystals, the reciprocal rod direction is normal to the plane of the same index. Therefore, the angle between directions is also the angle between planes.

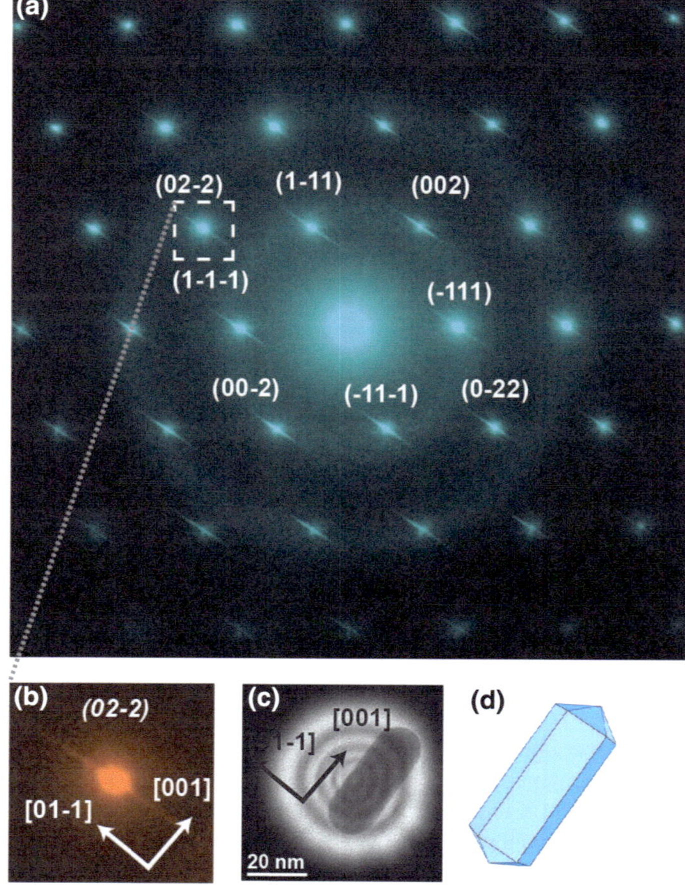

Fig. 17.4 a NAED pattern of a nanorod oriented to the [110] zone axis. Well-defined streaks point in the [0 1 −1] direction. **b** Magnified view of (0 2 −2) reflection showing streaks pointing to the [0 1 −1] direction. **c** Image of the nanoprobe on the nanorod. **d** A computer-generated image of the nanorod [reprinted with the permission from Shah et al. (2013)]

We measure the direction that the streaks were pointing toward by tracing a line from each streak until it intersects a Bragg reflection. In general, we select higher order reflections since the streaks typically point at reflections far from the origin. To determine the direction of these vectors from the origin, the angle between direction vectors for an fcc crystal is given by

$$\cos(\theta) = \frac{h_1 * h_2 + k_1 * k_2 + l_1 * l_2}{\sqrt{h_1^2 + k_1^2 + l_1^2}\sqrt{h_2^2 + k_2^2 + l_2^2}}$$

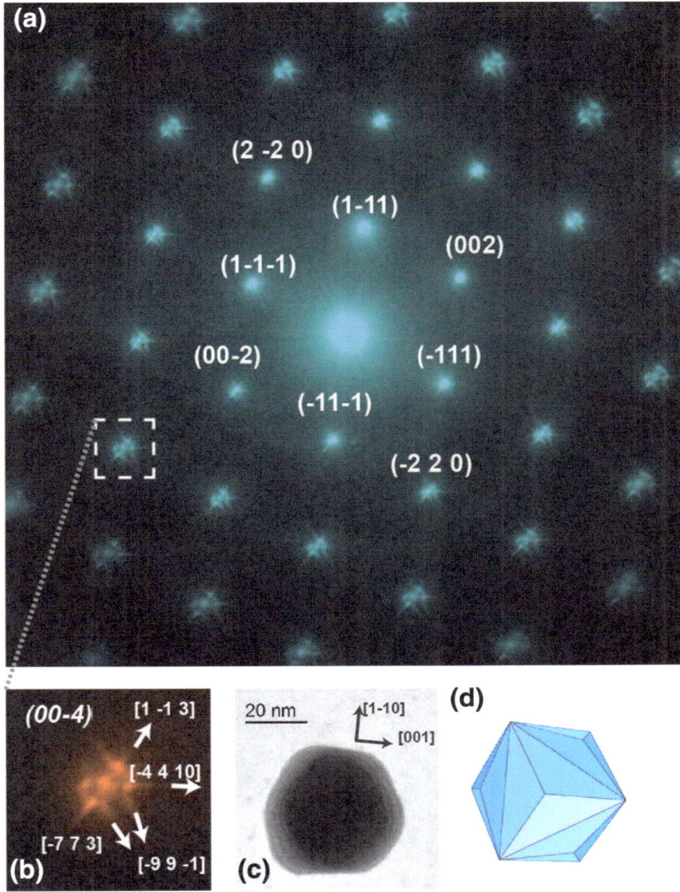

Fig. 17.5   a NED pattern of a 44-nm TOH crystal oriented to the [110] zone axis. The streaks are normal to the planes. **b** Magnified view of (0 0 −4) reflection showing streaks measured from this and other reflections. **c** TEM image of the TOH crystal. **d** Computer-generated image of the TOH crystal [reprinted with the permission from Shah et al. (2013)]

For example, using the reflection of (3 −3 −1), two streaks pointing to [−1 1 9] and [−6 6 −2] are observed; the directions of these streaks are determined to be [−4 4 10] and [−9 9 −1] by a vector difference, and the angle between these directions is 65.0° (Shah). Diffraction patterns such as Figs. 17.5 and 17.6 contain multiple reflections; several reflections may be used to index the complete array of streaks.

Figure 17.6 displays the angles between edge tangential (white text) and plane normals (black text) measured for the image of a TOH nanocrystal. The facet directions were measured by determining the angle between edge normal and fundamental directions in the Fourier transform of the image. The error of measuring angles directly from the image is 4 % and is primarily due to the facet edges appearing

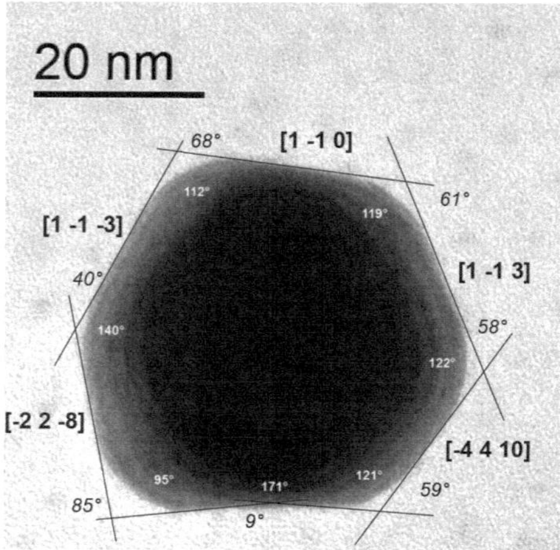

Fig. 17.6 TEM image of a 44-nm TOH gold nanoparticle. The tangents to the faces are drawn, and the angles between tangents are shown in *white text*. The angles between plane normals are 180° minus this angle and are shown in *black text*. The measurement error arises in drawing lines tangent to the shape edges. The facets are determined by tracing normals to the edges and measuring the angle from fundamental directions in the fast Fourier transform of the image [reprinted with the permission from Shah et al. (2013)]

curved in the TEM images. In comparison with the diffraction patterns, two facet indices match [(−4 4 10) and (1 −1 3)]. Streaks for the (1 −1 0) and (−2 2 −8) facets from the image are not observed in the diffraction pattern. We conclude that either these facets are small and not sharp or we are seeing a cliff-like edge in the transmission image where the nanocrystal is concave and the facet is not normal to the beam. The 44 nm TOH geometry is verified through the comparison of measured angles to a computer-generated 3D model with unit edge lengths 1 and $1 + 1/\sqrt{2}$ (Shah et al. 2013).

The measurement of streak length and width yields additional structural information about the roughness of each facet. According to Fig. 17.7, the roughness is characterized by two parameters, the average width (w) and height (H). They are related to the streak width and length, respectively. We can calculate the length of each streak in reciprocal space by solving for the camera constant in Eq. (17.5), λL, where R is the number of pixels from a reflection to the origin and d is a known d-spacing for Au for the particular reflection. We then measure the length of the streak R and define $1/R$ as a facet size parameter.

$$1/R = d/\lambda L \tag{17.5}$$

For example, in the rod diffraction pattern of Fig. 17.4, the streak facing [0 −2 2] measured from the reflection (0 0 −2) has a length of 119 pixels and the camera constant is 575 pixels * Å. The facet height parameter is 0.21 Å^{-1} or 4.8 Å.

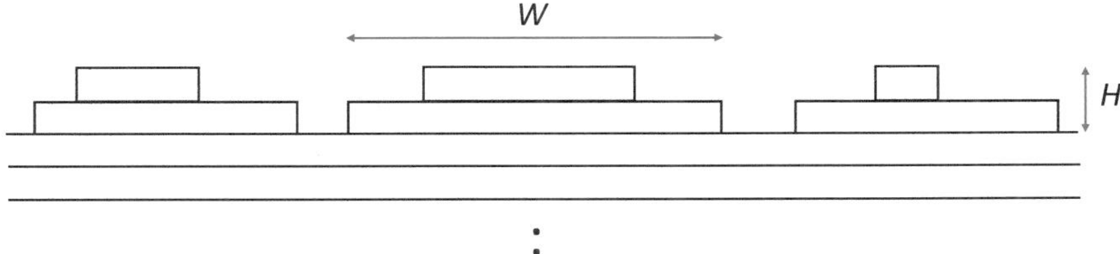

Fig. 17.7 Surface facet roughness model. The width (W) and height (H) define the roughness parallel and normal to the surface

The higher order reflections are expected to have a shorter streak length due to the small curvature of the Ewald sphere. Additionally, there can be errors in sample tilt, beam tilt, and projector lens distortion, which affect the length of the streak. A correction factor (c_{hkl}) can be expressed in Eq. (17.6) that accounts for these errors.

$$1/R_{hkl} = c_{hkl} * \frac{d}{\lambda L}. \tag{17.6}$$

An accurate measurement of beam tilt will thus improve the precision of measurement.

17.2.3   Identification of Planar Faults Using Coherent CBED

A useful effect was discovered from computations and observations of coherent CBED patterns from nanocrystals (Cowley and Spence 1981; Pan et al. 1989). This allows them to be used to find the fault vector which characterizes planar faults in crystals, in the spirit of the "g · b" analysis used in TEM imaging. Specifically, it was found that CBED disks that would normally be uniformly filled with intensity show annular rings of intensity instead, if the probe is placed near the edge of a crystal. Figure 17.8 shows the effect. We will refer to this effect loosely as "spot splitting." In subsequent work on planar faults, it was found that if the probe was situated over a fault (with the beam in the plane of the fault), then only those reflections for which $\vec{g} \cdot \vec{R} \neq 0$ were split. Here, $\vec{R}$ is the fault vector describing the translation needed to bring the crystal on one side of the fault into coincidence with that on the other. (Such a vector may not always be defined.) Thus, by noting which disks are split, $\vec{R}$ may be deduced. For planar faults, the splitting is normal to the

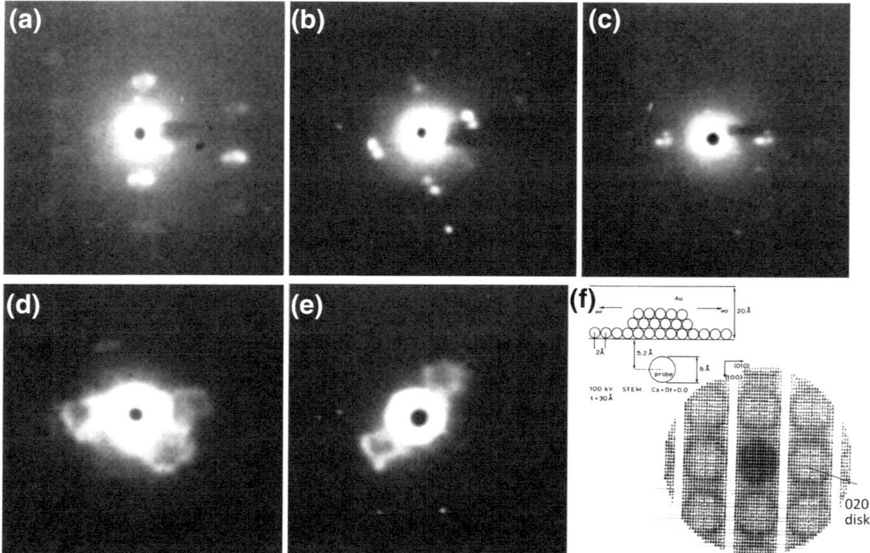

Fig. 17.8 Nanodiffraction patterns from small gold crystallites with the incident beam at the edge of the crystal. The beam convergence angle is about 3×10^{-3} rad (beam diameter at specimen 1.5–2 nm) for (**a–c**) and 2×10^{-2} rad for (**d, e**). (**f**) Is a computed CBED pattern for a probe diameter of 0.5 nm placed 0.52 nm outside a gold nanowire of 2 nm wide and 3 nm thick as shown *inset*

plane of the fault. Examples of this general approach can be found as follows: For a study of antiphase boundaries in Cu_3Au, see Zhu and Cowley (1982). Here, it was found possible to deduce the nearest-neighbor atomic coordination across a boundary from a study of coherent CBED patterns. For similar work on stacking faults and twins in steel, see Zhu and Cowley (1983).

By comparison with coherent electron diffraction or conventional dark-field TEM imaging, this approach has two important advantages—first the $\vec{g} \cdot \vec{R}$ conditions for many reflections may be obtained from a single diffraction pattern, and second, the method may readily be combined with STEM imaging. GP zones (thin precipitates) in Al-4 %Cu have also been studied by this method, providing support for a particular atomic model of the precipitate (Zhu and Cowley 1985). Microdiffraction studies on catalyst particles have proved informative. For work on Rh particles (about 2–3 nm in diameter) on Ce oxide substrates, see Pan et al. (1987). Epitaxial relationships between particle and substrate, twinning and oxidation, may all be investigated.

17.2.4 Nanocrystal Surface Reconstruction

The surface chemical bonds are different from the bulk; the unbalanced forces acting on surface atoms lead to surface relaxations, including bond length contraction or expansion or bond reconstructions, by forming a new surface structure (Bohnen and Ho 1993). On two-dimensional surfaces, surface reconstruction can be readily probed by STM (Besenbacher 1996). Surface reconstruction can also be studied by low-energy electron diffraction (LEED) (Van Hove et al. 1986) or reflection high-energy electron diffraction (RHEED) (Ichimiya and Cohen 2004) or surface ion scattering (Duke 1996). For example, the Au (110) surface has been known to reconstruct to exhibit the missing row structure in the form of the (1×2) reconstruction (Besenbacher 1996). Theoretically, on an extended two-dimensional (2D) surface of a simple metal, atoms contract to the underlying layer to lower their energy (Bohnen and Ho 1993). A simple model is that an electrostatic dipole moment is created on the surface due to the pullback of surface electrons, which leads to surface atom contraction. This is known as Smoluchowski smoothing effect (Smoluchowski 1941). Bond contraction with reduced coordination number for a surface atom was also suggested by Pauling (1947). In quantum mechanics, surface contraction is a result of the competition between the pairwise potential and the multiatom potential which takes into account the contribution from the electron gas in metals (Heine and Marks 1986).

Surface atomic contraction was first observed experimentally by ion scattering (Rieder et al. 1983) and LEED (Davis and Noonan 1983). For example, on the Cu(110) surface, an 8.5 % (Adams et al. 1983), or 5.3 % (Stensgaard et al. 1983), contraction was found between the 1st and the 2nd layers, while 1.2 % contraction was found on the (100) surface (Jiang et al. 1991). Similarly, a 8.5 % contraction was detected for the Al(110) surface by LEED (Davis and Noonan 1983; Noonan and Davis 1984).

Conventional surface scattering techniques using X-rays or ions or low-energy electrons require large and flat surfaces. To see how coherent diffraction patterns recorded from individual nanocrystals can be used to study surface relaxation, let us consider the diffraction intensity of a monoatomic nanocrystal under the kinematical approximation

$$I(\vec{S}) = f(\vec{S}) \left| \sum_i \exp(-2\pi i \vec{S} \cdot (\vec{R}_i + \delta \vec{r}_i)) \right|^2 \tag{17.7}$$

where $\vec{S}$ is the scattering vector, $f(\vec{S})$ denotes the atomic scattering factor including the Debye–Waller factor, $\vec{R}_i$, and $\delta \vec{r}_i$ denote the lattice site for the ith atom and the displacement from the perfect lattice position. The summation is over all atoms in the nanocrystal. In a perfect crystal, where all atoms are at the exact lattice sites (unrelaxed), $\delta \vec{r}_i = 0$, and Eq. (17.7) can be rewritten as follows:

$$I(\vec{S}) = f^2(\vec{S}) \left| \int s(\vec{r}) \cdot \left[\sum_i \delta(\vec{r} - \vec{R}_i) \right] \cdot \exp(-2\pi i \vec{S} \cdot \vec{r}) \mathrm{d}\vec{r} \right|^2$$
$$= f^2(\vec{S}) \left| s(\vec{S}) \otimes \sum_g \delta(\vec{S} - \vec{g}) \right|^2 , \qquad (17.8)$$

which is a reformulation of Eq. (17.3) for a monoatomic nanocrystal. The resulting diffraction pattern is thus a convolution between the diffraction of an infinite crystal and the Fourier transform of the shape of the nanocrystal, $s(\vec{S})$, which will be copied at each Bragg peak. The shape factor is centrosymmetric about the Bragg peak. When strain or surface reconstruction is present in the nanocrystal, $\delta\vec{r}_i$ is not zero. In this case, the shape factor can be rewritten as follows:

$$s_{\text{strained}}(\vec{r}) = s(\vec{S}, \vec{r}) \exp(-2\pi i \vec{S} \cdot \delta\vec{r})$$
$$s_{\text{strained}}(\vec{S}) = \int s_{\text{strained}}(\vec{S}, \vec{r}) \exp(-2\pi i \vec{S} \cdot \vec{r}) \mathrm{d}\vec{r} \qquad (17.9)$$

That is, the shape function of a strained nanocrystal can be represented as a complex function, with the phase corresponding to the strain field $\delta\vec{r}$ (Robinson et al. 2001). Note that in Eq. (17.9), the shape function is also a function of the scattering vector $\vec{S}$. Thus, different Bragg peaks now have a different shape factor. Because of the complex strain field, a non-centrosymmetric shape factor is expected from a strained particle (Robinson et al. 2001; Robinson and Vartanyants 2001).

The scattering vectors ($\vec{S}$), recorded by high-energy electrons, fall approximately onto a plane normal to the incident beam. Thus, a section of the amplitude of the complex 3D shape function is recorded. Figure 17.9 illustrates this principle using an 11-nm Au nanocrystal as an example. By tilting the incident beam using dark-field tilt, several sections were recorded from the 3D shape function at 5 mrad increments in a JEOL FEG TEM. Similar diffraction patterns can be obtained from larger nanocrystals by coherent X-ray diffraction (Robinson and Vartanyants 2001).

The effects of surface relaxation manifest in diffraction as an asymmetric shape factor around Bragg peaks. To illustrate how diffraction intensity around a Bragg peak depends on the sign and amount of surface relaxation, here we write out an analytic form of the kinematical intensity for a simple one-dimensional lattice model (Fig. 17.10). An atomic chain consists of $N - 2$ atoms with a regular bond length a and two atoms located at both ends with a contracted bond length $a - \gamma$. The scattered wave at the far field is the sum of scattered waves from all atoms in the chain:

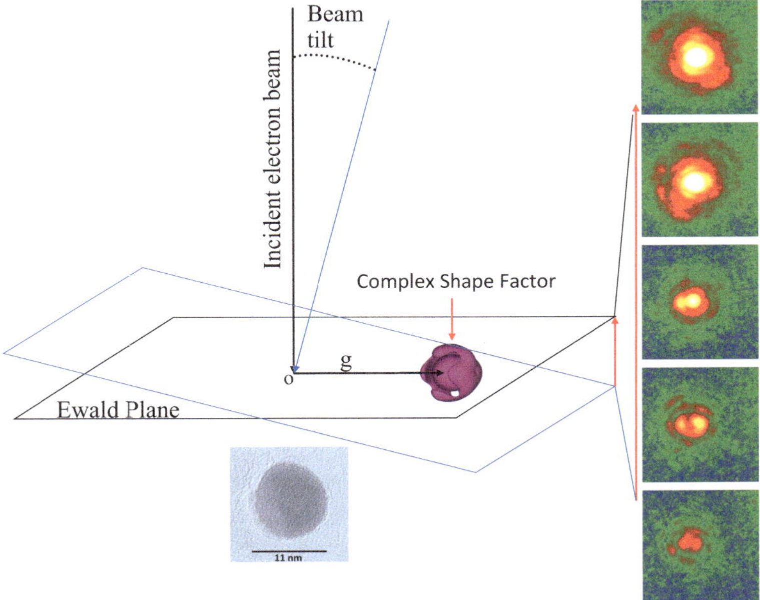

Fig. 17.9 Recording of 3D shape function for the reflection g by electron diffraction. The diffraction patterns on the right are from an Au nanocrystal of 11 nm in diameter (from Ji Li, University of Illinois)

$$\varphi(k)/f = \exp[-ik \cdot (\gamma - a)] + \exp[-ik \cdot ((N - 2)a + (a - \gamma))]$$
$$+ \sum_{n=0}^{N-2} \exp(-ik \cdot na) \tag{17.10}$$

where $k = 2\pi S$ is along the chain direction. The third term in Eq. (17.10) gives the Bragg peak for the unrelaxed "core." The first and the second terms are the scattered waves from the two contracted "surface" atoms. The intensity distribution in the proximity of a Bragg peak ($k = 2\pi/a$) in the diffraction pattern, specifically at the first-order local maxima where

$$k = \frac{2\pi}{a} \pm \frac{2\pi}{Na},$$

can be evaluated to the first order of γ as follows:

$$I\left(\frac{2\pi}{a} \pm \frac{2\pi}{Na}\right) = \varphi\left(\frac{2\pi}{a} \pm \frac{2\pi}{Na}\right) \cdot \varphi^*\left(\frac{2\pi}{a} \pm \frac{2\pi}{Na}\right) \approx A \pm \frac{B}{a} \cdot f^2 \cdot \gamma \tag{17.11}$$

where constants A and B depend on the chain length N. The last step in Eq. (17.11) is based on two assumptions: First, the total chain length N is large enough that the first-order local maxima are in the proximity of the Bragg peak of interest, i.e., $2\pi/a \gg 2\pi/(Na)$; secondly, the amount of contraction γ is much smaller compared to the ideal bond length a, i.e., the product of $|g| \cdot \gamma$ should be small, where g is the reciprocal of a. The normalized difference between the two first-order maxima,

$$\left[I\left(\frac{2\pi}{a} - \frac{2\pi}{Na}\right) - I\left(\frac{2\pi}{a} + \frac{2\pi}{Na}\right) \right] \Big/ \left[I\left(\frac{2\pi}{a} - \frac{2\pi}{Na}\right) + I\left(\frac{2\pi}{a} + \frac{2\pi}{Na}\right) \right],$$

is therefore proportional to $|g| \cdot \gamma$ for a given chain length N. The calculated diffraction pattern from this surface-contracted 1D lattice is shown in Fig. 17.10 (black line), where the patterns from an unrelaxed ($\gamma = 0$, red line) and an expanded lattice ($\gamma < 0$, green line) are superimposed for comparison. It is clear that the bond contraction of the two end atoms breaks the local symmetry of the shape factor around the Bragg peak. Specifically, the intensity of the first maximum on the side farther away from the direct beam ($k = 0$) is raised by $B|F|^2\gamma/a$, while the first maximum on the other side is reduced by the same amount. The degree of this local asymmetry in the shape factor, or the difference between these two maxima intensities normalized to the average intensity of the two maxima, is linearly proportional to the bond contraction γ for small γ. Reversing the sign of γ results in an expanded chain and a horizontally flipped diffraction pattern (green line). Therefore, the intensity distribution in the shape factor is extremely sensitive to both the sign and the amount of the end-atom contraction in the above 1D model. Local asymmetry around the Bragg peaks in the X-ray diffraction patterns from a microcrystal has been related to the strain in the crystal by Robinson and coworkers (Robinson and Vartanyants 2001).

The above diffraction model suggests that coherent diffraction from a nanocrystal can be used to determine its surface structure. This was demonstrated by Huang et al. (2008). Figure 17.11 shows a single-particle diffraction pattern of a 3.5-nm-diameter Au nanocrystal from their study. The Au nanocrystal was supported by graphene as the image in the inset shows. The diffraction pattern indicates

(1) the Au nanocrystal is single crystalline and is oriented approximately along its [110] zone axis on the substrate,

(2) the diffraction peak consists of a strong peak, surrounded by ring-like secondary peaks of first and second orders,

(3) for the three selected Bragg peaks, $(-11-1)$, $(00-2)$, and $(1-1-3)$ as indexed in Fig. 17.11, none is centrosymmetric and they are all different, suggesting a complex, reflection-dependent shape factor corresponding to a relaxed surface structure,

(4) the average Au–Au bond length is determined as 2.885 ± 0.005 Å from the Bragg peak positions using the supporting graphene lattice as calibration. This value is very close to the bond distance of bulk Au crystals, thus indicating that overall the nanocrystal is not strained by the graphene substrate.

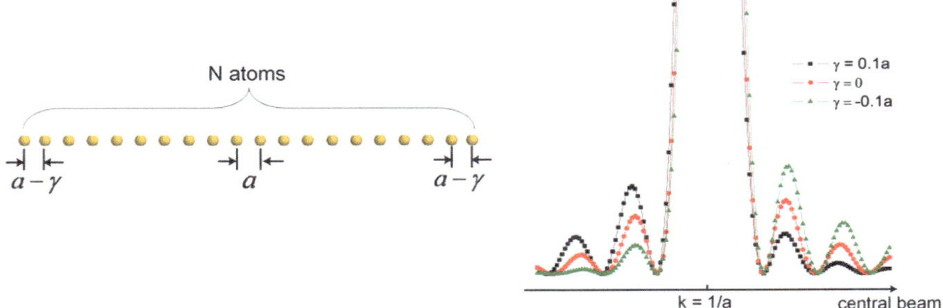

Fig. 17.10 1D lattice chain. *Left* a schematic model showing a 1D lattice chain with a regular lattice spacing a in the core and $a - \gamma$ for the two end atoms; *Right* calculated kinematical diffraction patterns in the proximity of the Bragg peak ($k = 1/a$) for $\gamma = 0.1a$ (*black*), $\gamma = 0$ (*red*), and $\gamma = -0.1a$ (*green*) to represent a contracted, unrelaxed, and expanded surface (Weijie Huang, Ph.D. Thesis, University of Illinois)

The basic features observed in the diffraction pattern of Fig. 17.11 can be explained using a coordination-dependent radial surface contraction model (Huang et al. 2008). In this model, the surface atoms are selected by their coordination numbers and are displaced toward the center of the nanocrystal by a distance Δr proportional to the natural log of the coordination number following the formulation of Pauling (1947):

$$\Delta r = \alpha \cdot \kappa(h, k, l) \cdot r \cdot \ln(12/n)/r_0 \tag{17.12}$$

where r is the distance to the center, r_0 the average radius of the nanocrystal, n the coordination number (defined by the number of nearest-neighboring atoms), α the scaling parameter, and $\kappa(h, k, l)$ a ratio used to describe the facet-dependent contraction. Both α and $\kappa(h, k, l)$ can be determined by matching simulations with the experimental pattern as shown in Fig. 17.12. The asymmetry observed in Fig. 17.12 rules out the possibility of unrelaxed surface or surface expansion for the Au nanocrystal, which would result in an intensity shift around Bragg peaks toward the central beam. The calculated diffraction patterns (marked as Model) for three Bragg peaks, $(-11-1)$, $(00-2)$, and $(1-1-3)$, were obtained using $\kappa(1, 0, 0) = 1$, $\kappa(1, 1, 1) \approx 0.35$, and $\alpha = 0.37$. A key feature is that the direction of the asymmetry (dashed arrows) is pointing along the $\langle 200 \rangle$ direction, which is reproduced only when $\kappa(100)/\kappa(111)$ is much larger than one (the solid arrows give the directions toward the central beam in the diffraction pattern).

For the 3.5-nm Au nanocrystal, the bond length distribution obtained from the best-fit radial contraction model is shown in Fig. 17.13. The atomic bond lengths at the surface are found to reduce by 1–8 % with respect to the interior of the particle. The amount of contraction in the model increases with a decreasing coordination number (Fig. 17.13b). The edge and the vertex atoms (with coordination number ≤ 7)

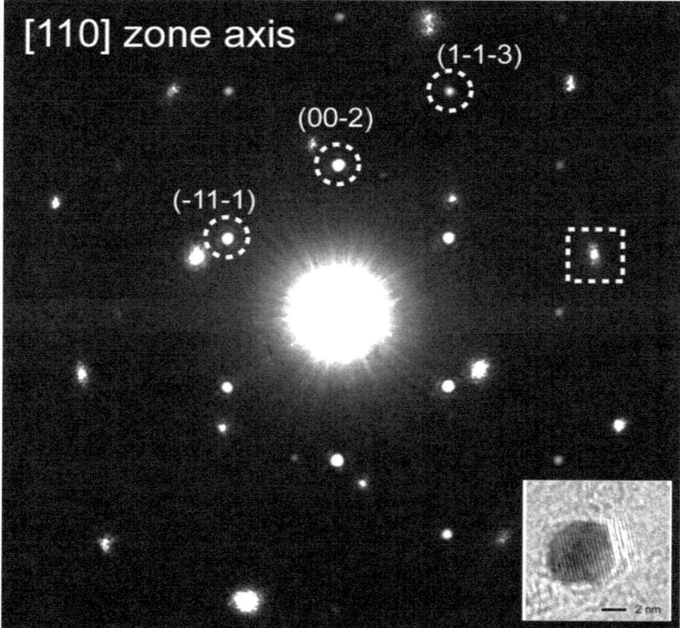

Fig. 17.11 A diffraction pattern recorded from a 3.5-nm Au nanocrystal supported on graphene along its [110] zone axis. *Three highlighted Bragg peaks* are analyzed for the surface relaxation. The *square* denotes the graphene reflection

contract by the largest amount (0.22 Å in average). The model also predicts an average out-of-plane bond length of 2.76 Å (a 0.13-Å contraction compared to the bulk Au–Au bond length of 2.885 Å) for {100} and 2.83 Å (0.05 Å with a distribution ranging from 0.03 to 0.07 Å) for {111} surface atoms.

Experimentally, coordination-dependent surface contraction was evidenced on Au nanocrystals of ∼3 to 5 nm. At this size range, coherent NAED records the strong first-order ring and sometimes the weaker second-order ring as well.

This coordination-dependent surface contraction is further supported by molecular dynamics (MD) simulations, which were performed by relaxing the models of Au nanocrystals at a simulated temperature $T = 500$ K using the embedded atom method (EAM) potential (Daw et al. 1993). Diffraction patterns were simulated using the atomic coordinates obtained from the MD simulation. The results are shown in Fig. 17.12 for the three Bragg peaks of $(-11-1)$, $(00-2)$, and $(1-1-3)$. Good agreement was obtained between the experimental patterns and the simulated patterns from MD; both the shapes of the central maxima and main features of the intensity distributions of the first-order oscillations are reproduced by MD. In particular, the intensity around $(00-2)$ in the experimental diffraction pattern shifts away from the central beam, while around $(-11-1)$ and $(1-1-3)$, the asymmetry directions are along $\langle 200 \rangle$. These trends are also observed in the

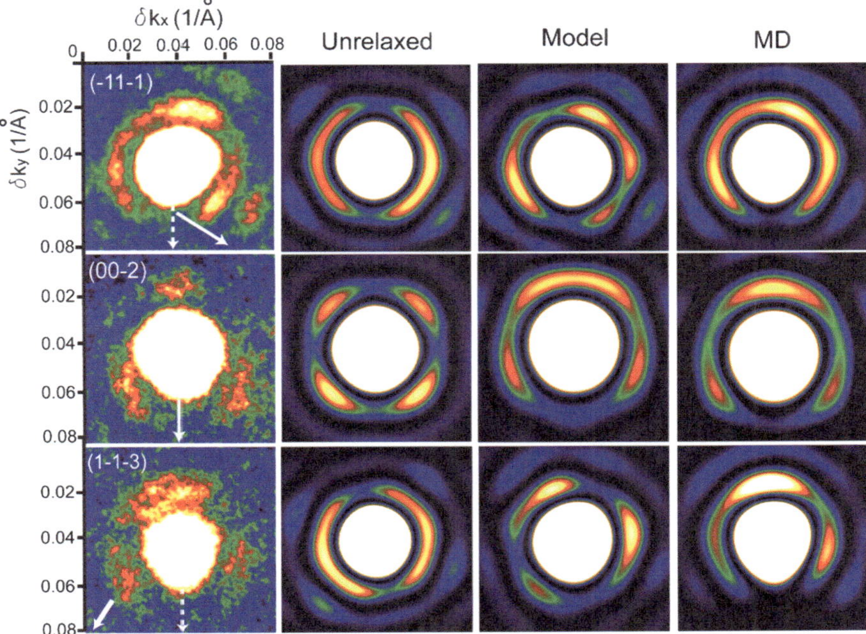

Fig. 17.12 Experimental diffraction intensities of an Au nanocrystal and their modeling. Experimental diffraction patterns are shown in the left column for the $(-11-1)$, $(00-2)$, and $(1-1-3)$ Bragg peaks (*upper*, *middle*, and *lower* panel). The *solid white arrows* in the pattern point toward the central beam in the experiment. The *dashed white arrows* indicate the directions of $\langle 200 \rangle$ in the reciprocal space. Simulated diffraction patterns are generated from a nanocrystal with surface atoms remaining in their original bulk position (unrelaxed), surface atoms relaxed according to Eq. (17.12), and surface atoms relaxed by molecular dynamics

contraction model of Eq. (17.12). Overall, the MD simulation improves the agreement with experiment, since it takes into account forces between atoms.

The MD simulations also provide further details on the bond deformation. Figure 17.14 plots the atomic displacements in 3D vector maps, with both the magnitudes and directions of the surface atom displacements represented. The displacements for atoms on the $\{100\}$ surfaces (including step atoms on $\{100\}$ facets) and edges are on average larger than those on the $\{111\}$, except for those close to the edge of the $\{111\}$ facets. The magnitude of displacements of the $\{100\}$ atoms varies from ~ 0.09 to 0.21 Å, while the displacements of $\{111\}$ atoms vary from 0.14 Å for atoms close to the edges, to 0.02 Å for atoms near the center of the facets. The atoms on the vicinal facets (between $\{111\}$ and $\{100\}$) have the largest magnitude of contraction (the largest is ~ 0.21 Å). There are much smaller contractions for $\{111\}$ surface atoms. While the vectors of the $\{100\}$ atoms predominantly point perpendicular to the surfaces, vectors of the $\{111\}$ atoms have larger in-plane components. The farther the atoms are located away from the center of the $\{111\}$ facets, the larger is the magnitude of the in-plane components. This suggests

Fig. 17.13 Bond length
distributions in the Au
nanocrystal obtained by
modeling the experimental
diffraction pattern. **a** Average
bond lengths versus the radial
distance to the center for the
best-fit radial contraction
model (*cross*) and molecular
dynamics simulation (*circle*).
b The out-of-plane bond
length (bond length between
atoms in different atomic
layers) is plotted as a function
of the atomic coordination
number

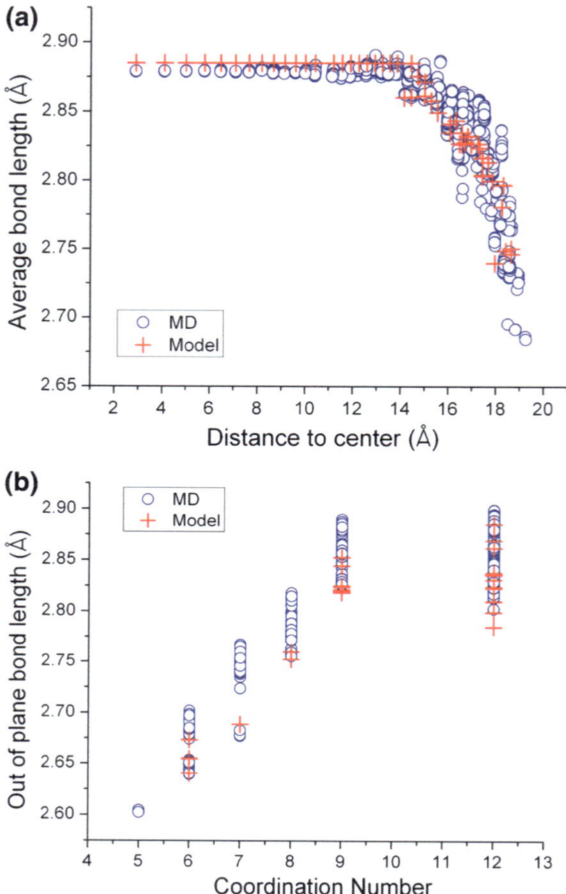

that surface atoms on different facets must accommodate each other, which is not
included in the radial contraction model. From the MD simulations, atomic displacements on the {111} surfaces appear to be partially forced by the neighboring
{100} and vicinal facets. Such structural relaxation dynamics involving collective
atomic rearrangements with different coordination numbers is unique to nanocrystal
surfaces and is not expected for bulk 2D surfaces.

17.2.5   *Surface Atoms of a Twinned Nanocrystal*

The above discussion demonstrates the characteristics of coherent diffraction
patterns from individual, single-crystalline nanocrystals, as well as surface effects.
The radial surface contraction model (Eq. 17.12) has provided the critical insight.

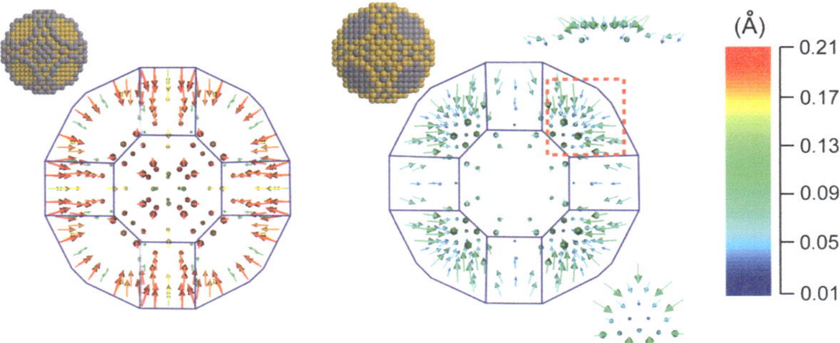

Fig. 17.14 Surface atom contraction obtained from the MD simulation. Surface atom displacements are shown here as vectors for atoms having a coordination number less than 9 (*left*, mostly are {100} atoms and neighboring vicinal facet atoms) and equal to 9 (*right*, mostly are {111} surface atoms). The *upper right inset* in *right* shows a magnified {111} facet viewed parallel to the facet, while the *lower right inset* shows the same facet viewed normal to the facet (which is tilted slightly to avoid atoms on the other facets). The magnitudes of the displacements are rendered using colors. The corresponding atoms in the nanocrystal model are shown as *inset at the top right corner*; the atoms whose displacements are shown are colored in *violet*. Both maps have the same display scale in the magnitude of displacement

In a nanocrystal with defects, the surface structure, however, could be modified by the presence of defects. Understanding the interplay between the surface atoms and the defects requires more sophisticated tools, especially the modeling of the defects, the surface relaxation, and their effects on diffraction patterns. Figure 17.15 shows an example. The experimental diffraction pattern was recorded from a twinned Au nanocrystal of 4 nm in diameter (see inset in Fig. 17.15a). Both the HREM image and the diffraction pattern show that the nanocrystal contains a single (111) twin boundary. Model crystals were constructed to simulate the effects of twins on the diffraction pattern. The approximate agreement between the experimental and simulated diffraction patterns (b and c) was obtained from a model containing two asymmetric fcc halves (d).

Two major features in the experimental patterns are reproduced by the modeling:

(1)   the Bragg peak pair consisting of the (111) and (002) Bragg peaks (inside the dashed boxes),
(2)   the intensity distribution around the Bragg peaks as shown in Fig. 17.15b, c.

The model from which the diffraction pattern was simulated was relaxed by MD. The good agreement between the experiment and the MD simulation suggests that the MD predicts the actual pattern of surface relaxation in this twinned nanoparticle. The surface displacement map of (111) surface atoms from the MD relaxation is shown in Fig. 17.15e. Most of the atomic displacements are pointing inward, as the case in single crystal. Interestingly, atoms near the twin boundary are displaced toward the boundary, instead of pointing to the center of the particle. This suggests that the presence of the twin boundary modifies the surface coordination environment locally, which influences the relaxation of nearby surface atoms.

17.2.6 *The Equilibrium Shape of Supported Nanocrystals*

A supported nanocrystal has an interface with the substrate. This interface modifies the equilibrium shape of the nanocrystal. The principle used to determine the equilibrium shape of the supported nanocrystal is based on the minimization of the free energy (see discussion below). Kaishew is credited with first modifying the Wulff theorem for predicting the equilibrium shape of supported crystals on a flat surface (Kaishew 1952). The excess free energy of a supported nanocrystal can be approximately expressed as Kern et al. (1979)

$$\Delta G(n) = -n\Delta\mu + \sum_{j\neq\text{Int}} \gamma_j A_j + (\gamma_{\text{Int}} - \gamma_S)A_{\text{Int}}, \tag{17.13}$$

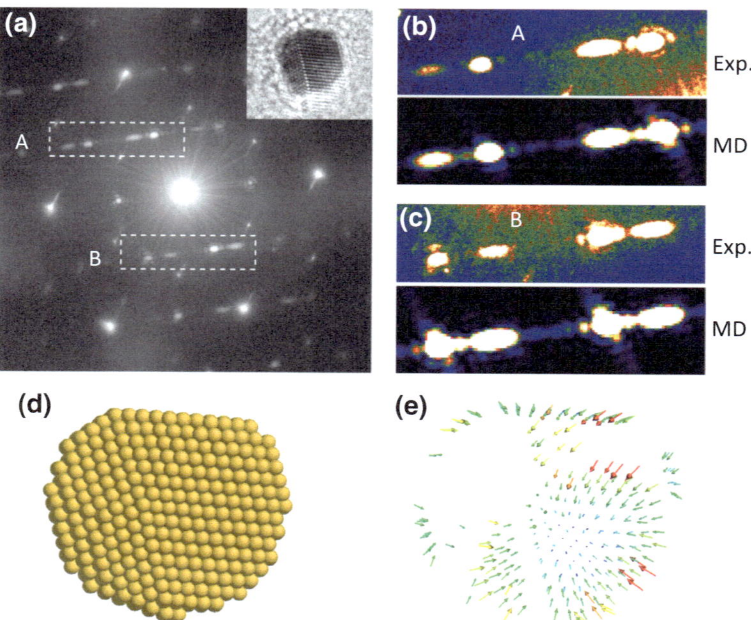

Fig. 17.15 Surface contraction from a single-twin Au nanocrystal. **a** Diffraction pattern recorded from a 4-nm single Au nanocrystal, inset: the HREM image of the nanocrystal; **b, c** comparisons between experimental (*upper*) and simulated (*lower*) diffraction features; the two selected areas are highlighted by *dashed box* in (**a**); **d** structural model of an asymmetric single-twin Au nanocrystal; **e** vector map for the displacements of (111) surface atoms in the model (from Weijie Huang, University of Illinois)

where γ_j and A_j are the nanocrystal surface energy and the surface area of the jth free surface, γ_{Int} and γ_S are the interfacial energy and the surface energy of the support, respectively, $\Delta\mu$ is the chemical potential difference between an atom in the gas phase and an atom in the nanocrystal, and n is the number of atoms in the nanocrystal. Noting that

$$V = nv$$

where V is the volume of the nanocrystal and v is the atomic volume, and

$$V = \frac{1}{2}\left(\sum_{j\neq\mathrm{Int}} h_j A_j + (h - \Delta h)A_{\mathrm{Int}}\right)$$

where h is the distance from the center of the crystal to the facets and Δh is the undercut (Fig. 17.16), and the derivative of ΔG can be written as follows:

$$\mathrm{d}\Delta G(n) = -\frac{\Delta\mu}{2v}\left(\sum_{j\neq\mathrm{Int}} h_j \mathrm{d}A_j + (h - \Delta h)\mathrm{d}A_{\mathrm{Int}}\right) + \sum_{j\neq\mathrm{Int}} \gamma_j \mathrm{d}A_j + (\gamma_{\mathrm{Int}} - \gamma_S)\mathrm{d}A_{\mathrm{Int}}$$

Using the partial equilibria conditions

$$\left(\frac{\partial\Delta G}{\partial A_{\mathrm{Int}}}\right)_{A_j,\dots,T,\Delta\mu} = 0 \quad\text{and}\quad \left(\frac{\partial\Delta G}{\partial A_j}\right)_{A_i,\dots,A_{\mathrm{Int}},T,\Delta\mu} = 0,$$

we obtain,

$$\frac{(\gamma_{\mathrm{Int}} - \gamma_S)}{h - \Delta h} = \frac{\Delta\mu}{2v}, \tag{17.14}$$

and

$$\frac{\gamma_h}{h} = \frac{\Delta\mu}{2v}, \tag{17.15}$$

where γ_h is the surface energy of the nanocrystal surface facet parallel to the interface. Combining Eqs. (17.14) and (17.15) gives the following equation:

$$\frac{(h - \Delta h)}{h} = \frac{(\gamma_{\mathrm{Int}} - \gamma_S)}{\gamma_h}. \tag{17.16}$$

It can be seen from the above equation that as the interfacial energy of an interface is lowered, the undercut (Δh) increases, so that the nanocrystal wets the

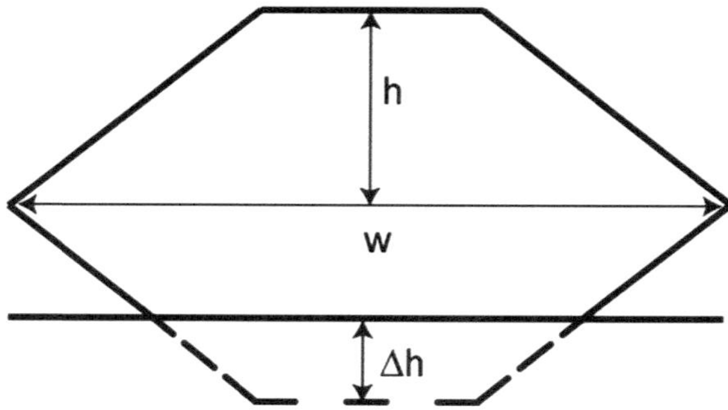

Fig. 17.16 Schematic illustration of a supported nanocrystal. h is the distance from the center of the crystal to the facets. "Δh" is the undercut. "w" is the width of the NC

support as the interfacial energy decreases. This is fundamentally similar to the effect of interfacial energy on the shape of supported liquid droplets. The wetting behavior is illustrated schematically in Fig. 17.17.

An alternative description of the interfacial energy uses the work of adhesion (W_{adh}) defined originally by Dupré (1869), where

$$W_{adh} = \gamma_h + \gamma_S - \gamma_{Int} \qquad (17.17)$$

which represents the net energy gained from forming an interface. A lower interfacial energy results in a larger work of adhesion (W_{adh}). Scaling the work of adhesion by the surface of energy of the nanocrystal interfacial facet gives the relative work of adhesion as follows:

$$\frac{\Delta h}{h} = \frac{W_{adh}}{\gamma_h}, \qquad (17.18)$$

which is a convenient measure of the adhesive strengths between disparate materials.

Using the Wulff–Kaishew principle, the work of adhesion and interfacial energies of nanocrystal interfaces can be determined. Since the shape provides only relative values of W_{adh} and γ_{int} (scaled by the surface energy), one has to know the surface energy a priori in order to find the absolute values of W_{adh} and γ_{int}. However, experimental determination of surface energies is difficult (Venables 2000), and consequently, theoretically calculated surface energies have been used in order to obtain absolute values for W_{adh} and γ_{int}.

The challenge of measuring the shapes of small (<10 nm) nanocrystals is that the atomic-resolution imaging of the interface and the facets of small nanocrystals is very challenging. Measurements of the equilibrium shapes have been previously reported on relatively large crystallites using HREM (Cosandey and Madey 2001; Sadan and Kaplan 2006) and SEM (Heyraud and Metois 1980; Sadan and Kaplan 2006). Other experimental observations were made using SPM (Hansen et al. 1999; Worren et al. 2001; Bonzel 2003; Koplitz et al. 2003) and grazing incidence

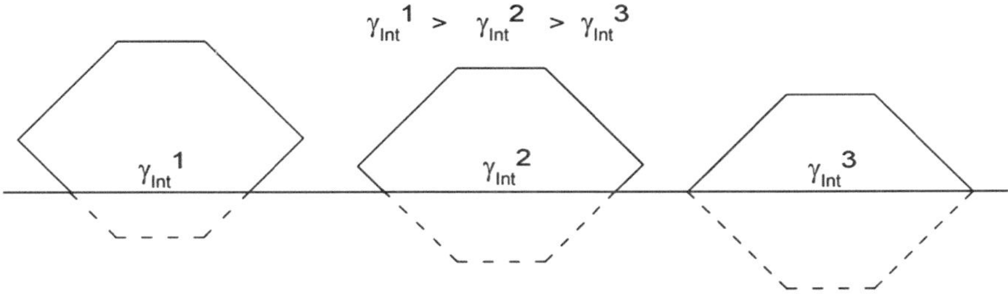

Fig. 17.17 Schematic illustration of wetting with decreasing interfacial free energy. The *dashed lines* are a guide to the eye and depict increasing undercut with decreasing interfacial energy

small-angle X-ray scattering (Renaud et al. 2003). Among these techniques, TEM and SPM are capable of measuring small (sub-10 nm) nanocrystals. Cross-sectional TEM is potentially helpful, but conventional cross-sectional sample preparation methods lead to modifications of the surface atomic structure of smaller nanocrystals. SPM has been successful in determining the shapes of small 2D islands (Hansen et al. 1999; Worren et al. 2001; Koplitz et al. 2003) but generally has difficulties with shape measurements of 3D nanocrystals that are less than 10 nm (Mitchell et al. 2001; Tanner et al. 2001; Pakarinen et al. 2006).

Figure 17.18 shows two examples of the equilibrium shape analysis of epitaxial Au nanocrystals supported on TiO_2 (110) using aberration-corrected STEM. The sample employed for the study was prepared by a special procedure, which involves making electron transparent (<100 nm) TiO_2 first, removing damaged surface by heating TiO_2 in an air furnace at 1000 °C, and the formation of Au nanocrystals by physical vapor deposition followed by annealing to promote the formation of the equilibrium shape, as well as epitaxy (for details, see Sivaramakrishnan et al. 2010).

The dimensions that can be measured directly from experimental images such as Fig. 17.18 are the total height ($2h - \Delta h$) and the width (w, which are related). For example, in Fig. 17.18c, it can be shown that $h = w/\sqrt{6}$. The interfacial energy in the case of Fig. 17.18 was calculated by using $\gamma_{Au(111)} = 1.283$ J/m^2 following Vitos et al. (1998) and $\gamma_{TiO_2(110)} = 0.33$ J/m^2 following Cosandey and Madey (2001). The average interfacial energy of Au nanocrystals similar to Fig. 17.18a was measured to be 0.61 ± 0.1 J/m^2 from a group of 23 nanocrystals (Sivaramakrishnan et al. 2010). In comparison, the interfacial energy of the nanocrystal in Fig. 17.18b and two others was measured at 1.12 ± 0.17 J/m^2 using $\gamma_{Au(100)} = 1.627$ J/m^2 (Vitos et al. 1998). The difference of 0.45 J/m^2 in interfacial energy between these two cases was attributed to the difference in the interfacial atomic structure. Experimentally, the reduced interfacial energy in the case of Fig. 17.18a is evidenced by their abundance among the annealed nanocrystals.

17.2.7   *Triple Junctions and Line Tension*

Our discussion of supported nanocrystals so far has neglected the triple junction at the interface of the support, the nanocrystal and vacuum. In general, triple junctions (TJs) are ubiquitous and their contribution to the free energy (TLE, triple line energy) is small in most materials. In the liquid–solid vapor TJs, the TLE is also commonly referred to as line tension (LT), which was first conceptually described by Gibbs (1839–1903). Later, Boruvka and Neumann (1977) generalized this concept using Young's equation for the contact angle of a liquid droplet on a solid surface, by incorporating LT as

$$\cos(\theta) = \frac{\gamma_{SV} - \gamma_{SL}}{\gamma_{LV}} - \frac{\tau}{\gamma_{LV} r} \tag{17.19}$$

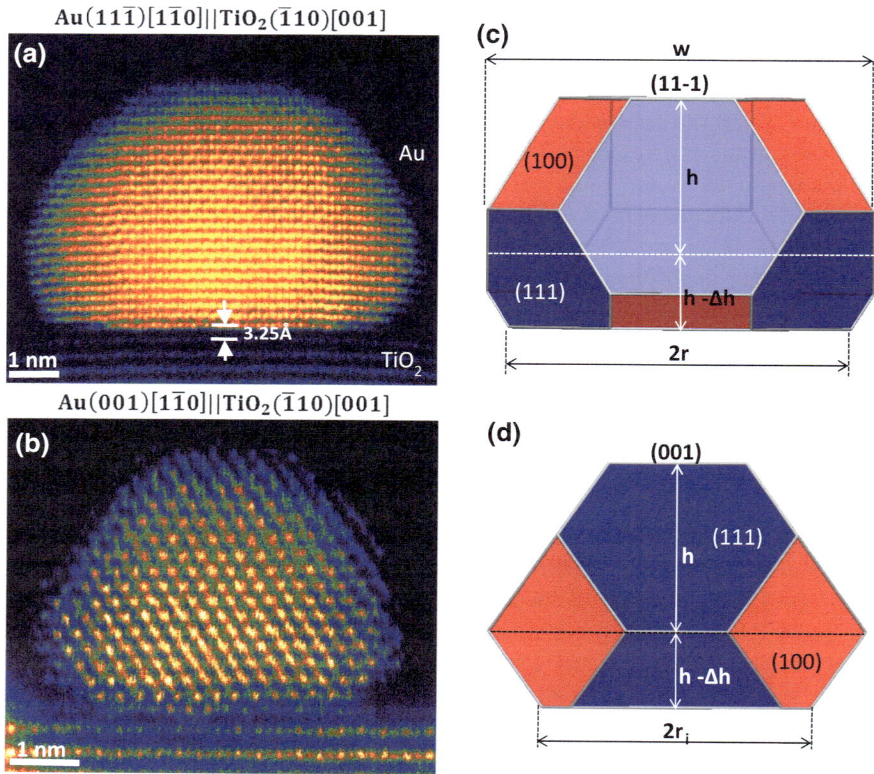

Fig. 17.18 **a**, **b** Atomic-resolution HAADF-STEM images of two Au nanocrystals of different epitaxy. **c**, **d** Models of the nanocrystal shapes in each case. The nanocrystals in (**a**) and (**b**) are viewed from the [112] and [1−10] zone axes, respectively (from Shankar Sivaramakrishnan, University of Illinois)

Here, θ is the contact angle of the droplet, γ_{SV} and γ_{LV} are the surface energies of the solid substrate and the liquid droplet, γ_{SL} is the interfacial energy, r is the radius of the droplet, and τ is the interfacial line tension. LT becomes important in determining the wetting angle (θ) of liquid droplets when the LT term in Eq. (17.19) becomes significant, i.e., as r reduces to the order of the characteristic length $l_c = \tau/\gamma$. From Eq. (17.19), it is clear that LT (τ) can be extracted from plots of $\cos(\theta)$ versus $1/r$ if all the other thermodynamic parameters are constant. Using this approach, LT has been measured in several liquids (Amirfazli et al. 1998; Pompe and Herminghaus 2000; Amirfazli and Neumann 2004). Theoretical estimates of LTs of liquid droplets are on the order 10^{-10} N, while experimental values range from 10^{-6} to 10^{-9} N (Amirfazli and Neumann 2004). For typical values of $\gamma_{LV} = 10^{-2}$ J/m^2 and LT $= 10^{-9}$ N, the effect of LT on liquid droplet shapes can be seen below 100 nm. By comparison with liquids, the surface energies of metallic solids are about two orders of magnitude larger. Thus, the characteristic length of metallic solids is expected to be the orders of magnitude smaller.

By taking into account the interfacial TLE (τ_{Int}) for supported nanocrystals, the Wulff–Kaishew principle can be modified as follows (Sivaramakrishnan et al. 2010):

$$\frac{\Delta h - h}{h} = \frac{\gamma_s - \gamma_{Int}}{\gamma_h} - \frac{\tau_{Int}}{\gamma_h}\frac{dl_{Int}}{dA_{Int}} \qquad (17.20)$$

where A_{Int} and l_{Int} are the interfacial area and interfacial perimeter, respectively. This expression for equilibrium shapes of solid nanocrystals with TLE is fundamentally similar to the effect of LT on liquid droplets (Boruvka and Neumann 1977) except for the "geometric factor" of dl_{Int}/dA_{Int}, which depends on the shape of the interface.

For epitaxial nanocrystals with the shape of a truncated cuboctahedron shown in Fig. 17.18a ($Au(11\bar{1})\|TiO_2(110)$), the measurement of the nanocrystal dimensions yields the quantities h, w, and $h - \Delta h$. The geometric factor is expressed in terms of the nanocrystal dimensions according to (Sivaramakrishnan et al. 2010)

$$\frac{dl_{Int}}{dA_{Int}} = \frac{2\sqrt{3}}{a_{Int}},$$

where

$$a_{Int} = w + \frac{2\Delta h}{1.633}$$

Figure 17.19c shows a plot of "$(\Delta h - h)/h$" versus "$1/a_{Int}$" for Au nanocrystals of $Au(11\bar{1})\|TiO_2(110)$ epitaxy. The data are fitted by a linear curve, which statistically give a better fit than for a horizontal line fit. From the slope of a direct-weighted linear fit to Fig. 17.19c, TLE was measured to be 0.28 ± 0.08 eV/Å ($4.53 \pm 1.27 \times 10^{-10}$ J/m). The error bar is about 24 %.

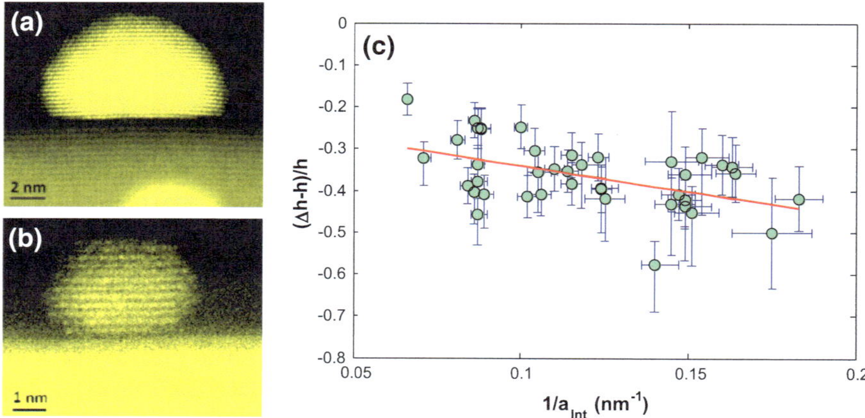

Fig. 17.19 Measurement of line tension. Two Au nanocrystals of (**a**) 10.9 nm and (**b**) 4.3 nm wide showing different wetting behavior. **c** Plot of $(dh - h)/h$ versus $1/a_{Int}$. Line tension and interfacial energy are measured from the slope and intercept, respectively

17.2.8   Interaction with Surface Steps

A major question in the study of supported nanoparticles or nanocrystals is how they interact with surface steps, which are present on any crystal surfaces. Interactions with surface steps are found in the early stage of thin-film growth and in supported catalytic nanoparticles. Preferential nucleation on surface steps is often observed in the studies of thin-film growth (for a review, see Kern et al. 1979). For metal nanoparticles, the interaction with surface steps can greatly influence the chemical activity and selectivity of the catalyst (Stevenson et al. 1987; Somorjai 1994; Liu 2003; Dvorak et al. 2016). Model systems that have been studied by surface science techniques are mostly 2D surfaces [for a review, see Somorjai (1994)]. The study of interactions with surface steps thus bridges the gap between the model and experimental systems by including faceted surfaces with terraces, atomic steps (or ledges), and kinks.

The parameters used to describe the nanocrystal orientation and interface are illustrated in Fig. 17.20b and labeled as:

(1)   the tilt axis marked as $\vec{n}_T$ for the tilt boundary, which is normal to the page,
(2)   the miscut angle, θ_v, which characterizes the orientation of the vicinal surface,
(3)   the normal to the "cusped" and vicinal surfaces, $\vec{n}_{cusp}$ and $\vec{n}_v$, respectively.
(4)   the nanocrystal rotation θ_N, which is measured relative to the normal of the cusped surface.

Additionally, the epitaxy and interfacial structure of nanocrystals on vicinal surfaces can be rationalized based on the concept of coincidence site lattice (CSL) cells (Balluffi et al. 1982; Loretto et al. 1989), following the CSL lattice

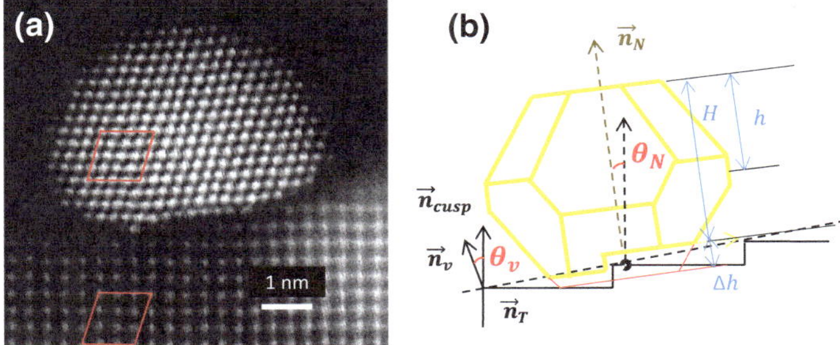

Fig. 17.20 Nanocrystal interaction with surface steps. (**a**) Atomic-resolution HAADF-STEM images of an Au nanocrystal in approximate epitaxial relation of Au(11−1)$_{[112]}$‖TiO$_2$(110)$_{[1−10]}$. The images were recorded along the TiO$_2$ [001], or Au [110], zone axis. The TiO$_2$ appears in a square lattice. The red boxes mark the identified CSL unit cells in Au and TiO$_2$ lattices, respectively. **b** A schematic illustration of an Au NC on vicinal surfaces. Parameters used to describe the vicinal surface structure are also marked in (**b**), including tilt axis $\vec{n}_T$, two angles here associated with surface miscut and nanocrystal rotation. The vectors $\vec{n}_{\mathrm{cusp}}$ and $\vec{n}_v$ denote the plane normals of the "cusped" and vicinal surfaces [from Gao et al. (2014)]

model for heterophase grain boundaries in crystalline solids. However, CSL is never perfect across a heterophase grain boundary. This is especially true in the case of nanocrystals (Bording et al. 2003). To describe the nanocrystal orientation relationship, we use the "fit–misfit" model developed by Mott (1948). In this model, the interface has partial lattice matching across the heterophase grain boundary and regions of "misfit" as boundary line defects possessing dislocation or boundary step character. Since nanocrystals have five degrees of freedom, the atoms in the boundary tend to fit as well as possible locally to assume the lowest possible energy (Bording et al. 2003), even though this often entails a small deviation from the orientation relationship of the bulk crystals. For the nanocrystal imaged in Fig. 17.19a, since 5 gold repeats along the (111) direction match 4 titanium columns in the TiO$_2$ (110) direction, from which this type of interface is calculated to be a $\sum 9$ boundary. The tilt axis $\vec{n}_T$ is along the TiO$_2$ (001) direction (which is also Au [110]), where $\vec{n}_{\mathrm{cusp}}$ represents the TiO$_2$ (110) surface normal. For a determination of the nanocrystal orientation, the rotation angles can be measured from the atomic-resolution image with the help of FFT. In Fig. 17.19a, the nanocrystal is rotated by 5.7° on a 9.7° miscut surface. The nanocrystal rotation angle is smaller than the surface miscut angle, which is true for other nanocrystals of $\sum 9$ boundary on the low-miscut TiO$_2$ (110) surfaces (Gao et al. 2014).

The interfacial atomic structure of a nanocrystal on a surface step is very different from the interface formed on a flat surface, as the examples in Figs. 17.19 and 17.17 show. Change in the interfacial structure impacts the interface energy and line tension, which in turn affects the shape of nanocrystals. A study of Au nanocrystals on vicinal rutile (110) surfaces with the Σ9 heterophase grain boundary (similar to the one shown in Fig. 17.19a) and a small rotation angle (<6°) shows that

(1) they have similar shapes in terms of the height and width ratio;
(2) the nanocrystals observed have a larger *h/w* ratio at 0.64, compared to the value of 0.49 for Au nanocrystals on TiO_2 (110) flat surfaces (Sivaramakrishnan et al. 2010).
(3) the triple line energy was measured to be $9.1 \pm 1.8 \times 10^{-10}$ J/m and the interfacial energy to be 0.48 ± 0.12 J/m^2. The interfacial energy is lower than the previous result for Au nanocrystals on flat surfaces at 0.61 ± 0.06 J/m^2 (Sivaramakrishnan et al. 2010). The triple line energy measured at $9.1 \pm 1.8 \times 10^{-10}$ J/m for Au NCs on stepped TiO_2 (110) surfaces is larger than the value of $4.5 \pm 1.3 \times 10^{-10}$ J/m reported earlier for Au NCs on flat TiO_2 (110) surfaces. (The same values of $\gamma_{Au(111)} = 1.283$ J/m^2 (Vitos et al. 1998) and $\gamma_{TiO_2(110)} = 0.33$ J/m^2 (Cosandey and Madey 2001) were used in the analysis).

Overall, the above case study shows that nanocrystals on low-miscut vicinal surfaces have a different interfacial atomic structure, equilibrium shape and surface, and interfacial energetics compared to a flat surface.

The interaction with surface steps also leads to interfacial strain. The atomic-resolution image of Fig. 17.20a shows deformation for the interfacial atomic columns near the surface steps. The interfacial displacements are reduced in atomic columns away from the interface, especially for these at 3 atomic layers away or more. The displacement of Ti atomic columns is also seen, but it is small compared to the shift of the Au atomic columns in the nanocrystal. Figure 17.21

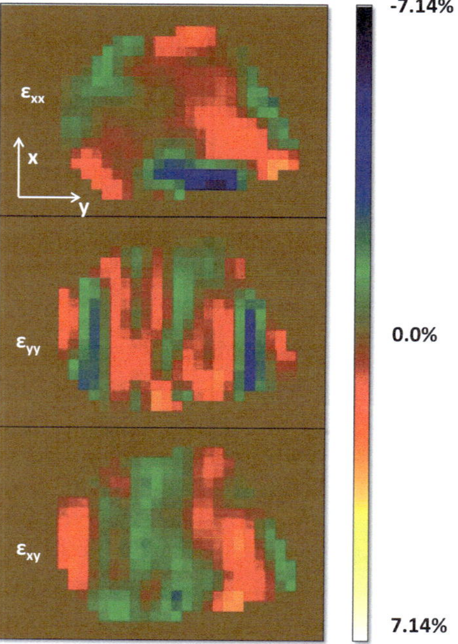

Fig. 17.21 Strain maps for the Au nanocrystal imaged in Fig. 17.19. These maps were calculated based on the measured displacements for normal stain in $x(\varepsilon_{xx})$ and y (ε_{yy}) direction and shear strain (ε_{xy}). The strain maps for the TiO_2 substrate are not shown here

shows the strain maps for ε_{xx}, ε_{yy}, and ε_{xy}, calculated using the method described in Sect. 16.5.2. In these strain maps, the Au nanocrystal is under compression in x direction (ε_{xx}) near the surface step and a small tensile strain is seen in regions near the top, bottom left, and bottom right corners of the nanocrystal. Surface Au atoms on the left and right sides are under a small compression. Along the y direction, the strain (ε_{yy}) is modulated by the interface, and expansion is seen where Au and TiO_2 terraces are in contact. The shear strain (ε_{xy}) map shows a change from positive to negative shear strain near the surface step. The large strain is seen next to the surface step close to -4.9 %, while the strain seen on the NC surface is only about 0.7 %.

17.3 Structure of Nanoclusters and Nanoparticles

The term "nanoparticles" generally refers to any nanometer-sized particle, except nanocrystals and nanoclusters. Nanocrystals as described in the previous section have interior atoms in the same atomic arrangements as the bulk crystal. Nanoclusters or simply clusters, on the other hand, refer to ultrafine particles consisted of several atoms to hundreds of atoms. Because of the small sizes, the constraint of lattice translational symmetry does not apply. Consequently, clusters can present non-crystallographic symmetries, icosahedra and decahedra being the best examples, in addition to crystalline clusters with the bulk lattice. Nanoclusters of certain particular size have increased stability, and they are observed with a much higher frequency than nanoclusters of other sizes. An example is Au55 (Au cluster with 55 atoms), which is much more stable than Au56 even though their difference is only one atom. Clusters like Au55 are often referred to as "magic number" clusters, as they assemble the number of atoms that equal exactly some "magic" number. A well-known example of magic clusters is the family of carbon fullerenes including C60, C70, and C84 (discussed later). The magic number is identified by mass spectrometry (Martin 1996), while the cluster structure is determined using a combination of diffraction and modeling. Nanoparticles of 3 nm or larger are mainly investigated by HREM and more recently using aberration-corrected STEM. However, electron imaging of small nanoparticles must be done with care under the conditions that minimize the electron beam and particle interaction, since small nanoparticles (including nanoclusters) can adopt several structural forms of similar energy (Iijima and Ichihashi 1986).

17.3.1 Diffraction by Free Clusters

Small nanoclusters are like molecules with several to tens of atoms. The structure of free molecules is determined by gas-phase electron diffraction, where a gas nozzle introduces a collimated beam of gas molecules, which are then probed by electron diffraction (Hargittai 2006). Such instruments have been adapted for the study of cluster beams. Instead of gas molecules, nanoclusters are first formed using an inert gas aggregation source and then introduced into the diffraction chamber. For example, the instrument described by Hall et al. (1991) uses a beam of 100 kV electrons from the electron gun and condenser lenses of a Philip EM300 TEM to record the diffraction pattern on a CCD camera.

At any given time, the fast electron sees the instantaneous atomic positions. There can be any number of clusters that pass through the electron beam. Each diffracting atom m is specified by a vector $\vec{r}_m$. Then, the instantaneous diffraction intensity $I(t)$ is given by the square of the amplitude of scattering from different atoms, which under the kinematic approximation can be written as follows:

$$I(t) = \sum_m f_m e^{2\pi i \vec{S} \cdot \vec{r}_m(t)} \sum_n f_n e^{-2\pi i \vec{S} \cdot \vec{r}_n(t)} \tag{17.21}$$

Both $\vec{r}_m(t)$ and $\vec{r}_n(t)$ change with time t, but the difference between these two is relatively constant for two atoms within the same cluster, while it can vary by a large amount for atoms belonging to different clusters. When averaged over time, only the atoms belonging to the same cluster contribute to the diffraction pattern. Taking this into account, we can simplify Eq. (17.21) by using the difference vector $\vec{r}_{mn} = \vec{r}_m(t) - \vec{r}_n(t)$:

$$I_c(t) = \sum_{m=1}^{N} \sum_{n=1}^{N} f_m f_n e^{2\pi i \vec{S} \cdot \vec{r}_{mn}}, \tag{17.22}$$

Here, c denotes a single cluster and N is the number of atoms in the cluster. The clusters are randomly oriented. To obtain the average intensity, we consider the clusters as rigid bodies, oriented with equal probability in any direction. The distribution of $\vec{r}_{mn}$ falls on a sphere, and by averaging the exponential term in Eq. (17.22), we then obtain

$$\left\langle e^{2\pi i \vec{S} \cdot \vec{r}_{mn}} \right\rangle = \frac{1}{4\pi r_{mn}^2} \int_0^{\pi} e^{i k r_{mn} \cos \phi} 2\pi r^2 \sin \phi \, d\phi = \frac{\sin k r_{mn}}{k r_{mn}}, \tag{17.23}$$

where $k = 2\pi S$. Substituting this result for the terms in Eq. (17.22), we obtain the averaged intensity

$$I_c = \langle I_c(t) \rangle = \sum_{m=1}^{N} \sum_{n=1}^{N} f_m f_n \frac{\sin k r_{mn}}{k r_{mn}}. \tag{17.24}$$

Equation (17.24) is known as the Debye scattering equation. It is general and important, since it applies to diffraction of all gases, within the limitations of the kinematical approximation.

As a simple example, we consider diffraction by the clusters of 4 atoms in a tetrahedron. Since all 4 atoms are the same, we can start with any one of the 4 atoms for m and then sum over n for all 4 atoms including the starting atom in the double sum of Eq. (17.24). The result gives

$$I_c/4 = f^2\left(1 + 3\,\frac{\sin kd}{kd}\right), \tag{17.25}$$

where d is the atomic bond distance. For larger clusters, additional terms are introduced with r_{mn} included for the second nearest neighbor, third nearest, etc. Also for clusters made of different types of atoms, the contribution by different atoms to the sum in Eq. (17.24) is weighted by the atomic scattering factor.

A cluster diffraction pattern is calculated by assuming a model structure for the clusters and applying the Debye formula to obtain a profile of diffracted intensity. For a monoatomic cluster, the Debye equation can be rewritten as follows:

$$I_c = Nf^2 + f^2 D^2 \sum_{m\neq n} \frac{\sin kr_{mn}}{kr_{mn}} \tag{17.26}$$

where the sum is over all pairs of atomic distances r_{mn} separating atom m from atom n. The Debye–Waller factor $D = \exp(-B\sin\theta/\lambda)$ is included to describe the attenuation of the atomic scattering factor, caused by thermal vibration. This simple model assumes that the displacement of all atoms in the cluster is random and isotropic about their equilibrium positions.

Figure 17.22 shows an example of the experimental intensity profile and the fit obtained from copper clusters reported by Reinhard et al. (1998). The fit was calculated using three model structure types for the copper clusters: face-centered cubic, decahedral, and icosahedral. Information about the composition and size ranges of clusters in the beam was extracted by fitting the experimental profile using a weighted combination of calculated diffraction patterns to the experimental data. For Fig. 17.22, the fit gives a population 11 % of *fcc*, 81 % icosahedral, and 8 % decahedral clusters with the average size of 3.4, 1.9, and 4.9 nm in diameters, respectively.

17.3.2 *Structure and Energetics of Metallic Nanoparticles*

Experimental data based on diffraction, electron microscopy, and theoretical studies (see Fig. 17.22 for an example) have narrowed down the pool of nanoparticle structures of fcc metals to a few candidate motifs: crystalline fcc lattice with the truncated octahedral morphology and its twinned variant and the non-crystalline structural motifs, such as Ino (1969) and Marks (1994) type of decahedron (Dh) and Mackay (1962b) type of icosahedron (Ih). Complex nanoparticle structures

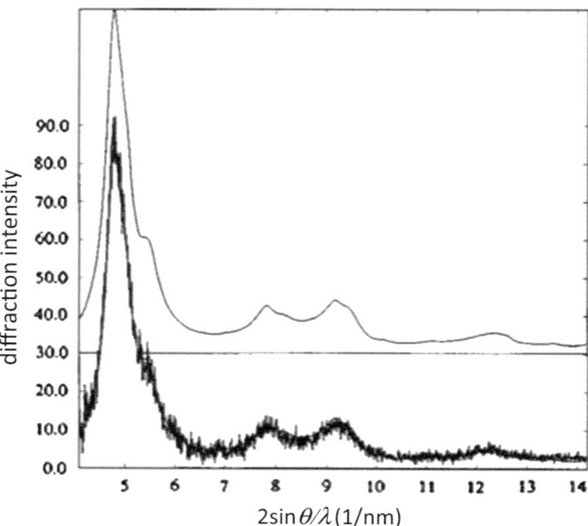

Fig. 17.22 Diffraction profile (*bottom*) obtained from copper clusters after removing scattering from carrier gas. *Top* is an example of a typical fit calculated with Eq. (17.26) (for details, see text), which is also superimposed on the experimental profile (reprinted from Reinhard et al. (1998) with the permission)

resulting from a combination of different structural motifs have also been reported Koga (2006). Evidence of different structural motifs in individual nanoparticles is provided by HREM (Marks 1994; Reinhard et al. 1997; Li and Zuo 2005) and electron diffraction (Fig. 17.23).

The models of crystalline fcc nanoparticles are constructed using the bulk metal bond length. The particle shape can be specified using two integers, N and M, which denote the center-to-vertex distance and center-to-truncated-facet distance, respectively, as shown in Fig. 17.24a. The use of the parameters N and M produces faceted particles with {111} and {100} facets. On the other hand, a round particle with a cutoff radius R would yield {111} and {100} facets as well as other vicinal surfaces.

An icosahedral nanoparticle has 12 fivefold axes and 20 triangular faces with *Ih* symmetry (Mackay 1962a; Yang et al. 1979). It consists of shells of atoms. At the core is an atom at the center surrounded by 12 atoms at the corner of an icosahedron. The next shell has 42 atoms, forming again a perfect icosahedron. An icosahedral nanoparticle with k completed shells has

$$N_{Ih}(k) = \frac{10}{3}k^3 - 5k^2 + \frac{1}{3}k - 1, \qquad (17.27)$$

number of atoms so that the series of magic numbers is 13, 55, 147, etc. Other nanoparticles with fivefold symmetry are Ino's and Marks's decahedron with the *Dh* symmetry as shown in Fig. 17.24d. They are also known together as multiply twinned particles (MTPs) (Ino 1966; Howie and Marks 1984; Marks 1984). The structure of MTPs consists of single-crystal tetrahedral subunits, which are twin-related on their adjoining faces. Icosahedrons and decahedrons are made of

Fig. 17.23 Electron diffraction patterns recorded from individual Au nanoclusters. **a** Single-crystalline diffraction pattern from Au nanocrystals of 1.2 nm viewed along its [110] zone axis. **b** Diffraction pattern from an icosahedral Au nanocluster of 3 nm with the direct beam subtracted off. The nanoclusters were supported on single-walled carbon nanotubes as shown to the right of the diffraction patterns. The icosahedral nanocluster model and simulated kinematic diffraction pattern are shown at the bottom. *Arrows* point to the features with good agreement between experiment and simulation (from Weijie Huang, University of Illinois)

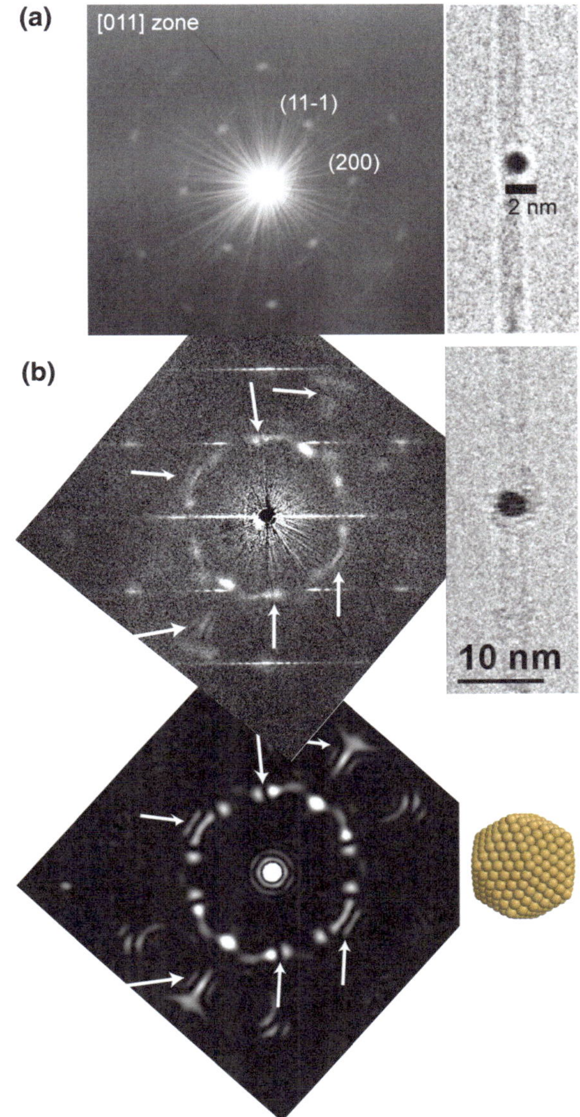

twenty and five constituent tetrahedral, respectively. The space filling of those subunits is not complete, leaving a large angular misfit. Consequently, MTPs are believed to be formed with disclination and related large strain (Howie and Marks 1984; Marks 1984; Johnson et al. 2008). Their respective constituent tetrahedra are deformed from the fcc structure as shown in Fig. 17.23b. For example, to form an icosahedron as proposed by Mackay (1962b), each of the tetrahedra has three coplanar edges stretched by 5 % (see Fig. 17.23b) and the close-packed plane

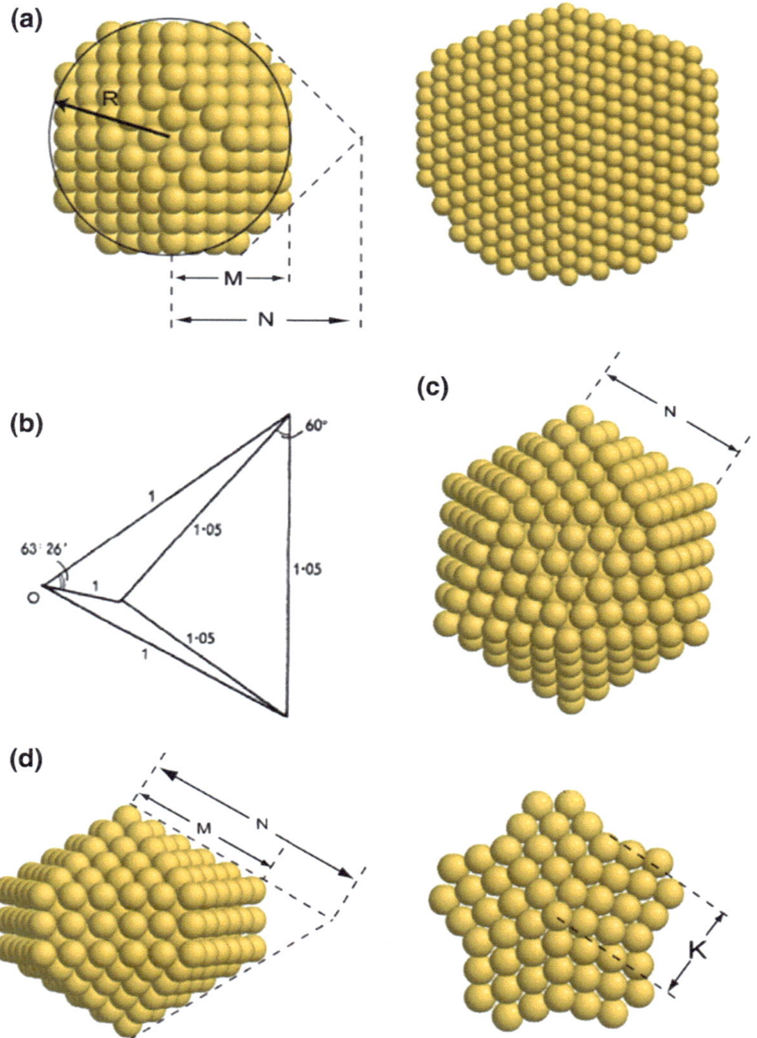

Fig. 17.24 Structural models for nanoclusters. **a** Single-crystal truncated octahedron (*left*) and its single-twin variant (*right*); **b** deformed constituent tetrahedron in the Mackey icosahedron model; **c** Mackey icosahedron; **d** Ino decahedron (*left*) and Marks decahedron (*right*)

defined by these three edges will be one of its twenty facets. To form a decahedron, each tetrahedron has one edge 2 % longer than the remaining five ones (Gryaznov et al. 1999) and that edge will serve as the fivefold axis. Two parameters, the center-to-vertex distance N and center-to-facet distant M, are needed to specify a decahedron, and an extra parameter K is required to describe the reentrant facets in the Marks decahedron. For icosahedra, only the center-to-vertex distance N is needed.

The binding energy E_B of a cluster of size N of a given structure can be approximately written in the following form (Baletto and Ferrando 2005):

$$E_B = aN + bN^{2/3} + cN^{1/3} + d \qquad (17.28)$$

where the first term corresponds to a volume contribution, while the others represent contributions from facets, edges, and vertices, respectively. The volume and surface contributions compete with each other. For example, theoretical calculations predict that fcc, single-crystalline nanoparticles will assume a truncated octahedron as their equilibrium thermodynamic shape. This structure is confined by the {111} and {100} families of planes (Henry 2005). The formation of such crystal shape, however, is energetically costly in small clusters due to the presence of under-coordinated vertex atoms (Negreiros et al. 2007). Conversely, particles of icosahedral or decahedral crystal morphology possess the lower energy {111} terminating facets at the cost of increased internal strain energy associated with the loss of lattice periodicity. The internal strain increases the volume energy term in Eq. (17.28). For noble metals, decahedral or icosahedral nanoparticles have been predicted to be energetically favored and observed when the size of the clusters is small, where surface energy wins over the internal cohesive energy.

The crossover from one energetically favorable structure to another has been studied by first principles and semiempirical simulations (Uppenbrink and Wales 1992; Cleveland et al. 1997; Baletto and Ferrando 2005). Generally, the results show that the icosahedron is found to be prevalent at very small sizes ($N < 100$); truncated octahedron is expected for larger sizes ($N > 1000$), and decahedron could be favored in the intermediate sizes. However, studies based on several fcc metallic systems (Baletto and Ferrando 2005) show that the crossover sizes from icosahedron to decahedron, or decahedron to truncated octahedron, are largely material dependent. Doye and Wales (1995) have proposed an intuitive principle for a quantitative understanding of the crossover size. Inter-atomic potentials of various materials can be grouped into two categories: soft and sticky interactions. The soft interaction has a wide potential well that can stabilize strain structures such as icosahedra or decahedra. The sticky interaction, on the other hand, is characterized by a narrower potential well and cannot easily accommodate strain, thus favoring crystalline structures such as truncated octahedra, even at very small sizes.

Despite the significant role thermodynamics plays in directing the geometric forms of nanoparticles, kinetics cannot be neglected and its role is often complex and sometimes acts as a competing factor. It has frequently been demonstrated, for example, that the adjustment of experimental parameters (i.e., capping agent concentration, choice of reducing agent, and temperature) can stimulate the adoption of specific morphological features (Narayanan and El-Sayed 2005; Lee et al. 2006; Yang et al. 2007; Zhu et al. 2007; Xia et al. 2009) and affect crystallite size during nanomaterial growth (Teranishi et al. 1997; Templeton et al. 1999; Frenkel et al. 2005; Narayanan and El-Sayed 2005; Xia et al. 2009). Then, the geometry of a nanoparticle is not solely determined by the minimization of its surface and strain energy (thermodynamic minimum). Instead, one must consider the possible formation of a metastable structure possessing a conversion potential well too deep to be easily overcome (Henry 2005). In this respect, the control of specific reaction conditions may result in the formation of thermodynamically unfavorable features (Tao et al. 2008). Large icosahedral nanoparticles are an example (Wu et al. 2012).

17.4 Carbon Nanostructures

Carbon materials, consisted mainly of carbon atoms, support a large industry. Graphite and graphitic materials, for example, are the basis for lubricants, brake linings, battery electrodes, etc. Diamond has many applications due to its extreme hardness, aside from its aesthetic appeal. Carbon blacks are used in catalysis, black inks and also as reinforcing fillers for tires in the motor industry. Activated carbons have wide applications as a green technology for water purification and deodorizing air. Nanostructured carbon has attracted tremendous interest starting with the experimental discovery of fullerenes in 1985 (for a review, see Avdeenkov et al. 2009), which was followed by carbon nanotubes (CNTs) (Iijima 1991) and more recently graphene Novoselov et al. (2004). Carbon nanotubes and the 2D graphene have become the major building blocks in nanotechnology, largely due to their structure-dependent and unique electronic and mechanical properties.

17.4.1 Carbon Allotropes and Bond Lengths

Elemental carbon can be present in different molecular configurations, which together with bulk diamond and graphite are known as carbon allotropes. Some examples of molecular carbon are shown in Fig. 17.25. The first three (C60, C540, and C70) are fullerenes in the form of a hollow sphere and icosahedral and ellipsoid shapes. The C60 molecule is the smallest stable fullerene. It has the carbon atoms forming 3 bonds like in graphite. The carbon bonds are connected as pentagons and hexagons, where the pentagons provide the curvature. Each fullerene contains 2 $(10 + M)$ carbon atoms with M hexagons and exactly 12 pentagons. The C60 has 20 hexagons, together with 12 pentagons, and it forms a perfect a ball (thus the nickname buckyballs).

Graphene is equivalent to a single atomic layer of graphite. Each carbon forms 3 bonds with other carbon atoms in a hexagonal network. Ideally, graphene is an atomically flat 2D structure. In practice, the presence of defects, such 7- and 5-membered rings (see Sect. 17.4.5), gives rise to local variations in atomic height. Graphene can be also manipulated into various folded or wrapped structures (Zhang et al. 2010; Kim et al. 2011), with applications for energy storage (Liu et al. 2012).

A single-walled carbon nanotube (SWCNT) can be regarded as a rectangularly shaped graphene wrapped into a seamless tube (Fig. 17.25e). The SWCNTs are stable at small diameters between ~ 1 and 3 nm approximately. The structure of a SWCNT is characterized by its diameter and chiral angle. Large-diameter carbon nanotubes (CNTs) exist in the form of double-walled CNTs (DWCNTs) or multiwall CNTs (MWCNTs). Two concentric SWCNTs make up a DWNT when they are attracted to each other. The diameter difference ΔD between the two tubes is on the order of the spacing between two graphene sheets in graphite (3.3 Å). The outer and inner tubes can have the same, or different, chiral angles. The structure is incommensurate if the two chiral angles are different from each other (Gao et al. 2006; Hirahara et al. 2006). A multiwall carbon nanotube has three or more shells of graphene of similar or different chiralities. The initial structural model proposed by Ijima consists of concentric tubes of similar spacing (Iijima 1991). Both SWCNTs and DWCNTs can form bundles made up of tubes of similar diameters. The bundles are formed by the van der Waals attraction between two or more nanotubes. The bundles can be highly ordered with the tubes close-packed and being parallel to each other. The structure of individual nanotubes in a bundle can be the same or different.

Hybrid carbon molecular structures are formed by combining different carbon nanoforms; for example, the peapods are formed by encapsulating fullerenes inside carbon nanotubes (Fig. 17.25f).

Other carbon nanoforms include disordered amorphous carbon and turbostratic carbon. Amorphous carbon has no long-range structure, and the carbon atoms form 4 bonds with tetrahedral symmetry. Turbostratic (or pyrolytic) carbon has a distorted, and irregular, lattice with rotational disorder about the graphite c-axis. The structure also has a significant amount of random unassociated carbon bonds. Coal consists of varying amounts of graphite, depending on its quality or grade (and transforms to diamond at high pressure and temperature), and the variability and richness of the compounds of carbon chemistry provide the basis for life.

The ground state electronic configuration of carbon atom is $(1s^2)$ $(2s^2 2p_x 2p_y)$. The inner electronic core is sufficiently compact to allow the outer valence electrons to hybridize with the electrons of neighboring atoms, so as to form linear bond (sp), or planar bonds (sp^2), or tetrahedral bonds (sp^3). The different forms of carbon bonds manifest themselves in a number of structural forms, for example, sp^3 σ bond formed in diamond and C_2H_6, sp^2 bonds in graphite and C_2H_4, and sp^1 bonds in carbyne and C_2H_6.

Carbon materials in general can be classified by the following two characteristics: the type of carbon hybridization and the characteristic size (Inagaki 2000). The sizes start from small organic molecules to small nanostructures of fullerenes, nanotubes, graphene, and nanodiamonds and then to larger structures such as multiwalled CNT (MWCNT) and the more complicated carbon architectures such as carbon black. At the macroscopic scale, we have diamond and graphite. The type of bonds in carbon materials is measured by the C–C bond length. The structure of

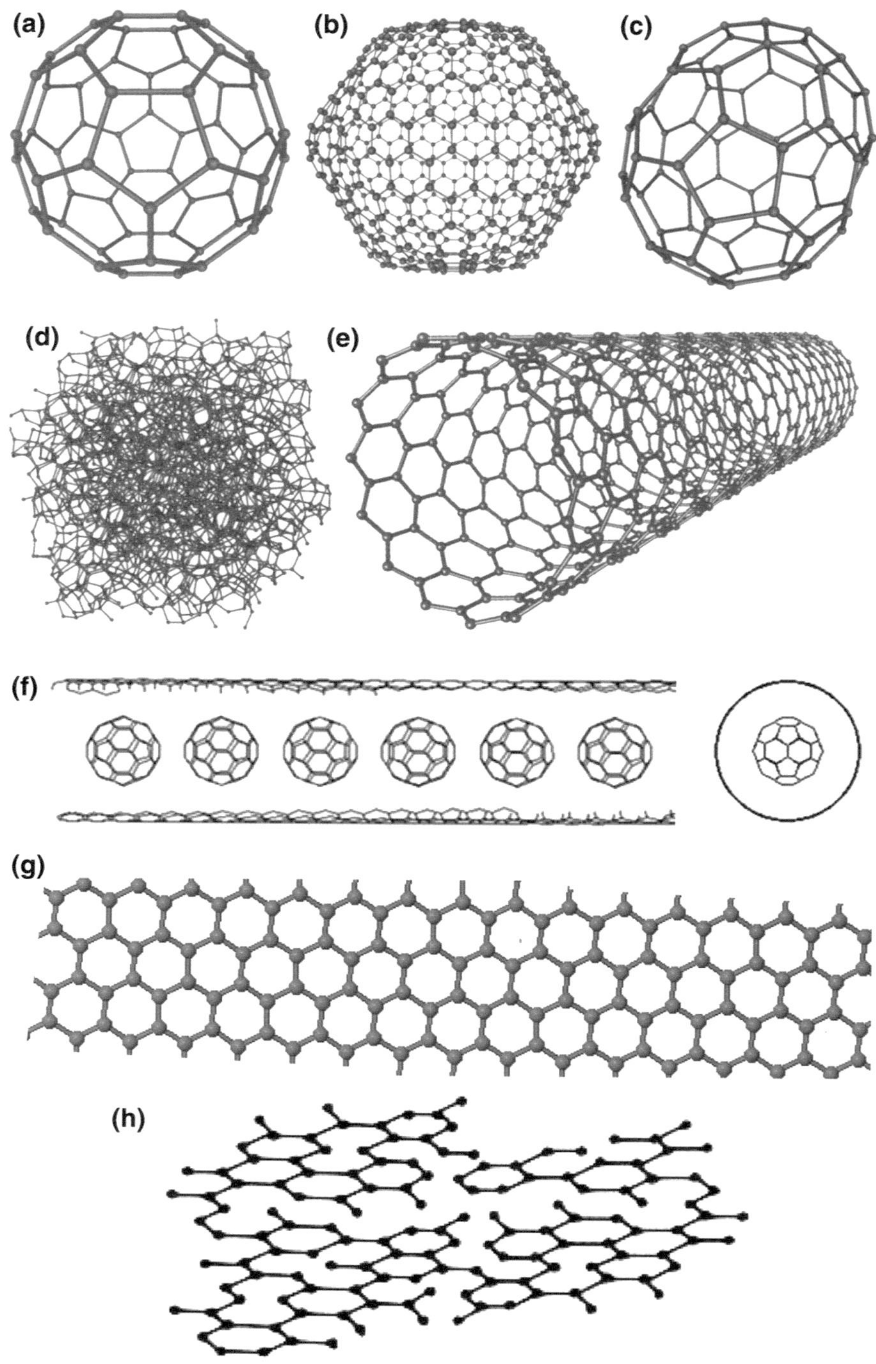

◄ **Fig. 17.25** Nanostructured carbon allotropes. **a** C60 buckminsterfullerene, **b** C540, fullerite **c** C70, **d** amorphous carbon, **e** single-walled carbon nanotube, **f** C60 encapsulated inside a single-walled carbon nanotube, known as peapod, **g** graphene and **h** pyrolytic or turbostratic carbon with distorted lattice structure and random unassociated carbon bonds. (Figures **a–e**) are from http://en.wikipedia.org/wiki/Allotropes_of_carbon)

graphite, for example, has an in-plane C–C bond length of 1.420–1.426 Å at room temperature as experimentally determined (Baskin and Meyer 1955; Lynch and Drickamer 1966; Trucano and Chen 1975) and 1.409–1.426 Å as predicted by theory (Hasegawa and Nishidate 2004). The C–C bond length of graphite is in between the bond lengths for a single (1.544 Å from diamond) and double (1.334 Å) bond. The carbon atoms in diamond have a coordination number of 4. The C–C single bond between two carbon atoms forming double bonds with other atoms has a shorter length at 1.504 Å (Pauling 1960). Pauling explained the bond length of graphite based on the so-called resonance model in which the extra electron is shared equally among the three carbon bonds to give a bond strength of 4/3 (Pauling 1960). The breakdown of equal bond sharing can lead to alternative graphene structures. In the so-called quinoid structure that Pauling proposed, two-thirds of the C–C bonds have a length of about 1.453 Å with 15 % double-bond character and one-third have a length of about 1.357 Å with 70 % double-bond character (Pauling 1966). Here, the extra electron resides predominantly in one of the three bonds. However, no experimental evidence of different bond lengths in graphite was found (Trucano and Chen 1975).

Two distinct bond lengths are observed in C60, and one is the so-called 6-6 bond at 1.387(3) Å and another 6-5 bond at 1.450(3) Å as measured by X-ray diffraction at 110 K. In CNTs, large curvatures are introduced when a graphene sheet is rolled into a tube, which breaks the symmetry of the three C–C bonds. According to the theoretical study of Kanamitsu and Saito, the difference between the two bonds has a strong tube diameter dependence that scales $1/D^2$ with D for tube diameter (Kanamitsu and Saito 2002). The change in bond length is about 0.3 % for 1-nm-diameter zigzag SWCNT. However, there are significant disagreements among theory about the C–C bond distance in SWCNTs. For example, the predicted C–C bond length around the tube ranges from 1.382 to 1.466 Å for the CNTs where the tube diameter is smaller than 10 Å in diameter (Robertson et al. 1992; Sanchez-Portal et al. 1999; Gulseren et al. 2002).

Within the carbon nanostructures, nanodiamonds, CNTs, and graphene are 0D, 1D, and 2D nanomaterials, respectively. Depending on the diameter and chirality, single-walled CNTs (SWCNT) can exhibit semiconducting or metallic behavior. Dissimilar carbon nanotubes may even be joined together to form molecular wires with interesting electronic, magnetic, optical, and mechanical properties (Dai 2006). SWCNTs are used as new materials for the development of small transistors (Tans et al. 1998; Bandaru et al. 2005; Delattre et al. 2009) and sensors (Perez et al. 2010; Salehi-Khojin et al. 2010; Somani 2010). Graphene is the thinnest known material in the universe. It is a zero band gap semiconductor with a linear energy–wavelength dispersion for both electrons and holes in the conduction and valence bands.

The properties of carbon nanostructures are dramatically influenced by the atomic structure, where carbon bonds show surprising diversity (Zhang and Zuo 2009; Dinadayalane and Leszczynski 2010). Structural characterization of complex carbon materials remains as a grand challenge in carbon nanoscience.

17.4.2   Electron Diffraction of Carbon Nanotubes

We start with diffraction by a SWNT. A structural model can be constructed by rolling up a single graphene sheet into a seamless cylindrical tube. At the seam, the sides can be connected in many ways by sliding one side against the other, which defines the chirality. The tube structure is then uniquely defined by the chiral vector $\vec{C} = n\vec{a} + m\vec{b}$, or (n, m), where $\vec{a}$ and $\vec{b}$ are the unit cell vectors of the graphene sheet (Fig. 17.27). In the rolled-up tube, the length of the chiral vector becomes the tube circumference. The angle between $\vec{a}$ and $\vec{b}$ is taken at $60°$ in the so-called solid-state convention (Saito et al. 1998). The tube chiral angle α is defined by the angle between the chiral vector $\vec{C}$ and the unit cell vector of $\vec{a}$.

The tube radius and chiral angle can be calculated from the tube chiral vector using:

$$r = \frac{|\vec{C}|}{2\pi} = \left|n\vec{a} + m\vec{b}\right|/2\pi = a\sqrt{n^2 + m^2 + nm}/2\pi \qquad (17.29)$$

and

$$\alpha = \cos^{-1}\left[\frac{2n + m}{2\sqrt{n^2 + m^2 + nm}}\right]. \qquad (17.30)$$

Diffraction of a cylindrical molecule, such as SWCNTs, is better described theoretically using the cylindrical coordinate (for computation, the Cartesian coordinate is both convenient and sufficient). The real space and corresponding reciprocal space coordinates are given by (r, ϕ, z) and (R, ψ, k_z), where r is the radius in the xy plane, ϕ is the angle between r and the x-axis, and z is the coordinate along the cylinder direction. Similarly, (R, ψ, k_z) are defined in the reciprocal space. The Fourier transform in the cylindrical coordinates is obtained from

$$F(R, \psi, k_z) = \int_0^\infty \int_0^{2\pi} \int_{-\infty}^\infty V(r, \phi, z) e^{2\pi i [R r \cos(\phi - \psi) + k_z z]} r\, dr\, d\phi\, dz \qquad (17.31)$$

The electron diffraction theory for carbon nanotubes has been developed based on the diffraction theory of helical structures (Lambin and Lucas 1997; Qin 2007 and references therein) based on the diffraction theory for helical structures in the kinematic approximation. The structure factor of a single helix in a single-walled carbon nanotube is given by

$$F_i(R, \psi, k_z) = Af(R, k_z) r J_N(2\pi R' r) \delta\left(k_z - \left(\frac{N}{P} + \frac{M}{p}\right)\right) e^{i\phi_i} \qquad (17.32)$$

where A is a tube-dependent phase factor, f is the atomic scattering factor, r is the tube radius, P is the helical pitch distance $(P = \left|\vec{C}\right| \tan\alpha = a\sqrt{n^2 + m^2 + nm}\sqrt{3}m/(2n + m))$, p is the distance between atoms on the helix $(p = a\sin\alpha = a\sqrt{3}m/2\sqrt{n^2 + m^2 + nm})$, and R and k_z are the components of the cylindrical reciprocal vector along the radial and tube axes, respectively. N and M are integers; each pair denotes a layer line of the helical diffraction pattern. The structure factor of a SWCNT is simply the sum of the structure factor of different helices that make up the tube:

$$F(R, \psi, \zeta) = \sum_i F_i(R, \psi, \zeta) \qquad (17.33)$$

The theory can be extended to DWCNTs or MWCNTs if the individual walls are cylindrical. The nanotube diffraction pattern consists of discrete lines (layer lines) from the delta function in Eq. (17.32) with

$$k_z = \left(\frac{N}{P} + \frac{M}{p}\right), \qquad (17.34)$$

which comes from the periodicity of the helix and the graphite structure. The layer line with $N = 0$ is called the equatorial line and goes through the origin of reciprocal space. The structure factor for the equatorial line averages over the atomic potential and is associated directly with the tube geometry.

The structural model of a single-walled carbon nanotube can be set up using the following procedure. First, a rectangular supercell is defined on the 2D graphene sheet (see Fig. 17.26) with the $\vec{a}'$-axis as the chiral vector:

$$\begin{aligned} \vec{a}' &= \vec{C} = n\vec{a} + m\vec{b} \\ \vec{b}' &= j\vec{a} + k\vec{b} \end{aligned} \qquad (17.35)$$

The $\vec{b}'$-axis is selected as the shortest graphene lattice vector perpendicular to $\vec{a}'$:

$$\vec{a}' \cdot \vec{b}' = 0 = nj + mk + (mj + nk)/2 \qquad (17.36)$$

This leads to

$$\frac{j}{k} = \frac{n+2m}{2n+m} \tag{17.37}$$

The values of j and k are selected as a pair, which have no common factors other than 1. The new supercell thus defines the smallest repeating unit of the graphene sheet along the tube axis. The carbon atomic positions in the supercell are obtained by transformation according to

$$\begin{pmatrix} x' \\ y' \end{pmatrix} = \frac{1}{nk-jm} \begin{pmatrix} k & -j \\ -m & n \end{pmatrix} \begin{pmatrix} x \\ y \end{pmatrix} \tag{17.38}$$

To make the tube, the supercell is transformed so that the b'-axis will be the tube axis and a' will be the circumference of the tube. The length of b' gives the period L along the tube axis. The coordinates of the nanotube are obtained by taking

$$z'' = y'$$
$$x'' = r\cos(2\pi x'/|\vec{C}|) \tag{17.39}$$
$$y'' = r\sin(2\pi x'/|\vec{C}|)$$

An example of transformation is shown in Fig. 17.26 for a chiral SWCNT.

Not all of helical layer lines appear in CNT diffraction patterns because of the periodic structure of the graphene sheet. To understand the extinction rules of CNT diffraction, let us consider a graphene sheet, where each reflection in the diffraction pattern is defined by the two-dimensional reciprocal lattice:

$$\vec{G} = h\vec{a}^* + k\vec{b}^* \tag{17.40}$$

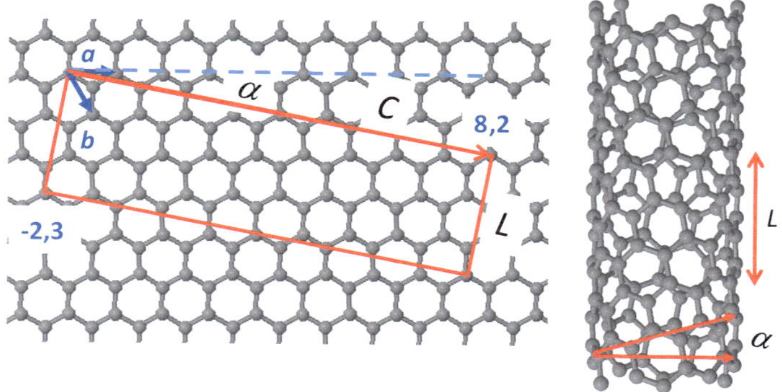

Fig. 17.26 Construction of a chiral single-walled carbon nanotube

with

$$\vec{a}^* = \frac{\vec{b} \times \hat{z}}{\vec{a} \times \vec{b} \cdot \hat{z}} = \frac{2}{3a^2}\left(2\vec{a} - \vec{b}\right)$$
$$\vec{b}^* = \frac{\hat{z} \times \vec{a}}{\vec{a} \times \vec{b} \cdot \hat{z}} = \frac{2}{3a^2}\left(-\vec{a} + 2\vec{b}\right) \tag{17.41}$$

The same reflection can be indexed in the reciprocal lattice, $\vec{a}^{*\prime}$ and $\vec{b}^{*\prime}$, of the supercell, which is a radial projection of the tube:

$$\vec{G} = h\vec{a}^* + k\vec{b}^* = h'\vec{a}^{*\prime} + k'\vec{b}^{*\prime} \tag{17.42}$$

There are two modifications to the reciprocal lattice of the supercell when it is transformed into a cylindrical tube. One is in the form of oscillations described by Bessel functions. The oscillations come from the curvature of graphene hexagons around the tube and are in the direction perpendicular to the tube. The second is the introduction of a second set of reflections that mirror the original ones. These two sets of reflections can be regarded as from the top and bottom half of the tube. In both cases, the reciprocal lattice vector along the tube, $\vec{b}^{*\prime}$, remains unchanged. Thus, the component of G along $\vec{b}^{*\prime}$ direction can be obtained by taking

$$G_{\perp} = \left(h\vec{a}^* + k\vec{b}^*\right) \cdot \frac{\hat{z} \times \vec{a}'}{|\vec{a}'|} = \frac{(2n - m)}{\sqrt{3}|\vec{C}|m}\left\{nh + mk - \frac{2(m^2 + n^2 + nm)}{(2n + m)}h\right\} \tag{17.43}$$

The relationship between the helical indices of (N, M) and the hexagonal indices of the graphene sheet, and (h, k) can be obtained by comparing the positions of layer lines predicted by two theories. For helical diffraction,

$$k_z = \frac{1}{P}\left(N + M\frac{P}{p}\right) = \frac{1}{P}\left(N + M\frac{2(n^2 + m^2 + nm)}{2n + m}\right) \tag{17.44}$$

Comparing this with Eq. (17.43) and using the definition of P, it can be shown that

$$G_{\perp} = \frac{1}{P}\left\{nh + mk - h\frac{P}{p}\right\} \tag{17.45}$$

Since $k_z = G_{\perp}$, the following extinction rules are obtained between the hexagonal and the helical indices of (h, k) and (N, M):

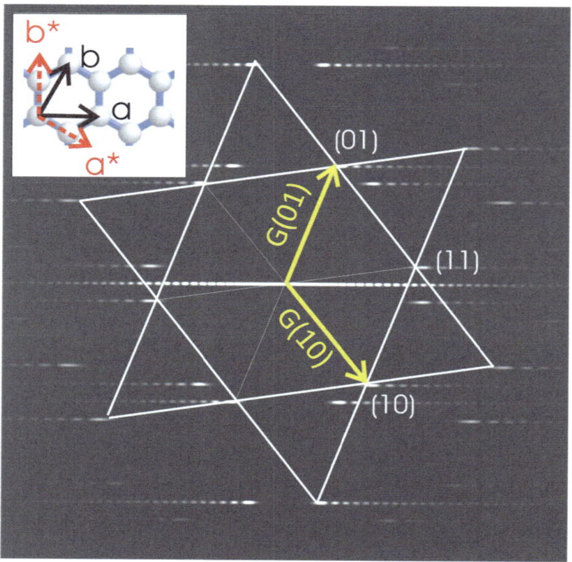

Fig. 17.27 Carbon nanotube diffraction pattern indexing and its relationship to the graphene structure. The diffraction pattern shown was simulated from a chiral, single-walled carbon nanotube. The *inset* shows the definition of real and reciprocal lattices of the graphene

$$N = nh + mk \quad \text{and} \quad M = -h \tag{17.46}$$

Thus, experimentally, carbon nanotube diffraction patterns can be first indexed based on the hexagonal lattice as shown in Fig. 17.27, and the order of Bessel functions can then be calculated with knowledge of the chiral vector (n, m).

The tube chiral angle can be measured from the layer line distances, $G_\perp(h, k)$. For example, the distances of three first-order graphene reflections, $G_\perp(1, 0)$, $G_\perp(0, 1)$, and $G_\perp(1, 1)$, can be used to obtain the chiral angle using the following equation:

$$\frac{1}{\sqrt{3}} \frac{G_\perp(1, 1) - G_\perp(1, 0)}{G_\perp(0, 1)} = \tan \alpha \tag{17.47}$$

The $G_\perp$ distances of the three reflections are obtained in the experimental pattern by measuring the distance of (11), (10), and (01) layer lines to the equatorial line (d_1, d_2, d_3). From Eq. (17.47), these distances are related to the chiral angle α by:

$$\alpha = a \tan\left(\frac{1}{\sqrt{3}} \frac{d_2 - d_1}{d_3}\right). \tag{17.48}$$

Both d_2 and d_3 correspond to the diffraction lines having relatively strong intensities and are further from the equatorial line. They can be measured to high accuracy.

The relationship defined in Eq. (17.48) is not affected if the tube is tilted relative to the incident electron beam. The distance from a diffraction line to the equatorial line in the case of a tilt of γ angle, d_γ, is enlarged compared to that for $\gamma = 0$, d_0, by:

$$d_\gamma = d_0 / \cos \gamma. \tag{17.49}$$

which can be used to calculate γ. In experimental analysis, γ is obtained by fitting the layer line intensities (see below).

To determine the structure of CNTs, the layer line intensity is fitted to $|f(R, k_z)J_N(\pi RD)|^2$, where f is the atomic scattering factor of carbon and R and k_z are the cylindrical reciprocal vectors along the radial and tube axes, respectively. To obtain the Bessel function of order N and the tube diameter from the experimental pattern, an intensity profile along the layer line (k_x) is taken and fitted. Matching the experimental profile thus requires the order of the Bessel function, N, and tube diameter, D, and the tube orientation angle γ. The angle γ takes into account the inclination between the tube axis and the plane normal to the electron beam. For a given N, the tube diameter D is calculated to match the positions of the maximum intensity for experiment (X_{max}) and Bessel function (x_N^{max}) using the relationship:

$$\pi R'D = \pi \frac{\sqrt{Z^2 \sin^2 \gamma + X_{\mathrm{max}}^2}}{L\lambda} D = x_N^{\mathrm{max}}$$

and

$$D = x_N^{\mathrm{max}} / \left(\pi \frac{\sqrt{Z^2 \sin^2 \gamma + X_{\mathrm{max}}^2}}{L\lambda} \right) \tag{17.50}$$

where X and Z are the experimental detector coordinates, $L\lambda$ is the camera constant, and R' is the scattering vector normal to the tube in the experimental coordinates. Figure 17.28 shows the examples of the fit. The Bessel orders that give consistent tube diameter and the tilt angle, γ, can be selected from the best fit.

17.4.3   Chirality and Diameters of Single-Walled Carbon Nanotubes

SWCNTs have excellent electric properties. Since their initial discovery in the early 1990s (Iijima and Ichihashi 1993), SWCNTs have attracted considerable interest for their applications in electronic devices and sensors (Baughman et al. 2002; Cao et al. 2008). Controlling the structure of SWCNTs through synthesis and processing is critical (Xie et al. 2015). The structure of a SWCNT is uniquely defined by its chiral vector (n, m), whose length and angle determine the tube diameter and

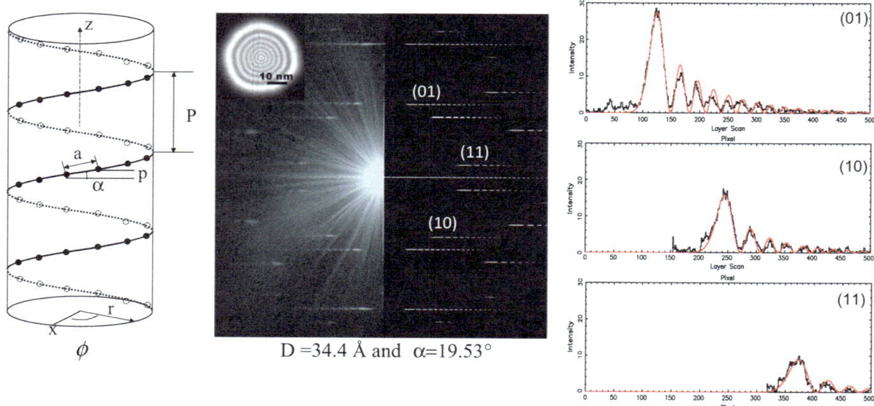

Fig. 17.28 Helical diffraction of SWCNT. *Left* the helical structure of a single C–C chain, middle, experimental, and simulated diffraction of a SWCNT, *right*

chirality as described in the last section. It has been shown by theory that a metallic tube is obtained if $(n - m) = 3l$ (l integer) or a semiconducting tube for $n - m \neq 3l$ (Saito et al. 1998). Semiconducting SWNTs are used for transistors, while metallic tubes are excellent conductors for interconnects.

SWCNTs can be synthesized by different techniques, including arc discharge. The most popular for device applications involves decomposition of carbon-containing gases on catalytic particles to provide carbon sources for growth, which can be controlled via deposition of catalytic particles (Kocabas et al. 2005; Matthews et al. 2006; Kang et al. 2007).

The structure of SWCNTs can be directly determined using electron diffraction. Electron diffraction patterns can be recorded from individual CNTs in the nanoarea electron diffraction (NAED) mode (Sect. 10.4.2) using a field emission gun TEM (Zuo et al. 2004). The small parallel beam allows the selection of an individual CNT and a reduction of background intensities in the recorded electron diffraction patterns from the surrounding materials. The NAED technique also improves the signal/noise ratio in the recorded patterns. Figure 17.29 shows an example from a SWCNT of 1.49 nm in diameter. Its structure was determined based on the fitting of the layer line intensities.

Figure 17.30 shows the chiral angles and diameters of 30 SWCNTs determined by electron diffraction. Because the handedness of the CNTs cannot be resolved from the diffraction pattern, we assume n ≥ m for the chiral vector, i.e., $0° \leq \alpha \leq 30°$. The carbon nanotubes studied here were grown by the following procedures. Catalyst (Fe:Mo:Al$_2$O$_3$) was prepared following the approach reported by Cassell et al. (1999) and Zhang et al. (2003). Catalyst was either spun or sprayed onto the TEM grids and then dried in a vacuum oven at 80 °C for 10 min before being loaded into the CVD reactor. The growth duration was 15 min with the reaction pressure at 10–100 Torr and the ratio of methane and hydrogen flow rate

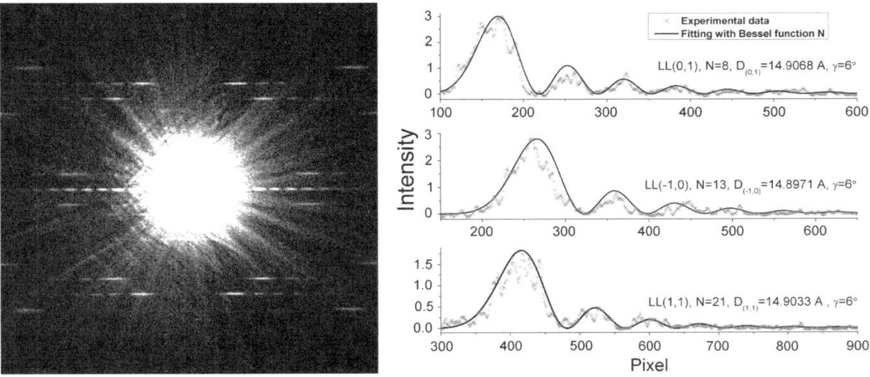

Fig. 17.29 An example of structure determination of SWCNTs. *Left* Experimental diffraction pattern recorded using nanoarea electron diffraction. *Right* Fitting of layer line intensities using kinematic theory. The chiral vector was determined as (13, 8) with the theoretical $D = 14.35$ Å and $\alpha = 22.17°$ and experimental $D = 14.90$ Å

Fig. 17.30 Statistic distribution of chiral angles and diameters determined from 30 SWCNTs

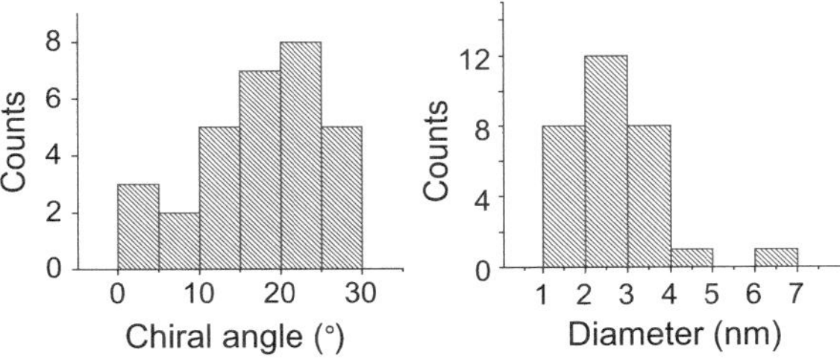

fixed at 60 and 40 cc/min, respectively. The CNTs produced depend strongly on the growth conditions and especially on the growth temperature. In most cases, we observe a mixture of SWNTs and DWNTs. The percentage of DWNTs and the number of tubes (yield) increase with temperature. The yield also increases under high-reaction gas pressure. By varying the temperature and reaction gas pressure, it is possible to grow CNTs of high density with dominantly SWNTs or DWNTs. For example, at 700 °C and 100 Torr of gas pressure, almost pure SWNTs were obtained, while at 800 °C, 10 Torr, more than 90 % tubes were DWNTs.

The tubes studied by diffraction were randomly selected. Among these tubes, 15 out of 30 (50 %) have chiral angles between 15° and 25°. The percentage of small chiral angle (0°–10°) was obviously suppressed. Two zigzag (26, 0), (37, 0) and one metallic armchair (13, 13) tubes were observed. The distribution observed here deviates significantly from the prediction geometrical model; if all (n, m) are equally possible, the chiral angle distribution should be approximately constant. The majority of measured SWCNTs have diameters between 1 and 4 nm, while very few are larger than 4 nm. The large-diameter SWCNTs are unstable, susceptible to collapse under the electron beam illumination.

Among the measured SWNTs, the ratio of metallic (including semimetallic) to semiconductive tubes is 10:19, which is very close to 1:2. The metallic (including semimetallic) and semiconductive tubes are distinguished according to the $n - m = 3l$ criterion mentioned earlier. The distribution indicates that the conductive property was not controlled by growth.

Carbon nanotubes grow by protruding from the catalyst particle at the tube–solid interface, where carbon atoms or molecules attach to the tube. The abundance of the chiral tubes of near-armchair configuration suggests a favored growth by hexagon addition with carbon dimers at the interface as illustrated in Fig. 17.30. For CNTs, the fraction of sites for dimer attachment increases from 0 to 100 % as α increases from $0°$ to $30°$. This is correlated with the increase in bond density (a zigzag tube has 0.406 carbon–metal bonds per Å compared to 0.515 for an armchair tube). The lack of armchair tubes can be explained based on growth energetics; in the classical growth theory, the interface process-controlled growth rate is proportional to $|\Delta G| \exp[-\Delta E/kT]/kT$, where ΔE is the interfacial energy barrier and $|\Delta G|$ is the driving force. The energy barrier ΔE includes a bond-breaking term for the carbon–metal bonds at the nanotube and catalyst interface. For a pure zigzag or armchair tube, all bonds have to be broken in order to grow an additional layer, while for a chiral tube, only a few bonds need to be broken (see Fig. 17.31). Thus, it has been proposed that carbon nanotubes grow by preferential attachment to steps that come

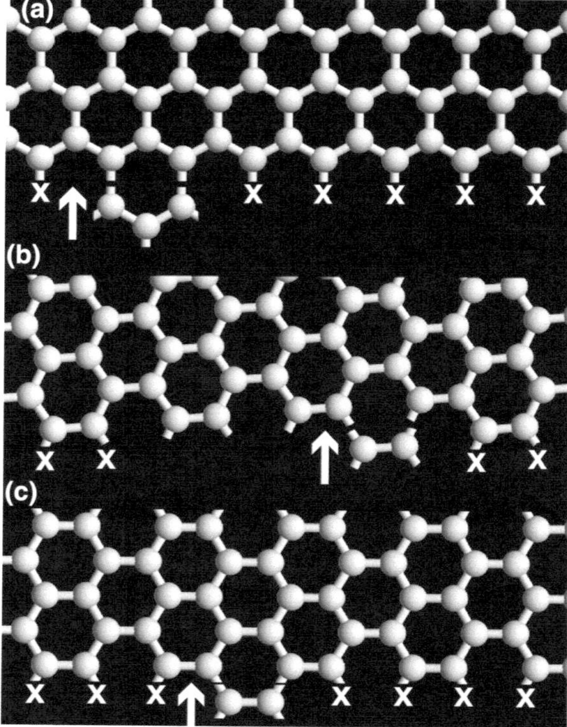

Fig. 17.31 Hexagon growth of (*left*) zigzag, (*middle*) chiral ($\alpha = 25°$), and (*right*) armchair carbon nanotubes with horizontal tube direction. Both the chiral and armchair nanotubes grow by the addition of carbon dimers, while the zigzag tubes require a combination of carbon trimers, dimers, and monomers. The X marks the carbon–metal bond at the nanotube and catalyst interface. The stepped tube–metal interface of a chiral tube requires few bond breakings and provides favored attachment sites as shown by the arrow (Gao et al. 2006)

with a chiral tube, and the growth is very much like the Frank's growth model for real crystals where steps formed by dislocations flow by preferential attachment (Gao et al. 2006).

17.4.4   Structure of Multiwalled Carbon Nanotubes

An ideal MWCNT consists of nested, cylindrical tubes (the "Russian doll" model), where the neighboring tubes are weakly bonded through van der Waals forces (Maniwa et al. 2001; Zuo et al. 2007). The MWCNT is incommensurate when each of its walls has its own chirality independent of other walls (Amelinckx et al. 1999; Zuo et al. 2007). In some cases, grouping of chiral angles was observed (Ruland et al. 2003; Koziol et al. 2005; Zhang et al. 2005) in MWCNTs grown by chemical vapor deposition (CVD). They can be distinguished by electron diffraction. Figure 17.32 shows two examples. The top and bottom diffraction patterns were recorded from two MWCNTs of approximately 20 and 10 nm in diameters, respectively. HREM inspection of the MWCNTs indicates that the smaller MWCNT is straight, without defects, while the larger MWCNT shows deviations from the straight- and parallel-walled tubes as indicated by the arrows. The two electron diffraction patterns are very different, with the larger MWCNT showing similar chiral angles from different walls of the MWCNT, while the smaller MWCNT has no correlation for the chiral angles.

The structure of MWCNTs is described by the structure of each wall. In the case of incommensurate MWCNTs, the structure can be treated the same as for nested SWCNTs. Experimental determination of an incommensurate MWCNT structure involves the measurement of the tube chiral angle and tube diameter or direct determination of the chiral vector (n, m) using the methods described in Sect. 17.4.2. The tube diameter measurements have to be performed in conjunction with the measurement of the tube inclination angle γ. Ignoring tube inclination can lead to large errors in the measured tube diameters. Both tube diameter and inclination angle γ can be measured by matching the oscillations of the layer intensities using Bessel functions. Overall, the structure determination problem of MWCNTs is overdetermined.

Figure 17.33 shows an electron diffraction pattern recorded from a 5-wall MWCNT. The pattern was recorded on imaging plates at the exposure time of 11 s. To determine the structure of the MWCNT, the same analysis procedure for SWCNT based on the layer line positions and intensities was used to determine the chiral vectors of different walls in the MWCNT. The challenging part is to sort out the layer lines into doublets (achiral tubes) or triplets (chiral tubes) corresponding to the graphene reflections of (1, 1), (1, 0), and (0, 1). This can be achieved using the following constraint to group the layer lines: $G_\perp(1,1) + G_\perp(1,0) = G_\perp(0,1)$, where $G_\perp(1,1)$, $G_\perp(1,0)$, and $G_\perp(0,1)$ are the axial component of the scattering factor of the corresponding graphene reflections (Gao et al. (2003)). The tube chiral angle is calculated from the layer line positions using Eq. (17.48). The Bessel

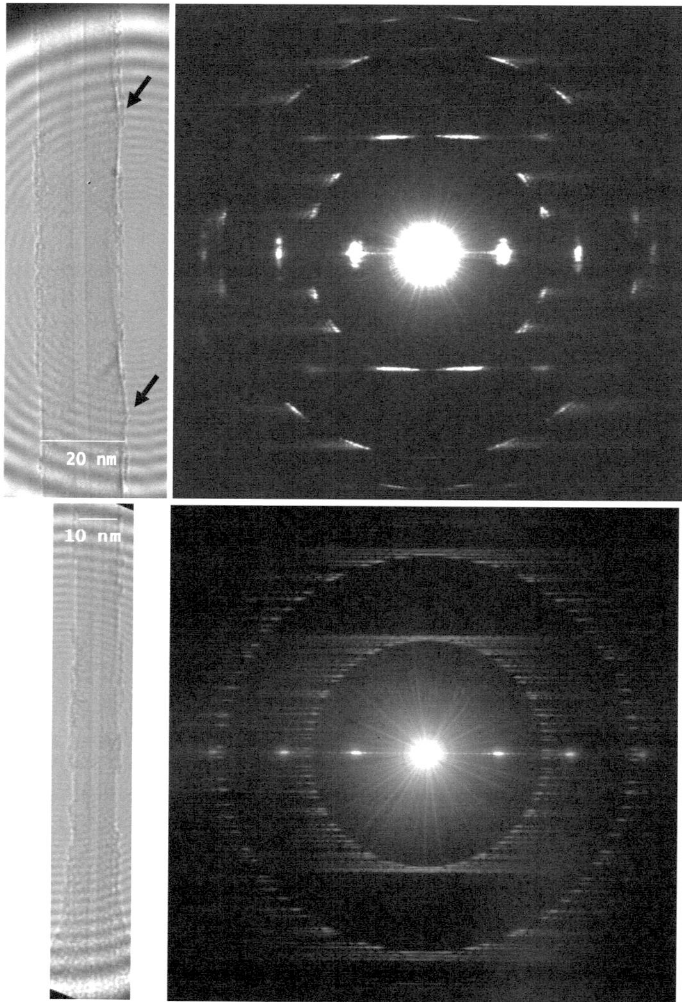

Fig. 17.32 Two types of MWCNTs with *Top* nearly commensurate and *Bottom* incommensurate structure

function orders of layer line (1, 0) and (0, 1) are directly related to the chiral indices (n, m) according to Eq. (17.46). The chiral indices, (n, m), can also be checked by comparison with the measured chiral angles and diameters. Based on these principles, the chiral indices for 5 walls are determined to be (14, 10), (25, 9), (39, 0), (29, 26), and (56, 4), with the experimentally determined diameters of 16.97, 24.32, 30.99, 37.61, and 45.79 Å, respectively. The structure of the MWCNT based on this structure determination is shown in Fig. 17.33c together with a simulated electron diffraction pattern from the model. There are significant differences in the

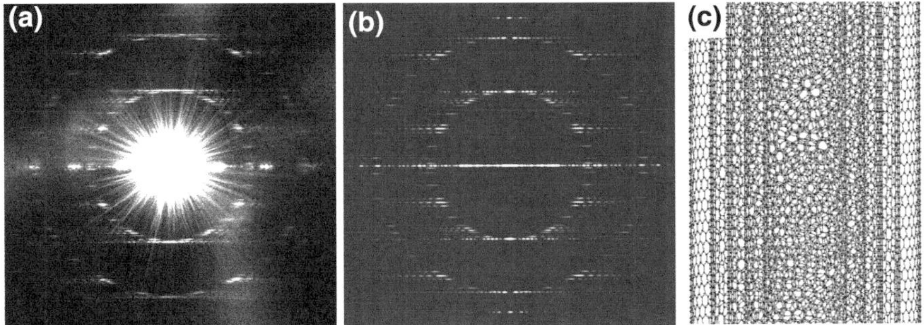

Fig. 17.33 Structure determination of MWCNTs. **a** An experimental ED pattern recorded from a 5-wall MWCNT. Atomic structure model (**c**) of ideal cylindrical tubes constructed based on the chiral vectors determined by electron diffraction and the simulated diffraction pattern (**b**) of model (**c**)

nanotube diameters compared with the ideal CNT structure model of the hexagonal graphene structure and C–C bond length of 1.421 Å. The largest difference is 3.8 ± 0.4 % for the innermost wall of ~ 17 Å in diameter. The difference is much smaller at 0.6 ± 0.2 % for the outermost wall of ~ 46 Å in diameter. Compared to the strong size dependence of the nanotube diameter difference, the axial periodicities of all 5 walls are almost the same as the ideal CNTs and no diameter dependence was observed. These experimental results indicate, on average, that there are two different bond lengths in each wall; the difference between the two bonds is most significant at close to 10 % for the small diameter walls (~ 3 nm or smaller) (Zhang and Zuo 2009).

17.4.5   Defects in Graphene and Carbon Nanotubes

Structural defects in the form of pentagons, heptagons, and octagons are responsible for curvature changes, as well as the introduction of dislocations and grain boundaries, in graphene and carbon nanotubes. Other types of defects involve bond rotations, Stone–Wales-type transformations, substitutional atoms (impurities, doping), vacancies, interstitials, and edge atoms.

The introduction of a single pentagon or heptagon is accompanied by a line defect that breaks the orientation symmetry, which is known as a disclination, from the study of liquid crystals. By comparison, a dislocation breaks translation symmetry. Disclinations are obtained in graphene by removing or adding a semi-infinite wedge of 60°, resulting in the core of a pentagon or heptagon embedded in the honeycomb lattice of graphene for negative ($s = -60°$) and positive ($s = 60°$) disclinations, respectively (Fig. 17.34). Here, s is the strength of the disclination, which measures how much the lattice rotates along a closed curve around the center of the pentagon or heptagon (the singularity point). Isolated disclinations are unlikely to be present in graphene because the large curvature (they inevitably

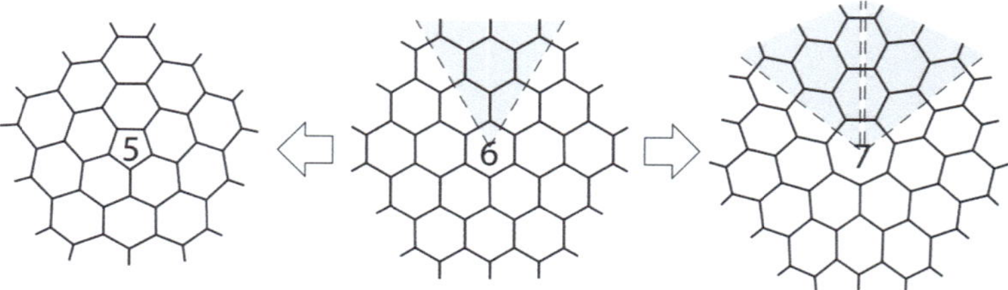

Fig. 17.34 Disclinations formed by removing and adding 60° segment from graphene (reprinted from Yazyev and Louie (2010) with the permission)

Fig. 17.35 *Top* Bond rotation leading to the 5–7–7–5 defect in the honeycomb lattice of graphene. From Suenaga et al. (2007). *Bottom* Atomic-resolution HAADF image of single-layer irradiated graphene showing the Stone–Wales defects [from Pan et al. (2014)]

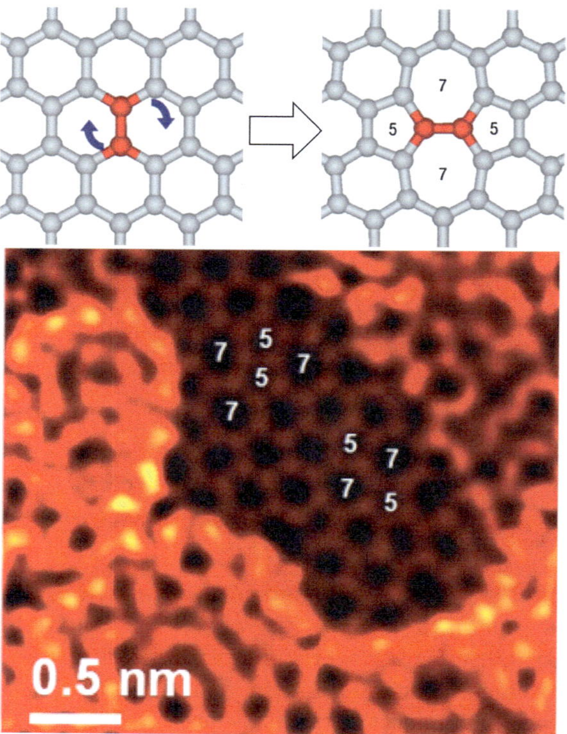

introduce) is incompatible with the planar structure of graphene. They are often observed in defective CNTs, where there is a change in the tube diameter or chirality or both. They are also observed in the end caps of CNTs.

A missing carbon atom or vacancy in graphene breaks three sp^2 covalent bonds at the cost of nearly 7.8 eV (Kaxiras and Pandey 1988). The likelihood of incorporating such defects during graphene or CNT growth is extremely small, even during high-temperature synthesis. Vacancy defects can be introduced by high-energy electron, ion, or neutron radiation through the knock-on damage process. Under an intense electron beam, these vacancies reorganize through bond rotations and eventually lead to amorphorization of the graphene (Kotakoski et al. 2011; Ran et al. 2012).

A single-bond rotation converts four adjacent hexagons into two pentagons and two heptagons (Fig. 17.35). This particular defect is known in the literature as a Stone–Wales (SW) defect Stone and Wales (1986). Compared with a single pentagon or heptagon, the SW defect can be incorporated into graphene without disturbing its 2D topology. The defect energy is approximately 3.5 eV. In CNTs, the SW defect is presumed to be as predominant as it is in graphite, even though the CNT synthesis, especially chemical vapor deposition (CVD) synthesis, proceeds at lower temperatures than typical graphitization. Once formed, the SW defects are relatively stable accompanied by a high-dissolution energy barrier. Other defects based on the combination of different numbers of pentagons and heptagons have also been observed in graphene and CNTs. The simplest is a single "5-7" pair, in which a pentagon adjacent to a heptagon is substituted for two hexagons. The 5-7 defect introduces a slight pucker to a graphene sheet. It is accompanied by an edge dislocation in graphene with the smallest possible Burgers vector equal to one lattice constant [$b = (1, 0)$]. The first TEM study of dislocations in graphite was reported in the early 1960s (Amelinckx and Delavignette 1960; Delavignette and Amelinckx 1962). Figure 17.36 shows an example of a pair of 5-7 defects separated at a distance, giving rise to a pair of positive and negative edge dislocations. The SW defect can be regarded as a pair of 5-7 defects having zero separation. Under strain, a SW defect can be separated into two 5-7 defects Walgraef (2007), accompanied by change in deformation that comes with dislocation glide (Warner et al. 2012) (Fig. 17.36).

In a cylindrical CNT, the dislocation introduced by a 5-7 defect is associated with a change in tube diameter and helicity (Fig. 17.37). The general rule is that the

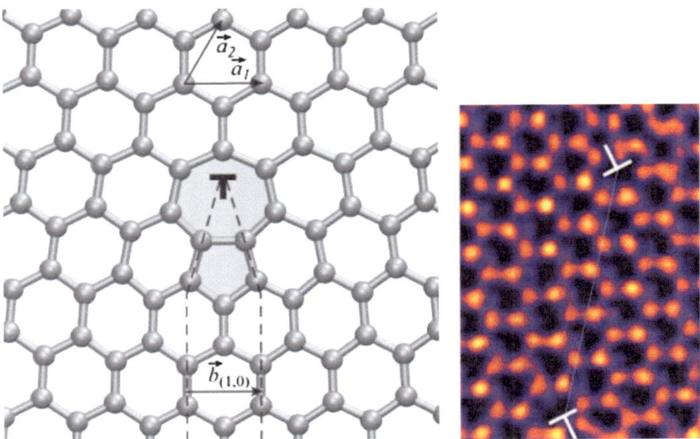

Fig. 17.36 *Left* Model of an edge dislocation in graphene [from Yazyev and Chen (2014)] and *right* HREM image showing an edge dislocation produced under electron beam irradiation (Warner et al. (2012), reproduced with the permission)

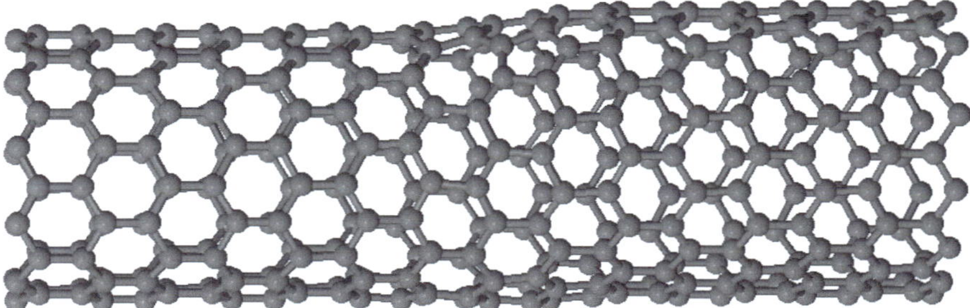

Fig. 17.37 Two zigzag nanotubes of chiral vectors (11, 0) and (12, 0) are connected with a pair of edge-sharing pentagon and heptagon oriented parallel to the tube axis

chiral vector (n, m) of a chiral CNT changes to $(n \pm 1, m \mp 1)$ with the incorporation of a 5-7 defect around a SWNT circumference (Chico et al. 1996). Since the SWNT electronic structure is dependent on helicity, the introduction of 5-7 defects creates a SWNT–SWNT heterojunction. Similar junctions can also be constructed from isolated pentagons, heptagons, or octagons. In these cases, the CNT forms a knee-like joint and the helicity change is accompanied by local concave or convex distortions to the SWNT sidewall. The additional curvature is allowed in the 0D and 1D topologies of fullerenes and CNTs, but forbidden in 2D, planar graphene (Collins 2009).

Graphene films of wafer scale produced by chemical vapor deposition (CVD) are polycrystalline. Two grains in a 2D film are rotated about their normal to form the tilt boundaries, and their 2D structure can be determined directly by HREM with the help of the improved resolution from aberration correction. Experimental results show that many grain boundaries (GBs) are covalently connected by chains of alternating pentagons and heptagons embedded in the honeycomb lattice of graphene (Fig. 17.38), in agreement with theory (Yazyev and Louie 2010). The GB structure can be generalized based on the Read–Shockley model of

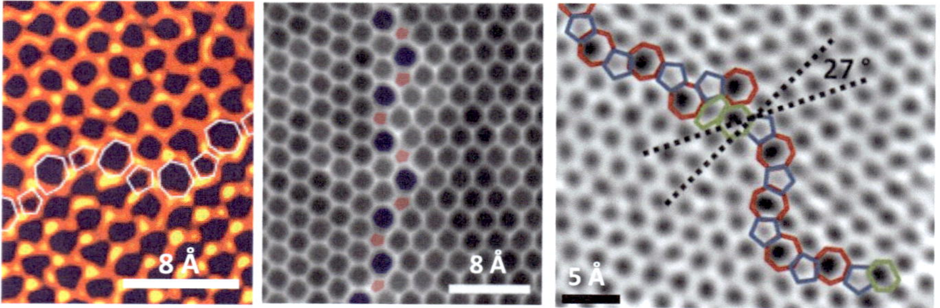

Fig. 17.38 Atomic structure of grain boundaries in polycrystalline graphene observed by HREM. From left to right, grain boundaries of 30°, 60° and a general character (from Rasool et al. (2013) and Huang et al. (2011), reproduced with the permission)

tilt GBs as arrays of edge dislocations. As discussed before, the core of edge dislocations in graphene is a pair of pentagons and heptagons. In such pairs, the coordination of all carbon atoms is conserved, thus automatically satisfying the bonding requirement. The presence of under-coordinated atoms is associated with high defect energy, and they are likely to be stabilized by adsorbates.

17.4.6   Van der Waals Forces and Molecular Interactions

The interaction between CNTs in a bundle or between two graphene sheets results from the same van der Waals (vdW) force between the graphite layers. The binding energy between CNTs can be determined from experiment, for example, by studying the interaction between two parallel CNTs. Figure 17.39 is a TEM image of such case. In this image, two CNTs grown from different catalysts merge and bond together to form a freestanding bundle of two parallel CNTs. The two CNTs have the same diameter $d = 4$ nm and lie essentially within the imaging plane of the microscope. The geometry of this pair of CNTs, resembling an opening crack, is a balance between the interface energy and strain energy, as illustrated by the geometrical model of tubes.

Each CNT can be modeled as an elastic beam since its length is much larger than the diameter. Two CNTs (beams) have the same hollow circular cross section with

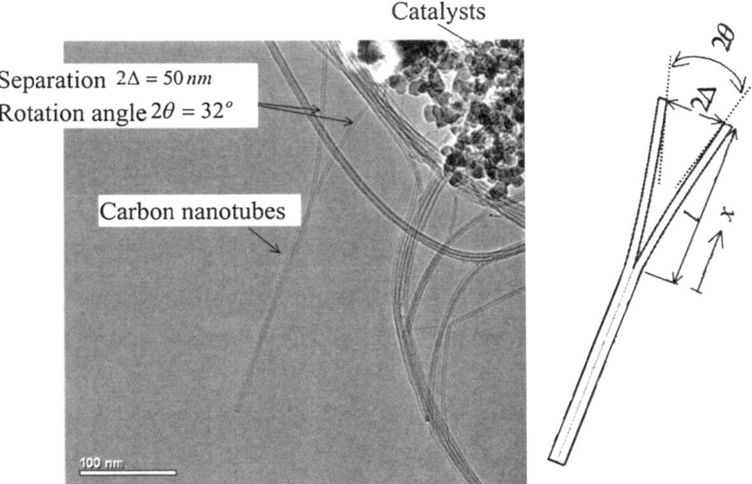

Fig. 17.39 Binding of two CNTs. *Left* TEM image of carbon nanotubes grown by chemical vapor deposition. Two carbon nanotubes connected to catalysts are sticking together and freestanding. They are both double-walled and have the same diameter. *Right* A double-beam model for two carbon nanotubes [provided by M. Gao, also see Chen et al. (2003)]

the outer diameter d, thickness t, and Young's modulus E. The bending stiffness of the beam is Timoshenko and Maccullough (1935)

$$EI = \frac{\pi}{64} E[d^4 - (d - 2t)^4].$$
(17.51)

A reasonable estimate of the Young's modulus of a CNT is the in-plane modulus of graphite, $E = 1.06\,\text{TPa}$ (Saito et al. 1998). For the double-walled CNT shown in Fig. 17.39, the thickness t is taken to be twice the interlayer spacing of graphite, i.e., $t = 2 \times 0.335\,\text{nm} = 0.67\,\text{nm}$. The bending stiffness in (17.51) becomes $EI = 1.07 \times 10^{-23}\,\text{Nm}^2$ (Chen et al. 2003).

As shown in Fig. 17.39, the left portion of the beams is bonded to represent the sticking part of the CNTs. The right portion of the beams is separated by a distance 2Δ over the length l. The rotation angle at the right ends is denoted by 2θ. For a separation length l, the surface energy in the system is then simply $\gamma\,l$, where γ is the aforementioned binding energy (interaction energy per unit tube length). The total energy in the system including both elastic and binding energies is given by Chen et al. (2003)

$$F = 2 * \frac{2EI}{l}\left(\theta^2 - 3\theta\frac{\Delta}{l} + 3\frac{\Delta^2}{l^2}\right) + \gamma l.$$
(17.52)

For very small or very large l, the elastic energy or binding energy becomes very large, respectively. Therefore, there exists a critical length l at which the total energy reaches the minimum. The CNTs in Fig. 17.39 are assumed to be at a state of minimum energy. This requires

$$\frac{\partial F}{\partial l} = 0,$$

which gives the binding energy γ,

$$\gamma = \frac{4EI}{l^2}\left(\theta - 3\frac{\Delta}{l}\right)^2.$$
(17.53)

Using EI estimated above and experimental values for l, Δ, and θ, we obtain from (3) an estimate of the binding energy of $\gamma = 0.36\,\text{nN}$.

The van der Waals interaction between atoms of the two CNTs is characterized by

$$V(r) = 4\varepsilon\left[\left(\frac{\sigma}{r}\right)^{12} - \left(\frac{\sigma}{r}\right)^{6}\right].$$
(17.54)

Two sets of σ and ε have been reported for carbon, namely $\sigma = 0.339\,\text{nm}$, $\varepsilon = 3.02\,\text{meV}$ (Quo et al. 1991), and $\sigma = 0.341\,\text{nm}$, $\varepsilon = 2.39\,\text{meV}$ (Girifalco et al.

Fig. 17.40 Structure model of folded graphene. **a** Schematic of folding a graphene sheet with a folding angle α. **b** The shape of the folded edge is determined by the minimization of energy [from Jiong Zhang, also see Zhang et al. (2010)]

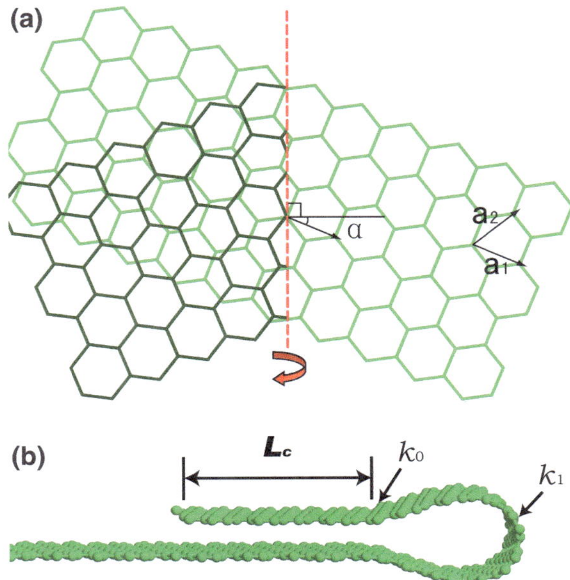

2000). This model gives binding energy (interaction energy per unit tube length) 0.39 nN and 0.32 nN for the above two sets of σ and ε, respectively. These values are rather close to the binding energy of 0.36 nN obtained from the experiment and the above mechanical model.

The molecular vdW force manifests itself in a number of phenomena in carbon nanostructures, such as interaction among carbon nanotubes in a bundle, collapsing of large-walled CNTs, packing of fullerenes, and folding (or origami) of graphene. Figure 17.41 shows the examples of folded sheets of 2 layers of graphene. Folded graphene of ~ 10 layers are also observed.

The structure of a folded graphene sheet is described by its folding angle and the balance between the attractive force between two graphene sheets and the tension cause by bending the graphene sheet (Fig. 17.40). The folding angle, α, is defined as the angle between the graphene lattice vector a_1 and the direction normal to the folding axis (Fig. 17.40a). The folded structure is characterized by a minimum length of flat, overlapped region, L_c, which provides the attractive force to balance the mechanical tension of the curved edge of the folded graphene. The shape of the curved edge and L_c can be determined by minimization of energy (Zhou et al. 2007). Assuming the energy of the flat, unfolded graphene to be zero, the total energy of a folded graphene U_{total} is composed of two parts, the bending energy in the curved edge U_b, and the adhesion energy of the flat region U_a due to vdW interaction. The bending energy U_b is obtained analytically as follows:

$$U_b = 2EI_G \int_0^{\sin^{-1}\left[\kappa_0^2/\left(\kappa_1^2-\kappa_0^2\right)\right]} \sqrt{\kappa_0^2 - (\kappa_1^2 - \kappa_0^2)\sin\theta}\, d\theta$$

$$+ EI_G \int_0^{\pi/2} \sqrt{\kappa_0^2 + (\kappa_1^2 - \kappa_0^2)\sin\theta}\, d\theta \tag{17.55}$$

where $EI_G = 0.69$ eV is the bending stiffness per unit length (nm) of graphene (Huang et al. 2006), κ_0 is the curvature at the connecting point between the curved edge and flat region, and κ_1 is the curvature at the right end point, as shown in Fig. 2b. The adhesion energy per unit length is defined by $U_a = -L_c\gamma$, where $\gamma = 1.45\,eV/\text{nm}^2$ (37 meV/atom) is the vdW interaction energy (Liu et al. 2004; Xiao et al. 2007). Energy minimization gives κ_0, κ_1 and the shape of the curved edge (Fig. 17.39b). From $U_{\text{total}} = 0$, the minimum length L_c to stabilize the fold is obtained as 1.62 nm.

Figure 17.41 shows diffraction patterns from folded graphene together with HREM images. The main features of the diffraction pattern include (1) two sets of honeycomb strong diffraction spots from top and bottom graphene layers; (2) elongated diffraction layer lines due to the curved structure of the folded edge; and (3) a relatively strong equatorial oscillation which is perpendicular to the edge

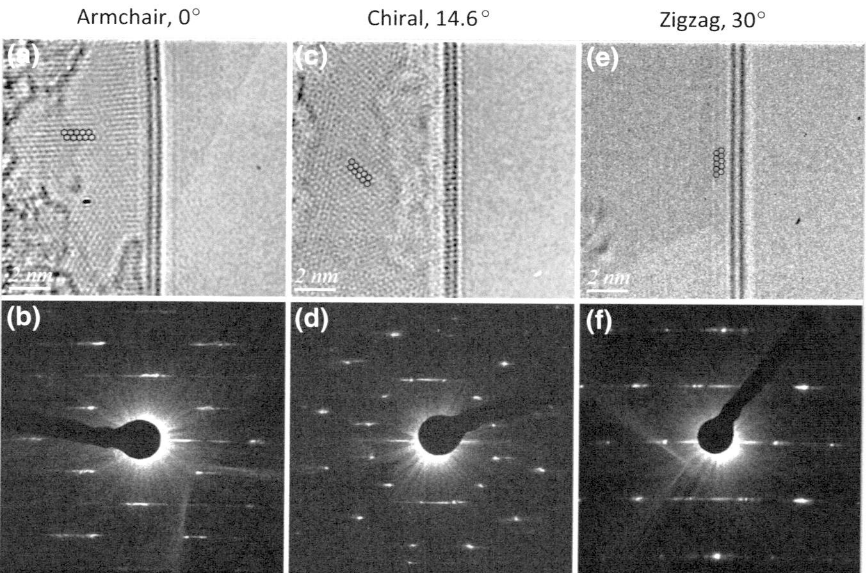

Fig. 17.41 Folded graphene. HREM images (**a**, **c**, **e**) and diffraction (**b**, **d**, **f**) from bilayer graphene folded edges of different types: (**a**, **b**) from armchair; (**c**, **d**) from a 14.6° chiral folded edge; (**e**, **f**) from zigzag [from Jiong Zhang, also see Zhang et al. (2010)]

direction. The graphene folding angle can be measured similarly to the chiral angle measurement of CNTs (Zhang et al. 2010).

The *vdW* force also plays a dominant role in SWCNT bundle formation, which is commonly observed in SWCNT synthesis (Thess et al. 1996). The attraction can result in a flattening of the CNT wall at the binding surface for nanotubes of both small and large diameters. The amount of flattening is small compared to large deformations, such as buckling and local indentations, seen by transmission electron microscopy (TEM) or atomic force microscopy (AFM) (Yu et al. 2000; Pantano et al. 2004). Flattening is also expected for a nanotube on a supporting substrate (Hertel et al. 1998). CNT deformations have a large effect on tube electron structures (Yang and Han 2000; Minot et al. 2003) and mechanical properties (Pantano et al. 2004). For example, experiments have shown tens of meV change in band gap per 1 % strain in SWCNTs (Yang and Han 2000; Minot et al. 2003).

The amount of tube flattening by vdW forces can be measured by electron diffraction by fitting the diffraction intensities using deformation models obtained from atomistic simulations (Jiang et al. 2008). Figure 17.42a shows the cross section of the deformed (17, 5) and (19, 11) CNT bundle. The radial deformation is plotted as a function of the angle θ as shown in Fig. 17.42b. For the (19, 11) and (17, 5) SWCNTs, the radial contraction at $\theta = 0°$ (flattening at the binding surface) is about 0.7 and 0.35 Å and the tube diameter change as viewed perpendicular to the binding surface is about 0.77 and 0.44 Å, respectively, at the 75 % of theoretically predicted deformation.

Fullerenes (C60) (Smith et al. 1998; Khlobystov et al. 2004) and metallo-fullerenes (Suenaga et al. 2003; Khlobystov et al. 2004) encapsulated inside SWCNTs, or the so-called peapods, are truly one-dimensional carbon-based molecular structures characterized by molecular vdW bonding. In bulk crystalline form, fullerenes are closely packed in a face-centered cubic (fcc) structure as

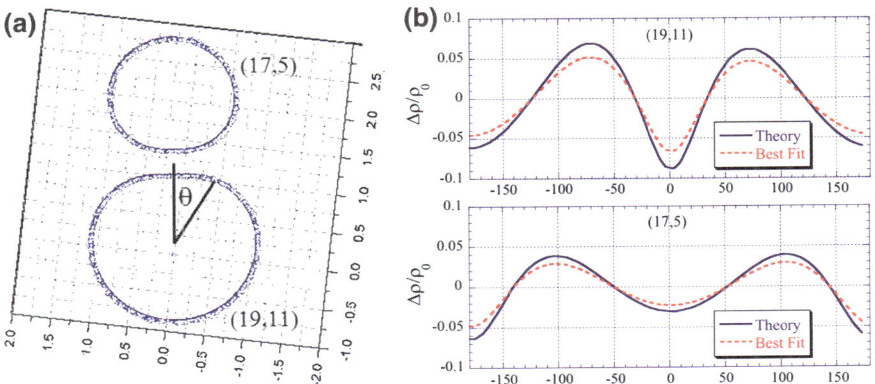

Fig. 17.42 Deformation of the (19, 11) and (17, 5) CNT bundle predicted by atomistic simulations and from the best-fit model with experiment. **a** The cross sections of the deformed tubes. **b** The radial deformation measured in the percentage of the original radius (ρ_o) plotted as a function of angle θ (in degrees) for the (19, 11) tube and *bottom* for the (17, 5) tube

topologically spherical molecules with weak vdW interactions between each other (David et al. 1991). The intermolecular spacing of adjacent C60s in the fcc structure is 10.02 Å (Hirahara et al. 2001). Packing of C60 molecules in peapod is reported to be different from that in three-dimensional fullerene crystals. Due to an inner tube pressure on the order of GPa generated by encapsulation of C60 molecules (Yoon et al. 2005), the fullerenes are arranged with a smaller inter-fullerene spacing by 3–4 % than in the bulk crystals. Based on X-ray diffraction, the distance between 2 C60 molecules was reported at 9.8 Å (Cambedouzou et al. 2004; Chorro et al. 2007) for peapod samples with an average tube diameter of 13.5 Å. The arrangement of C60 molecules inside the CNT is determined by interactions between molecules and with the CNT wall. Thus, a dependence on wall diameter is expected. Indeed, molecular dynamics (Michel et al. 2005; Troche et al. 2005; Verberck and Michel 2006; Verberck et al. 2006; Liu et al. 2008) and other theoretical simulations (Hodak and Girifalco 2003a, b) have predicted different types of ordered structures of C60s inside SWCNTs, which is largely determined by the diameter of the host SWCNT (Yoon et al. 2005; Verberck et al. 2006).

To determine the molecular structure of peapods requires a determination of the CNT structure, its orientation, and the arrangement of C60 molecules. This can be achieved by electron diffraction. Figure 17.43 shows an example. A section of the peapod was selected for electron diffraction at 80 kV by using a $\sim$25-nm-diameter parallel beam, as shown in the inset of Fig. 17.43a. Then, about 16 C60 molecules were included in electron diffraction. The electron beam has an electron current density of $\sim 10^5$ e/s nm^2. The diffraction pattern obtained is shown in Fig. 17.43b. The main features of this pattern consist of diffraction from the host SWCNT (Sect. 17.4.2), in the form of layer lines from the SWCNT helical structure, which are labeled by the arrows on the right and diffraction lines expected from the linear C60s chain. As marked by the arrows on the left side in Fig. 17.43b, the observed C60s diffraction intensities are distributed around the central area of the diffraction

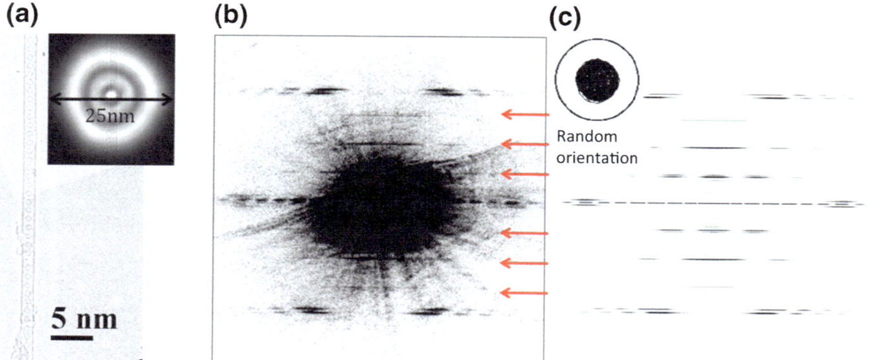

Fig. 17.43 **a** High-resolution TEM image of an isolated peapod. The *inset* shows the electron beam used for diffraction. **b** An experimental diffraction pattern recorded from the peapod shown in (**a**). **c** Simulated diffraction pattern based on the structural model shown in the *upper right inset*

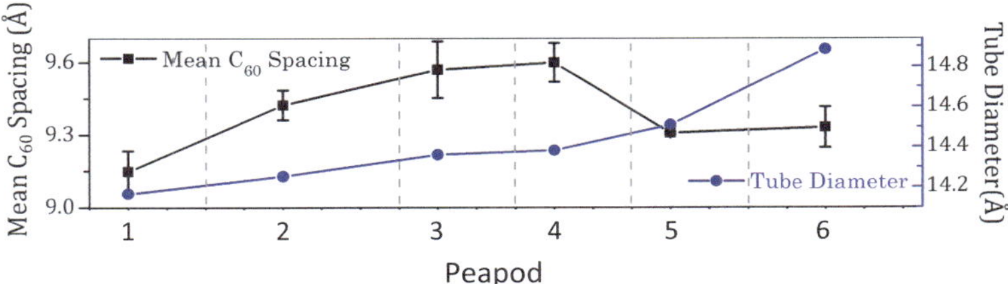

Fig. 17.44 Average axial spacing of C60 molecules in 6 peapods of different diameters, plotted together with the host nanotube diameter

pattern and appear as regularly spaced lines parallel to the SWCNT equatorial line. From the positions of these lines, the average spacing between C60 molecules along the tube axis direction could be measured with high accuracy (Ran et al. 2012).

Figure 17.44 shows the experimentally determined axial spacing of C60 molecules (Z_{C60}) in 6 peapods of different diameters, with the estimated error from diffraction patterns recorded at different locations of the peapods. The Z_{C60} increases in step with the tube diameter for the first 4 tubes, where the SWCNT diameters are smaller than 14.5 Å. This trend stops when the tube diameter exceeds 14.5 Å, in the case of tube 5-6, accompanied with a drop in Z_{C60} (difference between the 4th and 5th data point is larger than the respective error by a factor of 6.5 and 4.5). This result indicates that as D increases, the C60s packing changes from linear to zigzag as shown in Fig. 17.45a.

The linear-to-zigzag transition as indicated by the change in Z_{C60} can be understood based on the molecular interaction of a single encapsulated C60. By

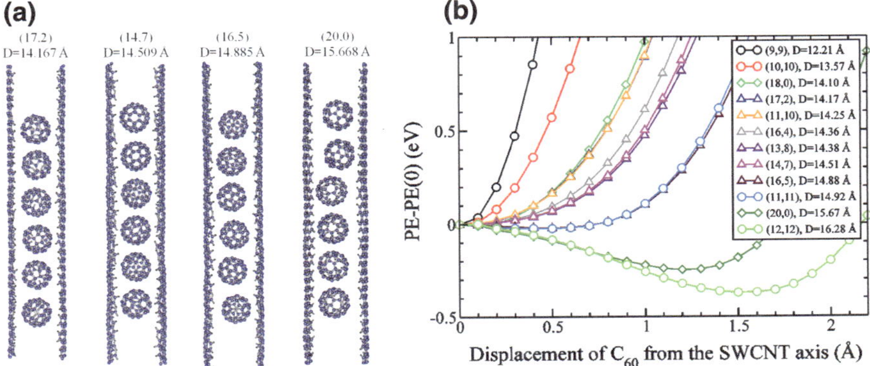

Fig. 17.45 Molecular configurations obtained by MD simulations. **a** Snapshots of the molecular configurations of 6 C60s encapsulated inside (17, 2), (14, 7), (16, 5), and (20, 0) SWCNT at 300 K during simulation, respectively. **b** The simulated single-C60 peapod system potential energy, PE, as a function of the C60's displacement from the CNT center axis. PE is shifted by PE(0), which denotes the potential energy when the C60 is located at the SWCNT center. The armchair, zigzag, and chiral nanotubes are represented by *circles*, *diamonds*, and *triangles*, respectively

calculating the minimum potential energy, during which the C60 can rotate and translate parallel to the tube axis freely, Fig. 17.45b shows the relative potential energy of the single-C60 peapod system as a function of the C60's displacement from the SWCNT axis. Diameters of the host SWCNTs range from 12.21 to 16.28 Å. For peapods with D smaller than 14.5 Å, the minimal system energy shows up at zero displacement (SWCNT centers). As D exceeds 14.5 Å, the energy minimum moves away from the tube axis. Therefore, the tube diameter of around 14.5 Å marks the transition of C60 packing from linear to zigzag. Importantly, a careful examination of Fig. 17.45 shows that peapods with D close to 14.5 Å exhibit a rather flat potential energy around their minima. This flat feature of the potential energy curve indicates that the encapsulated C60, previously subjected to 1D confinement along the SWCNT axis, is able to access space in 3D. Following the standard treatment of effective volume in a cell model for liquids (Hill 1956), the calculated effective area for a single C60 inside SWCNTs at 300 K gives

$$A_{\text{effective}} = \int_{0}^{\infty} 2\pi R \cdot e^{-\frac{U(R)-U_{\text{min}}}{kT}} dR. \tag{17.56}$$

Here, R is the C60 displacement off the SWCNT axis, $U(R)$ is the potential energy for the single-C60 peapod system at a displacement of R, and U_{min} is the potential energy minimum. The result shows that prior to the linear-to-zigzag transition point ($D \approx 14.5$ Å), $A_{\text{effective}}$ increases by 3–4 times as D increases from 13.6 to 14.2 Å. The sudden surge of accessible 3D space will lead to a 1D-to-3D transition of C60 configuration, as the C60s inside the narrower SWCNTs are still confined to 1D motion along the center axis (see Fig. 8). Undergoing this 1D-to-3D transition (significant expansion of accessible volume), the equilibrium configuration of C60s will move toward entropy-driven dissociation. This will likely contribute to the increase of Z_{C60} as the SWCNT diameter increases before the onset of the linear-to-zigzag packing transition (Ran et al. 2012).

References

Adams DL, Nielsen HB, Anderson JN (1983) Oscillatory relaxation of the Cu(110) surface. Surf Sci 128:294

Amelinckx S, Delavignette P (1960) Dislocation loops due to quenched-in point defects in graphite. Phys Rev Lett 5:50–51

Amelinckx S, Lucas A, Lambin P (1999) Electron diffraction and microscopy of nanotubes. Rep Prog Phys 62:1471–1524

Amirfazli A, Neumann AW (2004) Status of the three-phase line tension. Adv Colloid Interf 110:121–141

Amirfazli A, Kwok DY, Gaydos J, Neumann AW (1998) Line tension measurements through drop size dependence of contact angle. J Colloid Interf Sci 205:1–11

Avdeenkov AV, Bibikov AV, Bodrenko IV, Nikolaev AV, Taran MD, Tkalya EV (2009) Modified carbon nanostructures as materials for hydrogen storage. Russ Phys J 52:1235–1241

Baletto F, Ferrando R (2005) Structural properties of nanoclusters: Energetic, thermodynamic, and kinetic effects. Rev Mod Phys 77:371–423

Balluffi RW, Brokman A, King AH (1982) CSL/DSC lattice model for general crystal-crystal boundaries and their line defects. Acta Metall 30:1453–1470

Bandaru PR, Daraio C, Jin S, Rao AM (2005) Novel electrical switching behaviour and logic in carbon nanotube y-junctions. Nat Mater 4:663–666

Baskin Y, Meyer L (1955) Lattice constants of graphite at low temepratures. Phys Rev 100:544

Baughman RH, Zakhidov AA, de Heer WA (2002) Carbon nanotubes–the route toward applications. Science 297:787–792

Besenbacher F (1996) Scanning tunnelling microscopy studies of metal surfaces. Rep Prog Phys 59:1737–1802

Bohnen KP, Ho KM (1993) Structure and dynamics at metal surfaces. Surf Sci Rep 19:99–120

Bonzel HP (2003) 3d equilibrium crystal shapes in the new light of STM and AFM. Phys Rep 385:1–67

Bording JK, Li BQ, Shi YF, Zuo JM (2003) Size- and shape-dependent energetics of nanocrystal interfaces: experiment and simulation. Phys Rev Lett 90:226104

Boruvka L, Neumann AW (1977) Generalization of classical-theory of capillarity. J Chem Phy 66:5464–5476

Bratlie KM, Lee H, Komvopoulos K, Yang P, Somorjai GA (2007) Platinum nanoparticle shape effects on benzene hydrogenation selectivity. Nano Lett 7:3097–3101

Cambedouzou J, Pichot V, Rols S, Launois P, Petit P, Klement R, Kataura H, Almairac R (2004) On the diffraction pattern of C-60 peapods. Eur Phys J B 42:31–45

Cao Q, Kim H-S, Pimparkar N, Kulkarni JP, Wang C, Shim M, Roy K, Alam MA, Rogers JA (2008) Medium-scale carbon nanotube thin-film integrated circuits on flexible plastic substrates. Nature 454:495–500

Cassell AM, Raymakers JA, Kong J, Dai HJ (1999) Large scale CVD synthesis of single-walled carbon nanotubes. J Phys Chem B 103:6484–6492

Chen B, Gao M, Zuo JM, Qu S, Liu B, Huang Y (2003) Binding energy of parallel carbon nanotubes. Appl Phys Lett 83:3570–3571

Chen J, Lim B, Lee E, Xia Y (2009) Shape-controlled synthesis of platinum nanocrystals for catalyic and electrocatalytic applications. Nano Today 4:81–95

Chico L, Benedict LX, Louie SG, Cohen ML (1996) Quantum conductance of carbon nanotubes with defects. Phys Rev B 54:2600–2606

Chorro M, Cambedouzou J, Iwasiewicz-Wabnig A, Noe L, Rols S, Monthioux M, Sundqvist B, Launois P (2007a) Discriminated structural behaviour of C-60 and C-70 peapods under extreme conditions. EPL 79:56003

Chorro M, Delhey A, Noe L, Monthioux M, Launois P (2007b) Orientation of C-70 molecules in peapods as a function of the nanotube diameter. Phys Rev B 75:035416

Cleveland CL, Landman U, Schaaff TG, Shafigullin MN, Stephens PW, Whetten RL (1997) Structural evolution of smaller gold nanocrystals: the truncated decahedral motif. Phys Rev Lett 79:1873

Collins PG (2009) Defects and disorder in carbon nanotubes. In: Narlikar AV, Fu YY (eds) Oxford handbook of nanoscience and technology: frontiers and advances. Oxford University Press, Oxford

Cosandey F, Madey TE (2001) Growth, morphology, interfacial effects and catalytic properties of Au on TiO_2. Surf Rev Lett 8:73–93

Cowley JM, Spence JCH (1981) Convergent beam electron microdiffraction from small crystals. Ultramicroscopy 6:359–366

Dai L (2006) Carbon nanotechnology. Elsevier, Amsterdam

David WIF, Ibberson RM, Matthewman JC, Prassides K, Dennis TJS, Hare JP, Kroto HW, Taylor R, Walton DRM (1991) Crystal-structure and bonding of ordered C-60. Nature 353:147–149

Davis HL, Noonan JR (1983) Multilayer relaxation in metallic surfaces as demonstrated by LEED analysis. Surf Sci 126:245

Daw MS, Foiles SM, Baskes MI (1993) The embedded-atom method—a review of theory and applications. Mat Sci Rep 9:251–310

Delattre T, Feuillet-Palma C, Herrmann LG, Morfin P, Berroir JM, Feve G, Placais B, Glattli DC, Choi MS, Mora C, Kontos T (2009) Noisy Kondo impurities. Nat Phys 5:208–212

Delavignette P, Amelinckx S (1962) Dislocation patterns in graphite. J Nucl Mater 5(1):17–66

Dinadayalane TC, Leszczynski J (2010) Remarkable diversity of carbon–carbon bonds: structures and properties of fullerenes, carbon nanotubes, and graphene. Struct Chem 21:1155–1169

Doye JPK, Wales DJ (1995) Calculation of thermodynamic properties of small Lennard-Jones clusters incorporating anharmonicity. J Chem Phy 102:9659–9672

Duke CB (1996) Semiconductor surface reconstruction: the structural chemistry of two-dimensional surface compounds. Chem Rev 96:1237–1259

Dupré AM, Dupré P (1869) Théorie mécanique de la chaleur. Gauthier-Villars

Dvorak F, Farnesi Camellone M, Tovt A, Tran N-D, Negreiros FR, Vorokhta M, Skala T, Matolinova I, Myslivecek J, Matolin V, Fabris S (2016) Creating single-atom Pt-ceria catalysts by surface step decoration. Nat Commun 7:10801

Frenkel AI, Nemzer S, Pister I, Soussan L, Harris T, Sun Y, Rafailovich MH (2005) Size-controlled synthesis and characterization of thiol-stabalized gold nanoparticles. J Chem Phys 123:184701

Gao M, Zuo JM, Twesten RD, Petrov I, Nagahara LA, Zhang R (2003) Structure determination of individual single-wall carbon nanotubes by nanoarea electron diffraction. Appl Phys Lett 82:2703–2705

Gao M, Zuo JM, Zhang R, Nagahara LA (2006) Structure determinations of double-wall carbon nanotubes grown by catalytic chemical vapor deposition. J Mater Sci 41:4382–4388

Gao W, Choi AS, Zuo J-M (2014) Interaction of nanometer-sized gold nanocrystals with rutile (110) surface steps revealed at atomic resolution. Surf Sci 625:16–22

Girifalco LA, Hodak M, Lee RS (2000) Carbon nanotubes, buckyballs, ropes, and a universal graphitic potential. Phy. Rev B 62:13104–13110

Gleiter H (2000) Nanostructured materials: basic concepts and microstructure. Acta Mater 48:1–29

Gryaznov VG, Heydenreich J, Kaprelov AM, Nepijko SA, Romanov AE, Urban J (1999) Pentagonal symmetry and disclinations in small particles. Cryst Res Technol 34:1091–1119

Gülseren O, Yildirim T, Ciraci S (2002) Systematic ab initio study of curvature effects in carbon nanotubes. Phys Rev B 65:153405

Hall BD, Flueli M, Monot R, Borel JP (1991) Multiply twinned structures in unsupported ultrafine silver particles observed by electron-diffraction. Phys Rev B 43:3906–3917

Halperin WP (1986) Quantum size effects in metal particles. Rev Mod Phys 58:533–606

Hansen KH, Worren T, Stempel S, Lægsgaard E, Bäumer M, Freund HJ, Besenbacher F, Stensgaard I (1999) Palladium nanocrystals on Al_2O_3: structure and adhesion energy. Phys Rev Lett 83:4120

Hargittai I (2006) Gas-phase electron diffraction for molecular structure determination. In: Weirich TE, Lábár JL, Zou X (eds) Electron crystallography: novel approaches for structure determination of nanosized materials. Springer, Dordrecht

Hasegawa M, Nishidate K (2004) Semiempirical approach to the energetics of interlayer binding in graphite. Phys Rev B 70:205431

Heine V, Marks LD (1986) Competition between pairwise and multi-atom at noble metal surfaces. Surf Sci 165:65

Henry CR (2005) Morphology of supported nanoparticles. Prog Surf Sci 80:92–116

Hertel T, Walkup RE, Avouris P (1998) Deformation of carbon nanotubes by surface van der Waals forces. Phys Rev B 58:13870–13873

Heyraud JC, Metois JJ (1980) Equilibrium shape of gold crystallites on a graphite cleavage surface —surface energies and interfacial energy. Acta Metall 28:1789–1797

Hill TL (1956) Statistical mechanics: principles and selected applications. McGraw-Hill, New York

Hirahara K, Bandow S, Suenaga K, Kato H, Okazaki T, Shinohara H, Iijima S (2001) Electron diffraction study of one-dimensional crystals of fullerenes. Phys Rev B 64:115420

Hirahara K, Kociak M, Bandow S, Nakahira T, Itoh K, Saito Y, Iijima S (2006) Chirality correlation in double-wall carbon nanotubes as studied by electron diffraction. Phys Rev B 73:195420

Hodak M, Girifalco LA (2003a) Ordered phases of fullerene molecules formed inside carbon nanotubes. Phy. Rev B 67:075419

Hodak M, Girifalco LA (2003b) Systems of C-60 molecules inside (10, 10) and (15, 15) nanotube: a Monte Carlo study. Phys Rev B 68:085405

Howie A, Marks LD (1984) Elastic strains and the energy balance for multiply twinned particles. Philos Mag A 49:95–109

Huang Y, Wu J, Hwang KC (2006) Thickness of graphene and single-wall carbon nanotubes. Phys Rev B 74:245413

Huang WJ, Sun R, Tao J, Menard LD, Nuzzo RG, Zuo JM (2008) Coordination-dependent surface atomic contraction in nanocrystals revealed by coherent diffraction. Nat Mater 7:308–313

Huang PY, Ruiz-Vargas CS, van der Zande AM, Whitney WS, Levendorf MP, Kevek JW, Garg S, Alden JS, Hustedt CJ, Zhu Y, Park J, McEuen PL, Muller DA (2011) Grains and grain boundaries in single-layer graphene atomic patchwork quilts. Nature 469:389

Ichimiya A, Cohen PI (2004) Reflection high energy electron diffraction. Cambridge University Press, Cambridge

Iijima S (1991) Helical microtubules of graphitic carbon. Nature 354:56–58

Iijima S, Ichihashi T (1986) Structural instability of ultrafine particles of metals. Phys Rev Lett 56:616–619

Iijima S, Ichihashi T (1993) Single-shell carbon nanotubes of 1-nm diameter. Nature 363:603–605

Inagaki M (2000) New carbons control of structure and functions. Elsevier

Ino S (1966) Epitaxial growth of metals on rocksalt faces cleaved in vacuum. 2. Orientation and structure of gold particles formed in ultrahigh vacuum. J Phys Soc Jpn 21:346

Ino S (1969) Stability of multiply-twinned particles. J Phys Soc Jpn 27:941

Jiang QT, Fenter P, Gustafsson T (1991) Geometric structure and surface vibrations of Cu(001) determined by medium-energy ion-scattering. Phys Rev B 44:5773–5778

Jiang YY, Zhou W, Kim T, Huang Y, Zuo JM (2008) Measurement of radial deformation of single-wall carbon nanotubes induced by intertube van der Waals forces. Phys Rev B 77:153405

Johnson CL, Snoeck E, Ezcurdia M, Rodriguez-Gonzalez B, Pastoriza-Santos I, Liz-Marzan LM, Hytch MJ (2008) Effects of elastic anisotropy on strain distributions in decahedral gold nanoparticles. Nat Mater 7:120–124

Kaishew R (1952) Arbeitstagung fetsköper Phys., Dresden p. 81

Kanamitsu K, Saito S (2002) Geometries, electronic properties, and energetics of isolated single walled carbon nanotubes. J Phys Soc Jpn 71:483–486

Kang SJ, Kocabas C, Ozel T, Shim M, Pimparkar N, Alam MA, Rotkin SV, Rogers JA (2007) High-performance electronics using dense, perfectly aligned arrays of single-walled carbon nanotubes. Nat Nanotechnol 2:230–236

Kaxiras E, Pandey KC (1988) Energetics of defects and diffusion mechanisms in graphite. Phys Rev Lett 61:2693–2696

Kern R, Metois G (1979) Basic mechanisms in the early stage of epitaxy. Curr Top Mat Sci 3:135–419

Khlobystov AN, Britz DA, Ardavan A, Briggs GAD (2004) Observation of ordered phases of fullerenes in carbon nanotubes. Phys Rev Lett 92:245507

Kim K, Lee Z, Malone BD, Chan KT, Aleman B, Regan W, Gannett W, Crommie MF, Cohen ML, Zettl A (2011) Multiply folded graphene. Phys Rev B 83:245433

Kocabas C, Hur S-H, Gaur A, Meitl MA, Shim M, Rogers JA (2005) Guided growth of large-scale, horizontally aligned arrays of single-walled carbon nanotubes and their use in thin-film transistors. Small 1:1110–1116

Koga K (2006) Novel bidecahedral morphology in gold nanoparticles frozen from liquid. Phys Rev Lett 96:115501

Koplitz LV, Dulub O, Diebold U (2003) STM study of copper growth on ZnO (0001)-Zn and ZnO (0001)-O surfaces. J Phys Chem B 107:10583–10590

Kotakoski J, Meyer JC, Kurasch S, Santos-Cottin D, Kaiser U, Krasheninnikov AV (2011) Stone-Wales-type transformations in carbon nanostructures driven by electron irradiation. Phys Rev B 83:245420

Koziol K, Shaffer M, Windle A (2005) Three-dimensional internal order in multiwalled carbon nanotubes grown by chemical vapor deposition. Adv Mater 17:760–763

Lambin P, Lucas AA (1997) Quantitative theory of diffraction by carbon nanotubes. Phys Rev B 56:3571–3574

Lee H, Habas SE, Kweskin S, Butcher D, Somorjai GA, Yang P (2006) Morphological control of catalytically active platinum. Angew Chem Int Ed 45:7824–7828

Li BQ, Zuo JM (2005) Structure and shape transformation from multiply twinned particles to epitaxial nanocrystals: Importance of interface on the structure of ag nanoparticles. Phys Rev B 72:085434

Liu J (2003) Study of metal-support interactions in model nanocatalysts: anchoring of Pt metallic nanoparticles on alumina support. Micros Microanal 9(S02):290–291

Liu B, Yu MF, Huang YG (2004) Role of lattice registry in the full collapse and twist formation of carbon nanotubes. Phys Rev B 70:161402

Liu P, Zhang YW, Gao HJ, Lu C (2008) Energetics and stability of C-60 molecules encapsulated in carbon nanotubes. Carbon 46:649–655

Liu F, Song SY, Xue DF, Zhang HJ (2012) Folded structured graphene paper for high performance electrode materials. Adv Mater 24:1089–1094

Loretto D, Gibson JM, Yalisove SM (1989) Evidence for a dimer reconstruction at a metal-silicon interface. Phys Rev Lett 63:298–301

Lu XW, Gao WP, Zuo JM, Yuan JB (2015) Atomic resolution tomography reconstruction of tilt series based on a GPU accelerated hybrid input-output algorithm using polar fourier transform. Ultramicroscopy 149:64–73

Lynch RW, Drickamer HG (1966) Effect of high pressure on the lattice parameters of diamond, graphite, and hexagonal boron nitride. J Chem Phys 44:181–184

Maniwa Y, Fujiwara R, Kira H, Tou H, Nishibori E, Takata M, Sakata M, Fujiwara A, Zhao XL, Iijima S, Ando Y (2001) Multiwalled carbon nanotubes grown in hydrogen atmosphere: an X-ray diffraction study. Phys Rev B 64:073105

Marks LD (1984) Surface structure and energetics of multiply twinned particles. Philos Mag A 49:81–93

Marks LD (1994) Experimental studies of small particles structures. Rep Prog Phys 57:603

Marks LD, Peng L (2016) Nanoparticle shape, thermodynamics and kinetics. J Phys-Condens Matter 28:053001

Martin TP (1996) Shells of atoms. Phys Rep 273:199–241

Matthews KD, Lemaitre MG, Kim T, Chen H, Shim M, Zuo JM (2006) Growth modes of carbon nanotubes on metal substrates. J Appl Phys 100:044309

Miao JW, Charalambous P, Kirz J, Sayre D (1999) Extending the methodology of X-ray crystallography to allow imaging of micrometre-sized non-crystalline specimens. Nature 400:342–344

Michel KH, Verberck B, Nikolaev AV (2005) Nanotube field and one-dimensional fluctuations of C-60 molecules in carbon nanotubes. Eur Phys J B 48:113–124

Minot ED, Yaish Y, Sazonova V, Park JY, Brink M, McEuen PL (2003) Tuning carbon nanotube band gaps with strain. Phys Rev Lett 90:156401

Mitchell CEJ, Howard A, Carney M, Egdell RG (2001) Direct observation of behaviour of au nanoclusters on $TiO_2(1\ 1\ 0)$ at elevated temperatures. Surf Sci 490:196–210

Mott NF (1948) Slip at grain boundaries and grain growth in metals. Proc Phys Soc 60:391

Murphy CJ, San TK, Gole AM, Orendorff CJ, Gao JX, Gou L, Hunyadi SE, Li T (2005) Anisotropic metal nanoparticles: synthesis, assembly, and optical applications. J Phys Chem B 109:13857–13870

Narayanan R, El-Sayed MA (2005) Catalysis with transition metal nanoparticles in colloidal solution: nanoparticle shape dependence and stability. J Phys Chem B 109:12663–12676

Negreiros FR, Soares EA, de Carvalho VE (2007) Energetics of free pure metallic nanoclusters with different motifs by equivalent crystal theory. Phys Rev B 76:205429

Noonan JR, Davis HL (1984) Truncation-induced multilayer relaxation of the Al(110) surface. Phys Rev B 29:4349

Novoselov KS, Geim AK, Morozov SV, Jiang D, Zhang Y, Dubonos SV, Grigorieva IV, Firsov AA (2004) Electric field effect in atomically thin carbon films. Science 306:666–669

Pakarinen OH, Barth C, Foster AS, Nieminen RM, Henry CR (2006) High-resolution scanning force microscopy of gold nanoclusters on the KBr (001) surface. Phys Rev B 73:235428

Pan M, Cowley JM, Garcia R (1987) STEM and microdiffraction studies of Rh/CeO$_2$. Micron Microsc Acta 18:165–169

Pan M, Cowley JM, Barry JC (1989) Coherent electron microdiffraction from small metal particles. Ultramicroscopy 30:385–394

Pan CT, Hinks JA, Ramasse QM, Greaves G, Bangert U, Donnelly SE, Haigh SJ (2014) In-situ observation and atomic resolution imaging of the ion irradiation induced amorphisation of graphene. Sci Rep 4:6334

Pantano A, Parks DM, Boyce MC (2004) Mechanics of deformation of single- and multi-wall carbon nanotubes. J Mech Phys Solids 52:789–821

Pauling L (1947) Atomic radii and interatomic distances in metals. J Am Chem Soc 69:542–553

Pauling L (1960) The nature of the chemical bond (3rd). Cornell University Press, New York

Pauling L (1966) The structure and properties of graphite and boron nitride. Chemistry 56:1646–1652

Perez MS, Lerner B, Resasco DE, Obregon PDP, Julian PM, Mandolesi PS, Buffa FA, Boselli A, Lamagna A (2010) Carbon nanotube integration with a cmos process. Sensors 10(4):3857–3867

Petkov V, Wanjala BN, Loukrakpam R, Luo J, Yang L, Zhong C-J, Shastri S (2012) Pt-Au alloying at the nanoscale. Nano Lett 12:4289–4299

Pfeifer MA, Williams GJ, Vartanyants IA, Harder R, Robinson IK (2006) Three-dimensional mapping of a deformation field inside a nanocrystal. Nature 442:63

Pompe T, Herminghaus S (2000) Three-phase contact line energetics from nanoscale liquid surface topographies. Phys Rev Lett 85:1930–1933

Qin L-C (2007) Determination of the chiral indices (n, m) of carbon nanotubes by electron diffraction. Phys Chem Chem Phys 9:31–48

Quo Y, Karasawa N, Goddard WA (1991) Prediction of fullerene packing in C60 and C70 crystals. Nature 351:464–467

Ran K, Chen Q, Zuo JM (2012a) Fabrication and structure characterization of quasi-2-dimensional amorphous carbon structures. Acta Phys Chim Sin 28:1551–1555

Ran K, Mi X, Shi ZJ, Chen Q, Shi YF, Zuo JM (2012b) Molecular packing of fullerenes inside single-walled carbon nanotubes. Carbon 50:5450–5457

Rasool HI, Ophus C, Klug WS, Zettl A, Gimzewski JK (2013) Measurement of the intrinsic strength of crystalline and polycrystalline graphene. Nat Commun 4:2811

Reinhard D, Hall BD, Ugarte D, Monot R (1997) Size-independent fcc-to-icosahedral structural transition in unsupported silver clusters: an electron diffraction study of clusters produced by inert-gas aggregation. Phys Rev B 55:7868–7881

Reinhard D, Hall BD, Berthoud P, Valkealahti S, Monot R (1998) Unsupported nanometer-sized copper clusters studied by electron diffraction and molecular dynamics. Phys Rev B 58:4917–4926

Renaud G, Lazzari R, Revenant C, Barbier A, Noblet M, Ulrich O, Leroy F, Jupille J, Borensztein Y, Henry CR, Deville JP, Scheurer F, Mane-Mane J, Fruchart O (2003) Real-time monitoring of growing nanoparticles. Science 300:1416–1419

Rieder KH, Engel T, Swendsen RH, Manninen M (1983) A helium diffraction study of the reconstructed Au(100) surface. Surf Sci 127:223–242

Robertson DH, Brenner DW, Mintmire JW (1992) Energetics of nanoscale graphitic tubules. Phys Rev B 45:12592–12595

Robinson IK, Vartanyants IA (2001) Use of coherent X-ray diffraction to map strain fields in nanocrystals. Appl Surf Sci 182:186–191

Robinson IK, Vartanyants I, Williams GJ, Pfeifer MA, Pitney JA (2001) Reconstruction of shapes of gold nanocrystals using coherent X-ray diffraction. Phys Rev Lett 87:195505

Ruland W, Schaper AK, Hou H, Greiner A (2003) Multi-wall carbon nanotubes with uniform chirality: evidence for scroll structures. Carbon 41:423–427

Sadan H, Kaplan WD (2006) Au-sapphire (0001) solid-solid interfacial energy. J Mater Sci 41:5099–5107

Saito R, Dresselhaus G, Dresselhaus MS (1998) Physical properties of carbon nanotubes. Imperial College Press, London

Salehi-Khojin A, Lin KY, Field CR, Masel RI (2010) Nonthermal current-stimulated desorption of gases from carbon nanotubes. Science 329:1327–1330

Sanchez S, Small M, Zuo J-M, Nuzzo RG (2009a) Structural characterization of Pt-Pd and Pd-Pt core-shell nanoclusters at atomic resolution. J Am Chem Soc 131:8683–8689

Sanchez SI, Menard LD, Bram A, Kang JH, Small MW, Nuzzo RG, Frenkel AI (2009b) The emergence of nonbulk properties in supported metal clusters: negative thermal expansion and atomic disorder in Pt nanoclusters supported on γ-Al_2O_3. J Am Chem Soc 131:7040

Sanchez-Portal D, Artacho E, Soler JM, Rubio A, Ordejon P (1999) Ab initio structural, elastic, and vibrational properties of carbon nanotubes. Phys Rev B 59:12678–12688

Shah AB, Sivapalan ST, DeVetter BM, Yang TK, Wen JG, Bhargava R, Murphy CJ, Zuo JM (2013) High-index facets in gold nanocrystals elucidated by coherent electron diffraction. Nano Lett 13:1840–1846

Sivaramakrishnan S, Wen JG, Scarpelli ME, Pierce BJ, Zuo JM (2010) Equilibrium shapes and triple line energy of epitaxial gold nanocrystals supported on TiO_2 (110). Phys Rev B 82:195421

Smith BW, Monthioux M, Luzzi DE (1998) Encapsulated C-60 in carbon nanotubes. Nature 396:323–324

Smoluchowski R (1941) Anisotropy of the electronic work function of metals. Phys Rev 60:661

Somani PR (2010) Pressure sensitive multifunctional solar cells using carbon nanotubes. Appl Phys Lett 96:173504

Somorjai GA (1994) Introduction to surface chemistry and catalysis. Wiley, New York

Stensgaard I, Feidenhans'l R, Sorensen JE (1983) Surface relxation of Cu(110): An ion scattering investigation. Surf Sci 128:281

Stevenson SA, Dumesic JA, Baker RT (1987) Metal-support interactions in catalysis, sintering and redispersion. Van Norstrand Reinhold

Stone AJ, Wales DJ (1986) Theoretical-studies of icosahedral C60 and some related species. Chem Phys Lett 128(5–6):501–503

Suenaga K, Okazaki T, Hirahara K, Bandow S, Kato H, Taninaka A, Shinohara H, Iijima S (2003) High-resolution electron microscopy of individual metallofullerene molecules on the dipole orientations in peapods. Appl Phys A-Mater Sci Pro 76:445–447

Suenaga K, Wakabayashi H, Koshino M, Sato Y, Urita K, Iijima S (2007) Imaging active topological defects in carbon nanotubes. Nat Nanotechnol 2:358–360

Tanner RE, Goldfarb I, Castell MR, Briggs GAD (2001) The evolution of Ni nanoislands on the rutile TiO_2(110) surface with coverage, heating and oxygen treatment. Surf Sci 486:167–184

Tans SJ, Verschueren ARM, Dekker C (1998) Room-temperature transistor based on a single carbon nanotube. Nature 393:49–52

Tao AR, Habas S, Yang P (2008) Shape control of colloidal metal nanocrystals. Small 4:310–325

Templeton AC, Chen S, Gross SM, Murray RW (1999) Water-soluble, isolable gold clusters protected by tiopronin and coenzyme a monolayers. Langmuir 15:66–76

Teranishi T, Hosoe M, Miyake M (1997) Formation for monodispersed ultrafine platinum and their electrophoretic deposition on electrodes. Adv Mater 9:65–67

Thess A, Lee R, Nikolaev P, Dai HJ, Petit P, Robert J, Xu CH, Lee YH, Kim SG, Rinzler AG, Colbert DT, Scuseria GE, Tomanek D, Fischer JE, Smalley RE (1996) Crystalline ropes of metallic carbon nanotubes. Science 273:483–487

Timoshenko S, Maccullough GH (1935) Elements of strength of materials D. Van Nostrand Company, New York

Troche KS, Coluci VR, Braga SF, Chinellato DD, Sato F, Legoas SB, Rurali R, Galvao DS (2005) Prediction of ordered phases of encapsulated C-60, C-70, and C-78 inside carbon nanotubes. Nano Lett 5:349–355

Trucano P, Chen R (1975) Structure of graphite by neutron-diffraction. Nature 258:136–137

Uppenbrink J, Wales DJ (1992) Structure and energetics of model metal-clusters. J Chem Phys 96:8520–8534

Van Hove MA, Weinberg WH, Chan C-M (1986) Low-energy electron diffraction: experiment, theory and surface structure determination. Springer, New York

Venables JA (2000) Introduction to surface and thin film processes. Cambridge University Press, Cambridge

Verberck B, Michel KH (2006) Nanotube field of C-60 molecules in carbon nanotubes: Atomistic versus continuous approach. Phys Rev B 74:045421

Verberck B, Michel KH, Nikolaev AV (2006) The C-60 molecules in (C60)(n)@SWCNT peapods: crystal field, intermolecular interactions and dynamics. Fuller Nanotubes Carbon Nanostruct 14:171–178

Vitos L, Ruban AV, Skriver HL, Kollar J (1998) The surface energy of metals. Surf Sci 411: 186–202

Walgraef D (2007) On the mechanics of deformation instabilities in carbon nanotubes. Eur Phys J-Spec Top 146:443–457

Wanjala BN, Luo J, Loukrakpam R, Fang B, Mott D, Njoki PN, Engelhard M, Naslund HR, Wu JK, Wang LC, Malis O, Zhong CJ (2010) Nanoscale alloying, phase-segregation, and core-shell evolution of gold-platinum nanoparticles and their electrocatalytic effect on oxygen reduction reaction. Chem Mater 22:4282–4294

Warner JH, Margine ER, Mukai M, Robertson AW, Giustino F, Kirkland AI (2012) Dislocation-driven deformations in graphene. Science 337:209–212

Worren T, Hojrup Hansen K, Laegsgaard E, Besenbacher F, Stensgaard I (2001) Copper clusters on Al_2O_3/NiAl(1 1 0) studied with STM. Surf Sci 477:8–16

Wu J, Qi L, You H, Gross A, Li J, Yang H (2012) Icosahedral platinum alloy nanocrystals with enhanced electrocatalytic activities. J Am Chem Soc 134:11880–11883

Wulff G (1901) Z Kristallogr 34:449–530

Xia Y, Xiong Y, Lim B, Skrabalak SE (2009) Shape-controlled synthesis of metal nanocrystals: simple chemistry meets complex physics? Angew Chem Int Ed 48:60–103

Xiao J, Liu B, Huang Y, Zuo J, Hwang KC, Yu MF (2007) Collapse and stability of single- and multi-wall carbon nanotubes. Nanotechnology 18:395703

Xie X, Wahab MA, Li Y, Islam AE, Tomic B, Huang J, Burns B, Seabron E, Dunham SN, Du F, Lin J, Wilson WL, Song J, Huang Y, Alam MA, Rogers JA (2015) Direct current injection and thermocapillary flow for purification of aligned arrays of single-walled carbon nanotubes. J Appl Phys 117:134303

Xu R, Chen C-C, Wu L, Scott MC, Theis W, Ophus C, Bartels M, Yang Y, Ramezani-Dakhel H, Sawaya MR, Heinz H, Marks LD, Ercius P, Miao J (2015) Three-dimensional coordinates of individual atoms in materials revealed by electron tomography. Nat Mater 14:1099–1103

Yang L, Han J (2000) Electronic structure of deformed carbon nanotubes. Phys Rev Lett 85: 154–157

Yang CY, Yacaman MJ, Heinemann K (1979) Crystallography of decahedral and icosahedral particles. 2. High symmetry orientations. J Crystal Growth 47:283–290

Yang Y, Matsubara S, Xiong L, Hayakawa T, Nogami M (2007) Solvothermal synthesis of mulitple shapes of silver nanoparticles and their SERS properties. J Phys Chem C 111: 9095–9104

Yazyev OV, Louie SG (2010) Topological defects in graphene: dislocations and grain boundaries. Phys Rev B 81:195420

Yazyev OV, Chen YP (2014) Polycrystalline graphene and other two-dimensional materials. Nat Nanotechnol 9:755–767

Yoon M, Berber S, Tomanek D (2005) Energetics and packing of fullerenes in nanotube peapods. Phys Rev B 71:155406

Yu MF, Kowalewski T, Ruoff RS (2000) Investigation of the radial deformability of individual carbon nanotubes under controlled indentation force. Phys Rev Lett 85:1456–1459

Zhang J, Zuo JM (2009) Structure and diameter-dependent bond lengths of a multi-walled carbon nanotube revealed by electron diffraction. Carbon 47:3515–3528

Zhang R, Tsui RK, Tresek J, Rawlett AM, Amlani I, Hopson T, Fejes P (2003) Formation of single-walled carbon nanotubes via reduced-pressure thermal chemical vapor deposition. J Phys Chem B 107:3137–3140

Zhang GY, Bai XD, Wang EG, Guo Y, Guo WL (2005) Monochiral tubular graphite cones formed by radial layer-by-layer growth. Phys Rev B 71:113411

Zhang J, Xiao JL, Meng XH, Monroe C, Huang YG, Zuo JM (2010) Free folding of suspended graphene sheets by random mechanical stimulation. Phys Rev Lett 104:166805

Zhou W, Huang Y, Liu B, Hwang KC, Zuo JM, Buehler MJ, Gao H (2007) Self-folding of single- and multiwall carbon nanotubes. Appl Phys Lett 90:073107

Zhu J, Cowley JM (1982) Micro-diffraction from antiphase domain boundaries in Cu_3Au. Acta Cryst A38:718–724

Zhu J, Cowley JM (1983) Micro-diffraction from stacking-faults and twin boundaries in fcc crystals. J Appl Crystallogr 16:171–175

Zhu J, Cowley JM (1985) Study of early-stage precipitation in Al-4 %Cu by microdiffraction and STEM. Ultramicroscopy 18:419–426

Zhu J, Shen Y, Xie A, Qiu L, Zhang Q, Zhang S (2007) Photoinduced synthesis of anisotropic gold nanoparticles in room-temperature ionic liquid. J Phys Chem C 111:7629–7633

Zuo JM, Vartanyants I, Gao M, Zhang R, Nagahara LA (2003) Atomic resolution imaging of a carbon nanotube from diffraction intensities. Science 300:1419–1421

Zuo JM, Gao M, Tao J, Li BQ, Twesten R, Petrov I (2004) Coherent nano-area electron diffraction. Microsc Res Tech 64:347–355

Zuo JM, Kim T, Celik-Aktas A, Tao J (2007) Quantitative structural analysis of individual nanotubes by electron diffraction. Z Kristallogr 222:625–633

Appendix A
Useful Relationships in Electron Diffraction

The following is a compilation of frequently used results in electron diffraction.

$$E = mc^2 = \gamma m_e c^2 = m_e c^2 + e\Phi \tag{A.1}$$

$$\gamma = \left(1 - \beta^2\right)^{-1/2} = \frac{E}{m_e c^2}, \quad \text{and} \quad \beta = v/c \tag{A.2}$$

$$m = m_e/\sqrt{1 - v^2/c^2} = m_e/\sqrt{1 - \beta^2} = \gamma m_e \tag{A.3}$$

$$m = m_e\left(1 + \frac{e\Phi}{m_e c^2}\right) \tag{A.4}$$

$$v = c\sqrt{1 - \frac{1}{\gamma^2}} = \sqrt{1 - \frac{1}{\left(1 + \frac{e\Phi}{m_e c^2}\right)^2}} \tag{A.5}$$

$$p = \frac{1}{c}\sqrt{2m_e c^2 E + E^2} = \sqrt{2m_e E(1 + E/2m_e c^2)}, \quad \text{where} \quad E = e\Phi$$

$$\lambda \text{ (relativistic)} = \frac{h}{p} = h/\sqrt{2m_e E(1 + E/2m_e c^2)}, \qquad \text{where} \quad E = e\Phi \tag{A.6}$$
$$= hc/\sqrt{2m_e c^2 E + E^2}$$

$$U_g = \frac{2m_e}{h^2} V_g = \frac{K\sigma}{\pi} V_g = \frac{\gamma}{\pi} \frac{F_g^B}{V_c} \tag{A.7}$$

$$|U_g| = K/\xi_g \approx 1/\lambda\xi_g \tag{A.8}$$

$$\sigma = \frac{2\pi m_e \lambda}{h^2} \tag{A.9}$$

© Springer Science+Business Media New York 2017
J.M. Zuo and J.C.H. Spence, *Advanced Transmission Electron Microscopy*,
DOI 10.1007/978-1-4939-6607-3

$$V_g = \frac{h^2 F_g^B}{2\pi m_e e V_c} = 47.877647 \frac{F_g^B}{V_c}, \qquad (A.10)$$

if the cell volume V_c is given in Angstroms^3.

$$\left| V_g \right| = \frac{\pi}{\xi_g \sigma} \qquad (A.11)$$

$$k_o = \frac{1}{\lambda}$$

$K_o^2 = k_o^2 + U_o$, where K_o is wave number corrected for mean inner potential.

$$h^2/m_e = 274 \,\text{eV}\,\text{Å}^2$$

$$hc = 1.23984193 \,\text{eV}\,\mu\text{m}.$$

$h/m_e c = 0.02426\,\text{Å}$, Compton wavelength

$$m_e c^2 \approx 511 \,\text{keV}$$

$S_g = -\lambda g^2/2$ in the exact zone-axis orientation.

Appendix B
Electron Wavelengths, Physical Constants, and Atomic Scattering Factors

The fundamental physical constants used in this book are listed below (Source NIST).

$$c = 2.99792458 \times 10^8 \text{ ms}^{-1}$$

$$e = 1.6021766208(98) \times 10^{-19} \text{ C}$$

$$h = 6.626070040(81) \times 10^{-34} \text{ J s}$$

$$m_e = 9.10938356(11) \times 10^{-31} \text{ kg}$$

$$\varepsilon_0 = 8.854187817 \times 10^{-12} \text{ Fm}^{-1}$$

In addition,
$$1 \text{ eV} = 1.6021766208 \times 10^{-19} \text{ J}$$
$$h = 4.135667662 \times 10^{-15} \text{ eV s}$$
See Tables B.1 and B.2.

Table B.1 Electron wavelengths

Φ (V) (kV) (MeV)	λ (Å units)	λ^{-1} (Å^{-1})	$= \frac{m}{m_e}$	$= \frac{V}{c}$	$= \frac{V}{V_{100}}$
1	12.26	0.0815	1.0000020	0.0020	0.0036
10	3.878	0.2579	1.0000196	0.0063	0.0114
100	1.226	0.8154	1.0001957	0.0198	0.0361
500	0.5483	1.824	1.0009785	0.0442	0.0806
1	0.3876	2.580	1.00196	0.0625	0.1139
2	0.2740	3.650	1.00391	0.0882	0.1609
3	0.2236	4.473	1.00587	0.1079	0.1968
4	0.1935	5.167	1.00783	0.1244	0.2269
5	0.1730	5.780	1.00978	0.1389	0.2533
6	0.1579	6.335	1.01174	0.1519	0.2771

(continued)

© Springer Science+Business Media New York 2017
J.M. Zuo and J.C.H. Spence, *Advanced Transmission Electron Microscopy*,
DOI 10.1007/978-1-4939-6607-3

Table B.1 (continued)

Φ (V) (kV) (MeV)	λ (Å units)	λ^{-1} (Å^{-1})	$= \frac{m}{m_e}$	$= \frac{V}{c}$	$= \frac{V}{V_{100}}$
7	0.1461	6.845	1.01370	0.16638	0.2989
8	0.1366	7.322	1.01566	0.1749	0.3190
9	0.1287	7.770	1.01761	0.1852	0.3379
10	0.1220	8.194	1.01957	0.1950	0.3557
20	0.0859	11.64	1.0391	0.2719	0.4959
30	0.00698	14.33	1.0587	0.3284	0.5990
40	0.0602	16.623	1.0783	0.3741	0.6823
50	0.0536	18.67	1.078	0.4127	0.7528
60	0.0487	20.55	1.1174	0.4462	0.8139
70	0.0448	22.30	1.1370	0.4759	0.8680
80	0.0418	23.95	1.1566	0.5024	0.9164
90	0.0392	25.25	1.1761	0.5264	0.9602
100	0.0370	27.02	1.1957	0.5482	1.0000
200	0.0251	39.87	1.3914	0.6953	1.268
300	0.0197	50.80	1.5871	0.7765	1.416
400	0.0164	60.83	1.7828	0.8279	1.510
500	0.0142	70.336	1.9785	0.8629	1.574
600	0.0126	79.57	2.1742	0.8879	1.620
700	0.0118	88.56	2.3698	0.9066	1.654
800	0.0103	97.38	2.5655	0.9209	1.680
900	0.0094	106.1	2.7612	0.9321	1.700
1	0.0087	114.7	2.9569	0.9411	1.717
2	0.0050	198.3	4.9138	0.9791	1.786
4	0.0028	361.5	8.8277	0.9935	1.812
6	0.0019	523.5	12.742	0.9969	1.818
8	0.0015	685.2	16.655	0.9982	1.821
20	0.0012	846.8	20.569	0.9988	1.822

The following table gives as follows: functions of the accelerating voltage Φ, the relativistic electron wavelength λ, the wave number λ^{-1}, the relativistic factor $m/m_e = \left(1 - \beta^2\right)^{-1/2}$, the velocity $V = V/c$, and the correction factor V/V_{100} for ξ_g relative to 100 kV

Table B.2 Parameterized X-ray atomic scattering factors by Doyle and Turner (Acta Cryst. A24, 390, 1968)

Z	S	a_1	b_1	a_2	b_2	a_3	b_3	a_4	b_4	c
1	H	0.489918	20.6593	0.262003	7.74039	0.196767	49.5519	0.049879	2.20159	0.001305
1	H^{-1}	0.897661	53.1368	0.565616	15.187	0.415815	186.576	0.116973	3.56709	0.002389
2	He	0.8734	9.1037	0.6309	3.3568	0.3112	22.9276	0.178	0.9821	0.0064
3	Li	1.1282	3.9546	0.7508	1.0524	0.6175	85.3905	0.4653	168.261	0.0377
3	Li^{+1}	0.6968	4.6237	0.7888	1.9557	0.3414	0.6316	0.1563	10.0953	0.0167
4	Be	1.5919	43.6427	1.1278	1.8623	0.5391	103.483	0.7029	0.542	0.0385
4	Be^{+2}	6.2603	0.0027	0.8849	0.8313	0.7993	2.2758	0.1647	5.1146	−6.1092
5	B	2.0545	23.2185	1.3326	1.021	1.0979	60.3498	0.7068	0.1403	−0.1932
6	C	2.31	20.8439	1.02	10.2075	1.5886	0.5687	0.865	51.6512	0.2156
6	C^{VAL}	2.26069	22.6907	1.56165	0.656665	1.05075	9.75618	0.839259	55.5949	0.286977
7	N	12.2126	0.0057	3.1322	9.8933	2.0125	28.9975	1.1663	0.5826	−11.529
8	O	3.0485	13.2771	2.2868	5.7011	1.5463	0.3239	0.867	32.9089	0.2508
8	O^{-1}	4.1916	12.8573	1.63969	4.17236	1.52673	47.0179	−20.307	−0.01404	21.9412
8	O^{-2}	1.5774	55.4041	3.7626	19.1605	2.7959	6.6451	1.5084	0.385	0.3557
9	F	3.5392	10.2825	2.6412	4.2944	1.517	0.2615	1.0243	26.1476	0.2776
9	F^{-1}	3.6322	5.27756	3.51057	14.7353	1.26064	0.442258	0.940706	47.3437	0.653396
10	Ne	3.9553	8.4042	3.1125	3.4262	1.4546	0.2306	1.1251	21.7184	0.3515
11	Na	4.7626	3.285	3.1736	8.8422	1.2674	0.3136	1.1128	129.424	0.676
11	Na^{+1}	3.2565	2.6671	3.9362	6.1153	1.3998	0.2001	1.0032	14.039	0.404
12	Mg	5.4204	2.8275	2.1735	79.2611	1.2269	0.3808	2.3073	7.1937	0.8584
12	Mg^{+2}	3.4988	2.1676	3.8378	4.7542	1.3284	0.185	0.8497	10.1411	0.4853
13	Al	6.4202	3.0387	1.9002	0.7426	1.5936	31.5472	1.9646	85.0886	1.1151
13	Al^{+3}	4.17448	1.93816	3.3876	4.14553	1.20296	0.228753	0.528137	8.28524	0.706786
14	Si	6.2915	2.4386	3.0353	32.3337	1.9891	0.6785	1.541	81.6937	1.1407

(continued)

Table B.2 (continued)

Z	S	a_1	b_1	a_2	b_2	a_3	b_3	a_4	b_4	c
14	Si^{VAL}	5.66269	2.6652	3.07164	38.6634	2.62446	0.916946	1.3932	93.5458	1.24707
14	Si^{+4}	4.43918	1.64167	3.20345	8.43757	1.19453	0.2149	0.41653	6.65365	0.746297
15	P	6.4345	1.9067	4.1791	27.157	1.78	0.526	1.4908	68.1645	1.1149
16	S	6.9053	1.4679	5.2034	22.2151	1.4379	0.2536	1.5863	56.172	0.8669
17	Cl	11.4604	0.0104	7.1964	1.1662	6.2556	18.5194	1.6455	47.7784	−9.5574
17	Cl^{-1}	18.2915	0.0066	7.2084	1.1717	6.5337	19.5424	2.3386	60.4486	−16.378
18	Ar	7.4845	0.9072	6.7723	14.8407	0.6539	43.8983	1.6442	33.3929	1.4445
19	K	8.2186	12.7949	7.4398	0.7748	1.0519	213.187	0.8659	41.6841	1.4228
19	K^{+1}	7.9578	12.6331	7.4917	0.7674	6.359	−0.002	1.1915	31.9128	−4.9978
20	Ca	8.6266	10.4421	7.3873	0.6599	1.5899	85.7484	1.0211	178.437	1.3751
20	Ca^{+2}	15.6348	−0.0074	7.9518	0.6089	8.4372	10.3116	0.8537	25.9905	−14.875
21	Sc	9.189	9.0213	7.3679	0.5729	1.6409	136.108	1.468	51.3531	1.3329
21	Sc^{+3}	13.4008	0.29854	8.0273	7.9629	1.65943	−0.28604	1.57936	16.0662	−6.6667
22	Ti	9.7595	7.8508	7.3558	0.5	1.6991	35.6338	1.9021	116.105	1.2807
22	Ti^{+2}	9.11423	7.5243	7.62174	0.457585	2.2793	19.5361	0.087899	61.6558	0.897155
22	Ti^{+3}	17.7344	0.22061	8.73816	7.04716	5.25691	−0.15762	1.92134	15.9768	−14.652
22	Ti^{+4}	19.5114	0.178847	8.23473	6.67018	2.01341	−0.29263	1.5208	12.9464	−13.28
23	V	10.2971	6.8657	7.3511	0.4385	2.0703	26.8938	2.0571	102.478	1.2199
23	V^{+2}	10.106	6.8818	7.3541	0.4409	2.2884	20.3004	0.0223	115.122	1.2298
23	V^{+3}	9.43141	6.39535	7.7419	0.383349	2.15343	15.1908	0.016865	63.969	0.656565
23	V^{+5}	15.6887	0.679003	8.14208	5.40135	2.03081	9.97278	−9.576	0.940464	1.7143
24	Cr	10.6406	6.1038	7.3537	0.392	3.324	20.2626	1.4922	98.7399	1.1832
24	Cr^{+2}	9.54034	5.66078	7.7509	0.344261	3.58274	13.3075	0.509107	32.4224	0.616898
24	Cr^{+3}	9.6809	5.59463	7.81136	0.334393	2.87603	12.8288	0.113575	32.8761	0.518275

(continued)

Table B.2 (continued)

Z	S	a_1	b_1	a_2	b_2	a_3	b_3	a_4	b_4	c
25	Mn	11.2819	5.3409	7.3573	0.3432	3.0193	17.8674	2.2441	83.7543	1.0896
25	Mn^{+2}	10.8061	5.2796	7.362	0.3435	3.5268	14.343	0.2184	41.3235	1.0874
25	Mn^{+3}	9.84521	4.91797	7.87194	0.294393	3.56531	10.8171	0.323613	24.1281	0.393974
25	Mn^{+4}	9.96253	4.8485	7.97057	0.283303	2.76067	10.4852	0.054447	27.573	0.251877
26	Fe	11.7695	4.7611	7.3573	0.3072	3.5222	15.3535	2.3045	76.8805	1.0369
26	Fe^{+2}	11.0424	4.6538	7.374	0.3053	4.1346	12.0546	0.4399	31.2809	1.0097
26	Fe^{+3}	11.1764	4.6147	7.38639	0.3005	3.3948	11.6729	0.0724	38.5566	0.9707
27	Co	12.2841	4.2791	7.3409	0.2784	4.0034	13.5359	2.3488	71.1692	1.0118
27	Co^{+2}	11.2296	4.1231	7.3883	0.2726	4.7393	10.2443	0.7108	25.6466	0.9324
27	Co^{+3}	10.338	3.90969	7.88173	0.238668	4.76795	8.35583	0.725591	18.3491	0.286667
28	Ni	12.8376	3.8785	7.292	0.2565	4.4438	12.1763	2.38	66.3421	1.0341
28	Ni^{+2}	11.4166	3.6766	7.4005	0.2449	5.3442	8.873	0.9773	22.1626	0.8614
28	Ni^{+3}	10.7806	3.5477	7.75868	0.22314	5.22746	7.64468	0.847114	16.9673	0.386044
29	Cu	13.338	3.5828	7.1676	0.247	5.6158	11.3966	1.6735	64.8126	1.191
29	Cu^{+1}	11.9475	3.3669	7.3573	0.2274	6.2455	8.6625	1.5578	25.8487	0.89
29	Cu^{+2}	11.8168	3.37484	7.11181	0.244078	5.78135	7.9876	1.14523	19.897	1.14431
30	Zn	14.0743	3.2655	7.0318	0.2333	5.1652	10.3163	2.41	58.7097	1.3041
30	Zn^{+2}	11.9719	2.9946	7.3862	0.2031	6.4668	7.0826	1.394	18.0995	0.7807
31	Ga	15.2354	3.0669	6.7006	0.2412	4.3591	10.7805	2.9623	61.4135	1.7189
31	Ga^{+3}	12.692	2.81262	6.69883	0.22789	6.06692	6.36441	1.0066	14.4122	1.53545
32	Ge	16.0816	2.8509	6.3747	0.2516	3.7068	11.4468	3.683	54.7625	2.1313
32	Ge^{+4}	12.9172	2.53718	6.70003	0.205855	6.06791	5.47913	0.859041	11.603	1.45572
33	As	16.6723	2.6345	6.0701	0.2647	3.4313	12.9479	4.2779	47.7972	2.531
34	Se	17.0006	2.4098	5.8196	0.2726	3.9731	15.2372	4.3543	43.8163	2.8409

(continued)

Table B.2 (continued)

Z	S	a_1	b_1	a_2	b_2	a_3	b_3	a_4	b_4	c
35	Br	17.1789	2.1723	5.2358	16.5796	5.6377	0.2609	3.9851	41.4328	2.9557
35	Br^{-1}	17.1718	2.2059	6.3338	19.3345	5.5754	0.2871	3.7272	58.1535	3.1776
36	Kr	17.3555	1.9384	6.7286	16.5623	5.5493	0.2261	3.5375	39.3972	2.825
37	Rb	17.1784	1.7888	9.6435	17.3151	5.1399	0.2748	1.5292	164.934	3.4873
37	Rb^{+1}	17.5816	1.7139	7.6598	14.7957	5.8981	0.1603	2.7817	31.2087	2.0782
38	Sr	17.5663	1.5564	9.8184	14.0988	5.422	0.1664	2.6694	132.376	2.5064
38	Sr^{+2}	18.0874	1.4907	8.1373	12.6963	2.5654	24.5651	-34.193	-0.0138	41.4025
39	Y	17.776	1.4029	10.2946	12.8006	5.72629	0.125599	3.26588	104.354	1.91213
39	Y^{+3}	17.9268	1.35417	9.1531	11.2145	1.76795	22.6599	-33.108	-0.01319	40.2602
40	Zr	17.8765	1.27618	10.948	11.916	5.41732	0.117622	3.65721	87.6627	2.06929
40	Zr^{+4}	18.1668	1.2148	10.0562	10.1483	1.01118	21.6054	-2.6479	-0.10276	9.41454
41	Nb	17.6142	1.18865	12.0144	11.766	4.04183	0.204785	3.53346	69.7957	3.75591
41	Nb^{+3}	19.8812	0.019175	18.0653	1.13305	11.0177	10.1621	1.94715	28.3389	-12.912
41	Nb^{+5}	17.9163	1.12446	13.3417	0.028781	10.799	9.28206	0.337905	25.7228	-6.3934
42	Mo	3.7025	0.2772	17.2356	1.0958	12.8876	11.004	3.7429	61.6584	4.3875
42	Mo^{+3}	21.1664	0.014734	18.2017	1.03031	11.7423	9.53659	2.30951	26.6307	-14.421
42	Mo^{+5}	21.0149	0.014345	18.0992	1.02238	11.4632	8.78809	0.740625	23.3452	-14.316
42	Mo^{+6}	17.8871	1.03649	11.175	8.48061	6.57891	0.058881	0	0	0.34494
43	Tc	19.1301	0.864132	11.0948	8.14487	4.64901	21.5707	2.71263	86.8472	5.40421
44	Ru	19.2674	0.80852	12.9182	8.43467	4.86337	24.7997	1.56756	94.2928	5.37871
44	Ru^{+3}	18.5638	0.847329	13.2885	8.37164	9.32602	0.017662	3.00964	22.887	-3.1892
44	Ru^{+4}	18.5003	0.844582	13.1787	8.12534	4.71304	0.36495	2.18535	20.8504	1.4235
45	Rh	19.2957	0.751536	14.3501	8.21758	4.73425	25.8749	1.28918	98.6062	5.328
45	Rh^{+3}	18.8785	0.764252	14.1259	7.84438	3.32515	21.2487	-6.1989	-0.01036	11.8678

(continued)

Table B.2 (continued)

Z	S	a_1	b_1	a_2	b_2	a_3	b_3	a_4	b_4	c
45	Rh^{+4}	18.8545	0.760825	13.9806	7.62436	2.53464	19.3317	−5.6526	−0.0102	11.2835
46	Pd	19.3319	0.698655	15.5017	7.98929	5.29537	25.2052	0.605844	76.8986	5.26593
46	Pd^{+2}	19.1701	0.696219	15.2096	7.55573	4.32234	22.5057	0	0	5.2916
46	Pd^{+4}	19.2493	0.683839	14.79	7.14833	2.89289	17.9144	−7.9492	0.005127	13.0174
47	Ag	19.2808	0.6446	16.6885	7.4726	4.8045	24.6605	1.0463	99.8156	5.179
47	Ag^{+1}	19.1812	0.646179	15.9719	7.19123	5.27475	21.7326	0.357534	66.1147	5.21572
47	Ag^{+2}	19.1643	0.645643	16.2456	7.18544	4.3709	21.4072	0	0	5.21404
48	Cd	19.2214	0.5946	17.6444	6.9089	4.461	24.7008	1.6029	87.4825	5.0694
48	Cd^{+2}	19.1514	0.597922	17.2535	6.80639	4.47128	20.2521	0	0	5.11937
49	In	19.1624	0.5476	18.5596	6.3776	4.2948	25.8499	2.0396	92.8029	4.9391
49	In^{+3}	19.1045	0,551522	18.1108	6.3247	3.78897	17.3595	0	0	4.99635
50	Sn	19.1889	5.8303	19.1005	0.5031	4.4585	26.8909	2.4663	83.9571	4.7821
50	Sn^{+2}	19.1094	0.5036	19.0548	5.8378	4.5648	23.3752	0.487	62.2061	4.7861
50	Sn^{+4}	18.9333	5.764	19.7131	0.4655	3.4182	14.0049	0.0193	−0.7583	3.9182
51	Sb	19.6418	5.3034	19.0455	0.4607	5.0371	27.9074	2.6827	75.2825	4.5909
51	Sb^{+3}	18.9755	0.467196	18.933	5.22126	5.10789	19.5902	0.288753	55.5113	4.69626
51	Sb^{+5}	19.8685	5.44853	19.0302	0.467973	2.41253	14.1259	0	0	4.69263
52	Te	19.9644	4.81742	19.0138	0.420885	6.14487	28.5284	2.5239	70.8403	4.352
53	I	20.1472	4.347	18.9949	0.3814	7.5138	27.766	2.27353	66.8776	4.0712
53	I^{-1}	20.2332	4.3579	18.997	0.3815	7.8069	29.5259	2.8868	84.9304	4.0714
54	Xe	20.2933	3.9282	19.0298	0.344	8.9767	26.4659	1.99	64.2658	3.7118
55	Cs	20.3892	3.569	19.1062	0.3107	10.662	24.3879	1.4953	213.904	3.3352
55	Cs^{+1}	20.3524	3.552	19.1278	0.3086	10.2821	23.7128	0.9615	59.4565	3.2791
56	Ba	20.3361	3.216	19.297	0.2756	10.888	20.2073	2.6959	167.202	2.7731

(continued)

Table B.2 (continued)

Z	S	a_1	b_1	a_2	b_2	a_3	b_3	a_4	b_4	c
56	Ba^{+2}	20.1807	3.21367	19.1136	0.28331	10.9054	20.0558	0.77634	51.746	3.02902
57	La	20.578	2.94817	19.599	0.244475	11.3727	18.7726	3.28719	133.124	2.14678
57	La^{+3}	20.2489	2.9207	19.3763	0.250698	11.6323	17.8211	0.336048	54.9453	2.4086
58	Ce	21.1671	2.81219	19.7695	0.226836	11.8513	17.6083	3.33049	127.113	1.86264
58	Ce^{+3}	20.8036	2.77691	19.559	0.23154	11.9369	16.5408	0.612376	43.1692	2.09013
58	Ce^{+4}	20.3235	2.65941	19.8186	0.21885	12.1233	15.7992	0.144583	62.2355	1.5918
59	Pr	22.044	2.77393	19.6697	0.222087	12.3856	16.7669	2.82428	143.644	2.0583
59	Pr^{+3}	21.3727	2.6452	19.7491	0.214299	12.1329	15.323	0.97518	36.4065	1.77132
59	Pr^{+4}	20.9413	2.54467	20.0539	0.202481	12.4668	14.8137	0.296689	45.4643	1.24285
60	Nd	22.6845	2.66248	19.6847	0.210628	12.774	15.885	2.85137	137.903	1.98486
60	Nd^{+3}	21.961	2.52722	19.9339	0.199237	12.12	14.1783	1.51031	30.8717	1.47588
61	Pm	23.3405	2.5627	19.6095	0.202088	13.1235	15.1009	2.87516	132.721	2.02876
61	Pm^{+3}	22.5527	2.4174	20.1108	0.185769	12.0671	13.1275	2.07492	27.4491	1.19499
62	Sm	24.0042	2.47274	19.4258	0.196451	13.4396	14.3996	2.89604	128.007	2.20963
62	Sm^{+3}	23.1504	2.31641	20.2599	0.174081	11.9202	12.1571	2.71488	24.8242	0.954586
63	Eu	24.6274	2.3879	19.0886	0.1942	13.7603	13.7546	2.9227	123.174	2.5745
63	Eu^{+2}	24.0063	2.27783	19.9504	0.17353	11.8034	11.6096	3.87243	26.5156	1.36389
63	Eu^{+3}	23.7497	2.22258	20.3745	0.16394	11.8509	11.311	3.26503	22.9966	0.759344
64	Gd	25.0709	2.25341	19.0798	0.181951	13.8518	12.9331	3.54545	101.398	2.4196
64	Gd^{+3}	24.3466	2.13553	20.4208	0.155525	11.8708	10.5782	3.7149	21.7029	0.645089
65	Tb	25.8976	2.24256	18.2185	0.196143	14.3167	12.6648	2.95354	115.362	3.58324
65	Tb^{+3}	24.9559	2.05601	20.3271	0.149525	12.2471	10.0499	3.773	21.2773	0.691967
66	Dy	26.507	2.1802	17.6383	0.202172	14.5596	12.1899	2.96577	111.874	4.29728
66	Dy^{+3}	25.5395	1.9804	20.2861	0.143384	11.9812	9.34972	4.50073	19.581	0.68969

(continued)

Table B.2 (continued)

Z	S	a_1	b_1	a_2	b_2	a_3	b_3	a_4	b_4	c
67	Ho	26.9049	2.07051	17.294	0.19794	14.5583	11.4407	3.63837	92.6566	4.56796
67	Ho^{+3}	26.1296	1.91072	20.0994	0.139358	11.9788	8.80018	4.93676	18.5908	0.852795
68	Er	27.6563	2.07356	16.4285	0.223545	14.9779	11.3604	2.98233	105.703	5.92046
68	Er^{+3}	26.722	1.84659	19.7748	0.13729	12.1506	8.36225	5.17379	17.8974	1.17613
69	Tm	28.1819	2.02859	15.8851	0.238849	15.1542	10.9975	2.98706	102.961	6.75621
69	Tm^{+3}	27.3083	1.78711	19.332	0.136974	12.3339	7.96778	5.38348	17.2922	1.63929
70	Yb	28.6641	1.9889	15.4345	0.257119	15.3087	10.6647	2.98963	100.417	7.56672
70	Yb^{+2}	28.1209	1.78503	17.6817	0.15997	13.3335	8.18304	5.14657	20.39	3.70983
70	Yb^{+3}	27.8917	1.73272	18.7614	0.13879	12.6072	7.64412	5.47647	16.8153	2.26001
71	Lu	28.9476	1.90182	15.2208	9.98519	15.1	0.261033	3.71601	84.3298	7.97628
71	Lu^{+3}	28.4628	1.68216	18.121	0.142292	12.8429	7.33727	5.59415	16.3535	2.97573
72	Hf	29.144	1.83262	15.1726	9.5999	14.7586	0.275116	4.30013	72.029	8.58154
72	Hf^{+4}	28.8131	1.59136	18.4601	0.128903	12.7285	6.76232	5.59927	14.0366	2.39699
73	Ta	29.2024	1.77333	15.2293	9.37046	14.5135	0.295977	4.76492	63.3644	9.24354
73	Ta^{+5}	29.1587	1.50711	18.8407	0.116741	12.8268	6.31524	5.38695	12.4244	1.78555
74	W	29.0818	1.72029	15.43	9.2259	14.4327	0.321703	5.11982	57.056	9.8875
74	W^{+6}	29.4936	1.42755	19.3763	0.104621	13.0544	5.93667	5.06412	11.1972	1.01074
75	Re	28.7621	1.67191	15.7189	9.09227	14.5564	0.3505	5.44174	52.0861	10.472
76	Os	28.1894	1.62903	16.155	8.97948	14.9305	0.382661	5.67589	48.1647	11.0005
76	Os^{+4}	30.419	1.37113	15.2637	6.84706	14.7458	0.165191	5.06795	18.003	6.49804
77	Ir	27.3049	1.59279	16.7296	8.86553	15.6115	0.417916	5.88377	45.0011	11.4722
77	Ir^{+3}	30.4156	1.34323	15.862	7.10909	13.6145	0.204633	5.82008	20.3254	8.27903
77	Ir^{+4}	30.7058	1.30923	15.5512	6.71983	14.2326	0.167252	5.53672	17.4911	6.96824
78	Pt	27.0059	1.51293	17.7639	8.81174	15.7131	0.424593	5.7837	38.6103	11.6883

(continued)

Table B.2 (continued)

Z	S	a_1	b_1	a_2	b_2	a_3	b_3	a_4	b_4	c
78	Pt^{+2}	29.8429	1.32927	16.7224	7.38979	13.2153	0.263297	6.35234	22.9426	9.85329
78	Pt^{+4}	30.9612	1.24813	15.9829	6.60834	13.7348	0.16864	5.92034	16.9392	7.39534
79	Au	16.8819	0.4611	18.5913	8.6216	25.5582	1.4826	5.86	36.3956	12.0658
79	Au^{+1}	28.0109	1.35321	17.8204	7.7395	14.3359	0.356752	6.58077	26.4043	11.2299
79	Au^{+3}	30.6886	1.2199	16.9029	6.82872	12.7801	0.212867	6.52354	18.659	9.0968
80	Hg	20.6809	0.545	19.0417	8.4484	21.6575	1.5729	5.9676	38.3246	12.6089
80	Hg^{+1}	25.0853	1.39507	18.4973	7.65105	16.8883	0.443378	6.48216	28.2262	12.0205
80	Hg^{+2}	29.5641	1.21152	18.06	7.05639	12.8374	0.284738	6.89912	20.7482	10.6268
81	Tl	27.5446	0.65515	19.1584	8.70751	15.538	1.96347	5.52593	45.8149	13.1746
81	Tl^{+1}	21.3985	1.4711	20.4723	0.517394	18.7478	7.43463	6.82847	28.8482	12.5258
81	Tl^{+3}	30.8695	1.1008	18.3841	6.53852	11.9328	0.219074	7.00574	17.2114	9.8027
82	Pb	31.0617	0.6902	13.0637	2.3576	18.442	8.618	5.9696	47.2579	13.4118
82	Pb^{+2}	21.7886	1.3366	19.5682	0.488383	19.1406	6.7727	7.01107	23.8132	12.4734
82	Pb^{+4}	32.1244	1.00566	18.8003	6.10926	12.0175	0.147041	6.96886	14.714	8.08428
83	Bi	33.3689	0.704	12.951	2.9238	16.5877	8.7937	6.4692	48.0093	13.5782
83	Bi^{+3}	21.8053	1.2356	19.5026	6.24149	19.1053	0.469999	7.10295	20.3185	12.4711
83	Bi^{+5}	33.5364	0.91654	25.0946	0.39042	19.2497	5.71414	6.91555	12.8285	−6.7994
84	Po	34.6726	0.700999	15.4733	3.55078	13.1138	9.55642	7.02588	47.0045	13.677
85	At	35.3163	0.68587	19.0211	3.97458	9.49887	11.3824	7.42518	45.4715	13.7108
86	Rn	35.5631	0.6631	21.2816	4.0691	8.0037	14.0422	7.4433	44.2473	13.6905
87	Fr	35.9299	0.646453	23.0547	4.17619	12.1439	23.1052	2.11253	150.645	13.7247
88	Ra	35.763	0.616341	22.9064	3.87135	12.4739	19.9887	3.21097	142.325	13.6211
88	Ra^{+2}	35.215	0.604909	21.67	3.5767	7.91342	12.601	7.65078	29.8436	13.5431
89	Ac	35.6597	0.589092	23.1032	3.65155	12.5977	18.599	4.08655	117.02	13.5266

(continued)

Table B.2 (continued)

Z	S	a_1	b_1	a_2	b_2	a_3	b_3	a_4	b_4	c
89	Ac^{+3}	35.1736	0.579689	22.1112	3.41437	8.19216	12.9187	7.05545	25.9443	13.4637
90	Th	35.5645	0.563359	23.4219	3.46204	12.7473	17.8309	4.80703	99.1722	13.4314
90	Th^{+4}	35.1007	0.555054	22.4418	3.24498	9.78554	13.4661	5.29444	23.9533	13.376
91	Pa	35.8847	0.547751	23.2948	3.41519	14.1891	16.9235	4.17287	105.251	13.4287
92	U	36.0228	0.5293	23.4128	3.3253	14.9491	16.0927	4.188	100.613	13.3966
92	U^{+3}	35.5747	0.52048	22.5259	3.12293	12.2165	12.7148	5.37073	26.3394	13.3092
92	U^{+4}	35.3715	0.516598	22.5326	3.05053	12.0291	12.5723	4.7984	23.4582	13.2671
92	U^{+6}	34.8509	0.507079	22.7584	2.8903	14.0099	13.1767	1.21457	25.2017	13.1665
93	Np	36.1874	0.511929	23.5964	3.25396	15.6402	15.3622	4.1855	97.4908	13.3573
93	Np^{+3}	35.7074	0.502322	22.613	3.03807	12.9898	12.1449	5.43227	25.4928	13.2544
93	Np^{+4}	35.5103	0.498626	22.5787	2.96627	12.7766	11.9484	4.92159	22.7502	13.2116
93	Np^{+6}	35.0136	0.48981	22.7286	2.81099	14.3884	12.33	1.75669	22.6581	13.113
94	Pu	36.5254	0.499384	23.8083	3.26371	16.7707	14.9455	3.47947	105.98	13.3812
94	Pu^{+3}	35.84	0.484938	22.7169	2.96118	13.5807	11.5331	5.66016	24.3992	13.1991
94	Pu^{+4}	35.6493	0.481422	22.646	2.8902	13.3595	11.316	5.18831	21.8301	13.1555
94	Pu^{+6}	35.1736	0.473204	22.7181	2.73848	14.7635	11.553	2.28678	20.9303	13.0582
95	Am	36.6706	0.483629	24.0992	3.20647	17.3415	14.3136	3.49331	102.273	13.3592
96	Cm	36.6488	0.465154	24.4096	3.08997	17.399	13.4346	4.21665	88.4834	13.2887
97	Bk	36.7881	0.451018	24.7736	3.04619	17.8919	12.8946	4.23284	86.003	13.2754
98	Cf	36.9185	0.437533	25.1995	3.00775	18.3317	12.4044	4.24391	83.7881	13.2674

This table gives for different elements and charge states, and the nine parameters in the form $f^x(s) = \sum_{i=1}^{4} a_i e^{-b_i s^2} + c$

Appendix C
Crystallographic Data

See Table C.1.

Table C.1 The seven crystal systems

Crystal	Axial length and angles	Minimum symmetry elements
Cubic	Three equal axes at right angles: $a = b = c$ and $\alpha = \beta = \gamma = 90°$	Four, threefold rotation axes
Hexagonal	Two coplanar axes at 120° and third axis at right angles: $a = b \neq c$ and $\alpha = \gamma = 90°$, $\beta = 120°$	One, sixfold rotation (or rotation-inversion) axis
Trigonal (or rhombohedral)	Three equal axes equally inclined $a = b = c$ and $\alpha = \beta = \gamma \neq 90°$	One, threefold rotation (or rotation-inversion) axis
Tetragonal	Three axes at right angles: $a = b \neq c$ and $\alpha = \beta = \gamma = 90°$	One, fourfold rotation (or rotation-inversion) axis
Orthorhombic	Three orthogonal unequal axes: $a \neq b \neq c$ and $\alpha = \beta = \gamma = 90°$	Three, perpendicular twofold (or rotation-inversion) axes
Monoclinic	Three unequal axes and one pair not orthogonal $a \neq b \neq c$ and $\alpha = \gamma = 90°$, $\beta \neq 90°$;	One, twofold rotation (or rotation-inversion) axis
Triclinic	Three unequal axes and none at right angles: $a \neq b \neq c$, $\alpha \neq \beta \neq \gamma \neq 90°$	None

© Springer Science+Business Media New York 2017
J.M. Zuo and J.C.H. Spence, *Advanced Transmission Electron Microscopy*,
DOI 10.1007/978-1-4939-6607-3

C.1   Interplanar Spacings

The distance d between adjacent planes with Miller indices (hkl) may be found using the following expressions:

Cubic:

$$\frac{1}{d^2} = \frac{h^2 + k^2 + l^2}{a^2}$$

Tetragonal:

$$\frac{1}{d^2} = \frac{h^2 + k^2}{a^2} + \frac{l^2}{c^2}$$

Hexagonal:

$$\frac{1}{d^2} = \frac{4}{3}\left(\frac{h^2 + hk + k^2}{a^2}\right) + \frac{l^2}{c^2} \quad \text{(Miller indices)}$$

Rhombohedral:

$$\frac{1}{d^2} = \frac{(h^2 + k^2 + l^2)\sin^2\alpha + 2(hk + kl + hl)(\cos^2\alpha - \cos\alpha)}{a^2(1 - 3\cos^2\alpha + 2\cos^3\alpha)}$$

Orthorhombic:

$$\frac{1}{d^2} = \frac{h^2}{a^2} + \frac{k^2}{b^2} + \frac{l^2}{c^2}$$

Monoclinic:

$$\frac{1}{d^2} = \frac{1}{\sin^2\beta}\left(\frac{h^2}{a^2} + \frac{k^2\sin^2\beta}{b^2} + \frac{l^2}{c^2} - \frac{2hl\cos\beta}{ac}\right)$$

Triclinic:

$$\frac{1}{d^2} = \frac{1}{V^2}\left(S_{11}h^2 + S_{22}k^2 + S_{33}l^2 + 2S_{12}hk + 2S_{23}kl + 2S_{13}hl\right)$$

In the equation for triclinic crystals,
V = volume of unit cell (see Eq. A.9)

$$S_{11} = b^2 c^2 \sin^2 \alpha,$$

$$S_{22} = a^2 c^2 \sin^2 \beta,$$

$$S_{33} = a^2 c^2 \sin^2 \gamma$$

$$S_{12} = abc^2 (\cos \alpha \cos \beta - \cos \gamma)$$

$$S_{23} = a^2 bc (\cos \beta \cos \gamma - \cos \alpha)$$

$$S_{13} = ab^2 c (\cos \gamma \cos \alpha - \cos \beta)$$

C.2 Extinction Conditions Resulting from Screw and Glide Symmetry

See Table C.2.
   See Table C.3.

Table C.2 Absent reflections resulting from Bravais lattice

Lattice	Allowed reflections	Extinctions
Primitive	All	None
Body-centered	$(h + k + l)$ even	$(h + k + l)$ odd
Face-centered	h, k, l all odd or all even	$h + k + l$ mixed
Base-centered	h and k both odd or even	h and k mixed.

Additional extinctions may result from the presence of screw and/or glide symmetry elements
For the various Bravais lattices, the following extinction conditions are imposed on U_g

Table C.3 Extinctions due to screw and glide symmetry elements

Symmetry element		Affected reflection	Condition for systematic absence of reflection
Twofold screw (2_1)	a	$h00$	$h = 2n + 1 = $ odd
Fourfold screw (4_2) along	b	$0k0$	$k = 2n + 1$
Sixfold screw (6_3)	c	$00l$	$l = 2n + 1$
Threefold screw ($3_1, 3_2$) along c		$00l$	$l = 3n+, 3n + 2$
Sixfold screw ($6_2, 6_4$)	c		i.e., not evenly divisible by 3
Fourfold screw ($4_1, 4_3$) along	a	$h00$	$h = 4n + 1, 2,$ or 3
	b	$0k0$	$k = 4n + 1, 2,$ or 3
	c	$00l$	$l = 4n + 2, 3,$ or 4
Sixfold screw ($6_1, 6_5$)	c	$00l$	$l = 6n + 1, 2, 3, 4,$ or 5

(continued)

Table C.3 (continued)

Symmetry element		Affected reflection	Condition for systematic absence of reflection
Glide plane perpendicular to a			
Translation	$b/2$ (b glide)	$0kl$	$k = 2n + 1$
	$c/2$ (c glide)		$l = 2n + 1$
(n glide)	$b/2 + c/2$		$k = l = 2n + 1$
(d glide)	$b/4 + c/4$		$k + l = 4 + 1, 2,$ or 3
Glide plane perpendicular to b			
Translation	$a/2$ (a glide)	$h0l$	$h = 2n + 1$
	$c/2$ (c glide)		$l = 2n + 1$
(n glide)	$a/2 + c/2$		$h + l = 2n = + 1$
(d glide)	$b/4 + c/4$		$h + l = 4n + 1, 2,$ or 3
Glide plane perpendicular to c			
Translation	$a/2$ (a glide)	$hk0$	$h = 2n + 1$
	$b/2$ (c glide)		$k = 2n + 1$
(n glide)	$a/2 + b/2$		$h + k = 2n + 1$
(d glide)	$a/4 + b/4$		$h + k = 4n + 1, 2,$ or 3
A-centered lattice (A)		hkl	$k + l = 2n + 1$
B-centered line (B)			$h + l = 2n + 1$
C-centered lattice (C)			$h + k = 2n + 1$
Face-centered lattice (F)			$h + k = 2n + 1$ i.e., h, k, l not
			$h + l = 2n + 1$ all even or all
			$k + 1 = 2n + 1$ odd
Body-centered lattice (I)			$h + k + l = 2n + 1$

C.3 Symmetries in Zone-Axis CBED Patterns

See Table C.4.

Table C.4 The table shows the symmetries which will be observed in convergent beam diffraction patterns from crystals of the 32 point groups in the electron beam directions indicated (after Buxton et al. 1976)

		Electron beam direction						
Crystal system	Point group	<111>	<100>	<110>	<uv0>	<uuw>	[uvw]	
Cubic	$m3m$	$6_R mm_R$	$4mm1_R$	$2mm1_R$	$2_R mm_R$	$2_R mm_R$	2_R	
	$\bar{4}3m$	$3m$	$4_R mm_R$	$m1_R$	m_R	m	1	
	432	$3m_R$	$4m_R m_R$	$2m_R m_R$	m_R	m_R	1	
	$m3$	6_R	$2mm1_R$		$2_R mm_R$		2_R	
	23	3	$2m_R m_R$		m_R		1	
		[0001]	<11$\bar{2}$0>	<1$\bar{1}$00>	[uv.0]	[uu.	[u $\bar{u}$0w]	[uv.w]

(continued)

Table C.4 (continued)

Crystal system	Point group	Electron beam direction						
		$\langle111\rangle$	$\langle100\rangle$	$\langle110\rangle$	$\langle uv0\rangle$	$\langle uuw\rangle$	$[uvw]$	
Hexagonal	$6/mmm$	$6mm1_R$	$2mm1_R$	$2mm1_R$	2_Rmm_R	2_Rmm_R	2_Rmm_R	2_R
	$\bar{6}m2$	$3m1_R$	$m1_R$	$2mm$	m	m_R	m	1
	$6mm$	$6mm$	$m1_R$	$m1_R$	m_R	m	m	1
	622	$6m_Rm_R$	$2m_Rm_R$	$2m_Rm_R$	m_R	m_R	m_R	1
	$6/m$	61_R			2_Rmm_R			2_R
	$\bar{6}$	31_R			m			1
	6	6 $[0001]$	$\langle11\bar{2}0\rangle$		m_R		$[u\,\bar{u}0w]$	1 $[uv.w]$
Trigonal	$\bar{3}m$	6_Rmm_R	21_R				2_Rmm_R	2_R
	$3m$	$3m$	1_R				m	1
	32	$3m_R$	2				m_R	1
	$\bar{3}$	6_R						2_R
	3	3						1
Tetragonal	$4/mmm$	$4mm1_R$	$2mm_R$	$2mm1_R$	2_Rmm_R	2_Rmm_R	2_Rmm_R	2_R
	$\bar{4}2m$	4_Rmm_R	$2m_Rm_R$	$m1_R$	m_R	m_R	m	1
	$4mm$	$4mm$	$m1_R$	$m1_R$	m	m_R	m	1
	422	$4m_Rm_R$	$2m_Rm_R$	$2m_Rm_R$	m_R	m_R	m_R	1
	$4/m$	41_R				2_Rmm_R		2_R
	$\bar{4}$	4_R				m_R		1
	4	4				m_R		1
		$[001]$	$[100]$	$[u0w]$	$[uv0]$	$[uvw]$		
Orthorhombic	mmm	$2mm1_R$	$2mm1_R$	2_Rmm_R	2_Rmm_R	2_R		
	$mm2$	$2mm$	$m1_R$	m	m_R	1		
	222	$2m_Rm_R$	$2m_Rm_R$	m_R	m_R	1		
		$[010]$		$[u0w]$		$\langle uvw\rangle$		
Monoclinic	$2/m$	21_R		2_Rmm_R		2_R		
	m	1_R		m		1		
	2	2		m_R		1		
						$\langle uvw\rangle$		
Triclinic	$\bar{1}$					2_R		
	1					1		

C.4   Height H of HOLZ Planes for Zone (u, v, w)

Crystal class	H (in Eq. 3.24)
Monoclinic	$(u^2a^2 + v^2b^2 + w^2c^2 + 2uwac\cos\beta)^{-1/2}$
Orthorhombic	$(u^2a^2 + v^2b^2 + w^2c^2)^{-1/2}$
Hexagonal and rhombohedral {with $D^2 = (2/3)(c/a)^2$}	$[(3a^2/2)(u^2 + v^2 + t^2 + Dw^2)]^{-1/2}$
Tetragonal	$[a^2(u^2 + v^2) + w^2c^2]^{-1/2}$
Cubic	$a(u^2 + v^2 + w^2)^{-1/2}$

C.5   The Use of Metric Matrix for Crystallographic Calculations

The crystallographic quantities required in a computer program can most simply be derived from a metric matrix G defined (for any crystal system)

$$G = \begin{pmatrix} \vec{a} \cdot \vec{a} & \vec{a} \cdot \vec{b} & \vec{a} \cdot \vec{c} \\ \vec{b} \cdot \vec{a} & \vec{b} \cdot \vec{b} & \vec{b} \cdot \vec{c} \\ \vec{c} \cdot \vec{a} & \vec{c} \cdot \vec{b} & \vec{c} \cdot \vec{c} \end{pmatrix} = \begin{pmatrix} a^2 & ab\cos\gamma & ac\cos\beta \\ ab\cos\gamma & b^2 & bc\cos\alpha \\ ac\cos\beta & bc\cos\alpha & c^2 \end{pmatrix}. \quad (A.12)$$

The inverse metric matrix can be shown to be

$$G^{-1} = \begin{pmatrix} \vec{a}^* \cdot \vec{a}^* & \vec{a}^* \cdot \vec{b}^* & \vec{a}^* \cdot \vec{c}^* \\ \vec{b}^* \cdot \vec{a}^* & \vec{b}^* \cdot \vec{b}^* & \vec{b}^* \cdot \vec{c}^* \\ \vec{c}^* \cdot \vec{a}^* & \vec{c}^* \cdot \vec{b}^* & \vec{c}^* \cdot \vec{c}^* \end{pmatrix} \quad (A.13)$$

For example, the d-spacing of (hkl) planes is $d = 1/s$, where s is the length of reciprocal lattice vector $\vec{g} = h\vec{a}^* + k\vec{b}^* + l\vec{c}^*$. In terms of the metric matrix,

$$S^2 = (h, k, l)G^{-1} \begin{pmatrix} h \\ k \\ l \end{pmatrix} \quad (A.14)$$

The dot product of two real space vectors $\vec{r}_1 = n_1\vec{a} + m_1\vec{b} + p_1\vec{c}$ and $\vec{r}_2 = n_2\vec{a} + m_2\vec{b} + p_2\vec{c}$ is as follows:

$$\vec{r}_1 \cdot \vec{r}_2 = (n_1, m_1, p_1)G \begin{pmatrix} n_2 \\ m_2 \\ p_2 \end{pmatrix} \quad (A.15)$$

The dot product of two reciprocal space vectors $\vec{r}_1^* = h_1\vec{a}^* + k_1\vec{b}^* + l_1\vec{c}^*$ and $\vec{r}_2^* = h_2\vec{a}^* + k_2\vec{b}^* + l_2\vec{c}^*$ is as follows:

$$\vec{r}_1^* \cdot \vec{r}_2^* = (h_1, k_1, l_1)G^{-1}\begin{pmatrix} h_2 \\ k_2 \\ l_2 \end{pmatrix} \tag{A.16}$$

The transformation between real space and reciprocal space for the vector $\vec{r} = n\vec{a} + m\vec{b} + p\vec{c} = h\vec{a}^* + k\vec{b}^* + l\vec{c}^*$ is given by

$$\begin{pmatrix} h \\ k \\ l \end{pmatrix} = G\begin{pmatrix} n \\ m \\ p \end{pmatrix} \tag{A.17}$$

or

$$\begin{pmatrix} n \\ m \\ p \end{pmatrix} = G^{-1}\begin{pmatrix} h \\ k \\ l \end{pmatrix} \tag{A.18}$$

Volume of the unit cell

$$V = \sqrt{\det(G)} = abc\left(1 - \cos^2\alpha - \cos^2\beta - \cos^2\gamma + 2\cos\alpha\cos\beta\cos\gamma\right)^{1/2} = 1/V^* \tag{A.19}$$

Appendix D
Indexed Diffraction Patterns with HOLZ

Figures D.1, D.2, D.3, and D.4 show indexed diffraction patterns for commonly encountered structures. By successive addition of the basis vectors given (in three-dimensions), it is possible to construct the entire three-dimensional lattice as an aid to indexing lattice points in the outer HOLZ ring. The cross indicates one HOLZ point, from which the entire three-dimensional lattice may be constructed by translating the origin of the ZOLZ lattice to the cross.

© Springer Science+Business Media New York 2017
J.M. Zuo and J.C.H. Spence, *Advanced Transmission Electron Microscopy*,
DOI 10.1007/978-1-4939-6607-3

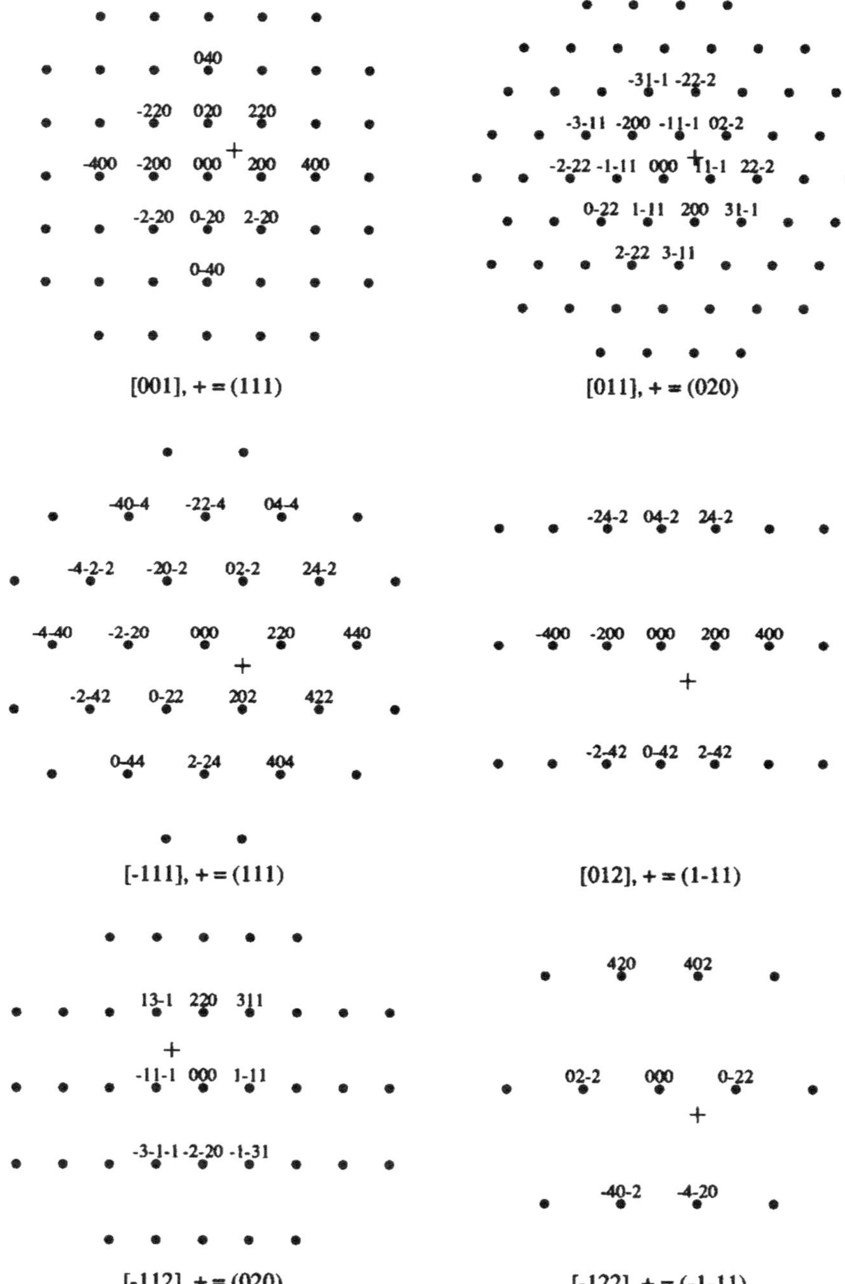

Fig. D.1 Twelve low-index zone-axis diffraction patterns for face-centered cubic crystals

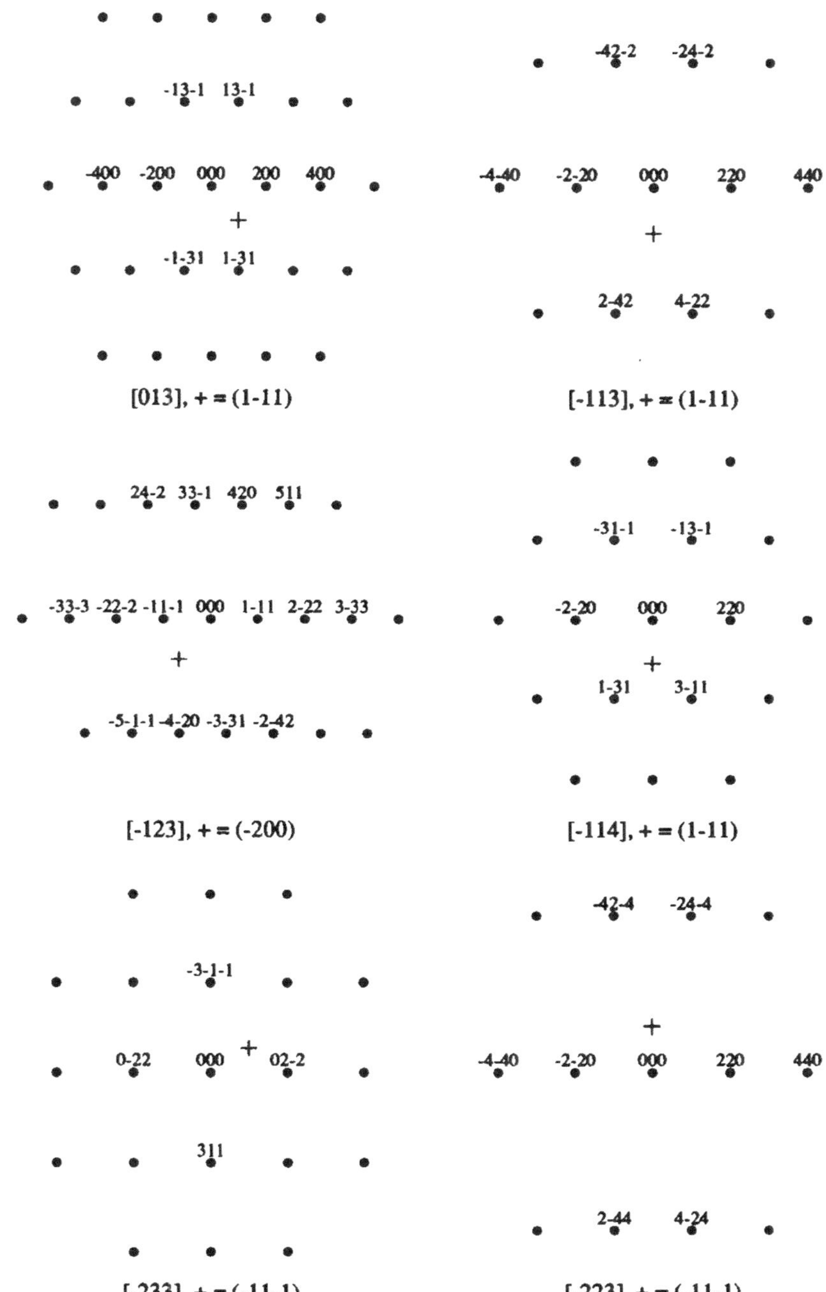

Fig. D.1 (continued)

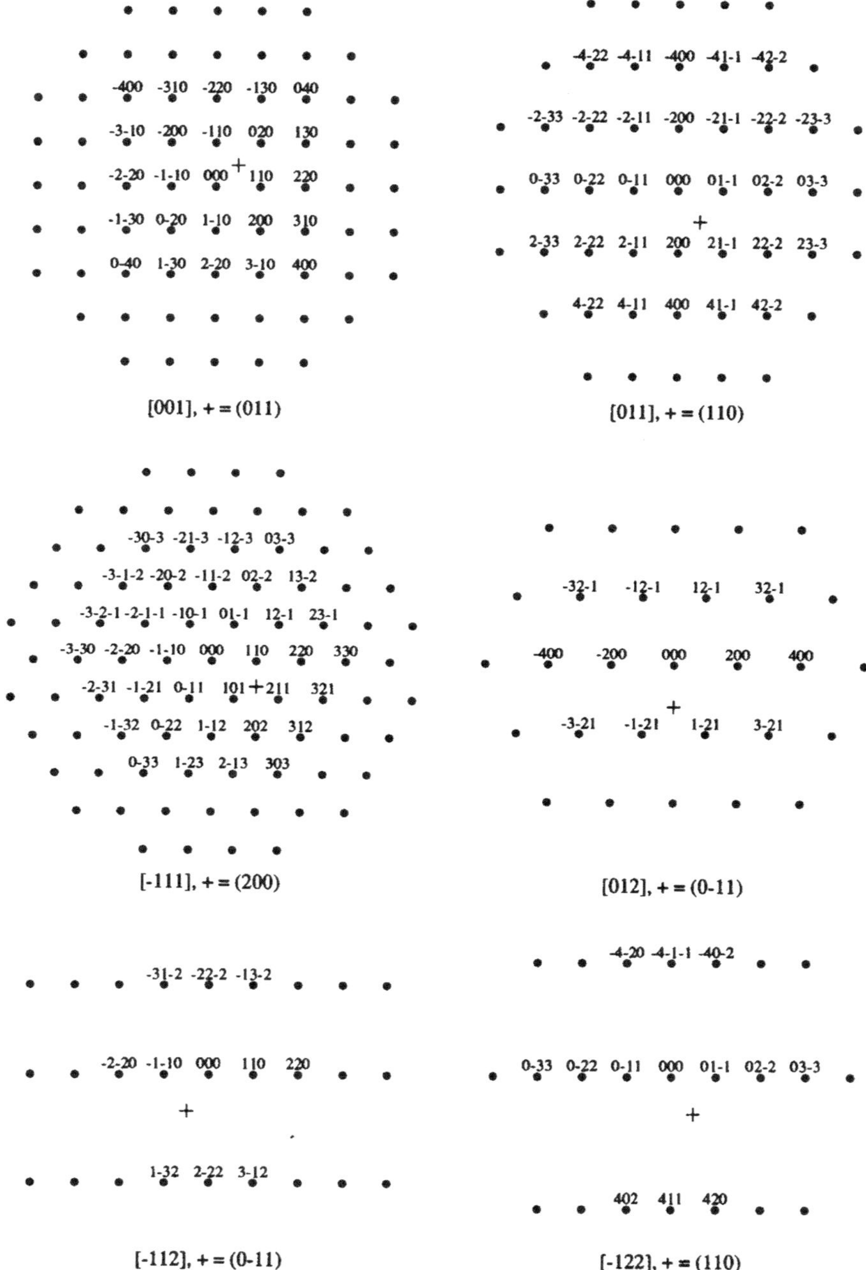

Fig. D.2 Twelve low-index zone-axis diffraction patterns for body-centered cubic crystals

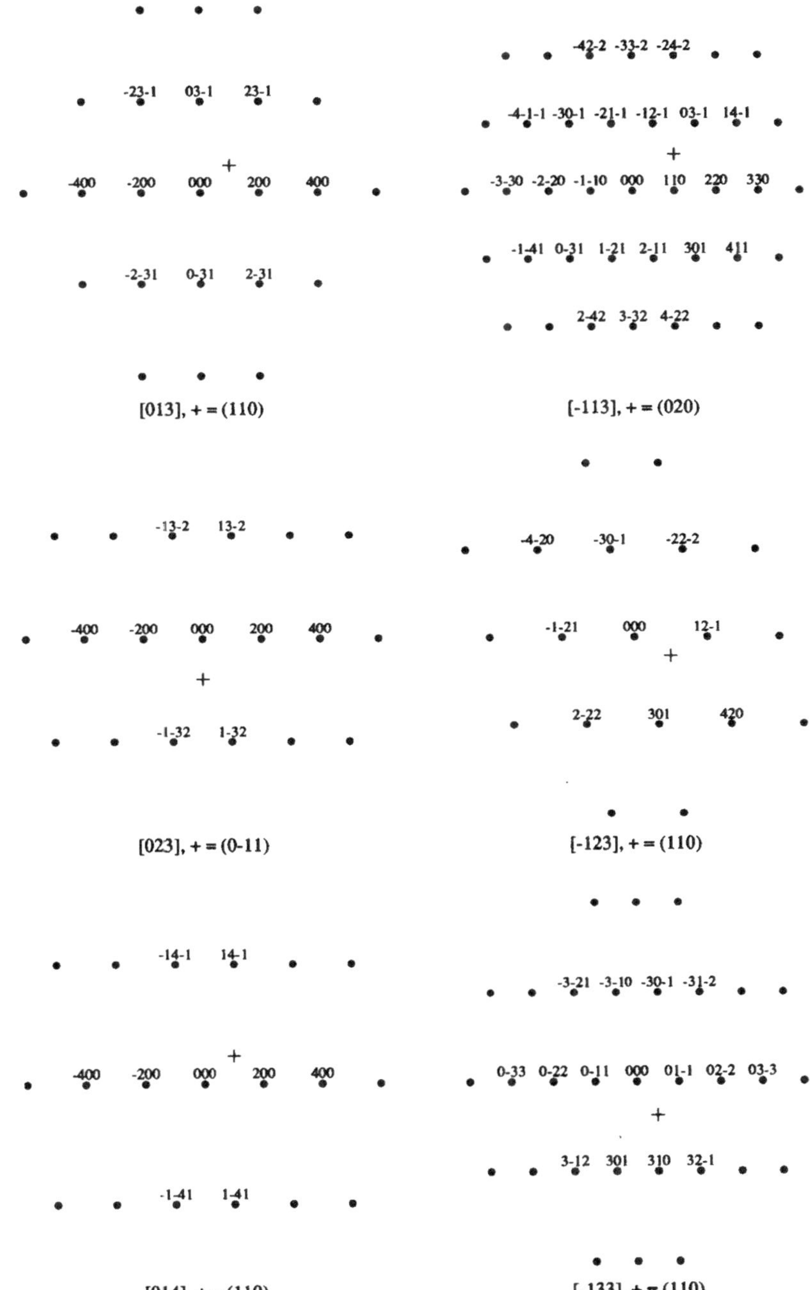

Fig. D.2 (continued)

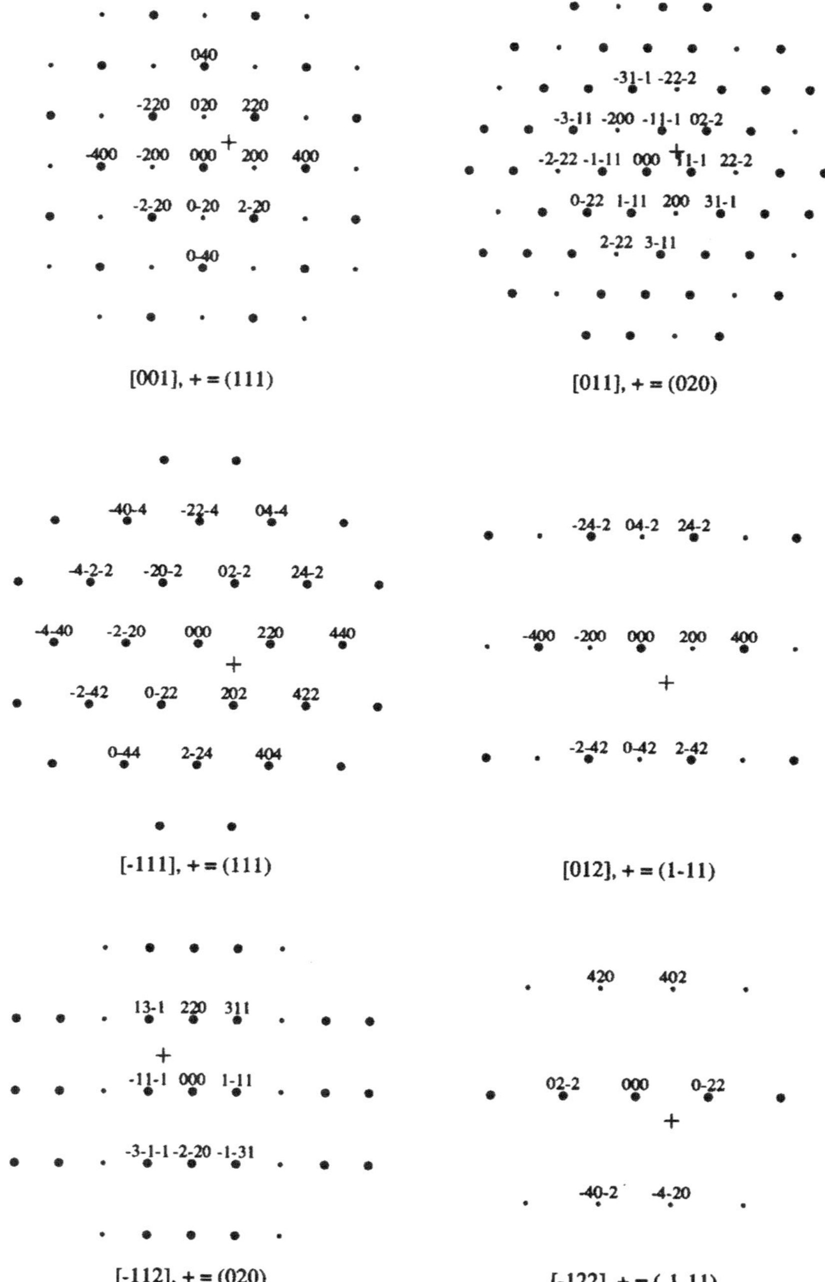

Fig. D.3 Twelve low-index zone-axis diffraction patterns for the diamond lattice. The reflections indicated by small dots are kinematically forbidden if the atoms are spherical

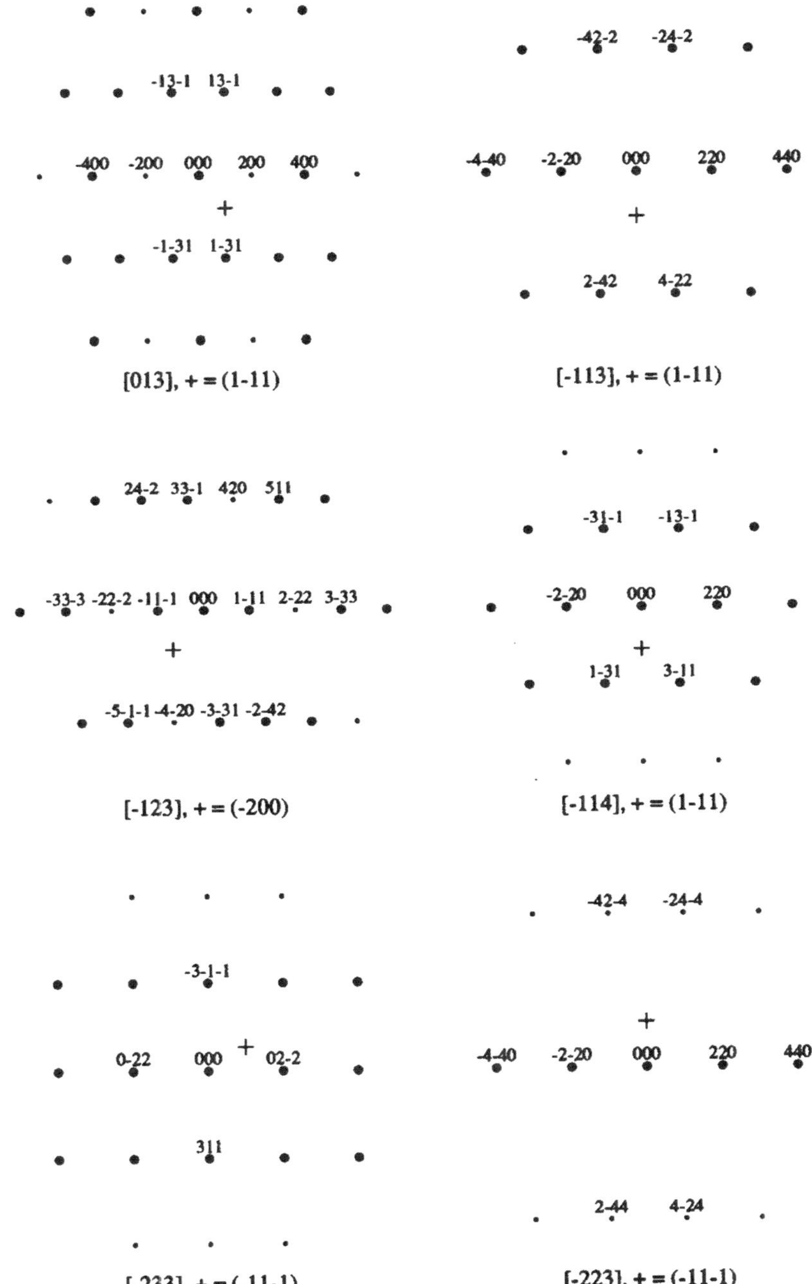

Fig. D.3 (continued)

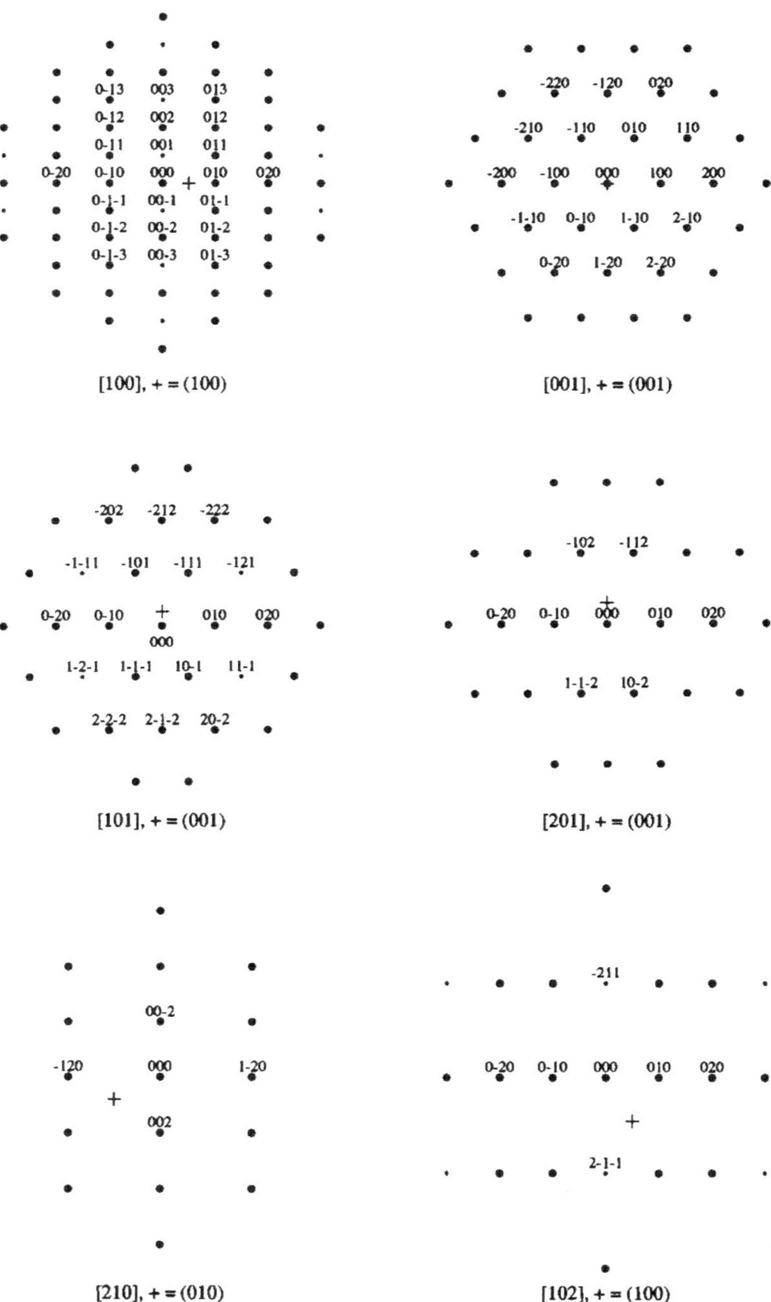

Fig. D.4 Twelve low-index zone-axis diffraction patterns for the hexagonal lattice with $c/a = 1.633$

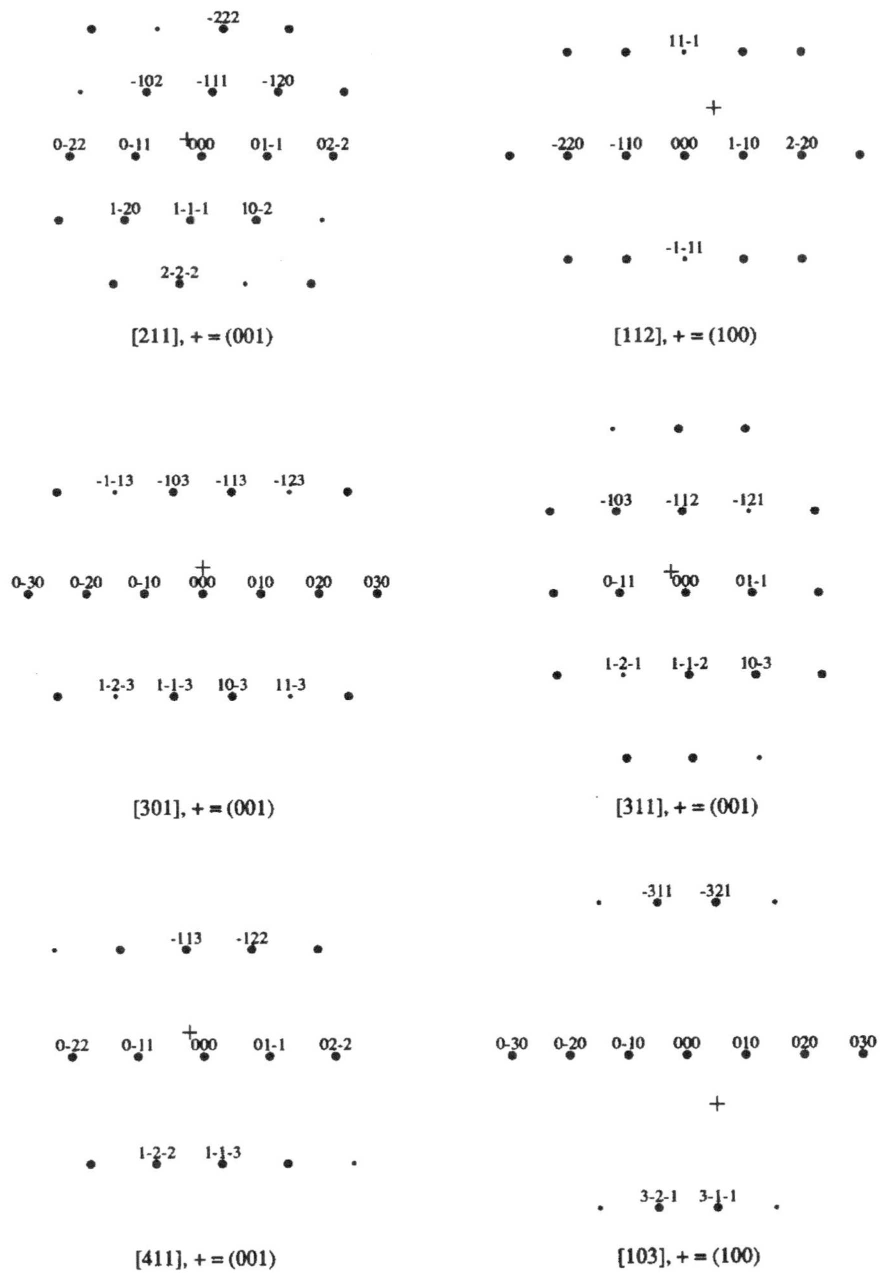

Fig. D.4 (continued)

Appendix E
Fourier Transforms, δ-Function, and Convolution

E.1 Fourier Transforms

The following type of integrals involving cosine and sine functions defines the continuous Fourier transform in one-dimensional Fourier analysis:

$$
\begin{aligned}
F_R(k) &= \int_{-\infty}^{\infty} f(x)\cos(2\pi kx)\mathrm{d}x \\
F_I(k) &= \int_{-\infty}^{\infty} f(x)\sin(2\pi kx)\mathrm{d}x
\end{aligned}
\tag{A.20}
$$

The combination of $F_R(k)$ and $F_I(k)$ in the complex form of $F_R(k) - iF_I(k)$ gives the following definition of complex Fourier transform

$$
F(k) = \int_{-\infty}^{\infty} f(x)\mathrm{e}^{-2\pi ikx}\mathrm{d}x,
\tag{A.21}
$$

where x is a coordinate in the real space and independent, and k is the transform variable in the Fourier space. In diffraction, k is a scattering vector along x in the reciprocal space.

The inverse Fourier transform is given by

$$
f(x) = \int_{-\infty}^{\infty} F(k)\mathrm{e}^{2\pi ikx}\mathrm{d}k.
\tag{A.22}
$$

© Springer Science+Business Media New York 2017
J.M. Zuo and J.C.H. Spence, *Advanced Transmission Electron Microscopy*,
DOI 10.1007/978-1-4939-6607-3

In 3D, x and k in Eqs. (A.21) and (A.22) are replaced by vectors and the integral is carried over 3D, which give the following

$$F(\vec{k}) = \int_{-\infty}^{\infty} f(\vec{r})e^{-2\pi i \vec{k}\cdot\vec{r}}d\vec{r}$$

$$f(\vec{r}) = \int_{-\infty}^{\infty} F(\vec{k})e^{2\pi i \vec{k}\cdot\vec{r}}d\vec{k}. \tag{A.23}$$

E.2 Parseval's Theorem and Plancherel's Theorem

The Plancherel theorem states that the integral of a function's squared modulus is equal to the integral of the squared modulus of its frequency spectrum, e.g.,

$$\int_{-\infty}^{\infty} |f(x)|^2 dx = \int_{-\infty}^{\infty} |F(k)|^2 dk. \tag{A.24}$$

The Fourier transform is unitary, and the Plancherel theorem is a consequence of this property. The unitarity of the Fourier transform can be proved by using Parseval's theorem, which gives

$$\int_{-\infty}^{\infty} f(x)g^*(x)dx = \int_{-\infty}^{\infty} F(k)G^*(k)dk, \tag{A.25}$$

where $G(k)$ is the Fourier transform of $g(x)$ and * denotes the complex conjugate of the two functions.

E.3   Properties of Fourier Transform

See Table E.1.

Table E.1 The properties of Fourier transform

Property	Function	Transform
Linearity	$af(x) + bg(x)$	$aF(k) + bG(k)$
Translation	$f(x - x_{\mathrm{o}})$	$F(k)\mathrm{e}^{-2\pi ikx_{\mathrm{o}}}$
Scaling	$f(ax), a \neq 0$	$\frac{1}{a}F(k/a)$
Inversion	$f(-x)$	$F(-k)$
Multiplication	$f(x)g(x)$	$F(k) \otimes G(k)$
Complex conjugate	$f^*(x)$	$F^*(-k)$
Differentiation	$\frac{\mathrm{d}}{\mathrm{d}x}f(x)$	$(2\pi ik)F(k)$
High-order differentiation	$\frac{\mathrm{d}^n}{\mathrm{d}x^n}f(x)$	$(2\pi ik)^n F(k)$

See Cowley, J. M. (1995), *Diffaction Physics (3rd)*, Elsevier Science, NL, for proofs of some of the listed properties

E.4 δ-Function

The δ-function was introduced by Dirac in his 1930 book on "The Principles of Quantum Mechanics." A starting point to define the δ-function is the rectangular function

$$\mathrm{rect}(x/a) = \begin{cases} 1/a & |x| \leq a/2 \\ 0 & |x| > a/2 \end{cases},$$

where a is the width of the rectangular function. It has the property of

$$\int_{-\infty}^{\infty} \mathrm{rect}(x/a)\mathrm{d}x = 1. \tag{A.26}$$

The Fourier transform of $\mathrm{rect}(x)$ gives

$$FT[\mathrm{rect}(x/a)] = \frac{1}{a} \int_{-a/2}^{a/2} \mathrm{e}^{-2\pi ikx}\mathrm{d}x = \frac{\sin(\pi ka)}{\pi ka} = \mathrm{sinc}(\pi ka).$$

The δ-function is obtained by taking the width a to the limit of $a = 0$ as

$$\delta(x) = \lim_{a->0} [\text{rect}(x/a)] \quad \text{and} \quad \delta(x) = 0 \quad \text{for} \quad x \neq 0, \tag{A.27}$$

and

$$\text{FT}[\delta(x)] = \lim_{a \to 0}[\text{sinc}(ka)] = 1 \tag{A.28}$$

The δ-function has even symmetry with

$$\delta(x) = \delta(-x). \tag{A.29}$$

Equations (A.27) and (A.28) are the basic properties of the δ-function, which can be met by a number of functions, including the followings

$$\delta(x) = \lim_{a \to 0}\left[\frac{1}{a\sqrt{\pi}}\exp\left(-\frac{x^2}{a^2}\right)\right], \tag{A.30}$$

$$\delta(x) = \lim_{a \to 0}\left[\frac{1}{a}\text{sinc}\left(\frac{x}{a}\right)\right]. \tag{A.31}$$

E.5   Convolution of two functions

The convolution of function $f(x)$ with function $g(x)$ is defined by

$$h(x) = \int_{-\infty}^{\infty} f(x')g(x - x')\mathrm{d}x', \tag{A.32}$$

which is denoted by $f(x) * g(x)$ or $f(x) \otimes g(x)$. The Fourier transform of the convoluted functions gives

$$\begin{aligned}
\text{FT}[h(x)] &= \int_{-\infty}^{\infty} \int_{-\infty}^{\infty} f(x')g(x - x')\mathrm{d}x' e^{-2\pi ikx}\mathrm{d}x \\
&= G(k) \int_{-\infty}^{\infty} f(x')e^{-2\pi ikx'}\mathrm{d}x' \\
&= F(k)G(k).
\end{aligned} \tag{A.33}$$

When $g(x) = \delta(x)$, the convolution gives

$$h(x) = \int_{-\infty}^{\infty} f(x')\delta(x - x')\mathrm{d}x' = \int_{-\infty}^{\infty} f(x')\delta(x' - x)\mathrm{d}x' = f(x).$$

E.6 Fourier Transform and Convolution of Gaussian Functions

The Gaussian function associated with the normal distribution in statistics has the standard form of

$$g(x) = \frac{1}{\sigma\sqrt{2\pi}}\mathrm{e}^{-x^2/2\sigma^2}, \tag{A.34}$$

which is a real and even function. It has a FWHM of $2\sqrt{2\ln 2}\,\sigma = 2.355\sigma$, where σ is standard deviation. The Fourier transform of $g(x)$ gives

$$\begin{aligned}
\mathrm{FT}[g(x)] &= 2\int_0^{\infty} g(x)\cos(2\pi kx)\mathrm{d}x \\
&= \frac{1}{\sigma\sqrt{2\pi}}\mathrm{Re}\left\{\int_0^{\infty} \mathrm{e}^{-x^2/2\sigma^2}\mathrm{e}^{2\pi ikx}\mathrm{d}x\right\},
\end{aligned}$$

Using the result of

$$\int_0^{\infty} \mathrm{e}^{-y^2}\mathrm{d}y = \sqrt{\pi}/2,$$

we obtain

$$\mathrm{FT}[g(x)] = \mathrm{e}^{-2\pi^2 k^2 \sigma^2}. \tag{A.35}$$

Convolution of two Gaussian functions with the standard deviations of σ_1 and σ_2 gives

$$FT[g_1(x) * g_2(x)] = e^{-2\pi^2 k^2 \left(\sigma_1^2 + \sigma_2^2\right)}$$

and

$$g_1(x) * g_2(x) = \frac{1}{\sqrt{2\pi\left(\sigma_1^2 + \sigma_2^2\right)}} e^{-x^2/2\left(\sigma_1^2 + \sigma_2^2\right)}.$$

Appendix F
Crystal Structure Data, Mean Inelastic Free Path, and Mean Potential

The following is a list of commonly occurring crystals, their structure, and their lattice constants. For more accurate data, refer, for example, Pearson (1967) (Tables F.1, F.2, F.3, F.4, F.5, F.6, F.7, F.8, F.9, and F.10).

*Anti-nickel arsenide structure

Table F.1 Crystal structures of the elements, room-temperature Debye-Waller factor (DWF) and inelastic mean free path (λ_i) of plasmon contribution (λ_p)

Z	S	Structure	Shortest distance between atoms (Å)	Cell constant (Å)	DWF $T = 293$ K (Theory) (B in Å^2)	Inelastic MFP λ_I, λ_p (nm)
1	H					
3	Li	b.c.c.	3.04	3.491	4.1, (4.81)	
4	Be	h.c.p.	2.24	2.27, 3.59	(0.423)	160, 169
5	B	Trigonal	1.71			123, 126
6	C	Graphite	1.42			
	C	Diamond	1.54	3.567	(0.144)	112, 116
	C	Amorphous				51.5**
8	O					
9	F					
11	Na	b.c.c.	3.71	4.225	7.9, (6.66)	
12	Mg	h.c.p.	3.21	3.21,5.21	(1.81)	150, 214
13	Al	f.c.c.	2.86	4.05	0.86, (0.781)	134, 160
14	Si	Diamond	2.35`	5.430	0.4668*, (0.485)	145, 168
15	P	Ortho (black)	2.17			160, 160
16	S	Ortho	2.12			200, 200
17	Cl					
19	K	b.c.c.	4.63	5.225	12, (10.7)	
20	Ca	f.c.c.	3.94	5.58	(2.00)	
21	Sc	h.c.p.	3.21	3.31, 5.21	(0.746)	

(continued)

© Springer Science+Business Media New York 2017
J.M. Zuo and J.C.H. Spence, *Advanced Transmission Electron Microscopy*,
DOI 10.1007/978-1-4939-6607-3

Table F.1 (continued)

Z	S	Structure	Shortest distance between atoms (Å)	Cell constant (Å)	DWF $T = 293$ K (Theory) (B in Å^2)	Inelastic MFP λ_I, λ_p (nm)
22	Ti	h.c.p.	2.95	2.95, 4.68	(0.514)	120, 202
23	V	b.c.c.	2.63	3.03	0.55, (0.572)	109, 158
24	Cr	b.c.c.	2.50	2.88	0.26, (0.251)	104, 149
25	Mn	Cubic	2.24			106, 146
26	Fe	f.c.c.			(0.558)	
26	Fe	b.c.c.	2.48	2.87	0.35, (0.325)	102, 121
27	Co	h.c.p.	2.51	2.51		98, 108
28	Ni	f.c.c.	2.49	3.52	0.37, (0.356)	98, 103
29	Cu	f.c.c.	2.56	3.61	0.57, (0.551)	100, 100
30	Zn	h.c.p.	2.66	2.66, 4.95	(1.14)	106, 106
31	Ga	Ortho	2.44	2.66, 4.95		110
32	Ge	Cubic	2.45	5.658	0.57, (0.600)	120, 126
33	As	Trigonal	2.90			
34	Se	Hexag.	2.32			130, 205
35	Br					
37	Rb	b.c.c.	4.90	5.585	(13.2)	
38	Sr	f.c.c.	4.30	6.08	(3.79)	
39	Y	h.c.p.	3.62	3.65, 5.73	(0.855)	124, 354
40	Zr	h.c.p.	3.20	3.23, 5.15	(0.569)	113, 268
41	Nb	b.c.c.	2.86	3.30	0.49, (0.450)	105, 194
42	Mo	b.c.c.	2.72	3.15	0.25, (0.216)	98, 163
43	Tc	h.c.p.	2.70	2.74, 4.40		
44	Ru	h.c.p.	2.65	2.71, 4.28		90, 134
45	Rh	f.c.c.	2.69	3.80		
46	Pd	f.c.c.	2.75	3.89	0.45, (0.446)	94, 118
47	Ag	f.c.c.	2.89	4.09	0.79, (0.734)	100, 125
48	Cd	h.c.p.	2.98	4.98, 5.62		107, 130
49	In	Tetrag.	3.25	3.25, 4.95		110, 129
50	Sn	Tetrag. (white)	3.02	6.49	(1.13)	115, 273
51	Sb	Trigonal	2.90			120, 234
52	Te	Hexag.	2.87			130, 216
53	I	Orthorhomb.	2.71			140, 233
55	Cs	b.c.c.	5.25	6.045	(17.5)	
56	Ba	b.c.c.	4.35	5.02	(3.06)	
57	La	h.c.p.	3.75	3.77	(1.86)	
58	Ce	f.c.c.	3.65	5.16		
59	Pr	Hexag.	3.64	3.67		

(continued)

Table F.1 (continued)

Z	S	Structure	Shortest distance between atoms (Å)	Cell constant (Å)	DWF $T = 293$ K (Theory) (B in Å^2)	Inelastic MFP λ_I, λ_p (nm)
60	Nd	Hexag	3.66	3.66		
61						
62	Sm					112, 280
63	Eu	b.c.c.	3.96	4.58		
64	Gd	h.c.p.	3.56	3.63, 5.78		110, 275
65	Tb	h.c.p.	3.53	3.60, 5.70	(1.01)	
66	Dy	h.c.p.	3.50	3.59, 5.65		118, 310
67	Ho	h.c.p.	3.52	3.58, 5.62	(0.839)	
68	Er	h.c.p.	3.55	3.56, 5.59		
69	Tm	h.c.p.	3.48	3.54, 5.56		
70	Yb	f.c.c.	3.87	5.48		110, 275
71	Lu	h.c.p.	3.46	3.50, 5.55		
72	Hf	h.c.p.	3.16	3.19, 5.05		95, 237
73	Ta	b.c.c.	2.86	3.30	0.32, (0.322)	88, 183
74	W	b.c.c.	2.74	3.16	0.18, (0.160)	82, 151
75	Re	h.c.p.	2.74	2.76, 4.46		78, 141
76	Os	h.c.p.	2.68	2.76, 4.32		
77	Ir	f.c.c.	2.71	3.86		78, 121
78	Pt	f.c.c.	2.88	4.08	0.32, (0.372)	82, 120
79	Au	f.c.c.	2.88	4.08	0.57, (0.620)	84, 120
80	Hg	Trigonal	3.00			
81	Tl	h.c.p.	3.41	3.46, 5.52		95, 135
82	Pb	f.c.c.	3.49	4.95	2.42, (2.13)	99, 141
83	Bi	Trigonal	3.11			105, 147
84	Po	Simple Cubic	3.35	3.34		
89	Ac	f.c.c.	3.75	5.31		
90	Th	f.c.c.	3.60	5.08	(0.730)	
91	Pa	b.c. tetrag..	3.21	3.92, 32.4		
92	U	Orthorhomb	2.77			
93	Np	Orthorhomb	2.60			
94	Pu	Monoclinic	3.28			

The inelastic mean free path data are taken from Iakoubovskii, K. and K. Mitsuishi, Phys. Rev. B 77, 104102 (2008), which were measured for 200 kV electrons. The sources of elemental DWFs are as follows: Peng, L.M., G. Ren, S.L. Dudarev and M.J. Whelan, Acta Cryst. A 52, 456 (1996), Sears, V.F. and S.A. Shelley, Acta Cryst., A47, 441 (1991)

*From Zuo, J.M., P. Blaha and K. Schwarz (1997). J. Phys.: Condens. Matter 9, 7541

**By electron holography, from https://www.ipen.br/biblioteca/cd/icem/1998/VOLUME1//V10259.PDF

The cell constants for c.p.h. structures are given as a, c in Angstroms

Table F.2 Crystals with the sodium chloride structure

Crystal	a	Crystal	a
Antimonides		*Carbides*	
CeSb	6.40	HfC	4.46
LaSb	6.47	NbC	4.47
ScSb	5.85	TaC	4.45
SnSb	6.13	TiC	4.32`
ThSb	6.32	UC	4.96
USb	6.19	VC	4.18
Arsenides		ZrC	4.68
CeAs	6.06	*Chlorides*	
LaAs	6.13	AgCl	5.55
ScAs	5.49	KCl	6.29
SnAs	5.68	LiCl	5.13
ThAs	5.97	NaCl	5.64
UAs	5.77	RbCl	6.58
Borides		*Cyanides*	
ZrB	4.65	KCN	6.53
Bromides		NaCN	5.89
AgBr	5.77	RbCN	6.82
KBr	6.60	*Phosphides*	
LiBr	5.50	CeP	5.90
RbBr	6.89	LaP	6.-01
Florides		ThP	5.82
AgF	4.92	UP	5.59
CsF	6.01	ZrP	5.27
KF	5.35	*Selenides*	
LiF	4.03	BaSe	6.60
NaF	4.62	CaSe	5.91
RbF	5.64	CeSe	5.98
Hydrides		LaSe	6.06
CsH	6.38	MgSe	5.45
KH	5.70	MnSe	5.45
LiH	4.08	PbSe	6.12
HaH	4.88	SnSe	6.02
RbH	6.04	SrSe	6.23
Iodides		ThSe	5.87
KI	7.07	*Sulfides*	
LiI	6.00	BaS	6.39
NH4I	7.26	CaS	5.69
NaI	6.47	CeS	5.78
RbI	7.34	LaS	5.84

(continued)

Table F.2 (continued)

Crystal	a	Crystal	a
Nitrides		MgS	5.20
CeN	5.01	MnS	4.45
CrN	4.14	PbS	5.94
LaN	5.39	SrS	6.02
NbN	4.70	ThS	5.68
ScN	4.44	US	5.48
TiN	4.23	ZrS	5.25
UN	4.88	*Tellurides*	
VN	4.113	BaTe	6.99
ZrN	4.61	BiTe	6.47
Oxides		CaTe	6.34
BaO	5.52	CeTe	6.35
CaO	4.81	LaTe	6.41
CdO	4.70	PbTe	6.45
CoO	4.27	SnTe	6.31
FeO	4.28-4.31	SrTe	6.47
MgO	4.21	UTe	6.16
MnO	4.45		
NbO	4.21		
NiO	4.17		
SrO	5.16		
TaO	4.42–4.44		
TiO	4.18		
UO	4.92		
ZrO	4.62		

Table F.3 Crystals with the cesium chloride structure

Crystal	a (Å)
AgCd	3.33
AgMg	3.28
CuBe	2.70
CuZn	2.94
CsBr	4.29
CsCl	4.12
CsCN	4.25
CsI	4.57
TlBr	3.97
TlCl	3.83
TlCN	3.82
TlI	4.20

Table F.4 Crystals with the sphalerite structure

Crystal	a (Å)	Crystal	a (Å)
AgI	6.47	GaSb	6.12
BN	3.62	HgS	5.85
BeS	4.85	HgSe	6.08
BeSe	5.07	HgTe	6.43
BeTe	5.54	InAs	6.04
CdS	5.82	InP	5.87
CuBr	5.69	InSb	6.48
CuCl	5.40	SiC	4.35
CuF	4.26	ZnS	5.41
CuI	6.04	ZnSe	5.67
GaAs	5.65	ZnTe	6.09
GaP	5.45		

Table F.5 Crystals with the wurtzite structure

Crystal	a (Å)	c (Å)
AgI	4.58	7.49
AlN	3.11	4.98
BeO	2.70	4.38
CdS	4.14	6.75
CdSe	4.30	7.02
NH_4F	4.39	7.02
SiC	3.08	5.05
TaN	3.05	4.94
ZnO	3.25	5.21
ZnS	3.81	6.23
ZnSe	3.98	6.53
ZnTe	4.27	6.99

Table F.6 Crystals with the nickel arsenide structure

Crystal	a (Å)	c (Å)	Crystal	a (Å)	c (Å)
CoS	3.37	5.16	NnTe	4.14	6.70
CoSb	3.87	5.19	NiAs	3.60	5.01
CoTe	3.89	5.36	NiSe	3.66	5.36
CrS	3.45	5.75	NiSn	4.05	5.12
CrSb	4.11	5.44	NiTe	3.96	5.35
CrSe	3.68	6.02	PtB*	3.36	4.06
FeS	3.44	5.88	PtBi	4.31	5.49

(continued)

Table F.6 (continued)

Crystal	a (Å)	c (Å)	Crystal	a (Å)	c (Å)
FeSb	4.06	5.13	PtSb	4.13	5.47
FeSe	3.64	5.96	PtSn	4.10	5.43
FeTe	3.80	5.65	VP	3.18	6.22
MnAs	3.71	5.69	VS	3.36	5.98
MnSb	4.12	5.78	VTe	3.94	6.13

Table F.7 Crystals with the fluorite structure

Crystal	a (Å)	Crystal	a (Å)
$BaCl_2$	7.34	Li_2S	5.71
BaF_2	6.20	Li_2Se	6.01
Be_2	4.67	Li_2Te	6.50
Be_2C	4.33	Na_2O	5.55
CaF_2	5.46	Na_2S	6.53
CdF_2	5.39	Na_2Se	6.81
CeH_2	5.59	Na_2Te	7.31
CeO_2	5.41	NbH_2	4.56
$CoSi_2$	5.36	$NiSi_2$	5.39
HfO_2	5.12	Rb_2O	6.71
HgF_2	5.54	Rb_2S	7.65
K_2O	6.44	$SrCl_2$	6.98
K_2S	7.39	SrF_2	5.80
K_2Se	7.68	ThO_2	5.60
K_2Te	8.15	UO_2	5.47
Li_2O	4.62	ZrO_2	5.07

Table F.8 Crystals with the rutile structure

Crystal	a (Å)	c (Å)	Crystal	a (Å)	c (Å)
CoF_2	4.70	3.18	NbO_2	4.77	2.96
FeF_2	4.70	3.31	PbO_2	4.95	3.38
GeO_2	4.40	2.86	SnO_2	4.74	3.19
MgF_2	4.62	3.05	TaO_2	4.71	3.06
MnF_2	4.87	3.31	TiO_2	4.59	2.96
MnO_2	4.40	2.87	WO_2	4.86	2.77
MoO_2	4.86	2.79	ZnF_2	4.70	3.13

Table F.9 Room-temperature Debye-Waller factors and inelastic mean free paths of selected binary crystals

Crystal (A_xB_y)	B_A (Å^2)	B_B (Å^2)	λ_i (nm)	Crystal (A_xB_y)	B_A (Å^2)	B_B (Å^2)	λ_i (nm)
Oxides				Zincblende			
BeO	0.355	0.28		AlAs			77(4)*
B_2O_3			120	GaP	(0.510)	(0.634)	
MgO	0.305	0.340	133	GaSb	(0.934)	(0.791)	
Al_2O_3	0.22	0.27	140	GaAs	(0.637)	(0.685)	67(4)*
SiO_2			155	InP	(0.882)	(0.578)	
CaO			130	InSb	(1.351)	(1.129)	
Sc_2O_3	1.3	1.4	125	ZnO	(0.537)	(0.460)	
TiO_2 (*Rutile*)	0.51	0.42	120	InAs	(0.946)	(0.641)	
V_2O_5			116	ZnS	(0.854)	(0.736)	
VO_2	0.42	0.59 0.78		ZnSe	(0.987)	(0.664)	
V_2O_3	0.31	0.51		ZnTe	(1.335)	(0.953)	46(2)*
CrO_3			118	CdTe	(1.915)	(1.285)	
Cr_2O_3	0.25	0.28		HgSe	(3.818)	(1.489)	
MnO	0.617	072		HgTe	(5.522)	(1.745)	
Fe_2O_3	0.32	0.40	116	CuCl	(3.625)	(2.344)	
CoO	0.509	0.67	115	CuBr	(2.018)	(2.877)	
NiO	0.414	0.61	115	CuI	(2.470)	(1.383)	
CuO	0.54	0.60		SiC	(0.218)	(0.226)	
Cu_2O	1.48	1.51					
ZnO	0.68	0.68	117				
Y_2O_3	0.60 0.55	0.66	122				
ZrO_2	0.33	0.55 0.46	115				
SnO_2	0.29	0.46	115				
TeO_2-α	0.59	0.92	128				
BaO			125				
La_2O_3	0.59	0.82 1.19	130				
Ce_2O_3	0.72**	0.68**	125				
TaO			110				
WO_3	0.7	1	110				
IrO_2	0.75	0.82	110				

(continued)

Table F.9 (continued)

Crystal (A_xB_y)	B_A ($\mathring{A}^2$)	B_B ($\mathring{A}^2$)	λ_i (nm)	Crystal (A_xB_y)	B_A ($\mathring{A}^2$)	B_B ($\mathring{A}^2$)	λ_i (nm)
Oxides				Zincblende			
PbO			122				
Bi_2O_3			125				

Data source The mean free paths of binary oxides are from Iakoubovskii, K. and K. Mitsuishi, Phys. Rev. B 77, 104102 (2008), which were obtained at 200 kV; theoretical DWFs of the zincblende structures are from J.S. Reid, Acta Cryst. A39, 1-13 (1983). The DWFs of binary oxides are experimental values from the published X-ray, neutron and electron diffraction data), λ_i, 200kV

* Obtained by electron holography, source: Chung, S., D. et al. (2007). Micros. and Microanal. 13, 329. Gan, Z. et al. (2015). Micros. and Microanal. 21, 1406

** T=313 K

Theoretical value is quoted in bracket ()

Table F.10 Crystal mean potential in volts. The theory is based on density functional theory calculations (see ref. b)

Crystal	Theory	Procrystal	Holography	RHEED
Mg	12.7	17.0	11.9–13.0	
MgO	12.64(2)	12.28 (ionic)	13.01(8)	
		12.85 (ionic)		
		17.94		
AlN	15.88	18.70		
	14.23	18.67		
AlP	11.39	13.34		
AlAs	12.34	13.93	12.1(7)	
AlSb	12.89	13.96		
Si	12.57	13.91	12.5(7)	12.0(4)
	12.21(2)	13.82	11.5(5)	
	12.48(6)		12.1(13)	
	12.38(2)			
	12.42(9)			
GaN	16.89	19.51		
	16.82	19.30		
GaP	13.63	14.88	14.38(12)	12.2
GaAs	14.19	15.26	14.0(6)	13.2
			14.53(17)	
GaSb	14.45	15.27	14.08(5)	
Ge	14.67	15.58	14.3(2)	
	14.12(2)		15.6(8)	
Ag			23.3(10)	
InN	18.90	19.25		
	17.35	19.57		
InP	13.90	15.06	14.53(7)	

Table F.10 (continued)

Crystal	Theory	Procrystal	Holography	RHEED
InAs	14.34	15.26	14.50(8)	
InSb	14.28	15.05		
ZnO	15.75	15.98		
ZnS	12.64	13.56		
ZnTe	13.77	13.50	13.7(6)	
	13.82			
Au	30.26	29.80	30.2	

Source of data (a) Spence JCH and M. O'Keeffe, Acta Cryst. A50, 33 (1994), (b) Kim M. et al. Phys. Stat. Sol. (a)166, 445 (1998), (c) Kruse, P. et al. Ultramic. 106, 105 (2006), (d) Chung S. et al. Microsc. Microanal. 13, 392 (2007), (e) Schowalter, M. et al. Appl. Phys. Lett. 88, 232108 (2016), (f) Menadue J. F., Acta Cryst. A28, 1 (1972). The mean potentials of vitrified ice and carbon foil were measured by Harscher A. and H. Lichte at 3.5 (12) and 10.7(1) volts respectively (reported at ICEM, 1998)

The procrystal is made of spherical atoms or ions (as specified)

Author Index

A

Adams, D.L., 593
Ade, G., 475
Adesida, I., 288
Aeschlimann, M., 195
Alam, M.A., 627, 628
Alden, J.S., 636
Aldred, P.J.E., 377
Aleman, B., 618
Alexander, H., 289, 543, 544
Allen, L.J., 262, 482, 572
Allport, P.P., 223
Allpress, J.G., 442
Almairac, R., 642
Amelinckx, S., 501, 631, 635
Amirfazli, A., 607
Amitani, K., 226
Amlani, I., 628
Amoros, J.L., 403
Anderson, J.N., 593
Ando, Y., 631
Aoki, T., 455
Ardavan, A., 641
Armigliato, A., 133, 556, 559, 567, 570
Aroyo, M.I., 311
Arslan, I., 542
Artacho, E., 621
Asaka, T., 557, 574
Authier, A., 311
Autrata, R., 215
Avdeenkov, A.V., 618
Avouris, P., 641
Ayer, R., 328, 329

B

Bacon, N.J., 276, 283
Bai, X.D., 628
Baker, R.T., 608
Balboni, R., 133, 556, 559, 567, 570

Baletto, F., 617
Balluffi, R.W., 501, 546, 548, 608
Bals, S., 573
Bandaru, P.R., 621
Bandow, S., 619, 642
Bangert, U., 634
Barbier, A., 604
Barge, D., 557
Barna, A., 559
Barnes, J.P., 556, 557, 575
Baroni, S., 380
Barreto, F.C.S., 340
Barrett, C.S., 390
Barry, J.C., 591
Bartels, M., 583
Barth, C., 605
Bartosik, M., 553
Bashir, R., 289
Basinski, Z.S., 529
Baskes, M.I., 598
Baskin, Y., 621
Batson, P.E., 472
Battaglia, M., 223
Baughman, R.H., 627
Baumann, F.H., 556
Bäumer, M., 604, 605
Baylac, E., 557
Bayle-Guillemaud, P., 505
Bean, D., 215
Bean, J.C., 556
Beche, A., 246, 556, 557, 560, 562, 563, 575
Becker, P.J., 368
Behan, G., 483
Beleggia, M., 45, 46, 575, 576
Bell, A.E., 199, 200
Bell, R., 223
Benedetti, A., 567
Benedict, L.X., 636
Benner, G., 236

© Springer Science+Business Media New York 2017
J.M. Zuo and J.C.H. Spence, *Advanced Transmission Electron Microscopy*,
DOI 10.1007/978-1-4939-6607-3

Subject Index

A

Abbe, Ernst, 441

Aberration
 coefficients, 162, 163, 168–170, 175, 176,
  183, 186, 281. 450, 455, 464
 correction, 16, 143, 145, 165, 177, 178,
  181, 184, 187. 188, 285, 288, 479, 481,
  542, 636
 correctors, 12, 143, 145, 177, 178, 190, 557
 function, 173, 174, 243, 244, 441, 452, 468,
  469, 471, 472, 495

Absorption, 97–99, 102–106, 110, 117–121,
 126, 209, 217, 313, 340, 362, 368, 372,
 428, 441, 475, 485, 486, 513

Absorptive potential, 494

Accelerating voltage, 13, 75, 111, 112,
 133–135, 137, 138, 149, 185, 204, 221,
 239, 240, 243, 270, 290, 291, 330, 332,
 333, 455, 479, 481, 566, 656

Acentric crystals, 102, 103, 105, 121, 122, 126,
 127, 132, 315, 323

Achroplanatic microscope, 190

ADF detector, 208, 210–215, 266, 442, 443,
 475, 476, 479, 494, 512, 562

Alloys, 227, 262, 353, 389, 411, 418, 419, 421,
 424, 540, 582

Aluminium oxide (Al_2O_3, corundum), 386, 387

Amorphous materials, 506

Amplitude contrast, 454, 455, 502

Amplitude-phase diagram, 518, 523

Angular momentum, 30

Anisotropic displacement parameters, 367

Annular bright field (ABF), 442

Annular dark field (ADF), 12, 208, 210–215,
 261, 266, 441–443, 463, 475, 476, 479,
 487, 494, 504, 512, 562, 574

Anomalous dispersion, 121, 312, 335

Anomalous transmission effect, 98

Antiphase domains in Cu_3Au, by CBED, 592

Astigmatism
 twofold, 174, 175, 178, 180, 189, 454, 455,
  458
 threefold, 174, 178, 181, 185, 189
 fivefold, 174
 sixfold, 174, 481

Asymmetric unit, 297, 310, 347, 348

Atom probe tomography, 540

Atomic absorption coefficient, 99

Atomic charges, 380, 392, 539

Atomic positions, finding, 350

Atomic resolution, 17, 186, 246, 249, 261, 289,
 334, 403, 406, 431, 441, 443, 475, 478,
 502, 504, 540, 542, 548, 550, 558, 570,
 581, 583, 604, 606, 609, 610, 634

Atomic scattering factor
 for electrons, 85
 for X-rays, 82–86, 95, 657

Axial aberrations, 171, 172, 174

B

Barium titanate ($BaTiO_3$), 339

Beam degeneracy, 203

Bent contours, 511

BeO, 210, 325, 326, 331, 334, 696, 698

Bessel functions, 428, 625, 626, 631

Bethe, H., 4, 13, 125, 375

Bethe potential, 13, 121, 125, 127, 130, 138

Bloch
 states, 485
 waves, 13, 101, 103, 104, 108, 120, 124,
  331, 484

Bloch wave method, 102, 139, 433, 437, 485,
 490, 491, 495, 531

Bond length, 363, 379, 380, 386, 387, 392,
 437, 506, 582, 593, 594, 596–598, 600,
 614, 618, 619, 621, 633

Born approximation, first order, 77–80, 82, 83

Born-Oppenheimer approximation, 366

© Springer Science+Business Media New York 2017

J.M. Zuo and J.C.H. Spence, *Advanced Transmission Electron Microscopy*,

DOI 10.1007/978-1-4939-6607-3

Möllenstedt, G., 8, 13, 574

Monochromator, 12, 28, 50, 165, 276, 283, 368, 455, 456

Moodie, A., 13, 14, 101, 132, 466, 468, 488

Morphotropic phase boundary (MPB), 341

Mott formula, 85

MTF, 218–222, 370

Multiple scattering, 4, 13, 15, 38, 47, 77, 79, 84, 122, 130, 133, 255, 259, 261, 330, 331, 369, 405, 406, 430, 432, 437, 443, 455, 461, 475, 484, 488, 491, 495, 516, 566, 567, 587

Multiple scattering via HOLZ, 331

Multiplicity, 145, 180, 189

Multipole
    aberrations, 145, 178, 180, 189
    fields, 145, 178
    lens, 145, 179

Multipole expansion of electron density, 374

Multislice method, 101, 431, 488, 489, 491–493, 495

N

Nanoarea electron diffraction (NAED), 15, 245, 250–252, 556, 586, 587, 628, 629

Nanobeam diffraction (NBD), 251, 252, 556, 560, 562–565

Nanoclusters, 581, 611, 612, 615, 616

Nanocrystallography, 581

Nanodiffraction, 15, 22, 231–233, 235, 236, 241, 242, 244, 246, 252, 261–263, 441, 492, 556, 562, 592

Nanoparticles, 77, 250, 252, 266, 459, 472, 495, 496, 581–583, 608, 611, 613, 614, 617, 618

Nanoprobes, 588

Nanostructure, critical length, 581, 582

Nanotips, aberrations of, 201

Nanowires, 581, 592

Neumann's principle, 297, 509

Newton's lens equation, 5, 158

Niggli cell, 329

Nodal point, 157

Noncentrosymmetric crystals, 15, 103, 120, 122, 124, 126, 132

Nonsystematics, three-beam, 122, 127, 128, 130, 132

Numerical aperture, 6, 237, 444, 482

Nyquist frequency, 493

Nyquist-Shannon theorem, 492

O

Object distance, 150, 170, 234

Objective aperture, 6, 12, 163, 239, 442, 448, 451, 453, 462, 464, 466, 475, 478, 479, 511, 575

Objective lens
    aberration, 162
    focus, 162
    prefield, 162
    symmetrical, 162

Objective minilens, 10

Object plane, 149, 150, 152, 158, 170–172, 184, 187, 232, 256, 270, 280, 285, 256, 288

Octupole-quadrupole corrector, 186

Off-axial aberrations, 171

Off-axial coma aberration, 171, 172, 187, 188, 190

Optical axis
    curved, 147

Optical potential, electron, 97–100

Over focus, 258

Overlapping CBED orders, consequences, 484

P

Parasitic aberrations, 171, 453

Paraxial approximation, 149, 445

Paraxial equations, 279, 281

Partially coherent, 27, 239, 511

Pauling, L., 360, 593, 597, 621

Perovskite oxides, 335

Perturbation theory
    for absorption, 117–121
    weak beams, 125

Phase contrast, 461, 474, 502, 510

Phase grating approximation, 489

Phase identification, 261, 335

Phase object, 77, 80, 81, 107, 449, 451, 457, 458, 468, 475, 488, 489, 491, 577

Phase problem
    solved by three-beam diffraction, 256

Phase velocity, 33

Phonon scattering, 97, 98, 99, 103, 269, 272, 495

Phosphors, 226

Photomultiplier tubes, 208, 210

Pixels, 215, 216, 217, 219, 220, 221, 223, 224, 226, 227, 264, 265, 271, 492, 562, 565, 573, 574, 577, 590

Planar faults, in CBED, 591

Plasmon scattering, 97, 269, 271, 431

If you have any concerns about our products,
you can contact us on
ProductSafety@springernature.com

In case Publisher is established outside the EU,
the EU authorized representative is:
Springer Nature Customer Service Center GmbH
Europaplatz 3, 69115 Heidelberg, Germany

Printed by Libri Plureos GmbH
in Hamburg, Germany